KU-429-108

Electrophilic Aromatic Substitution

Electrophilic Aromatic Substitution

R. TAYLOR

Reader in Chemistry, University of Sussex

JOHN WILEY & SONS

Chichester · New York · Brisbane · Toronto · Singapore

Other Wiley Editorial Offices

John Wiley & Sons, Inc., 605 Third Avenue,
New York, NY 10158-0012, USA

Jacaranda Wiley Ltd, G.P.O. Box 859, Brisbane,
Queensland 4001, Australia

John Wiley & Sons (Canada) Ltd, 22 Worcester Road,
Rexdale, Ontario M9W 1L1, Canada

John Wiley & Sons (SEA) Pte Ltd, 37 Jalan Pemimpin 05-04,
Block B, Union Industrial Building, Singapore 2057

Library of Congress Cataloging-in-Publication Data:
Taylor, R. (Roger)
Electrophilic aromatic substitution / by R. Taylor.
p. cm.
Includes bibliographical references.
ISBN 0 471 92482 2
1. Substitution reactions. 2. Aromatic compounds. I. Title.
QD281.S67T39 1990 89-22545
541.3′93—dc20 CIP

British Library Cataloguing in Publication Data:
Taylor, R. (Roger)
Electrophilic aromatic substitution.
1. Aromatic compounds. Electrophilic Substitution reactions
I. Title
547′.604593

ISBN 0 471 92482 2

Printed and bound by Courier International Ltd., Tiptree, Colchester.
Phototypesetting by Thomson Press (India) Limited, New Delhi.

Preface

Electrophilic aromatic substitution is one of the cornerstones of organic chemistry. It has a special significance in being the area where much of organic reaction mechanism theory, and in particular the electronic effects of substituents, was developed. No single field of organic chemistry has received so much attention as that of electrophilic substitution of benzene and its derivatives. One third of the worldwide organic chemical production involves aromatic compounds, leading to an ongoing interest in electrophilic aromatic substitution.

Twenty-five years ago the author wrote, with R. O. C. Norman, *Electrophilic Substitution in Benzenoid Compounds*. This was well received and is still referred to in current publications. However, it is now very much out of date, and the present book constitutes a major revision, the necessity for which is indicated by the fact that 60% of the 2700 references in it have been published since the original monograph was written; almost three times as many reactions are now known.

The coverage of the book is comprehensive, and the reader should be able to discover the salient particulars of a reaction, and especially the quantitative reactivity data, without recourse to the original literature. 'Aromatic' is taken to include not only benzenoid compounds, but also annulenes, metallocenes, and carboranes that undergo electrophilic substitution. Heteroaromatics are however excluded, since these are fully covered in a related volume that the author has recently completed (*Advances in Heterocyclic Chemistry*, Vol. 47, with A. R. Katritzky).

Chapters 1 and 2 provide enough background information to the subject so that any of Chapters 3–10 can then be understood and read in order to acquire more detailed information. Quantitative structure–reactivity relationships are described in Chapter 11. The sequencing of Chapters 3–10 is that of the electrophiles across the Periodic Table, and this sequence is also taken for leaving groups (for a given electrophile) in Chapters 4 and 10. Hydrogen exchange thus becomes the first of these chapters, and this is appropriate because it is the most typical of all electrophilic substitutions, is largely free from steric effects, and more quantitative data are available than for any other substitution. Substituent effects are therefore discussed in some detail in Chapter 3. The chapter on preparative aspects of electrophilic substitution, contained in the original book, has been deleted in deference to reviewer's comments and space limitations.

I would like to thank my wife Joyce for putting up with my extended absences from the family scene, and to Alan Sugar for marketing the budget-cost Amstrad word processor, without which I would surely have been unable to produce this text following the major cut-backs in British higher education.

ROGER TAYLOR
Sussex, April, 1989

Contents

CHAPTER 1

Introduction

1.1 THE STRUCTURE AND STABILITY OF AROMATIC COMPOUNDS

The most commonly found aromatic compound is benzene, first isolated by Faraday in 1825. The need to reconcile its molecular formula (C_6H_6), equivalence of the six carbon atoms, and sixfold symmetry, resulted in Kekulé's brilliant suggestion that benzene consisted of two structures (**1**) and (**2**) in dynamic

(**1**) ⇌ (**2**)

equilibrium.[1] This theory was able to account for the observed structural features of benzene, e.g. that there is only one *ortho*-disubstituted compound, $C_6H_4X_2$, rather than two. It could not account, however, for the differences in properties between benzenoid compounds and the formally similar alkenes, e.g. ethene reacts by *addition* to give 1,2-dibromoethane, whereas benzene reacts only in the presence of a catalyst such as iron fillings, and then undergoes *substitution* (eq. 1.1).

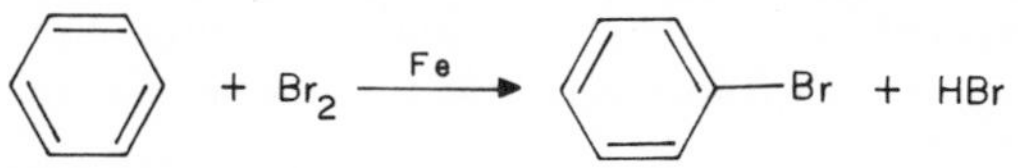

$$C_6H_6 + Br_2 \xrightarrow{Fe} C_6H_5Br + HBr \qquad (1.1)$$

The understanding of the comparative unreactivity of benzene and its tendency to undergo substitution rather than addition came with the development of wave-mechanical theory in the 1920s. These properties were shown to arise from the remarkable thermodynamic stability of the benzene nucleus, which in turn can be understood in terms of current theories of its structure.

Benzene has a very much greater thermodynamic stability than would be expected for a compound possessing three C—C, three C=C, and six C—H bonds. The stability is demonstrated by measurements of heats of hydrogenation: that of the double bond in cyclohexene is 120 kJ mol^{-1} and that of benzene, which formally contains three double bonds, is 209 kJ mol^{-1}.[2] The latter value is 151 kJ mol^{-1} less than three times the former, so this is the amount by which the heat of formation of benzene is greater than the value predicted by comparison with an aliphatic system. This stability may be understood in terms of both the Molecular Orbital and Valence Bond theories.[2,3]

According to the former, benzene consists of a planar ring of sp^2-hybridized carbons; because of the planarity of the molecule, the p-orbitals (each of which contains one electron and is perpendicular to this plane) can overlap laterally (Fig. 1.1) to give extended molecular orbitals in which the six p-electrons are situated in three bonding pairs. It is the delocalization of these pairs of electrons (termed π-electrons) that gives rise to the increased stability of benzene compared with a system containing isolated double bonds.

According to Valence Bond theory, the stability of benzene arises from resonance. Its structure is represented as a hybrid of several canonical structures of which the two (**1** and **2**) considered by Kekulé are the most important, the three Dewar structures of type **3** being less so. The structure of benzene is considered to lie 'in between' these extremes, and its stability is greater than that expected for any of them. The symbol ↔, which represents this intermediate character, does not indicate an oscillation or equilibrium between the structures as envisage by Kekulé. [However, there is very recent theoretical evidence that this may be incorrect, and that the π-electrons are partially localized[4] (as indeed they are for all other aromatics).]

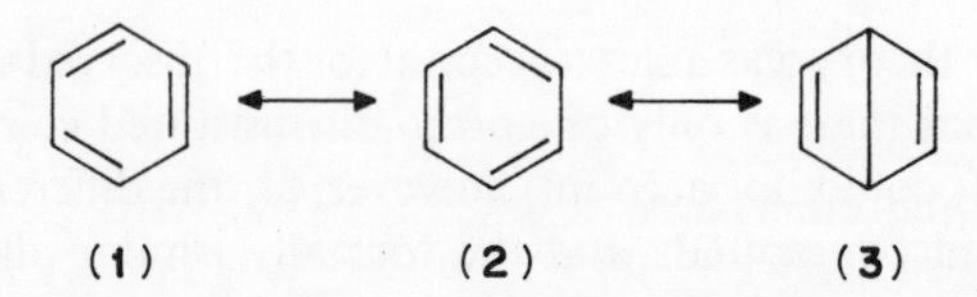

The molecular-orbital description of benzene has the advantage of giving an immediate pictorial idea of the source of the stability of the aromatic ring. On the other hand, it is easier to describe and discuss the reactions of benzenoid compounds in terms of Kekulé structures, and to account semi-quantitatively for the substitution patterns without recourse to complex calculations.

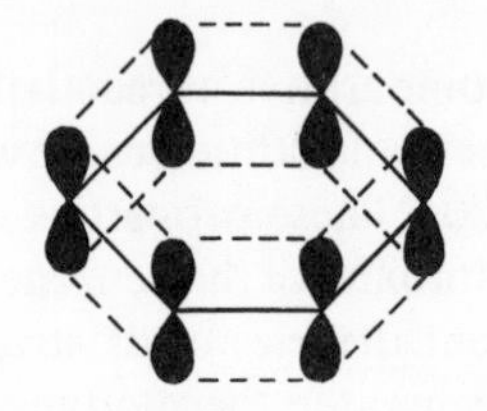

Figure 1.1 p-Orbital overlap in benzene.

The special thermodynamic stability possessed by benzene is termed *delocalization energy* or *resonance energy* and may be approximately equated with the experimentally derived value of 151 kJ mol^{-1}. It may, however, be better to regard the measured value as the *stabilization energy*, since it is not necessarily entirely attributable to resonance.[5] The reactions of benzene follow from the fact that it possesses this considerable stabilization energy.

Addition reactions, which result in the loss of this stability, are rendered energetically unfavourable relative to addition to alkenes. Thus, whereas the addition of 1 mol of hydrogen to benzene is endothermic, the reduction of ethene is exothermic by 137 kJ mol^{-1}.[2] Substitution in benzene does not, by contrast, result in the loss of the stabilization energy. The path of lowest energy for substitution involves first the addition of the reagent X to the nucleus, to give an intermediate which is represented as the hybrid of three canonical structures (**4**–**6**) (Section 2.2.2). Although the stabilization energy of the benzene ring is lost in this initial process (it is offset to some extent by the resonance energy of the resulting carbocation), it is subsequently regained when the proton is lost to give the substitution product.

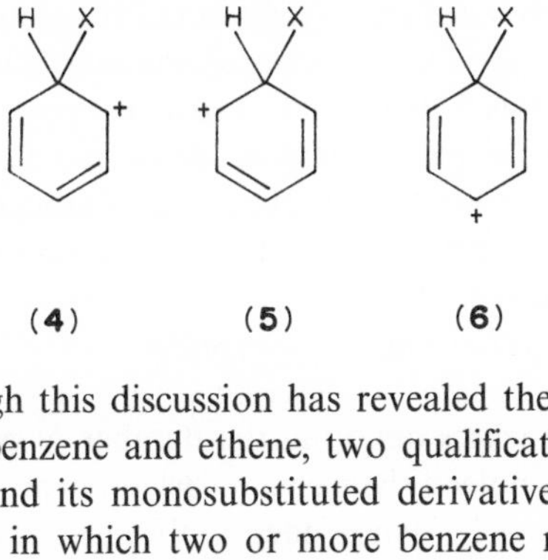

(**4**) (**5**) (**6**) 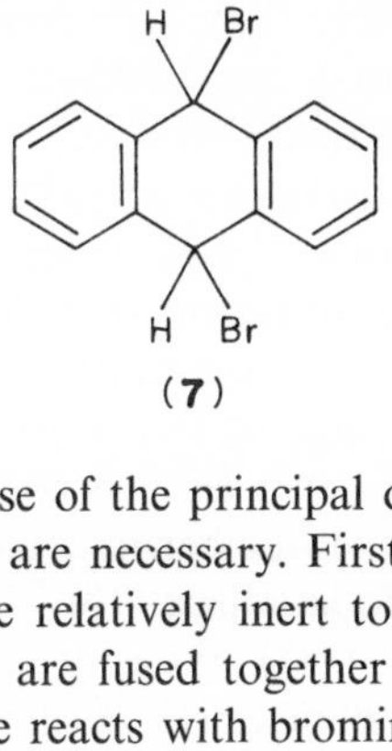 (**7**)

Although this discussion has revealed the cause of the principal differences between benzene and ethene, two qualifications are necessary. First, whereas benzene and its monosubstituted derivatives are relatively inert to addition, aromatics in which two or more benzene rings are fused together are often susceptible to addition. For example, anthracene reacts with bromine to give 9,10-dibromo-9,10-dihydroanthracene (7) which slowly loses hydrogen bromide at room temperature to form the substitution product, 9-bromoanthracene.[6] This is because the stabilization energy of anthracene, 390 kJ mol^{-1},[2] is much less than three times that of benzene: the adduct **7** has a stabilization energy of 302 kJ mol^{-1} derived from two benzene rings, so that the loss of stabilization energy is only about 48 kJ mol^{-1}, which is much less than that lost on addition to benzene.

Second, substituted benzenes are often further substituted very readily, e.g. acetanilide reacts with bromine in the absence of a catalyst to form mainly 4-bromoacetanilide, and aniline is so reactive that it gives 2,4,6-tribromoaniline immediately. This is a result of stabilization by the substituent of the unstable intermediate through which addition occurs. For acetanilide, this is represented by the contribution of **8** to the intermediate, so enabling the positive charge to be delocalized over nitrogen in addition to three of the nuclear carbon atom [cf. **4**–**6**]. Since the rate of formation of these intermediates is related to their stabilities (Section 2.2.2), acetanilide is brominated faster than benzene and no catalyst is required.

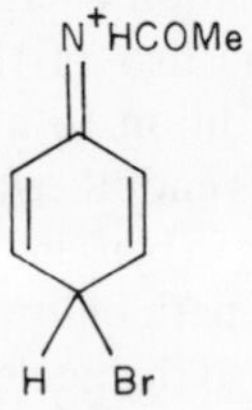

$N^{+}HCOMe$

H Br

(8) 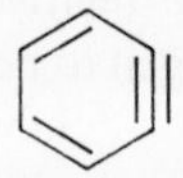 **(9)**

1.2 TYPES OF COMPOUNDS

It is appropriate to define the term 'aromatic' used in the title of this book. Unfortunately, there is no completely satisfactory definition,[7] but the one that is most generally applied describes a compound as being aromatic if it conforms to the Hückel rule. This states that *a planar monocyclic conjugated compound (one containing either sp- or sp^2-hybridized atoms) will be aromatic when the ring contains ($4n + 2$) π-electrons.* The number $4n + 2$ is known as a Hückel number, and the electrons must lie in the same plane. Thus, although benzyne (**9**) contains $4n$ π-electrons, it is nevertheless aromatic since two of these electrons occupy an orbital orthogonal to the orbitals containing the remainder.

The term *annulene* has come to be used to describe completely conjugated polyenes, the number in square brackets denoting the number of π-electrons. Thus, cyclobutadiene (**10**) is [4]annulene, benzene is [6]annulene, and both naphthalene (**11**) and azulene (**12**) are [10]annulenes. Only those annulenes in which this number is a Hückel number will have the particularly stable series of occupied π-molecular orbitals that makes them 'aromatic.' It is necessary, of course, for the adjacent p-orbitals to be able to overlap effectively, and if this is not possible owing to steric constraints (as in **13**) or because the ring is very floppy, i.e. the probability of concurrent overlap of each of the adjacent pairs of p-orbitals is low (as in **14**), then the energy level of the ground state will be relatively high. The molecule will then be both unstable and reactive towards addition. Work has been carried out in recent years to ascertain the extent to which departure from planarity affects the reactivities of annulenes, and benzenoid compounds such as cyclophanes and helicenes, especially in hydrogen exchange (Chapter 3).

Some organometallic compounds are also aromatic. Most notable are the metallocenes such as ferrocene (**15**), in which the outer shell of the metal has a

(10) 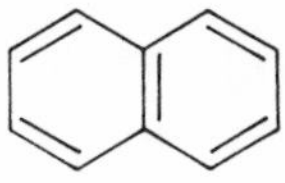 (11) 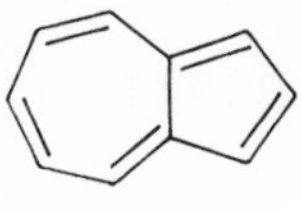 (12) 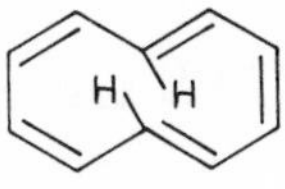 (13) 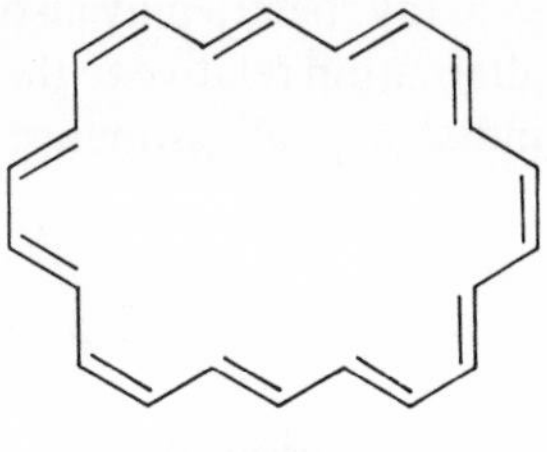 (14)

share of eighteen electrons, and each ring contains six π-electrons, shared over five carbons, making the compound very reactive. Compounds such as benzene chromiumtricarbonyl (**16**) also have a complete outer shell for the metal, with partial donation of electrons from the ring to the metal, so that they are less reactive than benzene.

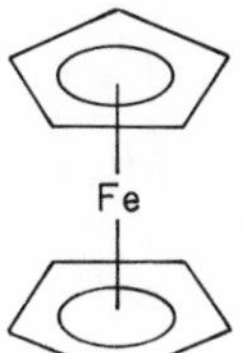

(15) 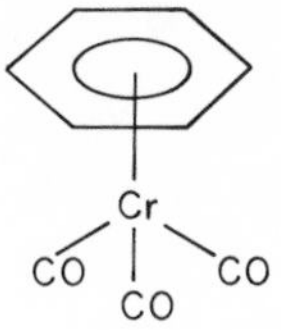 (16) 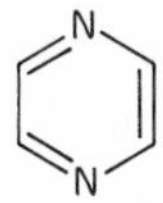 (17) 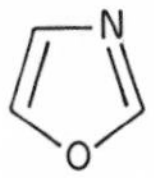 (18)

The definition of aromaticity encompasses a vast range of heterocyclic compounds such as pyrazine (**17**) and oxazole (**18**), but these are excluded from discussion here since their reactions are described fully elsewhere.[8]

Another commonly used index of aromaticity is the ability to exhibit a diamagnetic ring current under the influence of the magnetic field of an n.m.r. spectrometer. This gives rise to a secondary field which opposes the applied field within the aromatic ring and immediately above or below it, and reinforces it outside the ring (**19**). In the resulting spectrum, inner protons therefore appear upfield whereas outer protons appear downfield, as shown in the spectrum of [18]annulene (**20**) obtained at $-60\,°C$.[7]

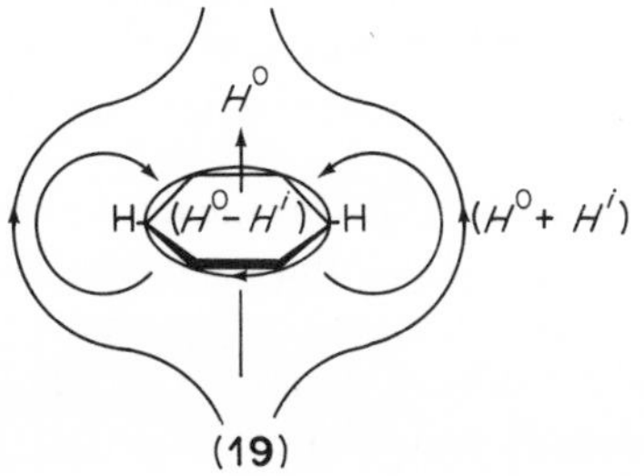

(19) 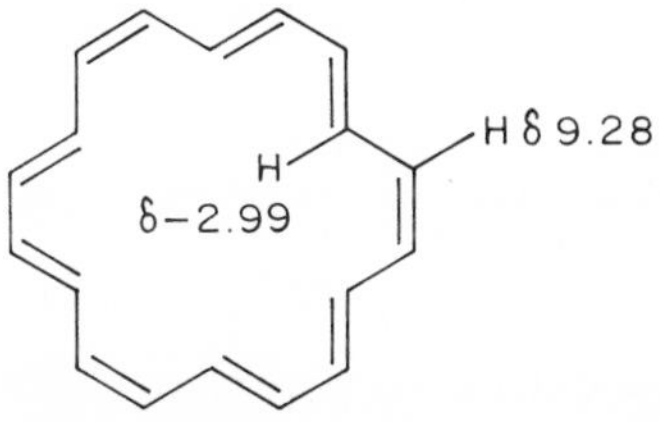 (20)

X-ray spectra may also be used, since a reduction in the amount of bond-length alternation relative to that expected from consideration of a single canonical form indicates that resonance is occurring.

1.3 TYPES OF SUBSTITUTION

Substitution in aromatics may be brought about by three types of reagent: electrophiles, nucleophiles, and free radicals. In electrophilic substitution, the two electrons which form the covalent bond between the aromatic and the reagent are both supplied by the former. Thus, in the nitration of benzene by the nitronium ion, NO_2^+ (Section 7.4), the aromatic ring supplies two electrons to form a covalent bond with nitrogen, causing a temporary electron deficiency, which is accommodated in the π-orbitals of the residual aromatic system (**21**), and reaction is completed by the removal of a proton from the nucleus (eq. 1.2). In this

H NO2 NO2

+ NO2+ ⟶ (+) —H+⟶ (1.2)

(**21**)

example, the reagent is clearly electron-deficient, i.e. electrophilic; it should be noted that examples are known in which the order of addition of the electrophile and removal of the proton is reversed, but the aromatic still provides the bonding electron pair.

Neutral species may also be electrophilic. For example, in the reaction of chlorine with a benzene ring, the new C—Cl bond is formed by the supply of two electrons from the aromatic (eq. 1.3), molecular chlorine behaving as the electrophile because of the ability of one chlorine atom to be displaced as chloride ion. This reaction could equally well be described as the nucleophilic displacement by the aromatic on chlorine, but it is customary to regard attack on carbon as the defining condition.

$$\mathrm{ArH}\ \ \mathrm{Cl{-}Cl} \longrightarrow \mathrm{Ar^{+}}\!\begin{smallmatrix}\mathrm{H}\\ \mathrm{Cl}\end{smallmatrix} \longrightarrow \mathrm{ArCl} + \mathrm{H^{+}} \qquad (1.3)$$

In nucleophilic substitution, two electrons are supplied for the formation of the new bond by the reagent, as in the formation of 2,4-dinitrophenol from 2,4-dinitrochlorobenzene. This proceeds via the intermediate **22** in which the nucleus accommodates a negative charge. Free-radical substitution, an example of which is the phenylation of benzene by the phenyl radical derived from thermal decomposition of dibenzoyl peroxide, proceeds via the intermediate **23**, in which the aromatic accommodates an unpaired electron provided by the attacking free radical.

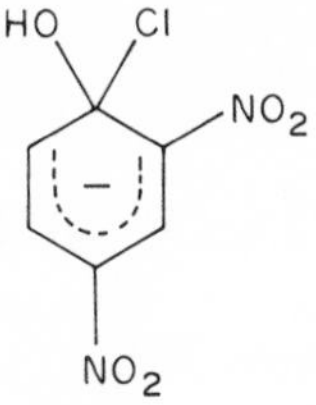

(22) 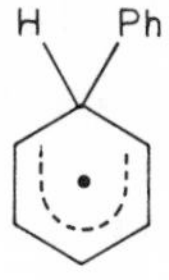 **(23)**

Because hydrogen provides an excellent positive leaving group and is the most common aromatic 'substituent,' electrophilic substitutions have been the most widely studied of the three classes of reaction; they have the advantage over free-radical reactions in being much more controllable. Moreover, reactions may be performed on the parent aromatic, so that electrophilic substitution is often the first reaction in a synthetic sequence involving aromatics.

Electrophilic substitutions in benzene may be generalized as in eq. 1.4, where Y may or may not carry a formal positive charge. Four classes of reaction may be distinguished:

(1) Both X and Y are hydrogen atoms but in different isotopic forms. For example, if X is tritium and Y—Z is an acid (Y = H) the process corresponds to acid-catalysed protiodetritiation.* Such reactions are generally known as hydrogen exchanges (Chapter 3).

(2) $X = H$, $Y \neq H$. This constitutes the substitutions of widest preparative applicability, including Friedel–Crafts reactions, nitration, sulphonation, and halogenation. Y (or Y—Z) may be any of a large number of species some of which are cations such as NO_2^+, ArN_2^+, Hg^{2+}, and others of which are covalent species such as Cl_2, Br_2, and SO_2.

(3) $X \neq H$, $Y = H$. Many substituents may be removed from the aromatic ring by protolytic displacements, e.g., $X = SiMe_3$ (protiodesilylation); these reactions are described in Chapter 4.

(4) $X \neq H$, $Y \neq H$. Since both X and Y may be one of a number of groups, there are a large number of reactions of this type, e.g. $X = B(OH)_2$, $Y—Z = Br_2$

*The method of naming substitution reactions used throughout this book is as follows. The name is composed of the following parts: the name of the incoming group, the syllable 'de,' the name of the departing group, and the suffix 'ation'; suitable elision of vowels or introduction of consonants is sometimes necessary for euphony. The protium isotope of hydrogen is described as 'protio' in reactions in which it is introduced or displaced, although its participation in reactions is often understood, the term 'protio' then being omitted. The term following the syllable 'de' usually identifies, in the leaving substituent, only the atom bonded to the aromatic ring; thus the removal of the $SiMe_3$ and $SiEt_3$ groups are both desilylation reactions.

(bromodeboronation). The known reactions are described in Chapter 10; many more are theoretically possible, but have yet to be investigated.

$$\mathrm{Ph{-}X} + \begin{cases} \mathrm{Y} \\ \mathrm{Y{-}Z} \end{cases} \longrightarrow \mathrm{Ph{-}Y} + \begin{cases} \mathrm{X} \\ \mathrm{X{-}Z} \end{cases} \tag{1.4}$$

1.4 THE POLAR EFFECTS OF SUBSTITUENTS

In reactions on monosubstituted benzenes, the substituent present usually influences strongly both the reactivity of the compound relative to benzene and the proportions of the three possible isomeric substitution products. For example, toluene is nitrated about 25 times faster than benzene, and gives *o*-, *m*-, and *p*-nitrotoluene in the relative proportions 15:1:10 [these figures vary slightly according to the conditions of nitration (Section 7.4)]. By contrast, chlorobenzene reacts about 100 times more slowly than benzene, giving the three nitrochlorobenzenes in the approximate proportions 30:1:70.

In many cases there is evidence that substitution at the *ortho* position is hindered or facilitated by the substituent that is already present, by a steric effect. However, steric effects cannot account for the marked variations in the ease of reaction at *meta* and *para* positions. Since the reactions involve polar reagents, the polar effects of substituents are of importance, and are examined to see how they play a part in determining aromatic reactivities.

Polar effects can be satisfactorily treated by dividing them into two main categories: (a) those which are transmitted either along σ-bonds (the inductive effect) or through space (the direct field effect), and (b) those which are transmitted through π-bonds (the resonance polar effect). This classification has the convenience that in aliphatic systems only the inductive (or field) effect can operate, so that it is possible to establish a scale of these effects and hence, from the total polar effect of given groups in systems which contain both σ- and π-bonds, to evaluate the importance of the resonance effect.

1.4.1 Inductive and Field Effects

The inductive effect depends on the substituent electronegativity whereas the field effect depends on the substituent dipole. They produce similar effects on molecular properties and so are difficult to distinguish from each other;[9,10] consequently, most workers do not attempt to do this. Where inductive effects are referred to subsequently in this book, a contribution from field effects (which may be substantial) is implied.

In a covalently bonded diatomic molecule formed by unlike atoms, e.g. HCl, the bonding electron pair is displaced towards the more electronegative atom.

The resulting non-coincidence of the centres of gravity of the electrons and protons in the molecule gives rise to a dipole moment. If one hydrogen in ethane is replaced by chlorine (which is more electronegative), the electron pair in the C—Cl bond is displaced further from carbon than in the C—H bond which it replaced. This carbon's nucleus is therefore less shielded than that of the carbon in ethane, so it becomes more electronegative. The electron pair of the C—C bond is accordingly displaced towards the carbon to which the chlorine is attached, so the second carbon is slightly affected. Thus the electronegativity of one atom results in permanent displacements of electrons throughout the molecule, the effect on successive atoms falling off with distance from the electronegative centre (**24**). The ability of chlorine to polarize the molecule in this way is described as its inductive effect. The reference point for the inductive effect is hydrogen, and substituents which are electron-attracting or -donating relative to this are designated as $-I$ and $+I$, respectively; the former are in the majority.

$$\overset{\delta\delta\delta+}{CH_2} \longrightarrow \overset{\delta\delta+}{CH_2} \longrightarrow \overset{\delta+}{CH_2} \longrightarrow \overset{\delta-}{Cl}$$

(**24**)

The field effect operates directly through space, and is considered to be of longer range than the inductive effect.[9] Attempts to distinguish the effects have been based on the fact that the field effect depends on the geometry of the molecule, whereas the inductive effect depends only on the nature of the bonds.[9,10] A typical example is the measurement of the pK_a values of **25** and **26**, which turn out to be 5.67 and 6.07, respectively;[11] if inductive effects alone operated, the values should be the same for each. Likewise, the differences in pK_a values and rates of esterification of *cis*- and *trans*-3-substituted propenoic acids imply a field effect.[12] Other experiments have been based on showing the electronic effect to be similar when different numbers of paths for transmission are available, as in **27**–**29**. However, these provide less convincing evidence for the field effect since the substituent will be able to satisfy its inductive requirements using electrons from each path, so the displacement of electrons in each is likely to be reduced in proportion to the number of paths available. Field effects should be of particular importance in the gas-phase owing to the

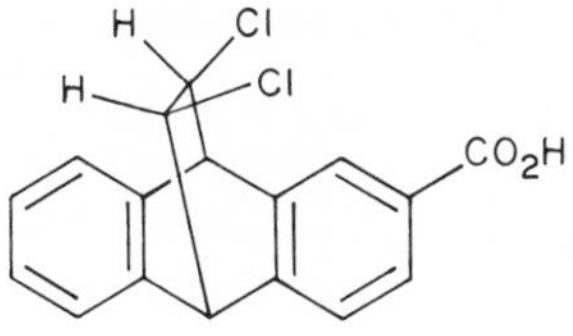

H Cl H Cl CO_2H

(**25**)

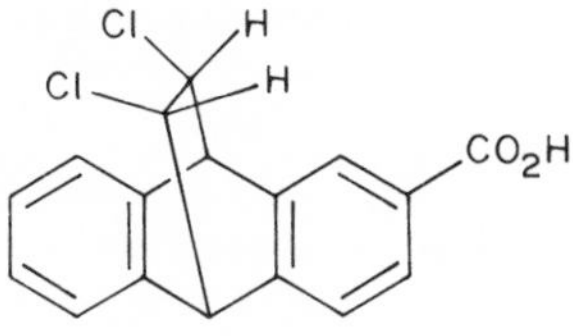

Cl H Cl H CO_2H

(**26**)

(27) (28) (29)

absence of solvent molecules intervening between the substituent and the reaction site.

Inductive (field) effects give rise to dipole moments. For example, MeF, MeOH, and $MeNH_2$ have moments of 1.81, 1.69, and 1.28 D, respectively.[13] Here the magnitudes of the moments are related to the electronegativities of fluorine, oxygen, and nitrogen. In some cases two or more moments can combine to give a zero or very small moment, e.g. *p*-dichlorobenzene has a zero moment, unlike its *ortho* or *meta* isomers, because the two C—Cl dipoles act in opposite directions. Nonetheless, each carbon atom in *p*-dichlorobenzene is positively polarized relative to those in benzene.

Inductive (field) effects are also revealed by examination of the relative strengths of aliphatic acids. The strength of an acid depends on the difference in free energies of the acid and its anion. An electron-attracting substituent near the carboxyl group enables the carboxylate anion to accommodate its negative charge more readily, i.e. it reduces the free energy of the anion relative to that of the acid, thereby increasing the ionization constant.

Some typical data for ionization constants of aliphatic acids at 25 °C are given in Table 1.1.[14,15] The following conclusions may be drawn:

(1) The halogens are electron-attracting, their $-I$ effects decreasing in the order F > Cl > Br > I. As the atomic number increases the nucleus is more strongly shielded by the inner shells of electrons.

(2) The inductive effect of a group falls off rapidly as its distance from the functional centre increases (and does so in free energy terms by a factor 2.5 ± 0.5 per intervening carbon atom).[16]

Table 1.1 Ionization constants ($\times 10^5$) of aliphatic acids $X—CO_2H$

X	CH_3	CH_2F	CH_2F	CH_2Cl	CH_2I
K_a	1.75	217	155	138	75
X	CH_3	CH_2Cl	$(CH_2)_2Cl$	$(CH_2)_3Cl$	
K_a	1.75	155	8.3	3.0	
X	H	CH_3	CH_2CH_3	$CH(CH_3)_2$	$C(CH_3)_3$
K_a	17.7	1.75	1.33	1.38	0.89

Table 1.2 σ_I values of substituents

Substituent	*t*-Bu	Me	H	Ph	NH_2	OMe	COMe	CO_2Me
σ_I	−0.07	−0.05	0	0.10	0.12	0.27	0.27	0.30
Substituent	I	CF_3	Br	Cl	F	CN	NO_2	NMe_3^+
σ_I	0.39	0.41	0.45	0.47	0.50	0.59	0.65	0.86

(3) Alkyl groups are inductively electron-releasing in the order *t*-Bu > *i*-Pr ≈ Et > Me, arising from the number of electropositive hydrogen atoms present; although the three hydrogens in the 'substituent' methyl are further removed from the carboxyl group than the hydrogen atom that this methyl replaces, their number more than compensates for the fall-off factor noted above.

The effects of some more commonly encountered substituents, essentially free from resonance effects and designated σ_I, have been evaluated by a procedure based on the Hammett equation (Section 1.4.4), and are given in Table 1.2. The values are selected from the compilation given in ref. 17; some reactions give slightly different values.

The effect of a given substituent is seen to be governed by the electronegativities of its constituent atoms. The effect of the phenyl substituent derives from the increasing electronegativity of hybridized carbon along the series $sp^3 > sp^2 > sp$.[18] Deuterium is electron-donating relative to protium.[19]

1.4.2 The Resonance Polar Effect

The ionization constants ($10^5 K$) of benzoic acid and its *meta* and *para* methoxy derivatives in water at 25 °C are 6.3, 8.1, and 3.4, respectively.[15] The *meta* derivative is a stronger acid than benzoic acid because OMe has a $-I$ effect, and consequently the *para* derivative should also be a stronger acid; that it is not is due to resonance.

The unshared electron pair in the p-orbital of the oxygen atom is able to overlap with the p-orbital of the *ipso* aromatic carbon, as represented in **30** and **31**. This electron pair acquires greater freedom of motion, with a consequent increase in delocalization energy, but in this process the π-electrons in the vicinity of the *ipso* carbon are repelled. The resulting displacement of these π-electrons (indicated by curved arrows in **31**) affects in turn the distribution of the remaining π-electrons of the aromatic system. Unlike inductive (field) displacement, which increases fairly uniformly with distance, this resonance polar effect gives rise to an *alternation* of polarized carbons, as shown in the valence bond structures **32**–**34**.

A group giving rise to a permanent displacement (polarization) of π-electrons in this way is described as having a resonance, conjugative, or *mesomeric* effect. The methoxy group, being electron-releasing as a result of the resonance effect,

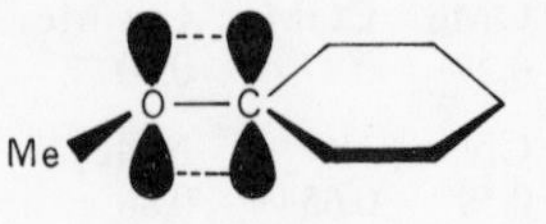

(30) 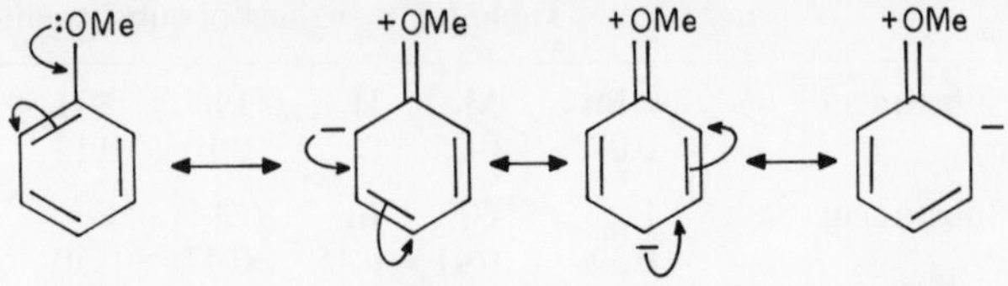 (31) (32) (33) (34)

is said to be a $+M$ substituent; its $+M$ effect opposes its $-I$ effect. The relative ionization constants of benzoic acid and 4-methoxybenzoic acid indicate that the former is more important than the latter, i.e. $+M > -I$. This is due in part to a direct mesomeric interaction between OMe and CO_2H (**35**), which provides an acid-weakening effect, but even when the carboxyl group is insulated from the benzene ring by an intervening methylene group so that such direct interaction is precluded, the $+M$ effect is still of greater consequence than the $-I$ effect, as 4-methoxyphenylacetic acid is weaker than phenylacetic acid.[20] (The conjugative effect is unlikely to be relayed through the CH_2 group here; more probably it raises the electron density at the aromatic 1-carbon and this is then transmitted inductively along the σ-bonds—see also **49**.)

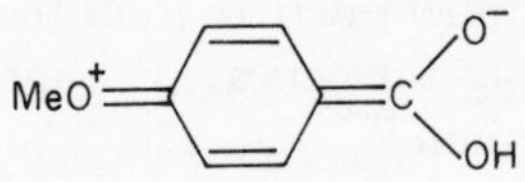

(35) 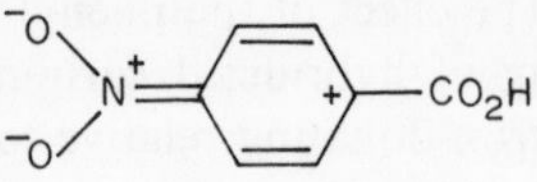 (36) 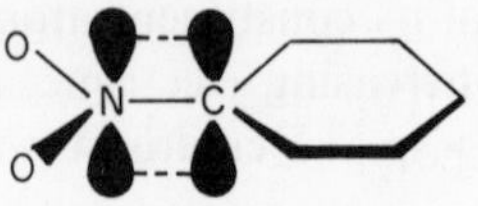 (37)

The operation of resonance may also result in electron withdrawal from the aromatic ring. For example, whereas *m*-NMe_3^+ increases the ionization constant of benzoic acid more than does *p*-NMe_3^+, as expected for a group which can only act inductively,[21] *p*-nitrobenzoic acid is *stronger* than its *meta* isomer.[22] This indicates that **36** contributes to the hybrid, a $-M$ effect here reinforcing the $-I$ effect of the nitro group. The corresponding representation in terms of π-orbital overlap is shown in **37**, and this also demonstrates a steric requirement for resonance to operate: for maximum overlap of the p-orbitals of nitrogen and the adjacent carbon, the nitro group and the benzene ring must be coplanar. When bulky substituents are present in the two *ortho* positions of the aromatic ring, the nitro group is twisted from this coplanar configuration and the mesomeric effect is reduced proportionally to $\cos^2\theta$, where θ is the angle of twist.[23] The *steric inhibition of resonance* is commonly revealed by dipole moment (μ) measurements, as illustrated by examination of **38**–**43**.[13] The mesomeric effect of bromine, represented by the contribution of **44** to the hybrid structure of bromobenzene, is not subject to steric influences, and the similarities of the dipole moments for bromobenzene and 3-bromodurene (**39**) show that the methyl substituents produce no overall effect. The markedly lower dipole

μ (D): 1.52 (38); 1.55 (39); 4.01 (40); 3.62 (41)

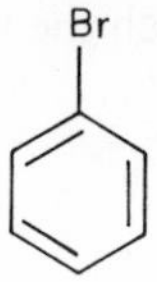

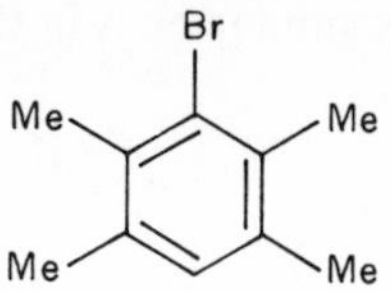

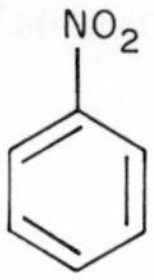

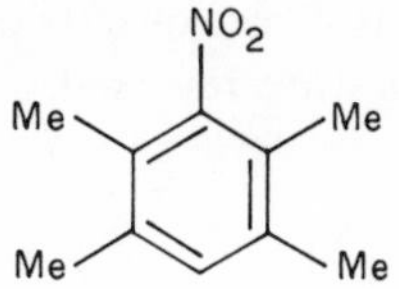

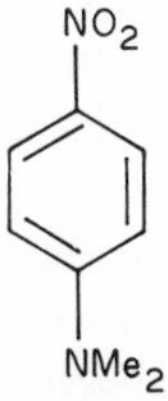

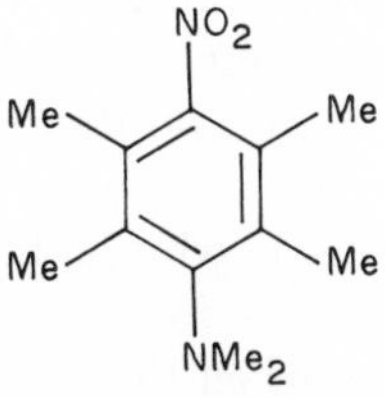

6.87 (42); 4.11 (43)

moment of nitrodurene (**41**) than that of nitrobenzene is therefore attributed to a reduction in the $-M$ effect of the nitro group because this is twisted from the plane. A more pronounced effect is apparent in the pair **42** and **43**. The larger dipole moment of the former is due in part to the contribution of **45** to the resonance hybrid.

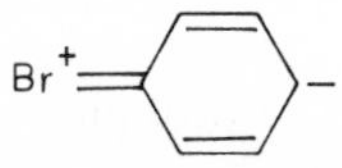

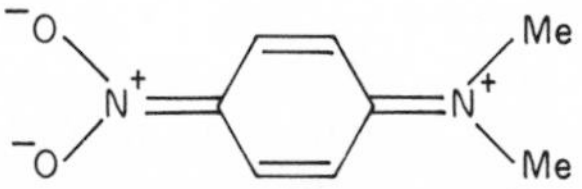

(44) (45)

It is commonly assumed that the nitro in nitrobenzene is coplanar with the ring, i.e. that there is no steric interaction between the oxygens of the nitro group and the adjacent hydrogens. This is unlikely to be correct, as demonstrated by the fact that the nitro group is exceptionally electron-withdrawing in five-membered heterocycles; in these compounds steric hindrance between adjacent groups is less than in benzene because the angles external to the ring are larger. The nitro group will therefore be able to become more coplanar with the ring thereby increasing its $-M$ effect.[24]

The occurrence of resonance does not necessarily result in the appearance of permanent polarization, as symmetrical molecules, e.g. biphenyl, may be resonance stabilized but have no resultant polarity. The molecule is a hybrid of structures such as **46** and other Kekulé forms, and **47** and other dipolar forms. There is no resultant polar effect from this resonance because structures such as **47** are exactly counterbalanced by similar ones in which the signs of the charges are reversed.

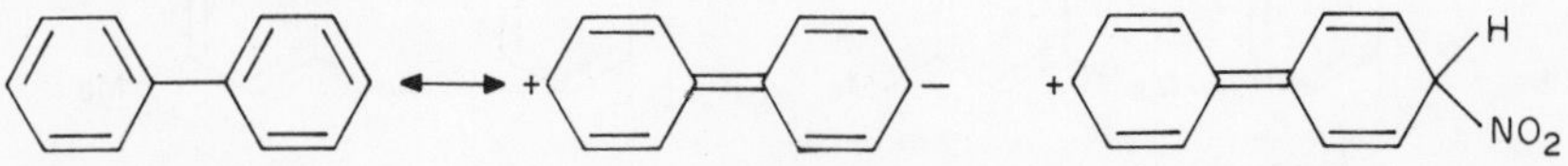

(**46**) (**47**) (**48**)

Although there is no resonance polar effect in biphenyl in the ground state, a polar effect can be brought into operation during the course of reaction. In electrophilic substitutions, the transition state resembles intermediates such as **21**, i.e. a positive charge is accommodated by the aromatic system. In a reaction on biphenyl (at the *ortho* or *para* position), this charge can be better accommodated than in the same reaction on benzene as it can be delocalized over two aromatic rings, e.g. **48**. This is why biphenyl is activated relative to benzene and is *o,p*-directing (see, e.g., Section 3.1.2.10). Since this property of the phenyl substituent is effective only when the reagent begins to bond to the ring, it is termed a *polarizability effect* to distinguish it from the permanent polarization discussed hitherto. The term *electromeric effect* (*E*) was introduced to distinguish such a mesomeric effect from that (*M*) which operates in the unexcited molecule, but is now rarely used; the expression 'time-variable resonance' is also to be found in the literature.

Although this discussion has emphasized the alternating character of the mesomeric effect by which the substituent effect is transmitted specifically to the *ortho* and *para* positions, there is evidence that a group with a powerful mesomeric effect, such as NMe_2, can affect the *meta* position by inductive relay, e.g. **49**.[25] The effect, which is small, may also be represented as arising from the canonical form **50** which, being of high energy, contributes only very little to the overall resonance hybrid.

(**49**) (**50**)

Quantitative values of the resonance polar effects of substituents have been determined in the same way, and on the same scale, as for σ_I values. However, owing to the variable polarizability effect, the values obtained depend on the model reaction chosen. Those obtained under conditions of 'average' polarizability are designated σ_R values, but these are less useful than those obtained under conditions whereby polarizability is minimal, designated $\sigma_R{}^0$; values selected from a recent compilation are given in Table 1.3.[16] The increase in the strength of the $+M$ effect in the order $F < OMe < NMe_2$ is the result of the decrease in the nuclear charge in that order. The order $F > Cl > Br > I$ arises

Table 1.3 σ_R^0 values of *para* substituents

Substituent	NMe_2	OMe	F	Cl	Br	I	*t*-Bu	Ph
	−0.54	−0.43	−0.32	−0.18	−0.16	−0.12	−0.13	−0.11
Substituent	Me	H	CN	CF_3	CO_2Me	COMe	NO_2	
	−0.10	0	0.08	0.10	0.17	0.19	0.19	

from the lateral interaction of p-orbitals that leads to the $+M$ effect, being greatest when the orbitals are of comparable size [see also Section 2.3.(iv)].

π-Electrons in the aromatic ring may be also be polarized by the substituent without actual π-electron transfer between the ring and the substituent. This is referred to as the π-inductive effect, but comprises both inductive and field interactions with the π-electrons of the ring.[10] It is difficult to distinguish this effect from normal resonance effects and there have been few, if any, attempts to do so in electrophilic aromatic substitution.

1.4.3 Hyperconjugation

Hyperconjugation is a mechanism of both electron release and electron withdrawal by certain substituents that far exceeds in importance that encompassed in its original proposal. Recent developments, and in particular those using gas-phase studies, have shown[26] that the traditional description of hyperconjugation and its experimental consequences is wrong, as are the accounts given in most existing textbooks. The following interpretation is consistent with all of the known data.

The $+I$ effects of alkyl groups increase with chain branching (Table 1.1). Thus, for example, the dipole moments of the alkylbenzenes, PhR, in benzene decrease in the order R = *t*-Bu (0.45) > *i*-Pr (0.43) > Et (0.39) > Me (0.37 D).[27] Further, the rates of the base-catalysed hydrolysis of ethyl *p*-alkylbenzoates in 56% aqueous acetone fall in the order H > Me > Et > *i*-Pr > *t*-Bu, as expected for a reaction which occurs through an electron-rich transition state and is therefore retarded by electron-releasing substituents.[28] However, contrary to expectation, the order for the hydrolysis rates of the same esters in 85% aqueous ethanol is H > *t*-Bu > *i*-Pr > Et > Me.[28]

This is one of a large number of examples, such as the solvolysis of phenyl(*p*-alkylphenyl)halomethanes,[29] in which the order of electron release for these alkyl groups is reversed. The reversal of the inductive order was first noted by Baker and Nathan in 1935 in their studies of the nucleophilic substitution of pyridine for bromine in *p*-alkylphenylbromomethanes,[30] and is usually referred to as the Baker–Nathan order. Since it nearly always occurs in the same situations as mesomeric electron release, namely when the substituent is in the *para* position, it was attributed to a form of mesomerism or conjugation in which a C—H bonding electron pair is delocalized.* The phenomenon, given the description *hyperconjugation*,[31] is represented for the *p*-methyl-substituted diarylhalomethane by the structure **51**, the contribution of which to the transition state lowers the free energy of activation for the solvolysis additional to that due to the inductive effect alone. Hyperconjugation is also sometimes referred to as *no-bond resonance* in view of the nature of structure **51**; the latter also shows the need for coplanarity of the C—H sigma bond, and the aryl p-orbital.

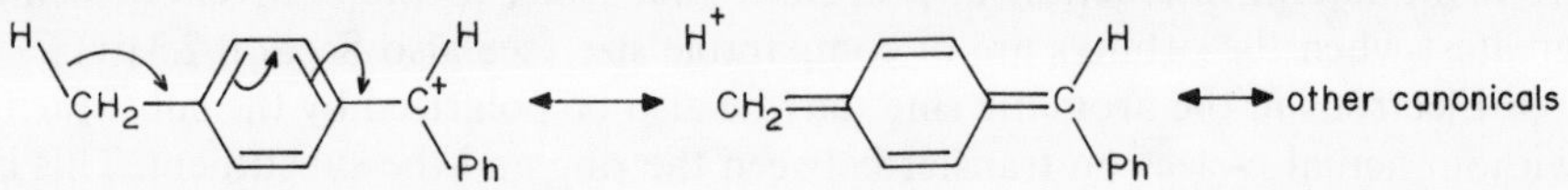

(**51**)

Baker and Nathan proposed (although with unclear reasoning) that as the hydrogen atoms of the methyl group were successively replaced by carbon atoms then the effect would diminish until, for the *tert*-butyl substituent, it disappeared. Subsequent work and calculations (e.g. refs 30 and 31) indicated that the latter assumption might not be correct. For example, the much greater activation of the *ortho* and *para* positions compared with the *meta* position in *tert*-butylbenzene (see, e.g., Table 3.2) showed that there must be substantial conjugative electron release from the *tert*-butyl group.[32] Some importance was therefore ascribed to C–C hyperconjugation which was, however, generally assumed to be less significant than C–H hyperconjugation, despite the fact that the C—H bond is stronger than the C—C bond. The overall electron-releasing order of the alkyl groups was thus believed to arise from the greater importance of C–H relative to C–C hyperconjugation. This is now known to be wrong.

Alternative theories were advanced to account for the occurrence of the Baker–Nathan effect (e.g. ref. 33), the most significant of which ascribed the reversal of the electron-releasing order of the alkyl groups to *steric hindrance to solvation*.[34] Thus the bulkier alkyl groups would more effectively hinder solvation of the charge that is delocalized to adjacent sites in the intermediate corresponding to the transition states for reaction at the *para* positions of the alkylaromatics (**52**); this would diminish the overall observed reactivity and hence the *apparent* electron release by the bulkier alkyl groups. Arguments then centred on whether hyperconjugation was a real phenomenon, or merely a manifestation of these solvation effects. It is now realized that both theories are partially correct, and it was the assumption that each of these alternatives was mutually exclusive that prevented a correct interpretation for four decades.

*The Baker–Nathan order has occasionally been observed when the alkyl substituent is in the *meta* position, e.g. in mercuriation (Table 5.2).

The concept of hyperconjugation is now supported overwhelmingly by evidence accumulated from widely differing areas of chemistry, the most conclusive and unambiguous of which is the conformational dependence of the electron-releasing ability of the substituent σ-bond (e.g. ref. 35). Moreover, the more reliable theoretical calculations now available have shown that the conjugative electron-releasing order of the alkyl groups should be the *same* as the inductive order,[36] and this has been confirmed by measurements of the electron-releasing abilities of alkyl groups in the gas phase.[37–40] Thus C–C hyperconjugation, as in structure **53**, is *more important* than C–H hyperconjugation (structure **51**); most texts contain the converse statement. It is thus

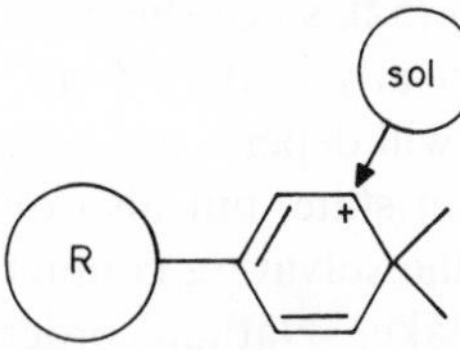

(**52a**) 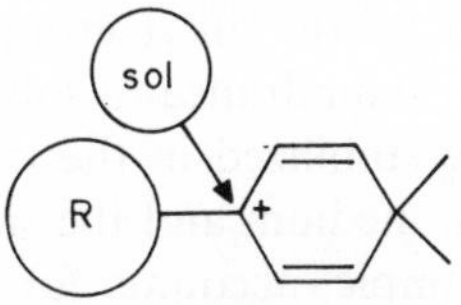 (**52b**) 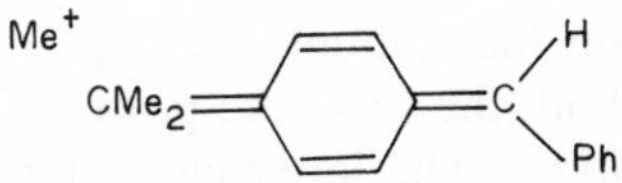 (**53**)

a paradox of physical–organic chemistry that Baker and Nathan proposed the important concept of hyperconjugation, but for the wrong reason, the reactivity order upon which they based their proposal being due merely to a solvent effect superimposed upon the *p*-*t*-Bu > *p*-Me electron-releasing order produced by the combined inductive and hyperconjugative mechanisms.

It should be noted that when alkyl groups are attached to unsaturated systems, ca 80% of their electron release is provided by hyperconjugation;[38] this contrasts with a lingering view that alkyl groups are mainly inductively electron releasing. The proportion may be even greater in situations where either steric strain exists within a molecule and is relieved through hyperconjugation,[40–42] or where the conformation of a substituent is such as to facilitate overlap between the appropriate C—H or C—C bond of the substituent and the adjacent p-orbital of the aromatic.[41,43] This *steric facilitation of hyperconjugation* has only recently been recognized as a factor producing enhanced electron release from some alkyl groups. Conversely, conjugative electron release may be diminished where overlap is sterically hindered (*steric hindrance to hyperconjugation*); both factors are discussed further in Section 3.1.2.1.

There are two other consequences of steric hindrance to solvation. The first is that for a given pair of alkyl groups, variation in *meta* reactivity ratios with change in conditions may be comparable to, or even greater than, the variation in *para* reactivity ratios.[39] Generally this has not been recognized because the inductive order is more marked at the *meta* position than at the *para* position, so trends towards a reversal of this order are less apparent. It arises because solvation of the intermediate corresponding to the transition state for *meta*

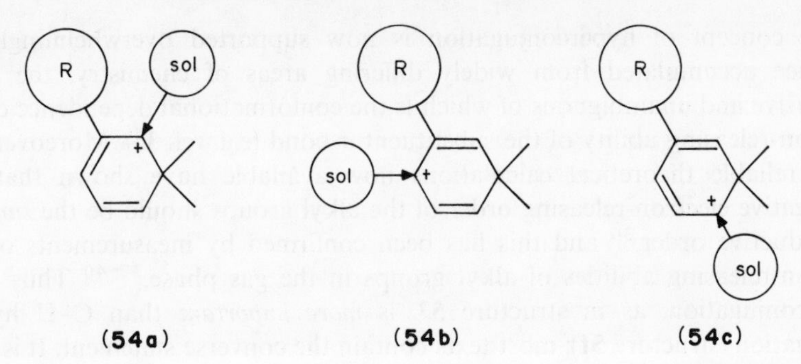

(**54a**) (**54b**) (**54c**)

substitution (**54**) can be more hindered than that for *para* (**52**), since there are two sites that can be hindered by the alkyl group, especially that in **54a**.[39] Secondly, the importance of steric hindrance to solvation will depend not only on the amount of charge to be stabilized in the transition state, but also on both the solvating power of the medium and the *size* of the solvating counter ion.[26] The latter effect, for example, accounts for the 'Baker–Nathan' order being obtained in a poorly solvating superacid medium, which contains a very large solvating counter ion, SbF_6^- [26] (cf. ref. 44).

The foregoing discussion describes how alkyl groups can be polarized to release electrons. A further mode of electronic behaviour should be mentioned, although it does not affect electrophilic aromatic substitution. When attached to *saturated* systems, alkyl groups appear to be able to polarize so as to *accept* electrons, the acceptance being greater the larger the group.[45]

Although hyperconjugation was originally proposed to account for conjugative electron release from the C—H bond, many other bonds, e.g. O—H, N—H, C—Si, and C—Hg, are able to release electrons by such a mechanism, and some of them do so extremely strongly;[46] in general, the weaker the bond the stronger is the electron release.

F CF_2—C_6H_4—CH_2^- ⟷ F^- CF_2=C_6H_4=CH_2 ⟷ other canonicals

(**55**)

It follows that since hyperconjugation occurs when carbon is bound to more electropositive elements of groups, the reverse process, *negative* or *reverse hyperconjugation*,[47] can occur when carbon is bound to very electronegative elements such as fluorine. Structure **55** illustrates the stabilization of a benzyltrifluoromethylbenzyl anion by C–F negative hyperconjugation.

Just as the order of electron release by hyperconjugation is C–Me > C–H, so the order of electron withdrawal by negative hyperconjugation are conformationally dependent,[48] the need for coplanarity of the sigma bond and the adjacent p-orbitals further confirming the nature of the process.

1.4.4 The Hammett Equation

The polar effects described above influence rates of reaction and it is in this context that they are of relevance to aromatic substitutions. Their influence on reaction rates of aromatic compounds are closely related to their effects in determining acid strengths, to the extent that there are quantitative relationships between structure and reactivity.

These relationships hold when the substituents are in the *meta* and *para* positions. Hammett discovered that a graph of the logarithms of the rate coefficients for the base-catalysed hydrolysis of the esters of *meta*- and *para*-substituted benzoic acids against the logarithms of the ionization constants of the corresponding acids is linear.[49] The straight line may be described by eq. 1.5, where k is the rate coefficient for the hydrolysis of a given ester, K is the ionization constant of the corresponding acid and ρ and A are constants. Straight-line plots are also obtained when rate coefficients for numerous other reaction series such as the acid-catalysed hydrolysis of benzamides, the solvolysis of benzoyl chlorides in ethanol, and the alkaline hydrolysis of ethyl 3-phenylpropenoates are plotted against each other or against the ionization constants in the manner of eq. 1.5.

$$\log k = \rho \log K + A \qquad (1.5)$$

$$\log k_o = \rho \log K_o + A \qquad (1.6)$$

$$\log (k/k_o) = \rho \log (K/K_o) = \rho\sigma \qquad (1.7)$$

The point for the unsubstituted compound is also fitted by the straight line according to eq. 1.6, where k_o is the rate coefficient for the unsubstituted ester and K_o is the ionization constant for benzoic acid. Subtraction gives eq. 1.7, where σ is defined as $\log (K/K_o)$. It is apparent that σ (Greek 's') depends only on the nature of the substituent, and is designated the *substituent constant*; ρ (Greek 'r') is the gradient of the linear plot, and varies with the nature of the reactions but is independent of the nature of the substituent, so is termed the *reaction constant*.* It is now possible to establish a scale of polar effects for *meta* and *para* substituents derived from the ionization constants of benzoic acids. Typical values† are given in Table 1.4 (several reviews of the Hammett equation have been published,[17,22,50] each of which contains compilations of σ values).

Equation 1.7, known as the *Hammett equation*, holds for a very large number of reactions of benzenoid compounds with *meta* or *para* substituents,[17,50] and it

*In eqs 1.5–1.7 ρ is strictly the ratio of the reaction constants for reaction of the ester to that for ionization of the acids. The latter is arbitrarily defined as 1.0 for dissociation in water at 25 °C, and so all derived ρ factors are relative to this standard value.

†It is emphasized that one must use $\log_{10}$ values and not $\log_e$ (ln) values when calculating rate data based on substituent constants.

Table 1.4 σ values of aromatic substituents

Constant	Substituent								
	Me	*t*-Bu	OMe	F	Cl	Br	I	CN	NO_2
σ_m	−0.07	−0.10	0.12	0.34	0.37	0.39	0.35	0.56	0.71
σ_p	−0.17	−0.20	−0.27	0.06	0.23	0.23	0.21	0.66	0.78

is instructive to examine the operation of the equation in one example. Consider the base-catalysed hydrolysis of ethyl benzoate and the ionization of benzoic acid. Withdrawal of electrons from the side-chain carbon atom stabilizes both the transition state for the hydrolysis (**56**) and the product of the ionization (**57**), and therefore an increase in k and K compared with k_o and K_o. This withdrawal may be brought about, for example, by a *m*-nitro group by virtue of its $-I$ effect, or by a *p*-nitro group by virtue of its $-I$ and $-M$ effects. The values for *m*-NO_2 (0.71) and *p*-NO_2 (0.78) indicate that the $-M$ effect of *p*-NO_2 makes it a more strongly electron-withdrawing substituent than *m*-NO_2.* In general, a positive σ-value indicates that the substituent is electron withdrawing, whereas a negative σ-value indicates that it is electron donating.

(**56**) (**57**)

In other reactions the reaction centre may become electron deficient in the transition state (or product if the process is an equilibrium). In that case, an electron-withdrawing group reduces k (or K) relative to k_o (or K_o) and this results in a Hammett plot of negative gradient (i.e. ρ is negative). Hence the sign of the ρ value of a reaction gives information about the electronic changes at

*σ Values can be dissected by means of eqs 1.8 and 1.9 into the parameters corresponding to the inductive and resonance components, i.e. σ_I and σ_R values, respectively.[51] The value of α is 0.33, showing the resonance effect to be relayed one third as effectively to the *meta* as to the *para* position.

$$\sigma_P = \sigma_I + \sigma_R \tag{1.8}$$

$$\sigma_m = \sigma_I + \alpha\sigma_R \tag{1.9}$$

Similar equations may be set up using σ_o values (Section 11.1.2) and, since these describe a smaller resonance contribution, the differential effect between *meta* and *para* positions becomes smaller, i.e. α is 0.5. Numerous variations on these equations have been proposed but it is inappropriate to consider them here.

the functional centre during the process, while the magnitude of ρ measures the sensitivity of the reaction to changes in the substituent. A positive ρ factor indicates that the reaction transition state is negatively charged, whereas a negative ρ factor indicates that the transition state is positively charged (as it is in electrophilic aromatic substitution).

The Hammett equation is usually not obeyed by *ortho*-substituted aromatics or, in its simple form, by substituted aliphatic compounds. Successful application of the equation to these systems would require each substituent to have a steric effect that was independent of the nature of the reaction, and would also require that the sensitivity of the reaction to these steric effects was proportional to its sensitivity to polar effects. Nevertheless, extensions of the Hammett equation have been examined which contain terms to allow for steric effects so that the overall effect of a substituent is divided into a polar and a steric component.[52]

In the form of eq. 1.7, the Hammett equation fails to correlate the rates of reactions in which a charge is developed adjacent to the aromatic ring in the transition state. This is because there can then be *direct* conjugation between the substituent and the charged centre, and thus the mesomeric component of the substituent is involved to a much greater extent than is possible in **56** or **57**. For example, in the ionization of phenols the negative charge on the phenate oxygen can conjugate directly with a *p*-nitro group, as in **58**. The nitro group therefore increases the acidity of phenols considerably more than is predicted from its effect on the ionization of benzoic acids. A special 'exalted' σ value (1.27) for *p*-NO_2 is then required in this reaction (and also in correlating the basicities of aromatic amines, where similar arguments apply). The set of values that are needed here (and also to correlate rates of nucleophilic aromatic substitutions for the same reason) are known as σ^- values; these differ significantly from σ values only for $-M$ substituents (a comprehensive compilation of σ^- values is given in ref. 17).

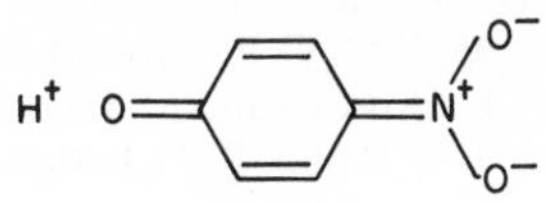

H^+ O=⟨ring⟩=$N^+(O^-)_2$ (58) MeO$^+$=⟨ring⟩=C(Me)Me Cl^- (59)

(58) 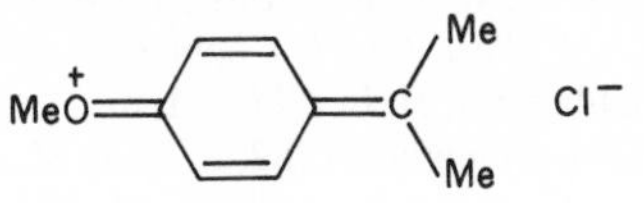 **(59)**

Conversely, $+M$ substituents fail to correlate the rates of reactions in which a positive charge is developed at the side chain α-position, because here direct conjugative electron release from the substituent to the reaction centre is possible. For example, in S_N1 nucleophilic substitution of 2-aryl-2-chloropropanes an 'exalted' σ value of -0.78 is required for *p*-OMe owing to the additional interaction giving rise to **59**. Electrophilic aromatic substitutions are correlated by these special values, designated σ^+, which differ substantially from σ values only for $+M$ substituents. This treatment is discussed fully in Chapter 11.

1.4.5 Conclusions

The foregoing discussion has shown, first, the ways in which a substituent attached to an aromatic compound can determine the electron density at the nuclear carbon atoms, and second, that the substituents can be placed in an order which describes their relative effects in governing the rates of certain reactions. A modified quantitative scale of substituent effects can be used to describe the relative rates of electrophilic substitution, and this is of great assistance in understanding the relative reactivities and orientations in these substitutions, which are discussed in the ensuing chapters.

REFERENCES

1. A. Kekulé, *Justus Liebigs Ann. Chem.*, 1872, **162**, 77.
2. G. W. Wheland, *Resonance in Organic Chemistry*, Wiley, New York, 1955.
3. C. A. Coulson, *Valence*, Oxford, London, 1952.
4. D. L. Cooper, J. Gerratt, and M. Raimondi, *Nature (London)*, 1986, **323**, 699.
5. M. J. S. Dewar and H. N. Schmeising, *Tetrahedron*, 1959, **5**, 166.
6. E. de B. Barnett and J. W. Cook, *J. Chem. Soc.*, 1924, 1084.
7. A full account of aromaticity and leading references may be found in P. J. Garrett, *Aromaticity*, Wiley, New York, 1986.
8. A. R. Katritzky and R. Taylor, *Adv. Heterocycl. Chem.*, Vol. 47, in press.
9. L. M. Stock, *J. Chem. Educ.*, 1972, **49**, 400, and references cited therein; A. R. Butler, *J. Chem. Soc. B*, 1970, 867; W. Adcock, P. D. Bettess, and S. Q. A. Rizvi, *Aust. J. Chem.*, 1970, **23**, 1921; R. Taylor, *J. Chem. Soc. B*, 1971, 622, 1450; T. W. Cole, C. J. Mayers, and L. M. Stock, *J. Am. Chem. Soc.*, 1974, **96**, 4555; J. H. Rees, J. H. Ridd, and A. Ricci, *J. Chem. Soc., Perkin Trans. 2*, 1976, 294; R. D. Topsom, *Prog. Phys. Org. Chem.*, 1976, **12**, 1; *J. Am. Chem. Soc.*, 1981; **103**, 39; C. A. Grob, A. Kaiser, and T. Schweizer, *Helv. Chim. Acta*, 1977, **60**, 391; J. H. P. Utley and S. O. Yeboah, *J. Chem. Soc., Perkin Trans. 2*, 1978, 766; A. J. Hofnagel, M. A. Hofnagel, and B. M. Wepster, *J. Org. Chem.*, 1978, **43**, 4720; W. F. Reynolds, *J. Chem. Soc., Perkin Trans. 2*, 1980, 985; O. Exner and P. Fiedler, *Collect. Czech. Chem. Commun.*, 1980, **45**, 1251; C. A. Grob, B. Schaub, and M. G. Schlageter, *Helv. Chim. Acta*, 1980, **63**, 57; K. Bowden and M. Hojatti, *J. Chem. Soc., Chem. Commun.*, 1982, 273; A. M. Aissani, J. C. Baum, R. F. Lengler, and J. L. Ginsberg, *Can. J. Chem.*, 1986, **64**, 532.
10. W. F. Reynolds, *Prog. Phys. Org. Chem.*, 1983, **14**, 165; R. W. Taft and R. D. Topsom, *Prog. Phys. Org. Chem.*, 1987, **16**, 1; R. D. Topsom, *Prog. Phys. Org. Chem.*, 1987, **16**, 125.
11. E. J. Grubbs, R. Fitzgerald, R. E. Phillips, and R. Petty, *Tetrahedron*, 1971, **27**, 935.
12. K. Bowden, *Can. J. Chem.*, 1965, **43**, 3354.
13. J. W. Smith, *Electric Dipole Moments*, Butterworths, London, 1955.
14. J. F. Dippy, *Chem. Rev.*, 1939, **25**, 151.
15. H. C. Brown, D. H. McDaniel, and O. Häflinger, *Determination of Organic Structures by Physical Methods*, Vol. 1, Academic Press, New York, 1956, p. 567.
16. R. W. Taft and I. C. Lewis, *J. Am. Chem. Soc.*, 1958, **80**, 2436.
17. O. Exner, in *Correlation Analysis in Organic Chemistry*, Eds. N. B. Chapman and J. Shorter, Plenum, New York, 1978, pp. 439–540.

18. H. A. Bent, *Chem. Rev.*, 1961, **61**, 275.
19. A. Streitwieser and H. S. Klein, *J. Am. Chem. Soc.*, 1963, **85**, 2759.
20. A. Fischer, B. R. Mann, and J. Vaughan, *J. Chem. Soc.*, 1961, 1093.
21. J. D. Roberts, R. A. Clement, and J. J. Drysdale, *J. Am. Chem. Soc.*, 1951, **73**, 2181.
22. D. H. McDaniel and H. C. Brown, *J. Org. Chem.*, 1958, **23**, 420.
23. M. J. S. Dewar, *J. Am. Chem. Soc.*, 1952, **74**, 3341, 3345.
24. R. Taylor, *Chemistry of Heterocyclic Compounds, Vol. 44, Thiophene and its Derivatives, Part 2*, Ed. S. Gronowitz, Wiley, New York, 1986, pp. 15, 26.
25. C. C. Price and D. C. Lincoln, *J. Am. Chem. Soc.*, 1951, **73**, 5838.
26. R. Taylor, *J. Chem. Res.* (*S*), 1985, 318.
27. T. L. Brown, *J. Am. Chem. Soc.*, 1959, **81**, 3232.
28. E. Berliner, M. C. Becketts, E. A. Blommers, and B. Newman, *J. Am. Chem. Soc.*, 1952, **74**, 4940.
29. E. D. Hughes, C. K. Ingold, and N. A. Taher, *J. Chem. Soc.*, 1940, 949.
30. J. W. Baker and W. S. Nathan, *J. Chem. Soc.*, 1935, 1844.
31. R. S. Mulliken, *J. Chem. Phys.*, 1939, **7**, 339; R. S. Mulliken, C. A. Rieke, and W. G. Brown, *J. Am. Chem. Soc.*, 1941, **63**, 41.
32. Conference on Hyperconjugation, Bloomington, IN, 1958, reported in *Tetrahedron*, 1959, **5**, 105.
33. E. Berliner and F. J. Bondus, *J. Am. Chem. Soc.*, 1948, **70**, 854.
33. W. M. Schubert and W. A. Sweeney, *J. Org. Chem.*, 1956, **21**, 119.
34. V. J. Shiner, *J. Am. Chem. Soc.*, 1960, **82**, 2655; C. G. Pitt, *J. Organomet. Chem.*, 1970, **C35**, 23; W. Hanstein, H. J. Berwin, and T. G. Traylor, *J. Am. Chem. Soc.*, 1970, **92**, 7476.
35. L. Radom, J. A. Pople, and P. von R. Schleyer, *J. Am. Chem. Soc.*, 1972, **94**, 5935; L. Radom, *Aust. J. Chem.*, 1974, **27**, 231; 1975, **28**, 1.
36. F. P. Lossing and G. Semeluk, *Can. J. Chem.*, 1970, **48**, 955.
37. W. J. Herre, R. T. McIver, J. A. Pople, and P. von R. Schleyer, *J. Am. Chem. Soc.*, 1974, **96**, 7162.
38. E. Glyde and R. Taylor, *J. Chem. Soc., Perkin Trans. 2*, 1977, 678.
39. W. J. Archer, M. A. Hossaini, and R. Taylor, *J. Chem. Soc., Perkin Trans. 2*, 1982, 181; M. A. Hossaini and R. Taylor, *J. Chem. Soc., Perkin Trans. 2*, 1982, 187.
40. H. V. Ansell and R. Taylor, *Tetrahedron Lett.*, 1971, 4915.
41. F. R. Jensen and B. E. Smart, *J. Am. Chem. Soc.*, 1969, **91**, 5686.
42. M. M. J. LeGuen and R. Taylor, *J. Chem. Soc., Perkin Trans. 2*, 1976, 559.
44. H. C. Brown, and M. Periasamy, and P. T. Perumal, *J. Org. Chem.*, 1984, **49**, 2754.
45. H. Kwart and T. Takeshita, *J. Am. Chem. Soc.*, 1964, **86**, 1161; R. C. Fort and P. von R. Schleyer, *J. Am. Chem. Soc.*, 1964, **86**, 4194; V. W. Laurie and J. S. Muenter, *J. Am. Chem. Soc.*, 1966, **88**, 2883; J. I. Brauman and L. K. Blair, *J. Am. Chem. Soc.*, 1971, **93**, 4315; J. E. Bartmess and R. T. McIver, *J. Am. Chem. Soc.* 1977, **99**, 6046; J. E. Bartmess, J. A. Scott, and R. T. McIver, *J. Am. Chem. Soc.*, 1979, **101**, 6046.
46. P. B. D. de la Mare, *Tetrahedron*, 1959, **5**, 107; *Pure Appl. Chem.*, 1984, **56**, 1755; C. Eaborn and S. H. Parker, *J. Chem. Soc.*, 1954, 939; C. Eaborn, *J. Chem. Soc.*, 1956, 4858; R. W. Bott, C. Eaborn, and P. M. Greasley, *J. Chem. Soc.*, 1964, 4804; W. Hanstein and T. G. Traylor, *Tetrahedron Lett.*, 1967, 4451.
47. L. O. Brockway, *J. Chem. Phys.*, 1937, **41**, 185, 747.
48. J. G. Stamper and R. Taylor, *J. Chem. Res.* (*S*), 1980, 128; (M) 2001.
49. L. P. Hammett, *J. Am. Chem. Soc.*, 1937, **59**, 96.
50. L. P. Hammett, *Physical Organic Chemistry*, 2nd edn, McGraw-Hill, New York, 1970; H. H. Jaffé, *Chem. Rev.*, 1953, **53**, 191; P. R. Wells, *Chem. Rev.*, 1963, **63**, 171; C. D. Johnson, *The Hammett Equation*, Cambridge University Press, Cambridge, 1973; J.

Shorter, *Correlation Analysis in Organic Chemistry*, Clarendon Press, Oxford, 1973.
51. R. W. Taft and I. C. Lewis, *J. Am. Chem. Soc.*, 1959, **80**, 2436.
52. R. W. Taft, in *Steric Effects in Organic Chemistry*, Ed. M. S. Newman, Wiley, New York, 1956, Ch. 13; R. Gallo, *Prog. Phys. Org. Chem.*, 1983, **14**, 115; S. H. Unger and C. Hansch, *Prog. Phys. Org. Chem.*, 1976, **12**, 91; M. Charton, *J. Am. Chem. Soc.*, 1975, **97**, 1552; *J. Org. Chem.*, 1976, **41**, 2217.

CHAPTER 2

The Mechanism of Electrophilic Aromatic Substitution

2.1 REACTION OF THE REAGENT WITH THE AROMATIC NUCLEUS

There are four ways in which an aromatic compound might react with an electrophile, differing with respect to whether the replaced group (commonly the proton) leaves either before, during, or after attack by the electrophile.

(1) If the proton leaves first, formation of the aromatic anion takes place in a rate-determining step and this is followed by rapid reaction of the anion with the electrophile (eq. 2.1). This mechanism is analogous to the S_N1 reaction of aliphatic compounds and is designated S_E1.

$$\mathrm{ArH} \underset{k_{-1}}{\overset{k_1}{\rightleftharpoons}} \mathrm{Ar^-} + \mathrm{H^+} \xrightarrow[k_2]{\mathrm{E^+}} \mathrm{ArE} \tag{2.1}$$

(2) A more probable alternative to mechanism (1) involves base catalysis of the rate-determining ionization which forms the aromatic anion, the latter reacting rapidly with the electrophile as before. This mechanism (eq. 2.2), is designated B-S_E1; groups other than a proton may be removed by the base.

$$\mathrm{B^-} + \mathrm{ArH} \underset{k_{-1}}{\overset{k_1}{\rightleftharpoons}} \mathrm{BH} + \mathrm{Ar^-} \xrightarrow[k_2]{\mathrm{E^+}} \mathrm{ArE} \tag{2.2}$$

(3) By synchronous formation of the C—E bond and cleavage of the C—H bond. In the transition state (**1**) both incoming and leaving groups would be partially bonded to the nucleus (eq. 2.3). The mechanism would correspond to the S_N2 reactions of aliphatic compounds, although it would necessarily have a different stereochemistry. In the event that a base would be required simultaneously to aid removal of the proton, the reaction would be termolecular and designated S_E3.

$$E^+ + ArH \longrightarrow Ar \begin{smallmatrix} \cdots H^{\delta+} \\ \cdots E^{\delta+} \end{smallmatrix} \longrightarrow ArE + H^+ \qquad (2.3)$$

(1)

(4) By addition of the electrophile to the nucleus followed by loss of a proton from the adduct (eq. 2.4), an intermediate σ-complex (**2**) being formed. In this two-step process either (a) the addition or (b) the expulsion step might be rate determining. The process is designated the S_E2 mechanism and in those reactions in which the electrophile is a proton the reaction is acid catalysed and is designated A-S_E2.

$$E^+ + ArH \underset{k_{-1}}{\overset{k_1}{\rightleftharpoons}} Ar^+ \begin{smallmatrix} H \\ E \end{smallmatrix} \xrightarrow{k_2(+B^-)} ArE + H^+(BH) \qquad (2.4)$$

(2)

In all cases for which evidence is available, the reaction follows mechanism 2.4. In addition there are six reactions, viz. hydrogen exchange (Chapter 3), protiodesilylation, protiodegermylation, and protiodestannylation (Chapter 4), lithiation (Chapter 5), and bromination (Chapter 9) which can also take place under base-catalysed conditions, and for these mechanism 2.2 applies (although in some cases the first and second steps are synchronous). The evidence for the mechanisms derives from two types of experiments: (a) the isolation and characterization of intermediates and (b) kinetic studies, including measurements of kinetic isotope effects.

2.1.1 Isolation of Intermediates

Trifluoromethylbenzene reacts with nitryl fluoride and boron trifluoride at low temperatures to give a coloured crystalline complex of mole ratio 1:1:1 which is stable below − 50 °C, and which above that temperature gives boron trifluoride, hydrogen fluoride, and *m*-nitrotrifluoromethylbenzene.[1] Structure **3** was assigned to the complex,[1] and since *m*-nitrotrifluoromethylbenzene is the product of the nitration of trifluoromethylbenzene under ordinary conditions, this is convincing evidence that the intermediate in the nitration is **3**. Analogous

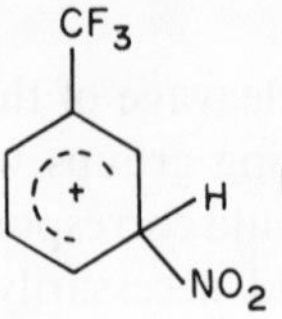

(**3**) 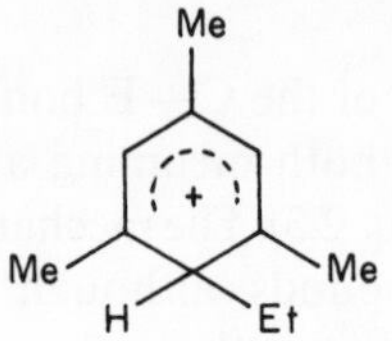 (**4**) 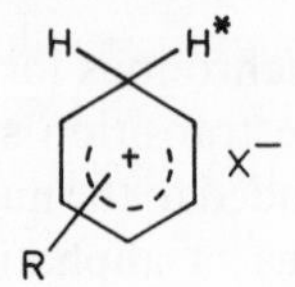 (**5**: **a**, H* = H; **b**, H* = D)

complexes have been isolated using other electrophilic reagents. For example, mesitylene, ethyl fluoride, and boron trifluoride give an orange conducting solid of m.p. $-15\,^{\circ}C$, which is considered to be **4**;[1] such complexes are the probable intermediates in Friedel–Crafts alkylations.

Complexes are formed by protonation of the aromatic nucleus. Benzene, toluene, and di- and polymethylated benzenes give coloured, conducting solutions in hydrofluoric acid.[2] The addition of boron trifluoride to the solutions increases the concentrations of the ions, and crystalline conducting complexes can be isolated from these solutions.[3a] The cation may be formulated as **5a**, where R represents one or more methyl groups, while the anion is F^- or, in the presence of boron trifluoride, BF_4^- [complexes completely stable at room temperature may be isolated when $R = 2,4,6\text{-}(NR_2)_3$, $X = ClO_4^-$[3b]]. These cations correspond to the intermediates in hydrogen exchange, further evidence for this being as follows:

(1) When hydrogen fluoride is replaced with deuterium fluoride, regeneration of the aromatic compound by heating the ionic complex gives both the original compound and its deuteriated derivative, consistent with the occurrence of exchange through **5b** which can eliminate either H^+ or D^+ on decomposition.[3]

(2) The electronic spectrum of a solution of anthracene in sulphuric acid is similar to that of the Ph_2CH^+. This is to be expected if anthracene is protonated at one *meso* carbon atom to give **6**.[4]

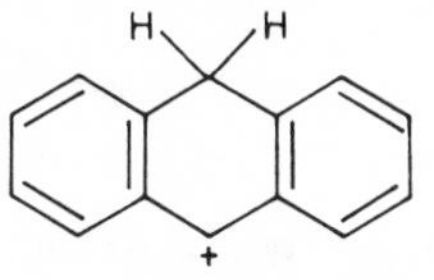

H H
+
(**6**)

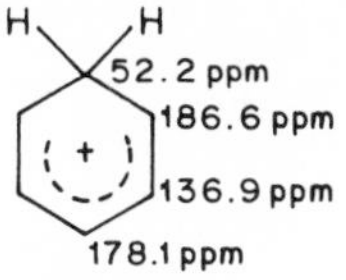

H H
52.2 ppm
186.6 ppm
+
136.9 ppm
178.1 ppm
(**7**)

(3) The ^1H n.m.r. spectra of various aromatics in trifluoroacetic acid–boron trifluoride indicate the presence of a methylene group, such as **5** and **6** contain.[5] This has been confirmed using superacids such as $HF–SbF_5$ [which produce even the benzenonium ion (**5a**, R = H)],[6] and ^{13}C n.m.r., which gives the chemical shifts shown in **7**.[7] These confirm that the positive charge is delocalized mainly to the *ortho* and *para* positions, since carbon bearing a charge has a greater chemical shift.

This accumulated evidence shows that aromatic compounds can form cations by the addition of an electrophile. Moreover, in the cases referred to above, the cations give rise directly to the normal products of aromatic substitution. The cations may also be captured by strong nucleophiles present to give isolable *adducts* or *addition complexes*, such as **8**, obtained in the nitration of *o*-xylene by nitric acid in acetic anhydride.[8] Decomposition of these adducts with acid results either in loss of nitrous acid giving **9** or, via 1,2-rearrangement and loss of acetic acid, 3-nitro-*o*-xylene (**10**). Thermal decomposition of the adducts results in 1,3-rearrangement and loss of acetic acid which, in the example shown, would produce 4-nitro-*o*-xylene. Adduct formation is discussed more fully in Sections 7.4.2(iii) and 9.2.2(1).

(**8**) (**9**) (**10**)

Evidence that the cations are intermediates in the substitutions is shown by the fact that a plot of log *k* for the chlorination of methylbenzenes[9] against log *K* for the equilibrium between the same compounds and the cations which they form with hydrogen fluoride[2] (see Table 2.1) is approximately linear (Fig. 2.1). This suggests that the chlorination rate is dependent on the ease of formation of a cation which is similar to that of the corresponding protonated aromatic.* The

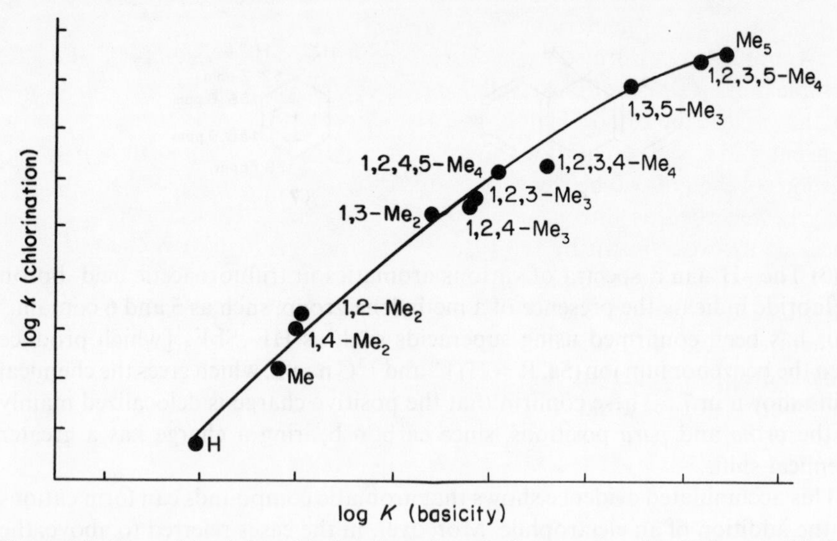

Figure 2.1. Plot of log *k* for chlorination of methylbenzenes vs log *K* for equilibrium between the same compounds and the cations which they form with hydrogen fluoride.

---

* A discussion of the observation that the factors which determine the position of an *equilibrium* also frequently govern the *rate* of formation of a product is given in Section 2.2.

methyl groups increase both k and K because they are electron releasing relative to hydrogen (Section 1.4); the plot is slightly curved because the rate-determining *transition state* for chlorination carries rather less charge than the cation.

2.1.2 Nature of the Intermediate Complexes

Complexes such as **3** and **4** are referred to as σ-complexes or, when they are described in the context of their intermediacy in aromatic substitutions, as Wheland intermediates.[10] Aromatics also form π-complexes both with ions such as Ag^{+}[11] and with neutral species such as iodine,[12] and hydrogen chloride.[9]

The structures and properties of the two types of complex are markedly different, as shown by comparisons of the addition compounds formed by aromatic compounds with hydrogen chloride (π-complexes) and with hydrogen chloride in the presence of aluminium chloride (σ-complexes).[9] The former have very low heats of formation and are largely dissociated in solution, are colourless, and do not conduct; when deuterium chloride is substituted for hydrogen chloride, the recovered aromatic compound does not contain deuterium. The latter, in contrast, can be obtained in much higher equilibrium concentrations, are coloured and conducting (and therefore ionic), and they provide a pathway for the exchange of deuterium for hydrogen.

π-Complexes are formed by a loose association of the reactants in which the complexing agent is not localized at a particular carbon atom, but is held near to the π-electron cloud. The crystal structure of the addition compound of benzene and silver perchlorate shows that the silver ion is equidistant from two carbons of each of two benzene rings, one on either side of it,[13] and the bonding may be regarded as three-centred. The aromatic ring is electron donating in these complexes, shown by the basicity of aromatic hydrocarbons (measured by their extent of π-complexing with HCl) increasing as methyl groups are added successively to the ring.[9] π-Complexes are therefore represented by structures such as **11**, the direction of the arrow indicating that the aromatic ring is electron donating, and its position relative to the ring indicating that the bond is not localized.

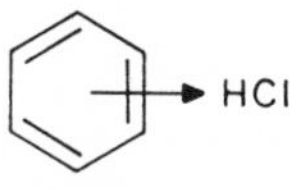

HCl

(**11**)

The formation of the σ-complex, and not the π-complex, is the rate-determining step in electrophilic substitution. This is shown, for example, by comparison of the relative stabilities of both π- and σ-complexes and halogen-

Table 2.1 Relative stabilities of π- and σ-complexes and relative rates of halogenation of alkylbenzenes

	Relative basicity		Relative reactivity	
Substituent	π-Complex (HCl)	σ-Complex (HF)	Cl_2	Br_2
H	0.61	0.09	0.005	0.004
Me	0.92	0.63	0.157	0.85
1,4-Me_2	1.00	1.00	1.00	1.00
1,2-Me_2	1.1	1.13	2.1	2.1
1,3-Me_2	1.26	26	200	204
1,2,4-Me_3	1.36	63	340[a]	600
1,2,3-Me_3	1.46	69	400[a]	660
1,2,4,5-Me_4		120	1 400[a]	—
1,2,3,4-Me_4	1.63	400	2 000	—
1,3,5-Me_3	1.59	2 800[b]	80 000	75000
1,2,3,5-Me_4	1.67	16 000	240 000	—
Me_5	—	29 000	360 000	—

[a]Calculated value, assuming additivity.
[b]Value obtained using HF–BF_3.[9]

ation rates for a number of alkylbenzenes (Table 2.1);[2,9,14] the rates evidently parallel more closely the stabilities of the corresponding σ-complexes (see Fig. 2.1). A specific feature of the results is that, whereas the relative stabilities of the π-complexes formed by the three xylenes increase steadily in the order $p < o < m$, in both σ-complex formation and halogenation the values for *o*- and *p*-xylene are comparable but those for *m*-xylene are enormously greater. This can be understood for the σ-complex stabilities by reference to the structure of the carbocations formed by protonation of the three compounds. That from *m*-xylene is a hybrid of **12–14**, i.e. it can be represented by **15**, in which both the electron-releasing methyl substituents can contribute to the stability of the ion by interaction with the adjacent positively polarized centres (hence protonation tends to occur *ortho* or *para*, and not *meta*, to the substituent). In the corresponding complexes from *o*- and *p*-xylene only one methyl group is directly bonded to a positively polarized carbon, so that the stabilization of the ion is

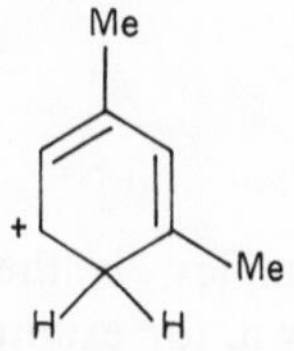

(**12**) 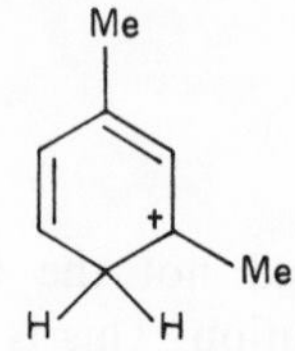 (**13**) 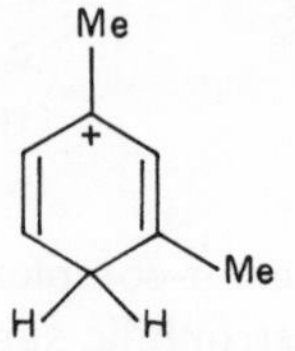 (**14**) 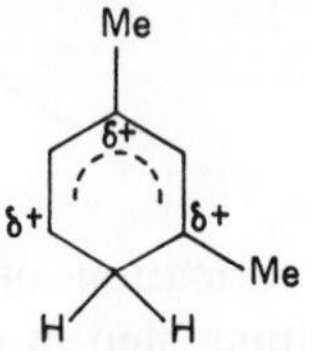 (**15**)

correspondingly reduced. The close parallelism of σ-complex stability and rates of halogenation indicates that the factors which govern the stabilities of the σ-complexes are reflected in the transition states which precede these complexes, and it is the formation of the σ-complex which constitutes the rate-determining step. The same effect is observed in the series 1,2,4-, 1,2,3-, and 1,3,5-trimethylbenzene.

Kinetic methods of examination can give information only about the maximum point on a free-energy profile of the reaction, and it is possible, and indeed likely, that π-complexes are formed in reactions in which formation of a σ-complex is rate determining. The most general representation of the kinetic pathway of an aromatic substitution in which σ-complex formation is rate determining is therefore that shown in Fig. 2.2, and the quantitative features of most electrophilic substitutions can be understood in terms of this pathway, i.e. one in which the transition state of the rate-determining step resembles the σ-complex (Section 2.2).

Some evidence appeared to indicate that π-complex formation provided the rate-determining step in some reactions,[15] the relative reactivities of compounds in these cases being markedly different from those generally observed [see, e.g., Sections 6.1.1(iii)(c), 7.4.1(vii), 9.2.2(iv) and 9.3.2(ii). However, it has subsequently been shown that these abnormalities arise from a combination of the high reactivity of the electrophile and insufficiently rapid mixing of the reagents (mixing control);[16] moreover, the rates do not correlate with π-complex stabilities.[17] In nitration there is evidence that an intermediate is formed prior to the formation of the σ-complex [Section 7.4.1(ii)]. However, this appears to consist of the electrophile and aromatic held in proximity by the solvent shell, without bonding between them of the kind found in π-complexes.

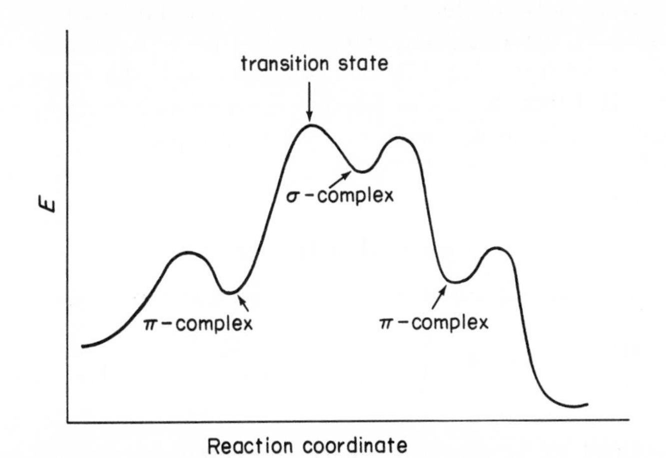

Figure 2.2. Energy profile for aromatic substitution pathway.

It is conceivable that the transfer of electrons from the aromatic to the electrophile may take place in a stepwise fashion.[18] This electron-transfer mechanism, which would produce a radical–radical cation pair $ArH^{+\cdot}\ E^{\cdot}$ as an intermediate, has been revived from time to time; the encounter pair referred to above could be of this kind. Support for the importance of such intermediates has been indicated by the formation of transient charge-transfer absorption bands between benzene derivatives, and halogens and mercury(II) trifluoroacetate. Moreover, the rate coefficients for the disappearance of these bands coincide with the rates of halogenation and mercuriation of the aromatic.[19] On the other hand, electrochemical generation of radical–radical cation pairs gives (for nitration) different relative reactivities to normal nitration.[20]

2.1.3 Kinetic Isotope Effects

The order of carbon—hydrogen bond strengths when carbon is attached to different isotopes (in the same structural environment) is C—T > C—D > C—H. Normally when a carbon—hydrogen bond is broken in the rate-determining step of a reaction a positive hydrogen isotope effect is observed, i.e. $k_H > k_D > k_T$. The absence of a positive isotope effect therefore indicates that the C—H bond is not ruptured during the rate-determining step.[21]

In many aromatic substitutions no isotope effect is observed. Thus deuteriobenzene and tritiobenzene are nitrated at the same rate as benzene, and similarly there is no kinetic isotope effect in the nitration of a number of other aromatics varying widely in reactivity, under nitrating conditions varying from pure nitric acid to nitric acid in oleum.[22] Again, there is no kinetic isotope effect in many halogenations, such as the iodine-catalysed bromination of toluene.[23]

In these cases it is concluded that the proton is not displaced from the aromatic ring in the rate-determining step. This is inconsistent with the synchronous mechanism (eq. 2.3), but consistent with the two-step process (eq. 2.4), provided that the rate of reaction is determined solely by the rate of the first step, which does not involve C—H bond rupture.

For the S_E2 reaction (eq. 2.4), the overall rate of formation of the product ArE is given by

$$\text{Rate} = k_2[\text{ArHE}^+][\text{B}^-] \tag{2.5}$$

The change in concentration of the intermediate with time is given by

$$\frac{\text{d}[\text{ArHE}^+]}{\text{d}t} = k_1[\text{ArH}][\text{E}^+] - k_{-1}[\text{ArHE}^+] - k_2[\text{ArHE}^+][\text{B}^-] \tag{2.6}$$

This is assumed to be zero (stationary concentration), hence $[\text{ArHE}^+] = k_1[\text{ArH}][\text{E}^+]/(k_{-1} + k_2[\text{B}^-])$. Substituting for this in eq. 2.5 gives eq. 2.7 as the rate equation:

$$\text{Rate} = \frac{k_1[\text{ArH}][\text{E}^+]k_2[\text{B}^-]}{k_{-1} + k_2(\text{B}^-]} \tag{2.7}$$

This reduces to eq. 2.8 if $k_2[\text{B}^-]$ is large, and to eq. 2.9 if $k_2[\text{B}^-]$ is small.

$$\text{Rate} = k_1[\text{ArH}][\text{E}^+] \tag{2.8}$$

$$\text{Rate} = k_1[\text{ArH}][\text{E}^+]k_2[\text{B}^-]/k_{-1} \tag{2.9}$$

It follows that if $k_2 \gg k_{-1}$, the overall rate is determined by the rate of the first step, i.e. eq. 2.8 applies. There will be no isotope effect and no base catalysis. If, however, the intermediate is partitioned between product formation and reversion to the starting materials, then since the rate at which the intermediate loses a proton is greater than that at which it loses D^+ or T^+ but the rate at which it reverts to starting material is independent of the nature of the isotope, the effect of partitioning would be to increase the rate of formation of the hydrogen-substituted product relative to the deuterium- or tritium-substituted products, i.e. a positive isotope effect would be observed.* The reaction rate will be governed by eq. 2.9 and base catalysis will be observed; there will be a linear dependence on base concentration if $k_2[\text{B}^-]/k_{-1} \ll 1$, and less than linear in all other cases. There are many electrophilic substitutions which show positive kinetic isotope effects. Examples are sulphonation,[22,25] diazonium coupling of phenols,[26] and the iodination of phenol and aniline and derivatives.[27] A positive value is consistent with either mechanism 2.3 or mechanism 2.4 in which k_2 is comparable to or less than k_{-1}. A distinction between these possibilities can only be made if further evidence such as the effect of base catalysis is available. For example, this has indicated that the two-step mechanism applies in iodination of phenol,[28] and in diazonium coupling (Section 7.2).

A further demonstration of the two-step mechanism is provided by the variation in isotope effect for a given reaction, according to substrate. Thus, whereas nitration usually gives no isotope effect, a value of k_H/k_D of 3.7 is obtained in nitration of 1-methyl-2,4,6-tri-*tert*-butylbenzene [Section 7.4.2(i)]. Here, steric hindrance to the electrophile increases k_{-1} and so alters the partitioning between k_{-1} and k_2, i.e. k_2 becomes relatively more rate determining; a number of examples of this effect are now known. Further, it is possible to decrease the isotope effects by either increasing k_2 through carrying out reactions in the presence of a higher concentration of base, or alternatively to decrease k_{-1} by using a more reactive electrophile [as in diazo coupling (Section 7.2)].

---

*Only the *primary* isotope effect has been considered here. *Secondary* isotope effects are to be expected in a reaction involving, as here, the conversion of an aromatic (sp^2) C—H bond into an aliphatic (sp^3) C—H bond, but the two most important secondary isotope effects should approximately nullify each other.[24]

The B-S_E1 reaction (eq. 2.2) is also base catalysed, and for hydrogen exchange (the main example of this reaction in which a proton is removed), a kinetic isotope effect is observed (Section 3.3). The unique feature of these reactions, however, is their electron-rich transition states, the consequence of which is that the substituent effects are reversed from those that are normally found in electrophilic substitutions. These reactions also show solvent isotope effects in the second step of the reaction [demonstrated in various demetallations (Sections 4.7.2 and 4.9.2)], and these vary according to the extent to which a free carbanion is produced in the transition state. The greater the charge on the carbanion, the less it will discriminate between solvents of differing isotopic composition.

2.2 THE THEORY OF REACTION RATES

The ultimate aim of a mechanistic treatment of an organic reaction is to calculate the reaction rate from first principles. This requires a detailed knowledge of the thermodynamic properties of the initial state and transition state of the process, for the reaction rate depends on the difference in free energies between them.[29] At present our knowledge of the structures of transition states is insufficiently advanced to enable much progress to be made towards this absolute end.

Next to the calculation of absolute reaction rates, it is desirable to obtain information about the *relative* rates of two or more compounds with a particular reagent, and this requires information only about the *differences* in free energies of activation of the compounds. This becomes a tractable problem in aromatic substitutions at *meta* and *para* positions, for then the free energy of activation is not likely to be much affected by changes in the kinetic contribution to the free energy as this should be relatively small. It is then possible to assess in a semi-quantitative manner the importance of the potential energy terms in the free energy of activation when the structure is modified.

Ignorance about the structure of the transition state necessitates a second simplification and has led to two methods of approach being made towards the problem of correlating structure with reactivity in aromatic substitutions: (i) the Isolated Molecule method and (ii) the Localization Energy method; these use the ground state and the intermediate, respectively, as models for the transition state.

2.2.1 The Isolated Molecule Method

This method assumes that an electrophile will bond to an aromatic carbon atom with an ease dependent upon the electron availability at that carbon. This is illustrated by comparison of the reactivities of the *meta* and *para* positions in nitrobenzene with that of one position in benzene. The σ-values of *m*- and *p*-NO_2 [0.71 and 0.78, respectively (Table 1.4)] indicate that the positions *meta* and *para* to the nitro group in nitrobenzene are both electron deficient with respect to any one position in benzene, the latter being the more deficient. An electrophile should therefore approach a carbon atom in benzene and the *meta* and *para* positions in nitrobenzene with an ease which decreases in that order, and this is consistent with the observed relative reactivities.

This approach is equivalent to obtaining information only about the Coulombic forces operating when the reagent begins its approach to the aromatic compound from a large distance. It takes no account of the changes that occur when the reactants mutually polarize each other as they approach more closely, i.e. the polarizability of the system is neglected. Often the polarizability at a given carbon atom parallels the permanent polarization at that position, and the success of the electrostatic method in correlating reactivities is probably due to this parallelism.[30]

The most general field in which the Isolated Molecule method is unsatisfactory is that of the alternant polycyclic hydrocarbons.[31,32] The electron densities are identical at all positions and equal to that at each position in benzene, yet there are marked differences in the reactivities of these compounds, e.g. a 2-position in naphthalene is nitrated 50 times faster than a position in benzene.[33]

A second example of a different type is the behaviour of the trimethylanilinium ion. The positive nitrogen pole withdraws electrons more strongly from the *meta* than from the *para* position, as indicated by the facts that the *m*-NMe_3^+ substituent increases the acidity of benzoic acid more than does *p*-NMe_3^+,[34] that in alkaline conditions the ester **16** is hydrolysed faster than **17**,[35] and similar evidence relating to reactivities influenced by the nitrogen pole.[34] The relative electron densities at the *meta* and *para* positions of the trimethylanilinium ion may therefore be represented by **18**, and this result is consistent both with the expectation that the $-I$ effect of a group should fall off with increasing distance and with theoretical calculations.[36]

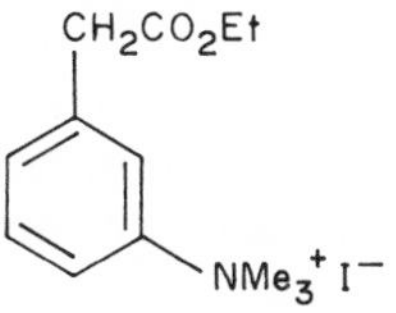

CH_2CO_2Et / $NMe_3^+ I^-$ (**16**)

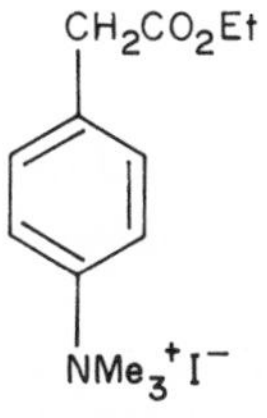

CH_2CO_2Et / $NMe_3^+ I^-$ (**17**)

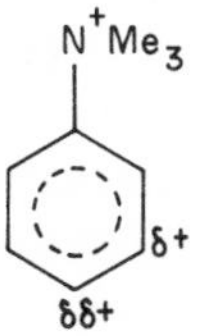

N^+Me_3 / $\delta+$ / $\delta\delta+$ (**18**)

These results predict that an electrophile would attack the *para* position of the trimethylanilinium ion more readily than the *meta* position, whereas in practice the *para* position is less reactive [e.g. in nitration, Section 7.4.3(xii)]. Evidently the ease with which a pair of electrons can be brought to the *meta* and *para* positions to form a covalent bond with the electrophile is not wholly related to the initial electron densities at these positions.

The reactivities both of the alternant polycyclic hydrocarbons and of the trimethylammonium ion are, however, satisfactorily predicted by the Localization Energy method.

2.2.2 The Localization Energy Method

This uses the Wheland intermediate[10] as a simplified model of the transition state since the structure can be written down, and certain conclusions can usually be drawn about its stability. The method clearly has its greatest value when the stabilities of the intermediate relative to the initial states can be derived theoretically. In practice, it is more common to derive *relative* values for the differences in energy between the intermediates and the initial states of a series of compounds reacting with a common electrophile.

For example, consider the Wheland intermediates **19** and **20**, formed by benzene and the 2-position of naphthalene with an electrophile, E^+ (each is, of course, a hybrid of several contributing structures). The formation of each may be regarded as occurring in two steps: (i) two π-electrons are isolated at one carbon of the nucleus; and (ii) a bond is formed between this carbon and the electrophile, the covalent pair being supplied by the former. Step (ii) is common to each molecule for reaction with a particular electrophile, so for comparative purposes it is only necessary to derive the energy required to reorganize the electrons in (i). This is known as the *localization energy*.

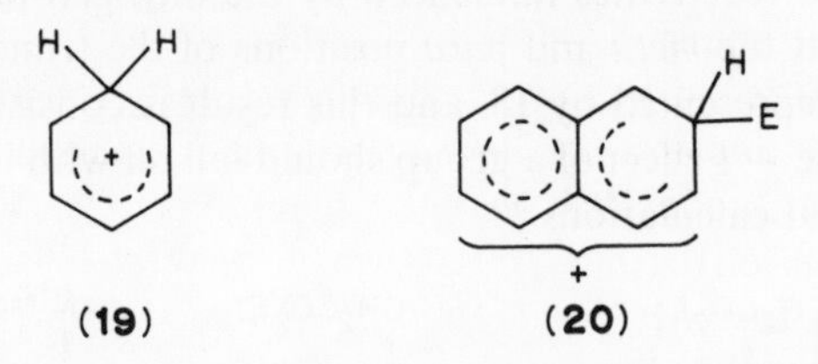

The energy levels may be calculated in terms of two energy units, the coulomb integral, α, and the resonance integral, β. The former is the amount of energy possessed by an isolated 2p orbital before overlap, and is the approximate measure of the electron-attracting power of the atom concerned. The latter is a measure of the degree of stabilization arising from π-orbital overlap, and is assumed to be negligble for non-adjacent atoms. For **19** the energy may be calculated to be $4\alpha + 5.46\beta$, whereas for the ground state of benzene it is $6\alpha + 8\beta$. The difference is $2\alpha + 2.54\beta$, and this is the localization energy for benzene.[37] This approach is referred to as the simple Hückel molecular orbital (HMO)

method, and is most easily applied (with very considerable success) to alternant polycyclic aromatic hydrocarbons since for each the 2α term effectively cancels, the reactivities being then predicted in terms of β-values only.

For compounds in which another atom is present in the nucleus or as a substituent, a greater number of approximations are involved in the calculations so the method is less successful, but nevertheless it can lead to useful results (see the discussion below of the nitration of $PhNMe_3^+$). Various other calculations have subsequently been introduced. The *self-consistent field* (SCF) method[38] overcomes the fact that electron–electron repulsions are either neglected or averaged out in the HMO method. It would be preferable, however, to include in the calculations *all* the electrons involved in interactions, and this has led, for example, to the *extended* Hückel method (EHMO),[39] and the all-electron version of SCF, known as *ab initio*.[40] However, the latter is an iterative method and so requires huge amounts of computer time even for molecules of only modest size, and is currently not feasible for most aromatics.

The suitability of the Wheland intermediate as a model for the transition state is shown by the facts that (i) for alternant polycyclics the localization energies are fairly linearly related to reactivities,[31] and (ii) $\log k$ values for typical electrophilic substitutions correlate with $\log K$ values,[31] as for example shown in Fig. 2.1. It is apparent, therefore, that the factors which determine the stability of σ-complexes, i.e. Wheland intermediates, also govern the stabilities of the transition states which lead to those complexes. A possible explanation of the observed parallelism between rate and equilibrium constants has been given by Hammond.[41] The formation of the intermediate in most aromatic substitutions is an endothermic process, and the free energy of the transition state is therefore closer to that of the intermediate than to that of the reactants (Fig. 2.2). Hence the major electronic reorganization should occur in the passage from the reactants to the transition state and a correspondingly smaller change in passage from the latter to the intermediate, so that a similarity in structure of the transition state and intermediate is reasonable. The intermediate should therefore be a better model than the reactants for the transition state, and this underlies the failure of the Isolated Molecule method to account for relative reactivities in some instances. Nevertheless, as a model for the transition state, the Wheland intermediate has certain deficiencies. First, the relative reactivities calculated on the basis of equating the two are invariably greater than those observed.[42,43] Second, the relative reactivities of a given group of aromatic compounds vary markedly with the nature of the reagent, as illustrated by the partial rate factors (see Section 2.3) for substitutions at the *meta* and *para* positions of toluene (Table 2.2). If the transition state was identical with the intermediate, the activation energies for a given group of compounds would differ from the localization energies by a constant amount, which would depend on the reagent, but the *relative* reactivities of members of the group towards different reagents would be the same.

Table 2.2 Partial rate factors for nitration and halogenation of toluene at 25 °C

	Nitration in Ac_2O	Chlorination in HOAc	Bromination in 85% HOAc
f_m	1.9	5.0	5.5
f_p	55.4	820	2420

These facts can be rationalized by the assumption that the transition state differs from the Wheland intermediate in one significant respect, viz. that the new σ-bond between a nuclear carbon and the reagent has not been completely formed in the transition state and, correspondingly, the π-electron system has not been completely reorganized.[42] It may be represented by **21**.

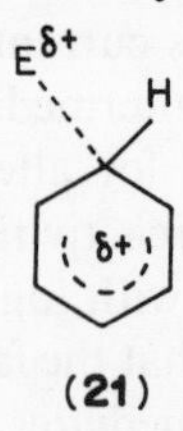

Hence the factors which stabilize one intermediate relative to another affect the corresponding transition states proportionally, giving rise to the observed correlations of rates and equilibria. However, the incomplete electronic reorganization in the transition state results in a reduction in the differential effects of substituents in determining the free energies of transition states compared with Wheland intermediates. The precise structure of the transition state will vary with the nature of the attacking reagent, resulting in changes in the differential effects of substituents with change of reagent and hence in the observed alteration of relative reactivities (Section 2.5). These ideas are both elaborated and evaluated in Chapter 11.

It is instructive to examine the nitration of the trimethylanilinium ion in the light of the preceding discussion. Because the *meta*-carbon is more positively polarized that the *para*-carbon (Section 2.2.1), an electrophile approaching from large separation is initially more strongly repelled from the *meta* than from the *para* position, and the energies corresponding to approach to these positions should rise as in the area A in Fig. 2.3. The Wheland intermediates corresponding to substitution at the two positions are hybrids of **22**–**24** and **25**–**27**, respectively, and calculations[34] based on the approximation of equal weighting of the three structures in each case have shown that the *para* is less stable than the *meta* intermediate by about 40 kJ mol^{-1}, primarily owing to the very high energy of structure **27**, which has adjacent positive charges. The relative energy contents of the two may be represented as in Fig. 2.3, and since it is known from

(22) (23) (24)

(25) (26) (27)

experiment that the transition state for *meta* substitution is of lower energy content than that for *para* substitution, the resulting energy profiles must therefore cross. The possibility of such crossing (see ref. 44) accounts for the failure of the Isolated Molecule method to predict correctly the relative reactivities in this case.

In summary, the Localization Energy method is theoretically more suitable than the Isolated Molecule method because it is more directly concerned with

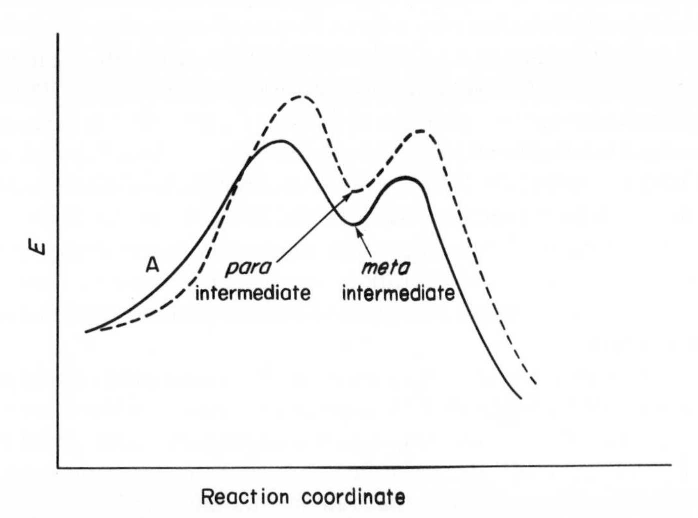

Figure 2.3. Energy profile for substitution at *meta* and *para* positions.

the transition state, and is empirically more widely applicable. Although the structure of the transition state is not identical with that of the Wheland intermediate on which the approach is based, nevertheless the canonicals which contribute to the intermediate are important contributors to the transition state hybrid; the resulting model gives a conceptually coherent picture of electrophilic substitutions. Subsequent discussions are therefore based on this model.

2.3 PARTIAL RATE FACTORS

The quantitative reactivities of aromatic compounds are commonly described in terms of *partial rate factors*, f, which measure the reactivity of the position concerned relative to that of one position in benzene. There are two methods of determining these factors, the *competition method* and the *direct kinetic method.*

In the competition method,[45] two aromatics (one of which is usually benzene) are allowed to react (in equal concentrations) with a deficiency of reagent for ca 10% of reaction. This latter condition ensures that the ratio of concentrations of the aromatics does not alter appreciably as reaction proceeds. From the overall amounts of each aromatic consumed and the isomer distribution within the compound under investigation, the partial rate factors may be calculated as illustrated by the nitration of toluene and benzene in acetic anhydride at 25 °C.[46] The yields of *ortho-*, *meta-*, and *para-*nitrotoluene were 63.0, 2.3, and 34.7%, respectively, and the overall relative reactivity of toluene to benzene was 26.6. The reactivity of the *ortho* position of toluene to benzene is clearly $(63.0/100) \times 26.6$ and since there are six positions available for nitration in benzene and only two *ortho* positions in toluene, the reactivity of the latter to a *single* position in benzene is $(6/2) \times (63.0/100) \times 26.6 = 50.3$. Likewise, the partial rate factor for the *meta* position in toluene is $(6/2) \times (2.3/100) \times 26.6 = 1.84$, and for the *para* position it is $6 \times (34.7/100) \times 26.6 = 55.4$. If the electrophile also replaces the substituent, then the relative rate at which this is removed is the *ipso* partial rate factor. If in the above example, y% of nitrobenzene were produced from toluene, the *ipso* factor would be $6 \times (y/100) \times 26.6$.

The competition method suffers from inaccuracy when the percentage of substitution at the least reactive position is very small, and indeed it is frequently impossible to calculate partial rate factors for these positions especially in more complex aromatics.

These difficulties are overcome in the direct kinetic method used in reactions involving a leaving group other than protium, e.g. T^+, $SiMe_3{}^+$. The sites in both benzene and the aromatic in question are thus prelabelled, and the relative rates of loss of the 'label' from benzene and from the aromatic is the partial rate factor. With this method partial rate factors differing by many orders of magnitude within a given molecule can be accurately determined.

2.4 THE ELECTRONIC EFFECTS OF SUBSTITUENTS

The main electronic effect of a substituent X on the reactivity of a benzene ring towards electrophiles may be considered under six headings:

(1) Certain substituents activate the nucleus. These include alkyl groups, ternary nitrogen substituents (e.g. $X = NH_2$, NMe_2, NHCOMe), and oxygen substituents (e.g. X = OH, OMe).

(2) Other substituents deactivate the nucleus. These include positive poles adjacent to the ring (e.g. $NMe_3{}^+$), carbonyl groups in various environments (e.g. CHO, COR, CO_2Et), and NO_2, CN, CF_3, and the halogens, Cl, Br, I.

(3) Substituents in group 1 cause substitution to occur predominantly in the *ortho* and *para* positions.

(4) Substituents in group 2, with some exceptions, cause substitution to occur mainly in the *meta* position. The most important exceptions are the halogens and —CH=CHY groups, where Y is an electron-attracting substituent.

(5) Some electrophilic reagents, such as the benzenediazonium cation, react only with phenols and amines and not with the less reactive compounds such as the alkylbenzenes and benzene itself.

(6) For both the *ortho,para*-directing and the *meta*-directing substituents the *ortho*:*para* ratio of substitution products is normally not equal to the statistical value of 2:1 but is sometimes greater and sometimes smaller than this.

A successful theory of aromatic substitution must account in the first instance for these six sets of observations.

The implications of the experimental data for one substituent, methyl, are first considered, and it will then be shown how these are consistent with describing the structure of the transition state in terms of that of the intermediate. This treatment will then be extended to other substituents.

In toluene the *para* position is more reactive than the *meta* position, and both are more reactive than a position in benzene. The results in Table 2.2 refer only to nitration and halogenation, but the *order* of reactivities is general although the values of the partial rate factors depend on the nature of the reagent (Section 2.7). Hence the free energies of activation are in the order toluene (p) < toluene (m) < benzene.

The free energies of activation for the *meta* and *para* positions of toluene are the differences in free energies of the two transition states and the initial state and, since the latter is common to both, the free energy of the *para* is less than that of the *meta* transition state. The observed relative reactivities of the *meta* and *para* positions of toluene and of one position in benzene indicate that the *difference* in free energies of transition and initial states is less for the *meta* and *para* positions in toluene than for benzene. These conclusions may be expressed schematically (Fig. 2.4) as follows: experimentally it is known that CE < CD < AB, and although the value of AC is not known, AC must be less than BD or BE. That is, (a) the methyl group has a greater effect in stabilizing the transition state

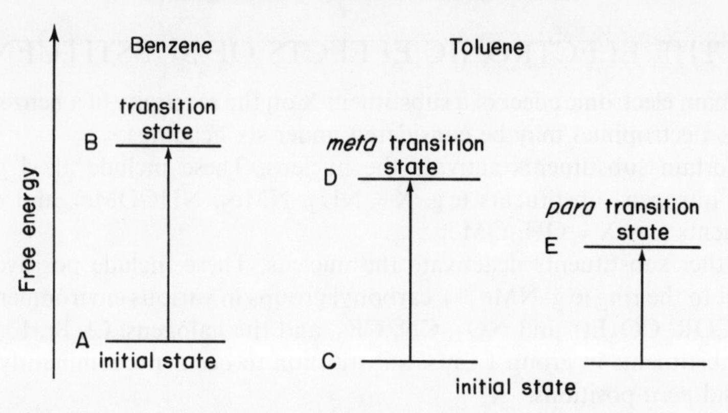

Figure 2.4. Representation of energy levels of initial and transition states.

when the reagent enters the *para* rather than the *meta* position, and (b) introduction of a methyl group into benzene affects the free energy of the transition state more than it affects that of the initial state.

These experimental facts are consistent with what is known about the transition state from discussion of the Localization Energy method. The Wheland intermediates formed by attack of a reagent on benzene and the *meta* and *para* positions of toluene are **28**, **29**, and **30**, respectively. First, in both **29** and **30** the methyl group provides stabilization by polarization (hyperconjugation) which is absent in **28**; such stabilization in the uncharged initial states is unlikely to be of little significance. Second, the free energy of **30** should be less than that of **29** because in the former the electron-releasing methyl group is bonded to a positively polarized carbon whereas in the latter its stabilizing effect must be relayed through one carbon atom. Thus to the extent that the factors which govern the relative stabilities of Wheland intermediates are reflected in the corresponding transition states, the observed order of reactivities is reasonable.

H E δ+ δ+ δ+ **(28)** H E δ+ δ+ δ+ CH_3 **(29)** H E δ+ δ+ δ+ CH_3 **(30)**

The effects of other substituents can now be discussed by this same procedure, i.e. to deduce their effects on the stability of this intermediate, from their known polar properties. One further qualification needs to be made. The structure of **28** suggests that a substituent should exert similar effects on *ortho* and *para* substitution rates by virtue of its polar effect. This is incorrect because **31** and **32** or **33** do not contribute equally to the hybrid represented by **28** (Section 11.2). In addition, the *ortho* position is subject to steric influences. For the moment, discussion is confined to *meta* and *para* reactivities, and the *ortho*:*para* ratio is discussed in detail later (Sections 2.5 and 11.2).

(**31**) (**32**) (**33**)

2.4.1 Alkyl Groups

From the above arguments, any group with a $+I$, $+M$ effect should activate both the *meta* and *para* positions, the latter more strongly. This is true for the alkylbenzenes, shown by the following partial rate factors for nitration in acetic anhydride at 0 °C[47] and chlorination in acetic acid at 25 °C.[48]

There are two features of interest in these data. First, the reactivities of the *meta* positions in both reactions are in the inductive order of electron release for the substituents, as are the reactivities of the *para* positions in nitration. In chlorination, however, the Baker–Nathan order (p-Me > p-t-Bu) is followed, and this is frequently observed in other reactions. It arises from the effect of steric hindrance to solvation of the transition state being superimposed upon the combined inductive and hyperconjugative electron-releasing effects of the

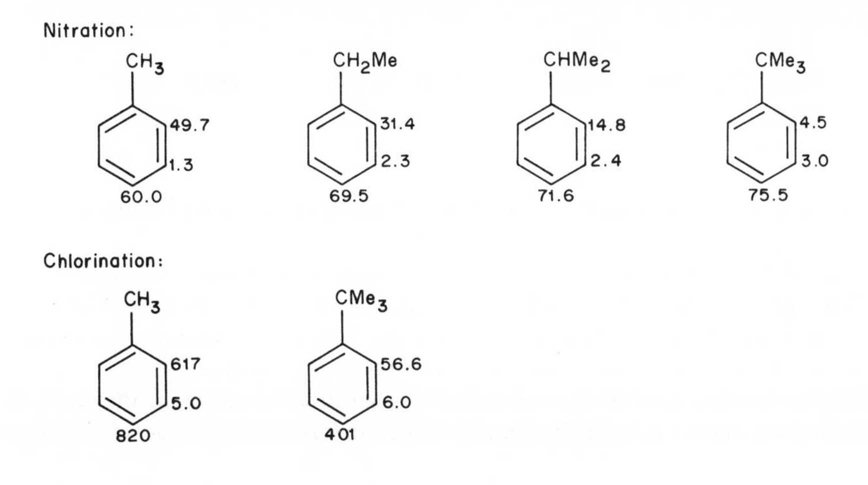

substituents, both of these latter being in the opposite (i.e. 'inductive') activation order. In consequence, the Baker–Nathan order is associated with reactions carried out in good solvating media, or in poorer solvating media where the solvating counter ion is very bulky (see also Sections 1.4.3 and 11.1.4).

Second, whereas in each reaction the *meta*:*para* ratio does not change very much as one alkyl group is replaced by another, indicating that the polar effects of the alkyl groups are closely similar, the *ortho*:*para* ratio decreases markedly in each series. This suggests that substitution at the *ortho* position becomes increasingly impeded by the substituent as this becomes larger.

2.4.2 Positively Charged Substituents

Substituents which contain an atom bearing a unit positive charge bonded to the benzene ring are predominantly *meta*-directing, and produce very marked deactivation of the nucleus. The groups SMe_2^+, $SeMe_2^+$, NMe_3^+, PMe_3^+, $AsMe_3^+$, and $SbMe_3^+$ give more than 80% of the *meta* nitro product in nitration.[49] This orientation is accompanied by very strong deactivation, e.g. the trimethylanilinium ion is 10^8 times less reactive than benzene.[50]

These substituents are of the $-I$ type and affect the transition states in the opposite way to the alkyl groups, i.e. stability decreases in the order **34** > **35** > **36** {the NMe_3^+ substituent also has a weak $+M$ effect, causing it to give proportionally more *para* substitution than do the other substituents [Section 7.4.3.(xii)]}.

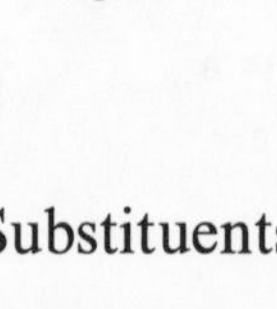

H E δ+ δ+ δ+ (34)

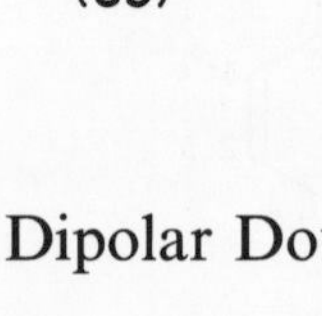

H E δ+ δ+ δ+ X^+ (35)

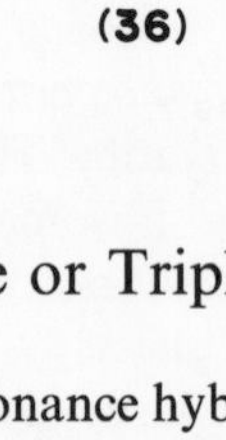

H E δ+ δ+ δ+ X^+ (36)

2.4.3 Substituents with Dipolar Double or Triple Bonds

These substituents are normally represented as resonance hybrids, e.g. **37**–**40**. Consequently, the atom bonded to benzene is positively polarized and the substituents therefore behave like positive poles in Section 2.4.2, causing deactivation of the nucleus with predominant *meta* orientation.

An interesting feature of these groups is that in some reactions they give an *ortho*:*para* ratio considerably greater than the statistical value of 2:1 [see

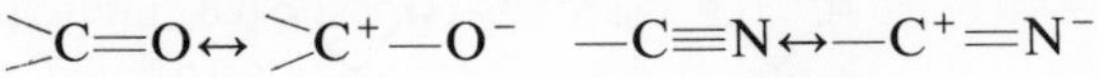

$$\gt C{=}O \leftrightarrow \gt C^{+}{-}O^{-} \qquad {-}C{\equiv}N \leftrightarrow {-}C^{+}{=}N^{-}$$

37 **38**

$$-\overset{+}{N}\begin{matrix} \nearrow O \\ \searrow O^{-} \end{matrix} \leftrightarrow -\overset{+}{N}\begin{matrix} \nearrow O^{-} \\ \searrow O \end{matrix} \qquad -\underset{\underset{O}{\|}}{\overset{\overset{O}{\|}}{S}}-R \leftrightarrow -\underset{\underset{O^{-}}{|}}{\overset{\overset{O^{-}}{|}}{S^{2+}}}-R$$

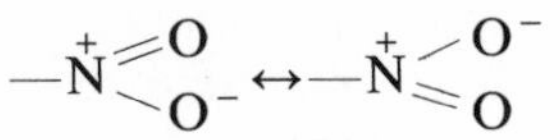

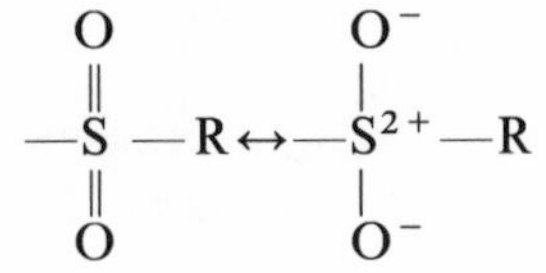

39 **40**

nitration (Table 7.15) and positive halogenation (Table 9.7)]. The explanation for this is given in Section 11.2.

The sulphonic acid group, $—SO_3H$, is also deactivating and *meta* orientating. This group is so strongly acidic that, even in the acidic media in which electrophilic substitutions are usually carried out, it must be considerably ionized and the anion, SO_3^-, might be expected to have the opposite effect to a cation, i.e. to be activating and *ortho,para* orientating. This is not so because the sulphur atom attached to the benzene ring is strongly electron deficient (see below), and this effect outweighs that of the negatively charged oxygens located further from the ring.

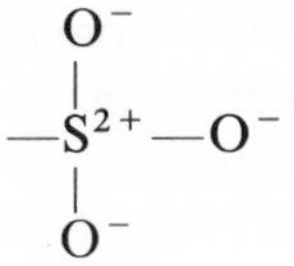

$$-\underset{\underset{O^{-}}{|}}{\overset{\overset{O^{-}}{|}}{S^{2+}}}-O^{-}$$

2.4.4 Substituents in which the Atom Bonded to Benzene has an Unshared Pair of Electrons

These substituents include those in which the atom bonded to the aromatic ring is ternary nitrogen, oxygen, sulphur, and the halogens. They are all predominantly *ortho,para*-directing groups, but they differ in that all except Cl, Br, and I are activating. The fluoro substituent is a borderline case, as it activates the *para* position in some reactions but deactivates it in others.

The lack of uniformity arises because the groups have opposing inductive and mesomeric effects ($-I$, $+M$). The latter results in further delocalization of the positive charge in the Wheland intermediate (**41**–**44**) when substitution occurs on the *para* (or *ortho*) position. The stabilization which results from this extra delocalization is, however, opposed by the $-I$ effect, which decreases the stability of the first three structures relative to the corresponding structures from benzene. The resulting effect for *para* (or *ortho*) substitution then depends on the relative importance of the inductive and conjugative effects.

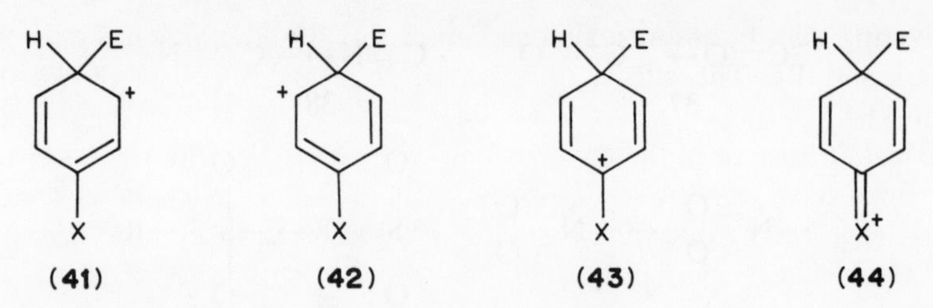

For the amino and hydroxyl substituents and their derivatives, the $+M$ effect is evidently dominant. For example, NMe_2 and OMe activate the *para* position in protiodesilylation [Section 4.7.1(ii)] by factors of about 10^7 and 10^3, respectively.[51] However, for Cl, Br, and I, the $-I$ effect dominates, e.g. in the same reaction the partial rate factors for *p*-Cl and *p*-Br are 0.13 and 0.10.[51]

Amino groups are the most powerfully activating of all neutral substituents. For example, in chlorination the *para* position of *N*,*N*-dimethylaniline is ca 10^{19} times as reactive as benzene,[52] and the reaction of *N*,*N*-dimethylaniline with benzene has an activation energy of almost zero.[53] The hydroxyl group is less strongly activating, but the O^- substituent is considerably more powerful than OH because first its $+M$ effect is greater and second it has a $+I$ effect. Thus phenols and anilines are able to take part in substitutions in which benzene and even fairly strongly activated aromatics are unreactive. The best known examples are diazonium coupling (Section 7.2), nitrosation (Section 7.3), and the Hoesch reaction (Section 6.9).

The acetamido group, NHCOMe, is less strongly activating than the amino or *N*,*N*-dimethylamino groups, because the unshared pair of electrons on nitrogen is already delocalized within the substituent ($-\ddot{N}H-\overset{|}{C}=O \leftrightarrow -\overset{+}{N}H=\overset{|}{C}-O^-$) and is not therefore as readily available as the pair in, say, $-\ddot{N}Me_2$ for π-orbital overlap with the electron-deficient transition state.

The resultant effects of substituents with unshared electron pairs on the reactivities of *para* positions are consistent with the relative values for the inductive and mesomeric effects of these groups. The $-I$ effects lie in the order $NMe_2 < OMe < F$ and the $+M$ effects in the order $NMe_2 > OMe > F$ (Section 1.4.2). Hence the *N*,*N*-dimethylamino group has the largest activating effect in aromatic substitutions; conversely, the fluoro substituent has the weakest activating effect, and indeed in many reactions it deactivates the *para* position[54] (see Section 11.1). The order of reactivities of the halobenzenes at their *para* positions arises because both $-I$ and $+M$ effects of the substituents are in the same order, $F > Cl > Br > I$, but these electronic effects do not fall off uniformly along the series (Sections 1.4.1 and 1.4.2). In addition, conjugation may occur between the π-electrons of the aromatic ring and the empty d-orbitals possessed by Cl, Br, and I, and this would also deactivate the aromatic ring. The observed order of electrophilic reactivities is $F > Cl > Br$, with I usually being more deactivating than F, but in some reactions more and in others less deactivating than Cl and Br. The effects of these substituents is considered further in Section 11.1.3.

The *meta* positions in these compounds are affected by the conjugative effect of the substituent only by second-order relay (Chapter 1, structures **49** and **50**), and so are normally deactivated by the $-I$ effect. Because of the operation of the $+M$ effect at the *para* position, the *meta*:*para* ratio in the direct substitution of a compound containing a substituent of the $-I$, $+M$ type is often too small to be measured accurately. For example, the yields of the *meta*-nitro derivative formed by nitration of chloro-, bromo-, and iodobenzene with acetyl nitrate in nitromethane were found, by isotope dilution analysis, to be 0.9, 1.2, and 1.8%, respectively,[55] whereas no *meta* derivative could be detected in the nitration of fluorobenzene with nitric acid in acetic anhydride.[47]

In such cases it is generally more accurate to evaluate *meta* reactivities by a kinetic investigation of the rate of displacement of a group in the *meta* position. Thus the relative rates of protiodetriethylgermylation of *meta*-substituted compounds *m*-$XC_6H_4GeEt_3$ are for X = OMe 0.51, F 0.032, Br 0.019, and Cl 0.019.[56] These results indicate, first, the deactivating effect of each *meta* substituent, second, that *m*-F is considerably more deactivating than *m*-OMe, consistent with the larger $-I$ effect of fluorine, and third, that for the fluoro substituent at least, a second-order relay of the $+M$ effect operates (**45**; cf. structure **50** in Chapter 1) to offset partly the deactivating $-I$ effect at the *meta* position, as consideration of the $-I$ effects alone would predict that *m*-F should be the most deactivating of the *m*-halo substituents.

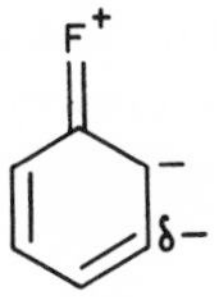

(**45**)

Finally, it is convenient to include in this group of substituents derivatives of styrene, PhCH=CHX, and of phenylacetylene, PhC≡CX, where X is electron attracting. Like the halobenzenes, these compounds are normally deactivated yet are substituted predominantly at the *ortho* and *para* positions (and this is true even for phenylacetylene itself,[57] where the deactivation comes from the electronegativity of the sp-hybridized carbon atoms). For example, ω-nitrostyrene (1-nitro-2-phenylethene) is nitrated at only one tenth the rate of benzene[58] but gives the nitro derivatives in yields of *o*- 31%, *m*- 2%, and *p*- 67%.[59] The substituents are of the $-I$, $+M$ type so the *meta* positions are deactivated, but the deactivation at the *para* positions is reduced (as in the halobenzenes) by the conjugative effect of the substituent which enables the positive charge on the intermediate to be further delocalized (**46**). Thus, a pair of electrons in the double (or triple) bond of the substituent act similarly to one of the unshared electron pairs on a halogen substituent.

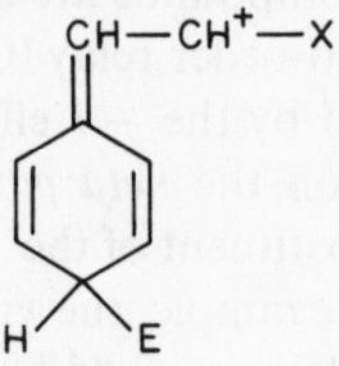

CH—CH$^+$—X

H E

(46)

2.4.5 Substituents Possessing a Vacant Atomic Orbital

Substituents possessing an empty p- or d-orbital that can conjugate with the π-electrons of the aromatic ring withdraw electrons by a $-M$ effect. In the case of electronegative elements such as the halogens (Section 2.4.4), the $-M$ effect is difficult to detect in the presence of the combined $+M$ and $-I$ effects. However, substituents containing electropositive elements such as boron or silicon have a $+I$ effect which activates the aromatic nucleus in all positions, and this is opposed by the $-M$ effect which reduces the reactivity of the *ortho* and *para* positions. Activation with *meta* orientation is therefore observed or, if the inductive effect is sufficiently large, the orientation becomes *ortho* > *meta* > *para*. Thus nitration of benzeneboronic acid with fuming HNO_3–H_2SO_4 gives the nitro derivatives in yields of *o*- 22%, *m*- 73%, and *p*- 5%,[60] and the partial rate factors for the $SiMe_3$ substituent in protiodesilylation are $f_o = 8.55$, $f_m = 1.63$, and $f_p = 1.25$;[61] steric acceleration in the latter reaction also enhances the reactivity of the *ortho* position.

2.4.6 Substituted Methyl Groups

Orientation and reactivity in compounds of the type $PhCH_2X$, $PhCHX_2$, and $PhCX_3$ depend on the electron-attracting capacity of X. When, in a series of compounds $PhCH_2X$, the electron-attracting power of X is increased, the overall $+I$, $+M$ effect of the group is reduced until it becomes a $-I$, $-M$ effect. The activating and *o,p*-directing nature of —CH_2X (X = H) is gradually changed into a deactivating and *meta*-directing effect (e.g. when X = NO_2 or $NMe_3{}^+$) (see nitration, Tables 7.6 and 7.13).

When one, two, or three electron-attracting substituents are substituted successively into the methyl group, a similar change is observed, e.g. the proportions of *m*-nitro product in nitration of $PhCH_2Cl$, $PhCHCl_2$, and $PhCCl_3$ are 13.9, 33.8, and 64.5, respectively.[62,63]

2.4.7 Biphenyl and Bi- and Polycyclic Systems

(1) Biphenyl is activated and *ortho,para*-orientating. For example, towards bromine the *para* position is about 3000 times as reactive as one position in benzene whereas the *meta* position is less reactive than a position in benzene.[64] The deactivation of the *meta* position results from the $-I$ effect of the phenyl substituent (Section 1.4.1). The activation of the *para* position derives from increased delocalization of the charge (to six possible positions) in the transition state (see, e.g., **48** in Chapter 1).

The introduction of an electron-attracting substituent into biphenyl decreases the reactivity and causes substitution to occur preferentially in the unsubstituted ring. If the electron withdrawal is sufficiently large, then the unsubstituted ring becomes deactivated towards substitution, yet may still give *ortho,para* orientation, i.e. the substituted ring behaves like a halogen substituent. Thus, hydrogen exchange of pentafluorobiphenyl gives partial rate factors for the pentafluoro substituent of $f_o = 0.0095$, $f_m = 0.0053$, and $f_p = 0.0158$.[65]

The introduction of an electron-releasing substituent, e.g. NHAc, into biphenyl increases the reactivity and causes substitution to occur preferentially in the substituted ring.

(2) The fusion together of two or more benzene rings increases the reactivity of the molecules and also introduces directing influences. The 1- and 2-positions of naphthalene, for example, have partial rate factors in nitration of 470 and 50, respectively.[66] Both results follow from examination of the transition states which, for 1- and 2-substitution in naphthalene, may be approximated as hybrids of **47**–**51** and **52**–**56**, respectively. For each, the delocalization of the positive

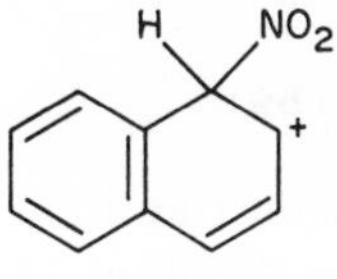

H NO2 (47) 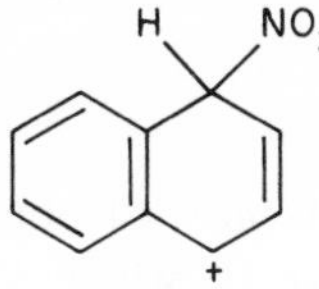 (48) 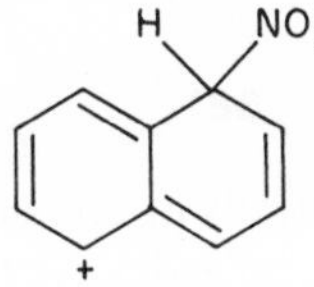 (49) 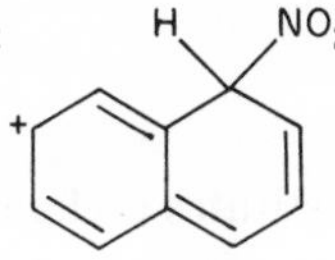 (50) 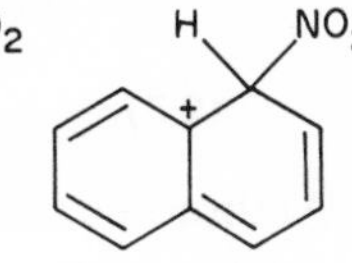 (51)

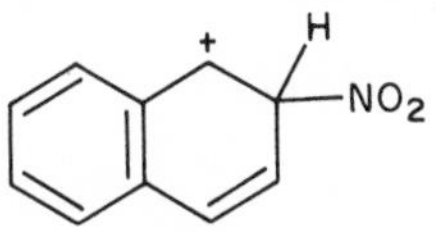

(52) 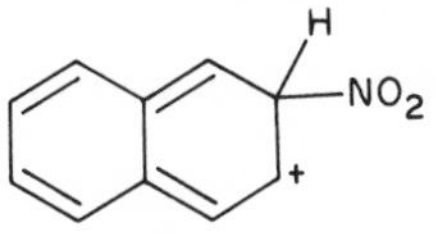 (53) 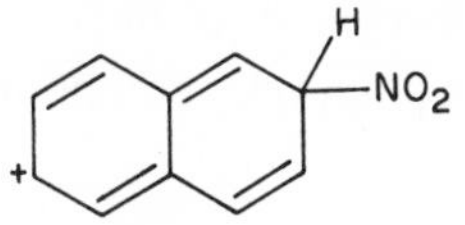 (54)

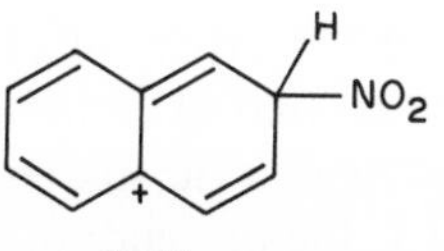

(55) 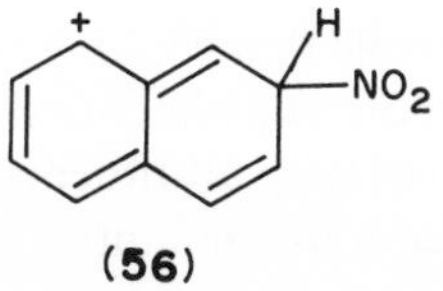 (56)

charge is considerably more extensive than in the transition state for substitution in benzene, leading to increased stability and decreased activation energy. Further, structures **47**, **48**, and **52** are benzenoid and therefore of lower energy content than the remainder, in which the benzenoid nature of the second ring has been interrupted. Since there are two low-energy contributors for 1-substitution, it is reasonable that the transition state for the former is of lower energy than that for the latter.

An electron-attracting substituent in one of the two aromatic rings causes deactivation of the molecule and the reagent reacts at the unsubstituted ring, predominantly at the 5- and 8-positions. For example, 1-nitronaphthalene gives 1,5- and 1,8-dinitronaphthalene on nitration,[67] and 2-nitronaphthalene gives mainly a mixture of 1,6- and 1,7-dinitro derivatives.[68]

An electron-releasing substituent causes activation and directs the reagent into the same ring as itself. If the substituent is in the 1-position, substitution occurs in the 2- and 4-positions, but a 2-substituent directs predominantly to the 1-position and not to the 3-position which is also *ortho* to the substituent. The reason is that stabilization by the substituent of the transition states, as 3- and 1-substitution interrupts the benzenoid character of the second ring in the former case (**57**), but not the latter (**58**). Expressed differently, **58** is a hybrid of two Kekulé structures. This is discussed in greater detail under hydrogen exchange (Section 3.1.2.13).

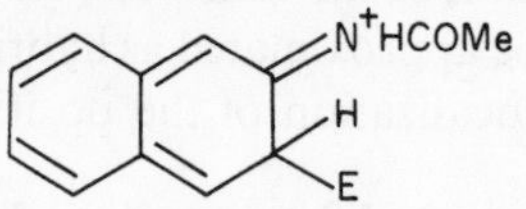

$N^{+}HCOMe$ H E $N^{+}HCOMe$ H E

(57) 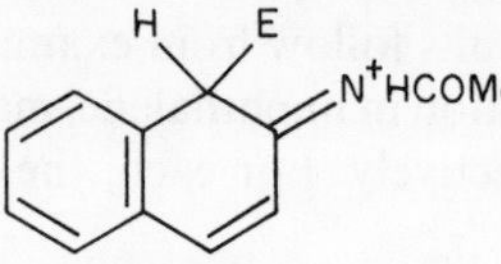 **(58)** 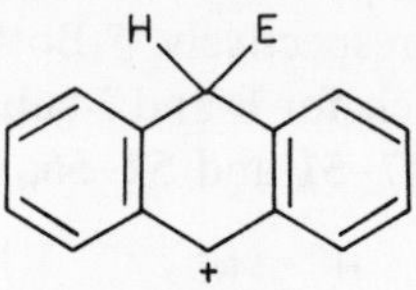 **(59)**

Anthracene substitutes preferentially at one of its *meso* carbons, the transition state having a considerably delocalized structure to which the dibenzenoid canonical **59** is the most important contributor. Anthracene also undergoes addition readily, the 9,10-dihydro adducts often undergoing elimination to give the 9-substituted products.

Other polycyclic compounds are discussed in later chapters.

2.4.8 Two or More Substituents

The simplest principle that could govern the reactivity of a benzenoid carbon affected by two or more substituents would be that of additivity. That is, if the introduction of each of two substituents was to alter the free energy of activation at a particular position by amounts x and y, the presence of both substituents would alter the free energy of activation by an amount $x + y$. Since $\log k \propto \Delta G^{\ddagger}$, the partial rate factor for substitution in the disubstituted compound would be equal to the product of the partial rate factors for the two monosubstituted compounds.

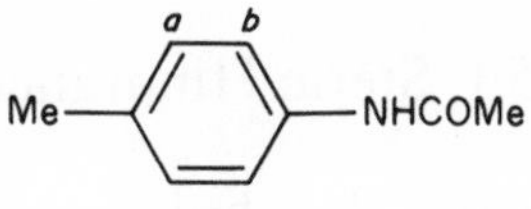

(60)

The principle is, to a first approximation, fairly successful in predicting substitution patterns, e.g. the partial rate factors for nitration of the three chlorotoluenes,[69] and more so for reactions, e.g. hydrogen exchange, where steric effects are less important. Nevertheless, discrepancies exist, and in general the principle underestimates substitution rates for aromatics containing two or more strongly deactivating substituents, and overestimates the reactivity of aromatics containing two or more activating substituents. The reasons for these deviations are considered more fully in Section 11.1.5.

The principle is of value in a semi-quantitative way. Consider the bromination of 4-acetamidotoluene (**60**). The acetamido group activates the *ortho* and *para* positions strongly and probably has a weak deactivating effect on the *meta* position, whereas the methyl group activates the *ortho* and *para* positions less powerfully than acetamido and also activates the *meta* position weakly. As a result, position *b* in **60**, which is *ortho* to acetamido and *meta* to methyl, is very much more reactive than position *a*, which is *ortho* to methyl and *meta* to acetamido. Consequently, reactions such as nitration and halogenation occur at *b*, and this provides a route to *meta*-substituted toluenes.

2.4.9 Summary

The effects of simple substituents may be classified on the basis of the above discussion as shown in Table 2.3.

Table 2.3 Classification of simple substituents

Substituent type	Effect on reactivity	Directing effect	Example
$+I, +M$	Activating	o, p	Alkyl groups, O^-
$+I, -M$	Activating	o, p	$SiMe_3$, CO_2^-
$-I, +M$[a]	Activating	o, p	NMe_2, NHCOMe, OMe, Ph
	Deactivating	o, p	Cl, Br, I, $CH{=}CHNO_2$
$-I$	Deactivating	m	Positive poles[b]
$-I, -M$	Deactivating	m	NO_2, CN, COR

[a] Effect on reactivity is determined by the relative importance of the two opposing polar effects.
[b] NR_3^+ groups also possess a very weak $+M$ effect.

2.5 THE STERIC EFFECTS OF SUBSTITUENTS AND ELECTROPHILES

2.5.1 Steric Hindrance

Since any substituent is larger than hydrogen, it might be expected that, compared with the *para* position, substitution at the *ortho* position would always be partially impeded by non-bonding repulsive forces. This is not necessarily so, for in some reactions there is evidence that the substituent and the reagent interact by covalent or coordinate bonding so that the reagent is held in a geometrically suitable position for transfer to the *ortho* position (see Section 11.2.3). Further, even when there is repulsion between substituent and reagent, the effect may be outweighed by a different steric phenomenon, namely steric acceleration (Section 2.5.2).

In these two cases, *ortho*:*para* ratios greater than the statistical value of 2.0 ($\log f_o : \log f_p > 1.0$) are commonly obtained, but even when values of less than 2.0 are found it is by no means because *ortho* substitution is sterically hindered, since electronic factors may be responsible (see Section 11.2.4).

The recognition of steric hindrance to *ortho* substitution is therefore difficult, for it is impossible to assess its importance when the *ortho*:*para* ratio is greater than 2.0, whereas in the remaining cases it is not often possible to separate steric from electronic factors. In the following examples, however, there is little doubt that steric hindrance is of considerable significance in determining the *ortho*:*para* ratio. The first example involves reaction with aromatics possessing substituents of differing bulk and the second involves reaction with electrophiles of differing bulk.

First, in the nitration of four alkylbenzenes, PhR (R = Me, Et, *i*-Pr, and *t*-Bu), $\log f_m : \log f_p$ is approximately constant, indicating that the electronic effects of the four alkyl substituents are closely similar, whereas the $\log f_o : \log f_p$ ratio decreases markedly as the alkyl group is made larger (Table 7.1). The decrease is attributed to the increasing importance of steric hindrance. A similar argument has revealed the occurrence of steric hindrance in halogenation of *tert*-butylbenzene (Table 9.4 and Fig. 9.1).

Second, whereas the $\log f_m : \log f_p$ ratios for acetylation and benzoylation of toluene are similar, the $\log f_o : \log f_p$ ratio in the former reaction is considerably less than that in the latter (Section 6.7.3), showing that the reagent involved in acetylation is larger than that in benzoylation. Similarly, examination of the data for methylation, ethylation, and isopropylation of toluene (Section 6.1.3) reveals that the alkylating agents have steric requirements which increase in that order.

2.5.2 Steric Acceleration

When a bulky substituent is being replaced in an aromatic substitution, steric compression between it and the *ortho* substituent(s) is relieved in passage from the initial (eclipsed) state to the transition (staggered) state, giving rise to steric acceleration. This effect is, of course, opposed by steric hindrance between the *ortho* substituent(s) and the attacking reagent, but if the reagent is small, steric acceleration can be of greater importance than steric hindrance and high *ortho*:*para* ratios then result. Steric acceleration is most commonly observed in protonolyses (Chapter 4) because steric hindrance in such reactions is small. It is undoubtedly important also in reactions in which a large substituent is replaced by reagents other than hydrogen, but diagnosis is more difficult here because steric hindrance is usually sufficient (and yet of unknown magnitude) to give small *ortho*:*para* ratios. Steric acceleration can precisely offset the effect of steric hindrance as in mercuridesilylation of 2-trimethylsilylbiphenyl (Section 10.31) and bromodesilylation of 1-trimethylsilylnaphthalene (Section 10.38). However, this is not always the case as results for bromodesilylation of 2-trimethylsilylbiphenyl show (Section 10.38).

2.6 THE EFFECT OF STRAIN

Strain can affect the reactivities of aromatic molecules in one of two ways. The first arises when the aromatic ring is deformed out of planarity as a result of some non-bonding interaction between adjacent atoms or groups, 4,5-dimethylphenanthrene (**61**) being a typical example (Table 3.25). The non-planarity results in a loss of ground-state stability, making the molecule more reactive than it should otherwise be.

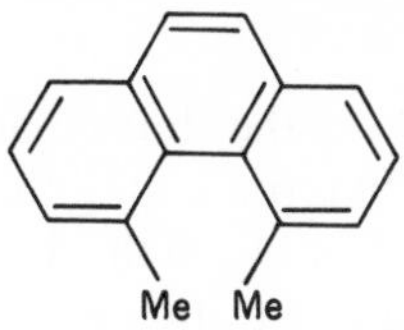

Me Me HO HO

(**61**) 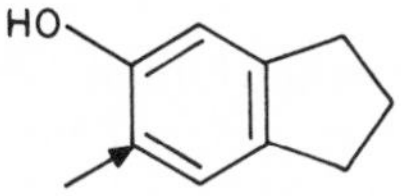 (**62**) 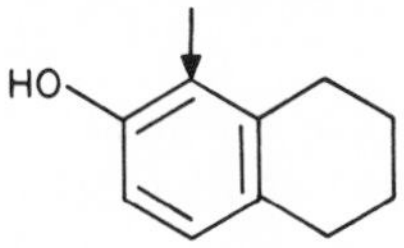 (**63**) 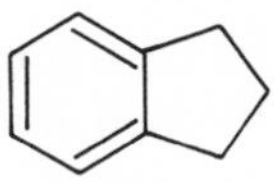 (**64**)

The second circumstance is that which arises when a strained cyclic group is attached to an aromatic ring, as for example in indane. It has long been recognized that sites adjacent to the five-membered ring are abnormally unreactive, so that for example β-hydroxyindane (**62**) preferentially brominated in the β'-position, whereas β-hydroxytetralin (**63**) preferentially brominated in the α-position (the Mills–Nixon effect).[70]

Earlier interpretations of this phenomenon were that the presence of the strained cyclic substituent produced bond fixation in the ground state, so that, for example, **64** would be the preferred canonical for indane. The resultant bond fixation would then account for the β'-position being more activated by the OH group than the α-position, since the conjugative effect of the OH substituent would be relayed more effectively across the bond of higher order.[70] Recent calculations of the bond orders in both benzcyclopropene and benzcyclobutene predict bond fixation in these very strained molecules[71] (although experimental data for benzo derivatives of dihydropyrenes appear to contradict this[72]). However, the magnitude of the coupling constants between adjacent hydrogens in the less strained indane, and in tetralin, determined in high-field n.m.r. experiments,[73] suggest that there is no bond fixation in these molecules, and this appears to be confirmed by ^{13}C n.m.r. measurements.[74]

A satisfactory explanation of the phenomenon is however obtained by considering the *change* in bond length, and accompanying change in strain in the cyclic substituent, on going to the transition state. Thus, in the resonance hybrid **65**–**67** representing the transition state for α-substitution, the bond common to both rings has double-bond character in two-thirds of the structures **65** and **66**, whereas in the ground state only half of the structures have a double bond in this position. Going from the ground state to the transition state therefore involves shortening of the common bond, and an increase in strain when the side-chain is a five-membered ring and a smaller decrease in strain when it is a six-membered ring.[75]

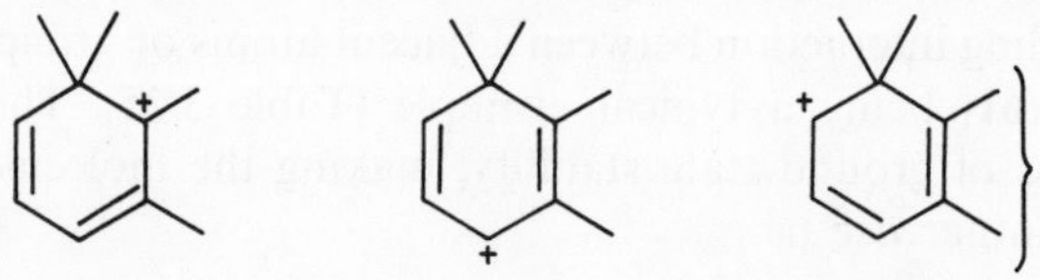

(65) (66) (67)

This argument was extended by the writer, for it is a corollary of the above that the bond common to both rings in the resonance hybrid **68**–**70** representing the transition state for β-substitution has double-bond character in only one third of the structures, **68**. Consequently, this bond is lengthened on going from ground state to transition state, and strain if present in the adjacent ring should be reduced, resulting in enhancement of reactivity at the β-position.[76] These

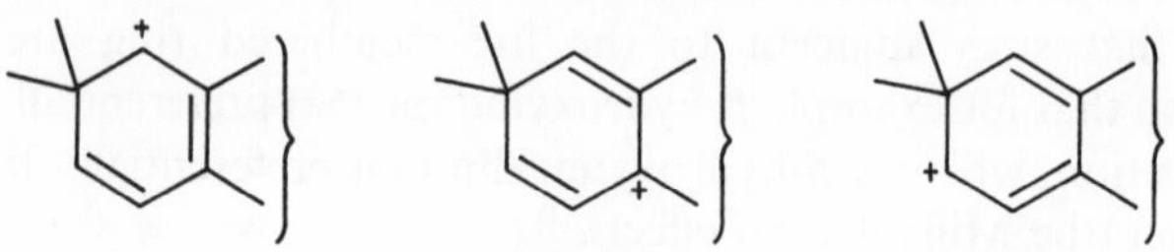

(68) (69) (70)

modifications to the positional reactivities have been shown to be observed in all molecules possessing strained cyclic side-chains[76] and are discussed in greater detail in Sections 3.1.2.3, 4.7.1(ii), and 11.3.

2.7 THE EFFECT OF ELECTROPHILE REACTIVITY ON ORIENTATION AND AROMATIC REACTIVITY

The discussion of orientation and reactivity has so far been mainly confined to the role of the substituent. However, whereas it is true that the substituent is responsible for determining whether substitution is predominantly *ortho*/*para*, or *meta*, and whether each of the three positions is more or less reactive than a carbon in benzene, the quantitative relationships between the reactivities of these positions are dependent also on the nature of the electrophilic reagent. This is illustrated by the data in Table 2.4.

Down this series, the *meta*:*para* ratio decreases and the reactivity of toluene relative to benzene increases. The changes can be understood as follows. Imagine a reagent of such high reactivity that it reacted at every collision with an aromatic molecule. It would not then discriminate between toluene and benzene, or between the *meta* and *para* positions in toluene; the relative reactivities of toluene to benzene and of the *meta* to *para* positions in toluene would be 5:6 and 2:1, respectively. This situation is most nearly approached, amongst the reactions in Table 2.3, by isopropylation, and it is reasonable that a carbocation should be so reactive. As the reactivity of the reagent is decreased there is an increasing demand for an electron pair to be supplied to the nuclear carbon to which the reagent is bonding in order that the activation energy barrier can be surmounted. Now the ease with which the electrons can be removed from the aromatic sextet and localized at one carbon is determined by the polar character and the position of the substituent, so that the greater the demand on the part of the reagent for an electron pair, the greater will be the differential effects of the substituents. It is

Table 2.4 Isomer distributions and relative rates for substitutions of toluene

Reaction	Isomer distribution/%		Relative reactivities of toluene and benzene
	m	*p*	
Isopropylation	25.9	46.2	1.8
Mercuriation	9.5	69.5	7.9
Nitration	3.0	38.0	>23
Benzoylation	1.5	89.3	110
Chlorination	0.5	39.7	350
Bromination	0.3	66.8	605

therefore understandable that the greatest differences in reactivities occur for the least reactive electrophiles such as the molecular halogens and that, in general, as the reactivity of the reagent decreases, its demands for an electron pair for bond formation at the transition state increases, resulting in an increase in the *selectivity* between *meta* and *para* positions and between different aromatic nuclei. This is an example of the *Reactivity–Selectivity Principle.*[77]

2.8 REVERSIBILITY AND REARRANGEMENT

Two features which may complicate electrophilic substitutions are the occurrence of the reverse reaction and of rearrangement, but both are relatively uncommon.

2.8.1 Reversibility

Although in principle all electrophilic substitutions are reversible, in practice the reverse reaction is too slow to be significant in most cases. Two reactions in which this is not completely true under commonly used conditions are sulphonation and Friedel–Crafts alkylation.

The sulphonic acid group may be removed from the aromatic nucleus in acidic conditions (Section 4.12). Use is made of this property in synthesis: the bulky sulphonic acid group may be introduced to block the most reactive position in a compound, a second substitution is then carried out on another position in the molecule, and finally the sulphonic acid group is removed.

The sulphonation of naphthalene at about 60 °C gives mainly the 1-sulphonic acid, but at about 160 °C the 2-sulphonic acid predominates. This is the result of the reversibility of sulphonation together with the greater thermodynamic stability of the 2-sulphonic acid, which arises from the repulsion between the bulky substituent and the *peri* hydrogen in the 1-isomer. At the lower temperature of sulphonation the 1-derivative is formed the faster of the two, as in other substitutions, whereas at the higher temperature not only is the rate of 2-sulphonation compared with 1-sulphonation increased (reactivity–selectivity effect), but also the 1-sulphonic acid undergoes desulphonation; hence there is a gradual build-up of the more stable 2-derivative.

Disproportionation and isomerization frequently occur in Friedel–Crafts alkylations as a result of protiodealkylation (Section 4.5).

2.8.2 Rearrangement

In the acid-catalysed equilibriation of the xylenes there is evidence for an intramolecular rearrangement:[78]

(2.10)

Analogous rearrangements occur during Friedel–Crafts alkylations, reaction on an alkylbenzene often giving an unusually large proportion of the *meta* derivative [Section 6.1.1(ii)].

2.9 MECHANISMS INVOLVING MIGRATION OF THE ELECTROPHILE FROM A SIDE-CHAIN

A few electrophilic substitutions take place via migration of the electrophile from a side-chain. Each reaction has a substitution counterpart involving the normal mechanism, each involves rearrangement from an OR or NHR side-chain (so that the activation of the aromatic ring is high), and there is evidence that both inter- and intramolecular processes are involved. The reactions are the Claisen rearrangement of allyl ethers, giving alkylation (Section 6.1.5), the Fries rearrangement of phenolic esters, giving acylation (Section 6.7.5), the nitramine rearrangement, giving nitration (Section 7.5), and the Orton rearrangement of *N*-chloroamides, giving chlorination [Section 9.2.2(iii)].

2.10 MICROSCOPIC AND MACROSCOPIC DIFFUSION CONTROL OF RELATIVE REACTIVITIES

In some reactions, the experimentally observed relative rates of two aromatics does not reflect the true difference in intrinsic reactivity; the reactivity of the more reactive compound is attenuated, leading to a reduced relative reactivity. Either of two factors may be responsible. The first is *microscopic diffusion control*, otherwise known as *encounter control*, and the other is *macroscopic diffusion control* or *mixing control*.

Encounter control occurs in homogeneous systems when only a small number of collisions occur between two entities before they react together. Under these conditions, the rate of reaction is governed only by the rate at which the two entities can diffuse together, which in turn depends on the viscosity of the medium. Since this constraint, when it applies, will clearly more seriously affect the aromatic compound that is intrinsically the more reactive, the reduced relative reactivity then follows.

Mixing control applies when reaction does not occur under truly homogeneous conditions. Consider the reaction under competition conditions of a cluster of electrophiles with a cluster of aromatic compounds A and B, of which A is the more reactive. In the aromatic cluster, A will become depleted, so that the remaining electrophiles are more likely to react with B than would be the case if the stoicheiometric ratio of A and B were maintained by proper mixing. The result is that the relative reactivities of A and B are attenuated. This problem affects substitutions involving very reactive electrophiles; with less reactive electrophiles, the stoicheiometric ratio between A and B has time to re-establish itself before further reaction takes place.

Examples of both kinds of diffusion control are common in nitration because of both the high reactivity of the electrophile, and the high viscosity of some of the nitration media (Section 7.4).

REFERENCES

1. G. A. Olah and S. J. Kuhn, *J. Am. Chem. Soc.*, 1958, **80**, 6541.
2. M. Kilpatrick and F. E. Luborsky, *J. Am. Chem. Soc.*, 1953, **75**, 577.
3. (a) G. A. Olah and S. J. Kuhn, *J. Am. Chem. Soc.*, 1958, **80**, 6535; (b) P. Menzel and F. Effenberger, *Angew. Chem., Int. Ed. Engl.*, 1972, **11**, 922.
4. V. Gold and F. L. Tye, *J. Chem. Soc.*, 1952, 2172.
5. C. MacLean, J. H. van der Waals, and E. L. Mackor, *Mol. Phys.*, 1958, **1**, 247.
6. G. A. Olah, *J. Am. Chem. Soc.*, 1965, **87**, 1103; G. A. Olah and T. E. Kiovsky, *J. Am. Chem. Soc.*, 1967, **89**, 5692; G. A. Olah, R. H. Schlosberg, R. D. Porter, Y. K. Mo, D. P. Kelly, and G. Mateescu, *J. Am. Chem. Soc.*, 1972, **94**, 2034.
7. G. A. Olah, J. S. Staral, G. Asencio, G. Liang, D. A. Forsyth, and G. Mateescu, *J. Am. Chem. Soc.*, 1978, **100**, 6299.
8. D. J. Blackstock, A. Fischer, K. E. Richards, J. Vaughan, and G. J. Wright, *J. Chem. Soc., Chem. Commun.*, 1970, 641.
9. H. C. Brown and D. J. Brady, *J. Am. Chem. Soc.*, 1952, **74**, 3570.
10. G. G. Wheland, *J. Am. Chem. Soc.*, 1942, **64**, 900.
11. R. M. Keefer and L. J. Andrews, *J. Am. Chem. Soc.*, 1952, **74**, 640.
12. R. M. Keefer and L. J. Andrews, *J. Am. Chem. Soc.*, 1955, **77**, 2164.
13. R. E. Rundle and J. H. Goring, *J. Am. Chem. Soc.*, 1950, **72**, 5337.
14. H. C. Brown and L. M. Stock, *J. Am. Chem. Soc.*, 1957, **79**, 1421.
15. G. A. Olah, S. J. Kuhn, and S. H. Flood, *J. Am. Chem. Soc.*, 1961, **83**, 4571; G. A. Olah and N. A. Overchuk, *Can. J. Chem.*, 1965, **43**, 3279.
16. W. S. Tolgyesi, *Can. J. Chem.*, 1965, **43**, 343; S. Y. Caille and R. J. P. Corriu, *Tetrahedron*, 1969, **25**, 2005; R. G. Coombes, R. B. Moodie and K. Schofield, *J. Chem. Soc. B*, 1968, 800; J. G. Hoggett, R. B. Moodie and K. Schofield, *J. Chem. Soc. B*, 1969, 1; P. F. Christy, J. H. Ridd, and N. D. Stears, *J. Chem. Soc. B*, 1970, 797; J. H. Ridd, *Acc. Chem. Res.*, 1971, **4**, 248; R. Taylor and T. J. Tewson, *J. Chem. Soc., Chem. Commun.*, 1973, 836; S. V. Naidenov, Yu. V. Guk, and E. L. Golod, *J. Org. Chem. USSR*, 1982, **18**, 1731.
17. P. Rys, P. Skrabal, and H. Zollinger, *Angew. Chem., Int. Ed. Engl.*, 1972, **11**, 874; F. P. DeHaan *et al.*, *J. Am. Chem. Soc.*, 1979, **101**, 1336; C. Santiago, K. N. Houk, and C. L. Perrin, *J. Am. Chem. Soc.*, 1979, **101**, 1337.
18. J. Kenner, *Nature* (*London*), 1945, **156**, 369.

19. S. Fukuzumi and J. K. Kochi, *J. Am. Chem. Soc.*, 1981, **103**, 7240.
20. L. Eberson and F. Radner, *Acc. Chem. Res.*, 1987, **20**, 53.
21. L. Melander, *The Use of Nuclides in the Determination of Organic Reaction Mechanisms*, University of Notre Dame Press, IN, 1955, p. 74.
22. L. Melander, *Acta Chem. Scand.*, 1949, **3**, 95; W. M. Lauer and W. E. Noland, *J. Am. Chem. Soc.*, 1953, **75**, 3689; T. G. Bonner, G. Bowyer, and G. Williams, *J. Chem. Soc.*, 1953, 2650.
23. L. Melander, *Ark. Kemi*, 1951, **2**, 211.
24. A. Streitwieser, R. H. Jagow, R. C. Fahey, and S. Suzuki, *J. Am. Chem. Soc.*, 1958, **80**, 2327.
25. J. C. D. Brand, A. W. P. Jarvie, and W. C. Horning, *J. Chem. Soc.*, 1959, 3844; U. Berglund-Larsson, *Ark. Kemi*, 1956, **10**, 549.
26. H. Zollinger, *Helv. Chim. Acta*, 1955, **38**, 1597, 1617; R. Ernst, O. A. Stamm, and H. Zollinger, *Helv. Chim. Acta*, 1958, **41**, 2274.
27. E. Grovenstein and D. C. Kilby, *J. Am. Chem. Soc.*, 1957, **79**, 2972; E. Shilov and F. Weinstein, *Nature* (*London*), 1958, **182**, 1300.
28. E. Grovenstein and U. V. Henderson, *J. Am. Chem. Soc.*, 1956, **78**, 569.
29. S. Glasstone, K. J. Laidler, and H. Eyring, *The Theory of Rate Processes*, McGraw-Hill, New York, 1941.
30. M. J. S. Dewar, *J. Am. Chem. Soc.*, 1952, **74**, 3355.
31. A. Streitwieser, *Molecular Orbital Theory for Organic Chemists*, Wiley, New York, 1961.
32. K. Yates, *Hückel Molecular Orbital Theory*, Academic Press, New York, 1978; C. A. Coulson, B. O'Leary, and R. B. Mallion, *Hückel Theory for Organic Chemists*, Academic Press, New York, 1978.
33. M. J. S. Dewar and T. Mole, *J. Chem. Soc.*, 1956, 1441; M. J. S. Dewar, T. Mole, and E. W. T. Warford, *J. Chem. Soc.*, 1956, 3576.
34. J. D. Roberts, R. A. Clement, and J. J. Drysdale, *J. Am. Chem. Soc.*, 1951, **73**, 2181.
35. R. O. C. Norman and P. D. Ralph, *J. Chem. Soc.*, 1963, 5431.
36. J. D. Roberts and D. A. Semenov, *J. Am. Chem. Soc.*, 1955, **77**, 3152.
37. F. H. Burkitt, C. A. Coulson, and H. C. Longuet-Higgins, *Trans. Faraday Soc.*, 1951, **47**, 553.
38. R. Pariser and R. G. Parr, *J. Chem. Phys.*, 1952, **21**, 466, 727; J. A. Pople, *Trans. Faraday Soc.*, 1953, **49**, 1375.
39. R. Hoffmann, *J. Chem. Phys.*, 1963, **39**, 1397.
40. M. J. S. Dewar, *The Molecular Orbital Theory of Organic Chemistry*, McGraw-Hill, New York, 1969.
41. G. S. Hammond, *J. Am. Chem. Soc.*, 1955, **77**, 334.
42. P. M. G. Bavin and M. J. S. Dewar, *J. Chem. Soc.*, 1956, 334.
43. M. J. S. Dewar and T. Mole, *J. Chem. Soc.*, 1957, 342.
44. R. D. Brown, *Q. Rev. Chem. Soc.*, 1952, **6**, 63.
45. C. K. Ingold, F. R. Shaw, and I. S. Wilson, *J. Chem. Soc.*, 1928, 1280.
46. S. Sotheeswaran and K. J. Toyne, *J. Chem. Soc., Perkin Trans. 2*, 1977, 2042.
47. J. R. Knowles, R. O. C. Norman, and G. K. Radda, *J. Chem. Soc.*, 1960, 4885.
48. H. C. Brown and L. M. Stock, *J. Am. Chem. Soc.*, 1957, **79**, 5175; 1959, **81**, 5615.
49. A. Gustaminza, T. A. Modro, J. H. Ridd, and J. H. P. Uttley, *J. Chem. Soc. B*, 1968, 534; H. M. Gilow, M. de Shazo, and W. C. van Cleave, *J. Org. Chem.*, 1971, **36**, 1745.
50. M. Brickman, J. H. P. Uttley, and J. H. Ridd, *J. Chem. Soc.*, 1965, 685.
51. C. Eaborn, *J. Chem. Soc.*, 1956, 4858.
52. P. W. Robertson, P. B. D. de la Mare, and B. E. Swedlund, *J. Chem. Soc.*, 1953, 782.
53. R. P. Bell and E. N. Ramsden, *J. Chem. Soc.*, 1958, 161.

54. L. M. Stock and H. C. Brown, *Adv. Phys. Org. Chem.*, 1963, **1**, 74.
55. J. D. Roberts *et al.*, *J. Am. Chem. Soc.*, 1954, **76**, 4525.
56. C. Eaborn and K. C. Pande, *J. Chem. Soc.*, 1961, 5082.
57. C. Eaborn, A. R. Thompson, and D. R. M. Walton, *J. Chem. Soc. B*, 1969, 859.
58. F. G. Bordwell and K. Rohde, *J. Am. Chem. Soc.*, 1948, **70**, 1191.
59. J. W. Baker and I. S. Wilson, *J. Chem. Soc.*, 1927, 842.
60. D. R. Harvey and R. O. C. Norman, *J. Chem. Soc.*, 1962, 3822.
61. C. Eaborn, *J. Chem. Soc.*, 1956, 4858; C. Eaborn, D. R. M. Walton, and D. J. Young, *J. Chem. Soc. B*, 1969, 15; C. Eaborn and P. M. Jackson, *J. Chem. Soc. B*, 1969, 21.
62. J. R. Knowles and R. O. C. Norman, *J. Chem. Soc.*, 1961, 2938.
63. A. F. Holleman, *Chem. Rev.*, 1925, **1**, 187.
64. H. C. Brown and L. M. Stock, *J. Am. Chem. Soc.*, 1962, **84**, 1238.
65. R. Taylor, *J. Chem. Soc., Perkin Trans. 2*, 1973, 253.
66. M. J. S. Dewar, T. Mole, and E. W. T. Warford, *J. Chem. Soc.*, 1956, 3581.
67. H. H. Hodgson and J. Walker, *J. Chem. Soc.*, 1933, 1346.
68. E. R. Ward and J. G. Hawkins, *J. Chem. Soc.*, 1954, 2975.
69. P. B. D. de la Mare and J. H. Ridd, *Aromatic Substitution*, Butterworths, London, 1959, p. 90.
70. W. H. Mills and I. G. Nixon, *J. Chem. Soc.*, 1930, 2510.
71. P. C. Hiberty, G. Ohanession, and F. Delbecgq, *J. Am. Chem. Soc.*, 1985, **107**, 3095.
72. R. H. Mitchell, P. D. Slowey, T. Kamada, R. V. Williams, and P. J. Garratt, *J. Am. Chem. Soc.*, 1984, **106**, 2431.
73. A. P. Laws and R. Taylor, unpublished work.
74. S. Sternhell, personal communication.
75. J. Vaughan, G. J. Welch, and G. J. Wright, *Tetrahedron*, 1965, **21**, 1665.
76. R. Taylor, *Chimia*, 1968, **22**, 1; R. Taylor, G. J. Wright, and A. J. Homes, *J. Chem. Soc. B*, 1967, 780; R. Taylor, *J. Chem. Soc. B*, 1968, 1402; 1559; 1971, 536.
77. For a recent account and leading references, see S. J. Formosinho, *J. Chem. Soc., Perkin Trans. 2*, 1988, 839.
78. D. A. McCaulay and A. P. Lien, *J. Am. Chem. Soc.*, 1952, **74**, 6246.

CHAPTER 3

Hydrogen Exchange

Hydrogen exchange is the general term used to describe reactions in which one isotope of hydrogen on an aromatic nucleus is replaced by another. The exchange reactions which have been studied are deuteriodeprotonation (deuteriation) and tritiodeprotonation (tritiation), in which hydrogen is replaced by deuterium and tritium, respectively, and protiodedeuteriation (dedeuteriation) and protiodetritriation (detritiation) in which deuterium and tritium, respectively, are replaced by hydrogen. The names by which these reactions are commonly known are shown in parentheses. Deuteriodetritiation has also been studied in connection with kinetic isotope-effect measurements. The reagents are acids or bases which transfer hydrogen cation (in one of its isotopic forms) to the aromatic compound, e.g. acid-catalysed dedeuteriation may be represented by

$$\mathrm{ArD + HA \rightleftharpoons ArH + DA} \tag{3.1}$$

For quantitative studies of the mechanism of aromatic substitutions, dedeuteriation and detritiation have a number of advantages over other aromatic substitutions:

(1) By a suitable choice of synthetic methods, the exact location of the labelled hydrogen may be known. The exchange rates of each of the specifically labelled isomers can then be measured, giving the partial rate factors directly. The advantages of this direct method over the competition method have been described in Section 2.3. (This advantage does not apply to deuteriation or tritiation, where it is necessary to presuppose the position of substitution during the reaction; generally more than one position will be substituted and analysis of the results may be complicated.)

(2) Hydrogen exchange can be carried out under a wide variety of conditions leading to an unparalleled $>10^{20}$ variation in reactivity of the electrophile. The accompanying change in selectivity makes it possible to obtain information, for example, about the way in which the activating or deactivating effect of a substituent is dependent upon the reactivity of the reagent.

(3) Steric effects are smaller in hydrogen exchange than in any other electrophilic substitution, and indeed are entirely absent in all but a very few

extremely hindered molecules. This makes the reaction particularly suitable for studying the electronic effects of *ortho* substituents.

(4) The reaction has a high sensitivity towards substituent effects and very accurate rate data may be obtained.

These advantages have encouraged the accumulation of more quantitative data (under a single condition) than for any other electrophilic substitution. The reaction is not merely of theoretical value, however. Because the reactivities of aromatic molecules in a given substitution are quantitatively related to those in another (Chapter 11), the information gained in hydrogen exchange may be used to predict the reactivity pattern in other substitutions. For example, kinetic studies in hydrogen exchange have shown that some isomer substitution patterns obtained in other reactions are in error.[1]

Aromatic hydrogens undergo exchange with those of acids, water, or the conjugate acids of bases, and these reagents usually act as solvents for the aromatics. Since the electrophile is therefore present in considerable excess, pseudo-first-order kinetics are obtained.[2] The kinetics may be followed in a number of ways: for deuterium exchange the deuterium content of the aromatic has been analysed by gravimetric analysis, density measurements, and infrared, mass, and nuclear magnetic resonance spectroscopy, and for tritium exchange the tritium content has been analysed by gas counting and scintillation counting. Of these methods, the last is the simplest and the most accurate, and has come to be that generally employed. However, it requires specific prelabelling of the aromatic, and for molecules where this involves difficult syntheses, the deuterium exchange–n.m.r. method may be preferable.

3.1 ACID-CATALYSED HYDROGEN EXCHANGE

The bulk of hydrogen-exchange kinetics have been carried out under conditions of acid catalysis, and a wide variety of acidic reagents have been employed, ranging from aqueous ammonium and biphosphate ions[3,4] to superacids.[5] The kinetic results obtained under each condition and the effects of co-solvents and catalysts have been described in detail.[2] The variation in the selectivity of the electrophilic hydrogen as conditions are changed may be illustrated by the spread in the partial rate factors for *p*-Me, for which values ranging from 175 (in 66.4 wt% H_2SO_4 at 65.75 °C)[6] to 6200 (in HBr at 20 °C)[7] have been recorded.*

The addition of small amounts of mineral acid to carboxylic acids raises the acidity considerably, thereby combining accessible exchange rates with the

*The lower value of 83 estimated for aqueous sulphuric acid[8] (quoted in ref. 11) has been shown to be too low by a factor of about 3.[9,10]

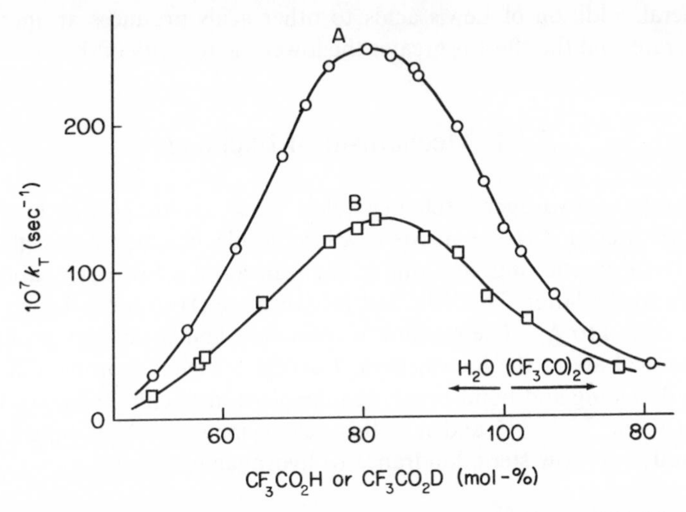

Figure 3.1. Plots of rate of detritiation of 4-tritio-*m*-xylene against mol-% CF_3CO_2H (A) and CF_3CO_2D (B).

property of a good solvent,[12] and such mixtures have been widely used for studying the exchange reaction. Unfortunately, other reactions (e.g. sulphonation) occur simultaneously in these and many of the other media employed for hydrogen exchange, and the extent of the side-reaction must be assessed if accurate values for the exchange rates are to be obtained.[10,13] By contrast, side-reactions are usually absent in hydrogen exchange in pure trifluoroacetic acid, and this feature, together with the convenient rates arising from the high acidity, have led to this being the acid most commonly used for exchange-rate measurements. The acid may be used over the temperature range from −10 to at least 180 °C,[14] so that a wide range of aromatic reactivities may be examined. The disadvantages of the acid are sensitivity to impurities and relatively high cost. Impurities may be removed by distilling from concentrated sulphuric acid and then from silver trifluoroacetate[15] (or more conveniently from silver oxide followed by sulphuric acid). The cost may be reduced by recovery and reuse (via neutralization, and distillation of the dried sodium salt with concentrated sulphuric acid[16]). Different batches of purified acid may give slightly different exchange rates, but this is inconsequential as rates can be measured relative to some standard (usually [4-^{3}H]toluene), the derived partial rate factors being independent of trivial differences in the acid.

Addition of aprotic solvent to trifluoroacetic acid generally produces a rate decrease.[2] Addition of water up to a concentration of ca 20 mol-% increases the rate, owing to increased solvation of the transition state. Thereafter the rate decreases because the decreasing acidity becomes the dominant factor governing the rate (Fig. 3.1).[15] This effect is also found with other polar co-solvents.[17]

In general, addition of Lewis acids to other acids produces an increase in exchange rate, and the effect is greater the lower the polarity of the protic acid.[2]

3.1.1 Mechanism of Exchange

Acid-catalysed hydrogen exchange takes place via the A-S_E2 mechanism (eq. 3.2) (see Section 2.1). It consists of a bimolecular reaction between an acid HA and the aromatic ring, resulting in the transfer of a hydrogen cation from the former to the latter. The Wheland intermediate (**1**) so formed then loses a hydrogen cation to A^-. The reaction is reversible, and the energy profile must be symmetrical about the intermediate **1** except for small isotopic differences. Thus bond-making and bond-breaking take place in essentially identical steps, as shown in eq. 3.2,[4,10,18] and it is this near symmetry which contributes to the generally very low steric hindrance to hydrogen exchange.

$$\text{HA} + C_6H_5H^* \underset{k_{-1}}{\overset{k_1}{\rightleftharpoons}} [C_6H_5(H)(H^*)]^+ \; (\mathbf{1}) \underset{k_{-2}}{\overset{k_2}{\rightleftharpoons}} C_6H_5H + H^*A \qquad (3.2)$$

(i) Nature of the acid catalysis

Earlier work had indicated that the reaction was specific-acid catalysed,[19–22] which led, by virtue of the symmetrical nature of the reaction pathway, to the proposal of a mechanism involving π-complexes. The experimental evidence upon which this was based was shown to be in error by Eaborn and Taylor,[10] who proposed the alternative A-S_E2 mechanism. This mechanism was subsequently confirmed by the observation of catalysis by a variety of acidic species (*general-acid catalysis*) by many workers.[3,4,23–35] Moreover, the assumption of the π-complex mechanism was based on the Zucker–Hammett hypothesis, viz. that a water molecule cannot be convalently bound in the transition state if the catalysing acid is H_3O^+.[36] However, this is now recognized to be an erroneous diagnosis of mechanism because similar bases show different protonation behaviour in a given acid.[15,18,37–44]

(ii) Intermediates

The A-S_E2 mechanism implies the existence of a Wheland intermediate (**1**). The presence of such an intermediate was first indicated by the observation that anthracene in H_2SO_4 gives a yellow species with an electronic absorption spectrum different from that of anthracene. Structure **2** was assigned to this species.[45] More recently, n.m.r. studies have given evidence for the presence of an aliphatic CH_2 group in this and related systems.[46–48] By the n.m.r. method it is possible to determine the charge distribution in these structures, and for the benzenonium ion itself (**3**).[48] This has important theoretical and experimental consequencies especially with regard to hydrogen exchange, and is discussed further below (Section 3.1.2).

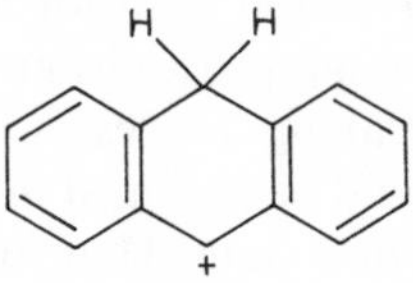

H H etc. H H 0.26 0.09 0.30

(**2**) (**3**)

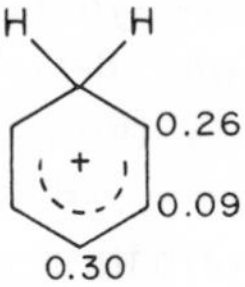

(iii) Extent of proton transfer

The transition state structure will differ from that of the intermediate in the extent of bond-breaking and bond-making, which will vary according to the reactivities of the aromatic and the catalysing acid. Consequently, work has been directed towards determining the extent to which the incoming proton (or isotope) has been transferred to the aromatic in the transition state. The extent of proton transfer is measured by the *Brønsted coefficient*, which is the slope of a plot of the logarithms of the exchange-rate coefficients of either a given aromatic base against the pK values of a range of catalysing acids (designated α) or of a range of aromatic bases against their pK values at a constant acidity (designated β). Thus the value of the coefficient should be 0 when no transfer has occurred and 1.0 when transfer is complete, as in the Wheland intermediate; a value between these extremes can be expected for the A-S_E2 mechanism. For example, detritiation of [1-^{3}H]azulene by anilinium ions, carboxylic acids, and dicarboxylic acid monoanions gave α-values of 0.61, 0.67, and 0.68, respectively.[49] As expected, the values become larger as the acid becomes weaker, corresponding to a later transition state in which the proton is more transferred. Likewise, the small (average) value of 0.54 for detritiation of [3-^{3}H]guaiazulene[49] accords with the expectation that the transition state for exchange of a more reactive aromatic should occur earlier along the reaction path. Detritiation of 1,3,5-trimethoxybenzene gave α-values of 0.56–0.71 (depending on the acid type)[4,50] and 3-labelled indoles gave β-values of 0.67–0.75 for detritiation by hydroxonium ion and acetic acid, respectively.[51] Challis and Millar[51] suggested that the Brønsted coefficients may not be a meaningful measure of the transition state structure because detritiation of [3-^{3}H]-2-methylindole gave α-values of 0.46[51] to 0.58,[52] i.e. different from the above β-values; evidence from aliphatic chemistry also casts some doubt on the validity of the coefficients.[53] On the other hand, one cannot expect precisely identical α- and β-values for reaction of a given aromatic. For reaction with a range of acids, a spectrum of transition states will be encountered so that the α-values determined will only represent the average transition-state structure. Indeed, a plot of log (rate coefficient) vs pK_a of the catalysing acid should be a curve; similar arguments apply to the interpretation of the β-values.

(iv) Isotope effects

The near symmetry symmetry of the reaction path for hydrogen exchange means that the isotope effects in bond-making processes are compensated to a substantial extent by those in bond-breaking processes. Consequently, the overall isotope effects are smaller than would be obtained in a substitution with comparable bond formation in the transition state. Thus, whereas the rates of dedeuteriation relative to deuteriation of 1,3,5-trimethoxybenzene in aqueous perchloric acid is 2.01, the relative rates of the steps k_2 in eq. 3.2 for these reactions is 8.1 (neglecting secondary isotope effects). The observed isotope effect is thus markedly reduced because in the first step of the reaction k_1 favours dedeuteriation by a factor of 4.02.[54] Similar results have been obtained in the hydrogen exchange of azulene.[32,34]

A further consequence of the symmetrical pathway is that the isotope effect should depend on the position of the transition state along the reaction coordinate, and should pass through a maximum. Some indication of this is given by the data in Table 3.1,[9,35,55] which refer to the isotope effect for the second step of the reaction. In Table 3.1 the reactivity of the aromatic is the function indicating the transition state structure, but the difference in pK_a between the aromatic and the catalysing acid may also be used. The isotope effect also shows a maximum against this parameter and, further, this maximum is obtained as expected at $\Delta pK_a = 0$, i.e. where $\alpha = 0.5$.[35] Challis and Millar[51] questioned these conclusions since they found little variation in the isotope effect with reactivity of the aromatic or the catalysing acid in exchange of substituted indoles. They suggested that the differences previously observed (e.g. in Table 3.1) could arise from differences in proton tunnelling rather than transition-state symmetry. An alternative explanation could be that the most (kinetically) basic site in indole is nitrogen rather than the 3-position, especially since earlier work showed that either site may be the more basic, depending on the proton-donating acid.[56]

The solvent isotope effect, k_{HX}/k_{DX}, in hydrogen exchange can be more or less than 1.0 depending on the strength of the catalysing acid and the reactivity of the aromatic base. For example, detritiation of [2-^{3}H]-*p*-cresol in aqueous HCl[57] and of [2-^{3}H]-1,3,5-trimethoxybenzene in aqueous $HClO_4$[24,50] and tritiation of *N*,*N*-dimethylaniline in aqueous H_2SO_4[58] gave values of 1.6–1.7. By contrast, reaction of 1,3,5-trimethoxybenzene with aqueous acetate buffers[24] and [4-^{3}H]-

Table 3.1 Correlation of $k_2(H)/k_2(D)$ with reactivity for protiodetritiation

Aromatic	Log $k_{rel.}$	$k_2(H)/k_2(D)$
[^{2}H]Benzene	0	3.4
[3-^{2}H]Toluene	0.8	3.4
[2-^{2}H]Toluene	2.6	4.6
[4-^{2}H]Toluene	2.6	5.5
[2-^{2}H]Anisole	4.3	7.2
[4-^{2}H]Anisole	4.8	6.7
[2-^{2}H]-1,3,5-Trimethoxybenzene	10.0	6.7
[1-^{2}H]Azulene	11.5	9.2
[3-^{2}H]Guaiazulene-2-sulphonate	12.2	7.4
[1-^{2}H]-4,6,8-Trimethylazulene	12.8	9.6
[3-^{2}H]Guaiazulene	13.0	6.0

m-xylene with aqueous trifluoroacetic acid (TFA) media[15] gave values between 0.5 and 0.75. The effect of the reaction-path symmetry in hydrogen exchange is shown by the much lower value of 0.16 obtained in protiodesilylation (Section 4.7) of 4-chlorophenyltrimethylsilane in these latter media. The difference in solvent isotope effects between the two reactions is a measure of the relative rates of reversion of the intermediate to starting materials in the exchange.[15,54]

(v) *Energy-barrier heights*

The heights of the energy barriers in the reaction pathway have been measured in protiodetritiation of [1-^{3}H]azulene in aqueous $HClO_4$.[58] The free energy of activation of the first step is 73.3 kJ mol^{-1}, an additional 6.3 kJ mol^{-1} being required to overcome the second barrier; the overall and relative heights of these barriers will, of course, depend on the reactivity of both the aromatic and the catalysing acid.

(vi) *Variation of rates with temperature*

A linear correlation of logarithms of reaction rates with reciprocal temperature (Arrhenius correlation) indicates generally that the reaction mechanism is constant throughout the temperature range. Hydrogen exchange is probably unique in that reaction rates have been measured over a 110 °C range, using TFA, and show some curvature in the Arrhenius plots, as did exchange rates determined over a 65 °C range using H_2SO_4.[59] It is unlikely, however, that this curvature is due to a change in mechanism, but rather that the acidity of acids generally decreases with increasing temperature.[60]

3.1.2 Substituent Effects

Partial rate factors, and the substituent effects which govern them, are discussed below. Three points are emphasized:

(1) The data obtained in some cases are approximate because no corrections were made for the principal side-reactions.

(2) The partial rate factors for the four hydrogen-exchange processes studied are considered together in order to simplify the discussion. Since dedeuteriation and detritiation involve the severence, in the formation of the second transition state, of a C—D and a C—T bond, respectively, partial rate factors for the two processes should be different.[9] Hence partial rate factors for the dedeuteriation of toluene in aqueous H_2SO_4 are greater than those for the corresponding detritiation.[9,38] The ratio of the ρ-factors is ca 1.03, which implies that the severence of the C—T bond is less advanced at the second transition state than severence of the C—D bond. The first transition state will be closely similar for both processes, with differences from secondary effects only. Since the reaction path for deuteriation must be the exact reverse of that for dedeuteriation, identical transition states must be encountered on each reaction path. Moreover, the differences in free energies of the ground states of a pair of reactants and products will not differ appreciably so that similar partial rate factors are expected for each reaction and the data in Table 3.2 show that this is so. The differences between the partial rate factors for dedeuteriation and detritiation are not significant enough to affect any of the conclusions regarding substituent effects described below.

(3) Because of the advantages of hydrogen exchange over other electrophilic substitutions, rate data obtained in the reaction are the most meaningful in terms of the interpretation of substituent effects. Substituent effects obtained in the reaction are therefore discussed fully and generally, discussion in other chapters being limited to those substituents which have either not been examined in hydrogen exchange or which show unique effects in the particular reaction.

3.1.2.1 Alkyl substituents

The partial rate factors and related data for hydrogen exchange of monoalkylated benzenes are given in Tables 3.2 and 3.3 and show the following features:

(1) The spread in rates depends markedly on the nature of the medium, so the necessity for stating the conditions of the exchange reaction is emphasized. It is also necessary to consider the temperature of the reaction because the wide range of temperatures at which rates can be measured means that the relationship governing the temperature dependence of Hammett ρ-factors:

$$\rho T = \rho' T' \tag{3.3}$$

must be taken into account. This produces a larger effect than may be realized,

Toluene					Ethylbenzene				Isopropylbenzene				*tert*-Butylbenzene					Conditions	Ref.
o	*m*	*p*	*o*:*p*	$\log f_o / \log f_p$	*o*	*m*	*p*	$\log f_o / \log f_p$	*o*	*m*	*p*	$\log f_o / \log f_p$	*o*	*m*	*p*	*o*:*p*	$\log f_o / \log f_p$		
		175													154			66.4 wt-% aq. H_2SO_4, 64.75 °C	6
283	6.3	235	1.20	1.03														80.66	
319[a]	6.2[a]	287[a]	1.11[a]															80.66	
			1.055															74.85	
			1.04															73.25	
																0.955	0.99	71.41 wt-% aq. H_2SO_4, 25 °C	10, 13, 61, 62
250	5	250	1.00	1.00									170		180			71.34	
			1.00															70.70	
			0.99															69.00	
			0.98															64.50	
330	7.2	313	1.055	1.01									393	23.7	387	1.015	1.00	CF_3CO_2H(92.04)–H_2O(5.45)–$HClO_4$(2.51)[b], 25 °C	13
305	5	313	0.975											9.3	230			CH_3CO_2H(26.0)–H_2O(34.0)–H_2SO_4(40.0)[b], 25 °C	13
234[c,d]	3.2[c,d]	347[c,d]	0.675	0.95											363[c,d]			CF_3CO_2H(97–75)–H_2SO_4(3–25), 25 °C	12
253	4[d]	421	0.60	0.91	257	6.9[d]	449	0.91	259		493	0.90	198	7[d]	490	0.40	0.85	aq. CF_3CO_2H,[e] 70 °C	63, 64, 65
						4.5[f,g]			250[f]	6.2[f,g]	483[f]	0.90							
219[h,i]	6.1	450[h,i]	0.49	0.88[i]					223		544	0.86	242	35.3	535	0.45	0.87	CF_3CO_2H, 70 °C	6, 62, 66, 67
			0.39													0.37		$C_3F_7CO_2H$, 70 °C	62
541	9.2	702	0.77										620	32	863	0.72		CF_3CO_2H(95.31)–H_2O(2.21)–H_2SO_4(2.48)[b], 25 °C	13, 62
			0.5															Liquid HI, 25 °C	69
1540	6.8	6000	0.26	0.84	1600	5.2	6600	0.84	1400	6.0	5800	0.84	1300	6.0	4800	0.84		Liquid HBr, 20 °C	7

[a] For dedeuteriation. The other values obtained in aqueous H_2SO_4 refer to detritiation.
[b] Medium composition in molar percentages (in parentheses).
[c] Uncorrected for errors due to accompanying sulphonation.
[d] Approximate values.
[e] This medium would have contained traces of HCl.[15]
[f] Values obtained for dedeuteriation; other data under these conditions are for deuteriation.
[g] Data obtained at 100 °C.

Table 3.3 Partial rate factors and $\log f_o$:$\log f_p$ ratios for hydrogen exchange of alkylbenzenes in trifluoroacetic acid at 70 °C[67,71,72]

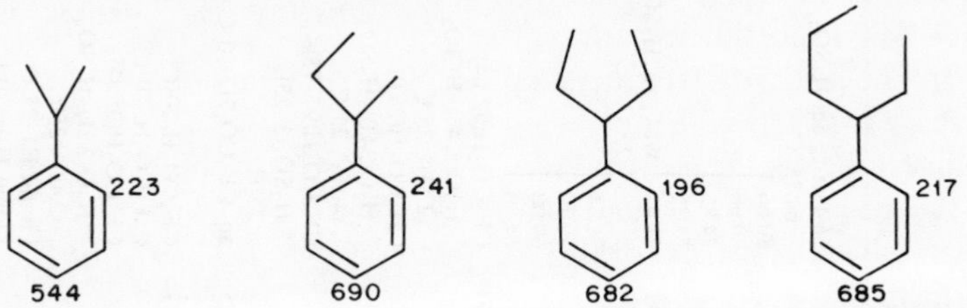

	223	241	196	217
	544	690	682	685
$\log f_o$:$\log f_p$	0.86	0.84	0.81	0.825
	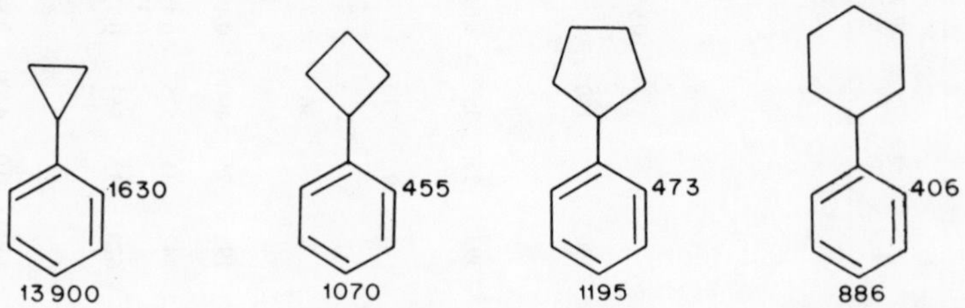 1630	455	473	406
	13 900	1070	1195	886
$\log f_o$:$\log f_p$	0.77	0.875	0.87	0.885
	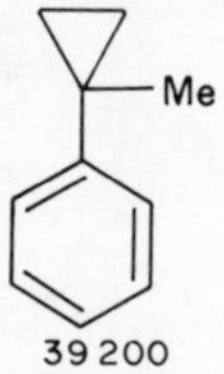 Me			
	39 200			
	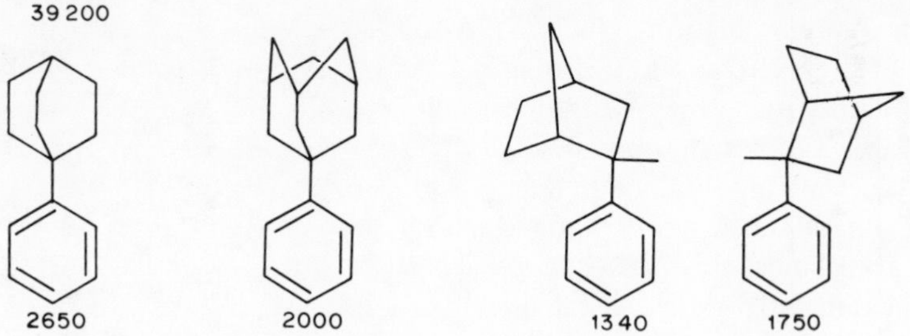 2650	2000	1340	1750

e.g. a partial rate factor of 250 at 25 °C is equivalent to a value of only 83 at 100 °C if the relationship holds perfectly. Since under most conditions hydrogen exchange is almost entirely unaffected by steric factors, it should adhere to this relationship better than any other electrophilic substitution. In practice, the variation is slightly less than predicted by the relationship, which may be due to the decrease in the acidity of the medium with increasing temperature.[60] Hence the value of f_p^{Me} of 250 at 25 °C (aqueous H_2SO_4) is predicted to be 129 at

65.75 °C compared with the observed value of 170, and the value of $f_m^{t\text{-Bu}}$ of 35.3 at 70 °C (TFA) is predicted to be 26.5 at 100 °C compared with 29.8 observed;[68] the change in partial rate factor is about two thirds of that predicted.

(2) Alkyl substituents are clearly *ortho*, *para*-directing and activating.

(3) The *meta* partial rate factors for the methyl and *tert*-butyl groups do not increase regularly with increase in the *para* values. In some cases this may be due to inaccuracy in the *meta* data as a result of side-reactions, e.g. sulphonation, or (in the case of deuteriation) to the *meta* reaction being less than 1% of the overall observed exchange. Nevertheless, discrepancies remain and these arise from steric hindrance to solvation (Sections 1.4.3 and 11.1.4); this causes *meta* partial rate factors obtained under good solvating conditions (e.g. HOAc–H_2O–H_2SO_4) to be diminished relative to those obtained under poor solvating conditions (e.g. TFA).[68]

(4) Media composed mainly of sulphuric acid give the Baker–Nathan activation order (*p*-Me > *p*-*t*-Bu), whereas media composed mainly of TFA give the inductive order.[13] Hydrogen exchange is one of two electrophilic substitutions known in which this order can be reversed merely by changing the solvent; the other is molecular bromination (Section 9.6.1). This result has been of great significance in unravelling the cause of the Baker–Nathan order, which had been commonly assumed to be hyperconjugation (Section 1.4.3). If this were the cause then the order should become more pronounced as the electron demand by the reagent (measured by the value of f_p^{Me}) increased.[73] The hydrogen-exchange results show that in fact the reverse is true, and that the occurrence of the order is dependent on some property of the solvent. Gas-phase data confirm this view[68,74] and, together with the hydrogen-exchange results, show that in the more polar (i.e. solvating) media the *tert*-butyl group, because of its size, prevents maximum solvation of the charge acquired by the aromatic ring in the transition state. This same effect also causes a reduction in all *meta* partial rate factors obtained under these conditions [and noted in (3) above], and it can be seen from Table 3.2 that the maximum $f_p^{t\text{-Bu}}:f_p^{Me}$ values and maximum (accurate) f_m values are all obtained in media composed mainly of TFA. Of all electrophilic aromatic substitutions, hydrogen exchange in anhydrous TFA gives results that parallel most closely those obtained in the gas phase, and are therefore the most meaningful for theoretical analysis. These aspects are considered further in Section 11.1.4.

(5) The bicyclo[2.2.2]octyl-, adamantyl-, *exo*-norborn-2-yl-, and *endo*-norborn-2-yl *para* substituents are all more activating in hydrogen exchange (and even more so in gas-phase elimination of 1-arylethyl acetates)[74] than in related reactions carried out in good solvating media.[72] This again shows the relative absence of steric hindrance to solvation in TFA. For the neopentyl substituent (*t*-$BuCH_2$—), f_p for detritiation in TFA at 70 °C is 472,[72] i.e. greater than for *p*-ethyl, whereas in molecular bromination employing a more polar solvent (Section 9.6.2) the converse is true; in gas-phase elimination the activation of *p*-neopentyl relative to *p*-ethyl is even greater.[74] The hydrogen-exchange (and

elimination) results again derive from the relative absence of steric hindrance to solvation and showed that the alternative explanation of the bromination results ('secondary hyperconjugation')[75] is incorrect.

The bicyclo[2.2.2]octyl substituent is significantly more electron releasing than the apparently similar adamantyl substituent. This has been ascribed to steric acceleration of carbon–carbon hyperconjugation in the former (4), which relieves eclipsing interactions (5) which are absent in the adamantyl substituent (6).[72]

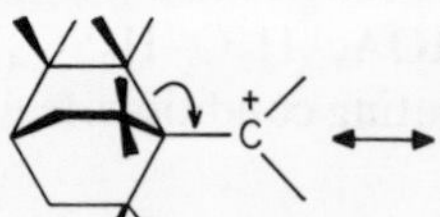

(4) 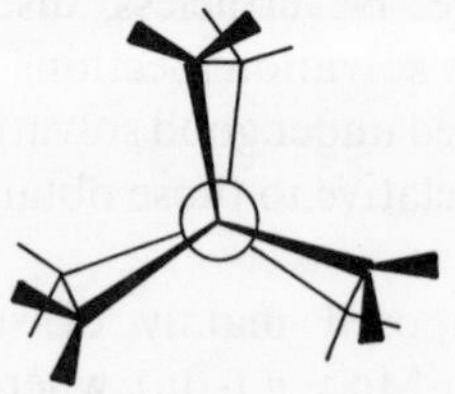 (5) 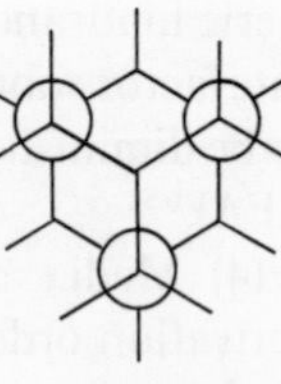 (6)

Steric hindrance to carbon–carbon hyperconjugation is believed to be the cause of the lower activation by the *endo*-norborn-2-yl substituent relative to the *exo* isomer. For maximum carbon–carbon hyperconjugation the aryl p orbitals must be able to lie coplanar with the electrons of the 1,2- and 2,3-bonds. There is no restriction on this in the *exo* isomer, whereas in the *endo* isomer there is an unfavourable interaction (7) between the aromatic *ortho* and *endo* hydrogens on C-6.[72,74] This view is confirmed by the occurrence of dealkylation alongside detritiation of the *endo* (but not the *exo*) isomer.[72] Dealkylation is known to occur when there is substituent–*ortho*-hydrogen interaction,[67] i.e. dealkylation becomes sterically accelerated.

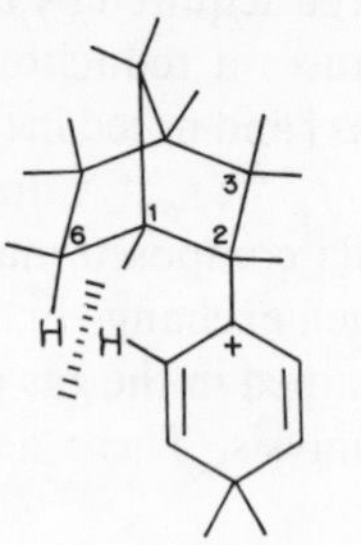

(7)

(6) The *ortho*:*para* ratios for toluene and *tert*-butylbenzene are similar, although in some media both are considerably less than the statistical value of 1.0. This argues against ascription of low ratios wholly or even mainly to steric effects, for if a low ratio for toluene was due to steric hindrance to *ortho* substitution, the ratio for *tert*-butylbenzene would be even lower.[62] One reason for the apparent decrease in ratio with increasing partial rate factor is simply a mathematical

consequence of the *Selectivity Principle* (Section 11.2.1), a proper measure of the electronic effects of the substituents being the log f_o:log f_p ratio.[76] These 'selectivity-corrected' values show a smaller variation in the ratio, but a genuine decrease remains. It has been argued that where steric hindrance is relatively unimportant the ratios are governed mainly by electronic effects which themselves depend on the extent of deformation of the aromatic sextet in the transition state.[77] This theory is developed in detail later (Section 11.2.4). One of its important conclusions is that the ratio for toluene should decrease as the electron demand of the reagent increases. The exchange results are consistent with this, for the ratios for both toluene and *tert*-butylbenzene decrease as the selectivity of the attacking species increases.[62]

(7) The cyclopropyl substituent is much more activating than other alkyl substituents. Severe strain in the ring weakens the bonds, thereby facilitating electron release by carbon–carbon hyperconjugation (**8**); cf. that described above for the bicyclo[2.2.2]octyl substituent. It follows that the electron release will be conformationally dependent[78] and this may account for the low log f_o:log f_p value, i.e. in this conformation (**8**), one *ortho* position is substantially shielded, although this explanation is not wholly satisfactory in view of the very low steric hindrance to hydrogen exchange. The results confirm that the high *ortho*:*para* product ratio obtained in nitration of cyclopropylbenzene by acetyl nitrate[79] is anomalous; a number of other aromatics give abnormally high ratios in nitration under these conditions [Section 7.4.3(iii)].

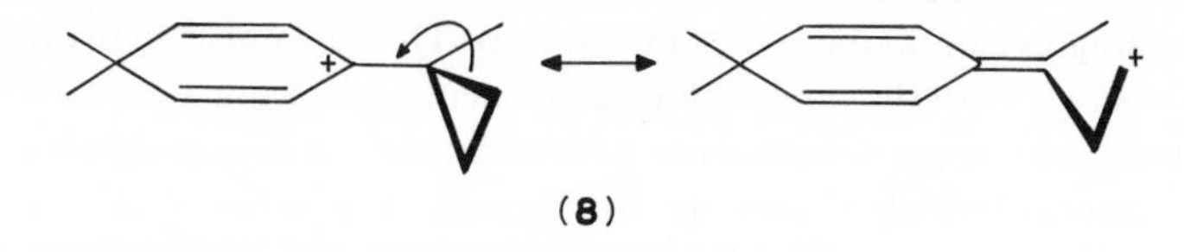

(**8**)

The *p*-1-methylcyclopropyl substituent is more activating than the *p*-cyclopropyl substituent (cf. *tert*-butyl vs. methyl) and this contrasts with molecular bromination, where steric hindrance to solvation again causes the order to be reversed.[80]

(8) The cycloalkyl groups activate in the order cyclopropyl > cyclopentyl > cyclobutyl > cyclohexyl, and the same order is obtained in solvolysis of 2-aryl-2-chloropropanes[81] and in nitration [Section 7.4.3.(iii)]. An important difference is that in the solvolysis the latter three groups are less activating than *p*-methyl. This result parallels the relative reactivity orders for *p*-methyl and *p*-*tert*-butyl in solvolysis[82] and in hydrogen exchange (in TFA), and show once again that bulkier aromatics do not exhibit their full reactivity under good solvating conditions, owing to steric hindrance to solvation.

The greater activation by the cyclopentyl group relative to the cyclohexyl and cyclobutyl groups (confirmed in nitration[83]) may arise because the C—H bonds of the cyclopentyl substituent are eclipsed. Interaction between adjacent hydrogens may thus cause elongation of the C—H bonds and thereby enhance inductive delocalization. Further, this eclipsing can be reduced by either C–C hyperconjugation (which produces ring opening, cf. **8** or by C–H hyperconjugation (**9**); this is thus a further example of *steric enhancement of hyperconjugation*. The C—H bonds are also eclipsed in the cyclobutyl substituent, but here they are further apart because of the greater strain in the ring, so steric facilitation of C–H hyperconjugation should be less. On the other hand, steric enhancement of C–C hyperconjugation should be greater and the latter may account for the greater activation by cyclobutyl relative to cyclohexyl.

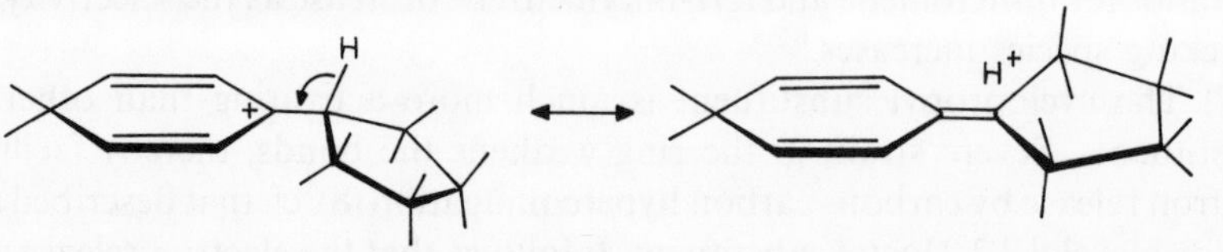

(**9**)

(9) The cycloalkyl substituents are all more electron-supplying than their open-chain counterparts, and this is confirmed in nitration.[83] Additional to the electronic explanations given above may be the fact that the transition states for reaction of the less bulky cyclo compounds may be better solvated.

(10) The log f_o:log f_p values for the secondary open-chain alkyl-substituted benzenes PhR (Table 3.3, R = but-2-yl, pent-3-yl, or hex-3-yl) are lower than those for the alkylbenzenes under the same conditions (Table 3.2) and the corresponding cycloalkylbenzenes (Table 3.3). The ratio also tends to decrease with increasing size of the open-chain substituent, indicating a steric effect, albeit a very small one; these substituents have a larger steric requirement than the *tert*-butyl group (which shows no hindrance). The ratio is of fundamental significance because not only do many other substituents in hydrogen exchange (and indeed in other electrophilic substitutions for which steric effects are minimal) give the same value,[84] but it is also predicted from the charge distribution in the Wheland intermediate (**3**). This means that in principle a linear free energy relationship can be derived for *ortho* substitution in the same way as for *meta* and *para* substitution (Section 11.2.4).

3.1.2.2 Polyalkyl substituents

Relative rates for deuteriation of polymethylbenzenes in aqueous TFA gave excellent agreement between observed and calculated reactivity,[95] and this has been widely quoted in support of the *Additivity Principle*. However, the data were only approximate,[86] and the more accurate partial rate factors for detritiation (Table 3.4,[62,66,70,87,88]) show that the observed values are up to ten times smaller than calculated. Moreover, this discrepancy increases regularly with increasing reactivity of the aromatic. This is consistent with current theories

Table 3.4 Observed and calculated partial rate factors for detritiation of polymethylbenzenes in trifluoroacetic acid

Position of methyl substituents	Position of tritium	Observed partial rate factor[a]	Calculated partial rate factor
1	2	210	
	3	6.12	
	4	439	
1,4	2[b]	1510	1285
1,2	3	1330	1285
	4	1900	2690
1,3	2[b]	39 100	44 100
	4	77 700	92 200
	5	45	37.5
1,2,4	3	165 000	270 000
	5	288 000	564 000
	6	8175	7865
1,2,3	4	253 000	564 000
	5	8560	16 450
1,3,5	2	7 100 000	19 400 000
1,2,3,4	5	1 190 000	3 450 000
1,2,4,5	3	776 000	1 650 000
1,2,3,5	4	25 750 000	118 000 000
1,2,3,4,5	6	73 800 000	711 000 000

[a]Calculated using the correct coefficient of $0.095 \times 10^{-7}\,s^{-1}$ for detritiation of benzene[14] (cf. ref. 70).
[b]In detritiation in aq. H_2SO_4 media, f(obs.) is substantially less than f(calc.), but this may be an artifact of the variation in f^{Me} values with acid concentration.[61]

of substituent effects, viz. that the more reactive an aromatic, the nearer to the ground state will be the transition state for reaction of it, and the smaller the observed effect of an additional substituent. This is one aspect of the *Reactivity–Selectivity Principle* which is discussed more fully in Section 11.1.5. Other features of these data are the following:

(1) The deviation noted above is small compared with the overall reactivity relative to benzene, and this provides further evidence that steric hindrance to hydrogen exchange is very small.

(2) The 3-position of *o*-xylene is *ortho* and *meta* to a methyl substituent whereas the 4-position is *para* and *meta* to a methyl substituent. The ratio of their reactivities should therefore be the same as that of the *ortho* and *para* positions in toluene, but this is not so, the respective $\log f_o{:}\log f_p$ values being 0.94 and 0.88.[87] The 4-position seems to be abnormally unreactive, and solvent effects are not responsible since the same result is observed in gas-phase studies of electrophilic reactivity.[89] Steric hindrance to hyperconjugation through

buttressing (**10**) may be the cause since conjugative effects are relayed more effectively to the *para* position.[70] This view is supported by the lower reactivity of the 3-position of *o*-xylene relative to that of the 2-position of *p*-xylene, the latter being also *ortho* and *meta* to methyl substituents. However, some other factor may be involved because, although this theory accounts for the 5-position of the highly buttressed 1,2,3-trimethylbenzene (p- $\times$ m^2-substituents) being hardly any more reactive than the 6-position of 1,2,4-trimethylbenzene (o- $\times$ m^2-substituents), it predicts that the difference in the reactivities of the 4-position of 1,2,3-trimethylbenzene and the 5-position of 1,2,4-trimethylbenzene (each with o- $\times$ m- $\times$ p-substituents) should be greater than between the 3-position of *o*-xylene and the 2-position of *p*-xylene; the converse is in fact observed.

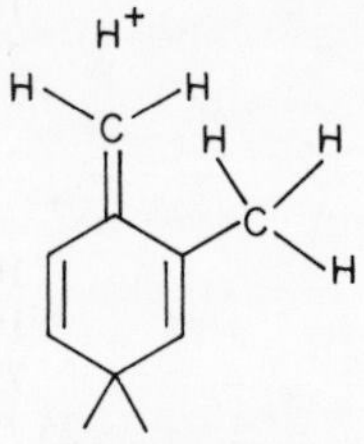

(**10**)

(3) The log f_o:log f_p ratio for 1,2,4-trimethylbenzene (0.895) is similar to that for *m*-xylene (0.885), toluene (0.88), and the value (0.87) predicted by the charge distribution in the Wheland intermediate (**3**) (Section 11.2.4). However, it is significantly different from that for *o*-xylene (0.94), yet the steric hindrance to hyperconjugation explanation should apply equally to each.

(4) An alternative explanation of these results is that there is some partial bond fixation in the polymethyl aromatics. Buttressing between adjacent methyl groups could cause elongation of the bonds between the aromatic carbons to which they are attached, resulting in lowering of the bond order and reduced transmission of substituent effects.

3.1.2.3 Cyclic alkyl substituents: the Mills–Nixon effect

Partial rate factors for detritiation in anhydrous TFA at 70 °C of aromatics bearing cyclic alkyl substituents, viz. indane (benzcyclopentene), tetralin (benzcyclohexene), and benzsuberane (benzcycloheptene), are shown in Table 3.5 together with those for *o*-xylene.[90,91] The data show that the α-position of indane is itself exceptionally unreactive, so accounting for the bromination results of Mills and Nixon (Section 2.6).[92] The explanation of this unreactivity by Vaughan *et al.*[93] was extended by Taylor and co-workers[76,88,94–96] (see Section 2.6) to predict that the reactivity of the β-position of indane should be

Table 3.5 Partial rate factors for detritiation in trifluoroacetic acid at 70 °C

	Me, Me; 1330; 1900	990; 4040	3750; 4050	1105; 3150
$\frac{\log f_o}{\log f_p}$	0.95	0.83	0.99	0.87

enhanced.* Thus, whereas the log f_o:logf_p ratio for benzsuberane is exactly that theoretically predicted by the charge distribution in the Wheland intermediate (**3**), the ratios for indane and tetralin are low and high, respectively (the ratio for *o*-xylene is, as noted above, abnormally high).

One other factor may contribute towards making indane and tetralin overall more reactive than *o*-xylene. Both of the former molecules have C—H bonds at the side-chain α-carbon atoms which are constrained almost perpendicular to the adjacent p-orbital, e.g. **11**, and this must considerably facilitate C–H hyperconjugation.[98,99] Deuteriation of benzcylooctene in aqueous TFA showed that it was less reactive than benzsuberane,[100] and so there appears to be a regular decrease in reactivity as the alkyl chain is lengthened. This follows not only from the loss of conformational advantage, but also the larger chain may hinder solvation of the aromatic ring in the transition state.[91]

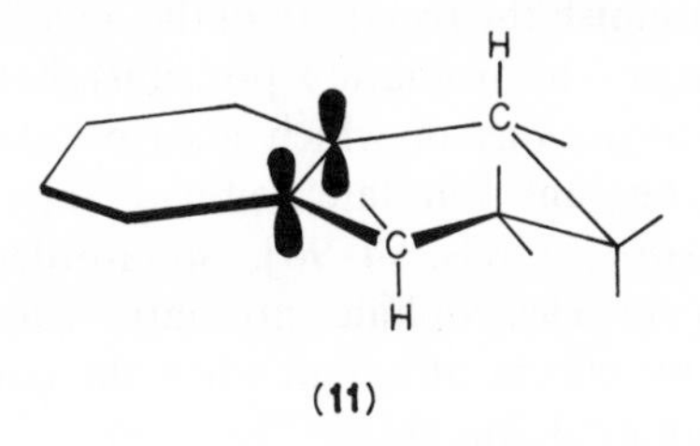

(**11**)

Taylor and co-workers showed that the bond-order theory explains why strained aromatic molecules *in general* are very unreactive at the α-position, and also that a parallel exists between the α/β-substitution ratio and the amount of strain. The cause of the abnormal substitution pattern in molecules such as benzcyclobutene, triptycene, biphenylene, etc., can therefore be understood;[76,88,94–96] further details are described in Sections 3.1.2.4 (9), 3.1.2.10 (11

*The β-position of indane should therefore be more reactive than that in tetralin as is found in protiodesilylation (Section 4.7) and solvolysis of 2-aryl-2-chloropropanes.[97] If the quoted experimental error[93] is taken into account, this order may also obtain in detritiation.

Table 3.6 Rate coefficients ($10^7 k/s^{-1}$) for detritiation in TFA at 20 °C

	HO, 28 000, Me, 79 000, Me, 35	HO, 44 200, 74 000, 71	HO, 98 000, 32 700, 60	HO, 3600, 4500
$\frac{k_\alpha}{k_{\alpha'}}$	800	623	1640	
	MeO, 62 000, Me, 105 000, Me, 54	MeO, 52 000, 129 000, 80	MeO, 280 000, 61 000, 79	
$\frac{k_\alpha}{k_{\alpha'}}$	1150	650	3540	

and 12), 3.1.2.12 (ii)(d), and 11.3. More recently, an attempt was made to explain the Mills–Nixon effect in terms of the strain in the bridgehead carbon atoms.[97,101] It was argued that this strain would modify their hybridization such that the orbitals which are used to form part of the aromatic ring have increased s character, and will therefore be more electronegative. Consequently the adjacent ring α-carbon atoms should be inductively deactivated. However, this effect should also diminish the reactivity of the β-carbon atoms, and further the reduction in reactivity of the aromatic α-positions should be less in reactions related to aromatic substitutions in which a carbocation is developed at a side-chain α-position in the transition state; neither of these predictions is fulfilled [Table 3.5; Section 4.7.1(ii)(17); refs. 94–96]. Bond-order effects are evidently the important factors in electrophilic aromatic substitutions which are dominated by conjugative effects, although when the latter are largely absent, electronegativity effects may dominate.[94,102]

Data for detritiation of oxy-substituted benzocycloalkenes in TFA at 20 °C (Table 3.6)[103] parallel the bromination results obtained by Mills and Nixon, but there are additional notable features. (Since oxy-substituents hydrogen bond in TFA,[104–106] which was not purified, and the exchange-rate coefficient for benzene is not known under this condition, partial rate factors are not calculated.) First, 7-hydroxybenzcycloheptene is less reactive than the other compounds, which confirms the results for the parent compounds (Table 3.5). Second, the ratios of the reactivities at the α- and α'-positions decreases along the series (β-oxy-)tetralin, *o*-xylene, indane. Since the α'-position is affected only by inductive effects of the oxy-substituent, the ratio provides a measure of the transmission of conjugative effects between the α- and β(oxy)-positions, and shows that even if there is no bond fixation in the ground state (Section 2.6), it must develop on going to the transition state.

Table 3.7 Rate coefficients ($10^7\,k/s^{-1}$) for hydrogen exchange[a] in TFA

				T/°C
	1,2-dimethoxybenzene: 36.1 (α), 10.4 (β) **(12)**	benzo-1,3-dioxole: 116 (α), 15.2 (β) **(13)**	benzo-1,4-dioxane: 49.0 (α), 18.6 (β) **(14)**	30
$\frac{\log f_\alpha}{\log f_\beta}$	0.885	0.835	0.915	
	4,5-dimethyl-1,2-dimethoxybenzene: 1300 **(15)**	5,6-dimethylbenzo-1,3-dioxole: 280 **(16)**	6,7-dimethylbenzo-1,4-dioxane: 2500 **(17)**	25
	MeO, Me, OMe substituted benzene: 21, 486 **(18)**	MeO-substituted dihydrobenzofuran: 3, 50 **(19)**	MeO-substituted chroman: 73, 25 **(20)**	20

[a]For reaction, see text.

Taylor[107] showed that the theory based on the change in bond lengths on going to the transition state also accounted for the relative positional reactivities of 1,2-dimethoxybenzene (**12**), benzo-1,3-dioxole (**13**), and benzo-1,4-dioxane (**14**); analysis[107] of deuteriation work of Serebryanskaya *et al.*[108] indicated that the reactivity of the α-position of **13** was considerably less than that of the other positions. This was confirmed by the more accurate rate coefficients (Table 3.7) for detritiation of **12**–**14**.[109] For benzene k may be calculated to be 6.3 $\times 10^{-11}\,s^{-1}$ under these conditions,[6] permitting calculation of partial rate factors and hence the $\log f_\alpha$:$\log f_\beta$ values shown in Table 3.7. These confirm that the ratios for **13** and **14** are low and high, respectively, as found for the carbocyclic analogues. More detailed analysis of the data is precluded by the hydrogen bonding of oxy-substituted aromatics which occurs in TFA [for example, whereas rate coefficients ($10^7\,k/s^{-1}$) for detritiation of the β-positions of **12** and **14** in TFA at 70 °C are 2320 and 2910 respectively, in a mixture (60:40, v/v) of TFA in HOAc the reactivity order is reversed, the respective coefficients being 52.6 and 50.6[110]]. Data for deuteriation of the dimethyl derivatives **15**–**17**[108] likewise show the diminished and enhanced reactivities, respectively, of the α-positions of **16** and **17** relative to their open-chain analogue.

The ring strain effects are also evident from the detritiation data for compounds **18**–**20**.[111] These also show the importance of hydrogen bonding because they are at least three orders of magnitude less reactive than their analogues in Table 3.6, whereas oxy substituents normally activate strongly relative to alkyl substituents.

For detritiation of hexahydrophenanthrylene (**21**) and hexahydroaceanthrylene (**22**) in TFA at 70 °C, the partial rate factors at the indicated positions were each 5.7×10^7. This is greater than the value which may be calculated from the effects of alkyl substituents in naphthalene (Section 3.1.2.13) and is attributed to steric facilitation of hyperconjugation.[112]

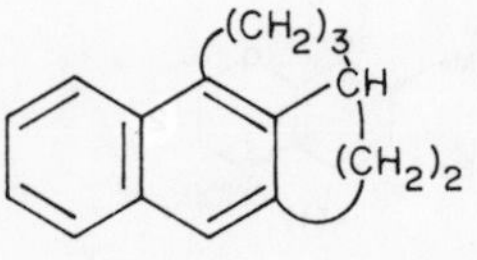

$(CH_2)_3$ CH $(CH_2)_2$ $(CH_2)_2$ CH $(CH_2)_3$

(**21**) 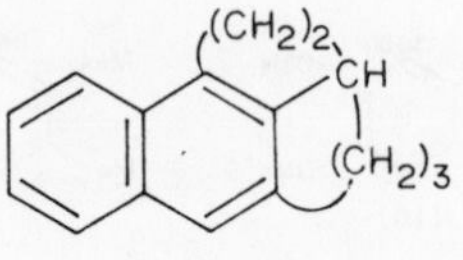 (**22**)

3.1.2.4 Substituted alkyl substituents

Partial rate factors for hydrogen exchange of derivatives of toluene under various conditions are shown in Table 3.8.[6,66,113–117] Notable features are the following:

(1) The CH_2SiMe_3 substituent is very strongly activating, showing that C–Si hyperconjugation (**23**) is greater than C–H hyperconjugation. This is because $SiMe_3$ is more electropositive than H, and the C—Si bond is weaker than the C—H bond.

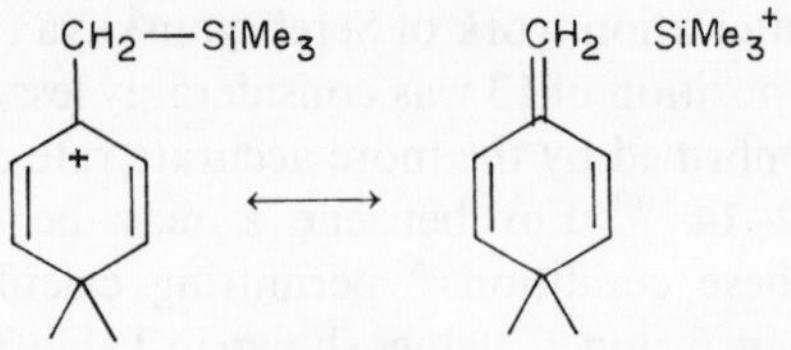

CH_2—$SiMe_3$ ⟷ CH_2 $SiMe_3^+$

(**23**)

(2) When the $SiMe_3$ group is conjugatively insulated from the ring a small activation (relative to methyl) remains. This confirms that the $SiMe_3$ substituent has a $+I$ effect.

(3) The lower *para* activation by $(CH_2)_3SiMe_3$ relative to the other $(CH_2)_nSiMe_3$ substituents is paralleled in protiodesilylation and nitration, [Section 7.4.3.(xi)]. An explanation for this anomaly is given in Section 4.7.1.(ii).

(4) With the exception of the $(CH_2)_2SiMe_3$ substituent, the other $SiMe_3$-

Table 3.8 Partial rate factors for substituted alkyl groups in hydrogen exchange

(i) Detritiation in TFA at 70 °C:

Substituent	f_o	f_m	f_p	$\log f_o / \log f_p$
CH_3	219	6.1	450	0.88
CH_2SiMe_3	9300		82 000	0.81
$(CH_2)_2SiMe_3$	450		810	0.91
$(CH_2)_3SiMe_3$	270		580	0.88
$(CH_2)_4SiMe_3$	270		690	0.86
CH_2Ph	47	3.8	120[a]	0.81
$CHPh_2$	14	2.3	42[b]	0.71
CPh_3	4.1	3.9	66	0.34
$(CH_2)_2Ph$	100c	5.7	200[d]	0.875

(ii) Deuteriation in liquid DBr at 25 °C[e]

Substituent	f_o	f_m	f_p	$\log f_o / \log f_p$
CH_3	1540		6000	0.84
CH_2SiCl_3	2.5	~0.03	56	0.82
$(CH_2)_2SiCl_3$	18		80	0.66
$(CH_2)_3SiCl_3$	90		380	0.76
$(CH_2)_4SiCl_3$	320		1300	0.80
$SiCl_3$	0.0003	0.03	0.00005	
CH_2Cl	0.29		1.14	

[a,b] Values of 1200 and 160 have been obtained for deuteriation in liquid DBr at 25 °C.[118]
[c,d] The corresponding values at 110 °C are 77 and 159, with $\log f_o{:}\log f_p = 0.86$.
[e] Data quoted as accurate to $\pm 20\%$.[116,117]

containing substituents give the constant $\log f_o{:}\log f_p$ ratio predicted by the charge distribution in the Wheland intermediate **(3)** (Section 11.2.4). The slightly lower value for the CH_2SiMe_3 substituent may reflect a small amount of steric hindrance, similar to that found with some *sec.*-alkyl substituents (Table 3.3).

(5) In contrast to the CH_2SiMe_3 substituent, CH_2SiCl_3 does not show strong activation because the inability of $SiCl_3$ to stabilize a positive charge makes C–Si hyperconjugation negligible. The results show $SiCl_3$ to have a $-I, -M$ effect, in contrast to $SiMe_3$, which is $+I, -M$. Deactivation by $SiCl_3$ at the *para* position is greater than at the *ortho* position because conjugative effects are relayed maximally to the *para* position, and this substituent gives a $\log f_o{:}\log f_p$

value (0.82) close to that (0.87) predicted theoretically. The ratios for some of the other $SiCl_3$-containing substituents are lower than expected, indicating some steric hindrance; these data and others below (Sections 3.1.2.5 and 3.1.2.6) suggest that the steric requirement of exchange in HBr is greater than under other conditions. For example, the relative rates of exchange at the 2- and 6-positions of 3-methyl(trichlorosilylmethyl)benzene[114] and 3-methylbenzyl chloride[115] (the 2-positions are more hindered in each case) is lower compared with that of the corresponding 2- and 4-positions in toluene.

(6) The ability to detect long-range substituent effects in hydrogen exchange (because of the high ρ-factor) is exemplified by the results for the $(CH_2)_4SiR_3$ substituents. Even through a four-carbon chain, a 5-fold deactivation by $SiCl_3$ and a 1.5-fold activation by $SiMe_3$ are observed.

(7) The *meta* and *para* partial rate factors for toluene, diphenylethane, diphenylmethane, and triphenylmethane show that phenyl has a $-I$ effect, and this also causes the $\log f_o{:}\log f_p$ ratio to decrease regularly on going to the latter two compounds. This ratio is however very much lower for tetraphenylmethane and it is probable that exchange at the *ortho* position of this highly crowded molecule is sterically hindered.

(8) The *meta* and *para* positions of tetraphenylmethane are more reactive than those in triphenylmethane. This is unlikely to be a solvation effect because almost the same rate ratio is obtained on adding a substantial quantity of a better solvent to TFA.[6] It is probable that steric crowding in tetraphenylmethane facilitates either neighbouring group stabilization of the transition state by one of the phenyl groups (**24**), or C–C hyperconjugation. In protiodesilylation, tetraphenylmethane is also more reactive than predicted, though less reactive than triphenylmethane,[6] which may reflect the smaller demand in this reaction for conjugative stabilization of the transition state.

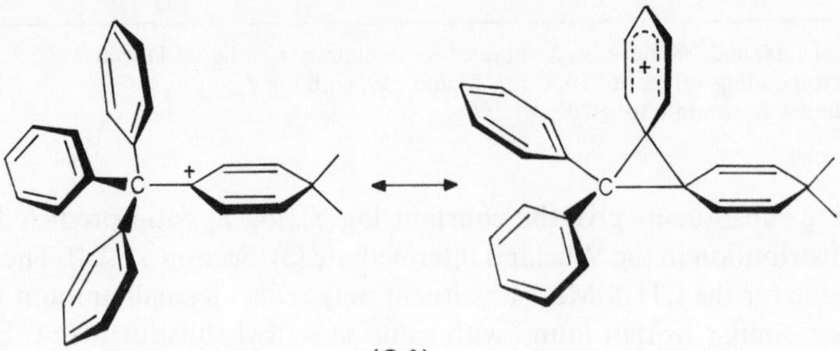

(**24**)

(9) Partial rate factors for detritiation of *o*-xylene, 9,10-dihydroanthracene, and triptycene (Table 3.9)[88] show that whereas the $\log f_\alpha{:}\log f_\beta$ ratio for 9,10-dihydroanthracene is the same as that for *o*-xylene, the value for triptycene (which has a strained side-chain) is much lower, as predicted by the theory based on bond-order effects (Section 3.1.2.3). Further, each position in 9,10-

Table 3.9 Partial rate factors for detritiation in trifluoroacetic acid at 70 °C

	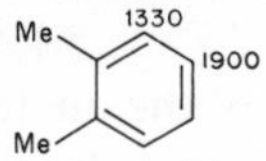 o-xylene (1330, 1900)	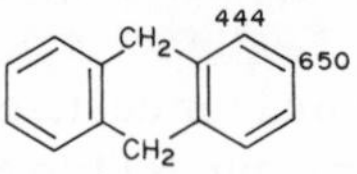 9,10-dihydroanthracene (444, 650)	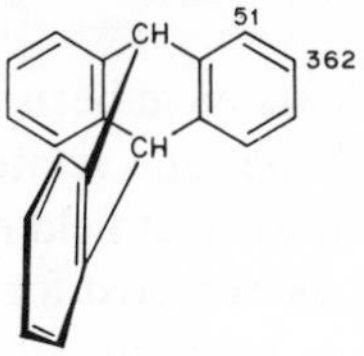 triptycene (51, 362)
$\log f_\alpha / \log f_\beta$	0.95	0.94	0.58

dihydroanthracene is reduced in reactivity by a factor of ca 3.0 relative to *o*-xylene, through the addition of the benzo substituent. Addition of a second benzo substituent should produce a similar rate reduction, the partial rate factors for triptycene being calculated to be 148 (α-position) and 221 (β-position). The observed α-reactivity is lower whereas that of the β-position is higher, as predicted by the bond-strain theory,[76] but not by the theory based on electronegativity effects.

9,10-Dihydroanthracene is ca twice as reactive as calculated by additivity based on the effect of the benzyl substituent (Table 3.8). This may reflect the reduced $-I$ effect of the phenyl group when acting through two chains, or it

Table 3.10 Rate coefficients for deuteriation in 96% sulphuric acid at 25 °C

(25) $R_3N^+(CH_2)_n-C_6H_4-(CH_2)_nNR_3^+$ (26)

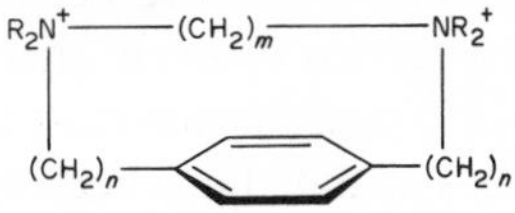

(25) n	(25) R	$10^5 k/s^{-1}$	(26) n	(26) R	(26) m
3	H	1.82×10^5			
3	Me	1.32×90^5			
2	H	725			
1	H	1.5×10^{-4}			
		2.10	3	Me	2
		0.132	2	Me	4

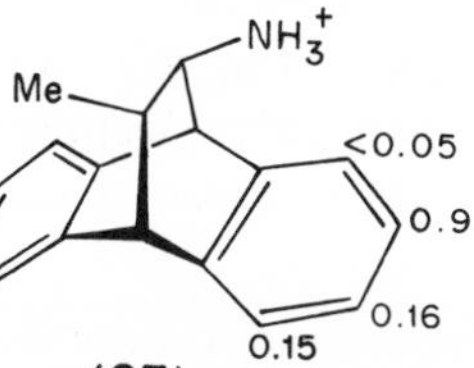

(27)

may be a further example of steric facilitation of hyperconjugation since one C—H bond on each methylene group is constrained parallel to the adjacent aryl p-orbital.

Rate coefficients have been measured for deuteriation of the quaternary salts of diazocyclophanes and related compounds (Table 3.10), in order to assess the importance of the direct field effect. The reactivities of compounds **25** decreases as *n* decreases, as expected for the operation of an inductive or field effect from the pole; the low reactivity of **26** relative to **25** for a given value of *n* suggested that a field effect was reponsible.[119] However, the partial rate factors for deuteriation of **27** in TFA–aqueous H_2SO_4, coupled with the fact that the exchange rates in each ring were identical, indicate that the NH_3^+ group exerts an inductive rather than a field effect in each ring.[120] Moreover, the compound with a *para* $(CH_2)_9$ bridge alone is very unreactive towards exchange,[121] probably owing to steric hindrance to solvation of one face of the aromatic ring, and it is this factor rather than a field effect which produces the low reactivity in compounds **26**.

Exchange between superacid and other aromatics possessing positively charged substituents has also been observed.[122]

3.1.2.5 Halogen substituents

Partial rate factors for exchange in the halobenzenes (Table 3.11)[5,7,110,123,124] show the following notable features:

(1) The data have been obtained using the widest variation in electrophilic reactivity for a given aromatic substitution, exchange in CF_3SO_3H being ca 2×10^{11} times faster than that in CF_3CO_2H at 70 °C.

(2) The data have been obtained over the largest temperature range employed for an electrophilic substitution, and therefore provide a valuable test of the Arrhenius relationship (eq. 3.3). Thus, although the rate factors depend on the conditions (shown also by the data for alkylbenzenes, Table 3.2), much of this variation arises from the temperature effect, e.g. the sixfold difference in partial rate factors for the *m*-Cl substituent at 55 and 180 °C can be shown to arise from the temperature difference. However, the observed deactivation at −40 °C is significantly less than calculated, probably owing to a much more reactive electrophile in CF_3SO_3H leading to a smaller ρ-factor.

(3) At lower temperatures the *o*- and *p*-chloro substituents are more deactivating than calculated from the partial rate factors at 55 °C, whereas at higher temperatures they are less deactivating than calculated.* This accords with a variation in the demand for conjugative electron release which will be strongest in the weakest acid (TFA) and weakest in the strongest acid (TFSA). This

*Comparison with the corresponding *meta* partial rate factors shows that this effect lies outside any possible failure to adhere rigidly to the Arrhenius relationship.

Table 3.11 Partial rate factors for hydrogen exchange of halobenzenes

Me	F			Cl			Br			I			
p	*o*	*m*	*p*	*o*	*m*	*p*	*o*	*m*	*p*	*o*	*m*	*p*	Conditions[a]
				0.005 (0.009	0.000345 0.000057	0.018 0.077)[b]							CF_3SO_3H, −40 °C
			1.73			0.127 (0.134)[b]			0.072			0.0865	CF_3CO_2H(73.40)–H_2O(1.98)–H_2SO_4(14.62), 25 °C
	0.136		1.79	0.035		0.161	0.027		0.098	0.043		0.112	CF_3CO_2H(96.77)–H_2O(0.82)–$HClO_4$(2.41), 55 °C
	0.126	~0.002		0.036	~0.001		0.027	~0.001	0.085	0.041	~0.003		CF_3CO_2H(89.50)–H_2O(4.55)–$HClO_4$(5.95), 55 °C
				0.047 (0.041	0.00115 0.00131	0.195 0.176)[b]							CF_3CO_2H(95.0)–CF_3SO_3H(5.0), 70 °C
				0.100 (0.088	0.00685 0.00662	0.44 0.27)[b]							CF_3CO_2H, 180 °C
240			2.18										76.29 wt-% aq. H_2SO_4, 25 °C
									0.073[c]				87.1 wt-% aq. H_2SO_4, 25 °C
313			1.7										CF_3CO_2H(92.04)–H_2O(5.45)–$HClO_4$(2.51), 25 °C
340			2.4										CF_3CO_2H(85.08)–H_2O(6.42)–H_2SO_4(8.50), 25 °C
6200			8.2	0.06		0.73							Liq. HBr, 20 °C

[a] Molar percentages in parentheses.
[b] Calculated for this temperature from the data obtained at 55 °C.
[c] Ref. 61.

effect is also apparent both from the data for each of the aqueous $HClO_4$–TFA media at 55 °C the deactivation being greatest in the strongest acid, and from exchange in liquid HBr which involves a weaker electrophile so producing less deactivation than does exchange in TFA.

(4) The small increase in ρ-factor (20%) on changing the exchange medium from TFA to trifluoromethane sulphonic acid (TFSA) shows that there is no tendency to change to a π-complex mechanism for the substitution, despite the enormous change in reactivity of the electrophile. This is very relevant to the evidence presented to support such a mechanism for nitration and alkylation under certain conditions [Section 7.4.1.(vii) and 6.1.1.(iii)(d)].

(5) The *p*-F substituent activates under each set of conditions as it does in other electrophilic substitutions in which the reagent is very selective and the demand for resonance stabilization of the transition state is large, e.g. molecular chlorination and bromination (Chapter 9). On the other hand, the f_p^{F} value does not exactly parallel the f_p^{Me} value, the latter usually being taken as a measure of the selectivity of the reagent (Section 11.1.1), and this may reflect differences in solvation between the media.

(6) For *meta* substituents, the deactivation order is $I < F < Cl = Br$, as it is for molecular bromination of polymethylbenzenes in nitromethane (Table 9.10). This shows that there is a secondary relay of resonance effects to the *meta* position, notably for fluorine. This is found in other reactions for which there is a strong demand for resonance stabilization of the transition states (Section 11.1) [cf. nitration, where *meta* halogens give the inductive deactivation order [Section 7.4.3.(viii)].

(7) For *para* substituents the deactivation order is $Cl < I < Br$, again as in molecular bromination in nitromethane, whilst in reactions of low electron demand such as protiodesilylation and protiodegermylation the order $Cl < Br \approx I$ obtains (Sections 4.7.1 and 4.8.1). This may be a further example of the dependence of the substituent effect on the nature of the reagent, and is discussed further in Chapter 11.

(8) The *ortho* halogens deactivate in the order $F < I < Cl < Br$, and when the halogen is moved from the *ortho* to the *para* position the increase in the partial rate factors is in the order $F > Cl > Br > I$. This demonstrates the increased effectiveness of the $-I$ effect at the *ortho* position. The significantly lower *ortho*:*para* ratio for exchange in chlorobenzene in liquid HBr may again indicate a small amount of steric hindrance in this medium. The electrophile under these conditions is considerably less reactive, and will be more closely associated with the acid anion in the transition state (see also Section 3.1.2.6).

3.1.2.6 Polysubstituted aromatics containing a halogen substituent

The deactivation factors produced by halogen substituents in detritiation of some polysubstituted aromatics in trifluoroacetic acid at 70 °C are shown in

Table 3.12 Exchange rate coefficients for detritiation in TFA at 70 °C and halogen deactivation factors

Compound (substituents as drawn)	Deactivation factor	$10^7 k/s^{-1}$
Cl	740	7×10^{-4}
Cl(Br), Me	562 (519)	3.7×10^{-2} (4.0×10^{-2})
Cl(Br), Me, Me	447 (442)	7.9 (8.6)
Cl, Me, Me	240	31
Br, Me, Me, Me	236	2870
Cl	25.0	2.4×10^{-3}
Cl, Me, Me	22.1	340
Cl, OMe	98	71
Cl, OMe	81	86
Cl, MeO	73.5	192
Cl, OMe	5.85	1190
Cl, MeO	5.16	3440

Table 3.12 together with the rates of exchange.[125] The following features are notable:

(1) All of the compounds are more reactive than calculated from additivity. This deviation is predicted by the *Reactivity–Selectivity Principle*[126] (Section 11.1.5); similar results are found in halogenation (Section 9.6.5).

(2) The deactivating effect of the halogens decreases with increased aromatic reactivity and similar results are found in deuteriation in a medium composed mainly of TFA.[127] A plot of the deactivating factors in Table 3.12 vs reactivity of the *m*-halogen-containing alkylbenzenes is approximately linear, and has a greater slope than that drawn through the corresponding points for the *o*-halogen-containing alkylbenzenes. This is reasonable since in an infinitely reactive aromatic the deactivating effects of both *ortho* and *meta* halogens will tend to zero.

(3) The methoxyaromatics show the same trends noted in (2), but the deactivating effects of the halogens appear to be much smaller. This is because the methoxy substituent is hydrogen bonded in trifluoroacetic acid[104–106,128] but in the presence of the electron-withdrawing halogens this bonding is reduced and the methoxy group becomes more activating. Comparison of the reactivities of the 2-methoxy-3-chloro- and 2-methoxy-5-chloro compounds indicates that in the former the methoxy group is twisted very slightly out of the plane of the aromatic ring.

(4) Further indication that exchange in liquid HBr is sterically hindered (cf. Sections 3.1.2.4 and 3.1.2.5) comes from the partial rate factors for the 2- and 4-positions of 3-chlorotoluene of 14 and 10, respectively.[7] This gives a value of $\log f_o{:}\log f_p$ of 0.53, compared with 0.84 in toluene itself (Table 3.2). In the less hindered 3-fluorotoluene, the ratio is 0.74.[124]

3.1.2.7 Oxygen- and sulphur-containing substituents

Partial rate factors for the OMe, SMe, OPh, and SPh substituents in detritiation in TFA at 70 °C are shown in Table 3.13.[104,129,130] Additional rate factors (for exchange in acetic acid containing 2 mol-% H_2SO_4 at 99.3 °C) are $f_o^{SMe} = 110$, $f_p^{SMe} = 390$, $f_p^{OMe} = 17\,900$[131,132] and $f_o^{OMe}{:}f_p^{OMe} = 0.33$–$0.52$ (depending on temperature and workers)[131,133,134]; f_m^{OMe} is 0.25 in 63 wt-% aqueous $HClO_4$.[135] Notable features are:

(1) Both *m*-OMe and *m*-OPh are deactivating, showing that secondary relay of the conjugative ($+M$) effect of oxygen to the *meta* position is not sufficient to offset the $-I$ effects of these groups.

(2) The $\log f_o{:}\log f_p$ ratios are closely similar and have the value predicted from the charge distribution in the Wheland intermediate (**3**) (see also Section 11.2.4).

(3) The order *p*-OMe > *p*-OPh shows that the oxygen in methoxyl is better able to supply electrons by a conjugative effect than that in phenoxyl. This follows from oxygen being able to interact mesomerically with both phenyl groups in diphenyl ether, and the fact that phenyl is electron withdrawing relative to methyl. The orders *p*-OMe > *p*-SMe and *p*-OPh > *p*-SPh reflect the greater $+M$ effect of oxygen compared with sulphur (cf. the order *p*-F > *p*-Cl); in each pair the smaller atom has the greater $+M$ effect when attached to unsaturated carbon, and in addition the larger atom may have a $-M$ effect arising from the empty d-orbitals.

(4) In TFA the *p*-OMe substituent is less activating than expected, owing to hydrogen bonding with the solvent,[105] and this is shown by the reduced σ^+ value of 0.60 needed to correlate its effect. This is also evident from the lower $\log f_p^{OMe}{:}\log f_p^{SMe}$ value of 1.08 compared with a value of 1.64 for exchange in

Table 3.13 Partial rate factors for detritiation of substituted benzenes in TFA at 70 °C

Substituent	*o*	*m*	*p*	$\text{Log } f_o{:}\log f_p$
OMe	7.30×10^4		1.9×10^5	0.92
SMe	2.15×10^4		8.2×10^4	0.88
OPh	6.90×10^3	~0.12	3.1×10^4	0.855
SPh	3.30×10^3		9.8×10^3	0.88

HOAc–H_2SO_4 (in which hydrogen bonding appears to be unimportant) and 1.68 for protiodesilylation [Section 4.7.1.(ii)]. Hydrogen bonding is also indicated by the higher activation energy for *para* exchange in anisole (78 kJ mol^{-1}) than for thioanisole (75 kJ mol^{-1}).[136] A second factor contributing to the differences in these ratios is the very high polarizability of the SMe substituent[129,137,138] which may arise from its ability to exhibit both $+M$ and $-M$ effects. The SMe substituent therefore activates relatively more in reactions with the largest rate spread. Similar arguments account for the differences in the log f_p^{OPh}:log f_p^{SPh} values for detritiation (1.12) and protiodesilylation (1.89).[129]

3.1.2.8 Substituted *N,N*-dialkylanilines

The rates of tritiation of *ortho*-, *meta*- and *para*-substituted *N,N*-dialkylanilines have been determined in tritiated ethanol containing enough H_2SO_4 to protonate the base fully.[139] The exchanging entity is the free base which exists in equilibrium with the protonated form, although where a strongly electron-supplying group, e.g. OMe, is *para* to the amino group, exchange takes place on the conjugate acid,[140,141] because the substituent increases the availability for protonation of the lone pair on the nitrogen. Such changes in the nature of the species undergoing reaction according to the substituent present are a common feature of aromatic substitutions.[142]

Most of the *ortho*-substituted compounds examined (especially those with bulky substituents) failed to undergo exchange. These groups probably inhibit attainment of the coplanarity necessary for resonance stabilization of the transition state **28**. This also shows that reaction occurs on the free base since the electronic effect of the protonated substituent would be independent of the conformation about the C—N bond. Moreover, the ρ-factor for the reaction (-3.54) is considerably less than that (ca -7.5) which applies for exchange of alkylbenzenes in aqueous H_2SO_4.[10] The transition state must therefore be substantially nearer the ground state and this would only be true if the substituent was the strongly electron-supplying NMe_2 rather than the electron-withdrawing NMe_2H^+. This provides further evidence for the *Reactivity–Selectivity Principle*, which is additionally confirmed by the ρ-factors for exchange in phenols (-4.9) and pyridinium ions (-12.9),[143] i.e. they parallel the reactivity of the aromatics.

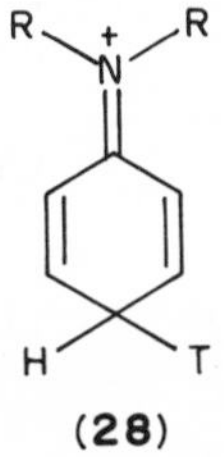

R R N + H T

(28)

3.1.2.9 Substituted acetophenones

Deuteriation of *para*-substituted acetophenones in sulphuric acid occurs on the free base,[144] as expected in view of the low reactivities of the aromatics. However, there was a marked difference in the log k vs acidity plots and those for benzene, toluene, or naphthalene.[10,145] The reason for this is not known; hydrogen bonding or increased steric hindrance to solvation with increasing acidity are possible explanations. In 75% acid the exchange rates were similar to that of benzene, yet the Brønsted α-coefficients for the acetophenones (ca 0.5) were markedly different, that for benzene being ca 1.0. This result was interpreted as evidence against the *Reactivity–Selectivity Principle.* However, errors in determining α can be very large (>0.3), especially if an insufficient acidity range is covered, as shown for example by data for naphthalene exchange in aqueous H_2SO_4 (Section 3.1.2.12).

3.1.2.10 Biphenyl and related compounds

Partial rate factors for exchange in biphenyl[1,7,69,146–149] and derivatives[1,66,104,105,115,148–152] are set out in Table 3.14 and 3.15. These data demonstrate the following:

(1) The deactivation, and activation by the *m*-Ph and *p*-Ph substituents respectively, show the phenyl group to have $-I$ and $+M$ effects (cf. structure **48** in Chapter 1 for the latter effect).

(2) Comparison of the f_p^{Ph} values, 52, 130, 163, and 2800, with the corresponding f_p^{Me} values, 313, 420, 450, and 3800, obtained under the same conditions (Table 3.2) shows that the ratio of the reactivities at the two positions decreases as

Table 3.14 Partial rate factors for hydrogen exchange in biphenyl

o	*m*	*p*	*o*:*p*	Log f_o:log f_p	Conditions[a]	Ref.
45	0.81	48	0.94	0.98	$CF_3CO_2H–H_2O–HClO_4$ (92.04:5.45:2.51), 70 °C	1
52	0.68	52	1.0	1.0	$CF_3CO_2H–H_2O–HClO_4$ (92.04:5.45:2.51), 25 °C	146
82	0.49	130	0.63	0.91	Aq. CF_3CO_2H, 70 °C	147
133	<1	143	0.93	0.99	$CF_3CO_2H–H_2O–H_2SO_4$ (95.31:2.21:2.48), 25 °C	146
85		163	0.6	0.87	CF_3CO_2H, 70 °C	148, 149
		132			CF_3CO_2H, 100 °C	149
			0.43		$C_3H_7CO_2H$, 70 °C	148
			0.25		Liq. HI, 25 °C	69
520	<0.4	2800	0.19	0.79	Liq. HBr, 25 °C	7

[a]Molar percentages in parentheses.

Table 3.15 Partial rate factors for detritiation of biphenyl and derivatives

(i) In TFA at 70 °C:

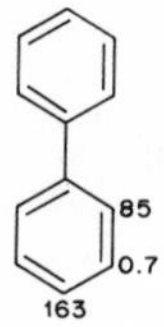

85
0.7
163
(29)

1970
189
100
2840
(30)

5500
125
21.4
16 800[a]
(31)

104
14100
(32)

	(29)	(30)	(31)	(32)
$\log f_o / \log f_p$	0.87	0.88 (1:3) 0.95 (4:2)	0.63 (1:3) 0.885 (4:2)	0.485

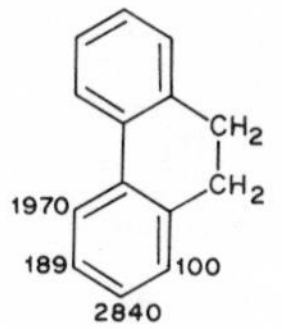

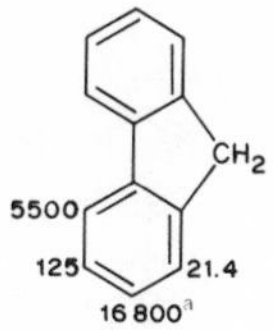

135
3670
160
313
(33)

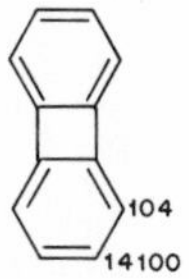

274
1830
362
430
(34)

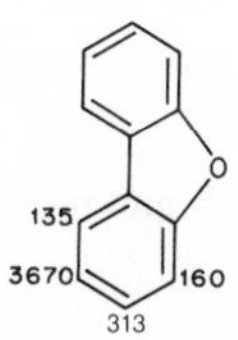

17 500[b]
(35)

20 000[c]
(36)

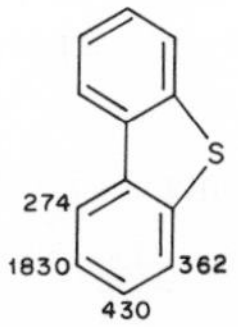

	(33)	(34)
$\log f_o / \log f_p$	0.62 (1:3) 0.85[d] (4:2)	0.785 (1:2) 0.92 (4:2)

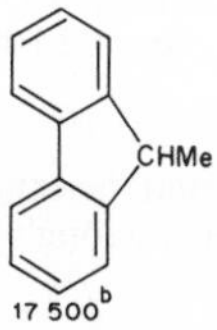

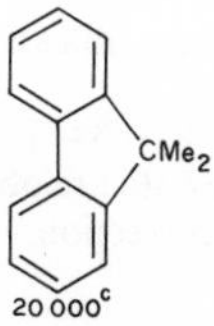

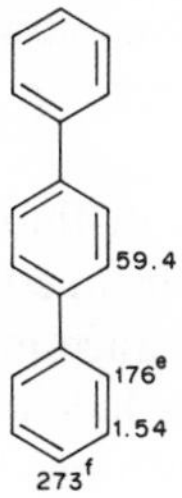

59.4
176[e]
1.54
273[f]
(37)

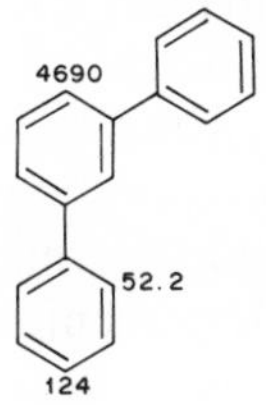

4690
52.2
124
(38)

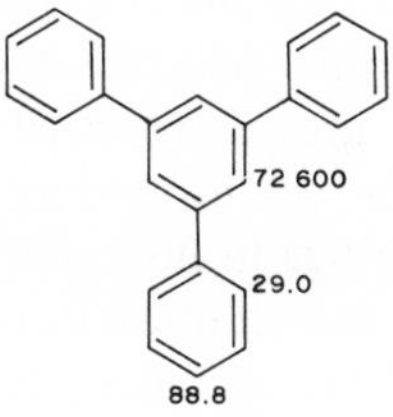

72 600
29.0
88.8
(39)

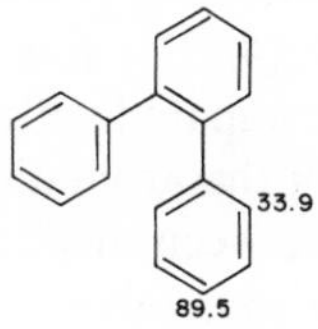

33.9
89.5
(40)

	(37)	(38)	(39)	(40)
$\log f_o / \log f_p$	0.92	0.82	0.78	0.75

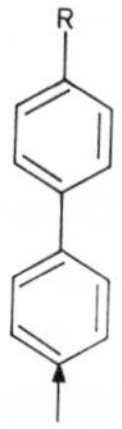

	Me	OMe	Cl[g]	Br	CO_2H	NO_2
R =	650	540	48	41	—	0.157
R′ =	65 000	51 000	4500	3600	345	6.3

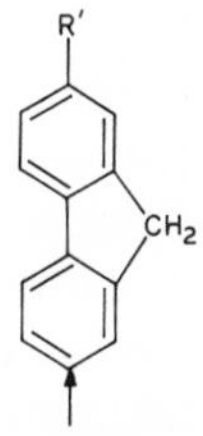

(continued)

Table 3.15 (*continued*)

(ii) In TFA containing ca. 5.5 moles-% H_2O and ca. 2.5 moles-% $HClO_4$ at 70 °C:

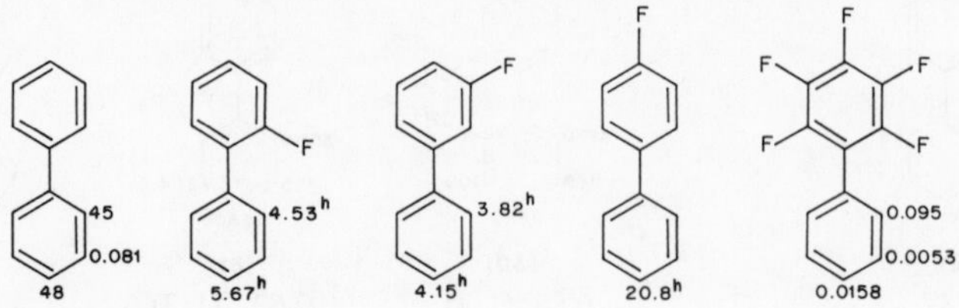

[a,b,c]These values become 3610, 3800, and 4870, respectively, in TFA(99.96)–H_2O(0.03)–$HClO_4$(0.01), and 2100, 2040, and 1930, respectively, in HOAc (20.83)–H_2O(34.43)–H_2SO_4(44.74) at 25 °C.
[d]This value was misreported as 0.43 in ref. 95.
[e,f]These values become 1200 and 14 000, respectively in liq. HBr at 25 °C,[7] giving a log f_o:log f_p value of 0.74.
[g]For R′ = 3′-Cl, $f = 20$.
[h]The data are obtained from the experimental data at 45 °C by means of the Arrhenius relationship; this correction was applied wrongly in ref. 153.

the reagent becomes more selective, owing to phenyl being more polarizable than methyl.

(3) Comparison for biphenyl of the log f_o:log f_p values with the *ortho*:*para* ratios shows that the large variation in the latter is mainly a selectivity effect. The variation in the former ratio is due to the change in structure of the transition state with reactivity of the electrophile (Section 11.2.4). Toluene shows similar behaviour (Table 3.2), and *p*-terphenyl (**37**) also shows the lowest log f_o:log f_p value under the most selective conditions.

(4) The much greater reactivity of the 2- and 4-positions in fluorene (**31**) compared with the corresponding positions in biphenyl arises from coplanarity of the aromatic rings in fluorene, which permits greater resonance interaction between them; a contribution from hyperconjugation in the methylene group is also involved.[113] The 2- and 4-positions in 9,10-dihydrophenanthrene (**30**) show intermediate reactivity, since its planarity lies between that of fluorene and biphenyl.

(5) Substituent effects are transmitted across the phenyl—phenyl bond but with a reduced magnitude, the ρ-factor being ca -2.1 compared with -8.75 in benzene. [The same 4-fold reduction factor is found in related gas-phase elimination reactions[154] and in protiodesilylation and other reactions [Section 4.7.1.(iii)]. Similar but slightly greater substituent effects are found in fluorene, the greater planarity of which facilitates the transmission of these effects. The lower reactivity of methoxybiphenyl and methoxyfluorene relative to their methyl counterparts shows again the effect of hydrogen bonding in TFA; addition of acetic acid causes the methoxy compounds to become increasingly reactive relative to the methyl compounds.[105] The reactivity of nitrobiphenyl may also be affected here by hydrogen bonding.[105]

(6) The reduced transmission of conjugative effects in biphenyl is also evident from the deactivation by *p*-F in contrast to its activation in benzene and by the *m*-F substituent deactivating *more* than the *m*-Cl substituent (cf. Table 3.11).

(7) Whereas changing the position of a fluoro substituent from *para* to *ortho* in one ring decreases the reactivity of the *para* position in the other ring by a factor of ca 4, keeping the position of the fluorine constant and changing the position of substitution from *ortho* to *para* in the other ring produces a rate change of only 20% (when the fluorine is *ortho*) and 6% (when the fluorine is *meta*). Since these rate changes are not proportional to the substituent–reaction site distance, this suggests (see also Section 3.1.2.4) that field effects are unimportant relative to inductive effects. The fluorophenyl substituent evidently produces its effect almost entirely by altering the electron density at the carbon by which it is attached to the ring undergoing substitution. Hence there is a strong parallel between **41** as a structure on the reaction path for biphenyl exchange and **42** which is on the path for benzene exchange; indeed, the difference between the *para*:*ortho* reactivity ratios in fluorobenzene (13.6) and 4-fluorobiphenyl (3.7) is very close to that predicted from the ρ-factors for exchange in the two systems.[1]

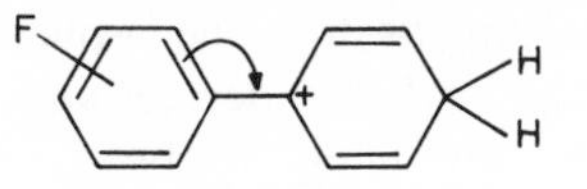

(**41**) 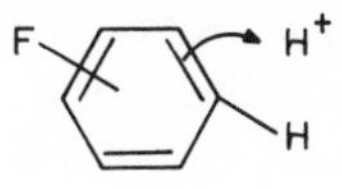 (**42**)

(8) The pentafluorophenyl substituent is a $-I$, $+M$ group producing deactivation with *ortho*,*para* orientation in hydrogen exchange. However, in reactions of low electron demand for resonance it produces greater electron withdrawal at the *ortho* and *para* positions, i.e. it can also exhibit a $-M$ effect.[1] This ability to change from $-M$ withdrawal to $+M$ supply according to electron demand is only shown by one other group, the ethynyl substituent. The pentafluorophenyl substituent is also less deactivating than predicted from additivity (2.2×10^{-3}) and this is paralleled by the greater reactivity than predicted of pentafluorobenzene towards demetallations (Sections 4.7 and 4.9) and in pyrolysis of 1-arylethyl acetates,[155] and of pentachlorobenzene towards nitration [Section 7.4.3.(ix)]. This may be attributed to enhanced mesomeric electron release under conditions of high electron demand[156] coupled with the inability of the halogens to exert their full $-I$ effect when acting in opposition.[155]

(9) The order of reactivities of the 2-position of fluorene (**31**), 9-methylfluorene (**35**), and 9,9-dimethylfluorene (**36**) can be reversed by changing the medium. Further, the media in which the order is **31** > **35** > **36** are those in which the *para* position in toluene is more reactive than that in *tert*-butylbenzene, and the converse is also true. Originally this was considered as evidence against ascribing the Baker–Nathan effect to steric hindrance to solvation (cf. Section 3.1.2.1), since the number of carbon atoms which share the positive charge of the electron-

deficient transition state is greater for the fluorenes than for the alkylbenzenes.[152] However, linear free-energy analysis of the data shows that steric hindrance to solvation is indeed less in the fluorene series; the difference between the ratios log f_2^{Fl}:log f_2^{DMFl} obtained in the TFA and aqueous H_2SO_4–HOAc media (0.029) is considerably less than the difference in the corresponding log f_p^{Me}:log $f_p^{t\text{-Bu}}$ ratios (0.077).

(10) The positions in the terminal rings of *p*- and *m*-terphenyl (**37** and **38**) are respectively more and less reactive than the corresponding positions in biphenyl, which follows from the expected electronic effects of the *p*- and *m*-phenyl substituents. By contrast, *o*-terphenyl (**40**) is less reactive than biphenyl because interaction between the *ortho* hydrogens of the terminal rings reduces the coplanarity and hence conjugation between the latter and the central ring.

The partial rate factor for the central ring position in *p*-terphenyl is correctly predicted by additivity (85 × 0.7 = 59.5). By contrast, the partial rate factor for the 4-position in *m*-terphenyl is three times less than predicted (85 × 163 = 13 900), indicating that *m*-terphenyl is slightly less planar than biphenyl. The central ring position of 1,3,5-triphenylbenzene (**39**) is 16 times less reactive than predicted, and since two *m*-phenyl substituents produce only twice the deactivation at the *para*-position of the terminal ring as does one *m*-phenyl substituent at the corresponding position in *m*-terphenyl, the degrees of coplanarity in *m*-terphenyl and 1,3,5-triphenylbenzene are likely to be similar. The low reactivity of the central ring of the latter is therefore probably one of the very few examples of steric hindrance to hydrogen exchange; few other electrophilic substitutions will take place at all at this position. Steric hindrance may slightly affect the reactivity of the *ortho*-position of the peripheral ring in 1,3,5-triphenylbenzene and, more probably, of the *ortho*-position of the terminal ring in *o*-terphenyl since the log f_o:log f_p ratios for exchange in these rings are significantly less than for biphenyl.

(11) The 1- and 4-positions of 9,10-dihydrophenanthrene (**30**), dibenzothiophene (**34**), dibenzofuran (**33**), and fluorene (**31**) are each alpha to a cyclic substituent, the strain in which increases along this series, and the 2- and 3-positions are beta positions. The 1- and 3-positions are *meta* to the phenyl group and respectively *ortho* and *para* to the other substituent. In the absence of other effects the log f_1:log f_3 ratio should be the same as the log f_o:log f_p ratios in the open-chain analogues PhXPh (Tables 3.8 and 3.13). The bond order–strain theory (Section 3.1.2.3) correctly predicts that the former ratio will decrease along the series so that the difference in the two sets of ratios will increase along the series, viz. X = $(CH_2)_2$, 0; S, 0.095; O, 0.165; CH_2, 0.18 (the ratios for diphenylethane and 9,10-dihydrophenanthrene are identical because bond shortening does not produce any increase in strain in the central ring of the latter). Similarly, the 4- and 2-positions are respectively *ortho* and *para* to phenyl and *meta* to the other substituent so that the log f_4:log f_2 ratio should be constant in the absence of other effects. Strain, however, causes it to fall along the series, viz.

X = $(CH_2)_2$, 0.95, S, 0.92; O, 0.85; CH_2, 0.885. The effect is less clear cut here because for fluorene the ratio is enhanced as a result of its reactivity being much higher than that of the other compounds.[95]

(12) The ratio of the reactivities of the α- and β-positions in biphenylene (**32**) is very low ($\log f_1 : \log f_2 = 0.485$). Since these positions are respectively *ortho* and *meta*, and *para* and *meta* to phenyl, the ratio should be the same as the $\log f_o : \log f_p$ ratio in biphenyl (0.87). The low value observed (less than for any other compound examined) arises from the increase in strain produced at the bond common to both rings on going to the transition state for α-substitution (see Section 3.1.2.3); since the central ring is only four-membered, the increase in strain is very large. Since both biphenylene and fluorene are planar, and the activation by *m*-methylene in hydrogen exchange will be ca 4-fold, the 2-position of biphenylene is predicted to be ca four times less reactive than the 2-position in fluorene, whereas it is only 16% less reactive. This is predicted by the bond strain theory (the reactivities of β-positions should be enhanced in strained aromatics) but not by the theory based on electronegativity effects.[95]

3.1.2.11 Atropisomers

Hexa-*o*-phenylene can exist as two stable conformers, or *atropisomers*, referred to as the helical (**43**) and crown (**44**) isomers. Partial rate factors have been determined (see **43**, **44**) for detritiation of these conformers in TFA at 70 °C along with those for tetra-*o*-phenylene (**45**) (which can exist only in the crown form) in

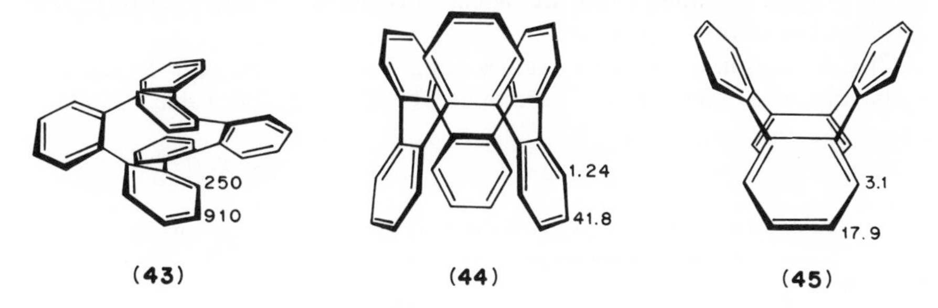

(**43**) (**44**) (**45**)

order to assess the effect of the different angles between the phenyl rings in the conformers on the reactivities. The main features of the results are:

(1) The helical atropisomer is much the more reactive because adjacent phenyl rings are more coplanar, thereby permitting greater resonance stabilization of the transition state.

(2) In both crown conformers, adjacent phenyl rings are virtually orthogonal. If the inter-ring angles were the same as in biphenyl then to a first approximation the partial rate factors for the 3- and 4-positions should be equal to *o*- × *m*-phenyl (**60**), and *p*- × *m*-phenyl (**114**), respectively. In fact, both are less owing to the orthogonality of the rings. Models show that some twisting towards coplanarity

is possible in **44** but not in **45**, which accounts for the greater partial rate factor at the most reactive site in the former. The very low reactivities of the '*ortho*' positions in both compounds may be due to the particular effectiveness of the $-I$ effect of the orthogonal phenyl groups.

(3) The log f_o:log f_p ratio for helical hexa-*o*-phenylene is 0.81, in good agreement with the value obtained for a wide range of substituents noted above, and also with theoretical prediction (Section 11.2.4).

(4) The 4-position of helical hexa-*o*-phenylene is (a) deactivated by a *m*-phenyl substituent (itself containing an *o*-phenyl substituent which will have only a small effect on the *m*-phenyl partial rate factor of 0.7), and (b) activated by a *p*-phenyl substituent (containing an *o*-phenyl substituent, calculated[157] to increase the value of f_p^{Ph} ca threefold. The calculated partial rate factor for the 4-position is thus ca $0.7 \times 163 \times 3 = 342$, and for the 3-position the value may be similarly calculated to be 178. The observed reactivities are greater, indicating that the coplanarity between adjacent phenyl rings is greater than in biphenyl and about the same as in 9,10-dihydrophenanthrene; models confirm this.

3.1.2.12 Polycyclic aromatics

The partial rate factors for hydrogen exchange of naphthalene (Table 3.16) show two features of interest:

(1) The 1-position is more reactive than the 2-position under all conditions, and the log f_1:log f_2 ratio is approximately constant (ca 1.5); this suggests that ratios, in other reactions, which are significantly less than 1.5 may reasonably be attributed to steric hindrance.[159]

(2) Comparison of the log f_2 values of 62, 151, and 1200 with the log f_p^{Me} values obtained under the same conditions, 313, 450, and 3800 (Table 3.2), shows

Table 3.16 Partial rate factors for hydrogen exchange in naphthalene

Position 1	Position 2	Log f_1:log f_2	Conditions[a]	Ref.
370	62	1.44	TFA(92.04)–H_2O(5.45)–$HClO_4$(2.51), 25 °C	146
1079	127	1.44	TFA(95.31)–H_2O(2.21)–H_2SO_4(2.48), 25 °C	146
1160	151	1.41	TFA, 70 °C	158
40 000–60 000[b]	1200	1.55–1.49	Liq. HBr, 25 °C	7

[a]Molar percentages in parentheses.
[b]Approximate value.

that the 2-position of naphthalene becomes more strongly activated relative to the *para* position of toluene as the selectivity of the reagent increases. This is due to the 3:4-benzo substituent being more polarizable than *p*-Me.[146]

Partial rate factors ($f_1 = 17\,800$, $f_2 = 266$) for detritiation of naphthalene in aqueous H_2SO_4 at pH = 0 and 100 °C have been calculated from exchange data at lower temperatures and higher acidities.[160] However, exchange at the 2-position showed an anomalous acidity dependence compared with the normal behaviour observed earlier under similar conditions by Stevens and Strickler.[145] Since the ability of sulphuric acid to dissolve aromatics decreases with decreasing acid concentration, anomalously low exchange rates can be observed particularly if the vapour space of the reaction vessel is not kept to the absolute minimum.[154] This accounts for the anomalous results reported in ref. 160; extrapolation of kinetic data obtained at 55 °C[160] would lead to the erroneous conclusion that the 1- and 2-position would be *deactivating* above acid concentrations given by $H_0 = -11$ and -8, respectively.

The excited states of naphthalene are more basic than the ground state, leading to faster exchange; in H_2SO_4 above ca 55 wt-% the 2-position is *more* reactive than the 1-position.[161,162] Although the first excited triplet is longer lived than the first excited singlet, the greater basicity of the latter (by a factor of 10^{13}) means that it is involved in the exchange process. By contrast, the first excited singlet and triplet states for anthracene have similar basicities so the greater lifetime of the latter results in it being the species involved in anthracene exchange under photochemical conditions.

Partial rate factors have been determined for exchange in a large number of polycyclic aromatics in anhydrous TFA at 70 °C (Table 3.17), and these have provided the following detailed insight into benzo substituent effects.

(i) Annelation

With only one exception, annelation increases the reactivity of the adjacent α- and β-positions, the greatest effect being at the former; Table 3.18 shows the annelation factors produced by the (dotted) benzo substituent at the positions shown. The reasons given in Chapter 2 to account for the higher reactivity of the 1- relative to the 2-position in naphthalene can evidently be applied to the other molecules. The magnitude of the annelation factor varies considerably, being much higher when a linear (anthracene-like) rather than an angled (phenanthrene-like) array of aromatic rings is produced. By contrast, the α/β rate ratios are reasonably constant.

The reactivity of a simple annelated molecule can be predicted by considering the principal canonical forms (those retaining benzenoid character) representing the transition states. For substitution at the 1- and 2-positions of anthracene there are more structures (**46** and **47**, respectively) than for substitution at the corresponding positions in naphthalene, and the ratio of the number of structures

Table 3.17 Partial rate factors for detritiation of polycyclics in TFA at 70 °C

Compound	Partial rate factors (as labelled on structure)
Naphthalene[158]	1160, 151
Anthracene[16]	127×10^5, 7900, 1135
Phenanthrene[163]	1630, 900, 173, 385, 810
Benzo[*c*]phenanthrene[164]	8680, 2050, 2465, 422, 1200, 1580
Dibenzo[*c*, *f*]phenanthrene[165]	18 400, 6650, 6930, 930, 2405, 2850, 10 200
Benzo[*c*]naphtho[1,2-*g*]-phenanthrene[166]	25 000, 8770, 6530, 905, 21600, 4400, 18 200, 10 250
Benz[*a*]anthracene[167]	3.5×10^6, 4485
Chrysene[168]	12 200, 975, 2790, 186, 307, 696
Benzo[*a*]naphtho[1,2-*h*]anthracene[166]	11100, 1190, 268, 3810
Fluoranthene[169]	4380, 247, 284, 70.5, 8860
Triphenylene[16]	620, 136
Pyrene[170]	8.5×10^5, 85[a], 1470[a]
Perylene[170]	1.47×10^6
Coronene[16]	7400

[a]Interpolated from data obtained in a slightly different medium.

Table 3.18 Annelating effects of benzo substituents

Structure	$f_{2:3\text{-benzo}}$	$f_{3:4\text{-benzo}}$	$f_{2:3}:f_{3:4}$
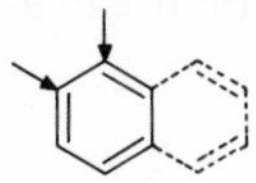	1160	151	7.7
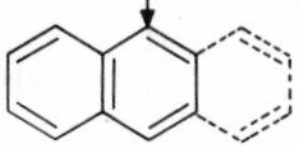	10 900	—	—
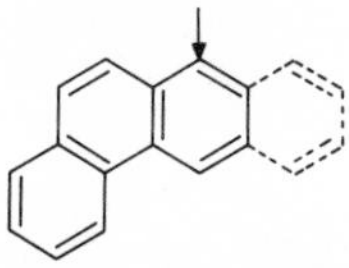	3900	—	—
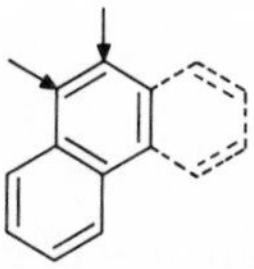	11	1.42	7.75
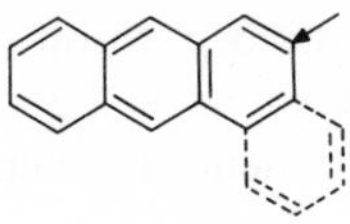	3.95	—	—
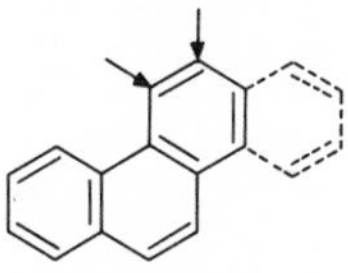	31.7	3.44	9.2
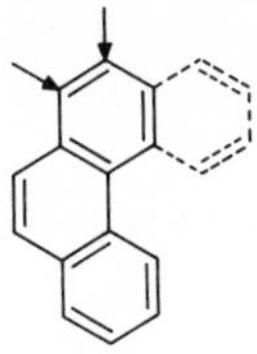	50	2.74	18
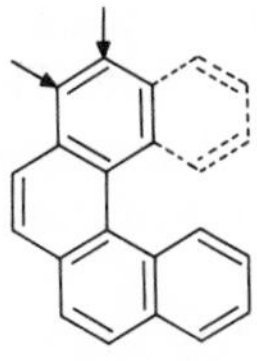	44	3.4	13
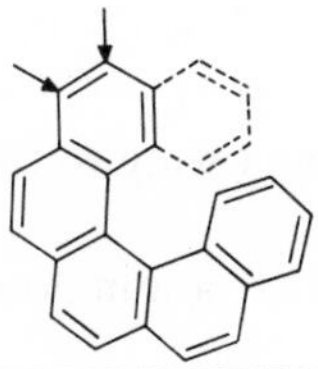	26.9	1.0	26.9

for 1- and 2-substitution is 1.5 and therefore less than for naphthalene (2.0). The 1- and 2-positions in anthracene should therefore be more reactive than the corresponding positions in naphthalene, and the difference in reactivity between them should be smaller; this is observed.[16] Molecular orbital calculations predict the same result.

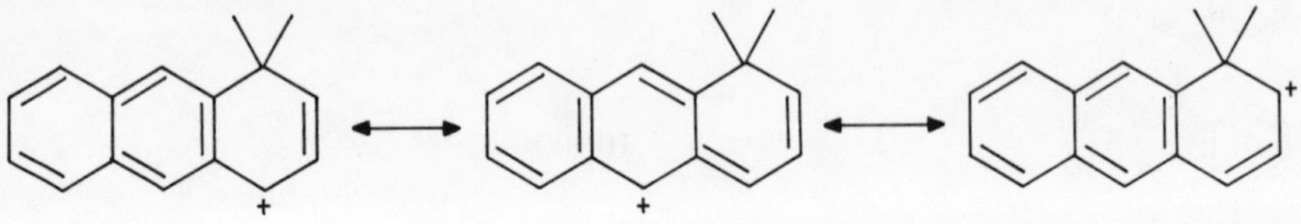

(46)

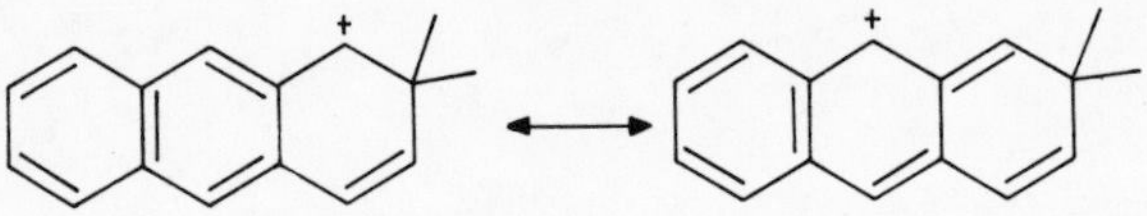

(47)

The effects of a remote benzo substituent (shown in Table 3.19) are described using the notation given in ref. 166. Thus in **48** the effect at the 5-position of the (dotted) ring annelated across the 1,2-bond is referred to as a 1,2–5-interaction; for simplicity, the notation is based on the naphthalene fragment

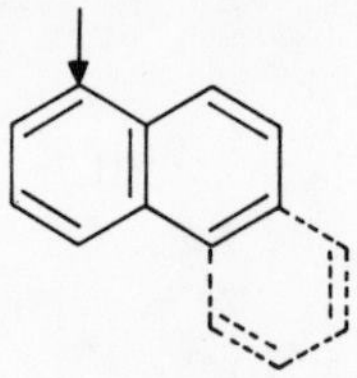

(48)

regardless of the molecule under consideration. The substituent effects are as follows:

(a) *The 1,2–5-annelation factor*. For planar molecules this produces a small deactivation, whereas in the non-planar helicenes a small activation results. The latter almost certainly arises from superimposition of the rate enhancement arising from non-coplanarity upon the deactivation that would otherwise apply (and is predicted by calculations).

(b) *The 1,2–6-annelation factor*. Except in the phenanthrene–triphenylene transformation (which is also anomalous with regard to the 1,2–7 factor that describes, through symmetry, the same position in triphenylene), this factor produces a moderate activation, largest in the helicenes, probably owing to the effect of non-coplanarity.

(c) *The 1,2–7-annelation factor*. With the exception of the anomalous

Table 3.19 Activating and deactivating effects of remote benzo substituents

Structure	Interaction			
	1,2–5	1,2–6	1,2–7	1,2–8
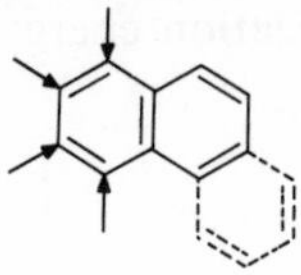	0.77	1.15	2.55	0.70
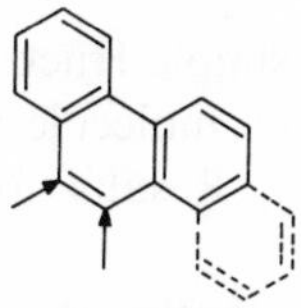			7.48	1.71
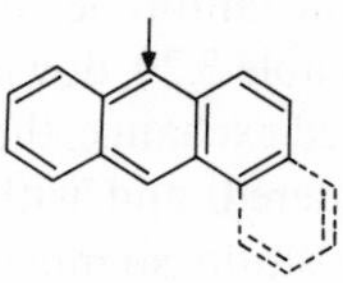	0.28			
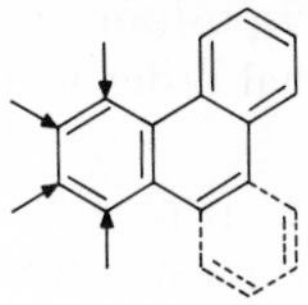	0.76	0.35	0.79	0.69
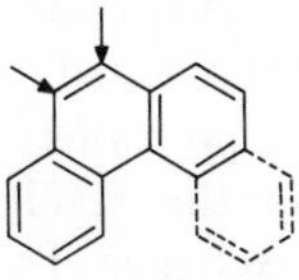	1.51	5.3		
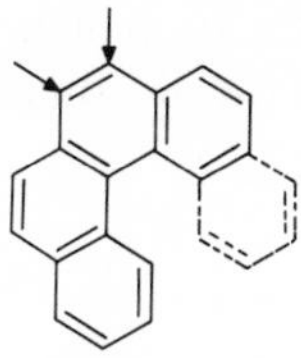	1.18	4.1		
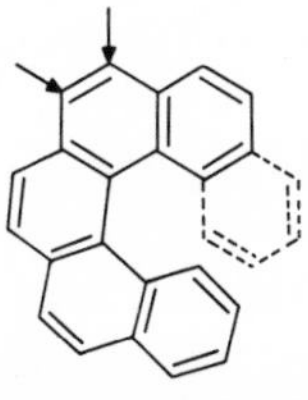	1.17	2.6		

phenanthrene–triphenylene transformation, this factor is the largest of the four interactions considered.

(d) *The 1,2–8-annelation factor.* This is slightly more deactivating than the 1,2–5 factor in the naphthalene–phenanthrene and phenanthrene–triphenylene transformations. However, activation is obtained in the phenanthrene–chrysene transformation which is both anomalous and contrary to localization energy predictions.

(ii) Positional reactivity orders

These have been calculated by various methods of which the simple Hückel method gives the best results. The positional order within a given molecule is predicted better than the order between different molecules. Results for individual molecules are as follows:

(a) *Phenanthrene.* The positional reactivity orders for phenanthrene in detritiation and other electrophilic substitutions are set out in Table 3.20, down which steric hindrance to substitution increases. In the unhindered exchange, the order is as calculated. The 4-position is clearly the most hindered, and with increasing size of electrophile soon becomes the least reactive position, so much so that no 4-substitution was detected in sulphonation. The 1- and 9-positions are about equally hindered and show a parallel decrease in positional order with increasing size of the electrophile.

(b) *The helicenes.* Observed and calculated reactivities for tetra-, penta-, and hexahelicene (Table 3.21) agree very well; only the reactivity of the most central position in each compound (the 6-, 7-, and 8-positions, respectively) is underestimated. Exchange at the 1-position in each compound is correctly predicted, so exchange at these sites is unlikely to be hindered. Models show these positions to be very crowded (especially in penta- and hexa-helicene), so the results demonstrate further the very low steric requirement of hydrogen exchange. The results also suggest that it is the 7-isomer which accompanies the formation of the 5-isomer in nitration, bromination, and acetylation, and not the 8-isomer reported in the literature.[174]

Table 3.20 Positional reactivity order for substitution in phenanthrene

Order	Reaction	Ref.
9 > 1 > 4 > 3 > 2	(Calculated)	175
9 > 1 > 4 > 3 > 2	Hydrogen exchange	163
9 > 1 > 3 > 2 > 4	Nitration	171
9 > 3 > 1 > 2 > 4	Benzoylation	172
3 > 9 > 1 > 2 > 4	Acetylation	172
3 > 2 > 9 > 1 > (4)	Sulphonation	173

Table 3.21 Observed and calculated reactivity orders for the helicenes

Compound	Observed	Calculated	Ref.
Tetrahelicene	5>6>4>1>2>3	5>4>6>1>2>3	164
Pentahelicene	5>7>6>4>1>2>3	5>4>6>7>1>2>3	165
Hexahelicene	5>7>8>1>4>6>2>3	5>7>4>8>6>1>2>3	166

(c) *Benzo[*a*]naphtho[1,2-*h*]anthracene and chrysene.* The order in benzo[*a*]naphtho[1,2-*h*]anthracene, 5>1>4>2, is as predicted[153] except that the reactivity of the 1-position is greater than that of the 4-position (as it is in the isomer hexahelicene). For chrysene the observed order 6>5>1>4>3>2 is close to that predicted, 6>1>5>4>3>2; as in the helicenes, the most central position in the molecule is more reactive than predicted.[168]

(d) *Fluoranthene.* The positional reactivity order (3>8 ≫ 1 > 7 > 2) differs from that calculated by various methods (3 > 8 > 7 > 1 ≫ 2) in that both the 1- and the 7-positions are much less reactive than expected.[169] This deviation is understandable in terms of bond-strain theory (Section 3.1.2.3) because both positions are α to a strained five-membered ring. Moreover, the ground-state strain in fluoranthene is greater than that in indane so the increase in strain on going to the transition state will be correspondingly larger.

(iii) The effect of benzene ring distortion

The addition of successive benzo substituents to naphthalene, giving the series phenanthrene (trihelicene), benzo[*c*]phenanthrene (tetrahelicene), pentahelicene, and hexahelicene, is accompanied by decreasing planarity as the molecules distort in order to achieve a minimum of 3 Å between the terminal carbon atoms. Calculations predict that the overall reactivity of each molecule, of which the average localization energy is an index, should remain constant. However, the reactivity increases regularly as the ring size is increased (Fig. 3.2). This is not a solvation effect (whereby poorer ground-state solvation of a large molecule would make it more reactive) because the less soluble benzo[*a*]naphtho[1,2-*h*]anthracene is much less reactive than the isomeric hexahelicene. Thus, whereas the sum of the localization energies for substitution at the 1-, 3-, 4-, and 5-positions in both benzo[*a*]naphtho[1,2-*h*]anthracene and hexahelicene is identical (-9.42β), the sum of the partial rate factors is 16 368 and 44 920, respectively. Moreover, coronene, which is the least soluble of all the aromatics shown in Table 3.17, has a partial rate factor which is significantly less than the average value per position for hexahelicene (11 950), yet the localization energy (-2.306β) predicts it to be more reactive than hexahelicene (for which the average value is -2.366β). The overall effect in hexahelicene of the distortion of the aromatic rings (which lowers the aromaticity and destabilizes the ground state towards

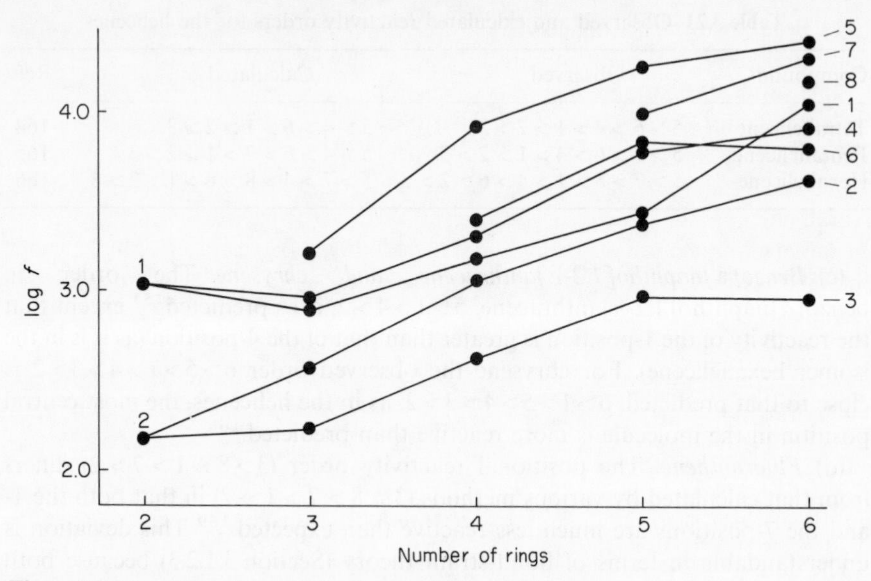

Figure 3.2. Variation in positional reactivities with number of rings in helicenes. 1–8 = Positions in each helicene.

substitution) appears to be a ca tenfold enhancement of reactivity, with a *ca* three to four-fold enhancement for the less distorted pentahelicene.

(iv) Silver-complex formation

The effect of the formation of silver complexes (Section 2.1.3) on the reactivity of an aromatic has been measured in hydrogen exchange.[176,177] The reactivities of ferrocene, phenanthrene (9-position), naphthalene (1-position), and *p*-xylene (2-position) were reduced by factors of ca 390, 13.7, 8.5 and 6.1, respectively, showing that the ease of formation parallels basicity; spectroscopic studies indicated that the [phenanthrene.Ag]$^+$ complex was more reactive than the [phenanthrene$_2$.Ag]$^+$ complex.

3.1.2.13 Substituted polycyclic aromatics

The study of substitution in benzene derivatives during the past 60 years has led to well developed theories accounting for benzene substituent effects; similar studies in polycyclic aromatics have been much more recent. Use of the hydrogen-exchange reaction has been particularly valuable in this respect because of the freedom from steric hindrance. Sufficient data have been amassed in detritiation (Table 3.22) to permit the identification of the basic factors governing these substituent effects. These are:

(1) Bond fixation, arising out of the need for the molecule to retain maximum benzenoid character, or to minimize bond strain.

(2) The number of sites over which charge can be delocalized.

(3) The inductive and mesomeric effects of the substituents, modified to take account of (1) and (2).

In considering the data in Table 3.22, it is important to distinguish between the overall partial rate factors relative to a single site in benzene (the values given in the table) and the effect of a substituent in a particular hydrocarbon; the latter is obtained by division of the partial rate factor for the position in question by the partial rate factor for the same position in the parent hydrocarbon. Thus the effect of a 2-methyl substituent at the 1- and 3-positions of naphthalene is 347 000/1160 and 541/151, i.e. 300 and 3.58, respectively. Failure to allow for the intrinsic reactivity of the appropriate position in the parent hydrocarbon can lead to very misleading conclusions.[98,178] Salient features of the data are as follows.

(i) Prediction of isomer yields in other reactions

The data in Table 3.22 permit the prediction of the yields of isomers in other reactions, e.g. 1-methylnaphthalene should give the order $4>2>5>8>7\approx 3>6$. Where the predicted orders are not obtained, steric hindrance is usually the cause, although in some cases discrepancies arise from incorrect analysis of products.

(ii) Conjugated vs non-conjugated sites

Electron-supplying substituents which are conjugated with a reaction site produce higher reactivity at that site than substituents that do not, and vice versa (see the data for the methylphenanthrenes, the starred sites being those conjugated with respect to the 9-position).

Relevant to this is the proposed involvement of covalent hydrates to account for 2-substitution in phenalone (last entry in Table 3.22).[187] It is notable that only the 2-, 5-, and 8-positions are *not* conjugated with the carbonyl group, so these will in any event be the most reactive. Moreover, substitution at the 5- and 8-positions involves loss of benzenoid character in the transition state, whereas 2-substitution does not. The 2,3-bond is also highly localized and will therefore have a high electron density. Consequently, 2-substitution is predicted; more complicated explanations seem unnecessary.

(iii) Bond fixation

The conjugative electron-supplying effect of a substituent across a 1,2-bond in naphthalene is greater than across a bond in benzene, and much greater than across a 2,3-bond in naphthalene, e.g. a 2-methoxy substituent activates the

Table 3.22 Partial rate factors for detritiation of substituted polycyclics in TFA at 70 °C

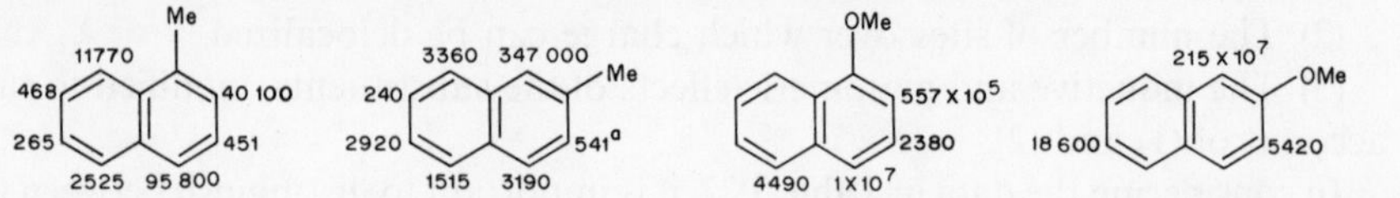

Ref. 158 158 158 158

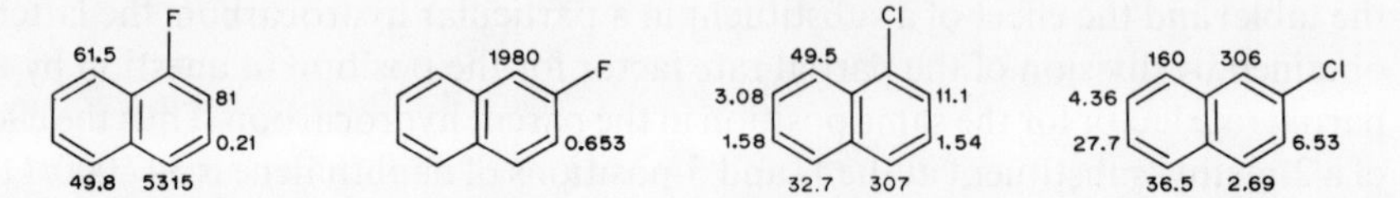

Ref. 158 158 158 158

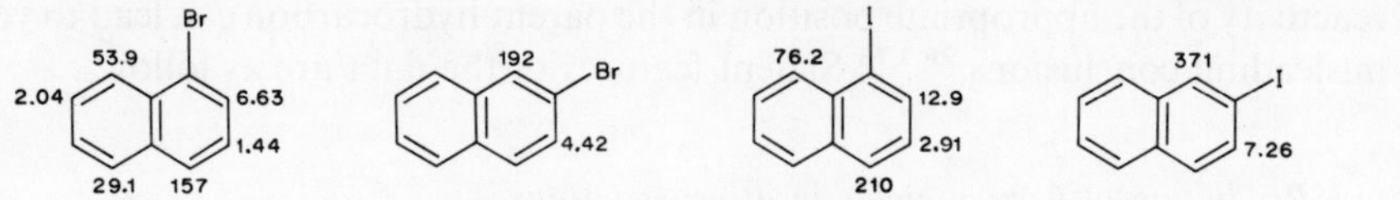

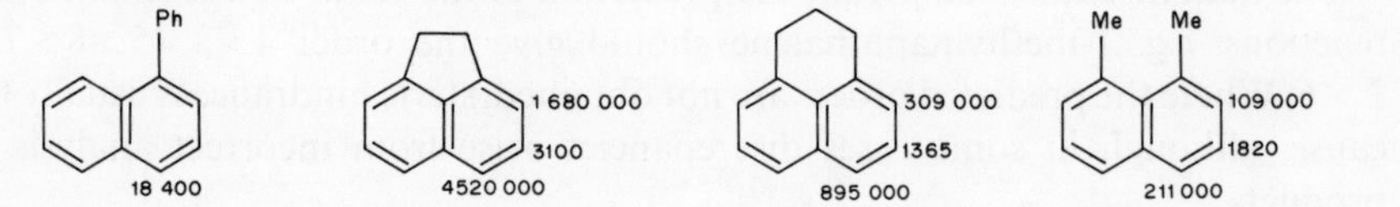

Ref. 158 158 158 158

Ref. 158 98,178 98, 178 179

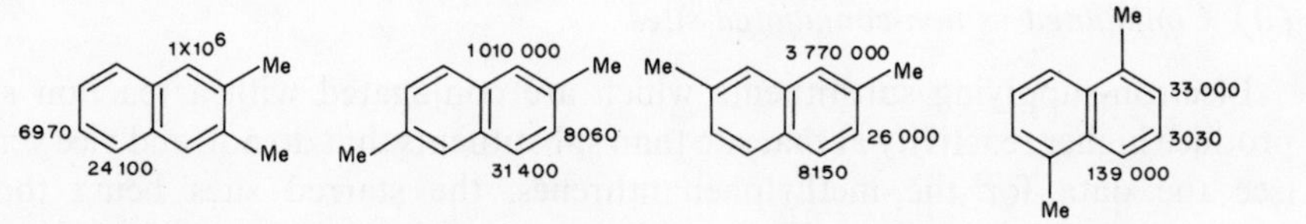

Ref. 179 179 179 179

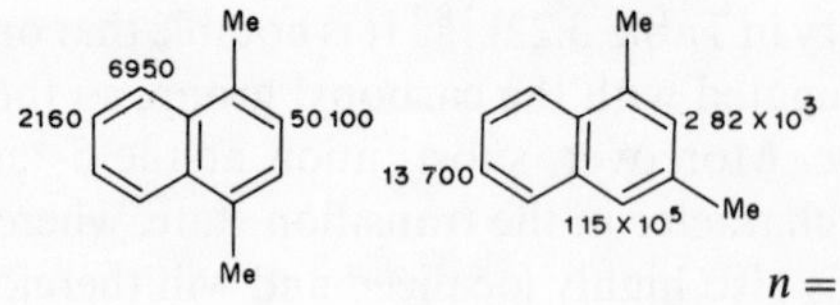

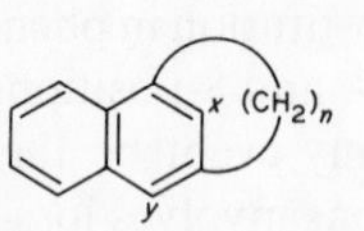

	$n =$	10	8	7
	$x =$	2.37×10^5	2.92×10^5	1.50×10^5
	$y =$	1.28×10^7	2.71×10^7	2.97×10^7

Ref. 179 179 180

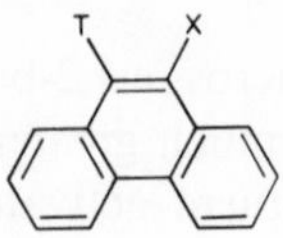

X =	Me	F	Cl	Br	I	CN
	297 000	673	30.4	17.7	22.6	0.66

Ref. 181

Table 3.22 (*continued*)

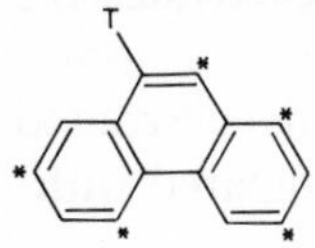

T

1-Me	2-Me	3-Me	4-Me	5-Me	6-Me	7-Me	8-Me
6990	4560	28 600	5050	3890	3210	4990	4290

Ref. 182

	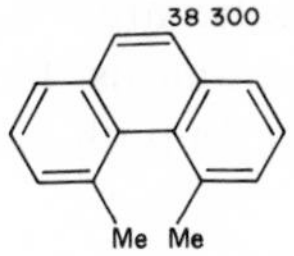 38 300 Me Me	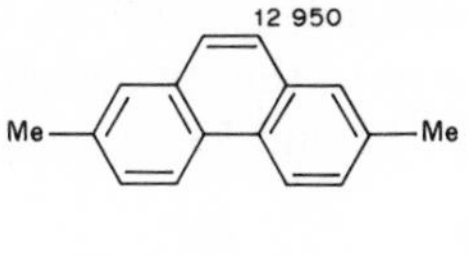 12 950 Me Me	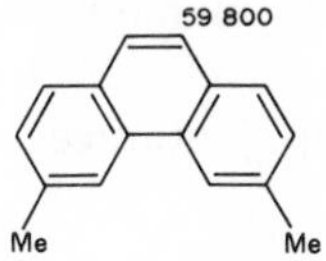 59 800 Me Me
Ref.	183	183	183
	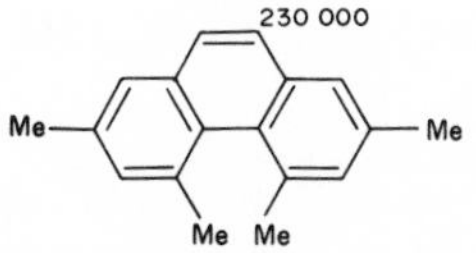 230 000 Me Me Me Me	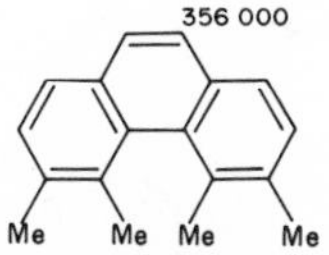 356 000 Me Me Me Me	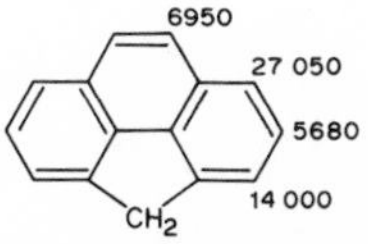 6950 27 050 5680 14 000 CH_2
Ref.	183	183	184
	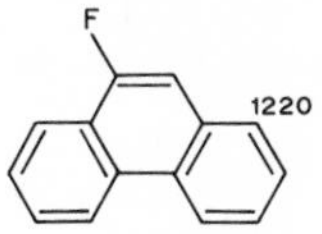 F 1220	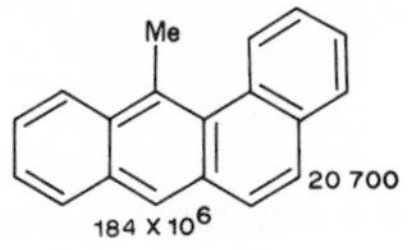 Me 20 700 184 X 10^6	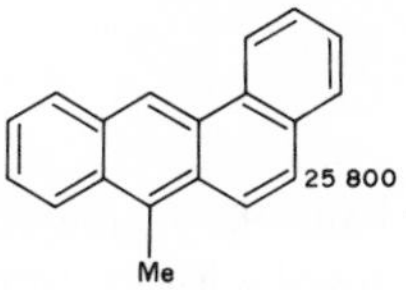 25 800 Me
Ref.	185	167	167
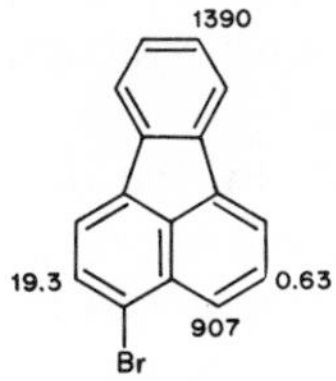 1390 19.3 0.63 907 Br	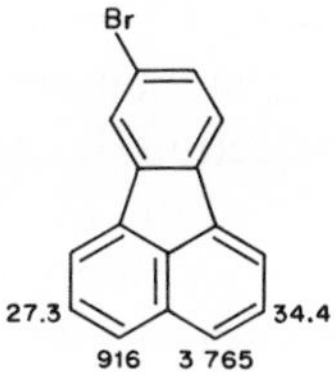 Br 27.3 34.4 916 3 765	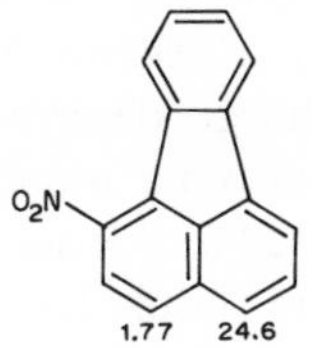 O_2N 1.77 24.6	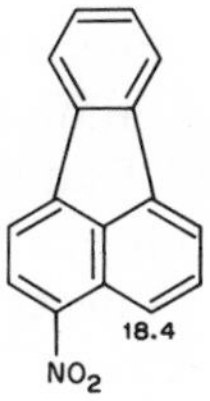 18.4 NO_2
Ref. 186	186	186	186
	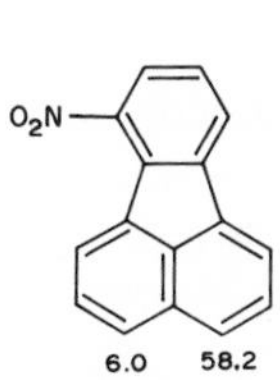 O_2N 6.0 58.2	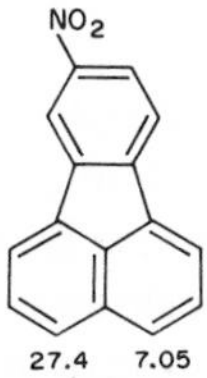 NO_2 27.4 7.05	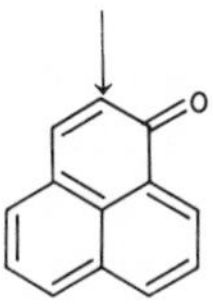 O
Ref.	186	186	187

[a]The different value given in ref. 158 is a typographical error.

adjacent 1- and 2-positions by factors of 3.7×10^5 and 36, respectively, compared with 7.3×10^4 in benzene. Likewise, a 2-fluoro substituent deactivates the 3-position by a factor of 0.044 (cf. 0.136 in benzene) but activates the 1-position 1.74 times. These results arise because there is bond fixation in naphthalene caused by the greater stability of the doubly benzenoid canonical **49** compared with the singly benzenoid canonicals **50** and **51**. Hence the bond orders for the 1,2- and 2,3-bonds in naphthalene and a bond in benzene are 1.725, 1.603, and 1.667, respectively.[188] Likewise, the preferred canonicals for phenanthrene and benz[*a*]anthracene are **52** and **53**, respectively. In both of these molecules bond

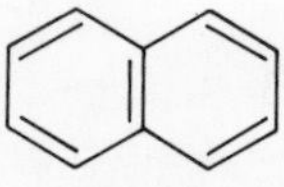

(49) 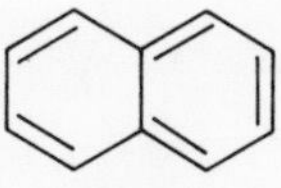 (50) 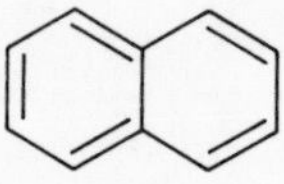 (51)

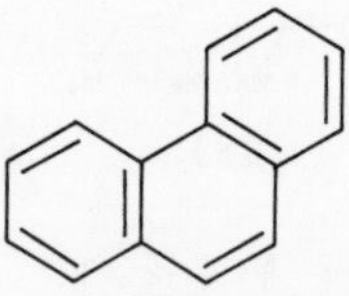

(52) 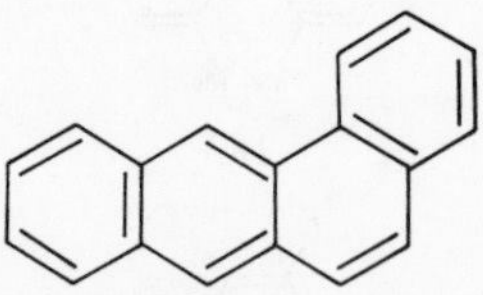 (53)

fixation is greater than in naphthalene, and consequently conjugative effects should, for example, be even greater than across the 9,10-bond in phenanthrene, confirmed by fluorine activating 4.1 times across this bond. However, the 182-fold activation by methyl is less than across the 1,2-bond in naphthalene because of the intervention of a second factor, the number of sites available for delocalization of the charge, described in Section (iv) below.

Bond fixation can also be induced by strain as shown by the data for fluoranthene.[115] The need to minimize strain and therefore avoid having double bonds in the central ring should lead to greater stability of structure **54** relative to structures **55** and **56**. Bond fixation should therefore be greater than in naphthalene. The resulting low order of the 3,16- and 4,16-bonds produces low transmission of the deactivating effect of bromine from the 3- to the 4-position, whereas the high order of the 2,3-bond facilitates transmission of the mesomeric effect of bromine from the 3- to the (conjugated) 2-position. Thus a 3-bromo substituent deactivates the 2- and 4-positions by much smaller factors (0.274 and 0.102, respectively) than those by which a 1-bromo substituent deactivates the corresponding 2- and 8-positions in naphthalene (0.047 and 0.044, respectively). Since the effect of the halogens is the product of a fine balance between the inductive and mesomeric effects, slight enhancement of one effect can lead to a large change in the observed overall result.

Formation of a double bond between the 12- and 13-positions of the central

(54) (55) (56)

ring of fluoranthene is also difficult because this would bring the benzene and naphthalene rings closer together, so making it harder for the 11,12-bond to bridge the C-13–C-15 distance. Conjugation between the benzene and naphthalene rings is therefore inhibited. Thus, whereas 4′-bromo and 4′-nitro substituents deactivate the 4-position of fluorene by 4.7 and 2670 times, respectively (Table 3.15), the analogous deactivation of the 3-position by the corresponding 8-substituents in fluoranthene is halved, being 2.35 and 1260 times, respectively. The greater deactivation by the 8-nitro substituent of the 3- relative to the 4-position in fluoranthene is due to resonance between the 8- and 4-positions requiring loss

Table 3.23 Calculated and observed substituent interactions in polycyclics

Aromatic	Interaction	$10^2 A_f$	Effect of CH_2 on localization energy/β	Activation by methyl
Benzene	1,2	16.7[a]	−0.80	220
	1,4	16.7[a]	−0.71	450
Naphthalene	1,2	18.2	−0.85	270
	2,1	18.2	−0.80	300
	1,4	14.5	−0.66	83.0
	1,5	3.6	−0.33	2.18
	1,7	7.3	−0.38	2.90
	2,8	7.3	−0.40	3.11
	2,3	10.9	−0.58	3.60
	2,6	10.9	−0.43	19.4
Phenanthrene	9,10	13.4	−0.66	182
	9,1	8.0	−0.415	4.24
	9,3	9.8	−0.425	17.4
	9,5	6.25	−0.22	2.37
	9,7	8.0	−0.25	3.05
	1,4	11.6	—	45[b]
	3,4	13.4	—	55[b]
Benz[a]anthracene	5,7	9.0	−0.44	5.79
	7,12	9.5	−0.56	24.0

[a] Because there is no bond fixation in benzene, these values are the averages of those obtained by using both Kekulé forms for the ground state.
[b] Calculated as 1.5 times the effect of methylene in 4*H*-cyclopenta[*def*]phenanthrene.

of benzenoid character of both rings in the naphthalene moiety, whereas resonance between the 8- and 3-positions requires loss of benzenoid character in only one ring.

(iv) Number of sites for delocalization of charge

Substituents at the 3- and 4-positions of naphthalene have a substantially smaller effect on the reactivity of the 1-position ($f_3^{Me} = 2.75, f_4^{Me} = 83$) than they do in benzene. The effect of benz[*a*]anthracene is even smaller, a 12-methyl substituent activating the 7-position only 53 times. This arises in part because the number of sites over which the transition state charge may be delocalized increases along the series benzene < naphthalene < phenanthrene < benz[*a*]anthracene, i.e. there will be less charge on any given site along this series. Calculations of the change in delocalization energy produced by introducing a CH_2^- substituent (i.e. an alkyl group with a $+M$ effect) at each point in the molecule (and which take account of the bond fixation factors) give a satisfactory indication of the substituent effect observed.[158] However, a much simpler method which requires no lengthy calculations is the following:[182]

(1) The number of C—C linkages which have the same bond order in the transition-state canonicals, e.g. **57**, **58**, and **59**, as in the most stable ground-state canonicals, e.g. **49** and **53**, are summed.

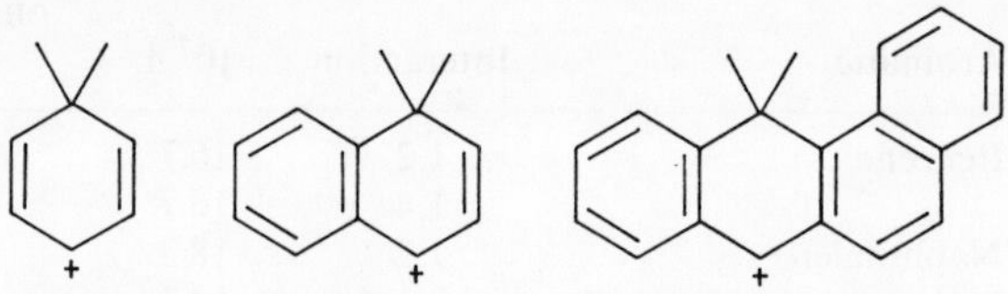

	(57)	(58)	(59)
No. of bonds in ground-state position:	3	8	18
Total no. of C—C linkages:	6	11	21
Sites available for charge delocalization:	3	5	9
$10^2 A_f$:	16.7	14.5	9.5

(2) This number is divided by the product of the total number of C—C linkages and the number of positions over which the transition state charge can be delocalized, i.e. factors for 18 (benzene), 55 (naphthalene), 11.2 (phenanthrene), 18.9 (benz[*a*]anthracene), etc., to give the *Substituent Interaction Factor*, A_f; the larger this number, the larger is the observed substituent effect. Some factors are listed in Table 3.23 together with the effect of the CH_2 substituent on the localization energy, and the observed methyl substituent effect.

Both methods incorrectly predict the relative activations by the *o*- and *p*-Me substituents in benzene, and the 2,3- relative to the 2,6-interaction in naphthalene, the localization energy method being worse in each case. The A_f factors correctly predict the following features: (a) the relative *para* interactions in benzene, naphthalene, phenanthrene, and benz[*a*]anthracene; (b) the poorer 1,5- relative to the 1,7-interaction in naphthalene; (c) the stronger interaction across the 1,2-bond in naphthalene than across the equivalent 3,4- and 9,10-bonds in phenanthrene; (d) the stronger 2,6-interaction in naphthalene than the corresponding 3,9-interaction in phenanthrene; and (e) the relative strengths of the interactions across the corresponding 2,8-, 1,9-, and 5,7-positions in naphthalene, phenanthrene, and benz[*a*]anthracene, respectively. Where they have been measured, the effects of other substituents are correctly predicted.

The effects of substituents in non-conjugated positions are incorrectly predicted by the localization energy method (which, for example, predicts deactivation by methyl in all positions). However, a modification of the A_f factor method can be applied which uses the average of the *ortho* values adjacent to the non-conjugated site, and these are designated A_f' values (Table 3.24).[182] They are not directly comparable to the A_f values but give good predictions of the relative reactivities: (a) activation in the polycyclics is predicted to be less than in benzene; (b) the activating effect of methyl substituents on the 1-position of naphthalene is correctly predicted to be $3 > 8 > 6$; (c) the activating effect of methyl substituents on the 2-position of naphthalene is predicted to be $4 = 5 > 7$ compared with $4 > 5 > 7$ observed; (d) the activating effect of methyl substituents on the 9-position of phenanthrene is predicted to be $4 > 2 = 8 > 6$ compared with the observed order of $6 > 4 > 2 \approx 8$; the activation from the 6-position appears to be too high but there is an important reason for this discrepancy and it is described under (viii) below.

Table 3.24 Calculated and observed interactions between non-conjugated sites in polycyclics

Aromatic	Site, substituent	A_f'	Activation by methyl
Benzene	1,3	16.7	6.1
Naphthalene	1,3	16.5	2.75
	1,6	5.4	1.31
	1,8	9.1	1.53
	2,4	12.7	3.0
	2,5	12.7	1.76
	2,7	9.1	1.67
Phenanthrene	9,2	8.9	2.72
	9,4	10.7	2.96
	9,6	7.1	3.16
	9,8	8,9	2.67

(v) The effect of i, j- vs j,i-substituent–site interactions

A substituent in a position *i* has approximately the same effect on the reactivity of a position *j* as does the same substituent in position *j* on the reactivity of position *i*. Most discrepancies, e.g. $f_1^{3\text{-Cl}} = 0.0023$, $f_2^{4\text{-Cl}} = 0.0103$; $f_1^{6\text{-Cl}} = 0.032$, $f_5^{2\text{-Cl}} = 0.0105$ can be accounted for in terms of the *Substituent Interaction Factors* noted in (iv) above. However, the effect of 2-halogens on the reactivity of the 1-position in naphthalene is three times greater than in the reverse direction and there is no obvious reason for this; the corresponding effects of methoxyl are 2.2×10^4 and 3.7×10^5.

Other theoretical methods are available for calculating substituent effects in polycyclics (Section 11.1.6), but these give no significant improvement over the above empirical methods. The substituent–reaction site interactions observed in detritiation are paralleled in the positional reactivities of the heterocycles quinoline and isoquinoline, measured in the gas phase,[189] so the substituent effects clearly describe the fundamental transmission properties of bicyclic aromatics.

(vi) Additivity of methyl substituent effects

Hydrogen exchange provides the only test of the *Additivity Principle* in polycyclic molecules. The observed polymethyl substituent effects for naphthalene and phenanthrene are shown in Table 3.25 together with calculated values derived from those in Table 3.22. In many cases there are substantial departures from additivity.

For 1,8-dimethylnaphthalene there is very good agreement at the 2- and 4-positions, showing that steric effects are unimportant. Hence the much higher reactivities of the corresponding positions in acenaphthene and perinaphthane (Table 3.22) do not arise from reduced steric hindrance, due to 'exposure' of these positions through strain in the bridge. This latter had been assumed to account for the high reactivities of the positions *ortho* and *para* to the 1,8-bridge in other substitutions of acenaphthene (Sections 6.7.3 and 9.6.7). Steric facilitation of C–H hyperconjugation **(60)**[98,190] and strain-enhanced C–C hyperconjugation account for the phenomenon. Not only are the side-chain C—H bonds aligned almost perfectly for C–H hyperconjugation in both acenaphthene and perinaphthane, but in the former molecule this process also removes the eclipsing

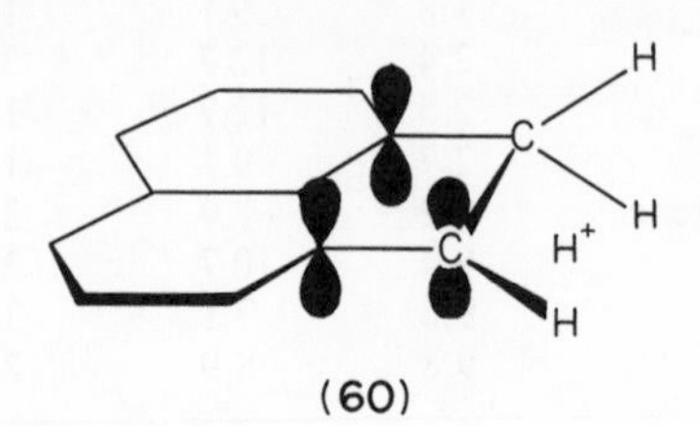

(60)

Table 3.25 Observed and calculated activation by polymethyl substituents in naphthalene and phenanthrene

Substituent	Position	Observed effect	Calculated effect	Obs./calc. activation
(a) *Naphthalene:*				
1,3-Me_2	2	1870	955	1.96
	4	10000	28400	0.403
	(5,7,8)	(37.7)[a]	(27.2)[a]	1.39
1,4-Me_2	2	332	801	0.414
	5	6.03	3.34	1.81
	6	14.3	5.44	2.63
1,5-Me_2	2	219	466	0.47
	3	20.1	9.36	2.15
	4	121	127	0.95
1,8-Me_2	2	722	827	0.873
	3	12.1	5.27	2.30
	4	184	180	1.02
2,3-Me_2	1	870	825	1.05
	5	21.0	3.80	5.53
	6	46.2	29.9	1.55
2,6-Me_2	1	878	393	2.23
	3	53.4	5.53	9.67
	4	27.3	7.98	3.42
2,7-Me_2	1	3280	870	3.77
	3	172	69.6	2.47
	4	7.09	3.60	1.97
(b) *Phenanthrene:*				
2,7-Me_2	9	7.94	8.40	0.95
3,6-Me_2	9	36.7	37.7	0.97
4,5-Me_2	9	23.5	6.97	3.37
2,4,5,7-Me_4	9	141	58.5	2.41
3,4,5,6-Me_4	9	218	263	0.83

[a] Average values.

of the C—H bonds, whilst C–C hyperconjugation relieves the ring strain; these two factors are much less important for perinaphthane which therefore has intermediate reactivity.

The deviations from additivity in the dimethylnaphthalenes indicate a decrease in bond fixation compared with the methylnaphthalenes, probably as a result of steric interactions causing a change in the bond lengths.[179] A decrease in bond fixation will produce the following effects:

(1) A decreased substituent interaction across the 1,2-bonds and an increased interaction across the 2,3-bonds. The latter will be of greater magnitude since the methylnaphthalene data show that relative to benzene, a decrease in bond order produces a much greater change in substituent effect than does a comparable

increase in bond order. This explains the deviations at the 2-positions of 1,3-, 1,4-, 1,5-, and 1,8-dimethylnaphthalene, the 4-position of 1,3-dimethylnaphthalene, and the 3-position of 2,6- and 2,7-dimethylnaphthalene.

(2) An increase in interactions between non-conjugated sites. These are considerably lower in naphthalene than in benzene (e.g. for methyl substituents, $\log f_4/\log f_3 = 4.4$ in naphthalene compared with 3.3 in benzene) because bond fixation causes relatively little of the resonance-delocalized transition state charge being relayed to an adjacent position. The increased interactions account for the deviations at the 3-positions of 1,5- and 1,8-dimethylnaphthalene, and at the 4-positions of 2,6- and 2,7-dimethylnaphthalenes; the reactivity of the 1-position of 2,3-dimethylnaphthalene results from compensation of the effects in (1) and (2).

(3) An increase in the inter-ring interactions except for the 2,6-interaction (which is high owing to bond fixation and will therefore be decreased). This accounts for all the remaining discrepancies and the magnitudes of some of the differences between them; some of the results are produced by a combination of this effect and that in (1).

Detritiation of 1,8- and 1,4-dimethylnaphthalene was accompanied by some sterically and electronically accelerated protiodealkylation, respectively.[179] This is described further in Section 4.5.

For the dimethylphenanthrenes, the reactivities of the 2,7- and 3,6-compounds are as predicted. In each of these the methyl substituents are in β-naphthalene-like positions and remote from each other so no ring distortion should occur, as the results confirm. For the 4,5-dimethyl- and the 2,4,5,7-tetramethyl-phenanthrenes, steric crowding between the 4- and 5-positions causes twisting of the phenanthrene nucleus, thereby destabilizing the ground state through loss of aromaticity. Enhanced exchange rates are therefore obtained. This factor must also apply to the 3,4,5,6-tetramethyl compound, but here there is severe crowding between the methyl groups which inhibits resonance (C–H hyper-conjugation) and this will be particularly important at the 3-position, which is strongly conjugated with the 9-position (Table 3.22); these two effects approximately cancel out.

(vii) Substitution at an enclosed site

The results for the cyclophanes (**61**) (Table 3.22) provide the only data for electrophilic substitution through a 'hole.'[112] Models indicate that access to the site X is prevented unless the alkyl chain lies to one side, and the required symmetry of the reaction path means that the process must occur as shown in **62**. Reduction in the size of the alkyl chain also causes buckling of the aromatic ring and consequent loss of ground-state stability. Position Y (*ortho* × *para* to the alkyl chain) is subject only to this latter effect, and with decreasing n there is a small increase in reactivity consistent with this loss of stability. Position X (*ortho*

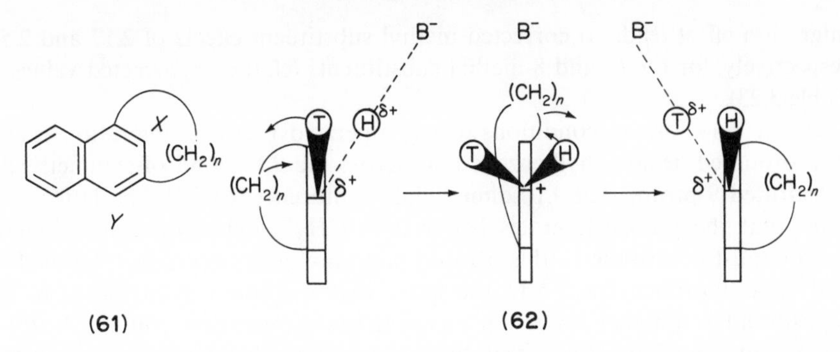

(61) (62)

× *ortho* to the alkyl chain) is subject to both effects. As n decreases there is initially a small increase in reactivity, but then a decrease as the effect of steric hindrance becomes more severe. The effect of steric hindrance is more clearly indicated by the ratio of the partial rate factors f_X/f_Y, which are 0.0245 (1,3-dimethylnaphthalene), 0.0185 (**61**, $n = 10$), 0.0108 (**61**, $n = 8$), and 0.0051 (**61**, $n = 7$).

(viii) Tritium migration during exchange

Kinetic runs with 3-methylphenanthrene showed a decrease in rate coefficient with time, whilst runs with the 6-methyl isomer gave anomalously high rates [Section 3.1.2.13(iv)]. This is due to tritium migration during exchange[182] (eq. 3.4), which is facilitated by the highly symmetrical nature of the exceptionally stable doubly benzenoid intermediate **63**; migration has not been observed in any other aromatic nucleus.

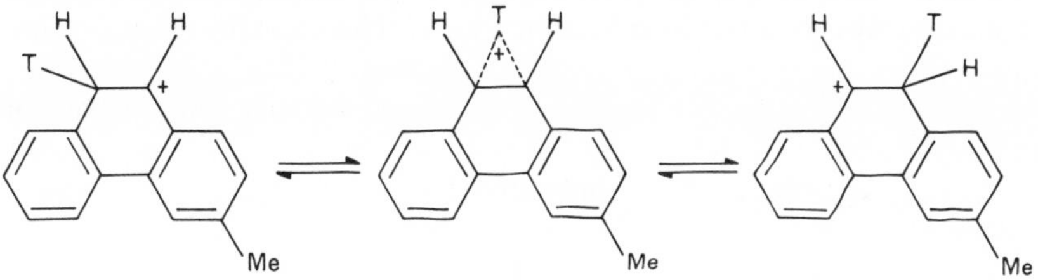

(3.4)

(63)

The reversibility of the reaction means that the 6-methyl isomer becomes contaminated with the (much more reactive) 3-isomer (leading to enhanced exchange rates) and the 3-methyl isomer becomes contaminated with the (much less reactive) 6-methyl isomer, leading to diminishing rate coefficients during a kinetic run. Rate coefficients for the 6-methyl isomer do not increase during runs because the more reactive 'impurity' exchanges rapidly. A similar factor affects the 1- and 8-isomers, but is less kinetically important since the former is only about 1.5 times as reactive as the latter. The reactivities of the other 'pairs' of isomers are so similar that no deviations are detectable. Correction for the migration effect leads to corrected methyl substituent effects of 2.17 and 2.56, respectively, for the 6- and 8-methyl substituents (cf. the uncorrected values in Table 3.23).

Under very forcing conditions (using superacids), proton migration occurs in protonated hexamethylbenzene and across the 9,10-bond of symmetrically substituted 9-protonated-9,10-dimethylphenanthrenes (**64**).[191] The latter indicates that the proton migrates better than CH_3^+ and steric and symmetry factors must contribute to this. Proton migration also occurs in 1,2-dimethyl- and 1,2-difluorobenzene.[192] In the latter case migration from the 3- to the 4-position has a higher activation energy than the migration from the 4- to the equivalent 5-position. Olah and Mo[192] proposed that benzonium ions (**65**) involving a 3-centre bond are intermediates in these rearrangements.

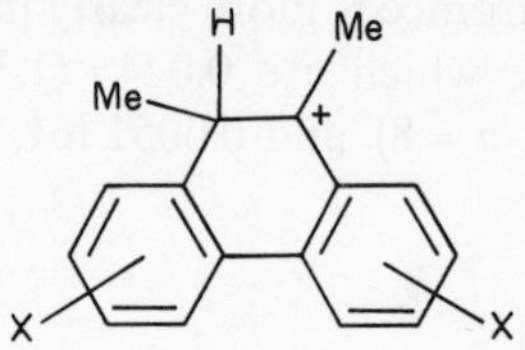

(**64**) 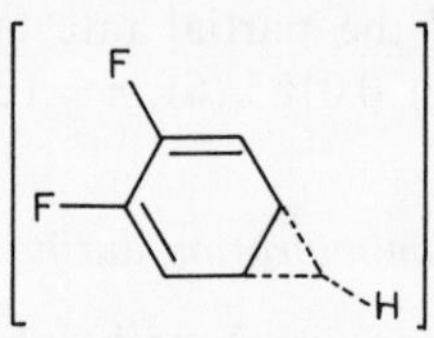 (**65**)

3.1.2.14 Long-range substituent effects

The high ρ-factor for hydrogen exchange makes possible the detection of long-range substituent effects (see also Section 3.1.2.4). This explains the $k_{rel.}$ values

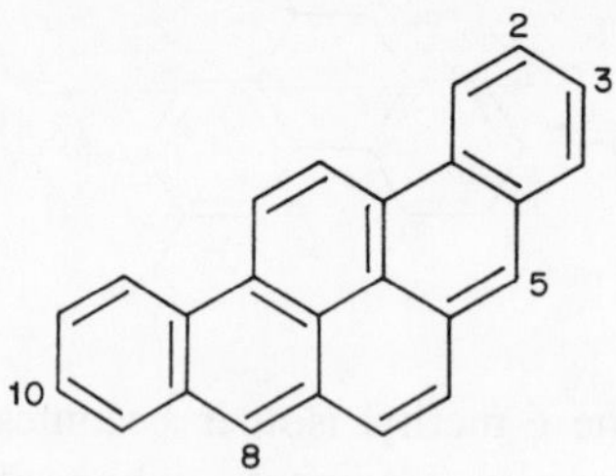

(**66**)

	Relative rates at position	
Substituent	5	8
H	1.0	1.0
2-F	0.7	1.0
3-F	0.87	0.44
2,10-F_2	0.38	0.58

obtained for deuteriation at the 5- and 8-positions of 2-, 3-, and 10-fluoro-substituted dibenzo[a,i]pyrenes (**66**).[193] The 5- (but not the 8-position) is conjugated with the 3-fluorine, whereas the 8- (but not the 5-position) is conjugated with the 2- and 10-fluorines. The observed relative reactivities are then accounted for by the balance of $-I$ and $+M$ effects (where the latter can operate).

3.1.2.15 Metal-containing substituents

The marked electron release by the CH_2SiMe_3 substituent due to C–Si hyperconjugation was noted in Section 3.1.2.4. Likewise, the very high activation of hydrogen exchange produced by the $CH_2Fe(CO)_2C_5H_5$ substituent[194] probably arises from C–Fe hyperconjugation; the corresponding molybdenum and tungsten compounds are three and six-times less activating, respectively, and methylene—metal bond cleavage accompanies the reaction.[194]

Ferrocene is $>10^8$ times more reactive than benzene towards exchange.[149] The deactivating factors produced by substituents in one ring upon deuteriation in the other (CF_3CO_2D, 140 °C) are 26 (Br), 65 (CO_2H), and 217 (CN).[195] In the same ring the effects are larger as expected: 69 (Br) and 256 (CO_2H) for the 2,5-(pseudo-*para*) interaction, similar values being obtained for the 3,4-(pseudo-*ortho*) interaction. A methyl substituent increases the exchange rate tenfold, whilst octamethyl- and hexamethylferrocenes are respectively 10^5 and 10^4 times as reactive as the parent. The conclusion that the methyl effects are not additive[196] is invalid, however, without the appropriate positional reactivity data. As a substituent, ferrocene produces marked activation with $f_p^{\text{ferr.}} \approx 10^5$.[149,197]

Apart from silver ion [Section 3.1.2.12(iv)], other metals from π-complexes to aromatics and the $Cr(CO)_3$ and $Cr(CO)_2PPh_3$ ligands produce weak and strong activation, respectively. The 1,3,5-trimethyl- and 1,3,5-trimethoxy substituents produced only trivial activation in the presence of these ligands, whereas the acetoxy-, carbomethoxy-, and protonated dimethylamino substituents produced a larger activation.[198] The overall result appears to be a delicate balance between two opposing effects, i.e. electron-supplying groups activate the ring, but weaken the bond to the electron-donating ligand, whereas electron-withdrawing groups produce the reverse effects. The $Mn(CO)_2PPh_3$ ligand is also much more activating than the $Mn(CO)_3$ ligand.[199]

3.1.2.16 Annulenes

The reactivities of these non-benzenoid aromatics are of interest because of their predicted aromaticity. Partial rate factors for detritiation of azulene (**67**),[200] 1,6-methano[10]annulene (**68**),[201] 11,11-difluoro-1,6-methano[10]annulene (**69**),[201] and 1,6:8,13-propane-1,3-diylidene[14]annulene (**70**)[202] show the following features:

(1) Azulene is extremely reactive and substitutes at the 1(3)-position. This is attributable to the formation of an aromatic 6π seven-membered ring (**71**) in the transition state. Exchange is not observed at any other position and consideration of the aromaticity of the respective transition states[200] predicts a positional reactivity order of $1>2>5>4>6$.

(2) The [10]- and [14]annulenes are very reactive and much more so than the

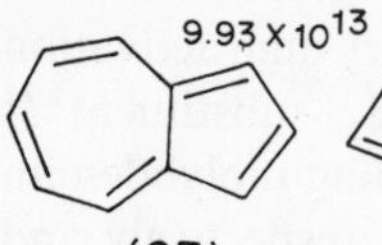

formally similar naphthalene and anthracene. This arises because their ground states are destabilized through non-planarity, and the greater reactivity of the [14]annulene accords with its calculated lower aromaticity.[203]

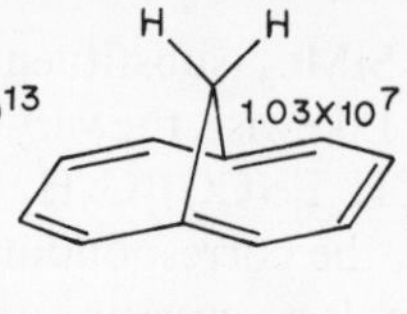

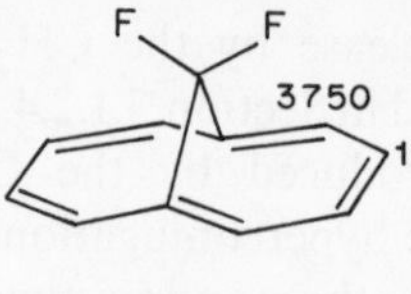

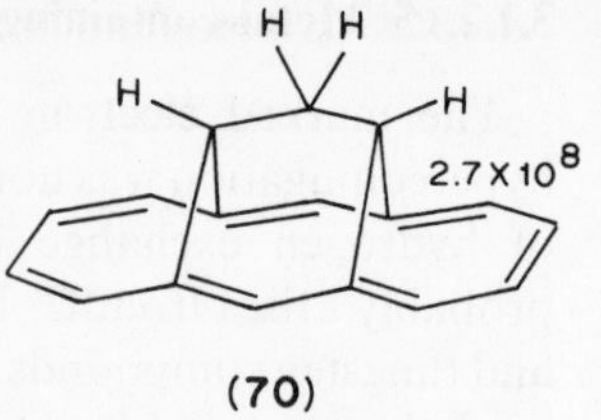

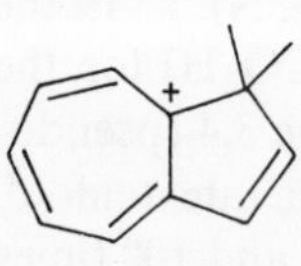

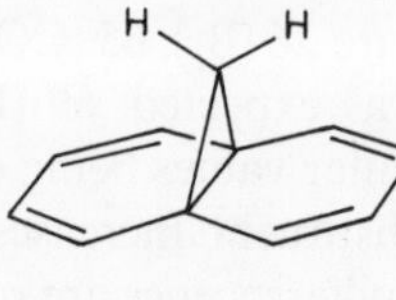

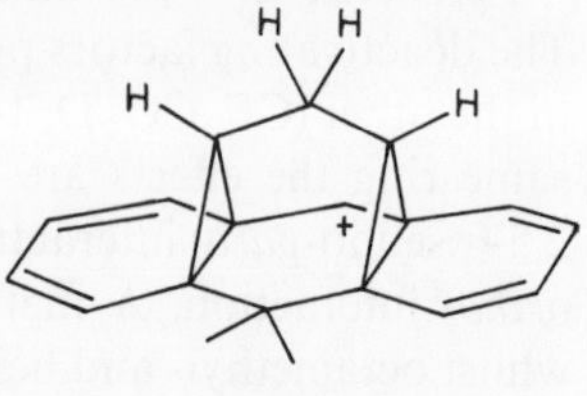

(3) The difference in the reactivities of the 2- and 3-positions in the [10]annulenes can be explained by assuming that transition states based on the norcaradiene structure (**72**) make a significant contribution to the overall resonance hybrid.[201] Spectroscopic studies have confirmed that in the ground state the bridgehead p-orbitals overlap to the extent of ca 40%.[204] The predominant substitution in the 2-position of the [14]annulene (rather than the 7-position corresponding to the 9-position in anthracene) also arises from the importance of the canonical **73** in which delocalization of the charge is not possible, thereby making the 7-position unreactive. By contrast, if **70** accurately described the molecule, delocalization over seven sites could occur.

(4) The difference in the reactivities at the 2-positions of the hydrogen- and fluorine-containing [10]annulenes is predicted by the difference in electronic effects of the m-CH_2 and m-CF_2 substituents.[201]

(5) From the exchange data and those for protiodesilylation [Section 4.7.1.(ii)] the differences in σ^+ and σ values may be calculated to be 0.365 for **68** and 0.31 for **70**, showing the latter to be slightly less polarizable; this may reflect its lower ground-state stability and hence higher polarization.

3.2 GAS-PHASE HYDROGEN EXCHANGE

In the gas phase, tritium from helium tritiide molecular ion HeT^+ exchanges into aromatics.[205] The distribution of tritium is fairly uniform in halogenobenzenes, mainly *ortho* in anisole, and *meta* in trifluoromethylbenzene, indicating the reaction to be an electrophilic substitution. The reaction is accompanied by aryl—substituent bond cleavage (as is also, to a slight extent, hydrogen exchange in solution) and is most marked with *tert*-butylbenzene, one of the compounds that cleaves in acid-catalysed exchange (Section 4.5). Isomerization of the intermediate arenonium ion occurs so that rearranged products are obtained; for halogenobenzenes the extent of rearrangement tends to be greater the weaker the aryl—halogen bond.

Deuterium tritiide, D_2T^+, also exchanges with aromatics. It is less reactive than helium tritiide, and shows higher intramolecular selectivity (as required by the *Reactivity–Selectivity Principle*). In the reaction with toluene the substitution pattern is predominantly *ortho,para*, although some exchange in the alkyl group also occurs.

3.3 BASE-CATALYSED HYDROGEN EXCHANGE

In the presence of strong bases, aromatic hydrogens exchange in a reaction which may be represented by eqs 3.5 and 3.6, where M is an alkali metal and B^- is a base, e.g. NH_2^-, O-*t*-Bu^-. This reaction is therefore an electrophilic substitution in which loss of a proton is the first rather than the last step, i.e. the mechanism is B-S_E1 (see Section 2.1). Rates of base-catalysed exchange correlate with energies of deprotonation of ArX, as do those for base-catalysed desilylation and destannylation, which involve the formation of a free carbanion in the rate-determining step (Sections 4.7.2 and 4.9.2).[206]

$$\mathrm{ArD + MB \underset{k_{-1}}{\overset{k_1}{\rightleftharpoons}} Ar^-M^+ + BD} \tag{3.5}$$

$$\mathrm{Ar^-M^+ + BH \underset{k_{-2}}{\overset{k_2}{\rightleftharpoons}} ArH + MH} \tag{3.6}$$

Since the ease of formation of the aromatic anion Ar^- governs the rate of reaction with a given base, it follows that electron supply decreases the rate of exchange and vice versa, as shown by the data for deuterium exchange given in Table 3.26.[101,207–212] These data (which are less accurate than those obtained in acid-catalysed exchange because of the greater experimental difficulties) show the following features:

(1) Inductive effects largely determine the exchange rates, and a correlation exists between *ortho* exchange rates and σ_I values.[213,214] Other data show that the *o*-nitro substituent activates more than the *p*-nitro substituent,[215] the *ortho*:*para* rate ratio being 2000.[216]

(2) In contrast to acid-catalysed exchange, base-catalysed exchange is sterically hindered, as shown for example by the data for mesitylene compared with toluene, and by the very low overall rate ratio (0.15) for *tert*-butylbenzene.[213] A

Table 3.26 Partial rate factors for deuterium exchange of substituted benzenes PhX and other aromatics with various bases

		Base			
X	Position	Potassamide	Lithium cyclohexylamide	Caesium cyclohexylamide	Sodium methoxide
Me	2	0.18	0.12	0.20	
	3	0.38	0.54	0.59	
	4	0.32	0.43	0.52	
1,4-Me_2	2	0.1			
1,3,5-Me_3	2	0.015		0.013	
1,2,4,5-Me_4	3	0.004			
Ph	2	3.7	1.2		
	3	3.2	3.7		
	4	3.0	2.3		
OMe	2	8000			500[a]
	3	10			1[a]
	4	0.1			0.5[a]
SMe	2	550			
	3	24			
	4	6.5			
OPh	3	41			
	4	3.7			
NHMe	2	33			
	3	2.9			
	4	1.3			
NMe_2	2	1.2			
	3	0.13			
	4	0.06			
F	2	$>4 \times 10^6$	6.3×10^5		1.8×10^5
	3	4000	107		107
	4	200	11.2		3.5
CF_3	2	6×10^5			
	3	1×10^4	580		
	4	1×10^4			
1,4-$(OMe)_2$	2	2330			
1,2-$(OMe)_2$	4	1.47			
1-OMe-4-Me	2	192			
1,2-$(OMe)_2$-4,5-Me_2	3	72.5			
2-OMe-1,3,5-Me_3	4	0.04			
1,4-$(NMe)_2$	2	0.835			
1-NMe_2-2-Me	4	0.945			
PMe_2	2	2.0			
	3	3.6			
	4	2.3			
PPh_2	2	9			
	3	18			
	4	17			

Table 3.26 (*continued*)

X	Position	Base			
		Potassamide	Lithium cyclo-hexylamide	Caesium cyclo-hexylamide	Sodium methoxide
NPh_2	2	29			
	3	9			
	4	4			
Aromatic:					
Naphthalene	1	9.8	6.5		
	2	4.4	4.1		
Phenathrene	9		12.9		
Anthracene	1		14.1		
	9		42(71)[b]		
Pyrene	1		26(32)[b]		
	2		(27)[b]		
	4		(40)[b]		
Biphenylene	1		(490)[b]		
	2		(7.0)[b]		
Benzo[*b*]biphenylene	3		(1865)[b]		
Triphenylene	1				20.8
	2				2.77
[2,2]Paracyclophane	2				0.45

[a]For reaction with sodium ethoxide.
[b]For tritium exchange.

possible explanation is that in acid-catalysed exchange the electrophile approaches the ring above or below the ring plane. By contrast, in base-catalysed exchange the base must approach in the plane of the ring and hence of the substituent and so it is reasonable to expect greater hindrance here.

(3) Steric hindrance is greater for reaction with the bulkier cyclohexylamides than for reaction with amide.

(4) Since the partial rate factors for reaction at the α- and β-positions of naphthalene are different, an addition–elimination mechanism is not involved.

(5) With tritiated benzene, caesium cyclohexylamide catalyses exchange 3300 times faster than does lithium cyclohexylamide. Since caesium is more electropositive than lithium, its nucleophilic counterion is more reactive. However, other factors are important because the isotope effects k_D/k_T are substantially different for reaction with the two reagents, being 1.3–1.5, and 2.3 for the lithium and caesium compounds, respectively.[207,208,217] This difference is ascribed to the greater stability of the intermediate Ar^-Cs^+ so that there is less tendency to undergo the reverse step k_{-1} in eq. 3.5 before diffusion permits unlabelled solvent

BD to be replaced by unlabelled BH. For the lithium cyclohexylamide-catalysed exchange, reaction k_{-1} is comparable in rate to k_2, leading to a low isotope effect.

(6) Substituents XPh are in general more activating than XMe, owing to the greater $-I$ effect of Ph relative to Me.

(7) Substituent effects are fairly additive, considering the semi-quantitative nature of some of the data.

(8) The difference in the positional reactivities of the phosphorus-containing aromatics have been interpreted in terms of pπ–dπ-bonding. This explanation may be incorrect since the rates of base-catalysed exchange are largely independent of conjugative effects. Ph_2PO is much more strongly *ortho*-activating than Ph_2P.[210]

(10) The aromatic positions α to a strained ring, i.e. the 1-positions of triptycene and biphenylene, are much more reactive than the 2-positions. This arises from the strain factors described previously (Section 3.1.2.3) and, since conjugative effects are relatively unimportant for base-catalysed exchange, electronegativity rather than bond-fixation factors are likely to be the cause. Similar factors may enhance the reactivity of [2,2]paracyclophane relative to *p*-xylene.

Substituents in the benzene ring of the highly reactive arenetricarbonylchromium compounds, $ArCr(CO)_3$, produce only very small effects on the rate (in exchange with NaOEt–EtOH),[218] and this is in accord with the *Reactivity–Selectivity Principle.* The cations $ArCr^+I^-$ exchange more readily than the diarylchromium compounds Ar_2Cr; methyl substituents in the ring of the former produce a small decrease in the exchange rate.[219]

Rate coefficients ($10^7\,k$/s^{-1}) for the deuterium exchange of some naphthalene derivatives with O-*t*-Bu$^-$/*t*-BuOD are shown in **74**–**77**.[220] Proper interpretation of these data requires the (unknown) $\alpha:\beta$ reactivity ratio for naphthalene under the same conditions, but this ratio is likely to be smaller than that (1.6) obtained from reaction catalysed by lithium cyclohexylamide. In general, the data indicate that exchange with this reagent is very sterically hindered (see, for example, the

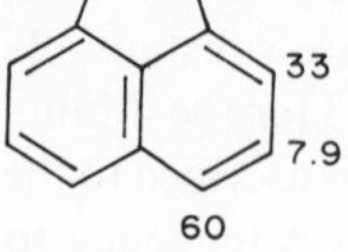

(74) 33, 7.9, 60 (75) 2.8, 26, 17 (76) 1.3, 71, 20 (77) 0.14, 3.9, 6.2 (78) CH_2 1.0, 29, 16

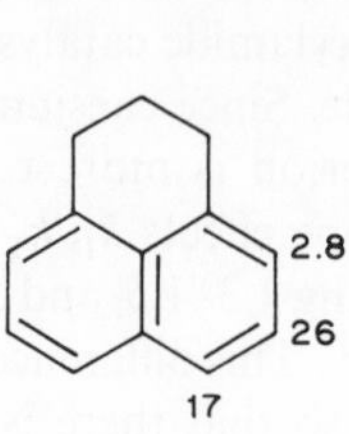

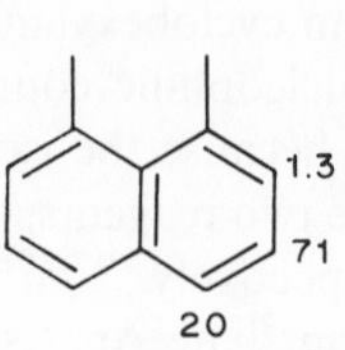

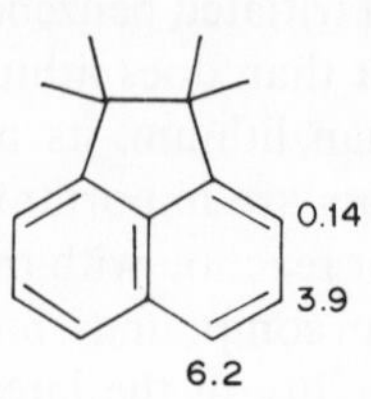

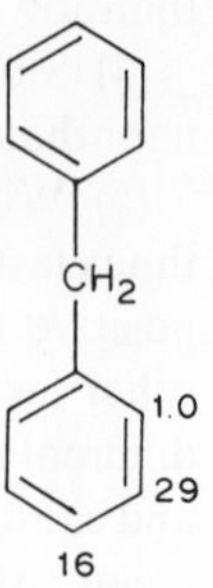

ortho:*para* ratio in diphenylmethane (**78**). Owing to the distortion of the naphthalene framework produced by the *peri* substituent, steric hindrance to exchange at the α'- and β-positions increases along the series **74**–**77**; these positions should be less hindered in acenaphthene (**74**) than in naphthalene and all the data support this interpretation. The data for β'-exchange have yet to be explained satisfactorily.

Detritiation of 1,3-dinitronaphthalene gave only 15% exchange in the 2-position and 85% in the 4-position,[221] which contrasts markedly with the exchange pattern in 1,3-dinitrobenzene.[216] This may be attributed to the low 2,3-bond order, which attenuates substituent effects as it does in acid-catalysed exchange. Thus the 2-position will be poorly activated by the 3-nitro substituent which will, by contrast, strongly activate the 4-position because of the high order of the 3,4-bond. In addition, the 1-nitro group will be less able to achieve coplanarity with the naphthalene ring than it will in benzene, owing to the *peri* hydrogen, and so electron withdrawal by it will be poorer. Taken together the effects account for the observed results.

REFERENCES

1. R. Taylor, *J. Chem. Soc., Perkin Trans. 2*, 1973, 253.
2. R. Taylor, *Comprehensive Chemical Kinetics*, Elsevier, Amsterdam, Vol. 13, 1972, pp. 194–277.
3. R. J. Thomas and F. A. Long, *J. Am. Chem. Soc.*, 1964, **86**, 4770.
4. A. J. Kresge and Y. Chang, *J. Am. Chem. Soc.*, 1959, **81**, 5509; 1961, **83**, 2877.
5. T. J. Tewson and R. Taylor, *J. Chem. Soc., Chem. Commun.*, 1973, 836.
6. H. V. Ansell and R. Taylor, *J. Chem. Soc., Perkin Trans. 2*, 1978, 751.
7. V. R. Kalinachenko *et al.*, *J. Org. Chem. USSR*, 1976, **12**, 93; P. P. Alikhanov, I. S. Temnova, V. R. Kalinachenko, and L. M. Yakimenko, *J. Org. Chem. USSR*, 1973, **43**, 152.
8. V. Gold and D. P. N. Satchell, *J. Chem. Soc.*, 1956, 2743.
9. S. Olsson, *Ark. Kemi*, 1960, **16**, 489.
10. C. Eaborn and R. Taylor, *J. Chem. Soc.*, 1960, 3301.
11. M. Liler, *Organic Mechanisms in Sulphuric Acid and Other Strong Acid Solutions*, Academic Press, New York, 1971.
12. E. L. Mackor, P. J. Smit, and J. H. van der Waals, *Trans. Faraday. Soc.*, 1957, **53**, 1309.
13. C. Eaborn and R. Taylor, *J. Chem. Soc.*, 1961, 247.
14. H. V. Ansell and R. Taylor, *J. Chem. Soc., Chem. Commun.*, 1973, 952.
15. C. Eaborn, P. M. Jackson, and R. Taylor, *J. Chem. Soc. B*, 1966, 653.
16. H. V. Ansell, M. M. Hirschler, and R. Taylor, *J. Chem. Soc., Perkin Trans. 2*, 1977, 353.
17. A. I. Serebryanskaya, P. A. Maksimova, and A. I. Shatenshtein, *J. Org. Chem. USSR*, 1977, **13**, 436.
18. L. Melander and S. Olsson, *Acta Chem. Scand.*, 1956, **10**, 879.
19. V. Gold and D. P. N. Satchell, *J. Chem. Soc.*, 1955, 3609.
20. V. Gold and D. P. N. Satchell, *J. Chem. Soc.*, 1955, 3622.
21. V. Gold and D. P. N. Satchell, *J. Chem. Soc.*, 1955, 3619.

22. D. P. N. Satchell, *J. Chem. Soc.*, 1956, 3911.
23. Ref. 2, pp. 196–208.
24. B. D. Batts and V. Gold, *J. Chem. Soc.*, 1964, 4284.
25. A. J. Kresge and Y. Chiang, *J. Am. Chem. Soc.*, 1972, **84**, 3976.
26. D. P. N. Satchell, *J. Chem. Soc.*, 1958, 3904.
27. A. J. Kresge and Y. Chiang, *Proc. Chem. Soc.*, 1961, 81.
28. F. A. Long and J. Schulze, *J. Am. Chem. Soc.*, 1961, **83**, 3340.
29. A. J. Kresge, Y. Chiang, and Y. Sato, *J. Am. Chem. Soc.*, 1967, **89**, 4418.
30. F. A. Long and J. Schulze, *J. Am. Chem. Soc.*, 1964, **86**, 327.
31. R. J. Thomas and F. A. Long, *J. Org. Chem.*, 1964, **29**, 3411.
32. J. Schulze and F. A. Long, *J. Am. Chem. Soc.*, 1964, **86**, 331.
33. J. Colapietro and F. A. Long, *Chem. Ind.* (*London*), 1960, 1056.
34. L. C. Gruen and F. A. Long, *J. Am. Chem. Soc.*, 1967, **89**, 1292.
35. J. A. Longridge and F. A. Long, *J. Am. Chem. Soc.*, 1967, **89**, 1292.
36. L. Zucker and L. P. Hammett, *J. Am. Chem. Soc.*, 1939, **61**, 2791.
37. N. C. Deno, J. Jarulzelski, and A. Shriesheim, *J. Am. Chem. Soc.*, 1955, **77**, 3044; N. C. Deno, P. T. Groves, and G. Staines, *J. Am. Chem. Soc.*, 1959, **81**, 5790; N. C. Deno, P. T. Groves, J. Jarulzelski, and M. Lugasch, *J. Am. Chem. Soc.*, 1960, **82**, 4729.
38. R. L. Hinman and J. Lang, *Tetrahedron Lett.*, 1960, **21**, 12.
39. R. W. Taft, *J. Am. Chem. Soc.*, 1960, **82**, 2964.
40. W. M. Schubert and R. H. Quacchia, *J. Am. Chem. Soc.*, 1962, **84**, 3778.
41. A. J. Kresge, G. W. Barry, K. R. Charles, and Y. Chiang, *J. Am. Chem. Soc.*, 1962, **84**, 4343.
42. A. R. Katritzky, A. J. Waring, and K. Yates, *Tetrahedron*, 1963, **19**, 465.
43. A. J. Kresge, R. A. More O'Ferral, L. E. Hakka, and V. P. Vitullo, *J. Chem. Soc., Chem. Commun.*, 1965, 46.
44. A. J. Kresge, S. G. Mylonakis, Y. Sato, and V. P. Vitullo, *J. Am. Chem. Soc.*, 1971, **93**, 6181.
45. V. Gold and F. L. Tye, *J. Chem. Soc.*, 1952, 2172, 2184.
46. C. MacLean, J. H. Van der Waals, and E. L. Mackor, *Mol. Phys.*, 1958, **1**, 247; G. Dallinga and G. Ter Maten, *Recl. Trav. Chim. Pays-Bas*, 1960, **79**, 737.
47. G. A. Olah and Y. K. Mo, *J. Org. Chem.*, 1973, **38**, 3212; G. A. Olah, H. C. Lin, and D. A. Forsyth, *J. Am. Chem. Soc.*, 1974, **96**, 6908; L. P. Kamshii and V. A. Koptyug, *Bull. Acad. Sci. USSR*, 1974, 232; I. B. Repinskaya, A. I. Rezvukhin, and V. A. Koptyug, *J. Org. Chem. USSR*, 1972, **8**, 1685, 1808; Yu. G. Erykalov, I. P. Beletskaya, I. S. Isaev, A. I. Rezvukhin, and V. A. Koptyug, *J. Org. Chem. USSR*, 1971, **7**, 1232.
48. G. A. Olah, R. H. Schlosberg, D. P. Kelly, and G. H. Mateescu, *J. Am. Chem. Soc.*, 1970, **92**, 2546.
49. R. J. Thomas and F. A. Long, *J. Am. Chem. Soc.*, 1964, **86**, 4770.
50. A. J. Kresge, S. Slae, and D. W. Taylor, *J. Am. Chem. Soc.*, 1970, **92**, 6309.
51. B. C. Challis and E. M. Millar, *J. Chem. Soc., Perkin Trans. 2*, 1972, 1618.
52. B. C. Challis and F. A. Long, *J. Am. Chem. Soc.*, 1963, **85**, 2524.
53. V. Gold and D. C. A. Waterman, *J. Chem. Soc. B*, 1968, 839, 849; F. G. Bordwell, W. J. Boyle, J. A. Hautala, and K. C. Lee, *J. Am. Chem. Soc.*, 1969, **91**, 4002; 1971, **93**, 511.
54. A. J. Kresge and Y. Chiang, *J. Am. Chem. Soc.*, 1967, **89**, 4411.
55. A. J. Kresge, *Discuss. Faraday Soc.*, 1965, **39**, 49.
56. R. L. Hinman and E. B. Whipple, *J. Am. Chem. Soc.*, 1962, **84**, 2534.
57. V. Gold, R. W. Lambert, and D. P. N. Satchell, *Chem. Ind.* (*London*), 1959, 1312; *J. Chem. Soc.*, 1960, 2461.
58. A. C. Ling and F. H. Kendall, *J. Chem. Soc. B*, 1967, 445.

59. B. C. Challis and F. A. Long, *J. Am. Chem. Soc.*, 1965, **87**, 1196.
60. C. D. Johnson, A. R. Katritzky, and S. A. Shapiro, *J. Am. Chem. Soc.*, 1969, **91**, 6654.
61. S. Olsson, *Ark. Kemi,* 1970, **32**, 89, 105.
62. R. Baker, C. Eaborn, and R. Taylor, *J. Chem. Soc.*, 1961, 4917.
63. W. M. Lauer, G. W. Matson, and G. Stedman, *J. Am. Chem. Soc.*, 1958, **80**, 6433.
64. W. M. Lauer, G. W. Matson, and G. Stedman, *J. Am. Chem. Soc.*, 1958, **80**, 6437.
65. W. D. Blackley, *Diss. Abstr.*, 1961, **20**, 1755.
66. K. C. C. Bancroft, R. W. Bott, and C. Eaborn, *J. Chem. Soc.*, 1964, 4806.
67. M. M. J. Le Guen and R. Taylor, *J. Chem. Soc., Perkin Trans. 2*, 1976, 559.
68. E. Glyde and R. Taylor, *J. Chem. Soc., Perkin Trans. 2*, 1977, 678.
69. A. I. Shatenshtein and P. P. Alikhanov, *Zh. Obshch. Khim.*, 1960, **30**, 992.
70. K. E. Richards, A. L. Wilkinson, and G. J. Wright, *Aust. J. Chem.*, 1972, **25**, 2369.
71. P. Fischer and R. Taylor, *J. Chem. Soc., Perkin Trans. 2*, 1980, 781.
72. W. J. Archer, M. A. Hossaini, and R. Taylor, *J. Chem. Soc., Perkin Trans. 2*, 1982, 181.
73. P. B. D. de la Mare and P. W. Robertson, *J. Chem. Soc.*, 1948, 100; P. B. D. de la Mare, *J. Chem. Soc.*, 1949, 2871; P. W. Robertson, P. B. D. de la Mare, and B. E. Swedlund, *J. Chem. Soc.*, 1953, 782; E. Berliner and F. Berliner, *J. Am. Chem. Soc.*, 1954, **76**, 6179; E. Berliner and M. M. Chen, *J. Am. Chem. Soc.*, 1956, **80**, 343; E. Berliner, *Tetrahedron*, 1959, **5**, 502; J. R. Knowles, R. O. C. Norman, and G. K. Radda, *J. Chem. Soc.*, 1960, 4885.
74. M. A. Hossaini and R. Taylor, *J. Chem. Soc., Perkin Trans. 2*, 1982, 187.
75. E. Berliner and F. Berliner, *J. Am. Chem. Soc.*, 1949, **71**, 1195.
76. R. Taylor, *Chimia*, 1968, **22**, 1.
77. R. O. C. Norman and G. K. Radda, *J. Chem. Soc.*, 1961, 3310.
78. W. van Dine and P. von R. Schleyer, *J. Am. Chem. Soc.*, 1966, **88**, 2321; P. von R. Schleyer and V. Buss, *J. Am. Chem. Soc.*, 1969, **91**, 5880.
79. R. Ketcham, R. Cavestri, and D. Jambotkar, *J. Org. Chem.*, 1963, **28**, 2139.
80. W. Kurtz, P. Fischer, and F. Effenberger, *Chem. Ber.*, 1973, **106**, 525.
81. R. C. Hahn, T. F. Corbin, and H. Schechter, *J. Am. Chem. Soc.*, 1968, **90**, 3404.
82. H. C. Brown and L. M. Stock, *J. Am. Chem. Soc.*, 1957, **79**, 1913.
83. J. M. A. Baas, and B. M. Wepster, *Recl. Trav. Chim. Pays-Bas*, 1972, **91**, 285.
84. H. V. Ansell, M. M. J. Le Guen, and R. Taylor, *Tetrahedron Lett.*, 1973, 13.
85. W. M. Lauer, G. W. Matson, and G. Stedman, *J. Am. Chem. Soc.*, 1958, **80**, 6439.
86. Ref. 2, p. 250.
87. H. V. Ansell and R. Taylor, *J. Chem. Soc. B*, 1968, 526.
88. R. Taylor, G. J. Wright, and A. J. Homes, *J. Chem. Soc. B*, 1967, 780.
89. E. Glyde and R. Taylor, *J. Chem. Soc., Perkin Trans. 2*, 1977, 1537.
90. J. Vaughan and G. J. Wright, *J. Org. Chem.*, 1968, **33**, 2580.
91. M. M. A. Stroud and R. Taylor, *J. Chem. Res. (S)*, 1978, 425.
92. W. H. Mills and I. G. Nixon, *J. Chem. Soc.*, 1930, 2510.
93. J. Vaughan, G. J. Welch, and G. J. Wright, *Tetrahedron*, 1965, **21**, 1665.
94. R. Taylor, *J. Chem. Soc. B*, 1968, 1402.
95. R. Taylor, *J. Chem. Soc. B*, 1968, 1559.
96. R. Taylor, *J. Chem. Soc. B*, 1971, 536.
97. H. Tanida and R. Muneyuki, *J. Am. Chem. Soc.*, 1965, **87**, 4794.
98. H. V. Ansell and R. Taylor, *Tetrahedron Lett.*, 1971, 4915.
99. F. R. Jensen and G. Maciel, *J. Org. Chem.*, 1960, **25**, 640.
100. T. J. Evenson, *Diss. Abstr.*, 1960, **20**, 3953.
101. A. Streitwieser, G. R. Ziegler, P. C. Mowery, A. Lewis, and R. G. Lawler, *J. Am. Chem. Soc.*, 1968, **90**, 1357.
102. A. R. Bassindale, C. Eaborn, and D. R. M. Walton, *J. Chem. Soc. B*, 1969, 12.

103. H. Selander and J. L. G. Nilsson, *Acta Chem. Scand.*, 1971, **25**, 1182; 1972, **26**, 3377.
104. R. Baker and C. Eaborn, *J. Chem. Soc.*, 1961, 5077.
105. R. Baker, R. W. Bott, C. Eaborn, and P. M. Greasley, *J. Chem. Soc.*, 1964, 627.
106. P. E. Peterson, D. M. Chevli, and K. A. Sipp, *J. Org. Chem.*, 1968, **33**, 972.
107. Ref. 2, pp. 243–248.
108. A. I. Serebryanskaya, A. V. Eltsov, and A. I. Shatenshtein, *J. Org. Chem. USSR*, 1967, **3**, 343.
109. J. H. Czernohorsky, K. E. Richards, and G. J. Wright, *Aust. J. Chem.*, 1972, **25**, 1459.
110. H. V. Ansell and R. Taylor, unpublished work.
111. H. Selander and J. L. G. Nilsson, *Acta Chem. Scand.*, 1972, **26**, 2433.
112. A. P. Laws, A. P. Neary, and R. Taylor, *J. Chem. Soc., Perkin Trans. 2*, 1987, 1033.
113. C. Eaborn, T. A. Emokpae, V. I. Sidorov, and R. Taylor, *J. Chem. Soc., Perkin Trans. 2*, 1974, 1454.
114. H. V. Ansell and R. Taylor, *J. Chem. Soc., Chem. Commun.*, 1973, 936.
115. H. V. Ansell and R. Taylor, *J. Chem. Soc., Perkin Trans. 2*, 1977, 866.
116. P. P. Alikhanov, V. R. Kalinachenko, G. V. Motsarev, I. S. Temnova, and L. M. Yakimento, *J. Gen. Chem. USSR*, 1977, **47**, 339; P. P. Alikhanov, T. G. Bogatskaya, V. R. Kalinachenko, G. V. Motsarev, and L. M. Yakimento, *J. Gen. Chem. USSR*, 1978, **48**, 550.
117. P. P. Alikhanov, V. R. Kalinachenko, T. S. Amamchan, V. R. Rozenberg, G. V. Motsarev, and L. M. Yakimento, *J. Org. Chem. USSR*, 1977, **13**, 691.
118. E. N. Yurigina *et al.*, *Zh. Fiz. Khim.*, 1960, **34**, 587.
119. R. Danielli, A. Ricci, and J. H. Ridd, *J. Chem. Soc., Perkin Trans. 2*, 1972, 2107.
120. A. J. Layton, J. H. Rees, and J. H. Ridd, *J. Chem. Soc., Chem. Commun.*, 1974, 518; J. H. Rees and J. H. Ridd, *J. Chem. Soc., Perkin Trans. 2*, 1976, 285.
121. L. M. Stock, *J. Chem. Educ.*, 1972, **49**, 400.
122. R. M. Pagni and R. J. Smith, *J. Am. Chem. Soc.*, 1979, **101**, 506.
123. C. Eaborn and R. Taylor, *J. Chem. Soc.*, 1961, 2388.
124. P. P. Alikhanov, T. S. Amamchan, T. G. Bogatskaya, O. N. Guve, V. R. Kalinachenko, G. V. Motsarev, and L. M. Yakimenko, *J. Org. Chem. USSR*, 1977, **13**, 515.
125. H. V. Ansell, K. C. C. Bancroft, C. Eaborn, R. E. Spillett, and R. Taylor, unpublished work.
126. R. Taylor, *Specialist Periodical Report on Aromatic and Heteroaromatic Chemistry*, Vol. 1, Chemical Society, London, 1973, p. 188.
127. P. P. Alikhanov, I. S. Temnova, V. R. Kalinachenko, and L. M. Yakimenko, *J. Gen. Chem. USSR*, 1973, **43**, 894.
128. R. W. Taft, E. Price, I. R. Fox, I. C. Anderson, and G. T. Davies, *J. Am. Chem. Soc.*, 1963, **85**, 3146; P. E. Peterson, *J. Org. Chem.*, 1966, **31**, 439; G. C. Levy, G. L. Nelson, and J. D. Cargioli, *J. Chem. Soc., Chem. Commun.*, 1971, 506.
129. F. P. Bailey and R. Taylor, *J. Chem. Soc. B*, 1971, 1446.
130. R. Baker, C. Eaborn, and R. Taylor, *J. Chem. Soc., Perkin Trans. 2*, 1972, 97.
131. A. I. Shatenshtein, E. A. Rabinovich, and V. A. Pavlov, *J. Gen. Chem. USSR*, 1964, **34**, 4050.
132. S. Oae, A. Ohno, and W. Tagaki, *Bull. Chem. Soc. Jpn.*, 1962, **35**, 681.
133. W. M. Lauer and J. T. Day, *J. Am. Chem. Soc.*, 1955, **77**, 1904.
134. S. Olsson and M. Russell, *Ark. Kemi*, 1970, **31**, 439, 455.
135. D. P. N. Satchell, *J. Chem. Soc.*, 1956, 3911.
136. Ref. 2, Table 159.
137. R. Taylor, *J. Chem. Soc. B*, 1971, 1450.
138. C. Eaborn and P. M. Jackson, *J. Chem. Soc. B*, 1969, 21.
139. B. B. P. Tice, I. Lee, and F. H. Kendall, *J. Am. Chem. Soc.*, 1963, **85**, 329; I. Lee and

F. H. Kendall, *J. Am. Chem. Soc.*, 1966, **88**, 3813; A. C. Ling and F. H. Kendall, *J. Chem. Soc. B*, 1967, 440.
140. J. R. Blackborow and J. H. Ridd, *J. Chem. Soc., Chem. Commun.*, 1967, 132.
141. G. P. Bean and A. R. Katritzky, *J. Chem. Soc. B*, 1968, 864.
142. Ref. 2, Ch. 1, Sections 3 and 8.
143. S. Clementi, A. R. Katritzky, and C. D. Johnson, *J. Chem. Soc., Perkin Trans. 2*, 1974, 1295.
144. T. J. Gilbert and C. D. Johnson, *J. Am. Chem. Soc.*, 1974, **96**, 5846.
145. C. G. Stevens and S. J. Strickler, *J. Am. Chem. Soc.*, 1973, **95**, 3918.
146. C. Eaborn and R. Taylor *J. Chem. Soc.*, 1961, 1012.
147. D. E. Rice, *Diss. Abstr.* 1961, **21**, 3961.
148. R. Baker, R. W. Bott, and C. Eaborn, *J. Chem. Soc.*, 1963, 2136.
149. Y. El-Din Shafig and R. Taylor, *J. Chem. Soc., Perkin Trans. 2*, 1978, 1263.
150. J. M. Blatchly and R. Taylor, *J. Chem. Soc.*, 1964, 4641.
151. H. V. Ansell, R. B. Clegg, and R. Taylor, *J. Chem. Soc., Perkin Trans. 2*, 1972, 766.
152. R. Baker, C. Eaborn, and J. A. Sperry, *J. Chem. Soc.*, 1962, 2382.
153. R. Taylor, *Specialist Periodical Report on Aromatic and Heteroaromatic Chemistry*, Vol. 2, Chemical Society, London, 1974, p. 226.
154. R. Taylor, unpublished work.
155. R. Taylor, *J. Chem. Soc. B*, 1971, 255.
156. C. Eaborn, J. A. Treverton, and D. R. M. Walton, *J. Organomet. Chem.*, 1967, **9**, 259.
157. M. M. Hirschler and R. Taylor, *J. Chem. Soc., Chem. Commun.*, 1980, 967.
158. C. Eaborn, P. Golborn, R. E. Spillett, and R. Taylor, *J. Chem. Soc. B*, 1968, 1112.
159. R. Taylor and G. G. Smith, *Tetrahedron*, 1963, **19**, 937.
160. J. Banger, C. D. Johnson, A. R. Katritzky, and B. R. O'Neill, *J. Chem. Soc., Perkin Trans. 2*, 1974, 394.
161. G. G. Stevens and S. J. Strickler, *J. Am. Chem. Soc.*, 1973, **95**, 3922.
162. M. G. Kuzmin, B. M. Uzhinov, G. S. Gyorgy, and I. V. Berezin, *J. Phys. Chem. USSR*, 1967, **41**, 400.
163. K. C. C. Bancroft, R. W. Bott, and C. Eaborn, *J. Chem. Soc., Perkin Trans. 2*, 1972, 95.
164. M. M. J. LeGuen and R. Taylor, *J. Chem. Soc., Perkin Trans. 2*, 1974, 1274.
165. M. M. J. LeGuen, Y. El-Din Shafig, and R. Taylor, *J. Chem. Soc., Perkin Trans. 2*, 1979, 803.
166. W. J. Archer, Y. El-Din Shafig, and R. Taylor, *J. Chem. Soc., Perkin Trans. 2*, 1981, 675.
167. H. V. Ansell, M. S. Newman, and R. Taylor, unpublished work.
168. W. J. Archer, R. Taylor, P. H. Gore, and F. S. Kamounah, *J. Chem. Soc., Perkin Trans. 2*, 1980, 1828.
169. K. C. C. Bancroft and G. R. Howe, *J. Chem. Soc. B*, 1970, 1541.
170. A. Streitwieser, A. Lewis, I. Schwager, R. W. Fish, and S. Labana, *J. Am. Chem. Soc.*, 1970, **92**, 6525.
171. M. J. S. Dewar and E. W. T. Warford, *J. Chem. Soc.*, 1956, 3570.
172. P. H. Gore, C. K. Thadani, S. Thorburn, and M. Yusuf, *J. Chem. Soc. C*, 1971, 2329.
173. O. I. Kachurin, E. S. Fedorchuk, and V. Ya. Vasilenko, *J. Org. Chem. USSR*, 1973, **9**, 1956, 1961.
174. P. M. op den Brouw and W. H. Laarhoven, *Recl. Trav. Chim. Pays-Bas*, 1978, **97**, 265.
175. A. Streitwieser, *Molecular Orbital Theory for Organic Chemists*, Wiley, New York, 1961, p. 336.

176. C. Eaborn, D. R. Killpack, J. N. Murrell, and R. J. Suffolk, *J. Chem. Soc., Perkin Trans. 2*, 1972, 432.
177. Y. El-Din Shafig and R. Taylor, unpublished work.
178. M. C. A. Opie, G. J. Wright, and J. Vaughan, *Aust. J. Chem.*, 1971, **24**, 1205.
179. A. P. Neary and R. Taylor, *J. Chem. Soc., Perkin Trans. 2*, 1983, 1233.
180. A. P. Neary, A. P. Laws, and R. Taylor, *J. Chem. Soc., Perkin Trans. 2*, 1987, 1033.
181. C. Eaborn, A. Fischer, and D. R. Killpack, *J. Chem. Soc. B*, 1971, 2142.
182. H. V. Ansell, P. J. Sheppard, C. F. Simpson, M. A. Stroud, and R. Taylor, *J. Chem. Soc., Chem. Commun.*, 1978, 586; *J. Chem. Soc., Perkin Trans. 2*, 1979, 381.
183. H. V. Ansell and R. Taylor, *J. Org. Chem.*, 1979, **44**, 4946.
184. W. J. Archer and R. Taylor, *J. Chem. Soc., Perkin Trans. 2*, 1981, 1153.
185. Calculated from D. R. Killpack, *D. Phil. Thesis*, Sussex University, 1969.
186. K. C. C. Bancroft and G. R. Howe, *J. Chem. Soc. B*, 1971, 400.
187. A. A. El-Anani, C. C. Greig, and C. D. Johnson, *J. Chem. Soc., Chem. Commun.*, 1970, 1024.
188. Ref. 175, p. 170.
189. E. Glyde and R. Taylor, *J. Chem. Soc., Perkin Trans. 2*, 1975, 1783.
190. A. Fischer, W. J. Mitchell, J. Packer, R. D. Topsom, and J. Vaughan, *J. Chem. Soc.*, 1963, 2892.
191. G. I. Borodkin, M. M. Shakirov, V. G. Shubin, and V. A. Koptyug, *J. Org. Chem. USSR*, 1978, **14**, 924.
192. G. A. Olah and Y. K. Mo, *J. Org. Chem.*, 1973, **38**, 3212.
193. D. J. Sardella, P. Mahathalong, H. A. Mariani, and E. Bohger, *J. Org. Chem.*, 1980, **45**, 2064.
194. S. N. Anderson, D. H. Ballard, and M. D. Johnson, *J. Chem. Soc., Chem. Commun.*, 1971, 779; D. N. Kursanov, V. N. Setkina *et al., Bull. Acad. Sci. USSR*, 1973, 199, 1602.
195. J. A. Mangravite and T. G. Traylor, *Tetrahedron Lett.*, 1967, 4457.
196. M. Sabatini, M. A. Franco, and R. Psaro, *Inorg. Chim. Acta*, 1980, **42**, 267.
197. A. I. Khatami, T. Kh. Kurbanov, I. R. Lyatifov, R. B. Materikova, and M. N. Nefedova, *Bull. Acad. Sci. USSR*, 1984, 2183.
198. D. N. Kursanov, V. N. Setkina, *et al., Proc. Acad. Sci. USSR*, 1970, **190**, 127; 1972, **202**, 75; *J. Gen. Chem. USSR*, 1971, **41**, 1345; *Bull. Acad. Sci. USSR*, 1974, 724.
199. F. S. Yakushin, V. N. Setkina, N. V. Kislyakov, D. N. Kursanov, and A. I. Shatenshtein, *Bull. Acad. Sci. USSR*, 1972, 270.
200. A. P. Laws and R. Taylor, *J. Chem. Soc., Perkin Trans. 6*, 1987, 591.
201. R. Taylor, *J. Chem. Soc., Perkin Trans. 2*, 1975, 1287.
202. A. P. Laws and R. Taylor, *J. Chem. Soc., Perkin Trans. 2*, 1987, 1692.
203. A. Sabljic and N. Trinajstic, *J. Org. Chem.*, 1981, **46**, 3457.
204. Leading references are given in ref. 202.
205. F. Cacace and G. Perez, *J. Chem. Soc. B*, 1971, 2086; F. Cacace, R. Cipollini, and G. Ciranni, *J. Chem. Soc. B*, 1971, 2089; F. Cacace, R. Cipollini, and G. Occhini, *J. Chem. Soc. B*, 1972, 84; F. Cacace and M. Speranza, *J. Am. Chem. Soc.*, 1976, **98**, 7305.
206. P. Dembech, G. Seconi, and C. Eaborn, *J. Chem. Soc., Perkin Trans. 2*, 1983, 301.
207. A. Streitwieser and R. G. Lawler, *J. Am. Chem. Soc.*, 1963, **85**, 2852; 1965, **87**, 5388; A. Streitwieser, R. G. Lawler, and C. L. Perrin, *J. Am. Chem. Soc.*, 1965, **87**, 5383.
208. A. Streitwieser and R. A. Caldwell, *J. Am. Chem. Soc.*, 1965, **87**, 5394.
209. A. Streitwieser, J. A. Hudson, and F. Mares, *J. Am. Chem. Soc.*, 1968, **90**, 648.
210. A. I. Shatenshtein *et al., Bull. Acad. Sci. USSR*, 1968, 1917; *J. Gen. Chem. USSR*, 1970, **40**, 1614; *Tetrahedron*, 1969, **25**, 1165.
211. A. Streitwieser and G. R. Ziegler, *Tetrahedron Lett.*, 1971, 415.

212. A. Streitwieser and F. Mares, *J. Am. Chem. Soc.*, 1968, **90**, 644.
213. G. E. Hall, E. M. Libby, and E. L. James, *J. Org. Chem.*, 1963, **28**, 311.
214. A. I. Shatenshtein, *Tetrahedron*, 1962, **18**, 95; *Adv. Phys. Org. Chem.*, 1963, **1**, 156.
215. N. A. Kozbulatova, E. A. Yakovleva, G. G. Isaeva, and Yu. S. Shabarov, *J. Org. Chem. USSR*, 1971, **7**, 2428; I. R. Bellabono and G. Sala, *J. Chem. Soc., Perkin Trans. 2*, 1972, 169.
216. E. Buncel, J. A. Elvidge, J. R. James, and K. T. Walkin, *J. Chem. Res. (S)*, 1980, 272.
217. A. Streitwieser, R. A. Caldwell, R. G. Lawler, and G. R. Ziegler, *J. Am. Chem. Soc.*, 1965, **87**, 5399.
218. M. Ashraf, *Can. J. Chem.*, 1970, **50**, 118; D. N. Kursanov, V. N. Setkina *et al.*, *Proc. Acad. Sci. USSR*, 1968, 1528; 1970, 429.
219. D. N. Kursanov, V. N. Setkina *et al.*, *J. Organomet. Chem.*, 1972, **37**, C35; *Bull. Acad. Sci. USSR*, 1974, 719.
220. D. H. Hunter and J. B. Stothers, *Can. J. Chem.*, 1973, **51**, 2884.
221. E. Buncel, A. R. Norris, J. A. Elvidge, J. R. Jones, and K. T. Walkin, *J. Chem. Res. (S)*, 1980, 326.

CHAPTER 4

The Replacement of a Substituent by Hydrogen

The reactions in this chapter include those which have been studied extensively with respect to both their mechanisms and the effects of substituents on the reaction rates, and others about which little is yet known. The former group has been especially useful in aiding our understanding of the mechanism of electrophilic aromatic substitution. In particular, the method of determining relative reactivities involves, as it does in dedeuteriation and detritiation, kinetic measurements on individual compounds which can be prepared in a high state of purity. Large differences in the rates of reaction of isomeric compounds can therefore be measured accurately.

Steric acceleration (Section 2.5.2) is encountered in these reactions because hydrogen replaces a bulkier group. Steric acceleration can occur even in the absence of large *ortho* groups, compression by the *ortho* hydrogens being sufficient to produce rate enhancement. This has been detected in protiodealkylation.

The rates at which various groups are cleaved relative to hydrogen are defined as the *ipso* partial rate factors.[1] For example, the relative rates of reactions 4.1 and 4.2 gives the *ipso* partial rate factor for the $SiMe_3$ group (although reaction 4.2

$$R\text{—}C_6H_4\text{—}SiMe_3 \xrightarrow{HA} R\text{—}C_6H_4\text{—}H + SiMe_3A \qquad (4.1)$$

$$R\text{—}C_6H_4\text{—}H \xrightarrow{HA} R\text{—}C_6H_4\text{—}H + HA \qquad (4.2)$$

cannot in fact be measured without isotopic labelling which alters the rate and this must be corrected for). Interpretation of the numerical meaning of *ipso* factors is complex because they comprise at least four effects:

(1) Ease of attack by the electrophile. This will be affected by electron supply to

C-1 by the substituent being cleaved and, in some cases, by the bulk of the substituent. It follows also that the *ipso* factor will depend upon electron supply to C-1 by other substituents in the ring, i.e. the relative rates of reactions 4.1 and 4.2 will depend on the nature of R.

(2) Ease of cleavage of the C-1–substituent bond. This will be affected by the bond strength, itself reduced if the substituent is very bulky. Under the latter circumstances, steric acceleration occurs.

(3) The ease with which the hyperconjugation structure **2** can be formed from the intermediate **1**;[2] this is related to condition (1). If C-1–substituent hyperconjugation is favourable, then the substituent will be a good leaving group.

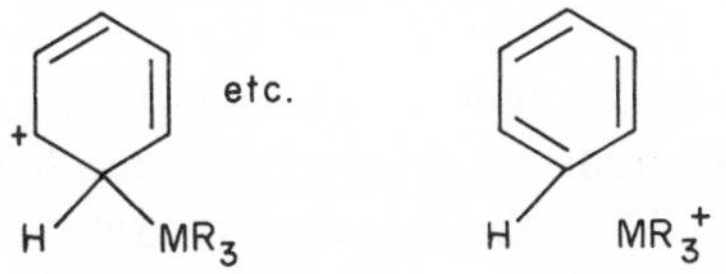

etc.

H MR$_3$ (**1**)

H MR$_3^+$ (**2**)

(4) The extent to which nucleophilic attack of A^- on the substituent determines the reaction rate.

It is therefore difficult to define a precise leaving-group ability. As an approximation, very reactive electrophiles, which have transition states for their substitutions near to the ground state with little or no kinetic isotope effect, will, by the principle of microscopic reversibility, be difficult to remove, and will have transition states near to products for the reverse reaction. Using this principle, a rough order of leaving group abilities has been defined[3] as $Cl^+ \approx NO_2^+ \approx R^+ < Br^+ < D^+ \approx ArN_2^+ \approx SO_3 \approx RCO^+ < NO^+ \approx H^+ \approx I^+ < Hg^{2+}$.

4.1 PROTIODEAURATION

Reaction of phenylauritriphenylphosphine with alcoholic hydrogen chloride produces benzene in a reaction (eq. 4.3) which is almost certainly an electrophilic substitution, although no further details are yet available.[4]

$$C_6H_5AuPPh_3 \xrightarrow{HCl} C_6H_6 + ClAuPPh_3 \qquad (4.3)$$

4.2 PROTIODEMAGNESIATION

The hydrolysis of arylmagnesium halides is an electrophilic substitution, but has not been examined kinetically in this context. However, the reaction between arylmagnesium bromides and hex-1-yne in diethyl ether, which results in the replacement of magnesium by hydrogen ion (eq. 4.4), has been studied.[5,6]

$$XC_6H_4MgBr + C_4H_9C{\equiv}CH \longrightarrow XC_6H_5 + C_4H_9C{\equiv}CMgBr \quad (4.4)$$

The reaction is first order in each reactant (although the order can apparently vary with conditions[6]), and the relative rates for substituents are X = 4-Me, 2.2; 4-Cl, 0.21; 3-Cl, 0.12; and 3-CF_3, 0.08. This is typical of an electrophilic process and a plot of $\log k_{\text{rel.}}$ against σ^+ values for the substituents (Section 11.1) is approximately linear, with $\rho = -1.7$.

4.3 PROTIODEMERCURIATION

Phenylmercury compounds undergo demercuriation in the presence of acids. The cleavage of diarylmercurials involves two mechanisms, and free-radical processes are not involved.[7,8] Cleavage by carboxylic acids gives rise to an S_E1 process in which the order in aromatic is 1.0 (eq. 4.5). However, this mechanism (in which the removal of mercury is aided by nucleophilic attack) may be a simplification since the order in carboxylic acid is approximately 3.0; the low $\log A$ factors for the reaction support the concerted process.

HgPh + HOAc ⟶ [cyclic transition state: Me–C(O)(O), H, HgPh] ⟶ H + PhHgOAc (4.5)

By contrast, the perchloric acid-catalysed reaction is strictly first order in both aromatic and acid, and is not catalysed by perchlorate ion and so is not aided by nucleophilic attack. This is consistent with the A-S_E2 process (eq. 4.6) and this also applies to the perchloric acid-catalysed reaction with acetic acid (in which the electrophile is believed to be H_2OAc^+). The reaction is accelerated by chloride ion (but only in the presence of acid), so this catalysis does not arise from nucleophilic attack on mercury to give the aryl carbanion, but probably from production of a small equilibrium concentration of hydrochloric acid, which gives a high specific cleavage rate.

HgPh + HA $\underset{k_{-1}}{\overset{k_1}{\rightleftharpoons}}$ [H, HgPh, +] + A^- $\xrightarrow[\text{fast}]{k_2}$ H + PhHgA (4.6)

Table 4.1 Partial rate factors for protiodemercuriation of arylmercury compounds RC_6H_4HgX by hydrochloric acid

R	X: Solvent: T/°C:	RC_6H_4 DMSO–dioxane 32	RC_6H_4 90% aq. dioxane 30	Cl 10% aq. ethanol 70
4-OEt			99.7	
4-OMe		28.4	80.3	148
2-Me			7.11	
4-Et			6.41	
4-Me			6.23	7.0
2-OMe			2.14	
3-Me			1.92	2.54
2:3-Benzo			1.92	
4-Ph		1.61		
H		1.0	1.0	1.0
4-F		0.84	0.875	
3-OMe			0.83	0.71
4-Cl		0.295	0.185	0.67
4-Br			0.127	
3-F			0.054[a]	
3-Cl			0.041[a]	0.26
2-OAc			0.0031[a]	
4-CO_2Me			0.028[a]	
2-CO_2Me			0.0155[a]	
2-Cl			0.0074	
3-NO_2		0.0086		

[a]Derived from data obtained at higher temperature.

The effects of substituents on the rate of demercuriation of symmetrical diarylmercurials by HCl in dimethyl sulphoxide–dioxane[9] or in 90% aqueous dioxane[10] are given in Table 4.1. Precise correlations with electrophilic substituent constants using the Yukawa–Tsuno equation[11] (Section 11.1.3) with $\rho = -2.8$ and -3.8, respectively ($r \approx 0.5$ in each case), confirm that the reaction is a typical electrophilic substitution. The relative rates of steps k_1 and k_2 in cleavage of bis-*o*-phenylenedimercury (**3**) under the former conditions is 6.6;[12] Dessy and Kim[12] argued that since the intermediate **4** gives **5** rather than

(**3**) —HCl, k_1→ (**4**) —HCl, k_2→ (**5**)

benzene and bis-*o*-chloromercuribenzene $C_6H_4(HgCl)_2$, the *o*-chloromercuri substituent must be electron supplying. However, Petrosyan and Reutov[13] believed the chloromercuri group to be electron withdrawing, the above results arising from steric hindrance to formation of bis-*o*-chloromercuribenzene, but this is incorrect since this product would be less hindered than **4**; a more reasonable explanation is that steric acceleration favours the formation of **5** from **4**.

For reaction of bis(pentafluorophenyl)mercury (with $Bu_4N^+I^-$ in DMF–H_2O) under base-catalysed conditions, an S_E1 mechanism has been proposed.[14] The ionization step for this mechanism will be favoured by the weak C—Hg bond, and the resulting carbanion will be stabilized by strong electron withdrawal in the aryl ring.

Protiodemercuriation of arylmercury(II) halides (in the absence of oxygen which affects the kinetics[15]) is first order in aromatic and acid, indicating an A-S_E2 mechanism analogous to eq. 4.6. The kinetics are complicated by complexing of the mercury(II) halide product with starting material, thereby causing a fall-off in rate (if sodium iodide is added to complex the product, so avoiding this problem, the arylmercury(II) chloride itself is also complexed to give $ArHgXI^-Na^+$ which is very much more reactive owing to the increased polarization of the C—Hg bond[16]). Removal of the mercury is evidently aided by nucleophilic attack because although the reaction of phenylmercury(II) bromide in DMF is faster with HCl than with HBr, added bromide ion is eight times as effective as chloride ion in catalysing the reaction.[17] Moreover, neither sulphuric nor perchloric acid will cleave arylmercury(II) chlorides, although reaction occurs on addition of chloride or bromide ions.[18] Partial rate factors in the latter work (Table 4.1) correlate with σ^+ values with $\rho = -2.45$ at 70 °C; the similarity of the ρ factor to that obtained with the diarylmercurials suggests a similar mechanism for each reaction.

4.4 PROTIODEBORONATION

The aryl—boron bond can be cleaved by water in an acid-catalysed reaction represented by eq. 4.7.[19] The reaction is first order in aromatic, is subject to general-acid catalysis, and is faster in a protium-containing than in a deuterium-containing solvent by factors of 3.7 and 1.7 for 4-methoxy- and 2,6-dimethoxy-benzeneboronic acid, respectively.[20] Since proton transfer to the aromatic takes place in a rate-determining step, the A-S_E2 mechanism is indicated and this is supported by the linear variation of rate with isotopic composition of the medium. (The Gross–Butler theory predicts a curve for the alternative A-1 mechanism in which protonation takes place in a rapid pre-equilibrium, the intermediate complex then rearranging in a rate-determining step.)

$$PhB(OH)_2 + H_2O \longrightarrow PhH + B(OH)_2 \qquad (4.7)$$

Plots of $\log k$ against the acidity function $-H_0$ in aqueous phosphoric and sulphuric acids are linear, the slopes being 1.9 and 1.15, respectively. Previously such linear plots were taken to indicate the occurrence of the A-1 mechanism, but the general-acid catalysis observed in deboronation showed this diagnosis to be unacceptable.

In aqueous sulphuric acid there is evidence of a second mechanism of protiodeboronation in which a pre-equilibrium involving attack of bisulphate ion on boron gives **6**. Rate-determining rearrangement via a six-membered cyclic transition state then gives the intermediate **7**, and thence the products (eq. 4.8). This (A-S_E1) mechanism appears to be more important the lower the reactivity of the aromatic and the lower the sulphuric acid concentration. Substituent effects for the reaction of ArX with 74.5% H_2SO_4 are for X = 4-OMe, 22 400; 4-Me, 66; 4-F, 2.1; 4-Br, 0.3; 3-F, 0.032; 3-Cl, 0.025; 3-NO_2, 0.00041; ferrocene is 7×10^6 times as reactive as benzene under these conditions.[21] These data correlate with σ^+ values with $\rho = -5.2$, confirming that protiodeboronation is a typical electrophilic substitution. However, data for reaction with substantially weaker sulphuric acid show a poorer correlation, probably owing to incursion of the first mechanism of deboronation.

$$PhB(OH)_2 + HSO_4^- \underset{}{\overset{fast}{\rightleftharpoons}} \mathbf{(6)} \xrightarrow{slow} \mathbf{(7)} \longrightarrow \text{products} \qquad (4.8)$$

A third mechanism of protiodeboronation has been detected in the reaction of arylboronic acids with malonate buffers in the pH range 2–6.7.[22] The rate coefficients pass through a minimum at ca pH = 5, indicating that in addition to the acid-catalysed reaction there is a base-catalysed reaction (eqs. 4.9 and 4.10).

$$ArB(OH)_2 + H_2O \overset{fast}{\rightleftharpoons} ArB(OH)_3^- + H^+ \qquad (4.9)$$

$$ArB(OH)_3^- + H_2O \overset{slow}{\rightleftharpoons} ArH + B(OH)_3 + OH^- \qquad (4.10)$$

Kinetic data here give an approximate correlation with σ^+ values using the Yukawa–Tsuno equation (Section 11.1.3) with $\rho = -2.3$ and $r = 0.5$.[21] The lower value of ρ than for the acid-catalysed reaction is consistent with the negative charge on boron facilitating addition of the electrophile, so that less electron donation by the aromatic ring is needed. The reaction gives very high *ortho*:*para* reactivity ratios, consistent with considerable steric acceleration which would accompany eq. 4.10. Cadmium ion catalyses the base- but not the acid-catalysed reaction, although the mechanism of this catalysis is not fully understood.[23]

4.5 PROTIODEALKYLATION

Friedel–Crafts alkylations are accompanied by isomerization and disproportionation [Section 6.1.1.(ii)]. Although some of these processes are intramolecular, protiodealkylation (eq. 4.11) is involved in many cases. The ease with which the reaction takes place depends on the stability of the leaving carbocation, hence de-*tert*-butylation is the most commonly observed dealkylation.

$$\mathrm{ArR + HA \longrightarrow ArH + RA} \qquad (4.11)$$

Most studies have involved cleavage by trifluoroacetic acid-containing media since no other observable substitution accompanies dealkylation under these conditions. Protiode-*tert*-butylation of *o*-methyl-*tert*-butylbenzene in TFA containing strong mineral acids is considerably faster than for the *p*-methyl isomer.[24]. Steric acceleration therefore contributes to the rate and this has been confirmed by other work, e.g. *o*-hydroxy- and *o*-methoxy-*tert*-butylbenzenes debutylate ca 50 times faster than their corresponding *para* isomers.[25,26] Likewise, 1,8-dimethylnaphthalene isomerizes to the 1,7-isomer during hydrogen exchange in TFA whereas the less hindered 2,3-isomer does not demethylate.[27] [This side reaction (ca 10% of the exchange rate in TFA) is the probable cause of the non-reproducibility of rate data for hydrogen exchange of the 3-position of 1,8-dimethylnaphthalene under various conditions.[28]] Dealkylation from α-positions of naphthalene is also aided by their higher intrinsic reactivity compared with the β-positions, and the electrophilic nature of the reaction is shown by the ready demethylation of 1,4-dimethylnaphthalene under conditions where 1-methylnaphthalene is unreactive.[27] A more comprehensive study with 1,4,5,8-tetramethylnaphthalene showed that one or two methyl groups migrate as shown in **8** and that in 1,8-dimethylnaphthalene a '*meta*' and a '*para*' methyl group activate the reaction 8- and 500-fold, respectively (after correction for symmetry effects).[29] The respective values for hydrogen exchange under these

Me Me Me Me → Me Me Me Me + Me Me Me Me + Me Me Me Me

(**8**)

conditions are 2.75 and 83 (Table 3.16 and 3.22), indicating a ρ-factor for the demethylation of ca -12.5.

Steric acceleration is an important factor even in the absence of *ortho* or *peri* substituents, since pent-3-yl and hex-3-ylbenzene dealkylate faster than *tert*-butylbenzene in anhydrous TFA,[30] even though the former compounds give carbocations less stable than the *tert*-butyl cation. In these compounds the alkyl groups on the side-chain can approach the *ortho* hydrogens more closely than can the methyl groups in the *tert*-butyl substituent, and to such an extent that these hydrogens themselves provide the steric acceleration. The alkyl cations produced in protiodealkylation will obviously tend to undergo the reverse reaction, alkylation. Since this should occur more readily on alkylbenzenes than on benzene, the overall result of protiodealkylation tends to be disproportionation.

Protiodealkylation appears to be the only acid-catalysed electrophilic substitution in which the second step of the reaction is rate-determining rather than the first, i.e. the mechanism of the reaction is A-1 (eqs 4.12 and 4.13) rather

$$\mathrm{ArR + H^+ \xrightleftharpoons{\text{fast}} ArRH^+} \tag{4.12}$$

$$\mathrm{ArRH^+ \xrightleftharpoons{\text{slow}} ArH + R^+} \tag{4.13}$$

than A-S_E2. This arises because in alkylbenzenes electron density on C-1 is higher than in any other aromatic, and this facilitates rapid attachment of the electrophile at C-1. Evidence for the A-1 mechanism is based on the following:

(1) Protiodealkylation is considerably slower than hydrogen exchange[30] (and ca five times slower even than sulphonation[31]) under the same conditions. Each reaction must involve the same electrophile, and since the electron *density at C*-1 (the point of attachment of the electrophile in dealkylation) should be greater than at C-2, -3, or -4 (the points of attachment of the electrophile in exchange), the first step in dealkylation should be faster than the first step in exchange. The lower overall reaction rate for dealkylation therefore implies that there is a slow step on the reaction pathway, which must be the second step.

(2) The reaction gives an inverse isotope effect[26] owing to the concentration of the substrate conjugate acid being higher in the deuteriated medium (if proton transfer was rate-determining then the reaction would be slower in the deuteriated medium). Moreover, under a given condition (60% perchloric acid), the value of k_H/k_D was larger for *o*-*tert*-butylphenol (0.36) than for *p*-*tert*-butylphenol (0.11). Loss of the *tert*-butyl cation in the former compound would be sterically accelerated so that the rate of the second step of the reaction would be comparable to that of the first, leading to a smaller isotope effect.[26] (This latter argument is not wholly satisfactory because an equally valid explanation is that the greater electron supply from the *para*- compared with the *ortho*-hydroxy

group would make attachment of the hydrogen ion faster for the *para* compound leading to the observed greater isotope effect.)

(3) The transition-state activity coefficients for debutylation of some *tert*-butylphenols and -anisoles showed insensitivity to medium composition over a wide acidity range, in marked contrast to the coefficients for hydrogen exchange in the same media.[26]

An unexpected conclusion from the above is that the alkyl cation must be a poorer leaving group than the proton, and this may be due to the greater ability of the proton to be solvated. Given that TFA is a very poor solvating medium, the difference between the rates of hydrogen exchange and protiodealkylation should be a minimum in this medium.

Although the mechanism of protiodebutylation is evidently different from that for hydrogen exchange and protiodeboronation, the rates of all three reactions show a similar dependence on the acidity of the medium[26] thereby providing further evidence (cf. ref. 32) that such dependences are poor criteria of mechanism. The A-1 mechanism is not general for debutylation, however, because in TFA both 2,4-di-*tert*-butyl- and 2,4,6-tri-*tert*-butylphenol readily debutylate (the latter almost instantaneously) and give a large solvent isotope effect, k_H/k_D, of 5.5. Since the corresponding anisoles were unreactive under the same conditions, a cyclic hydrogen-bonded transition state (**9**) may be involved here.[33]

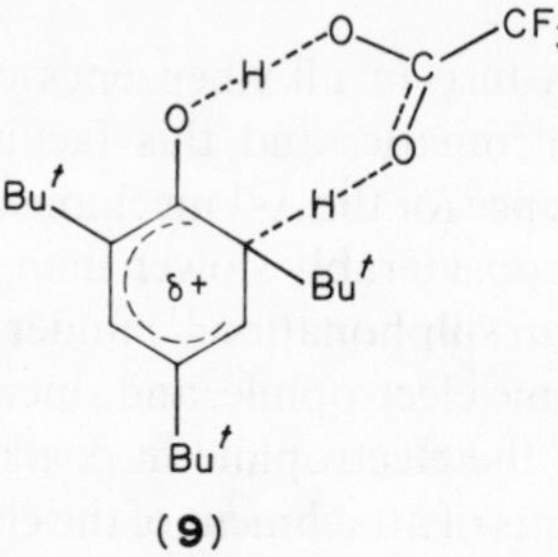

(**9**)

An anomalous result is the *decrease* in rate with *increasing* number of alkyl groups found in the protiodeisopropylation of the mono-, di-, and tri-isopropylbenzenes by HCl in the presence of $AlCl_3$;[34] formation of $AlCl_3$–aromatic complexes (progressively easier the more alkyl-substituted the aromatic) is the probable cause.

In contrast to protiodebutylation, protiodealkylation (or protiodetritylation) of triphenylmethylphenol (eq. 4.14) gives a solvent isotope effect, k_H/k_D, of 4.3, indicating that the A-S_E2 mechanism applies here.[26] This logically follows from the lower electron density on C-1 which will reduce the rate of proton addition, whereas the second step of the reaction should be sterically accelerated relative to that in debutylation. Steric acceleration and concomitant weakening of the C—aryl bond is also indicated by the ρ factor for the reaction (-3.9),[35] which is

$$HO-C_6H_4-CAr_3 + HA \longrightarrow Ar_3CA + HO-C_6H_5 \quad (4.14)$$

markedly lower than for desilylation which involves cleavage of a weak bond. However, the ρ factor was derived from the reaction of oxy-substituted compounds only and may therefore be too low because of hydrogen bonding to the substituents. (This reaction was incorrectly described as dearylation in ref. 36.)

4.6 PROTIODECARBONYLATION

Protiodecarbonylation reactions may be generally represented by eq. 4.15, where X is H, alkyl, or OH.

$$ArCOX + H^+ \longrightarrow ArH + XCO^+ \quad (4.15)$$

If X is H or alkyl the process is a deacylation, whereas when X is OH it is decarboxylation. Deacylation was discovered by Louise,[37] who found that reaction of benzoylmesitylene with hot phosphoric acid gave mesitylene and benzoic acid. This illustrates two requirements for deacylation, which are first that bulky groups must be adjacent to the acyl group so as to assist the reaction by steric acceleration, and second that electron-releasing substituents must be present in the ring to facilitate attachment of the proton to C-1. The decarboxylation reaction has similar requirements and thus two bulky *ortho* substituents were found necessary for decarboxylation of aromatic acids in hot phosphoric acid.[38]

4.6.1 Protiodeacylation

The sulphuric acid-catalysed protiodeacylation of 2,6-dimethyl- and 2,4,6-trimethylacetophenone gives the corresponding aromatic hydrocarbons and acetic acid.[39] An approximate correlation of log k with the acidity function H_0 led to the suggestion that the rate-determining step was a first-order decomposition of the conjugate acid BH^+, i.e. the A-1 mechanism was involved, but this was subsequently disproved by studies of the deacylation of aromatic aldehydes.[40] The rates of deacylation of various 2,4,6-trialkylbenzaldehydes in 50–100% H_2SO_4 each pass through a maximum at acid concentrations specific to each compound, and the decrease in rates above these maxima were shown to be accommodated satisfactorily only by the A-S_E2 mechanism (eq. 4.16). Three further aspects of these reactions are noteworthy. First, over the whole acid range in which a decrease in rate coefficient was observed, the aldehydes are present

$$\text{X-C}_6\text{H}_4\text{-CHO} + \text{HA} \underset{k_{-1}}{\overset{k_1}{\rightleftharpoons}} \text{X-C}_6\text{H}_5^{+}\text{(H)(CHO)} + \text{A}^- \xrightarrow[\text{(fast)}]{k_2} \text{X-C}_6\text{H}_5 + \text{CO} + \text{HA} \tag{4.16}$$

entirely as their (oxygen-protonated) conjugate acids $ArCHOH^+$, so that the formation of these less reactive species cannot be responsible for the rate decline. Second, added bisulphate ion increases the reaction rate so that k_2 is involved in determining the overall rate. Thirdly, the relative rates of deacylation of the 2,4,6-trialkylbenzaldehydes are Me 1.0, Et 4.5, and *i*-Pr 20.6, the increase along this series probably arising from steric acceleration of the protonation step.

Deacylation of 2,4,6-trialkylbenzaldehydes labelled with deuterium in the aldehyde group, and using deuteriated acid, indicated that the relative rates of steps k_1 and k_2 in eq. 4.16 depend on the size of the alkyl groups and the strength of the catalysing acid; the stronger the acid and the bulkier the alkyl groups, the more rate-determining k_1 becomes.[41] Thus, at acid concentrations above those at which oxygen is fully protonated, ArCHO reacts faster than ArCDO in sulphuric acid, which is compatible with eq. 4.16 provided that $k_{-1} > k_2$. The solvent isotope effect $[k(H_2SO_4)/k(D_2SO_4)]$ is greater in more concentrated acids. This is attributed to greater steric hindrance with the base HSO_4^- (more abundant in the stronger acid) than with the base H_2O, to the reverse of the first step in eq. 4.16. Consequently, 2,4,6-triisopropylbenzaldehyde (for which this steric hindrance would be greater) gives a large solvent isotope effect, showing that k_1 and k_2 are becoming more and less rate-determining, respectively.[42]

Deacylation of 2,4,6-trimethoxybenzaldehyde by mineral acids differs from that of the aldehydes above in that formic acid rather than carbon monoxide is quantitatively produced, and it has been proposed that here the base attacks the carbonyl carbon (eq. 4.17) instead of abstracting a proton (eq. 4.18) in the second step of the reaction.[43] The reason for this mechanistic difference is obscure, but a specific effect of the *o*-methoxyl group (e.g. hydrogen bonding with an attacking water molecule) may be responsible.

$$^{+}\text{Ar(H)-CH=O} + \text{A}^- \longrightarrow {}^{+}\text{Ar(H)-CH(A)-O}^- \longrightarrow \text{ArH} + \text{H-C(=O)-A} \tag{4.17}$$

$$^{+}\text{Ar(H)-CH=O} + \text{A}^- \longrightarrow \text{ArH} + \text{CO} + \text{HA} \tag{4.18}$$

Protiodeacylation is the first step in the (overall) sulphodeacylation of dimesityl ketone[44] (and other hindered aromatic ketones[45]). Mesitylene and mesitoic acid are the initial products, with mesitylene (some from decarboxylation of the acid) then sulphonating. This pathway must arise from steric hindrance to direct sulphodeacylation, coupled with the greater reactivity of the proton as an electrophile compared with SO_3 or HSO_3^+; hydrogen exchange is faster than sulphonation in most H_2SO_4 media. Substituent effects in the reaction of 4-X-2,6-dimethylphenyl methyl ketones with 89.8 wt-% H_2SO_4 at 25 °C have been determined as follows: X = Me, 35.1; *t*-Bu, 28.7; Ph, 23.4; F, 1.87, Cl, 0.47, I, 0.378; Br, 0.348. These data demonstrate the electrophilic nature of the reaction, and $\rho = -4.6$; this value supersedes an earlier one of ca -8.0 indicated by the partial rate factor for the 3-COMe substituent of 2.0×10^{-3} in protiodeacylation of acetylmesitylene. Deacetylation is 250 times faster than debenzoylation, which demonstrates steric acceleration well, since in the reverse reaction acetylation is more hindered than benzoylation (Section 6.7.3); deformylation is 5×10^4 times slower than deacetylation.[45] Steric acceleration also accounts for the greater reactivity in debenzoylation of *ortho*-substituted 9-benzoylanthracenes compared with the *para*-substituted isomers;[46] electron-supplying groups in the benzoyl group increased the reaction rate, consistent with the greater stability of the leaving benzoyl cation that is produced.

4.6.2 Protiodecarboxylation

The A-S_E2 mechanism for protiodecarboxylation (eq. 4.19) was proposed[47] as a result of studies with anthracene-9-carboxylic acid; this possesses the necessary structural features for protiodecarboxylation in that anthracene is very reactive towards electrophiles at the 9-position, and the carboxyl group is sterically compressed by the *peri* hydrogen atoms.

$$ArCO_2H + H^+(H_3O^+) \longrightarrow ArH + CO_2 + H^+(H_3O^+) \qquad (4.19)$$

The decarboxylation rates of alkyl-substituted benzoic acids pass through a maximum at acid concentrations which depend on the aromatic and acid. This is not due to reaction occurring on the acid anion (expected to be the more reactive species owing to the greater electron density on the *ipso* carbon[48]), the concentration of the anion decreasing with increasing acidity.[49] It is due to the free acid becoming converted to the conjugate acid (and hence $ArCO^+$ and water) at higher acidity.[49] Decarboxylation of 2,4,6-trihydroxybenzoic acid (which has extra sites for protonation) is less straightforward. Although the rate is negligible in water in which the $ArCOO^-$:ArCOOH ratio is high, suggesting that decarboxylation via the anion cannot be involved, the fall-off in rate with increasing acidity has been shown spectroscopically to correspond to conversion of the anion to the free acid, indicating that reaction occurs on the anion.[50]

The decarboxylation of mesitoic acid in 50–100% H_2SO_4 is first order in both carboxylic acid and hydronium ion[51] (although later work indicates that the acid catalysis is general rather than specific[52]). The rate of this last reaction is unaffected by the concentration of bisulphate ion,[51] so if the usual two-step A-S_E2 process is involved (eq. 4.20), the intermediate is converted into products much faster than its rate of reversion to reactants. This mechanism is supported by decarboxylation of 2,4-dihydroxybenzoic acid occurring 1.75 times faster in a medium containing H_2O than in one containing D_2O (after allowing for the differing strengths of the acids in the two media).[53] Since H_2O is a weaker base than D_2O, it follows that the hydrogen ion transfer in the slow step of eq. 4.20 will occur more readily in H_2O leading to an isotope effect $k(H_2O)/k(D_2O) > 1$, as observed. However, the decarboxylation of mesitoic acid in aqueous sulphuric acid gives carbon isotope effects C-12:C-13 = 1.038 and C-13:C-14 = 1.101,[54] which suggests that the C—C bond is broken in a step which is at least partially rate-determining; a similar isotope effect is obtained in the decarboxylation of 2,4-dihydroxybenzoic acid in acetate buffers.[55].

$$ArCO_2H + HA \underset{}{\overset{\text{slow}}{\rightleftharpoons}} Ar^{+}\langle{}^{H}_{CO_2H} \xrightarrow{\text{fast}} ArH + CO_2 + HX \quad (4.20)$$

Substituent effects in protiodecarboxylation are consistent with the view that the reaction is electrophilic. *Meta* partial rate factors obtained in deacarboxylation of substituted 2,4,6-trimethylbenzoic acids in 83 wt-% H_2SO_4 at 70 °C are 3-Et 4, 3-Me 3.2, and 3-OH 0.25; the 3-Br and 3-NO_2 compounds did not decarboxylate.[56]

The effect of steric acceleration of the reaction is evident from the fact that although 4-methyl-, 4-ethyl-, and 4-isopropyl substituents affect the rate of decarboxylation of 2,6-dimethoxybenzoic acid (**10**) approximately equally, the relative rates of decarboxylation of the trialkyl-substituted benzoic acids (**11**) are 2,4,6-triisopropyl- > 2,4,6-triethyl- > 2,4,6-trimethylbenzoic acid.[51,52]

Interpretation of the kinetics of decarboxylation of aminobenzoic acids[57] is complicated, as in the case of the hydroxy compounds, by the opportunity for protonation of the substituent, and also by the possibility of zwitterion formation; no definite conclusions regarding the mechanism have been reached.[58]

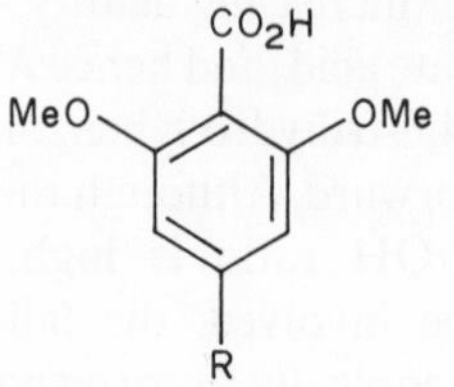

CO_2H, MeO, OMe, R

(**10**, R = Me, Et, *i*-Pr)

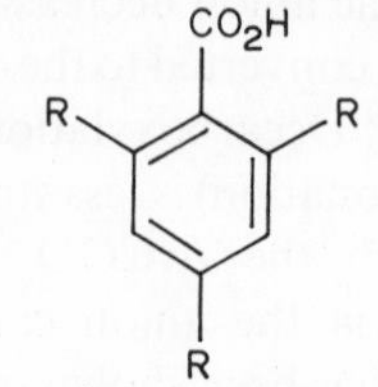

CO_2H, R, R, R

(**11**, R = Me, Et, *i*-Pr)

4.7 PROTIODESILYLATION

4.7.1 Acid-catalysed Protiodesilylation

The acid-catalysed solvolytic cleavage of aryl—silicon bonds in $ArSiR_3$ compounds is an electrophilic substitution in which a proton, initially solvated, is the attacking species.[59–61] Substituent effects in the reaction, and its mechanism have been extensively studied,[59,61–65,67,69–101] using in particular protiodetrimethylsilylation in which the $SiMe_3$ group is replaced by hydrogen.

(i) Mechanism

The features of the reaction are consistent with the A-S_E2 mechanism: First, in the presence of excess of acid the kinetics are first order in the silane.[59–61] Second, solvent isotope effects (k_H/k_D at 50 °C) of 1.53 and 6.22 have been obtained in protiodesilylation in a dioxane–HCl medium containing either H_2O or D_2O, and in CF_3CO_2H vs CF_3CO_2D media, respectively.[62–64] The magnitude of the isotope effect is governed by the balance of bond breaking in the acid HA and bond making of the aromatic C—H bond, and shows that the proton transfer is rather more than half complete at the transition state.[64] Third, the relative reactivities (k_{rel}) for cleavage of $ArMR_3$ compounds by $HClO_4$ in EtOH at 50 °C are[65] for M = Si, 1; Ge, 36; Sn, 3.5×10^5; and Pb, 2×10^8. Although the order of C—MR_3 bond strength decreases along this series,[65] this result is not due to the C—MR_3 bond being broken in the rate-determining step. It arises because it is also the order whereby the C—MR_3 bond releases electrons through hyperconjugation[66] and stabilizes the Wheland intermediate as shown in **1** and **2**.[2b,62,67] This results in large *ipso* factors and for the removal of $SiMe_3$ the value (aqueous H_2SO_4) is ca 10^4 (after correction for isotope effects).[65] The very high reactivities of the tin and lead compounds could arise from the nucleophilic attack by the hydroxylic solvent molecules on the metals, and facilitated by the availability of the vacant d-orbitals. This has been ruled out by the observation that the logarithms of the above relative cleavage rates correlate linearly with the absorption frequencies of the charge-transfer complexes between $PhCH_2MMe_3$ compounds and tetracyanoethene[68] (which depend on the electron release from the C—MR_3 bond), even though hydrocarbon solvents were used.[69]

$$C_6H_5SiR_3 + HA \underset{k_{-1}}{\overset{k_1}{\rightleftharpoons}} [C_6H_5(H)(SiR_3)]^+ + A^- \xrightarrow{k_2} C_6H_6 + SiR_3A \qquad (4.21)$$

The rate of protiodesilylation is also dependent on the nature of the groups attached to silicon, being greater when these groups are more electron releasing. Some relative rates for cleavage of the 4-anisyl—Si bond in 4-anisyl-$Si(C_6H_4R)_3$ compounds are R = 4-Me, 2.2; 3-Me, 1.3; H, 1.0; 4-Cl, 0.30; 3-Cl, 0.11; 2-Me, 0.005.[70] In these compounds (except the last) the group R is well removed from the reaction site but, when this is not the case (R = 2-Me) the *ipso* and *ortho* sites are crowded, which hinders both attachment of the electrophile and solvation of the transition state.[65,70] Steric hindrance is also evident from two studies which gave the approximate relative rates for $ArSiR_3$ compounds as follows: R = Me, 1.0; Et, 0.5; cyclohexyl, 0.1; *i*-Pr, 0.05; Ph, 0.018.[65,71] Further, the activating effects of *m*-alkyl groups in the aromatic ring increased as R was made larger. Steric hindrance produces a 'later' transition state in which a greater fraction of the charge from the electrophile is on the aromatic ring. Both this and the poorer solvation require greater stabilization by the ring substituent. In a related study of the cleavage of $PhSiR_3$ compounds in which the R groups consisted of a combination of Me and bulky $SiMe_3$ substituents, the $k_{rel.}$ values were $SiMe_3$, 1.0; $SiMe_2(SiMe_3)$, 2.1; $SiMe(SiMe_3)_2$, 0.35; and $Si(SiMe_3)_3$, 0.056.[72] The $SiMe_3$ group is more electron releasing than methyl when, as here, its $-M$ effect cannot operate, hence replacing one Me group by $SiMe_3$ causes a rate increase. However, when further groups are substituted, steric hindrance becomes dominant, so that subsequent replacements produce a rate decrease.

(ii) Substituent effects

The effects of a very large number of substituents have been determined in protiodetrimethylsilylation in methanol–perchloric acid and acetic acid–sulphuric acid media. The partial rate factors are given in Table 4.2.

The main features of these results are as follows:

(1) The log $k_{rel.}$ values give an approximately linear plot against σ^+ values ($\rho = -4.6$), showing that the reaction is an electrophilic substitution. A better correlation is obtained using the Yukawa–Tsuno equation (Section 11.1.3) with $\rho = -5.3, r = 0.65$. The r value shows that demands for resonance in the reaction are relatively small, consistent with the low ρ factor, indicative of an 'early' transition state.

(2) The slight activating in the 2,4,6-D_3 compound shows the $+I$ effect of deuterium and is the only example in electrophilic substitution.

(3) Because good solvating media are used, 4-*t*-Bu activates less than 4-Me as a result of steric hindrance to solvation being superimposed on the intrinsically greater $+I$ and $+M$ effects of *t*-Bu relative to Me. In HOAc–TFA (3:2, v/v) the corresponding partial rate factors are 27 and 32, and thus nearer to the solvation-free order. This is in keeping with the results for detritiation and bromination in trifluoroacetic acid, both of these reactions showing the *t*-Bu > Me activation order because TFA is a poor solvating medium. (In the gas phase, the largest

Table 4.2 Partial rate factors for protiodesilylation of arylsilicon compounds $XC_6H_4SiMe_3$ and $ArSiMe_3$ at 50 °C

X	f	X	f
In methanol–aqueous perchloric acid:			
2,4,6-$(OMe)_3$	$\sim 1.8 \times 10^8$	4-Et	19.5
4-NMe_2	3×10^7	4-*i*-Pr	17.2
4-OH	10 700	4-*t*-Bu	15.6
2-OH	3720	2-Me	17.8
4-OMe	1270[a]	4-CH_2Ph	7.9
2-OMe	335	2-CH_2Ph	3.75
4-OPh	88.5	4-$CHPh_2$	3.35
2-OPh	8.73	4-CPh_3	2.83
4-SMe	65.2	4-Ph	3.55
2-SMe	18.4	3-$C(SiMe_3)_3$	3.4
4-SH	11.3	3-$CH(SiMe_3)_2$	8.4
2-SH	4.42	3-$CHPrSiMe_3$	7.9
4-SPh	10.7	3-CH_2SiMe_3	6.4
2-SPh	1.30	3-$(CH_2)_2SiMe_3$	3.6
4-$C(SiMe_3)_3$	200	3-$(CH_2)_3SiMe_3$	3.8
4-$CH(SiMe_3)_2$	670	3-$(CH_2)_4SiMe_3$	3.6
4-CH_2SiMe_3	315	3-CH_2-*t*-Bu	4.05
4-$CHPrSiMe_3$	260	3-*t*-Bu	3.86
4-CH_2GeMe_3	490	3-Me	2.30
4-$(CH_2)_2SiMe_3$	28	3-$SiMe_3$	1.63
4-$(CH_2)_3SiMe_3$	22	4-$SiMe_3$	1.25
4-$(CH_2)_4SiMe_3$	24	2,4,6-D_3	1.265
2-CH_2SiMe_3	31	H	1.0
2-$(CH_2)_2SiMe_3$	17	3-OPh	0.36
2-$(CH_2)_3SiMe_3$	12	3-SMe	0.75
2-$(CH_2)_4SiMe_3$	13	4-F	0.75
4-Me	21.2	4-Cl	0.13
4-CH_2-*t*-Bu	20.7	4-Br	0.10
9-Anthracenyl	9.0×10^4	4-Dibenzothienyl	1.15
2-Anthracenyl	4.1	3-Dibenzothienyl	2.0
9-Phenanthrenyl	4.5	2-Dibenzothienyl	6.25
2-Phenanthrenyl	1.76	1-Dibenzothienyl	5.5
3-Phenanthrenyl	2.1	5-Tetralinyl	87
1-Naphthyl	8.1	6-Tetralinyl	67
2-Naphthyl	2.16	4-Indanyl	28.5
1-Pyrenyl	223	5-Indanyl	77.5
2-(9,10-Dihydro-phenanthrenyl)	12.8	3-Benzocyclobutyl	5.9
		4-Benzocyclobutyl	57
3-(9-Ethylcarbazolyl)	5.0×10^4	2-(1,6-Methano[10]annulenyl)	9270
1-Biphenylenyl	0.52		
1-Biphenylenyl	27.8	2-(11,11-Difluoro-1,6-methano[10]-annulenyl)	20.2
2-Fluorenyl	45.6		
2-(9-Methylfluorenyl)	42.2	2-(1,6:8,13-propane-1,3-diylidene[14]-annulenyl)	83 400
2-(9,9-Dimethyl-fluorenyl)	35.6		

Table 4.2 (*continued*)

X	f	X	f
4-Dibenzofuranyl	0.92		
3-Dibenzofuranyl	2.4		
2-Dibenzofuranyl	19.2		
1-Dibenzofuranyl	0.65		
In acetic acid–aqueous sulphuric acid:			
4-OMe	1010	4-CO_2H	0.00216[b]
4-CH_2SiMe_3	202	2-CO_2H	0.0052[b]
2-CH_2SiMe_3	31	4-CO_2Me	0.00188[c]
4-Me	18	4-SO_3H	0.00116[b]
2-$SiMe_3$	8.55	2-SO_3H	0.0026
2-*t*-Bu	8.0	4-SOMe	0.004[d]
4-Ph	2.8	4-NO_2	0.00014
2-Ph	6.0	2-NO_2	6.8×10^{-5}
4-CH_2OMe	1.27	4-Me_2NH^+	0.00118
4-CH_2Br	1.35	2-Me_2NH^+	6.8×10^{-5}
H	1.0	4-Me_3N^+	0.00030[e]
4-CH_2CO_2H	0.93	4-$CH_2Me_2NH^+$	0.0043
4-CH_2CO_2Me	0.84	4-$CH_2Me_3N^+$	0.0042
4-CH_2OEt	0.70	3-$CH_2Me_3N^+$	0.011
4-CH_2OH	0.64	4-Me_3As^+	0.00024
4-CH_2CN	0.20	4-Me_3P^+	7.5×10^{-5}[e]
4-$CH_2PO(OH)_2$	0.77	3-CO_2Me	0.0111
4-$CH_2PO(OEt)_2$	0.70	3-CO_2H	0.0088[b]
3-$CH_2PO(OH)_2$	0.32	3-CF_3	0.0023[e]
4-$CH_2PO(OEt)_2$	0.21	3-SO_3H	0.00145[c]
3-OMe	0.38	3-NO_2	0.00037
3-Ph	0.33	3-Me_3N^+	0.00021
4-F	0.95	3-Me_3P^+	0.00029
4-Cl	0.19	3-Me_3As^+	0.00058
4-Br	0.104	4-$OPO(OEt)_2$	0.35
4-I	0.101	3-$OPO(OEt)_2$	0.0012
2-F	0.073	4-$POMe_2$	5.9×10^{-5}
2-Cl	0.034	4-$POPh_2$	4.0×10^{-5}
2-Br	0.025	3-$POPh_2$	2.8×10^{-5}
2-I	0.038	4-$PO(OEt)_2$	12×10^{-5}
3-F	~0.0215	3-$PO(OEt)_2$	10×10^{-4}[b]
3-Cl	0.012	4-$PO(OH)_2$	21×10^{-5}
3-Br	0.0117	3-$PO(OH)_2$	13×10^{-4}
2,3,4,5,6-F_5	1.45×10^{-5}		

[a] This value was obtained by a more direct comparison[90] than the slightly higher value given in earlier publications.
[b] Lower values were obtained in more acidic media owing to hydrogen bonding.
[c] Value probably too low since the 4-CO_2H substituent gave a depressed value (0.00145) in the medium used for this determination.
[d] Probably refers to the protonated species (see chlorination, Section 9.6.3).
[e] Average value.

t-Bu:Me activation ratio is obtained[102]). Steric hindrance to solvation also causes the 4-CH_2CMe_3 substituent to be less electron releasing than 4-Me, and again this order is reversed in the gas phase[103] and in hydrogen exchange in TFA [Section 3.1.2.1(5)].

(4) The relative activation order by the 3- and 4-$SiMe_3$ substituents shows that the $-M$ effect [arising from $(p \rightarrow d)\pi$ bonding] for $SiMe_3$ is smaller than the $+I$ effect. However, steric hindrance to solvation produces less overall activation than is obtained in the gas phase (pyrolysis of 1-arylethyl acetates[104]). From the σ^+ values determined under the latter conditions, the rates in protiodesilylation can be calculated to be attenuated 3-fold, whereas in solvolysis of 1-aryl-1-methylethyl (*tert*-cumyl) chlorides, the effect of solvation is severe enough to cause both substituents to deactivate.[104]

The higher activation by the 2-$SiMe_3$ substituent is also found (but to a lesser extent) in the gas phase[104] and probably reflects a combination of the greater $+I$ effect and steric acceleration (which will be greater in desilylation). Other examples of steric acceleration due to adjacent $SiMe_3$ groups are given in paragraph (18).

(5) For the 2- and 4-$(CH_2)_nSiMe_3$ substituents a minimum in the activation produced occurs when $n=3$ (see also hydrogen exchange, Table 3.8). Models indicate a probable cause, for when $n=3$ the $SiMe_3$ group is able to sit directly over the aromatic π-cloud (**12**), thereby facilitating electron withdrawal through $(p \rightarrow d)\pi$ bonding. This would be less important for *m*-substitution, as the results indicate.

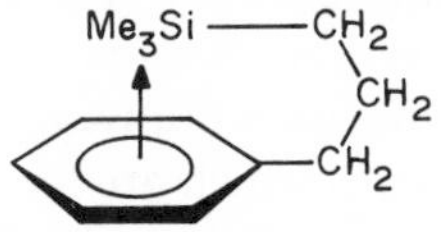

Me_3Si—CH_2—CH_2—CH_2

(**12**)

(6) The greater ease of C–Ge hyperconjugation relative to C–Si hyperconjugation is shown by the relative effects of the 4-CH_2SiMe_3 and 4-CH_2GeMe_3 substituents.

(7) The order of activation by the 4-$CH_{3-x}(SiMe_3)_x$ substituents is consistent with hyperconjugative electron release by the C—Si bond. This will be greater in $CH(SiMe_3)_2$ than in CH_2SiMe_3, but on going to $C(SiMe_3)_3$ conformational aspects dictate that one $SiMe_3$ group will lie in the plane of the aromatic ring and therefore be unable to participate in hyperconjugation. Moreover, there will be steric hindrance to bond shortening when three $SiMe_3$ groups are present, so there is a substantial decrease in activation relative to $CH(SiMe_3)_2$.[88]

(8) The data for CH_2X substituent effects are the largest collection available. The anomalous deactivation by CH_2OEt compared with activation by CH_2OMe may be due to greater steric hindrance to solvation in the former. This may also account for the greater deactivation by CH_2CO_2Me compared with CH_2CO_2H.

An alternative explanation is that hydrogen bonding occurs, and this should be greatest in the substituent containing the more electron-supplying alkyl group.

(9) The 4-CPh_3 substituent activates less than the 4-$CHPh_2$ substituent whereas in hydrogen exchange it activates more. In both reactions the decrease in activation on adding successive phenyl groups to CH_3 is not regular, and on this basis the 4-CPh_3 substituent is more activating than calculated. This enhanced activation can be accounted for in terms of steric acceleration of C–Ph hyperconjugation; in protiodesilylation the effect should be smaller in view of the lower demand for resonance stabilization of the transition state.[99]

(10) The deactivation order by *para*-halogens, $F < Cl < Br < I$, differs slightly from that in hydrogen exchange, $F < Cl < I < Br$ (Section 3.1.2.5), where the greater polarizability of iodine becomes apparent under the conditions of higher electron demand. At the *ortho* positions the order $F < I < Cl < Br$ is the same as in hydrogen exchange (polarizability effects are less important at the *ortho* position) and, as in that reaction, the increase in deactivation on moving the halogen from *para* to *ortho* ($F > Cl > Br > I$) parallels exactly the order of $-I$ effects. The smaller deactivation by *meta*-fluoro compared with the other halogens (as also in hydrogen exchange) confirms that there is secondary relay of resonance effects to the *meta* position.

(11) The pentafluorophenyl compound is about six times more reactive than predicted. This effect (also observed in the gas-phase pyrolysis of 1-arylethyl acetates[105]) has been attributed to the increased $+M$ effect of each fluorine when attached to a very electron-deficient molecule[85] and to a reduced $-I$ effect for the same reason.[105] The discrepancy between observed and calculated reactivity is even greater in protiodestannylation (Section 4.9).

(12) The results provide the first quantitative measure of the relative effects of a range of deactivating substituents, which are for both *meta* and *para* positions: $NO_2 > Me_3N^+ > SO_3H > CF_3 > CO_2H > CO_2Me > Cl$ (the SO_3H data may actually refer to the SO_3^- anion[106]). The charge distribution at the *ortho* and *para* positions in the Wheland intermediate predict that deactivating substituents will give high *ortho*:*para* ratios (Section 11.2.4). Whereas this is true for the CO_2H and SO_3H (SO_3^-) substituents, it is not true for NO_2 and Me_3N^+ substituents. The latter have the largest electron-withdrawing effects so the direct field effect may be responsible.[87] A discrepancy for such substituents is most likely to occur in a reaction with an 'early' transition state well removed from the Wheland intermediate. Such reactions will tend to have their isomer ratios governed rather more by inductive/field effects (Section 11.2.4).

The deactivating effect of the SOPh substituent is much greater than in molecular chlorination; hydrogen bonding in the present reaction may be responsible.[95]

(13) The 4-Me_3N^+ substituent deactivates more than the 4-Me_2NH^+ substituent as it does also in nitration. Steric hindrance to solvation may be partly responsible for this unexpected result, although the rate differences seem to be

too large to be accounted for solely by this. As in nitration (Table 7.13), 4-$CH_2NMe_3^+$ is more deactivating than 3-$CH_2NMe_3^+$.

The 3- and 4-Me_3N^+ substituents have very similar deactivating effects whereas in protiodegermylation and nitration the *meta* substituent deactivates most. Steric hindrance to solvation may again be partly responsible (a less solvating medium is used in degermylation).

The deactivation order by *para* positive poles is $Me_3P^+ > Me_3N^+ \approx Me_3As^+$. In the first substituent the $-M$ effect due to $(p \rightarrow d)\pi$ bonding will be greater. This will also apply for Me_3As^+, but here there is a weaker $-I$ effect so the overall result is about the same as for Me_3N^+. For the *meta* substituents conjugative effects are largely inoperative, so the inductive order $Me_3N^+ > Me_3P^+ > Me_3As^+$ is observed.

(14) The POX_2 substituents are believed to be fully protonated so that their deactivating effects are close to that of Me_3P^+. The CH_2POX_2 substituents, by contrast, are not very deactivating (cf. $CH_2NMe_3^+$), suggesting that they are not protonated, which may derive from the data being obtained in a much weaker acid medium.

(15) Although for sulphur-containing substituents the activation order is the expected one (SMe > SH > SPh) for both *ortho* and *para* series (this is the only determination of the effect of the SH substituent in electrophilic substitution), for the oxygen-containing compounds the order is OH > OMe > OPh. Since it has been argued that the oxygen anion is not involved,[72] and the difference in activation seems too great to be accounted for by steric hindrance to solvation, the result implies that OH hyperconjugation is better than OMe hyperconjugation, the reverse of the situation in the analogous carbon-containing substituents, which seems anomalous. It is possible that the O—H bond is weakened through hydrogen bonding with the solvent.

The $\log f_o$:$\log f_p$ ratios for the O- and S-containing substituents are low relative to those in hydrogen exchange,[90] because conjugative effects are much smaller in protiodesilylation, making the greater importance of the $-I$ effect at the *ortho* positions more evident. The same basic cause lies behind the anomalous ratios for electron-withdrawing substituents noted under (12) above.

(16) The low demand for resonance in protiodesilylation causes the reactivities of polycyclics to be very low compared with some other reactions, the most notable example being the reactivity of the 1-position of biphenylene which is deactivated in desilylation whereas it is activated in hydrogen exchange (Table 3.15). This is the only known example of a polycyclic behaving in this way, and compares with the behaviour of the 4-position of fluorobenzene which for the same reason can be slightly activated or deactivated (cf. hydrogen exchange and protiodesilylation, Tables 3.11 and 4.2).

Because of the different demands for resonance in the two reactions, comparison of the data on hydrogen exchange and desilylation can be used to determine σ-values for polycyclics (for which there are either few or no other data).[89] From the Hammett equations for each reaction[91] eq. 4.22 may be deduced. The σ values so obtained are susceptible to errors in the individual determinations, but nevertheless they are probably accurate to within 0.1 unit.

$$\sigma = (\log f_{detrit.} - 2.54 \log f_{desilyl.})/4.71 \qquad (4.22)$$

(17) The results for indane and tetralin parallel those in detritiation and the same explanation applies [Section 3.1.2.3]. The lower acidity of the reaction conditions in protiodesilylation also permitted examination of benzocyclobutene.[86] Strain theory[89,107,108] predicts that the ratios of the reactivities of the α- and β-positions should be even lower than in indane, and lower still in biphenylene (since the four-membered ring is here even more strained since it contains two double bonds). All of these conditions are fulfilled, the values being 0.36 (indane), 0.10 (benzcyclobutene), and 0.02 (biphenylene).[87] An even lower ratio of 0.002 has been reported for desilylation of 1,3-bis(trimethylsilyl)biphenylene.[109]

(18) Although the reactivities of the polycyclics are generally low in protiodesilylation, in four cases the partial rate factors are disproportionately large relative to those for other positions, owing to steric acceleration. The 9-position of anthracene is exceptionally reactive since the loss of $SiMe_3$ is accelerated by two *peri* hydrogens (the partial rate factor given in Table 4.2 is slightly lower than the literature value,[98] which was determined by comparison with an earlier value of 1510 for f_4^{OMe}, and the partial rate factor of 4.1 originally attributed to the 9-position[110] refers in fact to the 2-position[111]). The ratios of log f(hydrogen exchange):log f(desilylation) for the 1-positions of naphthalene (3.4) and pyrene (2.5) and the 9-position of phenanthrene (4.9) are significantly lower than for the unhindered 2-position of naphthalene (6.1) and the 2- and 3-positions of phenanthrene (9.1 and 8.0), indicating steric acceleration in the former positions (all of which are *peri* to a hydrogen atom). Moreover, for naphthalene, the $\log f_1{:}\log f_2$ ratio is much higher (2.72) than in other electrophilic substitutions.[112]

Steric acceleration is also evident in the acid-catalysed rearrangement of 1,8-bis(trimethylsilyl)naphthalene to the 1,7-isomer.[113] The corresponding

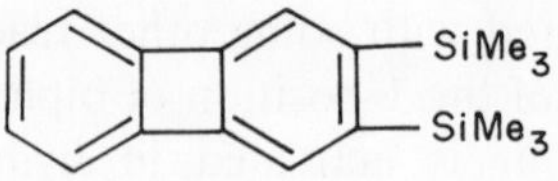

SiMe$_3$ SiMe$_3$

(13)

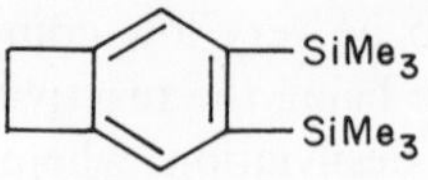

SiMe$_3$ SiMe$_3$

(14)

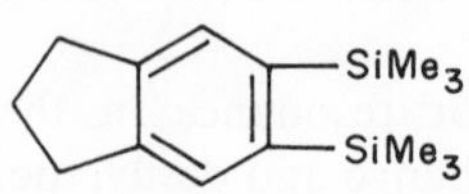

SiMe$_3$ SiMe$_3$

(15)

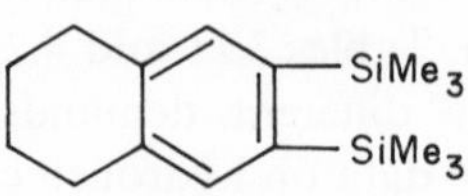

SiMe$_3$ SiMe$_3$

(16)

stannyl compound does not undergo this reaction because although Sn is larger than Si, the greater C—Sn distance makes steric acceleration lower. Steric acceleration provides the primary explanation for the observations[109,114] that the first $SiMe_3$ group is cleaved faster than the second in 2,3-bis(trimethylsilyl)-biphenylene (**13**), 4,5-bis(trimethylsilyl)benzcyclobutene (**14**), 5,6-bis(tri-methylsilyl)indane (**15**), and 6,7-bis(trimethylsilyl)tetralin (**16**), the rate ratios for the last three compounds being 38, 36 and 42, respectively.

(19) The relative positional reactivities of dibenzofuran and dibenzothiophene, originally thought to be anomalous,[77] are readily explained in terms of bond-strain theory[115] and this is discussed more fully in Section 11.3.

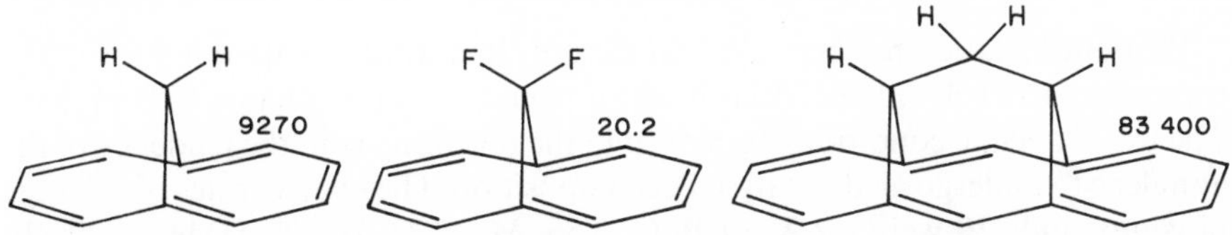

9270 (**17**) 20.2 (**18**) 83 400 (**19**)

(20) The protiodesilylation of 1,6-methano[10]annulene (**17**), its 11,11-difluoro derivative (**18**), and 1,6:8,13-propane-1,3-diylidene[14]annulene (**19**) provides, together with hydrogen exchange, the only quantitative data for electrophilic substitution of an annulene.[96,97] Assuming that there is no steric effect, σ values may be calculated as described in (16) above as -0.65, 0.05, and -0.865 respectively; if steric acceleration comparable to that found in naphthalene applies, these values become more positive by 0.215 σ units[96,97] (slightly different values were given in ref. 96 for **17** and **18** owing to the use of an older ρ value for desilylation). As in hydrogen exchange, the reactivity of the 2-position of **18** is approximately the product of the reactivity of **17** multiplied by two-thirds of the effect of the 3-CF_3 substituent ($9270 \times 0.0015 = 14$).

(21) Protiodesilylation of a few of the compounds described in Table 4.2, using HCl in aqueous HOAc, gave similar substituent effects.[71]

(iii) Substituent effects in biphenyl

The effects of substituents in the protiodesilylation of substituted biphenyls (Table 4.3)[80] are, in linear free energy terms, four times smaller than they are in benzene, i.e. ρ is fourfold less. This quantitatively parallels the corresponding effects in hydrogen exchange (Section 3.1.2.10), solvolysis of 1-aryl-1-chloroalkanes,[116] pyrolysis of 1-arylethyl acetates,[117] and esterification of carboxylic acids.[118] Similar results for the Me and Cl substituents were obtained by Benkeser and co-workers,[71] who found also that the 2′-Me and 2′-Cl substituents produced larger activation and deactivation, respectively, than their 4′-counterparts, presumably owing to the shorter substituent–reaction site

Table 4.3 Partial rate factors for protiodesilylation of 4-(4′-C_6H_4)$C_6H_4SiMe_3$ in aq. H_2SO_4–HOAc at 50 °C

X	OMe	Me	H	$SiMe_3$	Cl	Br	CO_2Me	CO_2H	NO_2
f	8.82	5.04	2.8	2.44	1.54	1.37	0.62	0.53	0.26

distance of (cf. hydrogen exchange). The resonance factor *r* in a Yukawa–Tsuno analysis (Section 11.1.3) is also attenuated ca 2-fold[80] as it is in other reactions.[116]

(iv) Substituent effects in 1,6-methano[10]annulene

Protiodesilylation has been used for the only quantitative determination of the transmission of substituent effects in an annulene, viz. 1,6-methano[10]annulene. The weakly acidic conditions employed in the reaction avoid the tendency of the annulene to undergo acid-catalysed decomposition. The relative rates of cleavage of compounds **20** were for X = *t*-BuO, 3.56; Me, 2.23; H, 1.0; $SiMe_3$, 0.90; Br, 0.055; CO_2Me, 0.037.[119] These data showed that there is very little conjugative interaction between the 2- and 7-positions, comparable in amount to that between the 1- and 5-positions in naphthalene [Section 3.1.2.13.(iv)]. This arises because the bridged structure **21** makes a significant contribution to the structure of the transition state. In **21**, no conjugation between the 2- and 7-positions is possible, whereas in the corresponding non-bridged structure, full conjugation could take place.

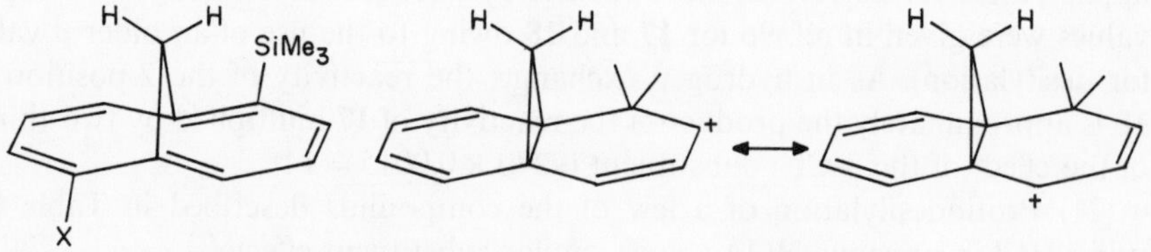

(**20**, X = O–*t*-Bu, Me, H, Br, CO_2Me) (**21**)

(v) Additivity effects

There have been three tests of the *Additivity Principle* in protiodesilylation. For polymethylbenzenes (Table 4.4),[76] additivity is excellent except for the 2,3-Me_2–, 2,4-Me_2–, and 2,4,6-Me_3-substituted compounds. In each case this is due to steric acceleration which, for the 2,3-Me_2 compound, arises from buttressing. Similar results have been obtained using *p*-toluenesulphonic acid or HCl in aqueous HOAc.[60,71] The ρ factors in these media are about 10% less, indicating a slightly earlier transition state, which may account for the lower steric acceleration found for the 2,3-Me_2 compound.

Table 4.4 Observed and calculated methyl substituent effects in protiodesilylation in aq. $HClO_4$–MeOH at 50 °C

Substituent	$f_{obs.}$[a]	$f_{calc.}$	Substituent	$f_{obs.}$[a]	$f_{calc.}$
2-Me	18.3		3,4-Me_2	56.1	54.3
3-Me	2.38		2,3-Me_2	71.9	43.6
4-Me	22.8		2,4-Me_2	422	417
3,5-Me_2	6.00	5.70	2,6-Me_2	3530	335
2,5-Me_2	42.9	43.6	2,4,6-Me_3	53 000	7640

[a] Slightly different values are obtained if the average data obtained over a range of acid concentrations are used (ref. 58, Table 233).

Table 4.5 Ratios of observed to calculated reactivities in protiodesilylation

	$SiMe_3$; MeO; X	$SiMe_3$; Me; NO_2	$SiMe_3$; MeO; CO_2H	$SiMe_3$; NO_2; Me
	X = F, Cl, Br, I			
$f_{obs.} : f_{calc.}$	~1.0	~0.75	~1.1[a]	0.36

[a] A greater discrepancy was reported in the original paper,[93] but the value of f_3^{COOH} appropriate to a much more acidic medium was incorrectly used (cf. footnote b to Table 4.2).

Steric acceleration appears not to affect the reactivity of 2,4,6-trimethoxyphenyltrimethylsilane, since the partial rate factor of ca 1.8×10^8 is very close to that calculated on the basis of additivity.[93]

Within the limits of the extrapolation needed, there is no significant departure from additivity for compounds containing an activating and a deactivating substituent (Table 4.5),[83,95] except in the case of 4-methyl-3-nitrophenyltrimethylsilane, where there may be interaction between both methyl and nitro groups. This would inhibit conjugation in both, but since nitro is *meta* to the reaction site, only activation by the methyl group would be significantly attenuated. The partial rate factor for 3-acetyl-2-methoxyphenyltrimethylsilane (2.4) was also reported in this study.[95]

(vi) Steric hindrance to protiodesilylation of alkylferrocenes

In protiodesilylation of 1′-alkyl-substituted ferrocenes (**22**), relative rates were R = Me 2.65, *i*-Pr 1.6, *t*-Bu 0.73.[120] This reduction in rate with increasing size of the alkyl group was attributed to hindrance of coordination of the attacking proton to iron (considered to be the initial step of this and related reactions).[121]

However, this is now considered to be unlikely in desilylation since the reactivity of ferrocene is very high (it has been semi-quantitatively determined as 10^5 times more reactive than benzene,[21] so σ^+ is ca -1.0) whereas the concentration of the protonated form is only 10^{-10} mol-%.[21] It is therefore probable that the proton attacks the 1′-carbon from the outside face of the molecule, and as the $SiEt_3$ group is displaced it interacts with bulky R groups, so diminishing the reaction rate.

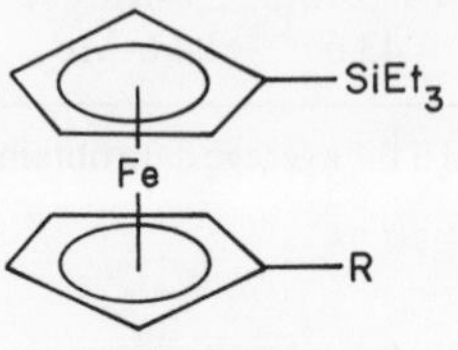

SiEt3
Fe
R

(22)

4.7.2 Base-catalysed Protiodesilylation

Base-catalysed hydrogen exchange involves attack of an electrophile on the aromatic anion, giving an electrophilic substitution in which the usual substituent effects are reversed, and this can also happen in protiodesilylation (eqs 4.23 and 4.24) (and in protiodegermylation, Section 4.8.2, and protiodestannylation, Section 4.9.2).

$$ArSiMe_3 + B^- \rightleftharpoons Ar^- + SiMe_3B \qquad (4.23)$$

$$Ar^- + HB \longrightarrow ArH + B^- \qquad (4.24)$$

Investigations of the mechanism have centred upon whether eq. 4.23 is synchronous with eq. 4.24, or whether a free carbanion is formed. A synchronous process was originally thought to be involved[122] (and also for the corresponding tin compounds), but later work has shown that whereas a synchronous mechanism applies to the tin compounds under most conditions, a virtually free carbanion is produced in the cleavage of the silicon compounds. In particular, the product isotope effect (the relative amounts of protonated to deuteriated products obtained on carrying out the cleavage in a 1:1 mixture of MeOH and MeOD), was approximately 1.2, indicating little discrimination by the intermediate species on reacting with the solvent; this must therefore be a virtually free carbanion,[123,124] and the same is true of base-catalysed protiodegermylation.[125] By contrast, the product isotope effect for protiodestannylation is 3.6–4.3, showing the absence of a free carbanion, and the synchronous mechanism[126] must be involved.

Because of the electrophilic assistance by the solvent in the tin cleavage, the relative reactivities of the silicon and tin compounds will be markedly solvent dependent. Thus the relative reactivities of $PhSnMe_3$ to $PhSiMe_3$ towards

Table 4.6 Partial rate factors for base-catalysed desilylation in aqueous DMSO[122,127,128]

Substituent	H_2O:DMSO ratio (v/v) 1:6 (40 °C)	1:9 (75 °C)	1:32.3 (70 °C)
3-NO_2	12960		
4-NO_2	10120		
3-Br	726		
3-CF_3	480		
3-Cl	400	180	
3-F	108		
4-Br	50.5		
4-Cl	34.5	30	
4-F	8.0	7.3	
3-Ph	3.84		
4-SMe		3.2	
4-Ph	2.49		
2,3-$(CH_2)_2$[a]			2.3
3-OMe	2.17	2.04	
H	1.0	1.0	1.0
2-Me			0.62
3-Me	0.45		0.55
4-Me	0.265		0.40
4-Et	0.40		
2,6-Me_2			0.50
2,5-Me_2			0.30
2,3-Me_2			0.26
2,4-Me_2[b]			0.225
3,4-Me_2			0.15
3,4-$(CH_2)_2$[c]			0.21
4-OMe	0.24	0.31	
3-NMe_2	0.131		
4-NMe_2	0.026		

[a]3-position of benzcyclobutene.
[b]Wrongly shown as the 2,5-compound in ref. 128.
[c]4-Position of benzcyclobutene.

cleavage by NaOH varies from $> 10^3$ in H_2O–MeOH to 0.1 in H_2O–Me_2SO (3:97, v/v); in the latter medium, the water protons are relatively unavailable because of hydrogen bonding, so that the tin cleavage becomes markedly slower.[125]

Electron-withdrawing substituents accelerate the reaction and vice versa. However, strong $+M$ substituents, e.g. 4-NH_2, 4-OMe, do not deactivate here very strongly, and an approximate correlation of rates with σ^0-constants and also with deprotonation energies is observed.[125]

Features of the substituent effect data (Table 4.6) are as follows:

(1) The 2, 6-dimethyl compound is more reactive than predicted on the basis of additivity, whereas all of the other dimethy compounds are slightly less reactive than predicted. Steric acceleration evidently applies as it does in acid-catalysed desilylation, confirmed by the partial rate factor for the 9-position of anthracene being at least 4 times greater than predicted.[98] However, as in the acid-catalysed reaction, there is a fine balance between steric acceleration and steric hindrance, and base-catalysed cleavage of the bulkier $SiEt_3$ group is slower than cleavage of $SiMe_3$.[124]

(2) Partial rate factors for some other polycyclics (1:6 aqueous DMSO at 70 °C) are[129] 1-naphthalene 12.6, 2-naphthalene 4.7, 9-phenanthrene 51, and 1-pyrene 71. These correlate approximately linearly with the corresponding data for base-catalysed hydrogen exchange, indicating a common mechanism for both.

(3) The α-position of benzocyclobutene is activated and is 8.8 times more reactive than the comparable position of 2, 3-dimethylbenzene and 10.6 times more reactive than the β-position. For these base-catalysed reactions, bond-strain theory cannot apply, and the results have been interpreted in terms of an enhanced $-I$ effect of the bridgehead carbon atom, arising from strain as proposed by Tanida and Muneyuki[130] (cf. ref. 131).

The relative rates of desilylation of compounds **23**–**25** (1:9 aqueous DMSO at 30 °C) appear to indicate that even the position β to a bridgehead can be activated by this effect.[132] This conclusion is less certain because substituents with lone pairs give exceptionally enhanced rates in base-catalysed desilylation as noted above (cf. ref. 122); in **24** the oxygen and aryl lone pair p-orbitals are held coplanar so that the exalted reactivity may arise from this effect alone.

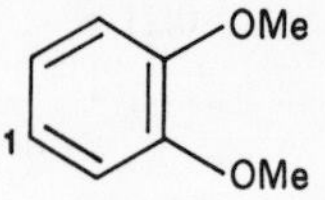

OMe
1
OMe
(**23**)

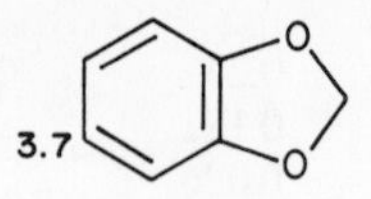

O
3.7
O
(**24**)

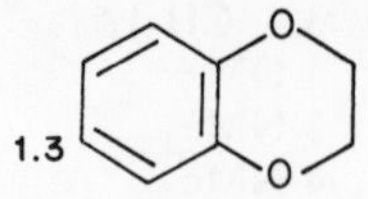

O
1.3
O
(**25**)

4.8 PROTIODEGERMYLATION

4.8.1 Acid-catalysed Protiodegermylation

This reaction (eq. 4.25) closely parallels protiodesilylation in every respect.[65,133,134] Cleavage of the tricyclohexylgermyl group is four times slower than cleavage of the triethylgermyl group, showing that, as in desilylation, bulky R groups hinder attachment of the proton to the *ipso* site. Protiodetriethylgermylation of the *para* position of anisole is 15.5 times faster than the corresponding protiodesilylation, and this factor becomes 36 for the same cleavage from benzene; the transition state for degermylation is therefore nearer to the ground state than is the transition state for desilylation. (This difference stresses the need to compare data for the same parent molecule only when devising *ipso* factors for various reactions, cf. ref. 135.) The solvent isotope effect (1.73) is slightly larger than in desilylation because the C—H bond is less formed in the transition state, and does not compensate to such an extent for the isotope effect arising from the O—H bond breaking.[62]

$$ArGeR_3 + HA \longrightarrow ArH + GeR_3A \quad (4.25)$$

Substituent effects are given in Table 4.7 and notable features are as follows:

(1) The log f values give an excellent correlation with the Yukawa–Tsuno equation (Section 11.1.3), with $\rho = -4.4$ and $r = 0.6$ for the aq. H_2SO_4–HOAc data. These values are both less than for desilylation, as expected for a reaction with a transition state nearer to the ground state.

(2) The reactivity of the 2,4,6-Me_3 compound is 1.24 times that predicted (cf. 1.215 in protiodesilylation). Thus degermylation is also sterically accelerated and to a slightly greater extent than in desilylation. Degermylation is the most sterically accelerated of the cleavages involving Group IVB elements; when the

Table 4.7 Partial rate factors for protiodetriethylgermylation of $ArGeEt_3$

	Reagent			Reagent	
Substituent	MeOH–aq. $HClO_4$	HOAc–aq. H_2SO_4	Substituent	MeOH–aq. $HClO_4$	HOAc–aq. H_2SO_4
4-NMe_2	3×10^6		H	1	1
2,4,6-Me_3	13,600		4-F	0.925	
4-OH	2,730		3-OMe	0.575	0.51
4-OMe	540	382	4-Cl	0.167	0.168
2-OMe	207	187.5	4-Br	0.130	0.133
4-CH_2SiMe_3	162		4-I	0.131	
4-OPh	36.8		3-F		0.032
2-Me	12.4		3-Br		0.019
4-Me	14.0	12.4	3-Cl	0.0165	0.0191
4-Et	13.0		3-CO_2H		0.0177
4-*i*-Pr	12.0		3-CF_3		5.4×10^{-3}
4-*t*-Bu	11.5		4-CO_2H		5.2×10^{-3}
2:3-Benzo	6.2		4-CF_3		2.6×10^{-3}
2-Ph	3.22		3-NMe_3^+		1.26×10^{-3}
4-Ph	2.69	2.43	4-NMe_3^+		1.06×10^{-3}
3:4-Benzo	1.79		3-NO_2		8.0×10^{-4}
3-*t*-Bu	3.3		4-NO_2		3.8×10^{-4}
3-Me	2.11	1.78			

metal is larger than germanium, the size of the metal removes the alkyl group further away from the rest of the aromatic ring so that the interactions which produce acceleration are reduced.

(3) The deactivating substituents confirm the pattern noted in protiodesilylation, except that 4-NMe_3^+ deactivates more than 3-NMe_3^+, as it does in nitration. For all other strongly deactivating substituents, deactivation is greatest from the *para* position, and the reason for this is given in Section 11.2.4.

4.8.2 Base-catalysed Protiodegermylation

The only study of this reaction concerned cleavage of the $GeMe_3$ group from the 2-position of benzo[*b*]thiophene. The reaction occurs 1300 times less readily than for cleavage of corresponding silicon compound[124] and the mechanism of the reaction is the same as for base-catalysed desilylation (Section 4.7.2).

4.9 PROTIODESTANNYLATION

4.9.1 Acid-catalysed Protiodestannylation

This reaction (eq. 4.26) resembles protiodesilylation (and protiodegermylation), but interesting differences arise from the greater ease of cleavage (by a factor of ca 3×10^5) of the aryl—tin bond. Much weaker acids may therefore be used to study the reaction, and this feature has also been exploited in a method for making deuterium-labelled benzene of high isotopic purity (through cleavage of $PhSnMe_3$ with CF_3CO_2D–D_2O), acid-catalysed hydrogen exchange being insignificant under the reaction conditions.[136]

$$ArSnMe_3 + HA \longrightarrow ArH + SnR_3A \tag{4.26}$$

Since cleavage occurs much more readily, the extent of aryl—hydrogen bond formation in the transition state is substantially less than in desilylation under given conditions. Consequently, a larger overall solvent isotope effect, k_H/k_D, is observed (2.55 for HCl in aqueous dioxane at 50 °C),[62] as there is substantially less compensation for the isotope effect arising from the O—H bond breaking in the solvent. A larger isotope effect (ca 5.2 at 50 °C)[64] is obtained in acetic acid (another report gave a value of ca 10),[137] and this is comparable to that (6.5 at 21 °C) found in desilylation in TFA. This shows that the difference in the acid strengths and the reactivities of the aromatic approximately cancel each other out here (to a greater extent than indicated since isotope effects increase with decreasing temperature), the transition states for each reaction under the given conditions occupying similar positions along the reaction coordinate.[64]

As in desilylation, the effects of R groups attached to a metal in cleavage of

$ArSnMe_3$ compounds show a combination of electronic and steric effects, a reactivity series being R = Me > *n*-Bu > *i*-Pr ≈ cyclohexyl ≫ Ph.[137,138] Inductive electron supply increases the reactivity (compare the effects of the latter pair of similarly sized groups), whereas larger alkyl groups activate less than smaller ones. Because the bulkier tin removes the R groups further away from the aryl ring, the steric effect is smaller here, as shown by the relative reactivities for the ethyl and cyclohexyl derivatives of 2.0 compared with 4.1 and 10.4 in protiodegermylation and protiodesilylation, respectively.[65]

Notable features of substituent effects (Table 4.8)[85,137–140] are as follows:

(1) For cleavage of the tricyclohexyl compounds the data correlate with the Yukawa–Tsuno equation (Section 11.1.3) with $\rho = -3.8$ and $r = 0.4$,[138] both factors showing that the transition state is nearer to the ground state than in desilylation, with less demand for resonance stabilization of the transition state. The spread of rates in the reaction is very dependent on the conditions, and for the cleavage of the trimethyl compounds by HCl, the ρ factor is only ca 2.2; the lower factor for a smaller R group again parallels the results for desilylation (Section 4.7.1.1). In protiodestannylation in $EtOH–HClO_4$ the ρ factor decreased with initial increase in water content of the medium, but thereafter remained constant; the rates passed through a minimum as the water content was increased. Both effects arise from the counteracting effects of solvation of the electrophile and of the transition state.[141]

(2) Because the transition state for the reaction has a relatively small amount of charge on the aromatic ring compared with most other electrophilic substitutions, solvation (and steric hindrance to it) is less important. Consequently, although the medium used is a fairly good solvating one, the 'inductive' order of electron release by alkyl groups is observed.

(3) The lower steric acceleration in the reaction compared with desilylation is shown by the smaller 2-Ph:4-Ph reactivity ratio of 1.2 (cf. Table 4.2). Likewise, 1,8-bis(trimethylstannyl)naphthalene does not rearrange to the 1,7-isomer, unlike the silicon analogue.[114]

(4) The very low acid concentrations needed for destannylation have permitted the only determination of the effects of the ethynyl and 1,3-butadienyl substituents in aromatic substitution. They have $-I$, $+M$ effects, the inductive effect being stronger for the latter substituent.[139,140]

(5) As in desilylation, the pentafluorophenyl compound is much more reactive than predicted on the basis of additivity.[85]

(6) The effects of bulky silicon-containing substituents demonstrate the $+I$ effect of the $SiMe_3$ group (and also perhaps Si–Si hyperconjugation). The deactivation by the ($+I$, $-M$) $SiMe_3$ substituent in destannylation (where steric hindrance to solvation is unimportant) in contrast to activation in desilylation constitutes one of the few pieces of evidence for the *inductomeric* (polarizability–inductive) effect.

Partial rate factors for cleavage of $SnMe_3$ by HOAc–dioxane at 27 °C from the

Table 4.8 Partial rate factors for protiodestannylation of $ArSnR_3$

Substituent	R = Medium = v/v = T/°C =	C_6H_{11} $HClO_4$– EtOH 2:50 50	Me HCl– MeOH — 25	Me $HClO_4$– EtOH 1:50 50	Me $HClO_4$– MeOH 2:5 50
4-NMe_2		~2×10^4			
4-OMe		63	48.8		
4-*t*-Bu		7.05			
4-*i*-Pr		6.95			
4-Et		5.3			
4-Me		5.6	5.74		
3-Me		1.845	1.73		
2-Ph		1.99			
4-Ph		1.77			
4-$Si(SiMe_3)_3$					1.76
4-$SiMe(SiMe_3)_2$					1.67
4-$SiMe_2(SiMe_3)$					1.31
H		1	1	1	1
4-$SiMe_3$					0.93
3-OMe		0.855			
4-F		0.62			0.93
2-F					0.41
4-Cl		0.187		0.37	
4-Br		0.145	0.55		
3-F					0.22
2,2,3,3,4-F_5					0.43
3-Cl		0.039		0.142	
3-Br				0.143	
4-C≡CH				0.425	
3-C≡CH				0.288	
4-$(C{\equiv}C)_2H$				0.263	
3-$(C{\equiv}C)_2H$				0.207	
4-CO_2H		0.030			
4-NMe_3^+		0.068			

1- and 2-positions of naphthalene, the 9-positions of phenanthrene and anthracene, and the 2-position of 1,6-methano[10]annulene have been determined as 2.61, 1.5, 2.8, 882, and 90, respectively;[142] these results parallel closely those in desilylation.

Other substituent effect measurements have shown that the relative rates of detrimethylstannylation of the 4-Me_3MCH_2-substituted benzenes are M = Si, 1; Ge, 1.35; Sn, 3.21.[143] Activation by these groups is largely by carbon–metal hyperconjugation, and since conjugative effects are less important in destannylation, the relative rates are lower than in desilylation (for which the Ge:Si reactivity ratio is 1.56).

Lastly, the ease of destannylation is shown by the ready cleavage by hot acetic acid of the $SnPh_3$ group from benzenetricarbonylchromium.[144]

4.9.2 Base-catalysed Protiodestannylation

This reaction differs from the corresponding desilylation in that the mechanism can vary from one in which a free carbanion is formed (cf. eqs 4.23 and 4.24) to one in which electrophilic attack by the solvent is synchronous with departure of the SnR_3 groups (eqs 4.27–4.29).

$$B^- + Me_3SnC_6H_4X \overset{\text{fast}}{\rightleftharpoons} BMe_3Sn^-C_6H_4X \tag{4.27}$$

$$BMe_3Sn^-C_6H_4X + BH \overset{\text{slow}}{\rightleftharpoons} BMe_3Sn^-C_6H_5{}^+X + B^- \tag{4.28}$$

$$BMe_3Sn^-C_6H_5{}^+X \xrightarrow{\text{fast}} BSnMe_3 + C_6H_5X \tag{4.29}$$

The synchronous mechanism is dominant in solvents, e.g. aqueous methanol, where electrophilic attack is favourable, and this gives rise to a small spread of rates (Table 4.9).[145,146] $+M$ substituents give rise to a higher reactivity than expected because they facilitate electrophilic attack by solvent molecules.[122,126] The reaction gives a product isotope effect, k_H/k_D, ranging from 3.8 (4-NO_2) to 5.2 (4-OMe),[124,147] confirming that a free carbanion is not produced under these conditions (cf. base-catalysed desilylation). However, the rate isotope effects (i.e. the rates in MeOH or MeOD) were halved, which is consistent with the mechanism given by eqs 4.27–4.29 in which a pentacoordinate species is formed in a pre-equilibrium. The position of this equilibrium will depend on the secondary isotope effect for desolvation of the methoxide ion. This should take place half as rapidly in the protium-containing as in the deuterium-containing solvent. Hence the product isotope effect arises in eq. 4.28 and the rate isotope effect is the multiple of the product and secondary isotope effects; the reaction rate and the composition of the products are evidently determined in the same step of the reaction.

Under conditions where electrophilic assistance is more difficult (aqueous DMSO, especially 3:100, v/v) cleavage occurs almost entirely by the free carbanion mechanism, as does base-catalysed desilylation, and there is a good linear correlation between the data in the two reactions.[146] The data also correlate with the energies of deprotonation of XC_6H_5, as do those for base-catalysed desilylation and hydrogen exchange (Section 3.3), confirming that a common mechanism applies for all three reactions.

The effect of the two different mechanisms is shown by the reactivity of $PhSnMe_3$ relative to $PhSiMe_3$ towards cleavage. This is greater than 10^2 in aqueous methanol because of the electrophilic assistance of the tin cleavage,

Table 4.9 Partial rate factors for base-catalysed protiodestannylation of $XC_6H_4SnMe_3$

X	Solvent		
	Aq. methanol	Aq. DMSO (1:9, v/v)	Aq. DMSO (3:100, v/v)
2-F	29		
4-NO_2		950	3370
4-NMe_3^+	10.3	192	530
3-CF_3	10.25	115	240
3-Br	9.1		
4-CF_3	8.2	70	144
3-Cl	8.1		
2-Cl	7.55		
2-CF_3	7.05		
3-F	5.45		
4-Cl	4.65	15.8	25
4-Br	4.6		
4-C≡CH	3.9		
3-C≡CH	3.8		
4-F	2.9		
4-SMe	2.4	3.2	3.4
4-Ph	1.8		
3-OMe	1.78		
3-Ph	1.69		
4-OMe	1.64	0.78	0.65
4-NMe_2	1.63	0.22	0.089
3-$SiMe_3$	1.39		
4-$SiMe_3$	1.13		
H	1	1	1
4-*t*-Bu	0.95		
2,3-C_4H_4	0.93		
4-Me	0.86	0.66	0.44
3-Me	0.83	0.59	0.58
3-NH_2	0.68		
2-OMe	0.61		
2-Me	0.25		

whereas in 3:100 (v/v) aqueous DMSO (where this does not apply) it becomes 0.1, and intermediate values are obtained in media with higher proportions of water.[123,125] The effect of the synchronous mechanism also shows up strongly in the higher reactivity of the 2-position of furan relative to that of thiophene (owing to the greater $+M$ effect of oxygen in the former) in destannylation in aqueous methanol, whereas in base-catalysed desilylation and hydrogen exchange furan is much less reactive than thiophene.[124]

The failure to observe a rate isotope effect in the destannylation of carboranylstannanes in ethanol (which gave the carborane reactivity order as *ortho* ≫ *meta* > *para*) led to the proposal that the free carbanion mechanism was involved.[149] However, in view of the above work it now appears more likely that the synchronous mechanism is involved, and that the product and secondary isotope effects cancel each other out.

Lastly, rates of base-catalysed destannylation of benzcylobutene have been measured, the results confirming those for base-catalysed desilylation.[128]

4.10 PROTIODEPLUMBYLATION

This reaction (eq. 4.30) is the most rapid of the cleavages involving Group IVB metals, and gives the largest solvent isotope effect, k_H/k_D, of 3.05 (for R = Et, X = Cl in aqueous dioxane[62]), since the transition state is nearest to the ground state. This is confirmed by the small ρ factor of -2.5 and r value of 0.4 found for correlation with the Yukawa–Tsuno equation (Section 11.1.2) of the partial rate factors (for cleavage of $ArPbMe_3$ by $HClO_4$–EtOH at 25 °C), viz. 4-OMe, 21.0; 4-Me, 3.4; 4-Cl, 0.32; 3-Cl, 0.125.[149]

$$ArPbR_3 \xrightarrow{HX} ArH + PbR_3X \tag{4.30}$$

4.11 PROTIODENITRATION

The strong resonance between the nitro group and the aromatic ring makes nitroaromatics particularly stable. However, nitration of the 9-position of anthracene gave a substrate isotope effect[150] which is uncharacteristic of nitration, indicating that the reaction here is reversible. This follows from the high electrophilic reactivity of the 9-position of anthracene and the presence of the *peri* hydrogens which not only sterically accelerate loss of the nitro group, but also weaken the C—N bond by forcing the nitro group out of coplanarity with the aromatic ring. 9-Nitroanthracene produced nitric acid on reaction with strong sulphuric acid in trichloroacetic acid, but anthracene itself was not detected.[151] The superacid fluorotantalic acid ($HTaF_5$) denitrates 9-nitroanthracene and nitropentamethylbenzene, and the displaced nitro group can be transferred to benzene, toluene, or mesitylene.[152] Toluene gave 95% *para* product, so a bulky denitrating/nitrating species was indicated. 6-Nitro-2,4-diisopropylacetanilide protiodenitrates on treatment with boiling ethanolic HCl;[153] here the denitration site is strongly activated towards electrophilic substitution.

The 1,3-rearrangement of the 2-nitro group to the 4- and 6-positions in 2,3-dinitroaniline, catalysed by concentrated H_2SO_4, involves protiodenitration. The migration of the nitro group is intramolecular and takes place in a rate-determining step; alkyl substituents have only a small effect on the migration rate.[154]

4.12 PROTIODESULPHONATION

This reaction (eq. 4.31) is of considerable preparative importance since the sulphonic acid group is used to block certain positions in the aromatic ring and, after further substitutions have been carried out, is removed by heating with aqueous acid.

$$ArSO_3H + HX \longrightarrow ArH + XSO_3H \tag{4.31}$$

There have been many mechanistic studies of the reaction, mostly involving H_2SO_4 as the catalysing acid.[155-169] Some were directed towards ascertaining whether the Jacobsen reaction (the isomerization of arylsulphonic acids in H_2SO_4) occurs intramolecularly or via protiodesulphonation followed by sulphonation; both processes appear to be involved.[168] There have been a number of mechanisms proposed, partly owing to the difficulties associated with studying a reaction in which a large number of species are present, and complications arising from the tendency for resulphonation to take place. The main features of the reaction are as follows:

(1) The activation energy is higher than for sulphonation, hence the relative ease with which resulphonation takes place.

(2) Desulphonation increases less readily with increasing acid concentration than does sulphonation,[159,163] and for many substrates the rates have been shown to pass through a maximum.[155,162,166]

(3) The reaction is most rapid for *ortho*-substituted sulphonic acids[165,168] and this steric acceleration is consistent with the large steric hindrance to the reverse reaction, sulphonation. Steric acceleration is so large that it overrides the normal electronic effects of the substituents. Thus, for benzenesulphonic acids possessing chloro, carboxyl, or sulphonic acid substituents in the *ortho*, *meta*, or *para* positions. The *ortho* isomer was the most reactive in each case.[165] Likewise, aniline-2, 3, 4, 6-tetrasulphonic acid loses the SO_3H group in the 3-position, the position least activated by the NH_2 group, but most crowded sterically.[170] Desulphonation of isomeric tetramethylbenzenedisulphonic acids takes place in the inverse order of the prediction based on additivity of substituent effects,[171] and steric acceleration is believed to be responsible. Likewise, 2-chloro-3, 5-dinitrobenzenesulphonic acid is more reactive than the 4-chloro isomer.[172]

Although earlier work indicated that reaction took place on the free sulphonic acid,[165,166] the balance of evidence is now strongly in favour of reaction occurring largely on the acid anion, as originally proposed by Lantz[157] to account for the maximum in rate with increasing acidity noted under (2) above. An extensive study of desulphonation using a technique which overcame the problem of resulphonation,[168] showed that the kinetic data were consistent with reaction on the acid anion, with the possibility of some reaction on the free acid at higher acidities.

Reaction on the acid anion would be favoured since there will be a high electron density on the *ipso* carbon, and indeed Baddeley *et al.*[160] proposed that the reaction between the electrophile and the anion would in consequence be a rapid step. However, the need (by the principle of microscopic reversibility) for the reaction mechanism to be the exact reverse of that for sulphonation has led to the mechanism given by eqs 4.32–4.35. The rapid pre-equilibrium to give the acid anion (eq. 4.32) is followed by a slow proton attachment to the *ipso* carbon (eq. 4.33) to give the intermediate. This then rapidly loses SO_3 to a base which will be either H_3O^+ in weaker acid media (eq. 4.34) or H_2SO_4 in stronger acid media (eq. 4.35); in other acids the free acid or acid anion would alternatively be involved in the last step.

$$ArSO_3H + H_2O \xrightleftharpoons{\text{fast}} ArSO_3^- + H_3O^+ \tag{4.32}$$

$$ArSO_3^- + H_2SO_4 \xrightleftharpoons{\text{slow}} Ar^+{<}^{H}_{SO_3^-} + HSO_4^- \tag{4.33}$$

$$Ar^+{<}^{H}_{SO_3^-} + H_3O^+ \xrightleftharpoons{\text{fast}} ArH + H_3SO_4^+ \tag{4.34}$$

$$Ar^+{<}^{H}_{SO_3^-} + H_2SO_4 \xrightleftharpoons{\text{slow}} ArH + SO_3\,(H_2SO_4) \tag{4.35}$$

The electrophilic nature of the reaction is shown by the increase in rate with increase in the number of alkyl substituents, so that a reactivity series is pentamethyl- > 1,2,3,5-tetramethyl- > 1,2,4,5-tetramethylbenzenesulphonic acid.[163] The effect of methyl substituents depends on temperature and acid concentration, but the values obtained in 79.3 wt-% H_2SO_4 at 120.1 °C, viz. $f_o = 152$, $f_m = 2.24$, $f_p = 69$ are representative[166] and also show the effect of steric acceleration. Both this and electronic effects combine to give the 1-:2-naphthalenesulphonic acid rate ratio of 50[165] and also the very high reactivity of anthracene-9-sulphonic acid which is readily cleaved by aqueous HCl at 100 °C.[173] Semi-quantitative results for a variety of other compounds also show that the reaction is electrophilic.[174]

4.13 PROTIODEHALOGENATION

Iodo, bromo, and chloro substituents can be displaced from aromatic rings in an acid-catalysed reaction (eq. 4.36), the ease of removal being I > Br > Cl. The reaction is sterically accelerated by large *ortho* substituents so that whereas 2,4,6-tri-*tert*-butylbromobenzene undergoes protiodebromination with strong acid, 2,4,6-trimethylbromobenzene is unreactive under the same conditions.[175] The steric strain in the former compound is shown by the fact that positive bromination of 1,3,5-tri-*tert*-butylbenzene gives a large kinetic isotope effect

[Section 9.3.1.(i)]. Thus the second step of the bromination, loss of the proton from the intermediate, must be partially rate-determining and therefore slow owing to the compression in the product; it follows that the first step of the reverse reaction, protiodebromination, will be accelerated.

$$\mathrm{ArHal} + \mathrm{HX} \longrightarrow \mathrm{ArH} + \mathrm{XHal} \tag{4.36}$$

Protiodehalogenation is electrophilic as shown by the effects of substituents in the HI-catalysed deiodination of iodobenzene, which are 2,4-$(OH)_2$ > 4-OH > 2-OH > 3-OH and 2-Me > 4-Me > 3-Me > H.[176] For the HI-catalysed deiodination of 4-hydroxyiodobenzenes in acetic acid, the partial rate factors are 2.4 (3-Me), 0.08 (3-Cl), 48 (2-Me), and 160 (2-Me × 5-*i*-Pr).[177] Both sets of work show steric acceleration by the *ortho*-methyl substituent, and the ρ factor is ca -3.0. For cleavages of other halogens, and from less activated substrates, higher ρ factors could be anticipated, but no data are available. The relative rates of the HCl–$SnCl_2$-catalysed deiodination of some amino-substituted iodoaromatics were also consistent with an electrophilic substitution.[178]

Choguill and Ridd[179] showed that the mechanism of protiodeiodination is that given in eq. 4.37. The reaction of 4-iodoaniline with aqueous mineral acids is first order with respect to the stoicheiometric concentration of aromatic, is independent of the concentrations of H^+ and I^-, and gives a solvent isotope effect, $k(H_2O)/k(D_2O)$, of 5.8. The latter feature follows from the principle of microscopic reversibility, since in the reverse reaction, iodination, a substrate kinetic isotope effect is observed, showing that loss of the proton is partly rate-limiting. The independence of the rate on the hydrogen ion concentration follows from the free amine (which must be the species which reacts with the proton) being present entirely as the conjugate acid in the acid range examined.

$$\mathrm{ArI} \underset{\text{slow}}{\overset{\mathrm{H}^+}{\rightleftharpoons}} \mathrm{Ar}^{+}\!\!<^{\mathrm{H}}_{\mathrm{I}} \underset{\text{fast}}{\overset{+2\mathrm{I}^-}{\rightleftharpoons}} \mathrm{ArH} + \mathrm{I}_3^- \tag{4.37}$$

Protiodebromination gives a useful route to *m*-bromophenols.[180] Bromination of methylphenols gives (via dienone intermediates) substitution at the sites most activated towards electrophilic substitution (i.e. *ortho* and *para* to hydroxy) and also at other sites. Subsequent reaction with HI removes the most labile bromines at the former positions. Thus, for example, 3-methylphenol gave 5-bromo-3-methylphenol and 4-methylphenol gave 3-bromo-4-methylphenol. Protiodebromination has also been observed in various bromothiophene derivatives[181] and with *n*-alkylanilines.[182]

Protiodechlorination is believed to be an intermediate step in the overall benzoyldechlorination of *p*-dichlorobenzene to give *p*-chlorobenzoylbenzene.[183] The formation of benzoylbenzene as a byproduct indicated that the reaction conditions were conducive to protiodechlorination (very strong acids, e.g. $HAlCl_4$, are present under Friedel–Crafts conditions), so that chlorobenzene is probably formed initially from the dichloro compound. The superacid $HAlCl_3OH$ (present in aqueous $AlCl_3$) causes chlorine migration via protiodechlorination in *o*-chlorofluorobenzene and *o*-dichlorobenzene; the latter gives mainly the *meta* isomer at equilibrium.[184]

4.14 PROTIODEPALLADIATION

In the palladium(II) acetate-catalysed formation of biaryls from iodobenzenes, protiodeiodination occurs and is accelerated by the presence of *ortho*-methyl substituents.[185] This is not due simply to steric acceleration since 4-methyl-iodobenzene underwent reaction more rapidly than 4-chloroiodobenzene, and an electrophilic substitution is indicated. The reaction is therefore believed to involve the intermediate formation of ArPdI, which is then cleaved by water:

$$\mathrm{ArPdI} + \mathrm{H_2O} \longrightarrow \mathrm{ArH} + \mathrm{PdI(OH)} \qquad (4.38)$$

REFERENCES

1. C. L. Perrin and G. A. Skinner, *J. Am. Chem. Soc.*, 1971, **93**, 3389.
2. (a) R. W. Bott, C. Eaborn, and P. M. Greasley, *J. Chem. Soc.*, 1964, 4804; (b) C. Eaborn, *J. Chem. Soc., Chem. Commun.*, 1972, 1255.
3. C. L. Perrin, *J. Org. Chem.*, 1971, **36**, 420.
4. E. G. Perevalova, T. U. Baukova, E. I. Gorimov, and K. I. Grundberg, *Bull. Acad. Sci. USSR*, 1970, 2031.
5. R. E. Dessy and R. M. Salinger, *J. Org. Chem.*, 1961, **26**, 3519.
6. L. V. Guild, C. A. Hollingworth, D. H. McDaniel, S. K. Podder, and J. H. Wotz, *J. Org. Chem.*, 1962, **27**, 762.
7. A. H. Corwin and M. A. Naylor, *J. Am. Chem. Soc.*, 1947, **69**, 1004; F. Kaufman and A. H. Corwin, *J. Am. Chem. Soc.*, 1955, **77**, 6280.
8. S. Winstein and T. G. Traylor, *J. Am. Chem. Soc.*, 1955, **77**, 3747.
9. R. E. Dessy and J.-Y. Kim, *J. Am. Chem. Soc.*, 1960, **82**, 686.
10. A. N. Nesmeyanov, A. E. Borisov, and I. S. Saveleva, *Proc. Acad. Sci. USSR*, 1964, **155**, 280.
11. Y. Yukawa and Y. Tsuno, *Bull. Chem. Soc. Jpn.*, 1959, **32**, 971.
12. R. E. Dessy and J.-Y. Kim. *J. Am. Chem. Soc.*, 1961, **83**, 1167.
13. V. S. Petrosyan and O. A. Reutov, *Proc. Acad. Sci. USSR*, 1968, **180**, 514.
14. I. P. Beletskaya, G. A. Artamkina, and O. A. Reutov, *Proc. Acad. Sci. USSR*, 1980, **251**, 124.
15. M. M. Kreevoy and R. L. Hansen, *J. Phys. Chem.*, 1961, **65**, 1055; F. R. Jensen and D. Heyman, *J. Am. Chem. Soc.*, 1966, **88**, 3438; B. F. Hegarty, W. Kitching, and P. R. Wells, *J. Am. Chem. Soc.*, 1967, **89**, 4816.
16. I. P. Beletskaya, A. L. Myshkin, and O. A. Reutov, *Bull. Acad. Sci. USSR*, 1965, 232.
17. I. P. Beletskaya, A. L. Kurts, and O. A. Reutov, *J. Org. Chem. USSR*, 1967, **3**, 1884.
18. R. D. Brown, A. S. Buchanan, and A. A. Humffray, *Aust. J. Chem.*, 1965, **18**, 1507.
19. A. D. Ainley and F. Challenger, *J. Chem. Soc.*, 1930, 2171.

20. H. G. Kuivila and K. V. Nahabedian, *Chem. Ind.* (*London*), 1959, 1120; *J. Am. Chem. Soc.*, 1961, **83**, 2159, 2164, 2167.
21. G. Cerichelli, B. Floris, G. Illuminati, and G. Ortiggi, *J. Org. Chem.*, 1974, **39**, 3948; B. Floris and G. Illuminati, *J. Organomet. Chem.*, 1978, **150**, 101.
22. H. G. Kuivila, J. F. Reuwer, and J. A. Mangravite, *Can. J. Chem.*, 1963, **41**, 3081.
23. H. G. Kuivila, J. F. Reuwer, and J. A. Mangravite, *J. Am. Chem. Soc.*, 1964, **86**, 2666.
24. G. Dallinga and G. ter Maten, *Recl. Trav. Chim. Pays-Bas*, 1960, **79**, 737.
25. K. C. C. Bancroft, *Ph.D. Thesis*, Leicester University, 1964; J. F. W. McOmie and J. A. Saleh, *Tetrahedron*, 1973, **29**, 4003.
26. T. A. Modro and K. Yates, *J. Am. Chem. Soc.*, 1976, **98**, 4247.
27. A. P. Neary and R. Taylor, *J. Chem. Soc., Perkin Trans. 2*, 1983, 1233.
28. H. V. Ansell and R. Taylor, *Tetrahedron Lett.*, 1971, 4515.
29. A. Oku and Y. Yuken, *J. Org. Chem.*, 1975, **40**, 3850.
30. M. M. J. LeGuen and R. Taylor, *J. Chem. Soc., Perkin Trans. 2*, 1976, 559.
31. H. Cerfontain, A. W. Kaandorp, and K. L. J. Sixma, *Recl. Trav. Chim. Pays-Bas*, 1963, **82**, 565.
32. R. O. C. Norman and R. Taylor, *Electrophilic Substitution in Benzenoid Compounds*, Elsevier, Amsterdam, 1965, pp. 205–207.
33. U. Svanholm and V. D. Parker, *J. Chem. Soc., Perkin Trans. 2*, 1973, 562.
34. E. P. Babin *et al.*, *Zh. Obshch. Khim.*, 1960, **30**, 430.
35. V. F. Lavrushin and Z. N. Tarakhno, *Proc. Acad. Sci. USSR*, 1971, **200**, 746.
36. R. Taylor, *Specialist Periodical Report on Aromatic and Heteroaromatic Chemistry*, Vol. 1, Chemical Society, London, 1973, p. 188.
37. E. Louise, *Ann. Chim. Phys.*, 1885, **6**, 206.
38. A. Klages and G. Lickroth, *Chem. Ber.*, 1899, **32**, 1549.
39. W. M. Schubert and H. K. Latourette, *J. Am. Chem. Soc.*, 1952, **74**, 1829.
40. W. M. Schubert and R. E. Zahler, *J. Am. Chem. Soc.*, 1954, **76**, 1.
41. W. M. Schubert and H. Burkett, *J. Am. Chem. Soc.*, 1956, **78**, 64.
42. W. M. Schubert and P. C. Myhre, *J. Am. Chem. Soc.*, 1958, **80**, 1755.
43. H. Burkett, W. M. Schubert, F. Schultz, R. B. Murphy and R. Talbott, *J. Am. Chem. Soc.*, 1959, **81**, 3923.
44. J. A. Farooqi, P. H. Gore, E. F. Saad, D. N. Waters, and G. F. Moxon, *J. Chem. Soc., Perkin Trans. 2*, 1980, 835.
45. J. A. Farooqi and P. H. Gore, *Tetrahedron Lett.*, 1977, 2983; P. H. Gore, E. F. Saad, D. N. Waters, and G. F. Moxon, *Int. J. Chem. Kinet.*, 1982, **14**, 55; P. H. Gore, A. M. G. Nasser, and E. F. Saad, *J. Chem. Soc., Perkin Trans. 2*, 1982, 983; J. Al-Kabi, J. A. Farooqi, P. H. Gore, A. M. G. Nasser, E. F. Saad, E. L. Short, and D. N. Waters, *J. Chem. Soc., Perkin Trans. 2*, 1988, 943.
46. A. I. Bukova and I. K. Buchina, *J. Org. Chem. USSR*, 1984, **20**, 1199.
47. H. Schenkel and M. Schenkel-Rudin, *Helv. Chim. Acta*, 1948, **31**, 514.
48. B. R. Brown, *Q. Rev. Chem. Soc.*, 1951, **5**, 131; B. R. Brown, D. L. Hammick, and J. B. Scholefield, *J. Chem. Soc.*, 1950, 778.
49. B. R. Brown, W. W. Elliott and D. L. Hammick, *J. Chem. Soc.*, 1951, 1184.
50. W. M. Schubert and J. D. Gardner, *J. Am. Chem. Soc.*, 1953, **75**, 1401.
51. W. M. Schubert, *J. Am. Chem. Soc.*, 1949, **71**, 2639.
52. W. M. Schubert, J. Donohue, and J. D. Gardner, *J. Am. Chem. Soc.*, 1954, **76**, 9.
53. A. V. Willi, *Z. Naturforsch., Teil A*, 1958, **132**, 997.
54. C. A. C. Bothner-By, and J. Bigeleisen, *J. Chem. Phys.*, 1959, **19**, 755; W. H. Stevens, J. M. Pepper, and M. Lonnsburg, *J. Chem. Phys.*, 1952, **20**, 192.
55. K.R. Lynn and A. N. Bourne, *Chem. Ind.* (*London*), 1963, 782.
56. F. M. Beringer and S. Sands, *J. Am. Chem. Soc.*, 1953, **75**, 3319.

57. A. V. Willi and J. F. Stocker, *Helv. Chim. Acta*, 1954, **37**, 1113; A. V. Willi, *Helv. Chim. Acta*, 1957, **40**, 1053; G. E. Dunn, P. Leggate, and I. E. Scheffler, *Can. J. Chem.*, 1965, **43**, 3080; J. M. Los, R. F. Rekker, and C. H. Tonsbeek, *Recl. Trav. Chim. Pays-Bas*, 1967, **86**, 622; A. V. Willi, C. M. Mon, and P. Vilk, *J. Phys. Chem.*, 1968, **72**, 3142.
58. R. Taylor, *Comprehensive Chemical Kinetics*, Elsevier, Amsterdam, Vol. 13, 1972, pp. 312–316.
59. C. Eaborn, *J. Chem. Soc.*, 1953, 3148, and references cited therein.
60. R. A. Benkeser and H. . Krysiak, *J. Am. Chem. Soc.*, 1954, **76**, 6353.
61. J. E. Baines and C. Eaborn, *J. Chem. Soc.*, 1956, 1436.
62. R. W. Bott, C. Eaborn, and P. M. Greasley, *J. Chem. Soc.*, 1964, 4804.
63. C. Eaborn, P. M. Jackson, and R. Taylor, *J. Chem. Soc. B*, 1966, 613.
64. C. Eaborn, I. D. Jenkins, and D. R. M. Walton, *J. Chem. Soc., Perkin Trans. 2*, 1974, 596.
65. C. Eaborn and K. C. Pande, *J. Chem. Soc.*, 1960, 1566.
66. C. G. Pitt, *J. Organomet. Chem.*, 1973, **61**, 49.
67. C. Eaborn, T. A. Emokpae, V. I. Sidorov, and R. Taylor, *J. Chem. Soc., Perkin Trans. 2*, 1974, 1454.
68. H. J. Berwin, *J. Chem. Soc., Chem. Commun.*, 1972, 237.
69. T. G. Traylor, W. Hanstein, and H. J. Berwin, *J. Am. Chem. Soc.*, 1970, **92**, 7476.
70. R. W. Bott, C. Eaborn, and P. M. Jackson, *J. Organomet. Chem.*, 1967, **7**, 79; R. C. Moore, *Ph.D. Thesis*, Leicester University, 1961.
71. R. A. Benkeser, W. Schroeder, and O. H. Thomas, *J. Am. Chem. Soc.*, 1958, **80**, 2283; R. A. Benkeser, R. A. Hickner, and D. I. Hoke, *J. Am. Chem. Soc.*, 1958, **80**, 2279; R. A. Benkeser and F. S. Clark, *J. Am. Chem. Soc.*, 1960, **82**, 4881.
72. M. A. Cook, C. Eaborn, and D. R. M. Walton, *J. Organomet. Chem.*, 1970, **25**, 85.
73. C. Eaborn, *J. Chem. Soc.*, 1956, 4858.
74. F. B. Deans and C. Eaborn, *J. Chem. Soc.*, 1959, 3031.
75. C. Eaborn, Z. Lasocki, and D. E. Webster, *J. Chem. Soc.*, 1959, 3034.
76. C. Eaborn and R. C. Moore, *J. Chem. Soc.*, 1959, 3640.
77. C. Eaborn and J. A. Sperry, *J. Chem. Soc.*, 1961, 4921.
78. R. Baker, C. Eaborn, and J. A. Sperry, *J. Chem. Soc.*, 1962, 2382.
79. R. W. Bott, C. Eaborn, K. C. Pande, and T. W. Swaddle, *J. Chem. Soc.*, 1962, 1217.
80. R. Baker, R. W. Bott, C. Eaborn, and P. M. Greasley, *J. Chem. Soc.*, 1964, 627.
81. R. W. Bott, C. Eaborn, and K. Leyshon, *J. Chem. Soc.*, 1964, 1971.
82. R. W. Bott, B. F. Dowden, and C. Eaborn, *J. Chem. Soc.*, 1965, 6306.
83. C. Eaborn and D. R. M. Walton, *J. Organomet. Chem.*, 1965, **3**, 169.
84. D. R. M. Walton, *J. Organomet. Chem.*, 1965, **3**, 438.
85. C. Eaborn, J. A. Treverton, and D. R. M. Walton, *J. Organomet. Chem.*, 1967, **9**, 259.
86. A. R. Bassindale, C. Eaborn, and D. R. M. Walton, *J. Chem. Soc. B*, 1969, 12.
87. C. Eaborn, D. R. M. Walton, and D. J. Young, *J. Chem. Soc. B*, 1969, 15.
88. A. R. Bassindale, C. Eaborn, D. R. M. Walton, and D. J. Young, *J. Organomet. Chem.*, 1969, **20**, 49.
89. R. Taylor, *J. Chem. Soc. B*, 1971, 536.
90. R. Taylor and F. P. Bailey, *J. Chem. Soc. B*, 1971, 1146.
91. C. Eaborn and J. F. R. Jaggard, *J. Chem. Soc. B*, 1969, 892.
92. C. Eaborn and P. M. Jackson, *J. Chem. Soc. B*, 1969, 21.
93. C. Eaborn, Z. S. Salih, and D. R. M. Walton, *J. Organomet. Chem.*, 1972, **36**, 41, 47.
94. C. Eaborn, Z. Lasocki, and J. A. Sperry, *J. Organomet. Chem.*, 1972, **35**, 245.
95. A. C. Boicelli, R. Danielli, A. Mangini, A. Ricci, and G. Pirazzini, *J. Chem. Soc., Perkin Trans. 2*, 1973, 1343.

96. R. Taylor, *J. Chem. Soc., Perkin Trans. 2*, 1975, 1287.
97. A. P. Laws and R. Taylor, *J. Chem. Soc., Perkin Trans. 2*, 1987, 1692.
98. C. Eaborn, R. Eidenschink, and D. R. M. Walton, *J. Organomet. Chem.*, 1975, **95**, 183.
99. H. V. Ansell and R. Taylor, *J. Chem. Soc., Perkin Trans. 2*, 1978, 751.
100. A. J. Cornish and C. Eaborn, *J. Chem. Soc., Perkin Trans. 2*, 1975, 874.
101. I. Szele, *Helv. Chim. Acta*, 1961, **64**, 2733.
102. E. Glyde and R. Taylor, *J. Chem. Soc., Perkin Tans. 2*, 1977, 678.
103. M. A. Hossaini and R. Taylor, *J. Chem. Soc., Perkin Trans. 2*, 1982, 187.
104. E. Glyde and R. Taylor, *J. Chem. Soc., Perkin Trans. 2*, 1973, 1632.
105. R. Taylor, *J. Chem. Soc. B*, 1971, 255.
106. T. A. Kortekaas, H. Cerfontain, and J. M. Gall, *J. Chem. Soc., Perkin Trans. 2*, 1978, 445.
107. J. M. Blatchly and R. Taylor, *J. Chem. Soc. B*, 1968, 1402.
108. R. Taylor, *Chimia*, 1968, **22**, 1.
109. R. C. Hillard and K. P. C. Vollhardt, *J. Am. Chem. Soc.*, 1977, **99**, 4058.
110. Ref. 32, p. 240.
111. M. M. Hirschler, H. V. Ansell, and R. Taylor, *J. Chem. Soc., Perkin Trans. 2*, 1977, 353.
112. R. Taylor and G. G. Smith, *Tetrahedron*, 1963, **19**, 937.
113. D. Seyferth and S. C. Vick, *J. Organomet. Chem.*, 1977, **141**, 173.
114. B. C. Berris, G. H. Hovakeemian, Y.-H. Lee, H. Mestdagh, and K. P. C. Vollhardt, *J. Am. Chem. Soc.*, 1985, **107**, 5670.
115. R. Taylor, *J. Chem. Soc. B*, 1968, 1559.
116. T. Inukii, *Bull. Chem. Soc. Jpn.*, 1962, **35**, 400; Y. Tsuno and W.-Y. Chong, T. Tawaka, M. Sawada, and Y. Yukawa, *Bull. Chem. Soc. Jpn.*, 1978, **51**, 596; R. Bolton and R. E. M. Burley, *J. Chem. Soc., Perkin Trans. 2*, 1977, 426.
117. R. Taylor, unpublished work.
118. P. Ananthakrishnanadar, N. Kannan, and G. Vorghesedharumaraj, *J. Chem. Soc., Perkin Trans. 2*, 1982, 505.
119. T. Suzuki, K. Takase, K. Takahashi, A. P. Laws, and R. Taylor, *J. Chem. Soc., Perkin Trans. 2*, 1988, 697.
120. R. A. Benkeser, Y. Nagai, and J. Hooz, *J. Am. Chem. Soc.*, 1964, **86**, 3742.
121. T. J. Curphey, J. O. Santer, M. Rosenblum, and J. H. Richards, *J. Am. Chem. Soc.*, 1960, **82**, 5249.
122. A. R. Bassindale, C. Eaborn, R. Taylor, A. R. Thompson, D. R. M. Walton, J. Cretney, and G. J. Wright, *J. Chem. Soc., Perkin Trans. 2*, 1971, 1155.
123. C. Eaborn, J. R. Jones, and G. Seconi, *J. Organomet. Chem.*, 1976, **116**, 83.
124. C. Eaborn and G. Seconi, *J. Chem. Soc., Perkin Trans. 2*, 1976, 925.
125. C. Eaborn, J. G. Stamper, and G. Seconi, *J. Organomet. Chem.*, 1981, **204**, 153, 204.
126. Ref. 58, pp. 347–348.
127. J. Cretney and G. J. Wright, *J. Organomet. Chem.*, 1971, **28**, 49.
128. C. Eaborn, A. A. Najam, and D. R. M. Walton, *J. Organomet. Chem.*, 1974, **46**, 255.
129. B. Bøe, C. Eaborn, and D. R. M. Walton, *J. Organomet. Chem.*, 1974, **82**, 13.
130. H. Tanida and R. Muneyuki, *J. Am. Chem. Soc.*, 1965, **87**, 4794.
131. A. Streitwieser, G. R. Ziegler, P. C. Mowery, A. Lewis, and R. G. Lawler, *J. Am. Chem. Soc.*, 1968, **90**, 1357.
132. L. J. Brocklehurst, K. E. Richards and G. J. Wright, *Aust. J. Chem.*, 1974, **27**, 895.
133. C. Eaborn and K. C. Pande, *J. Chem. Soc.*, 1961, 297.
134. C. Eaborn and K. C. Pande, *J. Chem. Soc.*, 1961, 5082.
135. S. R. Hartshorn, *Chem. Soc. Rev.*, 1974, **3**, 167.

136. W. A. Asomaning, C. Eaborn, and D. R. M. Walton, *J. Chem. Soc., Perkin Trans. 1*, 1973, 137.
137. J. Nasielski, O. Buchman, M. Grosjean, and M. Jauquet, *J. Organomet. Chem.*, 1969, **19**, 353; O. Buchman, J. Nasielski, and M. Planchon, *Bull. Soc. Chim. Belg.*, 1963, **72**, 286; *Helv. Chim. Acta*, 1964, **47**, 1695.
138. C. Eaborn and J. A. Waters, *J. Chem. Soc.*, 1961, 542.
139. C. Eaborn, A. R. Thompson, and D. R. M. Walton, *J. Chem. Soc. B*, 1969, 859.
140. C. Eaborn, A. R. Thompson, and D. R. M. Walton, *J. Chem. Soc. B*, 1970, 357.
141. C. Eaborn, A. R. Thompson, and D. R. M. Walton, *J. Organomet. Chem.*, 1971, **29**, 257.
142. W. Kitching, H. A. Olszowy, I. Schott, W. Adcock, and D. P. Cox, *J. Organomet. Chem.*, 1986, **310**, 269.
143. R. W. Bott, C. Eaborn, and D. R. M. Walton, *J. Organomet. Chem.*, 1964, **2**, 154.
144. G. K. Magomedov, V. G. Sirkin, A. S. Frenkel, A. V. Medvedeva, and L. V. Morozova, *J. Gen. Chem. USSR*, 1973, **43**, 803.
145. C. Eaborn, H. L. Hornfeld, and D. R. M. Walton, *J. Chem. Soc. B*, 1967, 1036.
146. P. Dembech, G. Seconi, and C. Eaborn, *J. Chem. Soc., Perkin Trans. 2*, 1983, 301.
147. R. Alexander, W. A. Asomaning, C. Eaborn, I. D. Jenkins, and D. R. M. Walton, *J. Chem. Soc., Perkin Trans. 2*, 1974, 304.
148. V. I. Stanko, T. V. Klimova, and I. P. Beletskaya, *J. Organomet. Chem.*, 1973, **61**, 191.
149. C. Eaborn and K. C. Pande, *J. Chem. Soc.*, 1961, 3715.
150. H. Cerfontain and A. Telder, *Recl. Trav. Chim. Pays-bas*, 1967, **86**, 3715.
151. P. H. Gore, *J. Chem. Soc.*, 1957, 1437.
152. G. A. Olah, S. C. Narang, R. Malhotra, and J. A. Olah, *J. Am. Chem. Soc.*, 1979, **101**, 1805.
153. K. M. Davies and J. Ellis, *Tetrahedron*, 1969, **25**, 1423.
154. J. T. Murphy and J. H. Ridd, *J. Chem. Soc., Perkin Trans. 1*, 1987, 1767.
155. J. Crafts, *Chem. Ber.*, 1901, **34**, 1350; *Bull. Soc. Chim. Fr.*, 1907, **1**, 917.
156. R. Lantz, *Bull. Soc. Chim. Fr.*, 1936, **2**, 2092.
157. R. Lantz, *Bull. Soc. Chim. Fr.*, 1945, **12**, 1004.
158. J. Pinnow, *Z. Elektrochem.*, 1915, **21**, 380; 1917, **23**, 240.
159. W. A. Cowdray and D. S. Davies, *J. Chem. Soc.*, 1949, 1871.
160. G. Baddeley, G. Holt, and J. Kenner, *Nature (London)*, 1944, **154**, 361.
161. V. Gold and D. P. N. Satchell, *J. Chem. Soc.*, 1956, 1635.
162. M. Kilpatrick, M. W. Meyer and M. L. Kilpatrick, *J. Phys. Chem.*, 1960, **64**, 1433.
163. M. Kilpatrick and M. W. Meyer, *J. Phys. Chem.*, 1961, **65**, 1312.
164. Y. Muramoto, *Chem. Abstr.*, 1960, **54**, 16416c.
165. A. A. Spryskov and N. A. Ovsyankina, *J. Gen. Chem. USSR*, 1950, **20**, 1083; 1951, **21**, 1649; 1954, **24**, 1777; *Chem Abstr.*, 1955, **49**, 6894e; O. I. Kachurin, A. A. Spryskov, and L. P. Melnikova, *Chem. Abstr.*, 1961, **55**, 2544; A. A. Spryskov and S. P. Starkov, *Zh. Obshch. Khim., SSSR*, 1956, **26**, 2862.
166. Y. I. Leitman and M. S. Pevsner, *J. Appl. Chem. USSR*, 1959, **32**, 2830; A. Koeberg-Telder, A. J. Prinsen, and H. Cerfontain, *J. Chem. Soc. B*, 1969. 1004.
167. H. Cerfontain and J. M. Arends, *Recl. Trav. Chim. Pays-Bas*, 1966, **85**, 358.
168. A. C. M. Wanders and H. Cerfontain, *Recl. Trav. Chim. Pays-Bas*, 1967, **86**, 1199.
169. I. R. Bellabono, *Chim. Ind. (Milan)*, 1959, 1079.
170. P. K. Maarsen, R. Bregman, and H. Cerfontain, *J. Chem. Soc., Perkin Trans. 2*, 1977, 1863.
171. A. Koeberg-Telder and H. Cerfontain, *Recl. Trav. Chim. Pays-Bas*, 1982, **101**, 41.
172. T. I. Potapova and A. A. Spryskov, *Chem. Abstr.*, 1963, **59**, 7337a.

173. P. H. Gore, *J. Org. Chem.*, 1957, **22**, 135.
174. B. I. Karavaev and A. A. Spryskov, *J. Gen. Chem. USSR*, 1956, **26**, 2231; A. G. Green and K. H. Vakil, *J. Chem. Soc.*, 1918, 35; G. Travagli, *Gazz. Chim. Ital.*, 1951, **81**, 668; K. Lauer, *J. Prakt. Chem.*, 1932, **135**, 182; V. V. Kozlov and A. A. Egorova, *J. Gen. Chem. USSR*, 1955, **25**, 963; V. V. Kozlov, *Dokl. Akad. Nauk SSSR*, 1948, **61**, 281; T. A. Kortekas and H. Cerfontain, *J. Chem. Soc., Perkin Trans. 2*, 1978, 742.
175. P. C. Myhre, *Acta Chem. Scand.*, 1960, **14**, 947.
176. J. B. Shoesmith, A. C. Hetherington, and R. H. Slater, *J. Chem. Soc.*, 1924, **125**, 1312, 2278.
177. V. Gold and M. Whittaker, *J. Chem. Soc.*, 1951, 1184.
178. B. H. Nicolet and J. R. Sampey, *J. Am. Chem. Soc.*, 1927, **49**, 1796; B. H. Nicolet and W. L. Ray, *J. Am. Chem. Soc.*, 1927, **49**, 1801.
179. H. S. Choguill and J. H. Ridd, *J. Chem. Soc.*, 1961, 822.
180. J. M. Brittain, P. B. D. de la Mare, N. S. Isaacs, and P. D. McIntyre, *J. Chem. Soc., Perkin Trans. 2*, 1979, 933; J. M. Brittain, P. B. D. de la Mare, and P. A. Newman, *J. Chem. Soc. Perkin Trans. 2*, 1981, 32.
181. R. M. Kellogg, A. P. Schaap, E. T. Harper, and H. Wynberg, *J. Org. Chem.*, 1968, **33**, 2902.
182. F. Effenberger and P. Menzel, *Angew. Chem., Int. Ed. Engl.*, 1971, **10**, 493.
183. M. Godfrey, P. Goodman, and P. H. Gore, *Tetrahedron*, 1976, **32**, 841.
184. G. A. Olah, W. J. Tolgyesi, and R. E. A. Dear, *J. Org. Chem.*, 1962, **27**, 3441, 2449.
185. F. R. S. Clark, R. O. C. Norman, and C. B. Thomas, *J. Chem. Soc., Perkin Trans. 2*, 1975, 121.

CHAPTER 5

Metallation

The replacement of aromatic hydrogens by metals constitutes an important class of electrophilic aromatic substitutions because of the synthetic usefulness of the organometallic intermediates. With the exception of mercuriation, these reactions are less well recognized as electrophilic substitutions, and in some cases detailed evidence for the electrophilic nature of the reactions has yet to be obtained. However, all of the reported metallations are included here in order to focus attention on the need for this additional work.

5.1 LITHIATION

This reaction is usually carried out with n-butyllithium (eq. 5.1) and is closely related to base-catalysed hydrogen exchange, the initial step in each reaction being removal of a proton from the aromatic by the base to give the aromatic anion, i.e. the mechanism is B-S_E1 (Section 2.1). In lithiation the greater stability of the product butane and its removal from solution means that the reaction here is irreversible, and the aryllithium is obtained as an intermediate. The electrophilic process is thus substitution of Li^+ into the aromatic anion which will be very fast, and may be synchronous with the first step as depicted in **1**.[1] As in the case of base-catalysed hydrogen exchange, lithiation differs from the conventional electrophilic substitutions (S_E2) in that the most acidic hydrogen is substituted (the π-cloud being largely uninvolved). This may take place at the site conventionally substituted, an example being lithiation of naphthothiophenes which occurs at the 2-position, as does nitration etc.[2,3]

$$\text{ArH} + \text{n-BuLi} \longrightarrow \text{ArLi} + \text{butane} \qquad (5.1)$$

$H^{\delta+}$ $R^{\delta-}$ $Li^{\delta+}$

(**1**)

Table 5.1 Yields (%)[a] in lithiation of substituted benzenes ArX

X	Position of lithium					Ref.
	2	3	4	5	6	
OMe	100					4
OEt	99	0.5	0.5			4
NMe_2	71	19				5
1,2-$(NMe)_2$		50–60				6
CH_2NMe_2	80–100					7–9
$(CH_2)_2NMe_2$	Mainly					10, 11
1,3-$(OMe)_2$	97		3			4
1-OMe-3-Me	40				60	4
Ph	8	58	34			4
CF_3	73–80	20–27				4, 12
1,3-$(CF_3)_2$	40		60			13
$CONR_2$	70–99					14
1-$CONR_2$-3-Me	(1)[b]				(2)[b]	14
SO_2NR_2	80					15
SO_2Ar	42–67					16
1-SO_3Li-4-Me(H)	100					17
1,2-$(CH_2)_2$		19–30				18

[a] In some cases the yields are based only on the amount of derivative recovered, so that the extent of lithiation at the site may be greater.
[b] Ratio of observed products.

However, for the most part lithiation gives the isomer (usually *ortho* to the substituent) which is not that principally obtained in other substitutions, and this is very useful synthetically (Table 5.1). The preference for substitution *ortho* to an electron-withdrawing group is also found in ferrocene, the 1-carboxamide substituent giving lithiation at the 2-position.[19]

There are substantial indications that lithium coordinates with substituents able to donate lone pairs, e.g. as in **2** (in which the Li—*t*-Bu bond may already be completely or partially broken). This may account for the exceptionally high *ortho* substitution produced by alkoxy and amino substituents, etc., and the preference for 8-substitution in 1-methoxy-[20] and 1-aminonaphthalene,[21] the

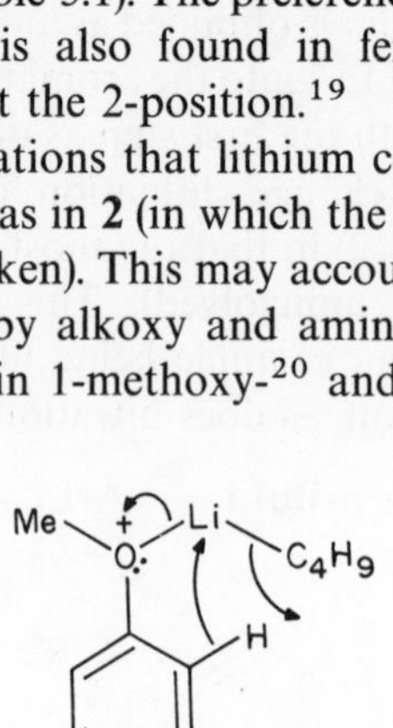

(2)

latter providing a useful route to 1,8-disubstituted naphthalenes; in the case of 1-methoxynaphthalene, 2-lithiation also occurs, the proportion of this increasing with time.[4] Coordination with the NMe_2 and OMe groups also explains the preference for 2-substitution in 1,1′-di(*N*,*N*-dimethylamino)ferrocene (disubstitution also occurs here, of course), *N*,*N*-dimethylaminoferrocene,[11] methoxyferrocene,[22] and *N*,*N*-dimethylaminoethylferrocene.[11] Coordination is absent in alkylbenzenes and the partial rate factors for isopropylbenzene,[1] viz. $f_o = 0.10$, $f_m = 0.38$, $f_p = 0.43$ reflect mainly the inductive effect of the alkyl group.

Because lithium does not have a large directing effect, but will activate through its $+I$ effect, polylithiation is feasible and has been observed with anthracene, biphenyl, fluorene, and indene, using n-butyllithium and N^4-tetramethylethylenediamine in hydrocarbon solvents, up to ten lithium atoms being introduced.[23]

The size of the alkyllithium oligomer varies according to the solvent and an order of steric requirement is n-BuLi/hydrocarbon (hexamer) > *t*-BuLi/hydrocarbon > BuLi/ether (tetramer) > n-BuLi (N^4-TED).[24] This is shown by the percentages of lithiation of the 6-position of 2-methylanisole which are 33% (n-BuLi/hydrocarbon), 42% (*t*-BuLi/hydrocarbon) and 75% (n-BuLi/N^4-TED),[25] the corresponding percentages for 2-*tert*-butylanisole under the former and latter conditions being 7.5 and 30%, respectively.[25, 26] Likewise, the relative percentages of 2- and 8-lithiation of 1-methoxynaphthalene were found to be 73:27 (n-BuLi/ether–hexane); 3:97 (*t*-BuLi/hydrocarbon); 99:1 (n-BuLi/N^4-TED)[27].

5.2 AURATION

The reaction of anhydrous gold(III) chloride with benzene produces a red solution which decomposes fairly rapidly to give gold(I) chloride and chlorobenzene in 98% yield; if a deficiency of benzene is used then 1,2,4,5-tetrachlorobenzene is the product. The rates of auration of a range of aromatics gave a reactivity order typical of an electrophilic substitution.[28] The intermediate arylauric dichlorides, formed according to eq. 5.2, are thermally unstable but may be isolated.[28,29] The *ortho*:*para* ratios, viz. 0.23 (toluene), 0.12 (anisole), and 0.11 (chlorobenzene), are consistent with a bulky electrophile.[30]

$$ArH + AuCl_3 \rightleftharpoons ArAuCl_2 + HCl \tag{5.2}$$

5.3 MERCURIATION

The group HgX may be introduced into the aromatic nucleus by reaction with a mercuric compound, usually the acetate or chloride, e.g. eq. 5.3. In a non-polar medium the distribution of reaction products is comparatively random, and the reaction is at least in part homolytic. In a polar medium or in the presence of catalysts such as perchloric acid, mercuriation takes place more readily giving the isomer distribution expected for an electrophilic process. The reaction has been studied under both homogeneous and heterogeneous conditions.

$$\mathrm{ArH + Hg(OAc)_2 \rightleftharpoons ArHgOAc + HOAc} \tag{5.3}$$

5.3.1 Mechanism

Homogeneous mercuriation of benzene with mercury(II) acetate in acetic acid, catalysed by mineral acids and by salts, is first order in both benzene and mercury(II) ion, indicating that the attacking entity is mercury(II) ion complexed with the anion of the catalysing species.[31] The catalysing effect of perchloric acid was attributed to the formation of a more polar and hence more reactive electrophile according to eq. 5.4,[32] A more comprehensive kinetic study of

$$\mathrm{Hg(OAc)_2 + HClO_4 \rightleftharpoons Hg(OAC)^+ClO_4^- + HOAc} \tag{5.4}$$

mercuriation in acetic acid by mercury(II) acetate and by mercury(II) perchlorate revealed distinct differences in behaviour,[33] including the following:

(1) The rate coefficient decreased on increasing the initial concentration of mercury(II) acetate, with mercury(II) perchlorate producing the opposite effect; these effects were more pronounced the lower the concentration of perchloric acid, and remained even in its absence with the perchlorate.

(2) Water retards the reaction, and more so with mercury(II) perchlorate in the absence of added perchloric acid than for mercury(II) acetate in the presence of perchloric acid.

(3) Added sodium perchlorate accelerated mercuriation with mercury(II) acetate in anhydrous acetic acid, and retarded mercuriation with mercury(II) perchlorate in acetic acid containing 0.2 mol l^{-1} water.

These observations arise because in acetic acid, mercury(II) perchlorate is partially converted to mercury(II) acetate and perchloric acid, the first step being eq. 5.5 and the second step being the reverse of eq. 5.4; the perchloric acid generated catalyses the reaction. These equilibria also accounted for the departure from second-order behaviour in media of low acidity. The retardation by water is due to its protonation by perchloric acid, which is then unavailable for catalysis, and the acceleration by sodium perchlorate of mercury(II) acetate mercuriation is due to formation of perchloric acid according to eq. 5.6.

$$\mathrm{Hg(ClO_4)_2 + HOAc \rightleftharpoons Hg(OAc)^+ClO_4^- + HClO_4} \tag{5.5}$$

$$\mathrm{NaClO_4 + HOAc \rightleftharpoons NaOAc + HClO_4} \tag{5.6}$$

Homogeneous mercuriation with mercury(II) trifluoroacetate in TFA gives pure second-order kinetics; there is no reaction of the mercury(II) salt with the solvent and little isomerization of the reaction products.[34] The reaction takes place much more readily (6.9×10^5 for benzene mercuriation at 25 °C)[34] because electron withdrawal from mercury by the trifluoroacetate ligand creates a greater positive charge on the electrophile. For the same reason, mercury trifluoromethanesulphonate is even more reactive.[35] Thus very unreactive aromatics can be substituted, e.g. tetrafluorobenzene with mercury(II) trifluoracetate[36] and pentafluorobenzene with mercury(II) trifluoromethanesulphonate.[35]

Heterogeneous mercuriation of benzene by mercury perchlorate in aqueous perchloric acid is strongly accelerated by perchloric acid and its neutral salts.[31,37] The rate of mercuriation here is an inverse function of the activity of water in the medium, so the acceleration by added salts is due to removal of water from the mercury(II) ion, resulting in a more reactive species.[38]

The carbon–mercury bond is weak, which has a number of consequences. One is the reversibility of mercuriation, which involves protiodemercuriation and thus occurs in the presence of acids. For example, bis(acetoxymercuri)benzenes are demercuriated by hydrochloric acid,[39] and arylmercury(II) chlorides can be similarly demercuriated.[40] Mercuriation of benzene with mercury(II) nitrate in nitric acid is reversible,[37] as also is mercuriation of phenylmercury(II) perchlorate with mercury(II) perchlorate in perchloric acid.[41] A second consequence is that large primary kinetic isotope effects are observed in mercuriation, values of k_H/k_D between 4.7 and 6.8 having been obtained, depending on the conditions [mercury(II) acetate in acetic acid,[42] mercury(II) perchlorate in perchloric acid,[38] mercury(II) trifluoroacetate in TFA].[43] Thus, in the simplified S_E2 scheme (eq. 2.4) k_{-1} becomes larger than k_2 and an isotope effect results. With mercury(II) trifluoroacetate, evidence was obtained that a π-complex is initially formed (this fast step has only very small isotope effects) and the isotope effect derived for the slow step was greater for toluene (7.5) than for benzene (3.7).[43] This contradicts the usual assumption that isotope effects should be greater for less reactive compounds since these will have a 'later' transition state. Uncatalysed mercuriation (of benzene) gives a *smaller* isotope effect (3.2 at 25 °C),[44] even though under these conditions the electrophile should be less reactive (and indeed the rates are lower), giving rise to a later transition state. This discrepancy has been attributed to a reduction in the concentration of the base responsible for the removal of the proton in the second step of the reaction, owing to the presence of the acid.[34]

The reversibility of mercuriation gives rise to isomerization. For example, the isomer distribution of tolylmercury(II) acetates formed from the perchloric acid-catalysed reaction of toluene with mercury(II) acetate was initially *ortho* 21.0, *meta* 9.5 and *para* 69.5%, but became 36, 31, and 33%, respectively, after a reaction time of 28 days.[45] This time-variable distribution was attributed to preferential demercuriation of the *para*-mercuriated derivative, since the observed relative reactivities of toluene to benzene also decreased with time.[45] However, this is difficult to reconcile with the expectation that protiodemercuriation will be sterically accelerated so that the rate of loss of the acetoxymercuri group from the *ortho* position should be comparable to that from the *para* position. It is possible that dimercuriation occurs followed by demercuriation, since the formation of *meta*- and *para*-dimercuriated benzenes from benzene and mercury(II) perchlorate involves demercuriation of the intermediate 1,2,4-tri(perchloratomercuri)benzene.[41] In this work the ready demercuriation of *ortho* bismercurials was shown by the fact that phenylmercury(II) acetate and mercury(II) acetate gave no *ortho*-disubstituted derivative if perchloric acid was present, yet a 65% yield was obtained in its absence.

Analysis of the products of aromatic mercuriation is rendered easier by replacing the HgX group by bromine (bromodemercuriation),[39] which occurs without isomerization for various substituted phenylmercury(II) compounds.[40]

5.3.2 Substituent Effects

Partial rate factors for uncatalysed mercuriation of some monosubstituted benzenes with mercury(II) acetate in acetic acid and with mercury(II) trifluoroacetate in trifluoroacetic acid, each at 25 °C are given in Table 5.2.[46,47] The main feature of these results are as follows:

(1) The logarithms of the partial rate factors correlate with σ^+ values (Section 11.1) giving values of ρ of -4.0 [mercury(II) acetate] and -5.7

Table 5.2 Partial rate factors for mercuriation of monosubstituted benzenes at 25 °C

	$Hg(OAc)_2$ in HOAc[a]			$Hg(OCOCF_3)_2$ in TFA		
Substituent	f_o	f_m	f_p	f_o	f_m	f_p
OMe	188		2310			
NHAc			277			
OPh			194			
Me	5.77	2.26	23.2	3.62	2.55	46.9
Et				1.97	2.37	42.8
i-Pr				0.533	2.18	39.0
t-Bu		3.41	17.2		1.97	32.2
Ph	0.081	0.77	6.42[b]			
F	0.60	0.038	2.92	0.0983	0.00687	1.51
Cl	0.075	0.054	0.34	0.0168	0.00820	0.232
Br	0.072	0.060	0.275	0.0126	0.00954	0.194

[a] For the less reactive compounds, rates and isomer distributions at 25 °C were extrapolated from data obtained at higher temperatures. For some of these compounds side-reactions of mercury(II) acetate with the solvent accounted for over 80% of reaction, and a competitive method had to be employed.
[b] For fluorene $f_2 = 122$.

Table 5.3 Partial rate factors for the catalysed and uncatalysed mercuriation of toluene at 50 °C

Conditions	f_o	f_m	f_p
$Hg(OAc)_2$ in HOAc + $HClO_4$	4.20	2.41	28.8
$Hg(OAc)_2$ in HOAc	4.61	1.98	16.8

[mercury(II) trifluoroacetate] (more recent work has indicated that the latter value may be as high as -6.4).[43] Thus the more reactive electrophile is more selective and the same feature is evident in perchloric acid-catalysed mercuriation which requires a higher ρ factor than uncatalysed mercuriation, as may be seen from the data in Table 5.3.[45] These contradictions of the *Reactivity–Selectivity Principle* evidently arise because the second step of mercuriation is rate-limiting. The rate-limiting transition state is thus beyond the Wheland intermediate but will be nearer to this the more reactive the electrophile. A greater charge to be stabilized by the substituent therefore resides on the aromatic ring.

(2) The *ortho*:*para* ratios are small, indicating that both reagents have large steric requirements, mercuriation with mercury(II) trifluoroacetate being the most hindered, as expected. Catalysed mercuriation appears to have a higher steric requirement than uncatalysed mercuriation (Table 5.3).

(3) Activation by the *p*-F substituent in this reaction of low selectivity is surprising, since there should be little call on conjugative ($+M$) effects to outweigh the $-I$ effect here. The order of deactivation by the *meta*-halogen substituents, Cl < F, indicates that there is little or no secondary relay of conjugative effects to the *meta* position, as is usual when the reagent is of low selectivity [e.g. in nitration, Section 7.4.3(viii)] and contrary to the order obtained when the reagent is of high selectivity (e.g. in hydrogen exchange, Section 3.1.2.5).

(4) A further anomaly is that the *p*-Me > *p*-*t*-Bu activation order applies even in the poorly solvating trifluoroacetic acid; no other reaction gives this in this medium (see Section 1.4.3). If the experimental result is correct (cf. thalliation, Section 5.5), it implies that the reaction requires exceptional solvation of the transition state.

(5) Although mercuriation is sterically hindered, the *Additivity Principle* holds remarkably well (Table 5.4).

Relative rates of perchloric acid-catalysed mercuriation of substituted benzenes at 25 °C have also been determined (Table 5.5)[33,42,45] and these confirm that the selectivity of the reaction is higher than for the uncatalysed reaction. Perchloric acid-catalysed mercuriation of toluene and toluene-$\alpha\alpha\alpha$-d^3 gave a negligible kinetic isotope effect, $k_H/k_D = 1.00 \pm 0.03$.[48]

Azobenzene mercuriates in the 2,2′- and 2,6-positions, but 2-methylazobenzene mercuriates in the 6-position whilst 2-iodoazobenzene mercuriates in the 2′-

Table 5.4 Observed and calculated relative rates in uncatalysed mercuriation of polymethylbenzenes at 50 °C

Positions of methyl groups	$k_{rel.}$ (observed)	$k_{rel.}$ (calculated)
None	1	—
1-	5.0	—
1,4-	8.2	9.1
1,2-	16.0	14.1
1,3-	34.5	30.1
1,2,4-	49	35.5
1,2,3-	68	62
1,3,5-	209	178
1,2,4,5-	30.0	27.8
1,2,3,4-	126	101
1,2,3,5-	257	235
1,2,3,4,5-	224	233

Table 5.5 Relative rates for catalysed mercuriation of substituted benzenes at 25 °C

Substituent	$k_{rel.}$	Substituent	$k_{rel.}$	Substituent	$k_{rel.}$
H	1	1,2–Me_2	32	NO_2	0.00041
Me	9.1	1,3-Me_2	90	Ph	4.2
Et	7.7	1,4-Me_2	12.6	C_4H_4[a]	12.3
i-Pr	7.1	F	0.38	HgOAc	0.69
t-Bu	6.4	Cl	0.053		
n-Pr	7.3	Br	0.047		
i-Bu	7.8	I	0.045		
s-Bu	6.8				

[a] Naphthalene.

and 6-positions; these results indicate an *ortho* coordination mechanism.[49] Acylferrocenes give the 1,2- and 1,1′-derivatives,[50] the former being facilitated relative to *ortho* disubstitution in benzene because of the larger distance between *ortho* positions. Mercuriation of ferrocene with $Hg(OAc)_2$–TFA in a 1:1:1 molar ratio gives a high yield (86%) of the monoacetoxymercuri product, but the 1,1′-disubstituted product is obtained if the molar ratio is changed to 1:2:2.[51] Triptycene is more reactive than *o*-xylene towards mercury(II) trifluoroacetate,[43] paralleling results for detritiation (Section 3.1.2.4); at 25 °C the rates relative to benzene are 15.4 and 39.5, respectively.

Lastly, acetate ions reduce the rate of mercuriation of anisole and the mercuridesilylation of *p*-tolyltrimethylsilane; consequently, the electrophile is believed to be a mixture of unionized mercury(II) acetate and acetoxymercury ion.[52] The silane/anisole reactivity ratios for reaction with these electrophiles were 2 and 15, respectively, so this appears to be another example of the breakdown of the *Reactivity–Selectivity Principle*, since the ratio should be higher with the less reactive electrophile. The reason may be that the first step of the reaction is probably rate-limiting for mercuridesilylation, whereas it is the second step in mercuriation.

5.4 BORONATION

The reaction between boron tribromide and benzene in the presence of aluminium tribromide gives phenylboron dichloride (eq. 5.7), and the electrophilic nature of the reaction was indicated by boronation of the more reactive diphenyl ether in the absence of the catalyst.[53]

$$PhH + BCl_3 \xrightarrow{AlCl_3} PhBCl_2 + HCl \qquad (5.7)$$

The reaction takes place with a range of aromatics and will occur (under pressure) with aluminium as catalyst, in which case hydrogen is produced; toluene gave a *para*:*meta* ratio of 4.6.[54]

5.5 THALLIATION

Thalliation is sometimes used as an alternative to mercuriation for introducing substituents into aromatic rings;[55] subsequent cleavage with potassium iodide, for example, gives the iodoaromatic. The main advantage of the reaction (which occurs 200–400 times less readily than mercuriation) is the relative absence of disubstitution (because it is more selective), although diarylthallium(III) compounds are readily formed.[56] Against this must be set the high cost of the reagents and the fact that they are very poisonous.

Thalliation by thallium triacetate is accelerated by the presence of perchloric and sulphuric acids and gives complex kinetics owing (as in mercuriation) to the establishment of equilibria between the various electrophiles, the reactivity order of which is $TlX_3 > TlOAcX_2 > Tl(OAc)_2X > Tl(OAc)_3$, ($X = ClO_4$ or HSO_4).[56,57] The reaction gives a kinetic isotope effect $k_H/k_D = 2.4$.[57]

As in the case of mercuriation, thalliation with the trifluoroacetate is faster (and most studies have been carried out under this condition); similarly, the trifluoromethanesulphonate is more reactive still and with this 1,3,5-trifluorobenzene can be thalliated.[58] With thallium tris(trifluoroacetate) substitution occurs mainly at the *para* position, although high yields of *meta* derivatives can be obtained under conditions of thermodynamic control; with some compounds, e.g. anisole, there is evidence of an *ortho*-coordination mechanism.[56,59] Thalliation of anisole is unusual in that at $-25\,^{\circ}C$ the *para* product is mainly obtained whereas at room temperature the *ortho* derivative is almost exclusively obtained, which indicates that the *ortho* derivative is thermodynamically the more stable, again indicative of an *ortho*-coordination mechanism (**3**).[60] Toluene gave the partial rate factors $f_o = 12.7$, $f_m = 4.5$, and $f_p = 226$, showing the reaction to be more selective than mercuriation, and $\rho \approx -7.0$;[61] toluene and *tert*-butylbenzene gave the expected *t*-Bu > Me activation order in this medium (trifluoroacetic acid),[62] in contrast to mercuriation, which behaves abnormally (Section 5.3).

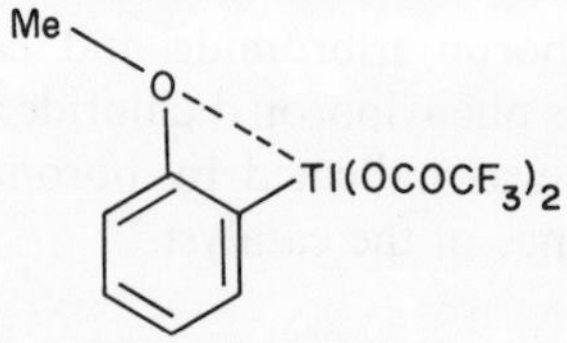

Me
O
$Tl(OCOCF_3)_2$

(**3**)

Thallium(I) impurities in the tris(trifluoroacetate) render kinetic data unreliable, but studies have been carried out with the triacetate sesquihydrate. These showed the rapid formation of a π-complex as in mercuriation, and a smaller kinetic isotope effect than for mercuriation, although that for the second step was greater. Use of σ^+ values gave a ρ factor which was larger (-8.3) than that for thalliation with the trifluoroacetate; a better correlation was obtained using the Yukawa–Tsuno method (Section 11.1.3), giving $\rho = -8.6$ and $r = 0.6$[63]

Chlorination in the presence of the reagents $TlCl_3 \cdot 4H_2O$, $Tl(OAc)_3$, TlCl, and TlOAc gave toluene:benzene rate ratios which increased along the series (from 43 to 133), whilst the *ortho*:*para* ratio increased from 1.1 and 2.5. The reaction appears to involve cleavage by chlorine of the arylthallium intermediate, whilst thalliation with thallium(I) is more selective and less hindered than the reaction with thallium(III).[64]

5.6 SILYLATION

The reaction between diphenyldichlorosilane with ferrocene in the presence of aluminium chloride gives, after hydrolytic work-up, the silanol $FerSi(OH)Ph_2$, in a reaction believed to be an electrophilic substitution.[65]

5.7 PLUMBYLATION

Reaction of aromatic ethers and lead tetraacetate in benzene or acetic acid at elevated temperatures leads to the *para*-aryllead triacetate (eq. 5.8).[66] As in mercuriation and thalliation, the reaction rate can be increased by using ligands from stronger carboxylic acids, e.g. halobenzenes may be plumbylated with lead tetraacetate in TFA, giving the *para*-halogenophenyllead tris(trifluoroacetates) in 3–34% yields.[67] Plumbylation of 1,3-dimethoxybenzene with lead tetraacetate gives a 78% yield of 2,4-dimethoxyphenyllead triacetate,[66] yet in the presence of monohalogenoacetic acids the yield becomes 99% and the reactions are faster.[68] Likewise, ligand exchange between lead tetraacetate and trichloroacetic acid results in rapid plumbylation of halogeno ethers by the resultant lead tetrakis(trichloroacetate).[69] The products of plumbylation are readily hydrolysed to oligomeric plumboxanes $[(ArPbX_2)_2O]_n$, but these may be converted back to the aryllead tricarboxylates by treatment with acid.[70]

$$ArH + Pb(OAc)_4 \rightleftharpoons ArPb(OAc)_3 + HOAc \qquad (5.8)$$

Plumbylation is more selective than either mercuriation or thalliation and the *ortho*, *meta*, and *para* partial rate factors for toluene are 9.6, 3.2, and 331, respectively, in dichloroacetic acid, and 61.2, 10.8, and 540 in TFA.[71] The low *ortho*:*para* ratios are consistent with the very bulky nature of the electrophile and the respective sizes of the ligands. It follows that protiodeplumbylation of *ortho*-substituted aryllead compounds will be rapid owing to steric acceleration.

5.8 ANTIMONATION

The reaction between aromatics and antimony pentafluoride gives stibines such as $Ar_2SbF_3.H_2O$ and Ar_3SbF_2. The high substrate selectivity and the predominance of *para* substitution are consistent with the proposed mechanism shown in Scheme 5.1.[72]

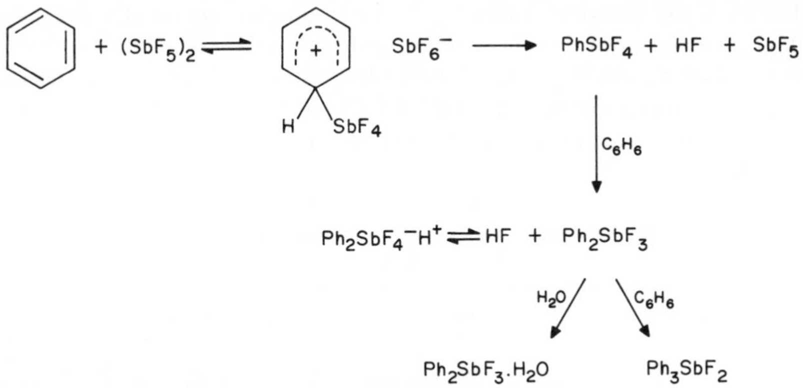

$$C_6H_6 + (SbF_5)_2 \rightleftharpoons [C_6H_6(H)(SbF_4)]^+ \; SbF_6^- \longrightarrow PhSbF_4 + HF + SbF_5$$

$$PhSbF_4 \xrightarrow{C_6H_6} Ph_2SbF_4^-H^+ \rightleftharpoons HF + Ph_2SbF_3$$

$$Ph_2SbF_3 \xrightarrow{H_2O} Ph_2SbF_3.H_2O \qquad Ph_2SbF_3 \xrightarrow{C_6H_6} Ph_3SbF_2$$

Scheme 5.1

5.9 TELLURIATION

The reaction of tellurium tetrachloride with activated aromatics, e.g. anisole, gives mainly the *p*-aryltellurium trichloride together with some bis(aryl)tellurium dichlorides.[73] Less reactive aromatics require catalysis by Lewis acids and an electrophilic substitution is indicated.

5.10 RHODIATION

Reaction of (octaethylporphyrinate)rhodium(III) chloride with aromatics in the presence of either silver perchlorate or silver tetrafluoroborate gives the corresponding phenyl–rhodium(III) complex. Anisole, toluene, and chlorobenzene each substitute exclusively *para*, and methyl benzoate gives a *meta–para* mixture in the ratio 92:8. These results confirm the bulky nature of the electrophile, $[(OEP)Rh^{III}]^+$ generated from anion exchange of the chloride with the silver salts.[74] The slow step is reaction of the electrophile with the aromatic; the data correlate with the Yukawa–Tsuno equation (Section 11.1.3), with $\rho = -4.5$, $r = 0.3$ (cf. ref. 74).

5.11 PALLADIATION

Fluorinated azobenzenes react with π-cyclopentadienylpalladium to give palladiation *ortho* to nitrogen indicating that, as in mercuriation, coordination to nitrogen is the initial step.[75]

In the presence of added oxidant, palladium diacetate reacts with benzene to give phenylpalladium acetate and thence phenyl acetate.[76,77] The isomer distributions for the reaction of the diacetate with some aromatics in acetic acid at 90 °C are shown in Table 5.6.[76] The results for *tert*-butylbenzene show the high steric hindrance to the reaction and the partial rate factors for toluene were $f_o = 1.09$, $f_m = 1.51$, $f_p = 3.19$, so the reaction is evidently very unselective. The reaction gives a kinetic isotope effect k_H/k_D of 4.5 for both the acetoxylation and the arylation which occurs in the absence of oxidant, indicating a common intermediate.[76]

Table 5.6 Isomer yields (%) in palladiation of aromatics ArX

X	*ortho*	*meta*	*para*
Cl	28	23	35
Me	26	36	38
t-Bu	0	58	42

REFERENCES

1. D. Bryce-Smith, *J. Chem. Soc.*, 1954, 1079.
2. K. Clarke, G. Rawson, and R. M. Scrowston, *J. Chem. Soc. A*, 1969, 537.
3. K. Clarke, D. N. Gregory, and R. M. Scrowston, *J. Chem. Soc., Perkin Trans. 2*, 1973, 2956.
4. D. A. Shirley, J. R. Johnson, and J. P. Hendrix, *J. Organomet. Chem.*, 1968, **11**, 209.
5. A. R. Lepley, W. A. Khan, A. B. Giumanini, and A. G. Giumanini, *J. Org. Chem.*, 1979, **42**, 237.
6. A. Cheminat, *J. Org. Chem.*, 1979, **42**, 237.
7. F. N. Jones and C. R. Hauser, *J. Org. Chem.*, 1962, **27**, 701.
8. F. N. Jones, H. F. Zinn, and C. R. Hauser, *J. Org. Chem.*, 1963, **28**, 663.
9. F. N. Jones, R. L. Vaulx, and C. R. Hauser, *J. Org. Chem.*, 1963, **28**, 3461.
10. N. S. Narasimbana and A. C. Ranade, *Tetrahedron Lett.*, 1966, 603.
11. D. W. Slocum, T. R. Engelmann, and C. A. Jennings, *Aust. J. Chem.*, 1968, **21**, 2319.
12. J. D. Roberts and D. W. Curtin, *J. Am. Chem. Soc.*, 1946, **68**, 1658.
13. P. Aeberli and W. H. Houlihan, *J. Organomet. Chem.*, 1974, **67**, 321.
14. P. Beak and W. A. Brown, *J. Org. Chem.*, 1977. **42**, 1823; 1979, **44**, 4463.
15. H. Watanabe, R. A. Schwartz, C. R. Hauser, J. Lewis, and D. W. Slocum, *Can. J. Chem.*, 1969, **47**, 1543.
16. H. Gilman and D. L. Esmay, *J. Am. Chem. Soc.*, 1953, **75**, 278; V. N. Drozd and L. A. Nikonova, *J. Org. Chem. USSR*, 1959, **5**, 313.
17. G. D. Figuly and J. C. Martin, *J. Org. Chem.*, 1980, **45**, 3728.
18. C. Eaborn, A. A. Najam, and D. R. M. Walton, *J. Chem. Soc., Chem. Commun.*, 1972, 840.
19. D. W. Slocum and F. E. Stonemark, *J. Org. Chem.*, 1973, **38**, 1675.
20. K. A. Barnes and L. J. Nehmsmann, *J. Org. Chem.*, 1962, **27**, 1939.
21. C. Eaborn, P. Golborn, and R. Taylor, *J. Organomet. Chem.*, 1967, **10**, 171.
22. D. W. Slocum, B. P. Koonsvitsky, and C. R. Ernst, *J. Organomet. Chem.*, 1972, **38**, 125.
23. A. F. Halason, *J. Organomet. Chem.*, 1971, **31**, 369.
24. D. A. Shirley, T. E. Harman, and C. F. Cheng, *J. Organomet. Chem.*, 1974, **69**, 327.
25. T. E. Harman and D. A. Shirley, *J. Org. Chem.*, 1974, **39**, 3164.
26. D. W. Slocum and B. P. Koonvitsky, *J. Org. Chem.*, 1973, **38**, 1675.
27. D. A. Shirley and C. F. Cheng, *J. Organomet. Chem.*, 1969, **20**, 251.
28. M. S. Kharasch and H. S. Isbell, *J. Am. Chem. Soc.*, 1931, **53**, 3053.
29. P. W. J. de Graaf, J. Boersma, and G. J. M. van der Kerk, *J. Organomet. Chem.*, 1976, **105**, 397.
30. R. O. C. Norman, W. J. E. Parr, and C. B. Thomas, *J. Chem. Soc., Perkin Trans. 1*, 1976, 1983.
31. R. M. Schramm, W. Klapproth, and F. H. Westheimer, *J. Phys. Chem.*, 1951, **55**, 843.
32. H. C. Brown and C. W. McGary, *J. Am. Chem. Soc.*, 1955, **77**, 2306.
33. A. J. Kresge, M. Dubeck, and H. C. Brown, *J. Org. Chem.*, 1967, **32**, 745.
34. H. C. Brown and R. A. Wirkkala, *J. Am. Chem. Soc.*, 1966, **88**, 1447, 1453, 1456.
35. G. B. Deacon and D. Tunaley, *J. Organomet. Chem.*, 1978, **156**, 403.
36. H. B. Albrecht and G. B. Deacon, *J. Organomet. Chem.*, 1973, **57**, 77.
37. F. H. Westheimer, E. Segel, and R. Schramm, *J. Am. Chem. Soc.*, 1947, **69**, 773.
38. C. Perrin and F. H. Westheimer, *J. Am. Chem. Soc.*, 1963, **85**, 2773.
39. O. Dimroth, *Chem. Ber.*, 1899, **32**, 748; 1902, **35**, 2853.
40. M. S. Kharasch and L. Chalkley, *J. Am. Chem. Soc.*, 1921, **43**, 607.
41. M. Malaiyandi, H. Sawatzky, and G. F. Wright, *Can. J. Chem.*, 1961, **39**, 1827.
42. A. J. Kresge and J. F. Brennan, *Proc. Chem. Soc.*, 1963, 215; *J. Org. Chem.*, 1967, **32**, 752.

43. C. W. Fung, M. Khorramdel-Vahed, R. J. Ranson, and R. M. G. Roberts, *J. Chem. Soc., Perkin Trans. 2*, 1980, 267.
44. C. W. McGary and G. Goldman, *PhD Theses*, Purdue University, 1955, 1961.
45. H. C. Brown and C. W. McGary, *J. Am. Chem. Soc.*, 1955, **77**, 2300, 2306.
46. H. C. Brown and G. Goldman, *J. Am. Chem. Soc.*, 1962, **84**, 1650.
47. H. C. Brown and R. A. Wirkkala, *J. Am. Chem. Soc.*, 1966, **88**, 1447, 1453, 1456.
48. C. G. Swain, T. E. C. Knee, and A. J. Kresge, *J. Am. Chem. Soc.*, 1957, **79**, 505.
49. P. V. Roling, J. L. Dill, and M. D. Rausch, *J. Organomet. Chem.*, 1974, **69**, C33; P. V. Roling, D. D. Kurt, J. L. Dill, S. Hall, and C. Hollstrom, *J. Organomet. Chem.*, 1976, **116**, 39.
50. R. F. Kovar and M. D. Rausch, *J. Organomet. Chem.*, 1972, **35**, 351.
51. V. I. Boev and A. V. Dombrovski, *J. Gen. Chem. USSR*, 1985, **55**, 790.
52. J. R. Chipperfield, G. D. France, and D. E. Webster, *J. Chem. Soc., Perkin Trans. 2*, 1972, 405.
53. Z. J. Bujwid, W. Gerrard, and M. F. Lappert, *Chem. Ind. (London)*, 1959, 1091.
54. E. L. Mutterties, *J. Am. Chem. Soc.*, 1959, **81**, 2597; 1960, **82**, 4163.
55. V. P. Glushkova and K. A. Kocheshkov, *Dokl. Akad. Nauk SSSR*, 1957, **116**, 233; *Izv. Akad. Nauk SSSR*, 1957, 1186, 1193.
56. P. M. Henry, *J. Org. Chem.*, 1970, **35**, 3083.
57. J. M. Briody and R. A. Moore, *Chem. Ind.* (*London*), 1970, 803; *J. Chem. Soc., Perkin Trans. 2*, 1972, 179.
58. G. B. Deacon and D. Tunaley, *Aust. J. Chem.*, 1979, **32**, 737.
59. A. McKillop, J. D. Hunt, M. Z. Zelesko, J. S. Fowler, E. C. Taylor, G. McGillivray, and F. Kienzle, *J. Am. Chem. Soc.*, 1971, **73**, 4841; E. C. Taylor, F. Kienzle, R. L. Robey, A. McKillop, and J. D. Hunt, *J. Am. Chem. Soc.*, 4845.
60. G. B. Deacon, R. N. M. Smith, and D. Tunaley, *J. Organomet. Chem.*, 1976, **114**, C1.
61. P. W. Kwok, L. M. Stock, and T. L. Wright, *J. Org. Chem.*, 1979, **44**, 2309.
62. R. M. G. Roberts, *Tetrahedron*, 1980, **36**, 3281.
63. S. F. Al-Azzawi and R. M. G. Roberts, *J. Chem. Soc., Perkin Trans. 2*, 1982, 677.
64. S. Uemura, O. Sasaki, and M. Okano, *J. Chem. Soc., Perkin Trans. 1*, 1972, 2268.
65. G. P. Sollott and W. R. Peterson, *J. Am. Chem. Soc.*, 1967, **89**, 5054.
66. G. W. K. Caviel and D. H. Soloman, *J. Chem. Soc.*, 1955, 1404; F. R. Preuss and I. Janshen, *Arch. Pharm.*, 1960, **293**, 933; D. R. Harvey and R. O. C. Norman, *J. Chem. Soc.*, 1964, 4860; L. C. Willemsens, D. de Vos, J. Spierenburg, and J. Wolters, *J. Organomet. Chem.*, 1972, **39**, C61.
67. D. de Vos, J. Wolters, and A. van der Gen, *Recl. Trav. Chim. Pays-Bas*, 1973, **92**, 701; H. C. Bell, J. R. Kalman, J. T. Pinhey, and S. Sternhell, *Tetrahedron Lett.*, 1974, 853.
68. D. de Vos, J. Spierenburg, and J. Wolters, *Recl. Trav. Chim. Pays-Bas*, 1972, **91**, 1465.
69. D. de Vos, F. E. H. Boschman, J. Wolters, and A. van der Gen, *Recl. Trav. Chim. Pays-Bas*, 1973, **92**, 467.
70. D. de Vos, W. A. A. van Barneveld, D. C. van Beelen, H. O. van der Kooi, J. Wolters and A. van der Gen, *Recl. Trav. Chim. Pays-Bas*, 1975, **94**, 97.
71. L. M. Stock and T. L. Wright, *J. Org. Chem.*, 1980, **45**, 4645.
72. G. A. Olah, P. Schilling, and I. M. Gross, *J. Am. Chem. Soc.*, 1974, **96**, 876.
73. J. Bergman, *Tetrahedron*, 1972, **28**, 3323.
74. Y. Aoyama, T. Yoshida, K. Sakurai, and H. Ogoshi, *Organometallics*, 1986, **5**, 168.
75. M. I. Bruce, B. I. Googall, and F. G. A. Stone, *J. Chem. Soc., Chem. Commun.*, 1973, 558.
76. L. M. Stock, K. Tse, L. J. Vornick, and S. A. Walstrum, *J. Org. Chem.*, 1981, **46**, 1757.
77. F. R. S. Clark, R. O. C. Norman, C. B. Thomas, and J. S. Willson, *J. Chem. Soc., Perkin Trans. 1*, 1974, 1289; K. Ichikawa, S. Uemura, and T. Okada, *J. Chem. Soc. Jpn.*, 1969, **90**, 212.

CHAPTER 6

Reactions Involving Electrophilic Carbon

Carbon bonded to an electronegative element or group X is positively polarized and therefore electrophilic, although not normally sufficiently so to react with benzenoid compounds. Its electrophilicity can, however, be increased by the addition of a species which can accept electrons from X and further polarize the C—X bond. The electron-accepting species may be a Lewis acid such as $AlCl_3$, $FeCl_3$, $SnCl_4$, and BF_3, the catalyst used depending on the type and reactivity of the electrophilic carbon compound and the aromatic. The carbon compounds which may be used include alkyl halides, alcohols, esters, and ethers. Electrophilic carbon may also be generated through protonation of an alkene or cyclopropyl groups. The products from these reactions are usually arylalkanes ArR (though diaryls ArAr′ have also been obtained from substitution by the

Table 6.1 Substitution of ArH by carbon electrophiles

Reaction	Reagents	Products
Aminoalkylation	e.g. $CHRO + R'_2NH$	$ArCHRNR'_2$
Hydroxyalkylation	e.g. $CHRO + H_2SO_4$	ArCHROH
Alkylthioalkylation	Dicylohexylcarbodiimide + Me_2SO	$ArCH_2SMe$
Haloalkylation	e.g. $CH_2O + HCl\text{–}ZnCl_2$	$ArCH_2Cl$
Gattermann–Koch reaction	$CO + HCl\text{–}AlCl_3$, CuCl	ArCHO
Vilsmeier–Haack reaction	$HCONMe_2 + POCl_3$	ArCHO
Gattermann reaction	$HCN + HCl\ (H_2O)$	ArCHO
Formylation	$Cl_2CHOMe + AlCl_3$	ArCHO
Hoesch reaction	RCN + HCl	ArCOR
Carboxylation	$COCl_2 + AlCl_3$	ArCOOH
Amidation	$RNCO + AlCl_3$	ArCONHR
Cyanation	$CNCCl_3 + HCl$	ArCN
Kolbe–Schmitt reaction[a]	CO_2 + pressure	$ArCOO^-$
Reimer–Tiemann reaction	$HCCl_3 + OH^-$	ArCHO

[a] Reaction with Ar^-.

phenyl cation). If the carbon compound is an acid, acid anhydride, amide, or acyl halide the product is an aryl ketone, ArCOR. These substitutions (which have been comprehensively reviewed)[1] are known as Friedel–Crafts reactions.[2]

There are some closely related processes involving electrophilic carbon and a Lewis or protic acid, and also a few which possess special features and therefore do not require either of these catalysts. These reactions, summarized in Table 6.1, are described in more detail below.

6.1 ALKYLATION

Alkylation is described by either eq. 6.1 or 6.2. The former involves catalysis by Lewis acids, the order of catalytic power being $AlBr_3 > AlCl_3 > GaCl_3 > FeCl_3 > SbCl_5 > TiCl_4 > ZnCl_2 > SnCl_4$, $ZrCl_4 > BCl_3$, BF_3, $SbCl_3$[3] (although an earlier report placed boron trifluoride midway along this series).[4] The catalyst chosen is ideally the one which is just active enough to effect alkylation at a convenient rate; the more reactive catalysts often bring about side-reactions which are described below [Section 6.1.1(ii)]. Tertiary compounds (halides, esters, alcohols, or ethers) generally require milder catalysts than primary compounds, and react mostly by the initial formation of a carbocation. By contrast, primary compounds form carbocations with extreme difficulty and reaction proceeds instead by displacement by the aromatic on a polarized alkyl compound–Lewis acid complex [Section 6.1.1 (iii)]. Alkylation by alkenes, or by cycloalkyl compounds with readily ruptured rings, may be brought about either with a strong proton acid or with a Lewis acid in the presence of a small quantity of water or a hydrogen halide [Sections 6.1.2(ii) and (iii)].

$$\mathrm{ArH + RX \xrightarrow{\text{Lewis acid (L.A.)}} ArR + HX} \tag{6.1}$$

$$\mathrm{ArH + alkene \xrightarrow{\text{HX or L.A.–}H_2O} ArR + HX} \tag{6.2}$$

Rearrangement within the alkyl group is a major complication for alkylation with primary and secondary compounds, and also some tertiary compounds (but not in *tert*-butylation). Another, general, difficulty is that introduction of one alkyl group into the benzene ring activates the nucleus so that further alkylation is favoured, and increasingly so with each additional substitution. The activation produced is small because alkylation is a very unselective electrophilic substitution, the more important feature being that the alkyl derivative is more soluble than the starting material in the catalyst layer (where reaction occurs).[5] The products of alkylation may also isomerize or disproportionate, so that often the orientation is thermodynamically rather than kinetically controlled. Di- or trihalides continue to react until all of the halogen is replaced by aromatic, e.g. $CHCl_3$ and benzene give Ph_3CH.

6.1.1 Mechanism of Alkylation by Alkyl Halides

Alkylation by alkyl halides has been the most widely studied and is therefore described in appropriate detail.

(i) Formation of complexes

The components of these alkylations form various complexes, some of which play a part in the substitution process:

(1) Whereas aluminium chloride is barely soluble in aromatic hydrocarbons, aluminium bromide dissolves readily to give coloured and highly polar solutions.[6] The composition of the complexes corresponds to $ArH \cdot Al_2Br_6$ ($ArH \cdot AlBr_3$ where the aromatic is more strongly basic) and they are of low stability. For a series of methyl-substituted benzenes the extent of complex formation increases with the number of methyl groups, indicating that the interaction is of the π-complex type in which the aromatic nucleus is the electron donor and aluminium is the electron acceptor, cf. the π-complexes formed between aromatics and acceptors such as hydrogen fluoride (Section 2.1.2). There is no evidence that these complexes are involved in alkylation, and indeed it would be improbable since the aromatic ring in the complexes would have a lower electron density and therefore be less reactive than the free aromatic. This appears to be the cause of the poor yields obtained in alkylation of polycyclic aromatics which form these complexes more readily.

(2) More important are the complexes formed by the chlorides and bromides of aluminium and gallium with alkyl halides.[7,8] For example, aluminium bromide dissolves in methyl bromide to form a 1:1 complex, and a solid of composition $CH_3Br \cdot AlBr_3$ has been isolated at $-78\,°C$. The complexes are polar but only feebly conducting, indicating that they are essentially ion pairs, e.g. $R^+AlBr_4^-$, which are very slightly dissociated into the free ions.[9] The magnitude of the dissociation is indicated by the 0.1% ionization of triphenylmethyl bromide–tin(IV) bromide in benzene, despite the triphenyl carbocation being particularly stable[9] [although ionization would here be less favoured owing to the use of tin(IV) bromide]. The formation of ion pairs is indicated by bromine exchange which takes place between alkyl bromides and aluminium bromide containing radioactive bromine, consistent with $AlBr_4^-$ as the exchange medium. Moreover, amongst alkyl halides the exchange reactivity order is tertiary > secondary > primary, consistent with formation of a carbocation intermediate.[10] With other catalysts the mechanism of the exchange is less certain. For the ^{36}Cl exchange between CH_3Cl and $GaCl_3$ (second order in catalyst and first order in methyl chloride), rate-determining attack by methyl chloride on 1:1 addition compounds or ion-pair dimers, with formation of dimethylchloronium ion, $MeCl^+Me$, as an unstable intermediate, has been proposed.[11] Chlorine exchange then occurs in a series of subsequent and rapid

reactions. The difficulty of generalizing a mechanism is demonstrated by the chlorine exchange between methyl chloride and antimony pentachloride being only first order in catalyst.[12] The relative exchange rates with catalysts were 170 ($GaCl_3$), 1 ($SbCl_5$), and <0.001 ($SbCl_3$).[12] In gallium chloride-catalysed methylation the methylation rate exceeded the chlorine exchange rate, showing that the same intermediate cannot be involved in both reactions.[11] Despite the uncertainties, it is clear that ionization of these complexes is slight so that a mechanism for alkylation by *primary* halides and which involves the generation of a free carbocation from the halide followed by reaction of this with the aromatic may be ruled out.

This conclusion does not apply if a primary alkyl fluoride is used with antimony pentafluoride as catalyst in liquid sulphur dioxide. The stability of the SbF_6^- ion is so high that free primary methyl or ethyl cations are formed and have been detected spectroscopically (n.m.r., laser Raman). This formation is confirmed by the alkylating power of the medium being unmatched by that observed previously for any methylating or ethylating reagent.[13] Isotope-labelling studies indicate that ethylation with ethyl fluoride and boron trifluoride under some conditions may involve the free primary carbocation.[14]

(3) Most pertinent of all to the mechanism of Friedel–Crafts alkylation is the formation of σ-complexes which are isolable at low temperature. For example, that from mesitylene, ethyl fluoride, and boron trifluoride is believed to be (**1**), its salt-like character being substantiated by its ionic conductance at the melting point.[15] The complex is evidently the intermediate in the alkylation of mesitylene, since on heating it decomposes to the alkylated product (**2**) together with hydrogen fluoride and boron trifluoride.

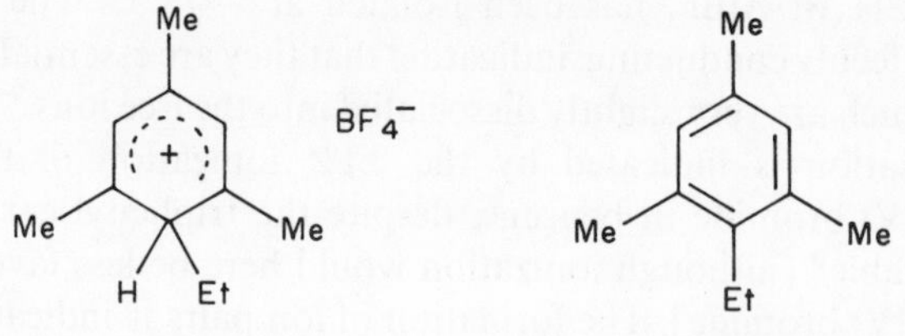

Me
Me Me
H Et
BF_4^-
(**1**)

Me
Me Me
Et
(**2**)

(4) Coloured ionic species are formed slowly and with high activation energies from aromatic hydrocarbons, hydrogen halides, and Lewis acids.[16] They are also σ-complexes and **3** is obtained from toluene, aluminium chloride, and hydrogen chloride.[17] These complexes, and those containing the ions $Al_2Cl_7^-$, $Al_3Cl_{10}^-$, etc., which can be formed from the hydrogen halide generated during a Friedel–Crafts alkylation, may facilitate alkylation by increasing the polarity of the medium.[18]

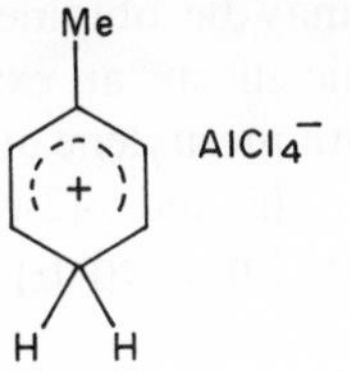

(3)

(ii) Rearrangements

Two types of rearrangement occur in alkylations: (1) a nuclear rearrangement in which an alkyl group attached to the nucleus migrates to a different nuclear position and may rearrange in the process; and (2) an aliphatic rearrangement in which the alkylating group itself rearranges.

(1) Nuclear rearrangements may be either intramolecular, intermolecular, or a mixture of both. Methyl groups participate only in intramolecular rearrangements, and an example is the reaction of *p*-xylene with $AlCl_3$–HCl to give a solid complex of *m*-xylene and the two inorganic compounds.[19,20] The isomerization may be represented as simple Wagner–Meerwein transformations (eq. 6.3).[21] Methyl groups migrate around aromatic rings in a series of successive 1,2-shifts, the rates of which appear to be dependent upon the π-density at the cationic centre.[22] Another example of isomerization is found in the $AlCl_3$-catalysed reaction of 2-chloropropane with *p*-xylene at 50 °C, which gives 62% of 2-(2-, propyl)-*p*-xylene and 38% of 5-(2-propyl)-*m*-xylene.[23] These migrations depend on the conditions: the $AlCl_3$-catalysed reaction of 1-chloropropane with *p*-xylene at low temperature gives 73% of 2-(1-propyl)-*p*-xylene together with 27% of the 2-propyl isomer, yet at high temperatures the product becomes exclusively 5-(2-

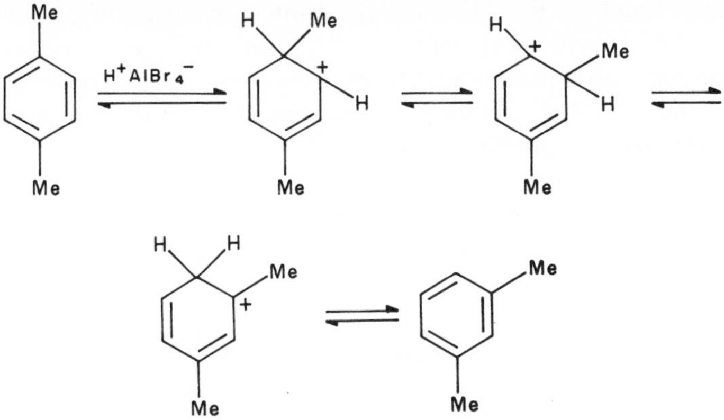

(4)

(6.3)

propyl)-*m*-xylene.[23] Products may be obtained which are the result of steric hindrance rather than electronic effects, an example being the $AlCl_3$-catalysed reaction of 1-chloropropane with mesitylene at 30 °C. The main products are 5-(2-propyl)-1,2,4-trimethylbenzene (**5**) and 5-(2-propyl)-1,2,3-trimethylbenzene (**6**) with little formation of the more hindered 1-propylmesitylene (**7**).[23]

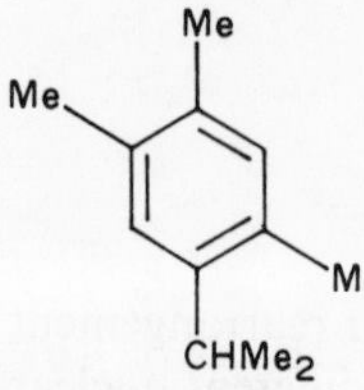

(**5**) 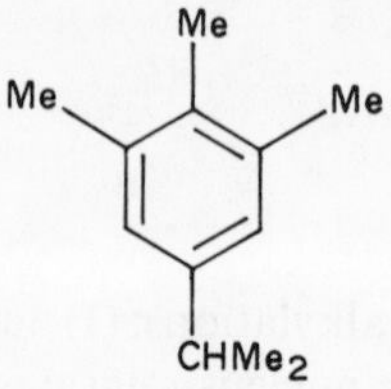 (**6**) 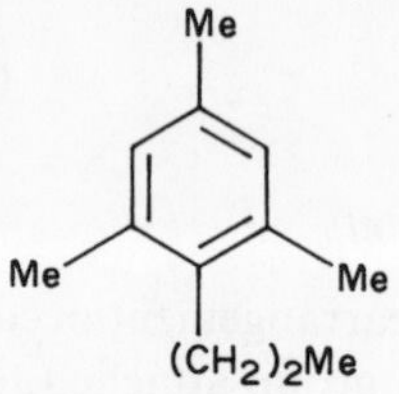 (**7**) 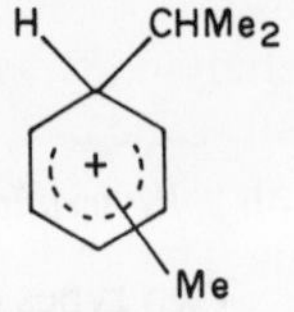 (**8**)

When one of the alkyl groups is branched at the α-carbon atom, intermolecular isomerizations become significant. For example, the main path in the isomerization of the methyl-2-propylbenzenes is an intermolecular rearrangement involving the separation of the 2-propyl cation from the arenium ion **8**.[24] The difference between this case and that of the xylene isomerization lies in the greater stability of Me_2CH^+ compared with, $CH_3{}^+$. It follows that *tert*-butyl groups migrate most readily, and mainly by an intermolecular mechanism. Ethyltoluenes isomerize by both inter- and intramolecular mechanisms,[25,26]

When the alkyl group can itself isomerize the rearrangement may involve migration of both H^- and Me^- within a carbocation produced from the alkyl group as a result of initial abstraction of hydride ion by other carbocations present in low concentration in the reaction medium. This explains the $AlCl_3$-catalysed interchange of the α- and β-carbon atoms in 1-propylbenzene,[27,28] and for the $AlCl_3$-catalysed interconversion of 2-methylpropylbenzene and 1-methylpropylbenzene[27,29] (eq. 6.4).[30] It accounts for the formation of mainly 2-methylpropylbenzene from the reaction of 2-chlorobutane with benzene at 80 °C,[31] and for a variety of products in the reaction of propyl- and butylbenzenes with aluminium chloride.[32]

$$\mathrm{PhCH(Me)CH_2Me} \xrightleftharpoons{-H^-} \mathrm{PhC^+(Me)CH_2Me} \xrightleftharpoons{\sim H^-} \mathrm{PhCH(Me)CH^+Me} \xrightleftharpoons{\sim Me^-}$$

$$\mathrm{PhCH^+CHMe_2} \xrightleftharpoons{+H^-} \mathrm{PhCH_2CHMe_2} \qquad (6.4)$$

The ready interconversion of these carbocations provides a pathway for both the interconversions of isomeric alkylbenzenes and also for isomerizations during alkylation. The ratios of the products isolated in an alkylation may therefore reflect differences in the rates of alkylation of different nuclear positions or, at least in part, the results of isomerizations in which equilibria are set up amongst the isomers. The importance of isomerization during alkylation also depends on the conditions. For example, in the formation of xylenes from toluene, $AlCl_3$, and MeCl, 27% of *m*-xylene is formed at 0 °C but 98% at 106 °C,[33] showing that isomerization increases with increasing temperature. The position of equilibrium also depends on the amount of Lewis acid present and its nature. In the BF_3-catalysed equilibration of xylenes, 60% of the *meta* isomer is present when the BF_3 concentration is low, but approaches 100% as the concentration is raised. At low concentrations, the equilibrium position is determined by the relative free energies of the three xylenes, which indicate that 57% of *m*-xylene should be present, whereas at high concentrations it is determined by the relative free energies of the xylene conjugate acids which are then the dominant species.[19] The conjugate acid of *m*-xylene (**4**) is more stable than those of *o*- or *p*-xylene because in **4** each methyl group interacts directly with positively charged carbons.

Alkylbenzenes also undergo disproportionation in the presence of Lewis acids and hydrogen halides.[34,35] For example, ethylbenzene in the presence of HF–BF_3 is converted into an equilibrium mixture of benzene (45%), ethylbenzene (10%), and *m*-diethylbenzene (45%).[34] These are intermolecular reactions as opposed to the isomerizations which are, in many cases, intramolecular; disproportionation and isomerization may compete with each other. For example, *p*-*tert*-butylbenzene and *o*-xylene react in the presence of HCl and $AlCl_3$ in nitromethane to give mainly 3,4-dimethyl-*tert*-butylbenzene by disproportionation (eq. 6.5), together with some *m*-*tert*-butyltoluene by isomerization.[25] The mechanism of disproportionation involves initial protiodealkylation (Section 4.1) and is facilitated here by the presence of the very strong acid $H^+AlCl_4^-$. For the reaction to proceed requires the presence of a more reactive aromatic which becomes alkylated; in the above example this is *o*-xylene. The ease with which the reaction takes place depends on the stability of the leaving carbocation and consequently de-*tert*-butylation is most commonly observed. This fact has been made use of in synthesis of *o*-benzoyltoluene; the *tert*-butyl group is used to block the reactive *para* position of toluene in acylation (Section 6.7), reaction then occurring *ortho* to the methyl group; the *tert*-butyl group is then removed by disproportionation with benzene (more reactive than *o*-benzoyltoluene) in the presence of aluminium chloride (eq. 6.6).[36]

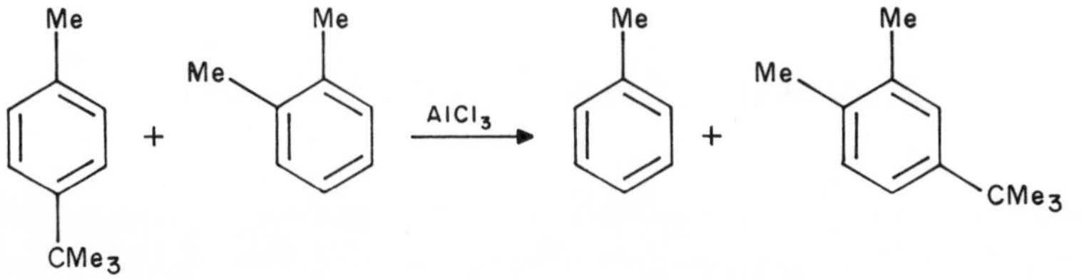

(6.5)

$$\text{4-}t\text{-Bu-toluene} \xrightarrow{PhCOCl,\ AlCl_3} \text{2-COPh-4-CMe}_3\text{-toluene} \xrightarrow{PhH,\ AlCl_3} \text{2-methylbenzophenone} \quad (6.6)$$

In studying the kinetics of Friedel–Crafts alkylations it is necessary to circumvent the complications which arise from the isomerization and disproportionation, by using for example the mildest possible reaction conditions, and short contact times between the reagents.[37]

Phenyl groups may also migrate, an example being $AlCl_3$-catalysed rearrangement of *o*-terphenyl into a mixture containing 7% *o*-, 70% *m*- and 23% *p*-terphenyl.[38]

(2) Friedel–Crafts alkylations with alkyl halides frequently result in isomerization of the alkylating group; e.g. the alkylation of benzene with 1-halopropanes and -butanes gives mixtures of 1- and 2-propylbenzene and 1- and 2-butylbenzene, respectively.[39] Whether isomerization occurs, and the extent to which it competes with direct alkylation, are dependent on the structure of the alkyl halide, the nature of the catalyst, the reactivity of the aromatic compound, and temperature.

First, isomerization is very common when primary alkyl halides are used. Hydride shifts occur particularly readily, e.g. **9** and **10**, as in the formation of 2-propylbenzene from 1-chloropropane and benzene and of *tert*-butylbenzene from 1-chloro-2-methylpropane and benzene.[21] Alkyl shifts also occur, as in the formation of 2-methyl-3-phenylbutane (**11**) from 1-chloro-2,2-dimethylpropane and benzene in the presence of $AlCl_3$. Both processes appear to be involved in the formation of various isomeric ethylcyclohexylbenzenes from benzene and chlorocyclooctane in the presence of $AlCl_3$.[40] (Rearrangement appears to be less severe if a primary alkyl group contains a terminal cyano group. For example, the $AlCl_3$-catalysed reaction of benzene with 6-bromohexanenitrile gives 85% of the primary 6-phenyl derivative and only 15% of the secondary 5-phenyl derivative.[41])

$$MeCH_2CH_2^+ \rightleftharpoons MeCH^+Me \quad \textbf{(9)} \qquad Me_2CHCH_2^+ \rightleftharpoons Me_3C^+ \quad \textbf{(10)}$$

$$MeC(Me)_2\text{—}CH_2Cl + PhH \xrightarrow{AlCl_3} MeCH(Me)\text{—}CH(Ph)Me \quad \textbf{(11)}$$

The products in these reactions can be considered to be derived from a carbocation more stable than that which could be formed by the original halide (although free carbocations are not necessarily present). However, when tertiary halides are used, products derived from a secondary carbocation are frequently formed.[39] For example, the $AlCl_3$-catalysed reaction of benzene with either 2-chloro-2-methylbutane or with 2-chloro-2,3-dimethylbutane gives mixtures consisting chiefly of 2-methyl-3-phenylbutane[21] and 2,2-dimethyl-3-phenylbutane,[39] respectively, arising from equilibria **12** and **13**. One explanation for this is that the greater stability of the tertiary ions also makes them less reactive, since the positive charge is more effectively delocalized away from the electrophilic centre. Thus, although the secondary ion may only be present in small concentration compared with the tertiary ion as a result of the equilibrium between them, its greater reactivity leads to a preponderance of the secondary substitution product. An alternative explanation is that the tertiary cation carries out the substitution, and this is followed by $AlCl_3$-catalysed rearrangement into the thermodynamically more stable secondary product.[42]

$$MeCH_2CM_2{}^+ \rightleftharpoons MeCH^+CHMe_2 \quad (12) \qquad Me_2CHCMe_2{}^+ \rightleftharpoons MeCH^+CMe_3 \quad (13)$$

Secondly, the extent of isomerization is dependent on the Lewis acid catalyst. Thus the rearrangement given by **13** does not occur if the corresponding alkylation is carried out in the presence of $FeCl_3$.[39] Iron(III) chloride is a less powerful catalyst than aluminium chloride and fails to break completely the alkyl—halogen bond necessary to form the alkyl cation through which isomerization occurs. It is nevertheless able to polarize this bond to the extent necessary for reaction to occur between the aromatic compound and the electrophilic carbon atom of the halide [see Section 6.1.1. (iii)].

Thirdly, the extent of isomerization decreases as the aromatic becomes more reactive. For example, while the alkylation of benzene with 1-chlorododecane in the presence of $AlBr_3$ gives only 40% of 1-dodecylbenzene, the remainder of the product being a mixture of isomeric derivatives, toluene and mesitylene under the same conditions give 68 and 88%, respectively, of the 1-alkyl product.[43] Similarly, in $GaBr_3$-catalysed alkylations, toluene gives less of the isomerized alkyl products than benzene.[44] Thus the more reactive aromatic compounds are able to react with the alkylating agent to a larger extent before the latter has had time to isomerize.[45]

Fourthly, the extent of isomerization can be minimized by using the lowest convenient temperature. For example, the $AlCl_3$-catalysed reaction of 1-chloropropane with benzene at room temperature gives 1- and 2-propylbenzenes in the ratio 60:40 at $-6\,^\circ C$ and 40:60 at $35\,^\circ C$;[46] similar results are obtained using $AlBr_3$ and 1-bromopropane.[43]

(iii) Kinetics

Kinetic investigations of alkylation have been mainly confined to the use of aluminium chloride, aluminium bromide, and gallium bromide as catalysts.

(a) *Kinetic effect of solvents.* The reaction involves polar transition states and is therefore more rapid in more polar solvents. Thus the relative rates of the reaction of *tert*-butyl chloride with benzene (Table 6.2) roughly parallel the dielectric constant of the medium.[47] It follows that either 1,2-dichloroethane or nitrobenzene is usually the solvent of choice for preparative use of the reaction; nitrobenzene is not reactive enough to undergo alkylation. Nitromethane has been used in kinetic studies, but this leads to some complications, as described below. Recently, mixtures of $AlCl_3$ and 1-methyl-3-ethylimidazolium chloride, which melt below room temperature and provide both solvent and catalyst, have been used to study both alkylation and acylation.[48] Initial rates correlated with the concentration of $Al_2Cl_7^-$, indicating this to be the catalysing species.

(b) *Studies in nitrobenzene and 1,2-dichloroethane.* The electrophile in at least some alkylations was shown not to be the free carbocation by the reactivity of butyl halides with benzene in CS_2 ($AlCl_3$-catalysed) being in the order RF > RCl > RBr > RI.[49] Subsequent investigations revealed fuller details of the mechanism. The $AlCl_3$-catalysed alkylation of benzene by benzyl halides in nitrobenzene is third order, rate = $k[PhH][ArCH_2Cl][AlCl_3]$. The rate is only moderately affected by marked changes in the polarity of the solvent and this, together with other evidence given below, indicates that the reaction does not involve the formation of a carbocation followed by rate-determining reaction between this and benzene, but involves instead nucleophilic attack of benzene on the polarized benzyl chloride–Lewis acid complex (eqns 6.7–6.9).[50] The evidence for this mechanism, which appears to be general for alkylation of

Table 6.2 Relative rates of *tert*-butylation of benzene in various solvents

Solvent	$k_{rel.}$	Dielectric constant/D
Carbon tetrachloride	1.0	2.24
Benzene	2.5	2.28
Carbon disulphide	4.0	2.64
1,1,2,2-Tetrachloroethane	30	8.0
1,2-Dichloroethane	110	10.5
Nitrobenzene	360	36.1

$$RCl + AlCl_3 \rightleftharpoons RCl \cdot AlCl_3 \quad (6.7)$$

$$PhH + RCl \cdot AlCl_3 \overset{\text{slow}}{\rightleftharpoons} RPhH^+ + AlCl_4^- \quad (6.8)$$

$$RPhH^+ + AlCl_4^- \rightleftharpoons PhR + HCl + AlCl_3 \quad (6.9)$$

primary alkyl halides in this solvent, is as follows:

(1) Alkylations in the presence of $AlCl_3$[50] and $AlBr_3$[51] are first order in aromatic and alkyl halide, and approximately so in catalyst. The aromatic is therefore present in the rate-determining step, but this cannot consist of a reaction between the aromatic and a free carbocation, because the reaction rate is slower with 4-nitrobenzyl chloride than with 3,4-dichlorobenzyl chloride. The greater electron withdrawal from the aromatic ring in the former reagent would produce a more positively charged carbocation, which would therefore be more reactive if it were involved. Likewise, electron-supplying substituents in the aryl ring of the benzyl chloride increase the rate.[52].

The $AlCl_3$-catalysed reaction of chlorocyclohexane with benzene in nitrobenzene and other solvents gave orders in catalyst that varied from 0.5 to as high as 2.1, and increased with decreasing polarity of the solvent.[53] This is consistent with the need for greater assistance of polarization of the alkyl halide in the media of lower dielectric.

(2) Both the relative reactivities of toluene and benzene and the isomer distribution of the products from toluene are different when the compounds are methylated with methyl bromide or with methyl iodide each in the presence of $AlBr_3$.[54] The electrophile therefore cannot be the methyl cation, as this would give the same reactivities for each reagent. Moreover, the *ortho*:*para* ratio is smaller when using methyl iodide, consistent with the bulkier iodine atom being present in the alkylating complex. The reaction is faster with methyl bromide, since bromine is a better donor than iodine for complexing with a Lewis acid.[54]

(3) Since the stabilities of the methyl, ethyl, and isopropyl ions increase in the order $CH_3^+ < MeCH_2^+ < Me_2CH^+$, the selectivities of the three cations, measured by the relative reactivities of toluene and benzene towards them, should also increase in this order. However, in $AlBr_3$-catalysed alkylation by the respective alkyl bromides, the selectivities lay in the opposite order, viz. 2.95 (methylation), 2.4 (ethylation), and 1.65 (isopropylation).[37] These results are consistent with rate-determining attack of the aromatic on the polarized alkyl halide–Lewis acid complex. The carbon—bromine bond will be least polarized in methyl bromide, resulting in greater discrimination between the aromatics by the cationic centre which will have the smallest positive charge. The results are not unambiguous, however, because the bulkier alkyl halide will give the least *ortho* substitution which will affect the overall rate ratios in the manner observed.[55] Indeed, measurements of selectivities in terms of *meta*:*para* ratios

indicated them to be the same for all three reagents, which led to the proposal that the differences in rate are due to a difference in the *concentration* of the polarized species.[56] The original interpretation appears to be confirmed by the relative alkylation rates (measured in 1,2,4-trichlorobenzene to reduce them to more accessible values), which are[51] 1:57:> 2500 (too fast to measure); the variation is greater than could arise from lack of any *ortho* substitution.

The mechanism is substantially the same if $GaBr_3$ is used as the Lewis acid, the only difference being that the reaction is second order in this weaker catalyst, indicating that two molecules of it are required to polarize the carbon—halogen bond successfully.[44] The relative reactivities of alkyl bromides were Me > Et > *n*-Pr > *i*-Pr > *t*-Bu; the rate difference between the last two halides was less than expected, but it is not clear whether this means that a free carbocation mechanism is involved for the *tert*-butyl compound. No kinetic isotope effect was found in the $GaCl_3$-catalysed reaction of methyl chloride with benzene and benzene-d_6, showing that proton loss in the step equivalent to eq. 6.9 is fast.[57]

(c) *Studies in nitromethane.* Alkylation with alkyl halides in nitromethane is fast and was characterized (in earlier work) by incompatible positional and substrate selectivities, the former being too high relative to the latter. Data for benzylation catalysed by the $AlCl_3$–nitromethane complex (Table 6.3) led to the proposal that substrate selectivities are determined by rate-determining formation of a π-complex between the aromatic and the polarized alkyl halide–catalyst complex, and that positional selectivity is determined in subsequent rearrangement to the σ-complex.[58] Isopropylation (Table 6.4)[59] and *tert*-butylation[60] gave substrate and positional selectivities more consistent with each other; the substrate selectivity in the latter reaction was raised to 16 by the use of less powerful

Table 6.3 Isomer distributions and relative reactivities in $AlCl_3$–$MeNO_2$-catalysed benzylation of PhX in nitromethane at 25 °C

X	Isomer distribution/%			Relative reactivity
	o	*m*	*p*	
H				1
Me	43.5	4.5	52.0	3.2
Et	42.4	5.0	52.6	2.45
Pr	39.6	8.1	52.3	2.22
Bu	39.1	8.6	52.3	2.08
F	14.7	0.2	85.1	0.46
Cl	33.0	0.6	66.4	0.24
Br	32.5	0.7	66.8	0.18
I	30.6	0.7	68.7	0.28

Table 6.4 Isomer distributions and relative reactivities in $AlCl_3$–$MeNO_2$-catalysed isopropylation of PhX with *i*-PrBr in nitromethane at 25 °C

X	Isomer distribution/%			Relative reactivity
	o	*m*	*p*	
Me	46.7	14.7	38.6	2.03
F	40.8	2.8	56.4	0.28
Cl	51.4	8.1	40.5	0.13
Br	51.8	11.6	36.6	0.11

catalysts, e.g. $SnCl_4$. In each alkylation small secondary kinetic isotope effects, $k_H/k_D \approx 1.2$, were observed, attributed to hybridization changes on going to the transition state (Section 2.1.3).

An alternative explanation of the benzylation results is that non-competitive conditions are responsible, i.e. the reagent is of such high reactivity and concentration that it reacts with each aromatic molecule on encounter (Section 2.10), i.e. the competition method becomes invalid. In the above work, a 16-fold variation in the ratio of the aromatic reagents produced only a small change in the substrate selectivity, leading to the belief that truly competitive conditions applied. This evidence is inconclusive, however, because non-competitive conditions may apply throughout the whole of the range of substrate ratios examined. Olah and Overchuck[61] attempted to resolve this problem by carrying out direct kinetic studies on the individual substrates. Good kinetic data were not obtained but the approximate rate data appeared to confirm the results of the competition experiments and moreover, the positional selectivities were similar under both conditions. However, recent work has shown that the kinetic order in aromatic is not first order, this latter being an essential pre-requisite for use of the competition method. Benzylation is zeroth order in aromatic for the $AlCl_3$-catalysed reaction with benzyl chlorides containing 4-Cl and 3,4-Cl_2 substituents, first order for reaction with 4-nitrobenzyl chloride, and zeroth order for the $TiCl_4$-catalysed reaction with the 4-methyl compound; under zeroth-order conditions formation of the electrophile is rate determining. Benzylation with the 4-methyl and 4-nitro compounds gives compatible positional and substrate selectivities, but the chloro-substituted compounds do not. These results indicate a change in mechanism from one involving the free cation (the formation of which is rate-determining) for the methyl and chloro compounds, to the nucleophilic displacement mechanism for the nitro compound. The 4-methylbenzyl cation gives normal selectivity since it is more stable and less reactive.[62]

A recent kinetic study of $AlCl_3$-catalysed isopropylation with isopropyl chloride gave selectivities substantially different from those given in Table 6.3, the values obtained being those theoretically predicted.[63] Reinvestigation of *tert*-butylation, using *tert*-butyl bromide also gave the theoretically expected results[63] whereas in the previous study this could only be obtained using weaker catalysts.[60] This work showed that the amount of water present in the reaction medium can very substantially alter yields and rate ratios.

(d) *Studies in non-polar solvents.* In the BF_3-catalysed ethylation of benzene and toluene by ethyl fluoride labelled at the 2-carbon, ca 50% of the label ended up in the α-position if the reaction was carried out in hexane or cyclohexane, but only 3.5% if nitromethane was the solvent. Hence the electrophile is evidently the free cation under the former conditions, the label being positionally equilibriated as in eq. 6.10, whereas in nitromethane it is the non-ionized halide–catalyst complex.[14] The toluene:benzene rate ratio of < 1 obtained in this work has been shown to be due to solubility effects.[64]

$$^{14}CH_3CH_2^+ \xrightarrow{\sim H^-} {}^{14}CH_2^+CH_3 \qquad (6.10)$$

In the $TiCl_4$-catalysed alkylation of anisole, toluene, and benzene with substituted benzyl chlorides, the isomer distribution in the former compounds tends towards statistical as the reactivity of the benzyl halide is increased, and this is accompanied by a decrease in the aromatic:benzene rate ratios (Table 6.5).[65] For anisole at least, the results were shown to be little affected by isomerization, and the selectivity change has been interpreted in terms of a change in the nature of the transition state from a σ-complex to a π-complex. It is improbable, however, that these small changes in electrophile reactivity would be sufficient to produce a change in the nature of the transition state. In hydrogen exchange a much greater increase in the reactivity of the electrophile (many orders of magnitude) is insufficient to produce such a change.[66] A more recent study of the $TiCl_4$-catalysed benzylation of toluene and benzene in excess aromatic showed

Table 6.5 Isomer distributions and reactivities relative to benzene in $TiCl_4$-catalysed benzylation of anisole and toluene with $XC_6H_4CH_2Cl$ at 25 °C

	Toluene				Anisole			
		Isomer distribution/%				Isomer distribution/%		
X	k_T/k_B	*o*	*m*	*p*	k_A/k_B	*o*	*m*	*p*
4-NO_2	2.5	59.6	6.2	34.2	7.0	54	< 0.5	46
H	6.3	40.5	4.3	55.2	14	47	< 0.3	53
4-OMe	97.0	28.6	1.5	69.9	15 500	30	—	70

the reaction to be first order in benzyl chloride and second order in catalyst. The effect of moisture present was shown to have a marked effect on the kinetics and with this carefully controlled, the toluene:benzene rate ratio of ca 20 is substantially higher than previously obtained.[67]

6.1.2 Mechanism of Alkylation by Other Reagents

Alkylations using sources of the alkyl group other than halides have not been extensively investigated, but the mechanisms do not appear to differ in any fundamental way from those with alkyl halides.

(i) Alkylation by alcohols

Electrophiles may be generated from alcohols either by the use of Lewis acid catalysts (BF_3 and $ZnCl_2$ are most commonly used) or by the addition of an acid. The former catalysts polarize the carbon—oxygen bond whereas the latter protonates oxygen to give, after loss of water, the carbocation (eqns 6.11 and 6.12). The catalyst chosen is that which will produce a sufficiently reactive electrophile yet cause the minimum of side-reactions. Rearrangement of the alkyl group is a general occurrence. The BF_3-catalysed alkylation of benzene with 1-propanol and 1-butanol gives good yields of 2-propylbenzene and 2-butylbenzene, respectively, and either 2- or 3-pentanol give mixtures of 2-pentylbenzene and 3-pentylbenzene in the same proportions (2.6:1).[68] This ratio is similar to that obtained in alkylation of benzene by ethylcyclopropane in the presence of HF, H_2SO_4, or $AlCl_3$ (1.7:1)[69] and by either 2-pentanol or 3-pentanol in the presence of $AlCl_3$ (1.5 and 3.2:1, respectively).[70] Since the alcohols do not interconvert under the reaction conditions, it is probable that the alkyl groups equilibriate in the environment of a complex with benzene, and the mechanism given by eq. 6.13 has been suggested.[70] In this series of reactions, step A leads to a π-complex of benzene and the carbocation through which rearrangement may occur (step B). Step B must be fast compared with the alkylation steps C and D to account for the formation of the same mixture of phenylpentanes from 2- and 3-pentanol. The direct formation of the carbocation, step E, probably occurs with tertiary alcohols.

$$ROH + H^+ \rightleftharpoons ROH_2^+ \tag{6.11}$$

$$ROH_2^+ \longrightarrow R^+ + H_2O \tag{6.12}$$

When the carbocation derived from the alkylating agent is a particularly stable one, e.g. Ph_2CH^+, reaction can occur by the prior formation of this ion in a fast equilibrium followed by rate-determining reaction of the ion with the aromatic. This has been established for alkylation of anisole by diphenylmethanol in the presence of sulphuric and acetic acids, $ZnCl_2$, and BF_3.[71] In each case the catalyst

$$ROH + BF_3 \xrightarrow[A]{PhH} C_6H_6\text{–}R^+ \underset{B}{\rightleftharpoons} C_6H_6\text{–}R'^+$$

E: $R^+ + BF_3OH^-$; C: arenium ion (H, R); D: arenium ion (H, R′) (6.13)

induces extensive ionization of the alkylating agent (eq. 6.14) to give the resonance-stabilized diphenylmethyl cation, which then reacts with the aromatic in the rate-determining step; the latter was confirmed by the absence of a kinetic isotope effect. The formation of ions from alcohols, e.g. Ph_3COH, in the presence of acids was first demonstrated by measurement of Van't Hoff *i*-factors.[72a].

$$Ph_2CHOH + H^+ \rightleftharpoons Ph_2CH^+ + H_2O \qquad (6.14)$$

(ii) Alkylation by alkenes

Carbocations are formed from alkenes through protonation (eq. 6.15), so a strong acid is necessary to produce alkylation from alkenes; CF_3SO_3H is particularly effective.[72b] Lewis-acid catalysts are inadequate unless traces of water, or an acid such as a hydrogen halide, are present, in which case very strong acids may be produced.

$$RCH{=}CH_2 + H^+ \rightleftharpoons RCH^+CH_3 \qquad (6.15)$$

Alkylation with alkenes shows the general features, e.g. isomerization, found for alkylation with alkyl halides. Isomerization follows the general pattern of being both temperature and catalyst dependent; some typical data are given in Table 6.6.[73]

Lewis acid-catalysed alkylation of anisole and 1,2-dihydroxybenzene with alkenes gives high *ortho* yields, coordination of the catalyst with the oxygen-

Table 6.6 Isomer ratios in alkylation of benzene with 3-methylbutene

Catalyst	Temperature/°C	Me_2CPhCH_2Me/%	$MeHCPhCMe_2$/%
$AlCl_3$–HCl	−40	100	—
$AlCl_3$–HCl	0	55	45
$AlCl_3$–HCl	21	11	87
HF	35	99	1
HF	100	70	30

containing substituent evidently being responsible.[74,75] Unusual alkenes which are activated by the electron-withdrawing group present are the pentadienyltricarbonyliron cation **14** and the cyclohexadienyltricarbonyliron cation **15**,[76,77] of which the former is the more reactive.

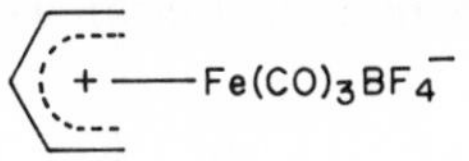

$Fe(CO)_3BF_4^-$ $Fe(CO)_3BF_4^-$

(**14**) 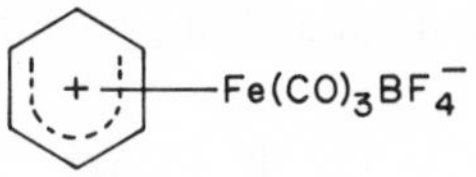 (**15**)

(iii) Alkylation by cycloalkanes

Small-ring cycloalkanes protonate readily and thus produce carbocations suitable for alkylation. The advantage of using cyclopropane is that high yields of n-alkylbenzene may be obtained. For example, the HCl–$AlCl_3$-catalysed reaction of cyclopropane with benzene at 0 °C gives 65–70% of 1-propylbenzene.[78,79] Alkylcyclopropanes give mixtures of secondary products owing to hydride shifts, but tertiary products are not possible. The amount of isopropylbenzene formed in the reaction of cyclopropane with benzene increases with increasing temperature.[80]

No comparative data are available but it would be expected that the reaction becomes more difficult with increasing size of the ring. Alkylation has been achieved with methylcyclobutane,[78] cyclopentane,[78] and indane,[81] but not with rings containing six or more carbons.

(iv) Alkylation by ethers

Ethers readily form complexes with BF_3, but these are ineffective alkylating agents unless traces of water are present[82] (it is probably true that this requirement, identified in alkylation by alkenes, applies generally to all Lewis acid-catalysed alkylations). Addition of strong acids greatly increases the alkylation rate, and the reaction is fast provided that the ratio of BF_3 to ether is greater than 1:1.[83] Carbocations are probably formed from the intermediate complex, especially when one of the alkyl groups is secondary or tertiary, and the most branched alkyl group will preferentially substitute. Both features are shown by formation of racemic 2-butylbenzene from the BF_3-catalysed alkylation of benzene with optically active 2-butyl methyl ether.[82] Alkylation with cyclic ethers resembles the reaction with cycloalkanes in giving mainly n-alkyl products, e.g. trimethylene oxide and benzene give a 50–70% yield of 1-hydroxy-3-phenylpropane.[84] They differ from alkylations with alkyl halides also in giving inversion of stereochemistry,[85] whereas the halides usually give racemization (although retention has been observed in one stereochemically favourable case).[86]

(v) Alkylation by esters, acids, acid halides, and sulphones

The Lewis acid-catalysed reaction of carboxylic acid esters with aromatics gives a mixture of the alkyl- and acylaromatics (eq. 6.16), arising essentially as a result of alkyl—oxygen and acyl–oxygen bond cleavage. The reaction is not therefore of much value as a preparative method. If electron withdrawal in the

$$\mathrm{R{-}C({=}O){-}O{-}R' + 2ArH \xrightarrow[\text{acid}]{\text{Lewis}} R{-}C({=}O){-}Ar + R'{-}Ar + H_2O} \tag{6.16}$$

group R can be increased, then alkyl—oxygen cleavage will become more favourable, as shown by increased rates of alkylation by halogeno-substituted esters.[87] Alkylation is also more rapid the more branched is R′.[87] Protonation of the carbonyl oxygen (**16**) greatly enhances alkyl—oxygen cleavage, and consequently TFA-catalysed alkylation by *tert*-butyl trifluoroacetate is both rapid and quantitative.[88] The much stronger trifluoromethanesulphonic acid may be used to catalyse the reactions of esters derived from weaker organic esters, 48–72% yields being obtained in benzylation by benzyl acetate.[89] Cyclic esters (lactones) may also be used, the products being carboxylic acids and here, as with cyclic ethers, the reaction is accompanied by inversion of configuration.[90]

$$\mathrm{R'{-}O{-}C({=}O){-}R \quad (+\ H^+\ \text{on carbonyl O})}$$

(16)

If an ester of an inorganic acid is employed, then the greater ability of the inorganic moiety to stabilize a negative charge means that alkyl—oxygen cleavage is favoured to the extent that only alkylation occurs. Alkyl sulphates (**17**), sulphinates (**18**), sulphonates (**19**), chloroformates (chlorocarbonates) (**20**), chlorosulphites (**21**), chlorosulphonates (chlorosulphates) (**22**), etc., have each been used, the ease of alkylation here being increased through the use of acid catalysis, e.g. CF_3SO_3H;[91] Lewis-acid catalysts may also be used.[92] The benzenesulphonic acid-catalysed benzylation of excess benzene by substituted benzyl benzenesulphonates (**19**, R = Ph, R′ = $PhCH_2$) is first order in ester and second order in acid.[93] Electron-supplying groups in the benzyl group increased the reaction rate and *vice versa*, and this led to an (incorrect) interpretation of the reaction mechanism in terms of rate-determining formation of a benzyl cation from an acid-solvated acid–ester ion pair. Electron withdrawal in the benzyl-cation would increase the reactivity of the cation, not the reverse, and the kinetic evidence points to the nucleophilic displacement mechanism [cf. Section 6.1.1. (iii)].

$R'OSO_2OR$ **(17)** $R'OSOR$ **(18)** $R'OSO_2R$ **(19)** $R'OCOCl$ **(20)** $R'OSOCl$ **(21)** $R'OSO_2Cl$ **(22)**

The alkyl group need only be attached to a strongly electron-withdrawing group. Thus alkylation has been observed with methanesulphonic acid (**23**),[94] alkylsulphonyl halides (**24**), and sulphones (**25**),[95] the last being less effective as expected. The kinetics of $AlCl_3$-catalysed benzylation of aromatics by phenylmethanesulphonyl chloride in nitromethane show the reaction to be first order in halide and catalyst but zeroth order in aromatic, so formation of the benzyl cation appears to be rate-determining.[96] The reaction gave incompatible substrate and positional selectivity, but these were both lower than in a previous report (which also gave incompatible selectivities);[95] the toluene:benzene rate ratio was 2.2, with 47.2, 9.0, and 43.8% *ortho*, *meta*, and *para* substitution, respectively. Alkylation in nitromethane appears to give anomalous results here as it does when the reagent is an alkyl halide.

$MeSO_2OH$ **(23)** $R'SO_2Cl$ **(24)** $R'SO_2R$ **(25)**

(vi) Alkylation by other reagents

The isomerization which accompanies the use of Friedel–Crafts catalysts has led to the investigation of the potential of reagents sufficiently polar to be able to alkylate without a catalyst. Reagents successfully employed so far include oxonium salts (**26**) and halonium salts (**27**), the latter being the more reactive owing to the greater electron-withdrawing ability of the halogen compared with oxygen; the presence of the positive pole makes both substantially more reactive than either esters or alkyl halides. They are formed by reaction of the aryl ether or halide with alkyl halide in the presence of SbF_5 (eq. 6.17). The oxonium salts will

$PhO^+Me_2\ SbF_6^-$ **(26)** $PhCl^+Me\ SbF_6^-$ **(27)**

$$PhOMe + MeF + SbF_5 \longrightarrow PhO^+Me_2\ SbF_6^- \quad (6.17)$$

$$C_6H_5\text{–}OMe + PhO^+Me_2\ SbF_6^- \longrightarrow Me\text{–}C_6H_4\text{–}OMe + PhOMe + HSbF_6 \quad (6.18)$$

react with compounds of the reactivity of anisole (eq. 6.18)[97] while the more reactive halonium salts react readily with toluene.[98] Side-reactions may also be minimized by using solid acid catalysts, such as perfluorinated alkanesulphonic acids, e.g. the acid form of the commercially available Nafion ion membrane resin. The activity of these catalysts can be increased by complexing with higher valency metal halides, e.g. SbF_5, and alkylation with alkyl halides, alcohols, alkenes, and esters occurs readily over these catalysts.[99] Chlorothiaformamidinium salts (**28**) have a cationic centre (resonance stabilized by the adjacent NR_2 and SR′ groups) and are therefore able to alkylate reactive aromatics without needing catalysts.[100]

Substituted alkenes tend to react with two molecules of aromatic, one replacing the substituent and the other adding to one end of the double bond, whilst substituted alkynes react with three molecules of aromatic in a similar manner.[101] However, by use of triaryl-substituted vinyl bromide and silver salts in the presence of a base, only the halogen is replaced, giving a vinyl-substituted aromatic in an electrophilic substitution of ρ factor -4.08.[102] α-Chloroenamines also react with only one molecule of (reactive) aromatic to give aminoalkenyl products (**29**).[103]

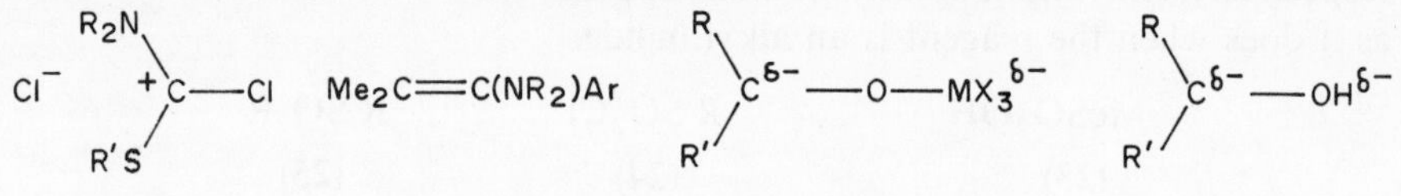

Cl^- $(R_2N)(R'S)\overset{+}{C}{-}Cl$ (**28**)

$Me_2C{=}C(NR_2)Ar$ (**29**)

$(R)(R')C^{\delta-}{-}O{-}MX_3^{\delta-}$ (**30**)

$(R)(R')C^{\delta-}{-}OH^{\delta-}$ (**31**)

In the presence of Lewis or protic acid catalysts, both aldehydes and ketones are able to alkylate aromatics; **30** and **31** are the probable electrophiles.[104] However, the formation mainly of anthracenes in the $AlCl_3$-catalysed reaction of benzaldehyde with benzene and other aromatics shows the mechanism to be less than straightforward,[104] and in the BF_3-catalysed reaction diarylmethanes are obtained in good yields instead. With toluene the product isomer distribution is *ortho* 42%, *meta* 8%, *para* 50%,[105] suggesting that a small electrophile is involved.

The reaction of n-butyllithium reagents with aromatics in ether solutions gives mainly the aryllithium derivative, although alkylaromatics are frequently observed as byproducts. This side-reaction can become dominant if a hydrocarbon solvent is used, the ease of reaction following the order tertiary > secondary > primary for the nature of R in RLi.[106] Since *tert*-butylation of naphthalene goes 95% into the 1-position, the reaction appears to be an electrophilic substitution, but the mechanism has not been investigated. Toluene and benzene have been alkylated by diazotization of amines with nitrosonium salts in nitromethane, and a normal isomer distribution in toluene is found.[107] The toluene:benzene rate ratios at 25 °C were 1.5 (ethylamine), 2.5 (isopropylamine), and 3.5 (benzylamine).

(vii) Alkylation in the gas phase

Alkylation may be carried out in the gas phase with alkyl cations, which may be produced in one of two ways. For example, fragmentation of dimethyl ether in an ion-cyclotron resonance spectrometer gives methylene cations, resulting in methylenation (eq. 6.19).

$$ArH + CH_2{}^+OCH_3 \longrightarrow ArCH_2{}^+ + CH_3OH \qquad (6.19)$$

The reaction gives almost statistical positional selectivity, accompanied by low substrate selectivity, i.e. 1.8 (toluene), 5 (ethyl- or isopropylbenzene), 0.1 (fluorobenzene) and < 0.05 (benzotrifluoride).[108]

Much more widely investigated is the reaction with radiolytically produced cations, the selectivities of which are in accordance with the expected stabilities of the cations. Methylation of toluene results in the formation mainly of *m*-xylene, the toluene:benzene reactivity ratio being 0.8. Use of the same radiolytically produced cations in solution gives, by contrast, a more normal rate ratio of 2.2 with the formation of mainly *o*- and *p*-xylene. Evidently the unsolvated ions in the gas phase are highly reactive, and produce a high-energy intermediate which rearranges to give *m*-xylene.[109] The ions MeX^+Me ($X =$ halogen) have been produced in the gas phase in various ways, and shown to discriminate between n- and π-type centres and react, for example, with phenol to give mainly *O*-methylated products, the extent of discrimination depending on the halogen.[110] Isopropylation is also very unselective, although the selectivity increases at higher reagent pressures because, in effect, the higher concentration of aromatic 'solvates' (or collisionally stabilizes) the ions and intermediate. Some partial rate factors have been determined for a range of methyl- and fluoro-substituted benzenes, but it is difficult to evaluate their meaningfulness in relation to solution data since, for example, *para*-methyl deactivates whereas *ortho*-fluoro activates substantially, the latter indicating an *ortho*-coordination mechanism,[111] and this evidently accounts for the enhanced *ortho* substitution in isopropylation of dihalobenzenes.[112a] Alkylation with $C_3H_7{}^+$ ions produced from protonation of cyclopropane and propene showed the former to give a much higher ratio of n-propyl to isopropyl aromatics, the ratio being higher for reaction with benzene than with toluene.[112b] This parallels behaviour in solution [see Section 6.1.2. (ii) and (iii)].

tert-Butylation differs from the foregoing gas-phase alkylations in giving fairly high selectivities, e.g. a toluene:benzene rate ratio which varies from 114 at 0 °C to 21 at 140 °C, the corresponding *para*:*meta* ratios varying from 13.3 to 3.8; no *ortho* substitution was found. These values are consistent with the lower reactivity of the much more stable *tert*-butyl cation.[113] Appreciable *ortho*-*tert*-butylation can occur at sufficiently activated sites, e.g. *m*-xylene gives 65% substitution at the 4-position.[114]

Collisionally induced decomposition spectra for, e.g., protonated isopropylbenzene and the isopropyl ion and benzene are identical, indicating that the resultant structure in each case must be the Wheland intermediate.[115] This indicates that the pathways for alkylation in solution and the gas phase are similar.

6.1.3 Substituent Effects

Nitrobenzene cannot be alkylated but, since the electrophile in alkylation appears to be highly reactive, and since nitrobenzene can be substituted in other electrophilic substitutions, it suggests that the lack of reactivity here arises from coordination of the Lewis acid catalyst with the nitro group, such coordination being well established for nitromethane. Coordination also occurs with OH, OR, NH_2, and NR_2 substituents, and consequently alkylation of aromatics containing these groups is difficult except by using reagents which do not require catalysts.

These difficulties mean that quantitative studies of the effects of substituents in determining orientation and reactivity have used only alkyl and halogen substituents. In deriving data it is necessary to ensure that the orientations observed are not due to thermodynamic control which would produce an increase in the proportion of *meta* isomer since this is thermodynamically the most stable. Partial rate factors derived from $GaBr_3$-catalysed alkylation of benzene and toluene in excess aromatic (Table 6.7)[44,116,117] were determined using minimum contact times in order to minimize isomerization. Even so, in

Table 6.7 Partial rate factors for $GaBr_3$-catalysed alkylation of toluene by RBr

R	f_o	f_m	f_p	Ref.
Me	9.51(8.3)[a]	1.70(1.8)[a]	11.8(9.4)[a]	44
Et	2.84	1.56	6.02	44
i-Pr	1.52	1.41	5.05	115
$PhCH_2$	4.9	2.3	9.2	116

[a] Data for $GaCl_3$-catalysed methylation.[119]

Table 6.8 Partial rate factors for $GaBr_3$-catalysed ethylation of PhX in 1,2-dichloroethane

X	f_o	f_m	f_p
Me	2.69	1.47	5.70
Ph	0.905	0.695	2.23
F	0.364	0.116	0.738
Cl	0.271	0.102	0.538
Br	0.096	0.087	0.433

Table 6.9 Partial rate factors for reaction of 1-(*p*-anisyl)-2,2-diphenylvinyl bromide with PhX

X	f_o	f_m	f_p
MeO	15.8	—	1230
Cl	0.036	—	0.48
Me	3.2	2.1	19.5

benzylation the isomer distribution was found to vary markedly with contact time and the true isomer distribution was therefore obtained by extrapolation to zero time of a plot of distribution against contact time. Despite these precautions, isomerization may not have been entirely eliminated.[59]

The low selectivity of alkylations is also shown by the data in Table 6.8 for $GaBr_3$-catalysed ethylation by ethyl bromide in 1,2-dichloroethane.[119] Low selectivity is also obtained in propylation in molten salts, the PhMe:PhH:PhCl rate ratios being 1.9:1:0.55.[48] Partial rate factors for the reaction of 1-(*p*-anisyl)-2,2-diphenylvinyl bromide (**32**) with substituted benzenes (Table 6.9)[102] show this reagent to be less selective, probably owing to (secondary) stabilization of the incipient cation by the *p*-methoxyphenyl group; its bulk is evident in the low *ortho*:*para* rate ratios.

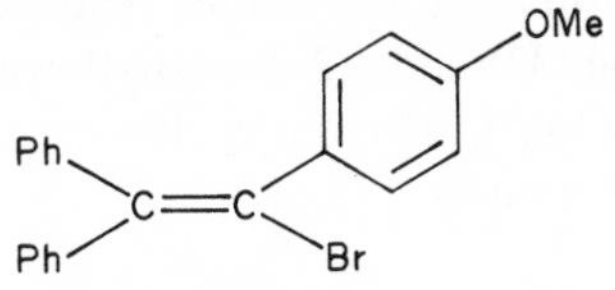

OMe

Ph

C=C

Ph Br

(**32**)

6.1.4 Cyclialkylation

This describes intramolecular alkylation which can take place with halides, alcohols, alkenes, and carbonyl compounds in which the functional group is at the end of an alkyl chain of a length sufficient to permit substitution in the aryl group at the other end of the chain, e.g. eq. 6.20. The reaction has been reviewed in

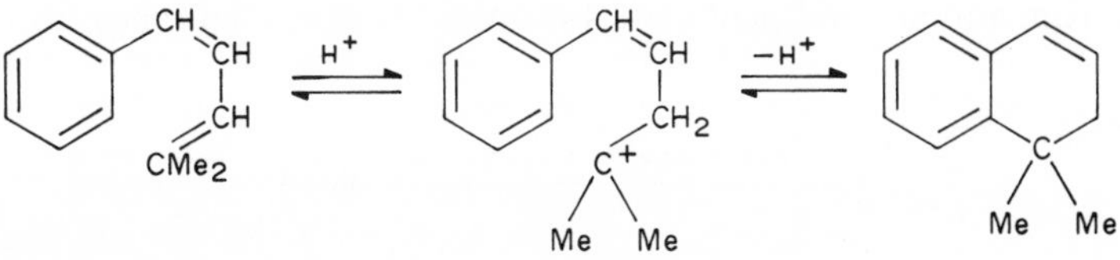

CH CH CH CMe_2 $\xrightleftharpoons{H^+}$ CH CH CH_2 C^+ Me Me $\xrightleftharpoons{-H^+}$ C Me Me (6.20)

considerable detail[120] and is of considerable synthetic utility, especially in the synthesis of heteroaromatics.

Cyclialkylation may be brought about by conditions similar to those used for alkylation, e.g. use of Lewis acid catalysts for alkyl halides and protic acids for alcohols, alkenes, and carbonyl compounds. Either six- or five-membered rings may be obtained (the former the more readily), and the reaction may be accompanied by rearrangement and hydride shifts. These are prevalent if the functional group is not in the position necessary to give rings of the required size. For example, the $AlCl_3$-catalysed cyclialkylation of α-bromoalkyl aryl ketones to indanones (eq. 6.21)[121] almost certainly occurs through a 1,2-hydride shift,

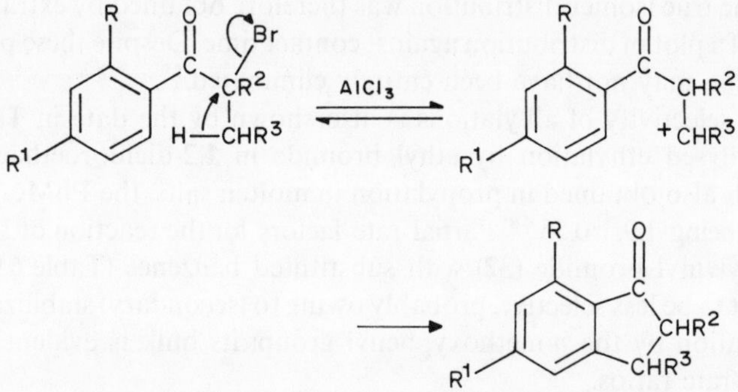

(6.21)

driven by the ability to form a stable product. Another example is the formation of 1-ethyl-1-methylindane (**33**) and 1,2-dimethyltetralin (**35**) from 2-hydroxy-3-methyl-1-phenylpentane (**34**).[122] These ring closures provide a major route to the synthesis of polycyclic aromatics.

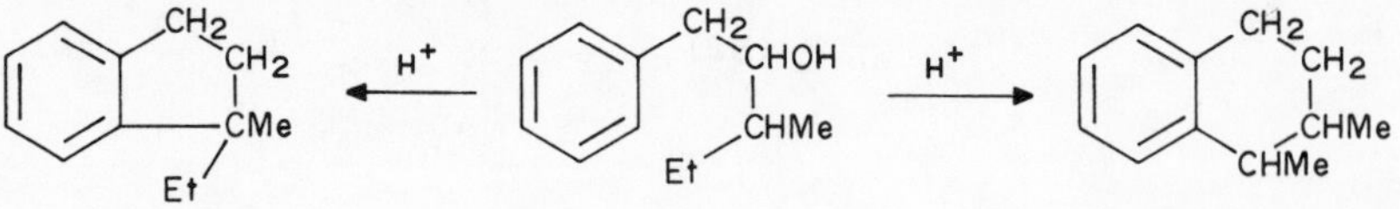

(**33**) (**34**) (**35**)

Ring closure is facilitated by the presence of electron-supplying groups *ortho* and *para* to the closure site. This feature is made use of in, for example, the synthesis of benzo[*b*]furans (eq. 6.22),[123] indoles,[124] and quinolines (eq. 6.23), the latter (Combes synthesis) involving acid-catalysed cyclodehydration of anils formed from anilines and pentane-2,4-diones. The side-chain may also contain

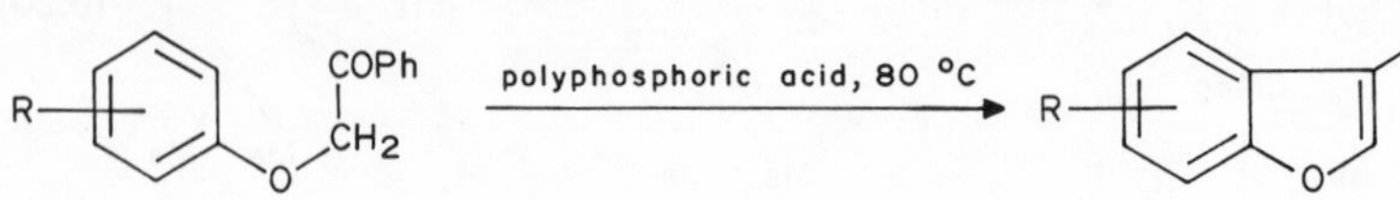

polyphosphoric acid, 80 °C (6.22)

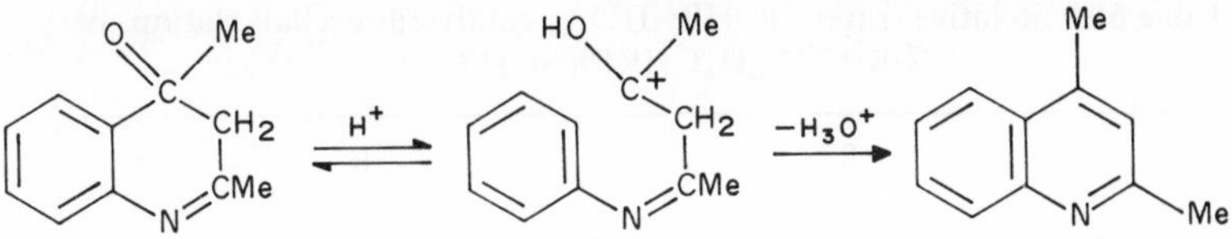

(6.23)

heteroatoms remote from the aromatic ring, this being the basis of the Bischler–Napieralski and Pictet–Spengler methods of preparing isoquinolines and derivatives, examples being given in eqs 6.24 and 6.25; numerous variations of the basic methods have been used.

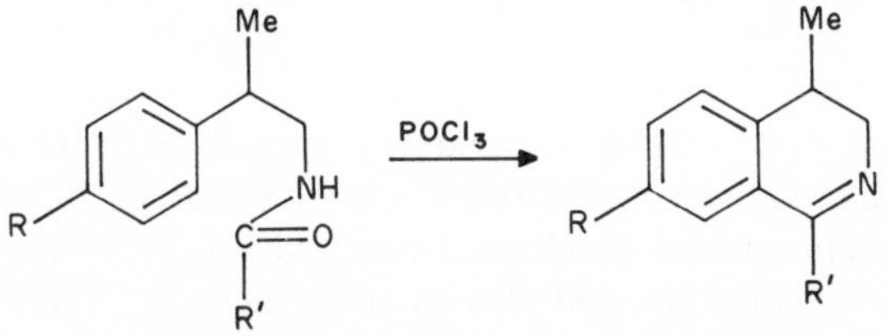

(6.24)

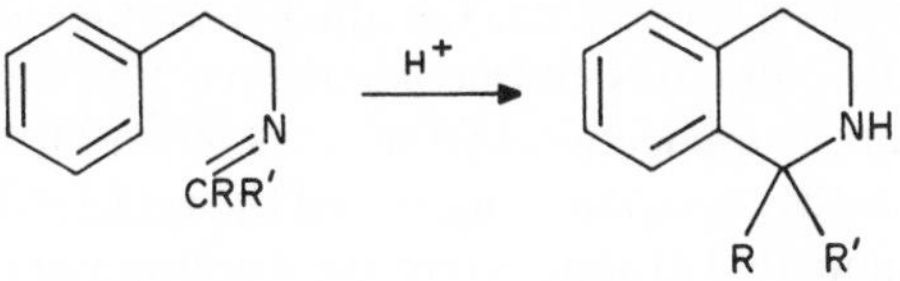

(6.25)

A kinetic study of the cyclialkylation in eq. 6.26 showed the reaction to be acid catalysed and first order in aromatic.[125] The electronic effects of substituents R^2 are fairly small because although electron supply facilitates protonation of the carbonyl group, at the same time it decreases the reactivity of the cation

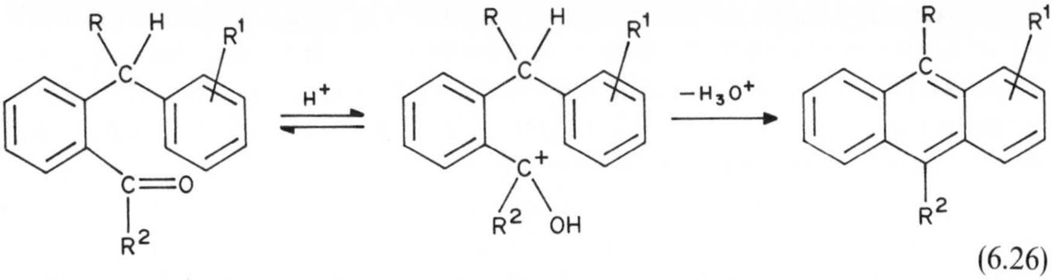

(6.26)

produced. This is evident from the indeterminate effects of *meta* and *para* substituents in benzyl groups R^2 (Table 6.10)[126] (and also in the formation of 9,9-diarylfluorenes from biphenyl-2-yldiarylmethanols).[127a] The tendency for *ortho* substituents in the benzyl groups to produce significantly greater deactivation, coupled with the much higher reactivity of the corresponding aldehyde ($R^2 = H$), suggests that the reaction is very sterically hindered. A clearer indication of the electronic effect has been obtained from the relative rates of cyclization of the

Table 6.10 Relative rates of HBr-HOAc-catalysed cyclialkylation of 2-$R^2COC_6H_4CHRPh$ at 117.5 °C

R^2	$k_{rel.}$	R^2	$k_{rel.}$	R^2	$k_{rel.}$
H	123	$C_6H_5CH_2$	5.24	3-$MeC_6H_4CH_2$	1.00
Me	15.9	4-$MeC_6H_4CH_2$	0.96	3-$FC_6H_4CH_2$	1.20
Et	6.8	4-$FC_6H_4CH_2$	0.62	3-$ClC_6H_4CH_2$	1.20
Ph	1.0	4-$ClC_6H_4CH_2$	1.15	3-$CF_3C_6H_4CH_2$	1.46
		4-$CF_3C_6H_4CH_2$	2.12	2-$MeC_6H_4CH_2$	0.17[a]

[a] From data obtained at 150 °C.

compounds with R, R^1 = H, R^2 = pyridyl or phenyl. Here the steric effects are constant, and the relative rates (at 100 °C were 3.8 (2-Py), 3.7 (3-Py), and 11.4 (4-Py).[127b] The reaction is also accelerated two fold when R = Me, as expected for an electrophilic substitution, and also by substituents R^1, the k_{rel} values being 3.5 (2-Me), 45.5 (3-Me), and 3.2 (4-Me).[128] (Note that here the 3-methyl substituent is *para* to the reaction site whereas the other substituents are both *meta*.) A 4-methyl substituent (i.e. *para* to N and *meta* to the reaction site) produced a similar (sixfold) acceleration in cyclization of the anils (eq. 6.23),[129] and in the same reaction the 2,3-dimethyl compound increased the rate 430-fold, from which the activating effect of a methyl group *para* to the reaction site may be calculated as 140-fold.[130] The reaction gives a small kinetic isotope effect $k_H/k_D = 1.5$[129], which may be a secondary effect (Section 2.1.3).

It appears from the above that substituent effects in the substituted ring are considerably greater in cyclialkylation than alkylation. Another recent example which also demonstrates this is the cyclization of 1- and 2-naphthylbutanol.[131] The latter substitutes normally in the 1-position to give the benzotetralin **36**, which is also produced (84%) from 1-naphthylbutanol. However, this compound also gave 16% of the spiro intermediate **37**, which then rearranged to **38**, as shown by the presence of deuterium at *both* side-chain α-positions, the deuterium being present only at the terminal position of the starting material. This is an example of *ipso* substitution, facilitated here because the 1-position of naphthalene is much more reactive than the 2-position. Moreover, when the starting material contained a 4-methoxy substituent (**38**), the increased activation of the 1-position resulted in 71% of reaction via *ipso* substitution. A spiro intermediate is also formed in one of two pathways leading to the formation of substituted tetralins in the trifluoroacetic acid-catalysed cyclialkylation of ω-arylalkenes $Ar(CH_2)_nCH{=}CH_2$.[132] The reaction here is notable in that the rates passed through a maximum for $n = 3$, indicating unique anchimeric assistance of protonation by the aryl group (eq. 6.27). Substituent effects indicated that the Ar_6 mechanism is preferred.

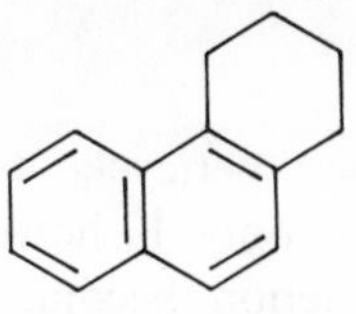

$(CH_2)_4OH$

OMe

(36) 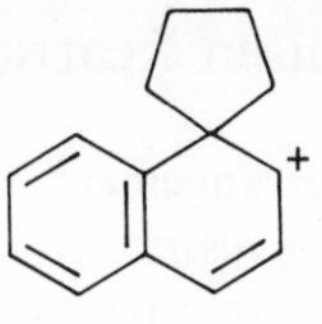 (37) 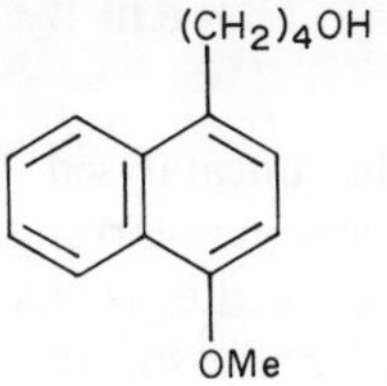 (38)

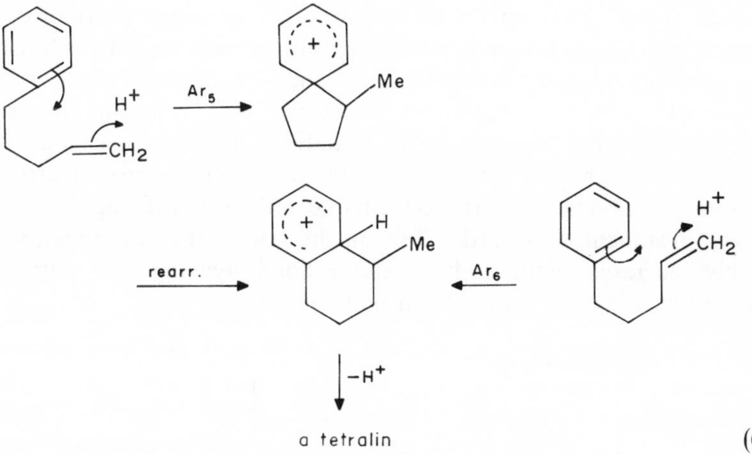

H^+ Ar_5 Me CH_2

rearr. H Me Ar_6 H^+ CH_2

$-H^+$

a tetralin (6.27)

Hydride-abstraction mechanisms which accompany normal alkylation [Section 6.1.1.(ii)] are also significant in cyclialkylation. An example of such a process and which appears to involve anchimeric assistance by a phenyl group occurs in cyclialkylation and bicyclialkylation of diphenylalkyl chlorides.[133] Apart from the main product 1-benzyl-3-methyltetralin (**39**), and others arising from hydride shifts followed by ring closure, the bicyclo derivative **40** was obtained if $AlCl_3$ was used as the catalyst (but not with $AlCl_3$–$MeNO_2$ or using the corresponding alcohols with H_3PO_4). Further, the proportion of **40** obtained depended markedly on the stereoisomer used for the reaction, indicating that the hydride abstraction from the 3-carbon of **39** which occurs (and leads to **40**) is anchimerically assisted by the phenyl group at the 1-position.

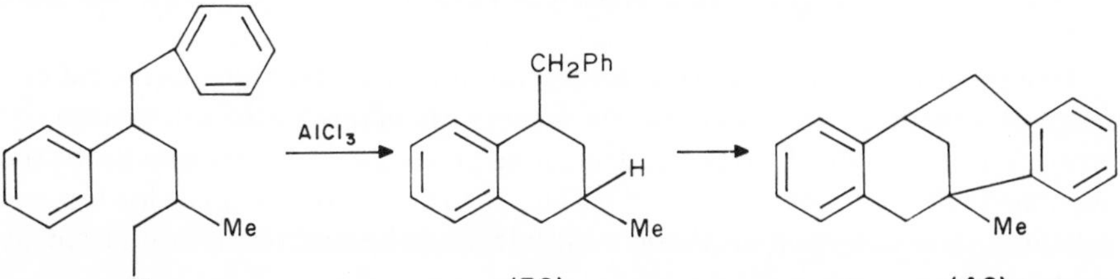

$AlCl_3$ CH_2Ph H Me Me Me Cl

(39) (40)

6.1.5 Claisen Rearrangement

The uncatalysed Claisen rearrangement is a concerted pericyclic [3, 3] sigmatropic rearrangement and gives rise to *ortho* products only. If, however, protic or Lewis acid catalysts are present then the reaction becomes an electrophilic substitution, is no longer concerted, and gives rise to *ortho*, *meta*, and *para* products (eq. 6.28). This is demonstrated, for example, by the reaction of 2-methoxyphenyl allyl ether with sulphuric acid. The ultimate location of the allyl group is shown, in percentage terms, in **41** and substitution also took place in the toluene solvent, showing that the reaction is at least in part intermolecular.[134] The reaction evidently proceeds via initial protonation at the carbon bearing the allyloxy substituent, and it is notable that in the TFA-catalysed reaction the rate variation with added water, lithium perchlorate, and trifluoroacetate salts[135] parallels exactly that observed in acid-catalysed hydrogen exchange.[136] The latter proceeds via rate-determining protonation and the correct interpretation of the rate variation in the rearrangement due to these added reagents is given in ref. 136. Substituents have little effect on the rate of rearrangement, which is probably the result of their effect on the bond-breaking step being largely cancelled by their effect on the bond-making step.

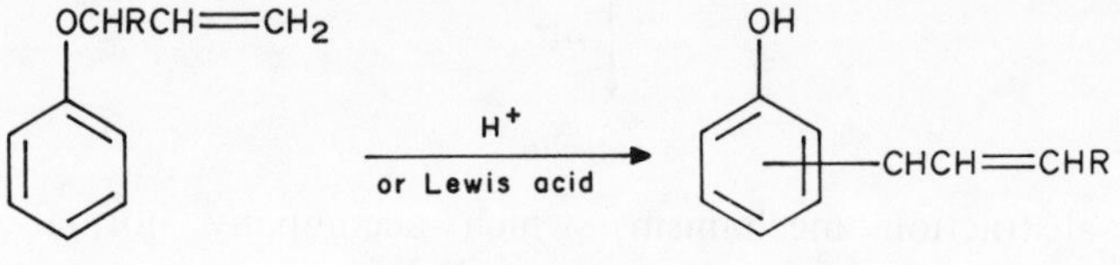

$$C_6H_5OCHRCH{=}CH_2 \xrightarrow[\text{or Lewis acid}]{H^+} HOC_6H_4{-}CHCH{=}CHR \qquad (6.28)$$

$OCH_2CH{=}CH_2$

OMe

44 %

14 %

2%

24%

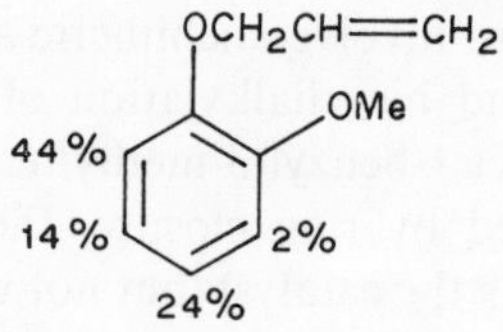

(41)

6.2 ARYLATION

Treatment of aromatics with a combination of a protic and Lewis acid can result in coupling of the aromatics. The mechanism of this the Scholl reaction[137a] is not clear, but it is likely to proceed by initial protonation of one aromatic by the superacid present to give a σ-complex. This then carries out normal electrophilic substitution of a second aromatic molecule, and the reaction is completed by dehydrogenation (eq. 6.29).

+ $HAlCl_4$ → PhH → $-H_2$, H^+ →

(6.29)

An incipient aryl cation is formed in the reaction between aryl halides and $AlBr_3$.[137b] The order of reactivity for the aryl halides is unusual, viz. PhF > PhCl > PhBr, and thus, for example, reaction of fluorobenzene gives 4-fluorobiphenyl. In the presence of water, halogen exchange also occurs. Yields of biaryls from the $AlCl_3$-catalysed reaction of 2-chlorothiophene with aromatics are toluene 3%, *m*-xylene 20%, anisole 44%, 1-methoxynaphthalene 83%.[137c]

Arylation may also be obtained by substitution with the aryl cation obtained from decomposition of aryldiazonium fluoroborates.[138] Partial rate factors for the reaction in acetonitrile or sulpholane (Table 6.11)[139] are typical of a very unselective electrophilic substitution. Other compounds which have also been used are phenylazo-*p*-tolylsulphone (**42**),[140] phenylazotriphenylcarbinol (**43**),[141] and phenylazotrifluoroacetate (**44**),[142] each of which gives similar partial rate factors. The ρ-factor for the reaction is ca -0.8 and the electrophile is thought to be the radical dication **45**, although the evidence for this (said to involve concerted uncoupling of a pair of π-electrons, one of which 'falls' into the vacant sp^2 σ-orbital) is not yet strong. There is also a difficulty in interpreting the

$PhNNSO_2$–C_6H_4–Me (**42**) $PhNNCPh_3$ (**43**) $PhNNOCOCF_3$ (**44**) (**45**)

Table 6.11 Partial rate factors for the reaction of phenyldiazonium tetrafluoroborate with PhX in acetonitrile at 40 °C

X	*ortho*	*meta*	*para*
OMe	3.14	0.68	3.53
Me	1.33	0.61	1.84
Cl	1.02	0.49	1.15
CO_2Me	0.42	0.40	0.23
NO_2	0.29	0.38	0.23

mechanism because, anomalously, *meta*-alkyl groups always deactivate. Partial rate factors provide evidence that the TFA-catalysed reaction of dibenzoyl peroxide with aromatics proceeds (at least in part) through a phenyl cation.[143]

Phenyl cations are also obtained in the reaction of aryl azides with trifluoroacetic and trifluoromethanesulphonic acids.[144] Arylnitrenium ions are formed as intermediates, resonance stabilization of which creates an aryl cation (**46**). Substitution can therefore give both diarylamines, or aminobiphenyls (eq. 6.30), and it is interesting that the nitrene formed from naphthalene (**47**) substitutes into benzene giving only 1-amino-4-phenylnaphthalene (**48**), whereas the nitrene formed from benzene (**46**) substitutes into naphthalene to give mainly *N*-phenyl-1-naphthylamine (**49**). This may be explained by the fact that stabilization of the positive charge from nitrogen into the ring will be more favourable for naphthalene because there is less loss of resonance energy than is the case in benzene. Thus, reaction as a carbon electrophile rather than a nitrogen electrophile is more favourable.

N_3 H^+ N_2^+ NH $-N_2$ NH^+ NH NH (**46**) ArX X NH o,p-NH_2 X

(6.30)

Substitution by a phenonium ion analogous to that shown in eq. 6.29 has also been observed in the decomposition by TFA of the adduct **50** (formed from 1-chloro-2,3-dimethylbenzene and acetyl nitrate).[145] Protonation and loss of acetic acid give a phenonium ion which substitutes into, e.g., mesitylene to give, after 1,4-loss of nitrous acid, the product **51**.

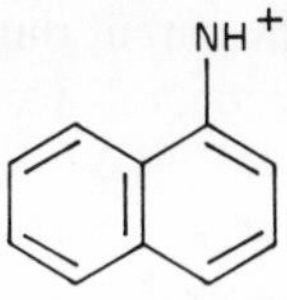

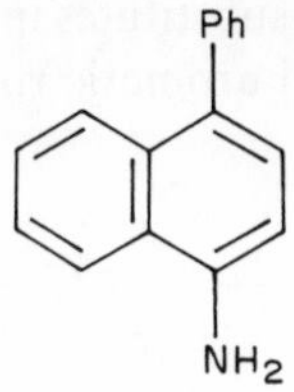

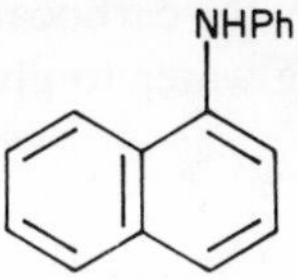

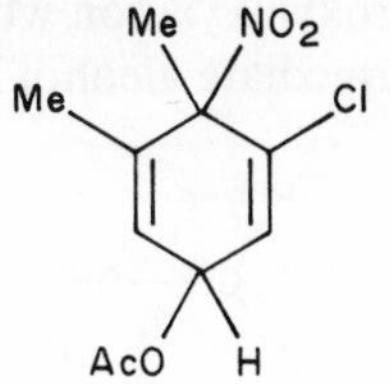

(47) (48) (49) (50)

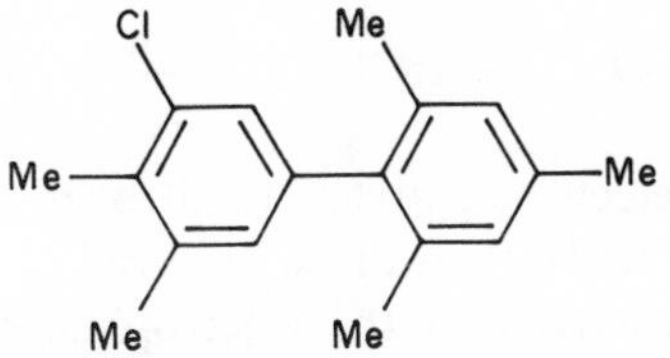

(51)

6.3 AMINOALKYLATION

Treatment of reactive aromatics with aldehydes, especially formaldehyde, and a secondary amine results in aminoalkylation, e.g. eq. 6.31.[146] The electrophile is an iminium derivative, e.g. $RN^{+}H{=}CH_2 \leftrightarrow RNH{-}CH_2{}^{+}$, and may also be prepared from aldehyde precursors (HCl–MeCN) and amines.[147] Electron-withdrawing substituents on nitrogen enhance the reactivity of the electrophile.[148]

$$Me_2N{-}C_6H_5 + HCHO + HN(CH_2)_4 \longrightarrow Me_2N{-}C_6H_4{-}CH_2N(CH_2)_4 + H_2O \quad (6.31)$$

6.4 HYDROXYALKYLATION

Aldehydes and ketones are protonated by protic acids and coordinated with Lewis acids to give the electrophiles **52** and **53**, respectively, and these substitute aromatics to give alcohols.[104] These may be further protonated, subsequent loss of water producing a second electrophile and substitution in a second molecule of aromatic (eq. 6.32). This leads to one drawback of the reaction, namely that the alkylated aromatics will have increased reactivity and polymeric products are often obtained, although this is used to advantage in the formation of phenol–formaldehyde resins. Some cyclialkylations (Section 6.1.4) involve a form of hydroxyalkylation whereby the carbocation substitutes into the parent ring, the intermediate alcohol losing water to give an aromatic ring.

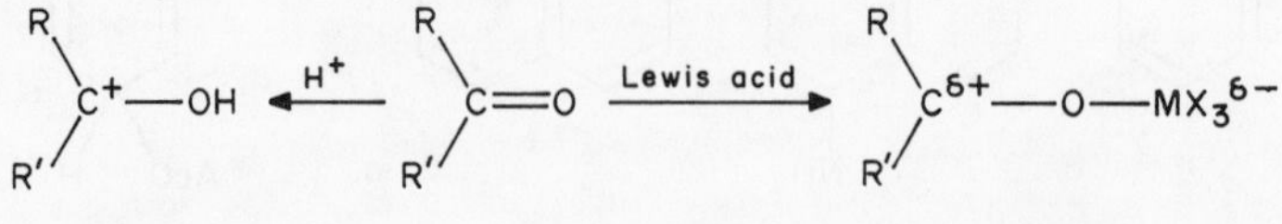

$$RR'C^{+}\text{—}OH \xleftarrow{H^+} RR'C{=}O \xrightarrow{\text{Lewis acid}} RR'C^{\delta+}\text{—}O\text{—}MX_3^{\delta-}$$

(52) (53)

$$ArH + RR'C{=}O \xrightarrow{H^+} ArCRR'OH \xrightarrow{H^+} ArCRR'Ar + H_2O \qquad (6.32)$$

Since oxygen is more electronegative than nitrogen, the electrophiles are, for a given structure, more reactive than those involved in aminomethylation. As in aminoalkylation, the reactivity of the electrophile can be enhanced by the presence of electron-withdrawing substituents. A well known example demonstrating these features is formation of DDT, in which the electrophile is sufficiently reactive to be able to substitute chlorobenzene (eq. 6.33).

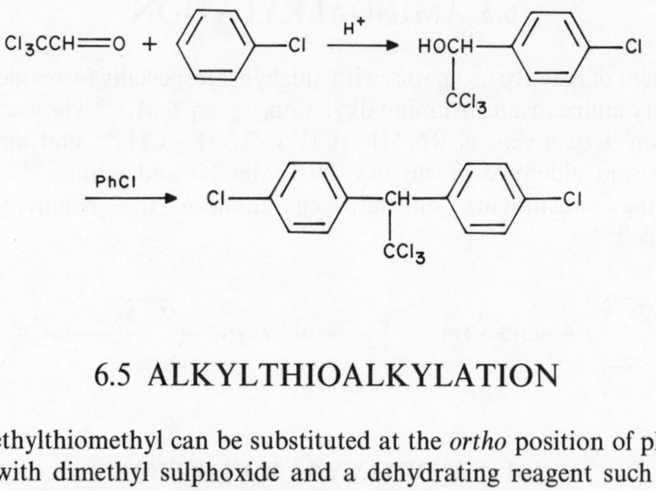

$$Cl_3CCH{=}O + C_6H_5Cl \xrightarrow{H^+} HOCH(CCl_3)C_6H_4Cl \xrightarrow{PhCl} Cl\text{—}C_6H_4\text{—}CH(CCl_3)\text{—}C_6H_4\text{—}Cl \qquad (6.33)$$

6.5 ALKYLTHIOALKYLATION

The methylthiomethyl can be substituted at the *ortho* position of phenols by reaction with dimethyl sulphoxide and a dehydrating reagent such as acetic anhydride, dicyclohexylcarbodiimide (DCC), or pyridine–SO_3, these reagents probably functioning also as polarizers of the S=O bond (eq. 6.34).[149] Coordination of the electrophile with the substituent evidently occurs.

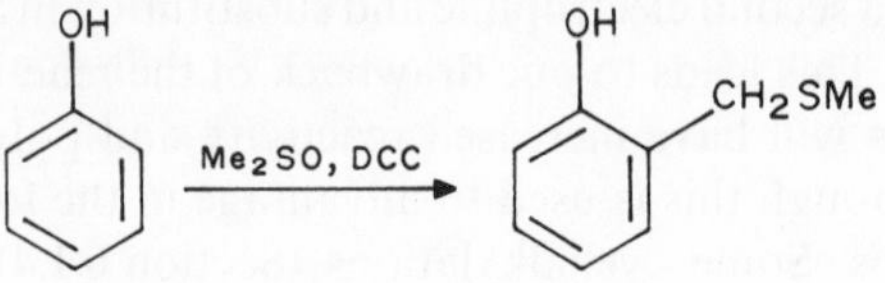

$$C_6H_5OH \xrightarrow{Me_2SO,\ DCC} 2\text{-}(CH_2SMe)C_6H_4OH \qquad (6.34)$$

6.6 HALOALKYLATION

The group $-CH_2$ hal can be introduced into aromatic rings by the reaction of the aromatic with formaldehyde and hydrogen halide in the presence of a protic or Lewis acid catalyst.[150] The simplest example is the preparation of benzyl chloride in 79% yield by heating benzene with paraformaldehyde and $ZnCl_2$ at 60 °C while HCl is passed in to the reaction mixture (eq. 6.35);[151] formalin may also be used as the source of formaldehyde.

$$PhH + CH_2O + HCl \xrightarrow{ZnCl_2} PhCH_2Cl + H_2O \qquad (6.35)$$

Haloalkylation shares features in common with hydroxyalkylation, including for example a tendency for diarylation (eq. 6.36), especially when the aromatic is very reactive, and/or polymerization. Indeed under some conditions the mechanism of both reactions is similar, differing only in the last step of the reaction. Thus, the reaction of mesitylene with formaldehyde–hydrochloric acid in acetic acid was first order in aromatic and aldehyde, the mechanism proposed being as in eqs. 6.37–6.39.[152] The halogen is only introduced in the fast step (eq. 6.39) and consistent with this is the observation that the bromomethylation of toluene, ethylbenzene, and isopropylbenzene gives the same *ortho*:*para* ratios as in chloromethylation, indicating that a common electrophile is involved.[153] Moreover, under the conditions used for chloromethylation, benzyl alcohol is completely converted to benzyl chloride.[153] The reaction is accelerated by the presence of halides, especially $ZnCl_2$, which may not necessarily act as a Lewis acid here but rather raise the acidity of the medium (by generating HCl).[154] The reaction succeeds with aromatics of the reactivity of chlorobenzene or greater.

$$ArH + ArCH_2Cl \xrightarrow[ZnCl_2]{ArH} ArCH_2Ar + HCl \qquad (6.36)$$

$$HCHO \overset{H^+}{\rightleftharpoons} CH_2OH \qquad \text{(fast)} \qquad (6.37)$$

$$ArH + CH_2{}^+OH \longrightarrow ArCH_2OH + H^+ \qquad \text{(slow)} \qquad (6.38)$$

$$ArCH_2OH \overset{HCl}{\rightleftharpoons} ArCH_2Cl \qquad \text{(fast)} \qquad (6.39)$$

A disadvantage with halomethylation under these conditions, and presumably therefore some hydroxyalkylations, is that the hydroxymethyl cation is carcinogenic. So too are some of the alternative and more reactive chloromethylating reagents, e.g. chloromethyl methyl ether ($ClCH_2OMe$),[155] bis(chloromethyl) ether [$(ClCH_2)_2O$],[156] methoxyacetyl chloride ($MeOCH_2COCl$),[157] and 1-chloro-4-(chloromethoxy)butane.[158] The last compound in particular is less volatile and potentially safer and, moreover, produces tetrahydrofuran as a byproduct, which then complexes with the acid catalyst and reduces side-reactions.

Bromomethylation may be carried out with bromomethyl methyl ether, which is about ten times more reactive than the chloro compound.[155] Iodomethylation has been carried out using chloromethyl methyl ether and HI[159] or HCHO–HI,[160] but fluoromethylation with either HCHO–HF or fluoromethyl methyl ether is unsuccessful since further reactions occur. Benzyl fluoride has been obtained by the $ZnCl_2$-catalysed reaction of fluoromethanol with benzene at low temperature.[161]

The relative reactivities of aromatics towards chloroalkylation have been measured but the results differ significantly according to the conditions employed, Table 6.12. Comparison of the sets of data suggests that the electrophile derived from chloromethyl methyl ether is bulkier, which seems reasonable. An earlier study using HCHO (paraformaldehyde)–HCl gave very small relative rates and deactivation by substituents which normally activate towards electrophilic substitution,[163] and these results are probably unreliable. Re-examination of the reactivity of toluene towards paraformaldehyde–HCl gave a toluene: benzene reactivity ratio of 112 and partial rate factors of $f_o = 117$, $f_m = 4.4$, $f_p = 430$ for the methyl substituent;[164] for reaction with chloromethyl methyl ether values of 45.6, 2.79, and 84.6 were obtained.[162] Under the latter conditions, the ρ factors have been quoted as -3.55 to -4.9.[162] Some isomer distributions which have been obtained in chloromethylation are given in Table 6.13. The *ortho*:*para* ratios for chloromethylation of toluene vary according to the catalyst concentration, which was attributed to a change in mechanism.[165] This is incorrect, however, since the $\log f_o$:$\log f_p$ values were constant (0.79) under all conditions (see Section 11.2.4).

From the data in Tables 6.12 and 6.13, and other reports, the following conclusions may be drawn.

(1) The electrophile in chloromethylation is less reactive and more selective than that in alkylation. For the HCHO–HCl reagent this is understandable since the hydroxymethyl cation should be a more stable reagent because of the resonance stabilization, $^+CH_2\text{—}\ddot{O}H \leftrightarrow CH_2\text{=}OH^+$.

(2) When chloromethyl methyl ether is used the electrophile is likely to be $^+CH_2Cl$ formed either by protonation of oxygen and loss of the stable species ROH, or by coordination of the Lewis acid with oxygen (more favourable than coordination with halogen). This species should be more electrophilic than $^+CH_2OH$, so accounting for the observed difference in selectivity. The higher reactivity in bromomethylation with bromomethyl methyl ether (and the predominance of secondary alkylation in fluoromethylation) can each arise from a differing balance of inductive and conjugative effects from the halogen in the corresponding electrophiles.

(3) The *ortho*:*para* ratios for the halogenobenzenes increase with increasing size of the halogen. This contrasts with, for example, the results for hydrogen

Table 6.12 Relative reactivities of ArH towards chloromethylation

ArH	$k_{rel.}$	
	HCHO–HCl, 65 °C[155]	$ClCH_2OMe$, 85 °C[162]
Benzene	1.0	1.0
Toluene	3.0	30.0
o-Xylene	—	46.5
p-Xylene	—	33.9
m-Xylene	24.0	1220
1,2,4-Trimethylbenzene	—	8010
Mesitylene	600	4800
1,2,3,4-Tetramethylbenzene	—	1500
Anisole	1300	—
2,5-Dimethylanisole	100 000	—
Chloromesitylene	2.0	—
Diphenyl ether	—	70.3

Table 6.13 Isomer distributions (%) in chloromethylation of PhX with HCHO–HCl

X	*ortho*	*meta*	*para*	Catalyst	Ref.
Me	45	1.1	54	—	165
Et	29	1.9	69	—	165
i-Pr	12	3.2	84	—	165
t-Bu	~0.1	5.8	94	—	165
CH_2CO_2Et	27	12.6	60.4	$ZnCl_2$ in CCl_4	153
F	11	—	89	$ZnCl_2$ in CCl_4	153
Cl	37	—	63	$ZnCl_2$ in CCl_4	153
Br	42	—	58	$ZnCl_2$ in CCl_4	153
I	47	—	53	$ZnCl_2$ in CCl_4	153

exchange (Section 3.1.2.5), and shows that demands for conjugative electron release here are not sufficient to outweigh the inductive effects; this is entirely consistent with the relative reactivities of the electrophiles in the two reactions.

(4) Although the electrophile in chloromethylation is less reactive than that in alkylation, the reaction is successful even with nitrobenzene,[150,166] which further suggests that the low reactivity of nitrobenzene in alkylation is due to coordination with the Lewis-acid catalysts (see Section 6.1.3).

6.7 FRIEDEL–CRAFTS ACYLATION

In acylation the —COR group is introduced into the benzene ring, usually by reaction of the aromatic with an acid halide or anhydride (less commonly with an acid or amide) in the presence of a Lewis acid, e.g. eq. 6.40. Carboxylic acids, catalysed by sulphuric acid, phosphoric acid, or hydrogen fluoride, may also be used. The reaction has been extensively reviewed.[1]

$$\mathrm{PhH + MeCOCl \xrightarrow{AlCl_3} PhCOMe + HCl} \tag{6.40}$$

Acylations are not subject to the same complications that beset alkylations. The acyl group which is introduced deactivates the nucleus to further substitution so that monoacyl derivatives are easily isolable. Rearrangements of the nuclear-substituted products or within the acyl group do not normally occur. The process is therefore of much greater value than alkylation and yields are usually high. Strongly deactivated nuclei such as that of nitrobenzene cannot, however, be acylated; this makes nitrobenzene a useful solvent for acylations.

6.7.1 Mechanism of Acylation by Acyl Halides

(i) Formation of σ-complexes

The treatment of alkylbenzenes with formyl fluoride and BF_3 at low temperatures gives 1:1:1 complexes which dissociate above their melting points to give aldehydes.[15] These are almost certainly σ-complexes such as **54** from mesitylene, and it is likely that similar intermediates are formed in other acylations.

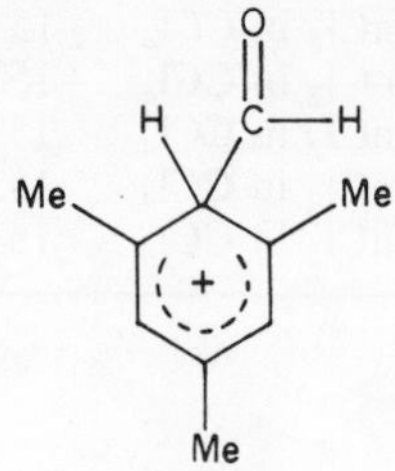

O, H, C—H, Me, Me, +, Me

(54)

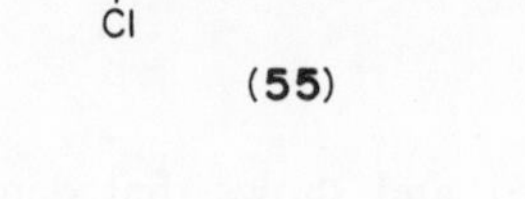

$$\mathrm{R{-}C(Cl){=}O^{\delta+}\cdots{}^{\delta-}MCl_n}$$

(55)

$$\mathrm{R{-}C^{+}{=}O}$$

(56)

$$\mathrm{R{-}C{=}O^{\delta+}\cdots{}^{\delta-}AlCl_3}$$
$$\mathrm{Cl\cdots AlCl_3^{\delta-}}$$

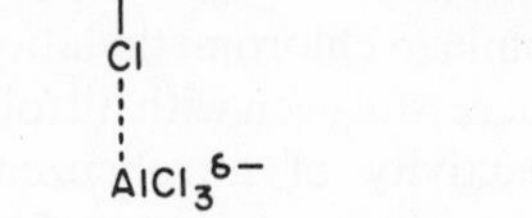

(57)

(ii) The electrophilic species

The Lewis acid polarizes the acylating agent to form either the polarized complex **55** or the acylium ion **56**. Either of these species may be primarily responsible for substitution, and that which is the more effective in a given case will depend on the nature of R, the aromatic, and the reaction medium. Evidence for these species is as follows:

(1) Many Lewis acids are known to form stable addition compounds with acyl halides, and these are mostly 1:1 donor–acceptor complexes.[167] Proton and fluorine n.m.r. studies of the addition complexes formed between acetyl, propanoyl, and benzoyl fluorides and SbF_5, AsF_5, PF_5, and BF_3 have shown them to be acylium salts, $RCO^+MF_{4(6)}^-$, in the solid state and strongly polarized complexes similar to **55** in solution.[168]

(2) Spectroscopic measurements show that the carbonyl oxygen of acid halides is coordinated to the metal in a halide–Lewis acid complex.[169] I.r. evidence shows that in a high dielectric medium, nitrobenzene, acetyl chloride and $AlCl_3$ form a mixture of both the polarized complex **55** and the acetylium ion **56**, but in chloroform only the former, and in the solid state only the latter are obtained. The existence of both species has been confirmed by X-ray spectroscopy,[170] which has in addition confirmed the proposal by Taylor[171] that the charge on the carbonyl carbon is better delocalized in the aromatic acyl halide–complex than in the aliphatic acyl halide–complex. Thus, acylation by the former will, relative to acylation by the latter, show a greater preference for the free acylium ion mechanism (see eq. 6.45) rather than the nucleophilic displacement mechanism (see eq. 6.46). This explains why benzoylation is less sterically hindered than acetylation (see below).

In less polar solvents, more than one molecule of Lewis acid is required to polarize the acyl halide, as shown by an i.r. study of the MeCOX–$AlCl_3$ system in 1,2-dichloroethane.[172] The structure of the 1:2 complex may be **57**, although coordination of catalyst dimer with oxygen is also possible.

Coordination of the catalyst with oxygen rather than halogen is also shown by the greater stability of the complexes formed between Lewis acids and ketones compared with those formed with acyl halides,[167,170] and this follows from the greater electron supply to oxygen in the former. Coordination with the ketone has kinetic consequences described below (Section iii).

(3) The conductance of a solution of $AlCl_3$ in benzoyl chloride shows that there is slight dissociation of the complex into free ions, while PhCOCl and molten $GaCl_3$ give a conducting medium in which ionization is extensive. The presence of the benzoyl cation is inferred in each case[173] (eq. 6.41).

$$PhCOCl \cdot AlCl_3 \rightleftharpoons PhCO^+ + AlCl_4^- \qquad (6.41)$$

(4) 2,4,6-Tribromobenzoyl chloride is rapidly equilibriated with the corresponding acid bromide in the presence of aluminium bromide in nitrobenzene. This indicates the intermediacy of the acylium ion **58** (eq. 6.42).[174]

COCl, Br, Br, Br $\xrightleftharpoons{AlBr_3}$ O, C+, Br, Br, Br, $AlBr_3Cl^-$ **(58)** $\xrightleftharpoons{-AlClBr_2}$ COBr, Br, Br, Br (6.42)

(5) Anisole reacts with pivaloyl chloride (Me_2CCOCl) in the presence of $AlCl_3$ to give *tert*-butyl 4-methoxyphenyl ketone, but under the same conditions benzene gives mainly *tert*-butylbenzene and carbon monoxide, while the monoalkylbenzenes give mixtures of ketone and aromatic hydrocarbon.[175] Both reaction products evidently derive from a common intermediate (**55** or **56**), which either acylates or, if the aromatic is less reactive, has time to lose carbon monoxide to give a carbocation which then alkylates the aromatic. The relative stability of a tertiary carbocation provides the driving force for the decarbonylation.

Acylations catalysed by proton acids are fairly rare. HF will catalyse acylation by acyl fluorides, through formation of the stable HF_2^- ion (eq. 6.43), and in nitromethane, in which the HCl_2^- ion is stable, HCl will catalyse acylation by acyl chlorides.[176] Trifluoromethanesulphonic acid is particularly useful for reactions involving nitro compounds, which give kinetic complications when Lewis acid catalysts are used. Its potential is indicated by the yields in reaction of benzoyl chloride with *p*-xylene, e.g. 82% (CF_3SO_3H), 28% (H_2SO_4), and 21% (CF_3CO_2H).[177]

$$RCOF + HF \rightarrow RCO^+HF_2^- \xrightarrow{ArH} ArCOR + 2HF \qquad (6.43)$$

(iii) Kinetics

These have been concerned with the reaction order, the effect of varying R in RCOX, the effects of solvents, and substituent effects.

(a) *The order in reagents and their relative reactivities.* There are two mechanisms for acylation, comparable to those which apply to alkylation, i.e. substitution by the free acylium ion into the aromatic, or nucleophilic displacement by the aromatic upon the polarized acyl halide–Lewis acid complex.[178]. The rate difference for reaction by the two mechanisms will diminish the more reactive the aromatic, and more reactive aromatics should tend to react by the latter mechanism (see, however, ref. 178). The reactivity order for acyl halides will differ for the two mechanisms: the reactivity of the acylium ion will be increased by electron withdrawal, whereas this will retard polarization of the C—halogen bond which takes place in a rate-determining step in the displacement mechanism.

Benzoyl halides tend to react by the acylium ion mechanism. Thus benzoylation of aromatics by 2,4,6-tribromobenzoyl halides gave a very small rate spread, consistent with reaction via the free acylium ion,[174] and the relative reactivities of substituted benzoyl chlorides is 4-NO_2 > 4-Cl > H > 4-Me for reaction with benzene ($\rho = 1.3$) or toluene,[179] and 4-Cl > 3-OMe > 4-OMe for reaction with fluorene.[180] However, for reaction with the very reactive 1,3-dimethoxybenzene the order becomes 4-OMe > H > 4-Cl,[181] suggesting that the nucleophilic displacement mechanism is intruding here. The latter mechanism accounts for electron-withdrawing groups Y in YCH_2COCl reducing the rate of the $AlCl_3$-catalysed acylation of reactive aromatics and increasing the rates of unreactive ones and *vice versa.*[179,182]

Reaction by the nucleophilic displacement mechanism should become more difficult as R in RCOX becomes larger. This appears to explain the relative reactivities for $AlCl_3$-catalysed acylation of mesitylene by a series of acyl halides which changed from X = I > Br > Cl > F for R = Me[183,184] to F > Cl > Br > I for R = *tert*-butyl; alkyl groups of intermediate size gave intermediate rate sequences.[184]

Early studies of the kinetics of acylation gave no examples of simple kinetics.[185] However, the $AlCl_3$-catalysed reaction of benzoyl chloride with benzene in benzoyl chloride as solvent gave straightforward second-order kinetics: rate = $k[PhH][PhCOCl{\cdot}AlCl_3]$; added HCl did not affect the rate.[186] These facts are consistent with reaction either by the free acylium ion (eqs 6.44 and 6.45), or by the nucleophilic displacement mechanism (eqs 6.44 and 6.46); an intermediate possibility would involve an ion pair. The kinetics departed from second order when 1 mol of benzophenone had been produced per mole of $AlCl_3$, owing to preferential complexing of the ketone with the catalyst.

$$\underset{\displaystyle\overset{|}{Cl}}{RC}{=}O + AlCl_3 \underset{}{\overset{\text{fast}}{\rightleftharpoons}} \underset{\displaystyle\overset{|}{Cl}}{RC}{=}O^+{-}AlCl_3^- \tag{6.44}$$

$$\left.\begin{array}{l} \underset{\displaystyle\overset{|}{Cl}}{RC}{=}O^+{-}AlCl_3^- \overset{\text{fast}}{\rightleftharpoons} RC{\equiv}O^+ + AlCl_4^- \\[2ex] RC{\equiv}O^+ + ArH \overset{\text{slow}}{\rightleftharpoons} {}^+ArHCOR \xrightarrow{\text{fast}} ArCOR + HCl \end{array}\right\} \tag{6.45}$$

$$\underset{\displaystyle\text{Cl}}{\text{RC}}{=}\text{O}^+{-}\text{AlCl}_3^- + \text{ArH} \underset{}{\overset{\text{slow}}{\rightleftharpoons}} \text{HAr}^+\underset{\displaystyle\text{Cl}}{\overset{\displaystyle\text{R}}{\text{C}}}\text{O}-\text{AlCl}_3^- \xrightarrow{\text{fast}} \text{ArCOR} + \text{HAlCl}_4 \quad (6.46)$$

The order in catalyst is doubled if weaker Lewis acids, e.g. $SbCl_5$, $FeCl_3$, or $GaCl_3$, are used, consistent with two molecules being required to polarize the acyl halide sufficiently for reaction to occur.[187] Complexes similar to **57** may well be involved here, especially since some of the catalyst have a greater affinity for chlorine than oxygen. The conclusion regarding the effectiveness of the catalysts contrasts with the relative rates they produce in benzoylation of toluene, benzene, or chlorobenzene with excess benzoyl chloride:[185] $SbCl_5$, 1300; $FeCl_3$, 570; $GaCl_3$, 500, $AlCl_3$, 1; $SnCl_4$, 0.003; BCl_3, 0.0006; $SbCl_3$, 'very small'. These rates, however, are probably valid only for initial rates of reaction since the kinetic orders differ according to the catalyst; it may be noted that the other works have found different relative reactivities, with $AlCl_3$ always the most effective.[188] $SnCl_4$ is useful for substrates which may be susceptible to decomposition by $AlCl_3$, and, since it coordinates less effectively to oxygen and in particular the ketone product, stoicheiometric concentrations of catalyst are not required. On the other hand its greater bulk means that steric hindrance to acylation under these conditions will be greater. For example, yields for reaction at the 4-position of 2-methyl-5-phenylthiophene decrease more with increasing size of R in RCOCl when $SnCl_4$ is used as the catalyst than when $AlCl_3$ is used.[189]

Nitrobenzene, although a very good solvent for acylations, results in complex kinetics, the order in $AlCl_3$ varying, in benzoylation, from 1.0 for benzene to 1.5 for toluene.[190] This appears to be due to complexing of the catalyst with nitrobenzene. Kinetic complications also arise if nitromethane is used as a solvent [see also alkylation, Section 6.1.1(iii)(c)], but it is possible to obtain reproducible kinetics provided that the solvent is carefully purified.[191] The kinetics are then third order but shown an inverse dependence on initial $AlCl_3$ concentration: rate $= [\text{ArH}][\text{RCOCl}][\text{AlCl}_3]/[\text{AlCl}_3]_0$. Partial rate factors for the methyl substituent, viz. $f_o = 119$, $f_m = 6.1$, $f_p = 2580$ indicate a surprisingly weak electrophile under these conditions, attributed to the inverse dependence on initial catalyst concentration.

(b) *The kinetic effect of solvents.* For the $AlCl_3$-catalysed benzoylation of toluene, the relative rates for different solvents are as follows: 1,2,4-trichlorobenzene 1.0, 1,2-dichlorobenzene 2.4, dichloromethane 3.5, 1,2-dichloroethane 4.1, excess benzoyl chloride 30.[192] Hence the use of excess acylating reagent appears to be easiest way of increasing reaction rates and, under these conditions, addition of a 1 or 2 molar excess of aluminium halide can bring about a 300–800-fold increase in the rate of acylation.[192,193]

6.7.2 Mechanism of Acylation by Other Reagents

(i) Acylation by carboxylic anhydrides, esters, amides, and acids

Acylation by these reagents has been reviewed.[176] Some have been the focus of recent interest because milder reaction conditions may be used, especially with anhydrides containing a good leaving roup. The main features are as follows:

(1) The addition of a Lewis acid to an anhydride probably converts it to the acyl halide, e.g. eq. 6.47, which carries out the acylation. This is even more likely for acylation by amides and esters since they will be less reactive.

$$RCOZ + AlCl_3 \longrightarrow RCOCl + AlCl_2Z \qquad (6.47)$$
$$(Z = OCOR, OR, \text{ or } NH_2)$$

(2) The acylium ion may be generated by addition of a proton acid to an anhydride, acid, or amide. Evidence for this comes from the freezing-point depression obtained on adding acid to the acylating reagent. For example, addition of H_2SO_4 to acetic anhydride indicates the presence of four species[194] (eq. 6.48) and five species on addition to a carboxylic acid (eq. 6.49).

$$(MeCO)_2O + 2H_2SO_4 \rightleftharpoons MeCO^+ + MeCO_2H_2{}^+ + 2HSO_4{}^- \qquad (6.48)$$

$$RCO_2H + H_2SO_4 \rightleftharpoons RCO^+ + 2H_3O^+ + 2HSO_4{}^- \qquad (6.49)$$

(3) If a mixed anhydride RCOOCOR′ is used, then two products may be obtained. If R and R′ are of similar electronic requirement then the ketone with the larger group predominates.[195] If R is more electron withdrawing than R′ then more of ArCOR′ will be obtained. Moreover, if R is very electron withdrawing then the anhydride will acylate without the need for any catalyst. Acetic trifluoroacetic anhydride will acetylate very reactive aromatics, whereas anhydrides from trifluoromethanesulphonic acid, $RCOOSO_2CF_3$, are more reactive than any other acylating reagent. For example, benzene may be benzoylated in 90% yield by merely warming with $PhCOOSO_2CF_3$.[196]

Other useful mixed anhydrides are the carboxylic dihalophosphoric anhydrides formed from the anhydrides of the constituent acids. Here electron withdrawal in the dihalophosphoric group makes it a very good leaving group, and good yields of ketones may be obtained under mild conditions (eq. 6.50).[197]

$$RCOOPO{\cdot}Cl_2 + ArH \xrightarrow{20\,°C,\,3\,h} ArCOR + Cl_2POOH \qquad (6.50)$$

(ii) Acylation by acylium salts

Acylium salts (prepared from the reaction between an acyl fluoride and SbF_5 in Freon at 0 °C) give high yields of acylaromatics, e.g. eq. 6.15.[198a] The reaction (first order in aromatic) is very sterically hindered, acylation by $EtCO^+SbF_6{}^-$

giving a *para*:*ortho* ratio of 37 with toluene, and ArH:PhH rate ratios of 92.2 (toluene), 13.7 (*p*-xylene), 204 (*m*-xylene), 652 (*o*-xylene), and 663 (mesitylene).[198b]

$$[BrCH_2C{=}O]^+SbF_5^- + PhH \xrightarrow{MeNO_2, 0°C} PhCOCH_2Br + HSbF_6 \quad (6.51)$$

(iii) Acylation by ketene

In the presence of $AlCl_3$, ketene and derivatives will acylate aromatics to give ketones (eq. 6.52). Toluene:benzene rate ratios are 47.2 and 174 for the reaction of dimethylketene and diphenylketene, respectively.[199] The former is smaller than for the corresponding reaction with acetyl chloride, which is consistent with the inability of the hydrogens of the CH_2 group to stabilize the incipient carbocation by hyperconjugation. Dimethylketene also gave lower *ortho*:*para* ratios than the reaction with 2-methylpropanoyl chloride showing the electrophile to be larger in the former case, attributable to its planarity whereas in the latter the alkyl groups are able to bend away from the aromatic ring in the transition state. The reaction of dimethylketene with benzene gave a kinetic isotope effect, $k_H/k_D = 1.06$.[199]

$$ArH + CR_2{=}C{-}O^{\delta+}{-}ArCl_3^{\delta-}$$

$$\longrightarrow Ar^+\!\begin{smallmatrix}H\\COCR_2\end{smallmatrix} AlCl_3^- \longrightarrow ArCOCHR_2 + AlCl_3 \quad (6.52)$$

(iv) Acylation with phosgene

The $AlCl_3$-catalysed reaction between phosgene and benzene first gives benzoyl chloride, and then, more slowly, benzophenone (eqs 6.53 and 6.54).[200]

$$PhH + COCl_2 \xrightarrow{AlCl_3} PhCOCl + HCl \quad (6.53)$$

$$PhCOCl + PhH \longrightarrow PhCOPh + HCl \quad (6.54)$$

This is consistent with the greater electron withdrawal from the carbon centre in phosgene. The strong withdrawal may account for the lack of acceleration by trifluoromethanesulphonic acid of acylation with phosgene.[201]

(v) Formylation

Formyl halides are, with the exception of the fluoride, unstable. Formylation may be carried out with this in the presence of BF_3,[202] but other methods of formylation (described below) are generally preferred.

(vi) Acylation in the gas phase

Gas-phase acetylation by ion-cyclotron resonance using acetone or biacetyl as precursors for the acetylium ion shows the general aromatic reactivity order expected for an electrophilic substitution, i.e. *o*-xylene > *m*-xylene > *p*-xylene > n-butylbenzene > ethylbenzene > toluene > benzene > halobenzenes.[203] The reactivity of *o*-xylene is anomalous, however, relative to the other isomers (cf. the order in hydrogen exchange, Section 3.1.2.2) and may reflect severe steric hindrance which acetylation also shows in solution (see below).

6.7.3. Substituent Effects

There have been many quantitative studies of acetylation and benzoylation; there are no data for highly deactivated compounds such as nitrobenzene, which cannot be acylated, or for anilines, which acylate at the substituent. Phenols acylate at both substituent and ring but the former can be minimized by careful choice of conditions.[204] Site selectivities can be very dependent on the conditions; in particular, the use of nitro solvents gives characteristics commensurate with increased steric hindrance of the electrophile, owing to complexing of this with the solvent.[205] By using various conditions it is possible to alter dramatically the isomeric composition of the products, a notable example being the acetylation of 1, 2, 3-trimethylbenzene.[206] In 1, 2-dichloroethane (excess of acetyl chloride) 91% of the 3-isomer and only 9% of the 2-isomer was obtained, whereas acetylation in carbon disulphide (slight excess of $AlCl_3$) gives 100% of the 2-isomer. Rearrangement of a kinetically formed product to a thermodynamically more stable product is rarer than in alkylation but may also occur. Thus acetylation of anthracene gives the hindered 9-isomer,[207] which under appropriate conditions gradually converts to the less-hindered 1- and 2-isomers. Likewise 1-acetyl-2-methyl- or methoxynaphthalenes rearrange into the 6-acetyl isomers and the α-acetyl groups in 1,5-diacetyl-2,6-dimethylnaphthalene rearrange into the alternative, less hindered α-positions.[209] Benzoylation of naphthalene with benzoic acid in polyphosphoric acid gives a 1:2 product ratio which is 15 at 70 °C but only 0.7 at 140 °C owing to acid-catalysed rearrangement (protiodebenzoylation) to the more stable (less hindered) 2-isomer.[210]

Partial rate factors for the aluminium chloride-catalysed reactions in 1,2-dichloroethane at 25 °C are set out in Tables 6.14[211,212] and 6.15.[213,214]

The principal features of the results are as follows:

(1) Acylation is very much more selective than alkylation (cf. Tables 6.7 and 6.8), indicating that it involves a much less reactive electrophile.

(2) Although acetylation and benzoylation have similar *meta* and *para* partial rate factors (indicating similar reagent selectivities), the *ortho* reactivities are much smaller in acetylation, showing the electrophile to have much larger steric

Table 6.14 Partial rate factors for acetylation of PhX

X	f_o	f_m	f_p
OMe			1.8×10^6
SMe	860		4.2×10^4
Me	4.5	4.8	749
Et	1.0	10.4	753
i-Pr		11.5	745
t-Bu		13.1	658
Ph		0.3	248
F			1.51
Cl		0.0003	0.125
Br			0.084

Table 6.15 Partial rate factors for benzoylation of PhX[a]

X	f_o	f_m	f_p
Me	32.6	4.9	626
Et	10.9	10.3	563
i-Pr	8.6	11.1	519
t-Bu		11.4	398
exo-Norborn-2-yl			1630
endo-Norborn-2-yl			1040
Ph		0.3	245

[a]Naphthalene: 1-position, 4660; 2-position, 1160.

requirements. This is because the electrophile in benzoylation is the acylium ion (or more nearly so), whereas in acetylation it is the bulkier acyl halide–catalyst complex [see Section 6.7.1(ii) and (iii)]. Steric effects may account for the preferential acetylation of 2-bromotoluene at the 5-position,[215] and for the reactivity order ($m > o > p$) in acetylation and benzoylation of iodotoluenes, in addition to the high *para* yields in these reactions with iodobenzene.[216]

(3) The *exo*-norborn-2-yl substituent is more activating than the *endo*-substituent, paralleling the result in both hydrogen exchange,[217] gas-phase pyrolysis of 1-arylethyl acetates,[218] and solvolysis of 2-aryl-2-chloropropanes.[219] The lower activation in the *endo* isomer has been attributed to steric hindrance to C–C hyperconjugation[217,218] (see Section 3.1.2.1).

Partial rate factors for acetylation of dichlorobenzenes in nitrobenzene at 25 °C using the Perrier procedure (premixing of catalyst and acyl halide) have also been determined (Table 6.16).[220] For the *ortho* and *para* compounds these differ from those calculated assuming additivity, the *ortho* isomer being more reactive than predicted whereas the *para* isomer was less reactive, cf. nitration [Section 7.4.3(ix)].

Table 6.16 Partial rate factors for acetylation of chlorobenzenes in $PhNO_2$

Substituent	Position	f	Substituent	Position	f
Cl	2	3.28×10^{-3}	1,3-Cl_2	2	1.23×10^{-4}
	3	1.60×10^{-4}		4	4.95×10^{-4}
	4	0.150		5	$<2 \times 10^{-6}$
1,2-Cl_2	3	1.48×10^{-5}	1,4-Cl_2	3	2.80×10^{-8}
	4	1.59×10^{-4}			

Table 6.17 Relative reactivities in acylation of naphthalene and derivatives

Acetylation in $CHCl_3$:

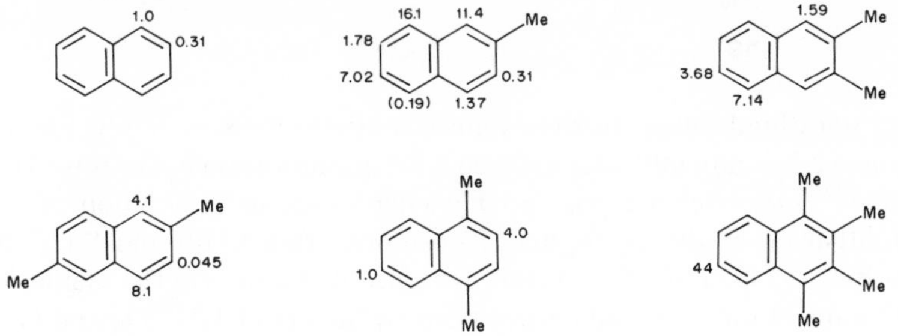

1.0 0.31

16.1 11.4 Me 1.78 7.02 0.31 (0.19) 1.37

1.59 Me 3.68 7.14 Me

4.1 Me 0.045 Me 8.1

Me 4.0 1.0 Me

Me Me 44 Me Me

Acetylation in $MeNO_2$:

1.0 8.3

42.3 6.9 Me 32.6 177 10.8 11.0 6.1

Benzoylation in $CHCl_3$:

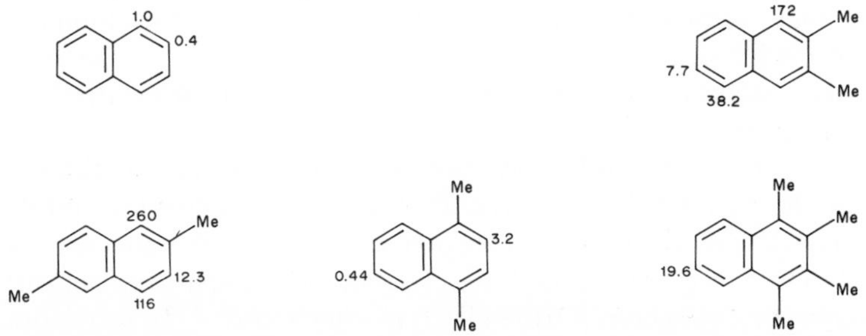

1.0 0.4

172 Me 7.7 38.2 Me

260 Me 12.3 Me 116

Me 3.2 0.44 Me

Me Me 19.6 Me Me

Data for acylation of methylnaphthalenes (Table 6.17)[208,209,221,222] demonstrate nicely the greater steric hindrance to each of α-substitution, acylation in nitro solvents, and acetylation compared with benzoylation. The proportion of β-substitution is generally greater under the more hindered conditions (the anomalous result for 5-acetylation of 2-methylnaphthalene in chloroform is almost certainly due to experimental error). The greater steric hindrance to acetylation compared to benzoylation is also shown by the diacylation of 2,7-dimethylnaphthalene in which **59** and **60** are the preferred products, respectively;[223] the 1- and 8-positions are much the most reactive sites in 2,7-dimethylnaphthalene, but they can both be substituted only in the less hindered benzoylation.

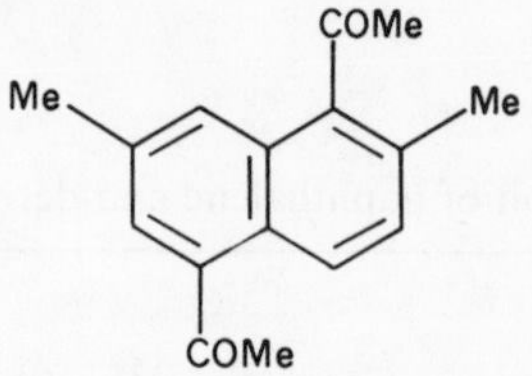

(59) 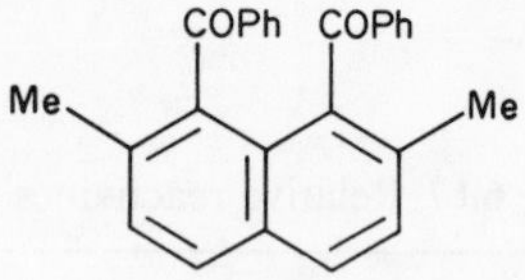 (60)

The higher steric hindrance in nitro solvents is shown by the 8-:6- product ratio in acetylation of 2-bromonaphthalene, which is 3.3 in chloroform but only 0.53 in nitrobenzene.[224] Likewise, the positional isomer yields in acetylation of 2-methoxynaphthalene at the 1-, 6-, and 7-positions are 9.5, 14, and 8.1% in chloroform, but 1.8, 43, and 0.6% in nitrobenzene; in chloroform the methoxy substituent activates the 1-, 6-, and 8-positions by factors of 1.72, 3.8, and 0.9, respectively.[208] An important feature of this work is the demonstration that the yield of a desired isomer can often be increased by careful choice of conditions. The results also confirm the findings from hydrogen exchange that substituent effects acting across a 1,2-bond are much greater than when acting across a 2,3-bond (Section 3.1.2.13).

There is some evidence that both second- and third-order terms are involved in acylation, arising from the involvement of a second molecule of catalyst.[225] This phenomenon may account for the time-variable α/β rate ratios found in acetylation of naphthalene, although the finding that reaction of the more reactive (and more sterically hindered) α-position is second-order in catalyst[226] is difficult to rationalize.

The lower steric hindrance to benzoylation is further confirmed by the 87% of 1-benzoylation of 2-methylnaphthalene (cf. acetylation, Table 6.17), and the 82% of 5-benzoylation of acenaphthene (**61**) compared with only 48% in acetylation.[227] The 5:3- product ratio for acetylation of acenaphthene can vary from 2.5 (catalysis by $ZnCl_2$ in Ac_2O) to 9 ($TiCl_4$ in white spirit).[228] In acetylation the (statistically corrected) relative reactivities of naphthalene, 1*H*-cyclobuta-

[*de*]naphthalene (**62**), perinaphthane (**63**), and acenaphthene (**61**) are 1.0:8.2: 21.3:55.8.[229] This first determination of the reactivity of **62** shows it to be less reactive than acenaphthene, owing, in the view of the writer, to the absence of steric acceleration (through eclipsing strain) of hyperconjugation (cf. acenaphthene) and the fact that hyperconjugative electron release from the substituent C—H bonds will increase the strain in the four-membered ring.

8 3 7 4 6 5

(**61**) (**62**) (**63**) (**64**)

Relative rates for acetylation in chloroform have also been determined (Table 6.18),[230] and the anomalous lower reactivity of 1,8-dimethylnaphthalene compared with 1-methylnaphthalene probably reflects increased steric hindrance at the *peri* positions due to compression between the methyl groups. The higher reactivity of the 5-position of acenaphthene may be partially due to relief of this steric compression, but is largely an electronic effect (see hydrogen exchange, Section 3.1.2.13). The inexplicable lower reactivity of the 3-position of 1-methylnaphthalene compared with the corresponding position in naphthalene indicates that other factors intervene to vitiate the results. The $AlCl_3$-catalysed acetylation of tetraphene (**64**) in CS_2 gives 47 and 53% of 3- and 5-substitution, respectively, the corresponding values being 72 and 28% if dichloromethane is used as solvent.[231] Acetylation of 5-acetylaminonaphthalene goes ca 50% in each of the 3- and 8-positions,[232] and the lack of 6-substitution can again be ascribed to steric hindrance. This may also account for the unexplained observation of 6- rather than 5-acetylation of 3-*tert*-butylacenaphthene (see **61** for numbering).[233] In this work, 3-acetyl-, bromo-, chloro-, and nitro-acenaphthene acetylated exclusively in the expected 6-position, 5-acetylace-naphthene acetylated in the 8-position (although 5-halogenoacenaphthenes gave mixtures of 3- and 8-acetyl products), and both 5-methyl- and 5-*tert*-butyl acenaphthenes acetylated in the 6-position, this latter being particularly surprising in view of the result for the 3-isomer.

Table 6.18 Relative reactivities in acetylation in chloroform at 20 °C

0.45
1.0

Me
0.18
0.10
62

Me Me
60

CH_2—CH_2
6.4
<0.2
94

Table 6.19 Relative rates of acylation of phenanthrene, triphenylene,[a] and chrysene[a]

Acetylation in $CHCl_3$ or 1,2-dichloroethane:

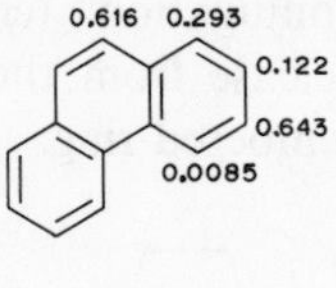

0.616 0.293 0.122 0.643 0.0085

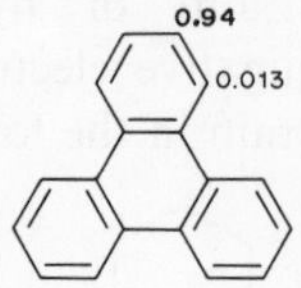

0.94 0.013

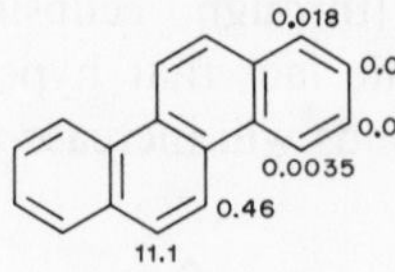

0.018 0.06 0.08 0.0035 0.46 11.1

Benzoylation in $CHCl_3$:

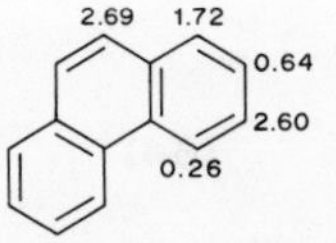

2.69 1.72 0.64 2.60 0.26

[a] Acetylation of naphthalene in 1,2-dichloroethane gave a time-dependent α/β reactivity ratio, the value of 0.95 given in ref. 235 differing from all previously measured values. The triphenylene and chrysene data are given relative to those for phenanthrene so the numerical values are 3.4-fold smaller than in ref. 235, in which naphthalene was used as a reference; the phenanthrene: naphthalene reactivity ratio differs 3.4-fold between refs. 234 and 235.

Relative rates for acylation of phenanthrene (Table 6.19)[234] also demonstrate the high steric hindrance to acylation and to acetylation in particular. Both reactions give the positional reactivity order $3 \approx 9 > 1 > 2 > 4$, in contrast to the order $9 > 1 > 4 > 3 > 2$ theoretically predicted and found in (unhindered) hydrogen exchange [Section 3.1.2.12(ii)], and the reactivity difference between acetylation and benzoylation is ca six times greater at the most hindered 4-position than at any other. The pattern of greater steric hindrance accompanying the use of nitro solvents is also apparent since these give a higher proportion of 2- and 3-substitution for both phenanthrene[234] and chrysene.[235] Likewise is acetylation of anthracene in nitrobenzene, the minor product is the 9-isomer,[236] whereas the 9-position is normally the most reactive. Relative rates for acetylation of triphenylene (Table 6.19)[235] give a much higher 2-:1- rate ratio of 74 than in hydrogen exchange (0.22, Table 3.17), where steric hindrance to 1-substitution does not apply.

Steric hindrance probably accounts for acetylation and formylation of hexamethyl-*trans*-15,16-dihydropyrene (**65**) occurring principally at the 2-position whereas nitration occurs at both 2- and 4-positions.[237] Both hydrogen exchange and trifluoroacetylation of azupyrene (**66**) take place preferentially at the 1-position, whereas nitration goes into the 4-position.[238] (Trifluoroacetylation may be less hindered than acetylation since it probably requires a lower degree of coordination between the anhydride and the BF_3–etherate catalyst.) Acetylation of 3-methylfluoranthene (**67**) gives a high 8-:4- isomer ratio of 23,[239] again owing to hindrance at the 4-position. Acylation of triptycene goes exclusively into the 2-position,[240] whereas hydrogen exchange predicts that a substantial amount of 1-substitution should be found (see Table 3.9). Likewise, acylation of fluorene gives 2- with only a small amount of 4-substitution,[241] in contrast to the substitution pattern in hydrogen exchange (Table 3.15).

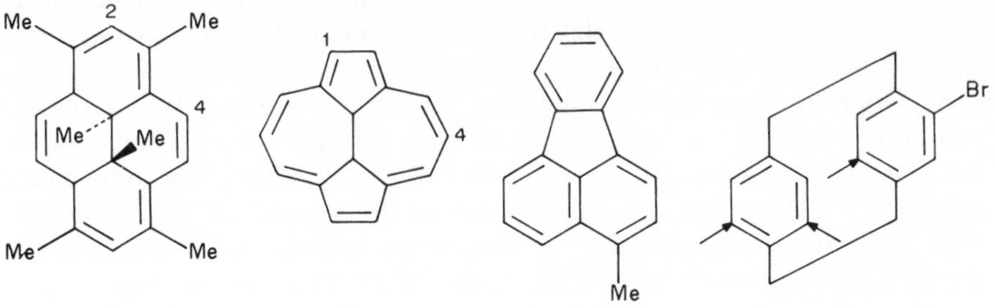

(65) (66) (67) (68)

Acetylation has been used to determine the effect of transannular interaction in *para*-cyclophanes. For compound **68**, $AlCl_3$-catalysed acetylation in 1,2-dichloroethane at 25 °C gave 17% substitution *para* to bromine and, in the other ring, respectively 6 and 41% 'pseudo' *ortho* and *para* to bromine.[242] These latter orientations led to the proposal that (in non-basic solvents) the neighbouring aromatic ring acts as an internal base in the product-determining step of the reaction. Similar results in bromination were confirmed by isotope studies but this was not possible for acetylation because of scrambling under the reaction conditions. Relative rates of acetylation of the cyclophanes **69** were >29 for $m = 2, n = 2$, 11.2 for $m = 3, n = 4$, 1.6 for $m = 4, n = 4$, and 1.0 for $m = 6, n = 6$.[243] The enhanced rates as the rings are brought closer together argues strongly for transannular stabilization as shown in **70**.

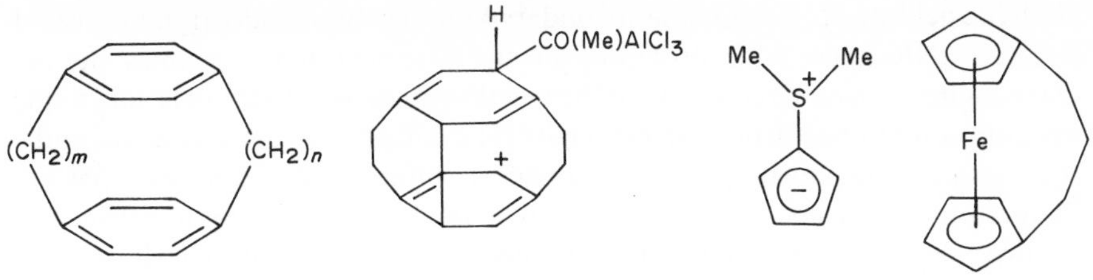

(69) (70) (71) (72)

Azulenes containing electron-withdrawing groups at the 1- and 3-positions acetylate (and halogenate) at the 5-position[244] as expected, since the 5-position is the most reactive in the seven-membered ring towards electrophilic substitution.[245] Acetylation, Vilsmeier formylation, ethylation, and sulphonylation of dimethylsulphonium cyclopentadienylide (**71**) at the 2-position demonstrate its aromatic character.[246] Both cyclopropylbenzene,[247] and cyclopropyl phenyl ether[248] acylate mainly at the 4-position. Benzoylation (and halogenation, nitration, and sulphonation) of benzocyclobutene occur largely at the 4-position.[249] This is largely an electronic effect (Section 3.1.2.3), the 3-position being relatively unhindered by the four-membered ring. Acetylation of hexahelicene gives mainly the 5-isomer together with what was assumed to be the 8-isomer,[250] but is probably the 7-isomer since hydrogen exchange shows the 5- and 8-positions to be most reactive [Section 3.1.2.12(ii)].

Acylation has been used in numerous studies of the reactivity of ferrocene and derivatives.[251] The high steric hindrance to the reaction is indicated by the substitution mainly into the unsubstituted ring of *N*-ferrocenylamides (and -imides),[252] and the failure to obtain either of the eclipsed 2,2′- or 3,3′-isomers from [3]ferrocenophanes (**72**).[253] Competitive acetylation of π-cyclopentadienyl metal complexes shows the order of reactivity to be $Fe(C_5H_5)_2 >$ anisole $> Ru(C_5H_5)_2 > C_5H_5Mn(CO)_3 > Os(C_5H_5)_2 > C_5H_5Cr(CO)_2NO > C_5H_5V(CO)_4 >$ benzene $> C_5H_5Re(CO)_3$.[254a] Acetylation has been used to show that the $Cr(CO)_3$ group is weakly deactivating in arenetricarbonylchromium complexes.[254b] This reinforces the results of base-catalysed hydrogen exchange (Section 3.3) where electron withdrawal by this ligand caused substantial rate acceleration.

6.7.4 Cycliacylation

The reaction of arylaliphatic acids possessing a side-chain of suitable length with both protic and Lewis acids results in intermolecular acylation (eq. 6.55, X = CH_2). Formation of six-membered rings is preferred when a choice of ring size exists; acid halides have also been used. The reaction is aided by electron-supplying groups at the *ortho* and *para* positions, and in consequence aryloxyaliphatic acids (6.55, X = O) readily undergo the reaction, as do arylamino acids (6.55, X = NR), these latter providing a route to substituted quinolines. Because of the higher collision frequency for the reaction compared with acylation, milder reaction conditions are needed, especially for cyclization of aryloxy acids, and in some cases phosphorus pentoxide alone is sufficient to effect the necessary dehydration. The acids may be generated *in situ* from reaction between an aromatic and a lactone, and is then termed intramolecular cycliacylation.[255]

As in the case of cyclialkylation, steric effects are very important and a *meta*-

$$C_6H_5XCH_2CH_2COOH \xrightarrow{H_2SO_4} \text{(cyclic ketone, X in ring)} \qquad (6.55)$$

Table 6.20 Relative rates of cyclization of benzoylbenzoic acids (**74**)

R′	3-OMe	4-Me	3-Me	4-Me	3,4-Me_2	H	4-Cl
	10 800	<0.0016	164	0.25	22	1.0	0.003

substituted acid will cyclize so that the substituent occupies mainly the position *para* to the entering group. This is evident in the partial rate factors for the effect of the methyl substituent in the HF-catalysed cycliacylation of 3-phenylpropanoic acids (**73**) to indan-1-ones.[256] The higher activation by *ortho*- relative to *para*-methyl suggests steric acceleration. For the corresponding *tert*-butyl compounds the *ortho, meta* and *para* partial rate factors were 1–10, 3, and 80, respectively, so steric hindrance is dominant here. In this work it was shown that the acylating species is the free acyl cation, formation of which is rate-determining at high cyclization rates. This masks any further activation of the ring, i.e. rates reach a maximum. Introduction of an α-methyl substituent caused a large increase in rate, attributed to a conformational effect whereby the $C^+(OH)_2$ group is forced next to the phenyl ring.

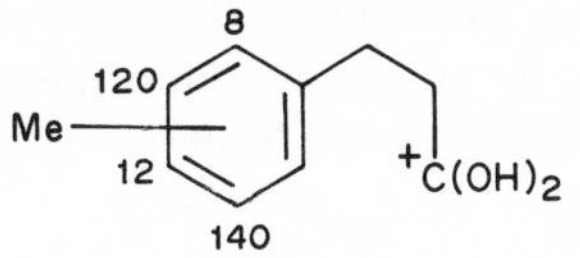

(**73**) 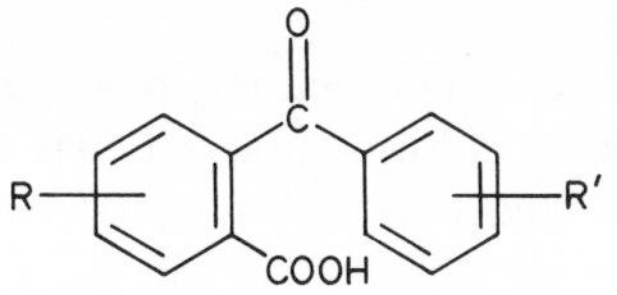 (**74**)

Cycliacylation will also take place with benzoylbenzoic acids (**74**) and the effects of substituents R′ for the H_2SO_4-catalysed reaction at 25 °C are given in Table 6.20.[257] The relative effects of *para* substituents R under conditions producing a constant concentration of acylium ion were OMe 0.09 and NO_2 4.1; the reaction gave a small kinetic isotope effect, $k_H/k_D = 1.25$.

6.7.5 Fries Rearrangement

This reaction is the acylation equivalent of the Claisen rearrangement. Phenolic esters, on heating with a Lewis-acid catalyst, rearrange to *o*- and *p*-acylphenols (eq. 6.56).[258] The *ortho*:*para* ratio depends on temperature, catalyst, and solvent, and in general higher temperature favours *ortho* substitution. The mechanism evidently involves a mixture of inter- and intramolecular reactions,[259] even though some experiments have indicated that either the former,[260] or the latter,[261] are exclusively formed; the formation of *para* products would seem to necessitate complete separation of the acyl group. The intermolecular reaction may be observed by carrying out the reaction in the presence of another aromatic which may be acylated. Whether or not this (crossover acylation) is observed is likely to depend on the relative reactivities of both aromatics. The balance between the two processes is temperature dependent,[259] owing to their differing activation energies. Intramolecular rearrangement to *meta* positions is possible if these are sufficiently activated by other substituents in the rings.[259]

$$C_6H_5OCOR \xrightarrow{AlCl_3} o\text{-}HOC_6H_4COR + p\text{-}HOC_6H_4COR \quad (6.56)$$

6.8 FORMYLATION

6.8.1 Gattermann–Koch Reaction

Treatment of moderately reactive aromatics with CO and HCl in the presence of copper(I) chloride and a Lewis acid (usually aluminium chloride) produced aldehydes (eq. 6.57).[262] If the reaction is carried out at high pressure, the CuCl is not required.

$$ArH + CO + HCl \xrightarrow[CuCl]{AlCl_3} ArCHO + HCl \quad (6.57)$$

It is improbable that the reaction proceeds via formation of formyl chloride, since this is unstable except at low temperatures,[263] and it has been suggested that the CO, HCl, and Lewis acid take part in a pre-equilibrium (eq. 6.58), which is followed by a rate-determining reaction between the formyl cation and the aromatic (eq. 6.59).[264] This is consistent with conductivity data and the reaction being first order in aromatic. Complex formation between the aldehyde and Lewis acid causes the reaction to be thermodynamically favourable, which it would otherwise not be.[264] Consequently, the Lewis acid is required in more than catalytic quantities, as shown by the increase in the yield of benzaldehyde from 21 to 65% when the $AlCl_3$ to benzene molar ratio is increased from 0.3 to 1.0.[265]

$$HCl + CO + AlBr_3 \rightleftharpoons HCO^+ + AlBr_3Cl^- \quad (6.58)$$

$$ArH + HCO^+ \xrightleftharpoons{slow} ArCHO + H^+ \quad (6.59)$$

Copper(I) chloride forms a complex with CO and its function may therefore be to accelerate the reaction with HCl. Other co-catalysts have been used, e.g. $TiCl_4$, but none is more efficient than CuCl. The reaction is very sterically hindered, as shown for example by the formation of 85% of 4-methylbenzaldehyde in the formylation of toluene.[266] Alkylbenzenes undergo rearrangement and disproportionation during the reaction (which has been comprehensively reviewed)[266,267a] owing to the presence of HCl and Lewis acid [Section 6.1.1(ii)]. A recent development has been to use catalysis by superacids such as $HF-BF_3$;[267b] trifluoromethanesulphonic acid at high pressure has also been used and, if cyclohexene is present, cyclohexyl ketones may also be obtained.[268]

6.8.2 Gattermann Reaction

Many aromatics can be formylated by treatment with HCN and HCl in the presence of a Lewis acid, followed by hydrolysis of the initial product (eq. 6.60). The initial product, which is not normally isolated, is evidently a derivative of the aldimine, ArCH=NH. The reaction is very similar to the Gattermann–Koch process, as CO and HCN are isoelectronic, but has the advantage that it can be successfully applied to phenols and to many heteroaromatics, although it fails with amines.[267a,269]

$$\mathrm{ArH + HCN + HCl \xrightarrow[\text{(ii) } H_2O]{\text{(i) Lewis acid}} ArCHO + NH_4Cl} \qquad (6.60)$$

As is generally the case for Lewis acid-catalysed reactions, the choice of catalyst depends on the reactivity of the aromatic. Thus, formylation of benzene requires $AlCl_3$,[270] whereas $ZnCl_2$ is adequate for phenols and aromatic ethers. Nowadays $Zn(CN)_2$ generally replaces HCN as the source of cyanide ion, and in the presence of HCl also generates the Lewis-acid catalyst.[271]

The mechanism of the reaction is not clearly understood and various complexes have been shown to be present.[267a,272] The reaction is probably best understood on the assumption that the ionic intermediate ($CH{=}NH^+ \leftrightarrow {}^+CH{=}NH$) is the effective electrophile. Reaction does not take place on the less reactive aromatics such as nitrobenzene and benzophenone, but chlorobenzene has been formylated in low yield. Yields from more reactive aromatics such as toluene can be almost quantitative,[270] although isomerization of alkyl substituents can occur as expected [see Section 6.1.1(ii)].

6.8.3 Vilsmeier–Haack Reaction

This is the most commonly used method of formylation and usually involves disubstituted formamides and $POCl_3$ as reagents (eq. 6.61).[267a,273] The reaction is limited to reactive aromatics and is successful with aromatic amines, in contrast to the two previous methods. The relative reactivities of the commonly used amides are $HCONMe_2 > HCONMePh > HCONHMe > HCONH_2$,[267a] and $COCl_2$ may also be used, in which case carbon dioxide is evolved during the reaction.

$$ArH + Me_2NCHO \xrightarrow{POCl_3} ArCHO + Me_2NH \quad (6.61)$$

The mechanism evidently involves the formation of $[R_2NCH^+Cl]PO_2Cl_2^-$ as the electrophile.[274] With phosgene the electrophile becomes $[R_2NCH^+Cl]Cl^-$, and since the negative charge is delocalized less effectively here, carbon carries a smaller positive charge and the reagent is less reactive. Thus, whereas the former reagent gave with a range of reactive aromatics (including heterocycles) kinetics which were either first or (with very reactive aromatics) zeroth order in aromatic, the latter reagent gave first-order kinetics throughout.[275] Zeroth-order kinetics are obtained when reaction with the aromatic is fast relative to the pre-equilibrium which forms the electrophile.

The view that electrophilic substitutions may, in part, proceed through the intermediacy of cation radicals has received support from the finding of high concentrations of cation radicals in Vilsmeier–Haack formylation of *N*-substituted dihydrophenazines.[276] The reaction has been used successfully to mono- and diformylate at the *meso* positions Cu^{II}, Ni^{II}, and Pd^{II} derivatives of octaethylporphyrin.[277]

6.8.4 Formylation with Dichloromethyl Alkyl Ethers

The Lewis acid-catalysed reaction of dichloromethyl alkyl ethers with aromatics gives aldehydes, after decomposition of the intermediate 1-alkoxy-1-arylethyl chlorides either by heating or by S_N1 hydrolysis (eq. 6.62);[278] $AlCl_3$ and $TiCl_4$ are the most frequently used catalysts. It may be noted that the electrophiles produced under these conditions and those employed in the Vilsmeier–Haack reaction (Section 6.8.3) differ essentially in that in the latter reaction they are ionized owing to the greater electron-supplying ability of NR_2 compared with OR.

$$ArH + Cl_2CHOR \xrightarrow{\text{Lewis acid}} ArCH(OR)Cl \begin{cases} \xrightarrow{\text{heat}} ArCHO + RCl \\ \xrightarrow{H_2O} ArCHO + ROH + HCl \end{cases} \quad (6.62)$$

There is evidence that $TiCl_4$ may coordinate with hydroxy substituents and produce substantial formylation (and other electrophilic substitutions) at the *ortho* position. Thus compounds **75** give substantial 3-substitution whereas

Table 6.21 Rate and product data for formylation of toluene and benzene

					Methylbenzaldehydes/%		
Reagent	Catalyst	Solvent	T/°C	k_T/k_B	*ortho*	*meta*	*para*
CO	HF–SbF_5	SO_2ClF	−95	1.6	45.2	2.7	52.1
HCOF	BF_3	xs. ArH	25	34.6	43.3	3.5	53.2
HCN–HCl	$AlCl_3$	xs. ArH	25	49.1	39.9	3.7	56.4
$Zn(CN)_2$–HCl	$AlCl_3$	$MeNO_2$	25	92.8	38.7	3.5	57.8
$Zn(CN)_2$–HCl	$AlCl_3$	xs. ArH	50	128	34.3	1.8	63.9
Cl_2CHOMe	$AlCl_3$	$MeNO_2$[a]	−27	200	42.3	0.5	57.2
Cl_2CHOMe	$AlCl_3$	$MeNO_2$	25	119	35.8	3.8	60.4
CO–HCl	$AlCl_3$·CuCl	xs. ArH	25	155	8.6	2.7	88.7
CO–HCl	$AlCl_3$	xs. ArH	0	319	6.6	0.8	92.6
CO–HF	BF_3	xs. ArH	0	860	3.5	0.5	96.0

[a] Ref. 280.

compounds **76** react exclusively at the 5-position.[279] Formylation of benzene and toluene in nitromethane at −27 °C is first order in aromatic, ether, and metal halide.[280] The rate and product data (Table 6.21) differ significantly from those obtained with the same reagents at higher temperature, possibly because high temperature causes dichloromethyl methyl ether to be partially decomposed by $AlCl_3$.[280]

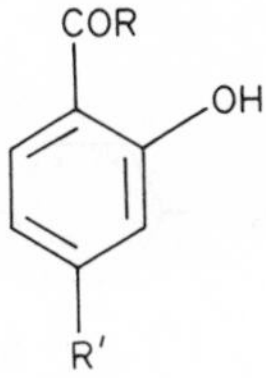

COR OH R′ (**75**)

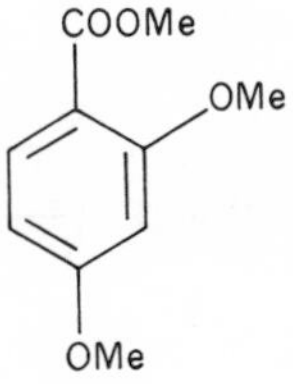

COOMe OMe OMe (**76**)

R = Me, OR;
R′ = OH, OMe, Me

6.8.5 Formylation with Formyl Fluoride

Unlike formyl chloride, formyl fluoride is stable and reacts with aromatics in the presence of BF_3 to give aldehydes.[202,267b] However, the relative difficulty of obtaining the fluoride negates the general use of this method.

6.8.6 Substituent Effects in Formylation

The isomer distribution and relative toluene to benzene rate ratios have been determined under a wide range of conditions (Table 6.21).[281] The very low selectivity in the reaction with CO–$HSbF_6$ may be assumed to be due to non-competitive conditions involving mixing and/or diffusion control (Section 2.10). The very low yields of 2-methylbenzaldehyde obtained with CO–Lewis-acid catalysts is not a selectivity effect since the $f_o : f_p$ ratios differ from the value of 0.87 theoretically required (Section 11.2.4) and indicate severe steric hindrance under these conditions (cf. ref. 281).

6.9 THE HOESCH REACTION

The reaction of nitriles and HCl with reactive aromatics such as phenols and phenolic ethers gives ketones.[282] A Lewis-acid catalyst (usually $ZnCl_2$) is required except when the aromatic is very reactive. An example of the reaction which has been reviewed[283] is the formation in 70–80% yield of 2,4,6-trihydroxyacetophenone (eq. 6.63).[284] Yields may be maximized by premixing the nitrile, HCl, and catalyst, and then adding the aromatic at 0 °C. In this way even aryl nitriles will react.[285] The reaction is particularly useful for preparing polyhydroxy- and polyalkoxyketones and hence (via Clemmensen or Wolff–Kishner reduction) the corresponding alkylphenols and their ether derivatives (these being of importance in natural product chemistry); moreover, substituents do not migrate during the reaction.

$$\text{1,3,5-}C_6H_3(OH)_3 + MeCN \xrightarrow[\text{(ii) } H_2O]{\text{(i) } ZnCl_2} \text{2,4,6-}(HO)_3C_6H_2COMe + NH_4Cl \qquad (6.63)$$

The reaction mechanism is complex, but can be understood in terms of formation of the imino chloride (eq. 6.64), which subsequently reacts with the aromatic to give the aromatic ketimine hydrochloride, followed by hydrolysis to the ketone (eq. 6.65). Consistent with this are the facts that (1) ketimine hydrochlorides are frequently isolated intermediates and (2) the imino chloride from chloromethyl cyanide has been isolated[286] and shown to react with *m*-dihydroxybenzene to give 2,4-dihydroxy-ω-chloroacetylbenzene.[287] However, imino chlorides are normally too unstable to exist, the above example being favoured by electron withdrawal in the alkyl group (which retards thermal elimination of HCl). The imonium chloride $RCCl{=}NH_2^+Cl^-$ has been suggested as a more probable intermediate, and indeed high yields can be obtained by pre-formation of the imonium chloride at low temperatures and reacting it with the aromatic in the presence of $ZnCl_2$.[285]

$$\mathrm{RCN + HCl \rightarrow R{-}\underset{\underset{\displaystyle NH}{\|}}{C}{-}Cl} \tag{6.64}$$

$$\mathrm{ArH + R{-}\underset{\underset{\displaystyle NH}{\|}}{C}{-}Cl \xrightarrow{ZnCl_2} Ar{-}\underset{\underset{\displaystyle NH_2{}^{+}Cl^{-}}{\|}}{C}{-}R \xrightarrow{H_2O} ArCOR + NH_4Cl} \tag{6.65}$$

6.10 CARBOXYLATION AND AMIDATION

The catalysed reaction of electrophiles of the general form $CO^{\delta+}\cdots X$ with aromatics gives carboxylic acids (in some cases after hydrolysis of the first-formed product). The simplest example is the formation of benzoic acid from benzene, CO_2, and $AlCl_3$,[288] which is facilitated by high pressure. The corresponding reaction with CS_2 gives dithiocarboxylic acids, ArCSSH.[289] It may be expected that a combination of CO_2 and superacid would also give carboxylation, but this reaction has so far not been reported.

The reaction takes place more readily with phosgene, as expected, but reaction of the intermediate acyl halide tends to produce diaryl ketones (eq. 6.66).[290] This can be largely avoided by carrying out the reaction in CS_2 in which the intermediate acyl halide–$AlCl_3$ complex is insoluble and therefore precipitates out of solution.[291] Oxaloyl chloride, $(COCl)_2$, has also been used as it is more convenient. The intermediate acyl halide, ArCOCOCl, readily loses CO to give ArCOCl and hence the acid after hydrolysis; benzil derivatives, ArCOCOAr, are also obtained as byproducts.[290]

$$\mathrm{ArH + COCl_2 \xrightarrow{AlCl_3} ArCOCl} \begin{cases} \xrightarrow{H_2O} \mathrm{ArCOOH} \\ \xrightarrow{ArH} \mathrm{ArCOAr} \end{cases} \tag{6.66}$$

Ketone formation can be avoided by using various reagents that produce intermediate proucts which may be hydrolysed to the carboxylic acid, but which are themselves insufficiently reactive as acylating reagents. For example, the dinitrile of carbonic acid, $CO(CN)_2$, may be used in place of phosgene, and this gives ArCOCN as intermediate; however, $CO(CN)_2$ is difficult to obtain.[292] More useful are amides, $ClCONR_2$, derived from chloroformic (chlorocarbonic) acid (these compounds are also the acyl chlorides of carbamic acid). The *N*-phenylamide (phenylcarbamoyl chloride) is readily produced from phenyl isocyanate and HCl, and reacts with the aromatic in the presence of $AlCl_3$ to give the *N*-phenylamide of the arylcarboxylic acid (eq. 6.67).[290,293] Cyanic acid itself (HC=N=O) may also be used in this reaction,[294] as well as other isocyanates, RN=C=O. If R is a strongly electron-withdrawing group such as SO_2X (X = alkyl, aryl, or Cl) then reaction with very reactive aromatics will take place without the need for a catalyst (eq. 6.68).[295] Carbamoyl chloride ($ClCONH_2$) and, more recently, the *N*,*N*-diethyl derivative,[296] have been used, as have alkyl chlorothiolformates (RSCOCl).[297] The latter reaction showed high steric hindrance and is fairly selective, giving a ρ factor of ca -11.7 (although using σ rather than σ^+ values, which exalts the ρ factor). For $ClCOSC_6H_4R$ reagents, increased electron withdrawal by R produced a decrease in selectivity between toluene and benzene, suggesting a free cation as the electrophile. Catalysts other than $AlCl_3$ give various byproducts, including alkylaromatics (which could arise from extrusion of COS from an intermediate $RSCO^+$ cation).

$$\text{Ph—N=C=O} + \text{HCl} \longrightarrow \text{PhNHCOCl} \xrightarrow{\text{ArH,AlCl}_3}$$

$$\text{ArCONHPh} + \text{HCl} \longrightarrow \text{ArCOOH} \qquad (6.67)$$

$$\text{R}_2\text{N—C}_6\text{H}_5 + \text{XSO}_2\text{—N=C=O} \longrightarrow \text{R}_2\text{N—C}_6\text{H}_4\text{—CONHSO}_2\text{X}$$

$$\xrightarrow{\text{H}^+,\text{H}_2\text{O}} \text{R}_2\text{N—C}_6\text{H}_4\text{—COOH} \qquad (6.68)$$

Alkylthiocarboxylation and -dithiocarboxylation have been achieved by reaction of aromatics with *S*-methyl thiocarboxonium fluoroantimonate (from MeF, SbF_5, and COS) and *S*-ethyl dithiocarboxonium fluoroantimonate (from EtF, SbF_5, and CS_2), respectively (eq. 6.69).[298] The former electrophile is more reactive as expected, and benzene gave a 72% yield, increased and decreased by electron-supplying and electron-withdrawing substituents, respectively, thus showing typical electrophilic properties for the reactions. The reactions show very high *para* selectivity, consistent with the high steric hindrance expected.

$$\text{ArH} + \text{RC(=X)S}^+\text{SbF}_6^- \longrightarrow \text{ArC(=X)SR} + \text{HSbF}_6 \qquad (6.69)$$

$$(\text{R} = \text{Me, Et; X} = \text{O, S})$$

Another method involves use of KOCN in aqueous HF; this forms the intermediate carbamoyl fluoride which reacts immediately with the aromatic to give the amide (eq. 6.70); use of KSCN gives the corresponding thioamide.[299]

$$KXCN + HF - H_2O \longrightarrow KOH + \left[\begin{array}{c} H_2NCF \\ \| \\ O \end{array}\right] \xrightarrow{ArH} \begin{array}{c} ArCNH_2 \\ \| \\ O \end{array} + HF \quad (6.70)$$

$$(X = O, S)$$

The chloro-oxocarbocation $COCl^+$ is believed to be formed in $CO–Br_2–SbCl_5$–liquid SO_2 at $-70\,°C$, and this reacts with benzene to give the intermediate $PhH{\cdot}COCl^+$, which hydrolyses to benzoic acid.[300]

Carboxylation accompanies chlorination by thallium(III) chloride tetrahydrate in CCl_4; the electrophile here is thought to be $CCl_3{}^+$.[301]

6.11 CYANATION

Aromatics of the reactivity of benzene or greater can by cyanated with Cl_3CCN, BrCN, or $Hg(ONC)_2$ (mercury fulminate).[302] In the case of trichloromethyl cyanide, the electrophile is probably $Cl_3C^+{=}NH$, produced by protonation (eq. 6.71).

$$Cl_3CCN \xrightarrow{HCl} Cl_3CC^+{=}NHCl^- \xrightarrow{ArH} Ar{-}\underset{\underset{NH_2{}^+Cl^-}{\|}}{C}{-}CCl_3 \xrightarrow{NaOH} ArCN \quad (6.71)$$

Cyanation may also be achieved using cyanogen bromide and aluminium chloride, and the yields for the reaction (Table 6.22)[303] indicate little steric hindrance to the reaction.

Table 6.22 Yields in cyanation using $CNBr–AlCl_3$ in CS_2

Compound	Position	Yield/%	Position	Yield/%
1,3,5-Trimethylbenzene	2	92		
Naphthalene	1	42	2	1
1-Methylnaphthalene	4	72		
2-Methylnaphthalene	1	85		
Anthracene	9	92		
Phenanthrene	9	83		
Chrysene	5	2	6	66
Triphenylene	1	32	2	48

6.12 THE KOLBE–SCHMITT AND REIMER–TIEMANN REACTIONS

Both of these reactions take advantage of the high reactivity of the phenoxide ion. Reaction of phenols under basic conditions therefore occurs with electrophiles of relatively low reactivity. The Kolbe–Schmitt reaction[304] (eq. 6.72) involves reaction with carbon dioxide and, if sodium phenoxide is used, substitution occurs mainly at the *ortho* position, indicating that some kind of coordination mechanism is involved.[305] However, potassium phenoxide is substituted mainly in the *para* position.[306]

$$C_6H_5O^- + CO_2 \longrightarrow o\text{-}(COOH)C_6H_4O^- \qquad (6.72)$$

The Reimer–Tiemann reaction produces aldehydes and uses as reagents chloroform and a base (eq. 6.73), substitution taking place mainly at the *ortho* position, unless both of these are blocked.[307] The reaction will also take place with reactive heterocycles such as pyrrole, and here insertion reactions accompany substitution, confirming that the electrophile is dichlorocarbene, $:CCl_2$.[308]

$$C_6H_5O^- + :CCl_2 \longrightarrow \text{(cyclohexadienone, H, } CCl_2^-) \xrightarrow{+H^+,\ -H^+} \text{(cyclohexadienone, } ^-CHCl_2) \longleftrightarrow o\text{-}(CHCl_2)C_6H_4O^- \longrightarrow o\text{-}(CHO)C_6H_4O^- \qquad (6.73)$$

REFERENCES

1. (a) G. A. Olah, *Friedel–Crafts and Related Reactions*, Interscience, New York, 1963–65; (b) G. A. Olah, *Friedel–Crafts Chemistry*, Wiley, New York, 1973.
2. C. Friedel and J. M. Crafts, *C. R. Acad. Sci.*, 1877, **84**, 1392, 1450.
3. G. A. Russell, *J. Am. Chem. Soc.*, 1959, **81**, 4834; see also ref. 65 for a more recent list of 126 Lewis acids in approximate order of catalytic activity.
4. K. L. Nelson and H. C. Brown, in *The Chemistry of Petroleum Compounds*, Eds. B. T. Brooks *et al.*, Vol. 3, Reinhold, New York, 1955, p. 465.

5. A. W. Francis, *Chem. Rev.*, 1948, **43**, 257.
6. H. C. Brown and W. J. Wallace, *J. Am. Chem. Soc.*, 1953, **75**, 6265.
7. A. Wohl and E. Wertyporoch, *Chem. Ber.*, 1931, **64B**, 1357; E. Wertyporoch, *Chem. Ber.*, 1931, **64B**, 1369; 1933, **66B**, 1232.
8. H. C. Brown, L. P. Eddy, and R. Wong, *J. Am. Chem. Soc.*, 1953, **75**, 6275.
9. F. Fairbrother, *J. Chem. Soc.*, 1945, 503; 1941, 293.
10. F. Fairbrother, *Trans. Faraday Soc.*, 1941, **37**, 763.
11. F. P. de Haan, H. C. Brown, D. C. Conway, and M. G. Gibby, *J. Am. Chem. Soc.*, 1969, **91**, 4854.
12. F. P. de Haan, M. G. Gibby, and D. R. Aebersold, *J. Am. Chem. Soc.*, 1969, **91**, 4860.
13. G. A. Olah, J. R. DeMember, R. H. Schlosberg, and Y. Halpern, *J. Am. Chem. Soc.*, 1972, **94**, 156.
14. R. Nakane, O. Kusihara, and A. Natsubori, *J. Am. Chem. Soc.*, 1969, **91**, 4528; A. Natsubori and R. Nakane, *J. Org. Chem.*, 1970, **35**, 3372; R. Nakane, O. Kusihara, and A. Takematsu, *J. Org. Chem.*, 1971, **36**, 2753; T. Oyama, T. Hamano, K. Nagumo, and R. Nakane, *Bull. Chem. Soc. Jpn.*, 1978, **51**, 1441.
15. G. A. Olah and S. J. Kuhn, *J. Am. Chem. Soc.*, 1958, **80**, 6541.
16. H. C. Brown and D. J. Brady, *J. Am. Chem. Soc.*, 1952, **74**, 3570.
17. H. C. Brown and H. W. Pearsall, *J. Am. Chem. Soc.*, 1952, **74**, 191.
18. H. C. Brown and W. J. Wallace, *J. Am. Chem. Soc.*, 1953, **75**, 6268.
19. D. A. McCaulay and A. P. Lien, *J. Am. Chem. Soc.*, 1952, **74**, 6246.
20. H. C. Brown and H. Jungk, *J. Am. Chem. Soc.*, 1955, **77**, 5579.
21. G. Baddeley, *Q. Rev. Chem. Soc.*, 1954, **8**, 355.
22. S. V. Mozorov, M. M. Shakirov, V. G. Shubin, and V. A. Koptyug, *J. Org. Chem. USSR*, **14**, 924.
23. R. M. Roberts and D. Shiengthong, *J. Am. Chem. Soc.*, 1964, **86**, 2851.
24. R. H. Allen, T. Alfrey, and L. D. Yats, *J. Am. Chem. Soc.*, 1959, **81**, 42.
25. R. H. Allen, *J. Am. Chem. Soc.*, 1960, **82**, 4856.
26. R. H. Allen, L. D. Yats, and D. S. Erley, *J. Am. Chem. Soc.*, 1960, **82**, 4853.
27 R. M. Roberts and S. G. Brandenberger, *J. Am. Chem. Soc.*, 1960, **82**, 4853.
28. R. M. Roberts and J. E. Douglas, *Chem. Ind.* (*London*), 1958, 1557.
29. R. M. Roberts, Y. W. Han, C. H. Schmidand, and D. A. Davis, *J. Am. Chem. Soc.*, 1959, **81**, 640.
30. D. A. McCaulay, in ref. 1a, Vol. II, Part 2, p. 1065.
31. R. M. Roberts and D. Shiengthong, *J. Am. Chem. Soc.*, 1960, **82**, 732.
32. R. M. Roberts, A. A. Khalaf, and R. N. Greene, *J. Am. Chem. Soc.*, 1964, **86**, 2846.
33. J. F. Norris and D. Rubinstein, *J. Am. Chem. Soc.*, 1939, **61**, 1163.
34. D. A. McCaulay and A. P. Lien, *J. Am. Chem. Soc.*, 1957, **79**, 5953.
35. D. A. McCaulay and A. P. Lien, *J. Am. Chem. Soc.*, 1953, **75**, 2407, 2411.
36. P. S. Hofman, D. J. Reiding, and W. Th. Nauta, *Recl. Trav. Chim. Pays-Bas*, 1960, **79**, 790.
37. H. C. Brown and H. Jungk, *J. Am. Chem. Soc.*, 1956, **78**, 2182.
38. G. A. Olah and M. W. Meyer, *J. Org. Chem.*, 1962, **27**, 3682.
39. L. Schmerling and J. P. West, *J. Am. Chem. Soc.*, 1954, **76**, 1917, and references cited therein.
40. V. M. Akhmedov, F. D. Alieva, and M. A. Mardanov, *J. Org. Chem. USSR*, 1973, **9**, 1676.
41. D. E. Butler, *Tetrahedron Lett.*, 1972, 1929.
42. A. A. Khalaf and R. M. Roberts, *J. Org. Chem.*, 1970, **35**, 3717.
43. S. H. Sharman, *J. Am. Chem. Soc.*, 1962, **84**, 2945.
44. C. R. Smoot and H. C. Brown, *J. Am. Chem. Soc.*, 1956, **78**, 6245, 6249, 6255.

45. H. C. Brown and W. J. Wallace, *J. Am. Chem. Soc.*, 1953, **75**, 6279.
46. V. N. Ipatieff, H. Pines, and L. Schmerling, *J. Org. Chem.*, 1940, **5**, 253.
47. N. N. Lebedev, *J. Gen. Chem. USSR*, 1958, **28**, 1211.
48. J. A. Boon, J. A. Levisty, J. L. Pflug, and J. S. Wilkes, *J. Org. Chem.*, 1986, **51**, 480.
49. N. O. Calloway, *J. Am. Chem. Soc.*, 1937, **59**, 1474.
50. H. C. Brown and M. Grayson, *J. Am. Chem. Soc.*, 1953, **75**, 6285.
51. H. Jungk, C. R. Smoot, and H. C. Brown, *J. Am. Chem. Soc.*, 1956, **78**, 2185.
52. N. N. Lebedev, *Chem. Abstr.*, 1959, **53**, 9106.
53. N. N. Lebedev, *J. Gen. Chem. USSR*, 1954, **24**, 673.
54. H. C. Brown and H. Jungk, *J. Am. Chem. Soc.*, 1955, **77**, 5584.
55. R. Taylor, *Comprehensive Chemical Kinetics*, Vol. 13, Elsevier, Amsterdam, 1972, p. 145.
56. R. H. Allen and L. D. Yats, *J. Am. Chem. Soc.*, 1961, **83**, 2799.
57. F. P. de Haan and H. C. Brown, *J. Am. Chem. Soc.*, 1969, **91**, 4844.
58. G. A. Olah, S. J. Kuhn, and S. H. Flood, *J. Am. Chem. Soc.*, 1962, **84**, 1688, 1695.
59. G. A. Olah, S. H. Flood, S. J. Kuhn, M. E. Moffat, and N. A. Overchuck, *J. Am. Chem. Soc.*, 1964, **86**, 1046.
60. G. A. Olah, S. H. Flood, and M. E. Moffatt, *J. Am. Chem. Soc.*, 1964, **86**, 1060, 1065.
61. G. A. Olah and N. A. Overchuck, *J. Am. Chem. Soc.*, 1965, **87**, 5786.
62. F. P. de Haan *et al.*, *J. Am. Chem. Soc.*, 1984, **106**, 7038.
63. F. P. de Haan *et al.*, *J. Org. Chem.*, 1986, **51**, 1587, 1591.
64. B. J. Carter, W. D. Covey, and F. P. de Haan, *J. Am. Chem. Soc.*, 1975, **97**, 4783.
65. G. A. Olah, M. Tashiro, and M. Kobayashi, *J. Am. Chem. Soc.*, 1970, **92**, 6369; G. A. Olah, S. Kobayashi, and M. Tashiro, *J. Am. Chem. Soc.*, 1972, **94**, 7448; G. A. Olah, J. A. Olah, and T. Ohyama, *J. Am. Chem. Soc.*, 1984, **106**, 5284.
66. T. J. Tewson and R. Taylor, *J. Chem. Soc., Chem. Commun.*, 1973, 836.
67. F. P. de Haan *et al.*, *J. Org. Chem.*, 1984, **49**, 3954.
68. A. Streitwieser, D. P. Stevenson, and W. D. Schaeffer, *J. Am. Chem. Soc.*, 1959, **81**, 1110.
69. H. Pines, W. D. Huntsman, and V. N. Ipatieff, *J. Am. Chem. Soc.*, 1951, **73**, 4343, 4483.
70. A. Streitwieser, W. D. Schaeffer, and S. Andreades, *J. Am. Chem. Soc.*, 1959, **81**, 1110.
71. D. Bethell and V. Gold, *J. Chem. Soc.*, 1958, 1905, 1930; V. Gold and T. Riley, *J. Chem. Soc.*, 1960, 2973.
72. (a) L. P. Hammett and A. Deyrup, *J. Am. Chem. Soc.*, 1933, **55**, 1900; (b) B. L. Booth, M. Al-Kinany, and K. Laali, *J. Chem. Soc., Perkin Trans. 1*, 1987, 2049.
73. B. S. Friedman and F. L. Moritz, *J. Am. Chem. Soc.*, 1956, **78**, 2000.
74. R. A. Kretchmer and M. B. McCloskey, *J. Org. Chem.*, 1972, **37**, 1989.
75. I. S. Belostotskaya, N. L. Komissarova, E. V. Dzharyyan, and V. V. Ershov, *Bull. Acad. Sci. USSR*, 1972, 1535.
76. T. G. Bonner, K. A. Holder, and P. Powell, *J. Organomet. Chem.*, 1974, **77**, C37.
77. L. A. P. Kane-Maguire and C. A. Mansfield, *J. Chem. Soc., Chem. Commun.*, 1973, 540.
78. A. V. Grosse and V. N. Ipatieff, *J. Org. Chem.*, 1937, **2**, 447.
79. L. Schmerling, *Ind. Eng. Chem.*, 1948, **40**, 2072.
80. V. N. Ipatieff, H. Pines, and L. Schmerling, *J. Org. Chem.*, 1940, **5**, 243; V. A. Isidorov, B. V. Ioffe, B. V. Stolyarov, and V. A. Valovoi, *Proc. Acad. Sci. USSR*, 1972, **205**, 595.
81. N. S. Namelkin, E. Sh. Finkelstein, E. B. Portuyk, and V. M. Vdovin, *Proc. Acad. Sci. USSR*, 1974, **214**, 110.
82. R. L. Burwell, L. M. Elkin, and A. D. Shields, *J. Am. Chem. Soc.*, 1952, **74**, 4567, 4570.
83. H. C. Brown and R. M. Adams, *J. Am. Chem. Soc.*, 1942, **64**, 2557.
84. S. Searles, *J. Am. Chem. Soc.*, 1959, **76**, 2313.

85. T. Nakajima, S. Suga, T. Sugita, and R. Ichikawa, *Bull. Chem. Soc. Jpn.*, 1967, **40**, 2980.
86. S. Musada, T. Nakajima, and S. Suga, *J. Chem. Soc., Chem. Commun.*, 1974, 954.
87. V. F. Traven and B. I. Stepanov, *J. Org. Chem. USSR*, 1971, **7**, 517.
88. U. Svanholm and V. D. Parker, *J. Chem. Soc., Perkin Trans. 1*, 1973, 562.
89. K. Nyberg, *Chem. Scr.*, 1974, **5**, 143.
90. J. I. Brauman and A. J. Paudell, *J. Am. Chem. Soc.*, 1967, **89**, 5421.
91. B. L. Booth, R. N. Haszeldine, and K. Laali, *J. Chem. Soc., Perkin Trans. 1*, 1980, 181.
92. G. A. Olah and J. Nishimura, *J. Am. Chem. Soc.*, 1974, **96**, 2214.
93. C. D. Nenitzescu, S. Titeica, and V. Ioan, *Bull. Soc. Chim. Fr.*, 1955, **20**, 1272, 1279; V. Ioan, D. Sandeulescu, S. Titeica, and C. D. Nenitzescu, *Tetrahedron*, 1963, **19**, 323.
94. E. G. Willard and H. Cerfontain, *Recl. Trav. Chim. Pays-Bas*, 1973, **92**, 739.
95. G. A. Olah, J. Nishimura, and Y. Yamada, *J. Org. Chem.*, 1974, **39**, 2430.
96. F. P. de Haan *et al.*, *J. Org. Chem.*, 1984, **49**, 3967.
97. G. A. Olah and E. G. Melby, *J. Am. Chem. Soc.*, 1973, **95**, 4971.
98. G. A. Olah, *Halonium Ions*, Wiley, New York, 1975; G. A. Olah and E. G. Melby, *J. Am. Chem. Soc.*, 1972, **94**, 6220.
99. J. Kaspi, D. D. Montgomery, and G. A. Olah, *J. Org. Chem.*, 1978, **43**, 3147; J. Kaspi and G. A. Olah, *J. Org. Chem.*, 1978, **43**, 3142.
100. R. L. N. Harris, *Aust. J. Chem.*, 1974, **27**, 2635.
101. R. Koncos and B. S. Friedman, and V. Franzen, in ref. 1a, Vol. II, Ch. 15 and 16.
102. T. Kitamura, S. Kobayashi, H. Taniguchi, and Z. Rappoport, *J. Org. Chem.*, 1982, **47**, 5003.
103. J. Marchand-Bryanaert and L. Ghosez, *J. Am. Chem. Soc.*, 1972, **94**, 2869.
104. J. E. Hofmann and A. Schriesheim, in ref. 1a, Vol. II, Ch. 19.
105. P. R. Stapp, *J. Org. Chem.*, 1974, **39**, 2466.
106. J. A. Dixon and D. H. Fishman, *J. Am. Chem. Soc.*, 1963, **85**, 1356.
107. G. A. Olah, N. A. Overchuk, and J. C. Lapierre, *J. Am. Chem. Soc.*, 1965, **87**, 5785.
108. R. C. Dunbar, J. Shen, E. Melby, and G. A. Olah, *J. Am. Chem. Soc.*, 1973, **95**, 7200.
109. F. Cacace and P. Giacomello, *J. Chem. Soc., Perkin Trans. 2*, 1978, 652.
110. M. Speranza, N. Pepe, and R. Cipollini, *J. Chem. Soc., Perkin Trans. 2*, 1979, 1179; N. Pepe and M. Speranza, *J. Chem. Soc., Perkin Trans. 2*, 1981, 1430; M. Colosimo and R. Bucci, *J. Chem. Soc., Perkin Trans. 2*, 1983, 933.
111. S. Takamuka, K. Iseda, and H. Sakurai, *J. Am. Chem. Soc.*, 1971, **93**, 2420; F. Cacace and E. Possagno, *J. Am. Chem. Soc.*, 1973, **95**, 3397; M. Attina, F. Cacace, G. Ciranni, and P. Giacomello, *J. Am. Chem. Soc.*, 1977, **99**, 2611; *J. Chem. Soc., Perkin Trans. 2*, 1979, 891; M. Attina and F. Cacace, *J. Am. Chem. Soc.*, 1983, **105**, 1122; M. Attina and P. Giacomello, *Tetrahedron Lett.*, 1977, 2373; M. Attina, G. de Petri, and P. Giacomello, *Tetrahedron Lett.*, 1982, 3525.
112. (a) B. Aliprandi, F. Cacace, and S. Farnarini, *Tetrahedron*, 1987, **43**, 2831; (b) M. Attina, F. Cacace, and P. Giacomello, *J. Am. Chem. Soc.*, 1980, **102**, 4768.
113. F. Cacace and P. Giacomello, *J. Am. Chem. Soc.*, 1973, **96**, 5851; F. Cacace and G. Ciranni, *J. Am. Chem. Soc.*, 1986, **108**, 887.
114. P. Giacomello and F. Cacace, *J. Chem. Soc., Chem. Commun.*, 1975, 379.
115. D. L. Miller, J. D. Lay, and M. L. Gross, *J. Chem. Soc., Chem. Commun.*, 1982, 970.
116. S. U. Choi and H. C. Brown, *J. Am. Chem. Soc.*, 1959, **81**, 3315.
117. H. C. Brown and B. A. Bolto, *J. Am. Chem. Soc.*, 1959, **81**, 3320.
118. H. C. Brown and A. H. Neyens, *J. Am. Chem. Soc.*, 1962, **84**, 1233, 1655.
119. F. P. de Haan, H. C. Brown, and J. C. Hill, *J. Am. Chem. Soc.*, 1969, **91**, 4850.
120. L. R. C. Barclay, in ref. 1a, Vol. II, Ch. 22; C. K. Bradsher, *Chem. Rev.*, 1987, **87**, 1277.
121. R. W. Layer and I. R. MacGregor, *J. Org. Chem.*, 1956, **21**, 1120.

122. R. O. Roblin, D. Davidson, and M. T. Bogert, *J. Am. Chem. Soc.*, 1935, **57**, 151.
123. W. Davies and S. Middleton, *J. Chem. Soc.*, 1958, 822.
124. P. L. Julian, E. W. Meyer, and H. C. Printy, in *Heterocyclic Compounds*, Ed. R. C. Elderfield, Vol. 3, Wiley, New York, 1952, p. 1.
125. E. Berliner, *J. Am. Chem. Soc.*, 1944, **66**, 553; L. K. Brice and R. D. Katstra, *J. Am. Chem. Soc.*, 1960, **82**, 2669.
126. C. K. Bradsher and F. A. Vingiello, *J. Am. Chem. Soc.*, 1949, **71**, 1434; F. A. Vingiello and J. G. van Oot, *J. Am. Chem. Soc.*, 1951, **73**, 5070; F. A. Vingiello, J. G. van Oot, and H. H. Hannabass, *J. Am. Chem. Soc.*, 1952, **74**, 4546.
127. (a) H. Hart and E. A. Sedor, *J. Am. Chem. Soc.*, 1967, **89**, 2342; (b) F. A. Vingiello and M. M. Schlechter, *J. Org. Chem.*, 1963, **28**, 2448.
128. F. A. Vingiello, M. O. L. Spangler, and J. E. Bondurant, *J. Org. Chem.*, 1960, **25**, 2091.
129. T. G. Bonner, M. P. Thorpe, and T. M. Williams, *J. Chem. Soc.*, 1955, 2351; T. G. Bonner and J. M. Watkins, *J. Chem. Soc.*, 1955, 2358; T. G. Bonner and M. Barnard, *J. Chem. Soc.*, 1958, 4176, 4181.
130. Ref. 55, p. 162.
131. A. H. Jackson, P. V. R. Shannon, and P. W. Taylor, *J. Chem. Soc., Perkin Trans. 2*, 1981, 286.
132. T. J. Mason and R. O. C. Norman, *J. Chem. Soc., Perkin Trans. 2*, 1973, 1840.
133. A. A. Khalaf and R. M. Roberts, *J. Org. Chem.*, 1973, **38**, 1388.
134. E. D. Laskina, T. A. Devitskaya, N. P. Koren, E. A. Simanovskaya, and T. A. Rudolfi, *J. Org. Chem. USSR*, 1972, **8**, 617.
135. U. Svanholm and V. D. Parker, *J. Chem. Soc., Perkin Trans. 2*, 1974, 169.
136. C. Eaborn, P. M. Jackson, and R. Taylor, *J. Chem. Soc. B*, 1966, 613.
137. (a) A. T. Balaban and A. Schriesheim, in ref. 1a, Vol. II, Ch. 23; (b) G. A. Olah, W. J. Tolgyesi, and R. E. A. Dear, *J. Org. Chem.*, 1962, **27**, 3441, 3449; (c) T. Sone, R. Yokoyama, Y. Okuyama, and K. Sato, *Bull. Chem. Soc. Jpn.*, 1986, **59**, 83.
138. A. N. Nesmeyanov, L. G. Makarova, and T. P. Tolustaya, *Tetrahedron*, 1957, **1**, 145.
139. R. A. Abramovitch and F. F. Gadallah, *J. Chem. Soc. B*, 1968, 497; M. Kobayashi, H. Minato, E. Yamada, and N. Kobori, *Bull. Chem. Soc. Jpn.*, 1970, **43**, 215, 223.
140. M. Kobayashi, H. Minato, and N. Kobori, *Bull. Chem. Soc. Jpn.*, 1970, **43**. 219.
141. N. Kamigata, M. Kobayashi, and H. Minato, *Bull. Chem. Soc. Jpn.*, 1972, **45**, 1231.
142. N. Kamigata, R. Hisada, H. Minato, and M. Kobayashi, *Bull. Chem. Soc. Jpn.*, 1973, **46**, 1016.
143. N. Kamigata, H. Minato, and M. Kobayashi, *Bull. Chem. Soc. Jpn.*, 1974, **47**, 894.
144. H. Takeuchi, K. Takano, and K. Koyama, *J. Chem. Soc., Chem. Commun.*, 1982, 1254; H. Takeuchi and K. Takano, *J. Chem. Soc., Chem. Commun.*, 1983, 447.
145. A. Fischer and C. C. Greig, *J. Chem. Soc., Chem. Commun.*, 1974, 50.
146. M. Miocque and J. Vierfond, *Bull. Soc. Chim. Fr.*, 1970, 1896, 1901, 1907.
147. J. Gloede, J. Freiberg, W. Bürger, G. Ollmann, and H. Gross, *Arch. Pharm. (Weinheim)*, 1949, **302**, 354; D. D. Reynolds and B. C. Cossar, *J. Heterocycl. Chem.*, 1971, **8**, 605.
148. H. E. Zaugg, *Synthesis*, 1970, 49; K. Nyberg, *Acta Chem. Scand., Ser. B*, 1974, **28**, 825; A. K. Sheinkman, E. N. Nelin, A. I. Kostin, and V. P. Kostin, *J. Org. Chem. USSR*, 1978, **14**, 1193.
149. M. G. Burdon and J. G. Moffatt, *J. Am. Chem. Soc.*, 1966, **88**, 5855; R. A. Olofson and G. Marino, *Tetrahedron*, 1971, **27**, 4195. P. Claus, *Monatsh. Chem.*, 1986, **99**, 1034; 1971, **102**, 913; P. Claus, N. Vavra, and P. Schilling, *Monatsh. Chem.*, 1972, **102**, 1072; Y. Hayashi and R. Oda, *J. Org. Chem.*, 1967, **32**, 457; G. H. Pettitt and T. H. Brown, *Can. J. Chem.*, 1967, **45**, 1306.

150. G. A. Olah and W. S. Tolgyesi, in ref. 1a, Vol. II, Ch. 21; L. I. Belenkii, Yu. B. Volkenstein, and I. B. Karmanova, *J. Gen. Chem. USSR*, 1977, **46**, 891.
151. R. C. Fuson and C. H. McKeever, *Org. React.*, 1942, **1**, 63.
152. Y. Ogata and M. Okano, *J. Am. Chem. Soc.*, 1956, **78**, 5423.
153. I. N. Nazarev and A. V. Semenosky, *Bull. Acad. Sci. USSR*, 1957, 103, 212, 861, 997.
154. M. M. Lyushin, S. V. Mekhtiev, and S. N. Guseinova, *J. Org. Chem. USSR*, 1970, **6**, 1445.
155. G. Vavon, J. Bolle, and J. Calin, *Bull. Soc. Chim. Fr.*, 1939, **6**, 1025.
156. H. Suzuki, *Bull. Chem. Soc. Jpn.*, 1970, **43**, 3299; M. E. Kuimova and B. M. Mikhailov, *J. Org. Chem. USSR*, 1971, **7**, 1485.
157. A. McKillop, F. A. Madjdabadi, and D. A. Long, *Tetrahedron Lett.*, 1983, **24**, 1933.
158. G. A. Olah, D. A. Beal, and J. A. Olah, *J. Org. Chem.*, 1976, **41**, 1627.
159. W. Badger, J. W. Barton, and J. F. W. McOmie, *J. Chem. Soc.*, 1958, 2666.
160. R. B. Sandin and L. F. Fieser, *J. Am. Chem. Soc.*, 1940, **62**, 3098.
161. G. A. Olah and A. Pavlath, *Acta Chim. Acad. Sci. Hung.*, 1953, **3**, 203, 425.
162. G. S. Mironov, I. V. Budnii, M. I. Farberov, V. D. Shein, and I. I. Bespalova, *J. Org. Chem. USSR*, 1966, **2**, 1615; 1970, **6**, 1231.
163. H. H. Szmant and J. Dudek, *J. Am. Chem. Soc.*, 1949, **71**, 3763.
164. H. C. Brown and K. L. Nelson, *J. Am. Chem. Soc.*, 1953, **75**, 6292.
165. G. A. Olah, D. A. Beal, and J. A. Olah, *J. Org. Chem.*, 1976, **41**, 1627.
166. T. Matsukawa and K. Shirakawa, *J. Pharm. Soc. Jpn.*, 1950, **70**, 25.
167. F. R. Jensen and G. Goldman, in ref. 1a, Vol. III, Ch. 36.
168. G. A. Olah, S. J. Kuhn, W. S. Tolgyesi, and E. B. Baker, *J. Am. Chem. Soc.*, 1962, **84**, 2733.
169. N. N. Lebedev, *J. Gen. Chem. USSR*, 1951, **21**, 1788; B. P. Susz and I. Cooke, *Helv. Chim. Acta*, 1954, **37**, 1273; I. Cooke, B. P. Susz, and C. Herschmann, *Helv. Chim. Acta*, 1954, **37**, 1280; D. Cook, *Can. J. Chem.*, 1959, **37**, 48.
170. S. E. Rasmussen and N. C. Broch, *Acta Chem. Scand.*, 1966, **20**, 1351; R. Chevrier and R. Weiss, *Angew. Chem., Int. Ed. Engl.*, 1974, **13**, 1.
171. Ref. 55, pp. 181–182.
172. R. Corriu, M. Dore, and R. Thomassin, *Tetrahedron*, 1971, **27**, 5819.
173. E. Wertyporoch and T. Firla, *Z. Phys. Chem.*, 1932, **162**, 398; N. N. Greenwood and K. Wade, *J. Chem. Soc.*, 1956, 1527.
174. G. Baddeley and D. Voss, *J. Chem. Soc.*, 1954, 418.
175. E. Rothstein and R. W. Saville, *J. Chem. Soc.*, 1949, 1950, 1954, 1959, 1961; M. E. Grundy, W. H. Hsü, and E. Rothstein, *J. Chem. Soc.*, 1958, 581.
176. D. P. N. Satchell, *Q. Rev. Chem. Soc.*, 1963, **17**, 160.
177. F. Effenberger and G. Epple, *Angew. Chem., Int. Ed. Engl.*, 1972, **11**, 300.
178. J. M. Tedder, *Chem. Ind. (London)*, 1954, 5320.
179. I. Hashimoto, T. Nojiro, and Y. Ogata, *Tetrahedron*, 1970, **26**, 4603; L. R. Pettiford, *J. Chem. Soc, Perkin Trans. 2*, 1972, 52; V. A. Ustinov, G. S. Mironov, and M. I. Farberov, *J. Org. Chem. USSR*, 1973, **9**, 754.
180. Á. I. Bokova and N. G. Sodorova, *J. Org. Chem. USSR*, 1971, **7**, 1075.
181. I. N. Zemzina and M. I. Inagamova, *J. Org. Chem. USSR*, 1972, **8**, 2419.
182. L. I. Belenkii, A. P. Yakubov, and Ya. L. Goldfarb, *J. Org. Chem. USSR*, 1970, **6**, 2531; I. Hashimoto, A. Kawasaki, and Y. Ogata, *Tetrahedron*, 1972, **28**, 217.
183. N. O. Calloway, *J. Am. Chem. Soc.*, 1937, **59**, 1474.
184. Y. Yamase, *Bull. Chem. Soc. Jpn.*, 1961, **34**, 480, 484.
185. Ref. 55, pp. 166–171.
186. H. C. Brown and F. R. Jensen, *J. Am. Chem. Soc.*, 1958, **80**, 2291.
187. F. R. Jensen and H. C. Brown, *J. Am. Chem. Soc.*, 1958, **80**, 3039.

188. O. C. Dermer, D. M. Wilson, F. M. Johnson, and V. H. Dermer, *J. Am. Chem. Soc.*, 1941, **63**, 2881; O. C. Dermer and R. A. Billmeier, *J. Am. Chem. Soc.*, 1942, **64**, 464.
189. G. Dana, P. Scribe, and J. P. Girault, *C. R. Acad. Sci. Ser. C*, 1972, **275**, 49; *Tetrahedron*, 1973, **29**, 413.
190. H. C. Brown and H. L. Young, *J. Org. Chem.*, 1957, **22**, 719.
191. F. P. de Haan *et al.*, *J. Org. Chem.*, 1984, **49**, 3900.
192. F. R. Jensen, G. Marino, and H. C. Brown, *J. Am. Chem. Soc.*, 1959, **81**, 3303.
193. S. C. J. Olivier, *Recl. Trav. Chim. Pays-Bas*, 1973, **92**, 739.
194. R. J. Gillespie, *J. Chem. Soc.*, 1950, 2997; H. P. Treffers and L. P. Hammett, *J. Am. Chem. Soc.*, 1937, **59**, 1708.
195. W. R. Edwards and E. C. Sibelle, *J. Org. Chem.*, 1963, **28**, 674.
196. F. Effenberger and G. Epple, *Angew. Chem., Int. Edn. Engl.*, 1972, **11**, 299.
197. F. Effenberger, G. König, and H. Klenk, *Angew. Chem., Int. Edn. Engl.*, 1978, **17**, 695.
198. (a) G. A. Olah and H. C. Lin, *Synthesis* 1974, 342, 895; (b) G. A. Olah, J. Lukas, and E. Lukas, *J. Am. Chem. Soc.*, 1969, **81**, 5319.
199. K. R. Fountain, P. Heinze, M. Sherwood, D. Maddox, and G. Gerhardt, *Can. J. Chem.*, 1980, **58**, 1198.
200. E. Ador and J. M. Crafts, *Chem. Ber.*, 1877, **10**, 2173; C. Friedel, J. M. Crafts, and E. Ador, *C. R. Acad. Sci. Ser. C*, 1877, **85**, 673.
201. I. R. Butler and J. O. Morley, *J. Chem. Res. (S)*, 1980, 358.
202. G. A. Olah and S. J. Kuhn, *J. Am. Chem. Soc.*, 1960, **82**, 2380.
203. R. C. Dumbar, J. Shen, and G. A. Olah, *J. Am. Chem. Soc.*, 1972, **94**, 6852; D. A. Chatfield and M. M. Bursey, *J. Chem. Soc., Faraday Trans. 1*, 1976, 417.
204. P. H. Gore, G. H. Smith, and S. Thorburn, *J. Chem. Soc. (C)*, 1971, 651.
205. G. E. Lewis, *J. Org. Chem.*, 1966, **31**, 749.
206. L. Friedman and R. J. Honour, *J. Am. Chem. Soc.*, 1969, **91**, 6344.
207. P. H. Gore, *Chem. Rev.*, 1955, **55**, 229; P. H. Gore and C. K. Thadani, *J. Chem. Soc. C*, 1966, 1729; 1967, 1498.
208. R. B. Girdler, P. H. Gore, and J. A. Hoskins, *J. Chem. Soc. C*, 1966, 181; P. H. Gore, A. S. Siddiquei, and S. Thorburn, *J. Chem. Soc., Perkin Trans. 1*, 1972, 1781.
209. P. H. Gore and M. Yusuf, *J. Chem. Soc. C*, 1971, 2586.
210. I. Agranant, Y. Shih, and Y. Benter, *J. Am. Chem. Soc.*, 1974, **96**, 1259.
211. S. Clementi and P. Linda, *Tetrahedron*, 1970, **26**, 2869.
212. L. M. Stock and H. C. Brown, *Adv. Phys. Org. Chem.*, 1963, **1**, 35.
213. H. C. Brown and G. Marino, *J. Am. Chem. Soc.*, 1959, **81**, 3308; 1962, **84**, 1236.
214. F. R. Jensen and B. E. Smart, *J. Am. Chem. Soc.*, 1969, **91**, 5686, 5688.
215. T. A. Elwood, W. R. Flack, K. J. Inman, and P. W. Rabideau, *Tetrahedron*, 1974, **30**, 535.
216. P. H. Gore, S. Thorburn, and D. J. Weyell, *J. Chem. Soc., Perkin Trans. 1*, 1973, 2940.
217. W. J. Archer, M. A. Hossaini, and R. Taylor, *J. Chem. Soc., Perkin Trans. 2*, 1982, 181.
218. M. A. Hossaini and R. Taylor, *J. Chem. Soc., Perkin Trans. 2*, 1982, 187.
219. H. C. Brown, B. N. Gnedin, K. Takeuchi, and E. N. Peters, *J. Am. Chem. Soc.*, 1975, **97**, 610.
220. P. A. Goodman and P. H. Gore, *J. Chem. Soc. C*, 1968, 2452.
221. P. H. Gore, C. K. Thadani, and S. Thorburn, *J. Chem. Soc. C*, 1968, 2502.
222. P. H. Gore and J. A. Hoskins, *J. Chem. Soc. C*, 1971, 3347.
223. P. H. Gore, A. Y. Miri, and A. S. Siddiquei, *J. Chem. Soc., Perkin Trans. 2*, 1973, 2936.
224. R. B. Girdler, P. H. Gore, and J. A. Hoskins, *J. Chem. Soc. C*, 1966, 518.
225. R. Corriu, M. Dore, and R. Thomassin, *Tetrahedron Lett.*, 1968,'2759.
226. A. D. Andreou, R. V. Bulbulian, P. H. Gore, F. S. Kamounah, A. H. Miri, and D. N. Waters, *J. Chem. Soc., Perkin Trans. 2*, 1981, 377.

227. I. P. Tsukervanik, Kh. Kim, and A. S. Rurbatova, *J. Gen. Chem. USSR*, 1963, **33**, 227.
228. P. P. Gnatyuk, S. E. Pokhila, A. M. Kuznetsov, V. A. Yakobi, V. L. Plakidin, and I. A. Rister, *J. Org. Chem. USSR*, 1973, **9**, 173.
229. F. E. Friedli and H. Schechter, *J. Org. Chem.*, 1985, **50**, 5710.
230. P. H. Gore and M. Jehangir, *J. Chem. Soc., Perkin Trans. 2*, 1979, 3007.
231. F. Perin, P. Jacquignon, and N. P. Buu-Hoi, *J. Chem. Soc., Chem. Commun.*, 1966, 592.
232. N. S. Dokunikhin, G. N. Vorozhtsov, and N. B. Feldblum, *J. Org. Chem. USSR*, 1972, **8**, 354.
233. L. W. Deady, P. M. Gray, and R. D. Topsom, *J. Org. Chem.*, 1972, **37**, 3335.
234. R. B. Girdler, P. H. Gore, and C. K. Thadani, *J. Chem. Soc. C*, 1967, 2619; P. H. Gore, C. K. Thadani, S. Thorburn, and M. Yusuf, *J. Chem. Soc. C*, 1971, 2329.
235. P. H. Gore, F. S. Kamounah, and A. Y. Miri, *J. Chem. Res. (S)*, 1980, 40.
236. P. H. Gore and C. K Thadani, *J. Chem. Soc. C*, 1967, 1498.
237. H. B. Renfroe, J. A. Gurney, and L. A. R. Hall, *J. Org. Chem.*, 1972, **37**, 3045.
238. A. G. Anderson, E. R. Davidson, E. D. Duags, L. G. Kao, R. L. Lindquist, and K. A. Quenemoen, *J. Am. Chem. Soc.*, 1985, **107**, 1896.
239. N. Campbell and N. H. Wilson, *J. Chem. Soc., Perkin Trans. 1*, 1972, 2739.
240. C. J. Paget and A. Burger, *J. Org. Chem.*, 1965, **30**, 1329; V. R. Skvarchenko, I. I. Brunovlenskaya, A. M. Novikov, and R. Ya. Levina, *J. Org. Chem. USSR*, 1970, **6**, 1514.
241. A. I. Bukova and N. G. Sidorova, *J. Org. Chem. USSR*, 1970, **6**, 1716.
242. H. J. Reich and D. J. Cram, *J. Am. Chem. Soc.*, 1969, **91**, 3505.
243. D. J. Cram, W. J. Wechter, and R. W. Kierstead, *J. Am. Chem. Soc.*, 1958, **80**, 3126.
244. A. G. Anderson and L. L. Replogle, *J. Org. Chem.*, 1963, **28**, 2578.
245. A. P. Laws and R. Taylor, *J. Chem. Soc., Perkin Trans. 2*, 1987, 591.
246. Z. Yoshida, S. Yoneda, and M. Hazama, *J. Org. Chem.*, 1972, **37**, 1364.
247. H. Hart, R. M. Schlosbergh, and R. K. Murray, *J. Org. Chem.*, 1968, **33**, 3800.
248. A. A. Petinskii, S. M. Shostakovskii, and E. I. Kositsyna, *Bull. Acad. Sci. USSR*, 1972, 1720.
249. J. B. F. Lloyd and P. A. Ongley, *Tetrahedron*, 1964, **20**, 2185; 1965, **21**, 245.
250. P. M. op den Brouw and W. H. Laarhoven, *Recl. Trav. Chim. Pays-Bas*, 1978, **97**, 265.
251. M. Rosenblum and W. G. Howells, *J. Am. Chem. Soc.*, 1962, **84**, 1167; M. Rosenblum and F. W. Abbate, *J. Am. Chem. Soc.*, 1966, **88**, 4178; H. L. Letzner and W. E. Watts, *Tetrahedron*, 1972, **28**, 121; K. Yamakawa, M. Hisatome, Y. Sato, and S. Ichida, *J. Organomet. Chem.*, 1975, **93**, 219.
252. A. N. Nesmeyanov, V. N. Drozd, and V. A. Sazonova, *Bull. Acad. Sci. USSR*, 1965, 1169.
253. J. A. Winstead, R. R. McGuire, R. E. Cochoy, A. D. Brown, and G. J. Ganthier, *J. Org. Chem.*, 1972, **37**, 2055.
254. (a) E. O. Fischer, M. von Foerster, C. G. Kreiter, and K. E. Schwarzhaus, *J. Organomet. Chem.*, 1967, **7**, 113; (b) J. L. von Rosenberg and A. R. Pinder, *J. Chem. Soc., Perkin Trans. 1*, 1987, 747.
255. S. Sethna, in ref. 1a, Vol. III, Ch. 35.
256. D. M. Brouwer, J. A. van Doorn, A. A. Kiffen, and P. A. Kramer, *Recl. Trav. Chim. Pays-Bas*, 1974, **93**, 189.
257. D. S. Noyce and P. A. Kittle, *J. Org. Chem.*, 1967, **32**, 2459; D. S. Noyce, P. A. Kittle, and E. H. Banitt, *J. Org. Chem.*, 1968, **33**, 1500.
258. A. Gerecs, in ref. 1a, Vol. III, Ch. 33.
259. S. Munavelli, *Chem. Ind. (London)*, 1972, 293.

260. F. Krausz and R. Martin, *Bull. Soc. Chim. Fr.*, 1965, 2192; R. Martin, *Bull. Soc. Chim. Fr.*, 1974, 983; 1979, 373.
261. Y. Ogata and H. Tabuchi, *Tetrahedron*, 1964, **20**, 1661.
262. L. Gattermann and J. A. Koch, *Chem. Ber.*, 1897, **30**, 1622.
263. H. A. Staab and A. P. Datta, *Angew. Chem.*, 1963, **75**, 1203.
264. M. H. Dilke and D. D. Eley, *J. Chem. Soc.*, 1949, 2601, 2613.
265. J. H. Holloway and N. W. Krase, *Ind. Eng. Chem.*, 1933, **25**, 497.
266. N. N. Crounse, *Org. React.*, 1949, **5**, 290.
267. (a) G. A. Olah and S. J. Kuhn, in ref. 1a, Vol. III, Ch. 38; (b) G. A. Olah, L. Ohannesian, and M. Arvanaghi, *Chem. Rev.*, 1987, **87**, 671.
268. B. L. Booth, T. A. El-Fekky, and G. F. M. Noori, *J. Chem. Soc., Perkin Trans. 2*, 1980, 181.
269. W. E. Truce, *Org. React.*, 1957, **9**, 37.
270. L. E. Hinkel, E. A. Ayling, and W. H. Morgan, *J. Chem. Soc.*, 1932, 2793.
271. R. Adams and E. Levine, *J. Am. Chem. Soc.*, 1923, **45**, 2372; R. Adams and E. Montgomery, *J. Am. Chem. Soc.*, 1924, **46**, 1518.
272. L. E. Hinkel and R. T. Dunn, *J. Chem. Soc.*, 1930, 1834; 1931, 3343; F. B. Dains, *Chem. Ber.*, 1902, **35**, 2496.
273. A. Vilsmeier and A. Haack, *Chem. Ber.*, 1927, **60**, 119; M.-R. de Maheas, *Bull. Soc. Chim. Fr.*, 1962, 1989.
274. R. Wizinger, *J. Prakt. Chem.*, 1939, **154**, 25; H. Lorenz and R. Wizinger, *Helv. Chim. Acta*, 1945, **28**, 600.
275. S. Alumni, P. Linda, G. Marino, S. Santini, and G. Savelli, *J. Chem. Soc., Perkin Trans. 2*, 1972, 2070; S. Clementi, F. Fringuelli, P. Linda, G. Marino, G. Savelli, and A. Taticchi, *J. Chem. Soc., Perkin. Trans. 2*, 1973, 2097.
276. V. D. Pokhodenko, V. G. Koshenko, and A. N. Inozomstev, *J. Chem. Soc., Chem. Commun.*, 1985, 72.
277. E. Watanabe, S. Nishimura, H. Ogashi, and Z. Yoshida, *Tetrahedron*, 1975, **31**, 1385.
278. A. Rieche, H. Gross, and E. Höft, *Chem. Ber.*, 1960, **93**, 88.
279. T. M. Cresp, M. V. Sargent, J. A. Elix, and D. P. M. Murphy, *J. Chem. Soc., Perkin Trans. 1*, 1973, 340.
280. F. P. de Haan *et al.*, *J. Org. Chem.*, 1984, **49**, 3963.
281. G. A. Olah, F. Pelizza, S. Kobayashi, and J. A. Olah, *J. Am. Chem. Soc.*, 1976, **92**, 296.
282. K. Hoesch, *Chem. Ber.*, 1927, **60**, 389, 2537.
283. P. E. Spoerri and A. S. DuBois, *Org. React.*, 1949, **5**, 387; W. Ruske, in ref. 1a, Vol. III, Ch. 32.
284. K. C. Gulati, S. R. Seth, and K. Venkataraman, *Org. Synth., Coll. Vol. 2*, 1943, 522.
285. E. N. Zil'berman and N. A. Rybakova, *J. Gen. Chem. USSR*, 1960, **30**, 1972.
286. J. Tröger and O. Lüning, *J. Prakt. Chem.*, 1904, **69**, 347.
287. H. Stephen, *J. Chem. Soc.*, 1920, 1529.
288. C. Friedel and J. M. Crafts, *C. R. Acad. Sci.*, 1878, **86**, 1368; *Ann. Chim. Phys.*, 1883, **14**, 433.
289. H. Jorg, *Chem. Ber.*, 1889, **22**, 1215; S. R. Ramadas and P. S. Srinivasan, *J. Chem. Soc., Chem. Commun.*, 1972, 345.
290. G. A. Olah and J. A. Olah, in ref. 1a, Vol. III, Ch. 39.
291. J. F. Norris and E. W. Fuller, *US Pat.*, 1 542 264, 1925.
292. O. Achmatowicz and O. Achmatowicz, *Rocz. Chem.*, 1961, **35**, 813.
293. R. Leuckart, *Chem. Ber.*, 1885, **18**, 873; *J. Prakt. Chem.*, 1890, **41**, 306.
294. L. Gatterman, *Justus Liebigs Ann. Chem.*, 1890, **23**, 1190.
295. M. Seefelder, *Chem. Ber.*, 1963, **96**, 3243.
296. Yu. A. Naumov, A. P. Isakova, A. N. Kost, V. F. Zakharov, V. P. Zvolinskii, N. F. Moiseikina, and S. V. Nikeryasova, *J. Org. Chem. USSR*, 1975, **11**, 362.

297. G. A. Olah and P. Schilling, *Justus Liebigs Ann. Chem.*, 1972, **761**, 77.
298. G. A. Olah, M. R. Bruce, and F. L. Clouet, *J. Org. Chem.*, 1981, **46**, 438.
299. A. E. Feiring, *J. Org. Chem.*, 1976, **41**, 148.
300. M. Yoshimura, T. Namba, and T. Tokura, *Tetrahedron Lett.*, 1973, 2287.
301. S. Uemura, O. Sasaki, and M. Okano, *J. Chem. Soc., Perkin Trans. 1*, 1972, 2268.
302. G. A. Olah, in ref. 1a, Vol. I, pp. 119–120.
303. P. H. Gore, F. S. Kamounah, and A. Y. Miri, *Tetrahedron*, 1979, **35**, 2927.
304. A. S. Lindsey and H. Jeskey, *Chem. Rev.*, 1957, **57**, 583.
305. J. L. Hales, J. I. Jones, and A. S. Lindsey, *J. Chem. Soc.*, 1954, 3145.
306. H. J. Shine, *Aromatic Rearrangements*, Elsevier, New York, 1967; K. Ota, *Bull. Chem. Soc. Jpn.*, 1969, **47**, 2343.
307. H. Wynberg and E. W. Meijer, *Org. React.*, 1982, **28**, 1.
308. E. A. Robinson, *J. Chem. Soc.*, 1961, 1663; J. Hine and J. M. van der Veen, *J. Am. Chem. Soc.*, 1959, **81**, 6446.

CHAPTER 7

Reactions Involving Nitrogen Electrophiles

Reactions described in this chapter are those involving nitrogen electrophiles, nitrogen being attached to either aryl, alkyl, or hydrogen (amination), nitrogen (diazonium coupling), one oxygen atom (nitrosation), or two oxygen atoms (nitration). The reactions become easier as electron withdrawal from the nitrogen is increased by greater overall electronegativity of the attached atoms.

7.1 AMINATION

Amination may be brought about in a variety of ways, but some of these do not involve normal electrophilic substitution:

(1) The thermal decomposition of azides produces nitrenes which will react with aromatics, provided that X is sufficiently electron withdrawing. For example the reaction with *N,N*-dimethylaniline (eq. 7.1, X = CN) gives *o*- and *p*- semidines.[1]

$$X{-}C_6H_4{-}\ddot{N}: + Me_2N{-}C_6H_5 \xrightarrow{130\ ^\circ C} Me_2N{-}C_6H_4{-}NH{-}C_6H_4{-}X \qquad (7.1)$$

(2) Reaction of hydrazoic acid with $AlCl_3$ or H_2SO_4 gives nitrenium ions which react with aromatics forming primary amines in 10–65% yields.[2] The mechanism shown in eq. 7.2 is a simplification.

$$HN_3 \xrightarrow{H^+} H_2{}^+N_3 \longrightarrow N_2 + H_2N^+ \xrightarrow{ArH} ArNH_2 + H^+ \qquad (7.2)$$

(3) A variation on (2) uses aryl azides (ArN_3) and either CF_3CO_2H or CF_3SO_3H to generate the electrophile $ArNH^+$. This can then either aminate aromatics directly giving diarylamines or, through delocalization of the positive charge into the aromatic ring, produce arylation (Section 6.2). A ρ factor of

-4.5 was obtained for the formation of diarylamines from toluene, prop-2-ylbenzene, and chloro- and bromobenzenes (each of which substituted mainly *ortho,para*); naphthalene substituted mainly at the 1-position.[3] A similar reaction occurs between aromatics, aryl azides, and phenol at -60 °C, the toluene:benzene rate ratio being 20 (phenyl azide) or 30 (*p*-tolyl azide).[4]

(4) Replacement of hydrogen in hydrazoic acid (eq. 7.2) by an electron-withdrawing group facilitates reaction. Thus reaction of ethyl azidoformate with aromatics in the presence of TFA gives ethyl *N*-arylcarbamates (eq. 7.3). For reaction in nitrobenzene at 109 °C the ρ factor was -1.7, and partial rate factors for toluene were $f_o = 2.8$, $f_p = 3.0$.[5]

$$\mathrm{EtOC(=O)N_3} \xrightarrow{H^+} \left[\mathrm{EtOC(=O)\ddot{N}H^+} \leftrightarrow \mathrm{EtOC(OH){=}\ddot{N}^+}\right] \xrightarrow{ArH} \mathrm{ArNHCO_2Et} \qquad (7.3)$$

(5) Amines containing electron-withdrawing groups also provide suitable electrophiles. For example, tertiary aromatic amines have been obtained from the reaction of *N*-chlorodialkylamines, $ClNR_2$, with aromatics, in some cases in the presence of H_2SO_4 or $AlCl_3$ catalysts, the yields being 50–90%. For the H_2SO_4-catalysed reaction of toluene, *meta* products dominated (irrespective of irradiation), but with Lewis acids the yields were mainly *ortho*, *para*; under the former conditions a PhMe:PhH:PhCl rate ratio of 9.3:1:0.1 was obtained.[6] Both secondary and tertiary aromatic amines have also been obtained by reaction of *N*-chloroalkylamines or *N*-chlorodialkylamines with aromatics and H_2SO_4 with metal ions as catalysts, although here the attacking species is said to be a radical cation, $R_2NH^{+\cdot}$.[7] This is electrophilic and gives the usual orientation pattern, e.g. *ortho,para* orientation with anilides and ethers. Amination with some haloamines and with NCl_3 in the presence of $AlCl_3$ also gives mainly *meta* products. This is attributed to Cl^+NCl_2–$AlCl_3^-$ polarization, and an addition–elimination mechanism with Cl^+ substituting initially in the site most activated towards electrophilic attack.[8]

(6) Variations of (5) use *N*-alkyl- or *N,N*-dialkylhydroxylamines–$AlCl_3$[9] or *N*-arylhydroxylamines–TFA,[10] the latter making use of water as a leaving group (eq. 7.4); the selectivity is low under the former conditions.

$$\mathrm{Ar'NHOH} \xrightarrow{H^+} \mathrm{H_2O} + \mathrm{Ar'NH^+} \xrightarrow{ArH} \mathrm{Ar'NHAr} + \mathrm{H^+} \qquad (7.4)$$

Two other variations use derivatives of amides as reagents. One involves heating the aromatic with a hydroxamic acid in polyphosphoric acid (eq. 7.5) and forms amides, but works only with phenolic ethers.[11] The second involves *N*-chloro-*N*-methoxyamides in the presence of TFA and silver ion and has been used for intramolecular amination, the products being *N*-methoxyamides (eq. 7.6); yields are 87% for $n = 1$ and 2 and 60% for $n = 3$.[12] Cyclization from the 2- to either the 1- or 3-position of naphthalene takes place in 68% yield (for $n = 1$), and radicals are not involved in the reaction.

$$RCONHOH \xrightarrow{H^+} H_2O + RCONH^+ \xrightarrow{ArH} RCONHAr + H^+ \quad (7.5)$$

$$Ph(CH_2)_nC(=O)-N(Cl)-OMe \xrightarrow{Ag^+, TFA} \text{[benzo-fused lactam: } (CH_2)_n\text{, C=O, N–OMe]} + H^+, AgCl \quad (7.6)$$

(7) A range of hydroxylammonium salts, e.g. $HONH_3Cl$, in the presence of $AlCl_3$ have been used and these give characteristically low selectivities, e.g. a PhMe:PhH rate ratio of 2.9 for the reaction of the chloride salt.[13]

7.2 DIAZONIUM COUPLING

Diazonium coupling involves electrophilic substitution of aromatic diazonium ions $Ar—N^+{\equiv}N$ into aromatics (eq. 7.7).[14] Because the positive charge is readily delocalized into the aryl ring, the ions are weakly electrophilic and reactions are essentially limited to amines and phenols (the latter as phenoxide ion). The size of the electrophile results in *para* substitution unless the *para* position is blocked, in which case substitution will take place *ortho*. The standard test for an aromatic amine (diazotization followed by coupling with β-naphthol to give an azo dye) makes use of the intrinsic relatively high reactivity of the 1-position of naphthalene together with the very strong activation by the 2-hydroxy group across the high-order 1,2-bond.

$$ArN_2{}^+X^- + Ar'H \longrightarrow ArN{=}NAr' + HX \quad (7.7)$$

The pH of the medium is very important. Amines will react in neutral or mildly acidic media (the free base remains the active reagent in the latter), but the less reactive phenols must be coupled in weakly basic media to produce the more reactive phenoxide ion. If the medium is too basic, neither amines nor phenols will react because the diazonium ion becomes converted to the unreactive diazo hydroxide, $ArN{=}NOH$, i.e. the positive charge is neutralized by electron supply from the oxygen.

The reaction with phenols is first order in phenol and diazonium hydroxide,[15] but is consistent only with bimolecular attack of the diazonium ion on the phenoxide ion, rate $= k_2[ArN_2{}^+][Ar'O^-]$.[16] This rate expression differs from that which applies to the neutral reagents, rate $= k_2[ArN_2OH][Ar'OH]$, by the presence of one molecule of water in the former and, since the concentration of the water solvent is constant, the two expressions are kinetically identical. For coupling with amines the appropriate concentration terms were shown to be

$[ArN_2^+]$ and $[Ar'NR_2]$ since pH–rate profiles passed through a maximum.[17] As the pH is increased the concentration of free base relative to its conjugate acid increases, as does the rate. At high pH the rate decreases because the diazo ion is converted to the hydroxide, as noted above.

The electrophilicity of the benzenediazonium ion can be increased by the presence of an electron-withdrawing group in the aromatic ring, and conversely decreased by the presence of an electron-supplying group; these groups modify the concentration of positive charge carried by the nitrogen. This is illustrated by the following approximate relative rates for the substitution of *para*-substituted benzenediazonium ions into five phenols: *p*-X = NO_2, 1300; SO_3^-, 13; Br, 13; H, 1; Me, 0.4; OMe, 0.1.[18] The effect of substituents is such that 2,4,6-trinitrobenzenediazonium ion will substitute into mesitylene,[19] which is very much less reactive than phenoxide ion.

The ρ factors for the effects of substituents in the diazo ions have been determined as ca 4.0 for substitution into aniline.[20] 4-Methyl- and 4-methoxyphenol,[21] 1-naphthoate ion,[22] 4-hydroxynaphthalene-1-sulphonic acid,[22] and 6-amino- and 6-hydroxynaphthalene-2-sulphonic acid;[23a] later work gave smaller values for the latter[23b] and also ca 3.1 for substitution into acetanilide. These latter could be due to breakdown of additivity of substituent effects in the diazonium ions used, since non-linear Hammett plots were obtained. Encounter control (Section 2.10) and solvation effects may account for the constant rate spread for reaction of diazo ions with resorcinol and its monoanion and dianion,[24,25] the relative rates of which are $1:10^{7.5}:10^{13}$, and for the generally poor correlation of reactivities of diazo ions with ρ-factors.[26] 1-Naphthol and its anion differ in reactivity by 7–9 orders of magnitude, depending on the diazonium ion used.[22] For the substitution of diazotized sulphanilic acid into the 2-position of 4-substituted phenols the ρ-factor has been determined as ca -3.8,[27] although this assumes, incorrectly, that σ^+ values apply to *ortho* positions.

The possibility that diazonium coupling may proceed by an electron-transfer mechanism has received support from CIDNP studies[28] and it has been proposed that charge-transfer structures are intermediates in the substitution pathway.[29]

Diazo coupling is usually subject to strong base catalysis,[30] but the rate of such base-catalysed reactions is not linearly related to the base concentration, thereby ruling out a termolecular S_E3 mechanism (Chapter 2, eq. 2.3), and the two-step (S_E2) mechanism (eq. 7.8) must apply. In accordance with this, isotope effects are observed alongside base catalysis, e.g. the reaction of substituted benzenediazonium ions with 1-D-2-hydroxynaphthalene-6,8-disulphonic acid gives values of k_H/k_D of 6.55 (4-Cl), 5.48 (3-Cl), and 4.78 (4-NO_2). By contrast,

$$ArN_2^+ + Ar'H \underset{k_{-1}}{\overset{k_1}{\rightleftharpoons}} Ar'^{+}\begin{matrix} H \\ N_2Ar \end{matrix} \xrightarrow[k_2]{+B} ArN_2Ar' + BH^+ \qquad (7.8)$$

the reaction of 2-methoxybenzenediazonium ion with 2-D-1-hydroxy-4-sulphonic acid shows neither effect. The occurrence of base catalysis and isotope effects depends on the ratio of the rate coefficients k_2 and k_{-1} (Section 2.1.3). Greater electron withdrawal in the aryldiazonium ion reduces the value of k_{-1} and hence the isotope effect as shown by the effect of the substituents above. On the other hand, an increase in steric hindrance in the aromatic should increase k_{-1} relative to k_2, thereby leading to an increased isotope effect and more effective base catalysis. This too is observed, values of k_H/k_D of 1.04 and 3.12 being obtained for (1) and (2), respectively.

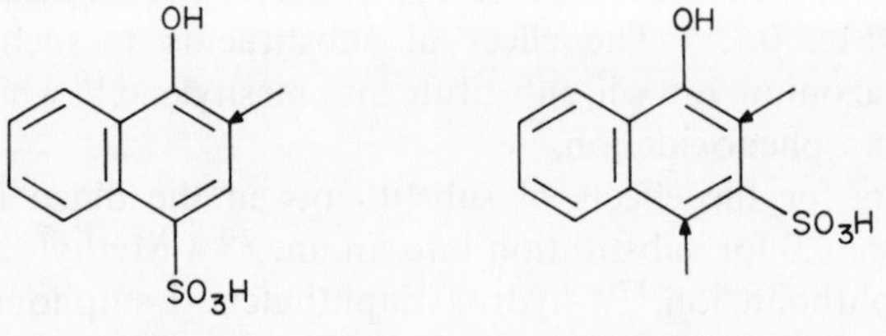

OH OH

SO_3H SO_3H

(1) **(2)**

Moreover, substitution at the 2-position of **2** occurs more readily than at the 4-position, so the 2:4-ratio becomes higher when deuterium is present. This is because reaction at the 4-position is more strongly base-catalysed, hence the isotope effect for 4-substitution is also larger.[31]

The magnitude of the isotope effect also depends on the pK_a of the catalysing base and passes through a maximum at $pK_a \approx 1$.[32] This parallels the results observed in hydrogen exchange [Section 3.1.1(iv)], showing that the difference in basicity between proton donor and proton acceptor is a measure of transition state symmetry.

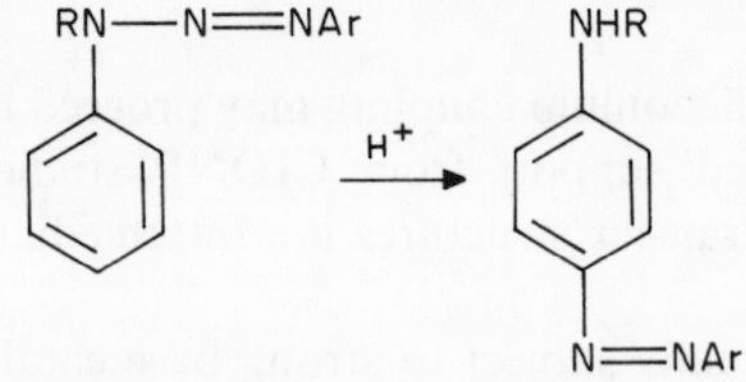

RN—N═NAr $\xrightarrow{H^+}$ NHR … N═NAr (7.9)

Substitution by the diazo group can also occur through rearrangement from a substituent. Thus, aryltriazines on treatment with acid give exclusively *para*-azo derivatives of secondary or primary aromatic amines (eq. 7.9);[33] the mechanism is not fully resolved, but appears to be largely intermolecular.[34a]

In one study, $ArCR{=}CR{=}CR{=}N_2^+$ ions have been used as electrophiles; here diazocyclization gives benzo-1,2-diazepine derivatives as products. Although the aromatic ring here is not strongly activated, entropy factors facilitate substitution.[34b]

7.3 NITROSATION

Nitrosation, which has recently been reviewed,[35] involves a relatively unreactive electrophile, but compounds of the reactivity of toluene or greater may be readily nitrosated. Although the nitrosonium ion (the most reactive nitrosating species) is 10^{14} times less reactive than the nitronium ion,[36] nitrosation frequently accompanies nitration with nitric acid because the concentration of the nitrosonium ion is greater; here only nitro compounds are obtained because the nitroso products are very readily oxidized by the acid.

7.3.1 Nitrosation with Nitrous Acid

Kinetic studies of nitrosation have usually been carried out in the context of nitration via nitrosation [Section 7.4.1(ix)]. Nitrous acid in the presence of an excess of nitric acid exists mainly in the form of dinitrogen tetroxide, and this in turn gives rise to equilibria 7.10–7.13. A number of entities are therefore present which may be effective nitrosating agents, i.e. the nitrosonium ion NO^+, and the carriers of this ion, N_2O_4, N_2O_3, HNO_2, and $H_2NO_2^+$. The nitrosonium ion should be the most effective of these, but if one of the carriers of the ion is present in high concentration it could be of comparable kinetic significance.

$$N_2O_4 \rightleftharpoons NO^+ + NO_3^- \qquad (7.10)$$

$$\left[N_2O_4 + \right] H_2O \rightleftharpoons N_2O_3 + 2HNO_3 \qquad (7.11)$$

$$N_2O_3 \rightleftharpoons NO + NO_2 \qquad (7.12)$$

$$N_2O_3 \rightleftharpoons NO^+ + NO_2^- \qquad (7.13)$$

For example, the kinetics of nitrosation during concurrent nitration and nitrosation in acetic acid obey the complex rate law rate $= (k + k'/[NO_3^-])[ArH][N_2O_4]$, indicating that NO^+ and N_2O_4 are both involved as nitrosating species. The former is generated from N_2O_4 via eq. 7.10 and its concentration is accordingly reduced by added nitrate ion. The value of k' is at least 10-fold greater than k, showing that NO^+ is a more reactive electrophile than N_2O_4,[37] but this numerical value is unlikely to be of real significance in view of the observation of kinetic isotope effects (below). The kinetics of *O*- and *N*-nitrosation also suggest the effectiveness of the carriers of the nitrosonium ion as nitrosating agents;[38] the nitrous acidium ion $H_2NO_2^+$ has been suggested as the electrophile in carboxylic acid media of pH 1–ca 5.[39]

Nitrosation shows substantial kinetic isotope effects, k_H/k_D, of 1.26–4.5,[40] confirming that the electrophile has low reactivity, and thus k_{-1}/k_2 in the normal S_E2 process is large.[40] Rate-determining decomposition of the Wheland intermediate then accounts for the fact that the ρ factor for nitrosation (-6.9) is small, despite the low reactivity of the electrophile, i.e. the rate-limiting transition state lies beyond the Wheland intermediate. Since the electrophile reacts with the aromatic in a pre-equilibrium, conclusions regarding the nature of the electrophile cannot be made from kinetic studies.[40] A recent theoretical investigation of the reaction mechanism suggests electron transfer from the frontier orbital of benzene to the π^*-orbital of the electrophile occurs to give, initially, two interconverting π-complexes of C_{2v} and C_1 symmetry (**3**).[41]

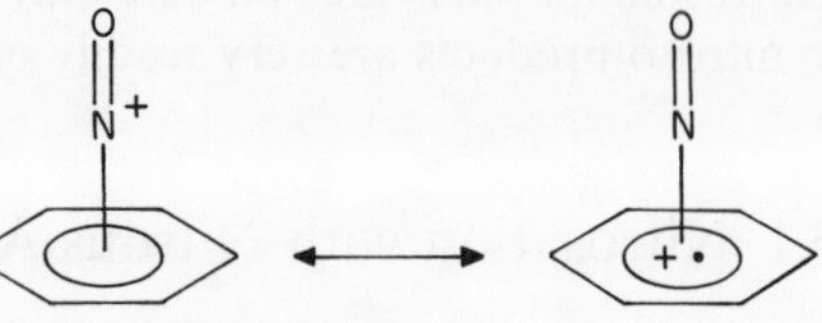

(3)

For *para* nitrosation of PhX by $NaNO_2$ in 10.4 M $HClO_4$ at 52.9 °C, partial rate factors are Me 102, PhO 3780, MeO 210 000, and HO 168 000, giving $\rho = -6.9$,[36] and for nitrosation *meta* to X in 4-X-phenols in 6.8 M $HClO_4$ at 0 °C, $\rho = -6.2$.[42] The reaction typically gives very low *ortho*:*para* ratios, e.g. with biphenyl.[43]

7.3.2 Nitrosation with Nitrosonium Compounds

The ease of nitrosation by NOX compounds could be expected to be greater the more electron-withdrawing is X, and this is the case. Thus the relative rates of nitrosation of 2-naphthol by NOCl, NOBr, and NOSCN are 3928:236:1.[44]

7.3.3 The Fischer–Hepp Rearrangement

Nitrosation may also occur via rearrangement of the electrophile from the substituent, and is known as the Fischer–Hepp rearrangement (eq. 7.14).[35,45] The reaction goes exclusively *para*, only works satisfactorily with HCl, and is particularly useful for preparing *C*-nitroso derivatives of secondary aromatic amines since these cannot be nitrosated directly. The mechanism of the reaction involves initial protonation of the *N*-nitrosoamine, which then decomposes

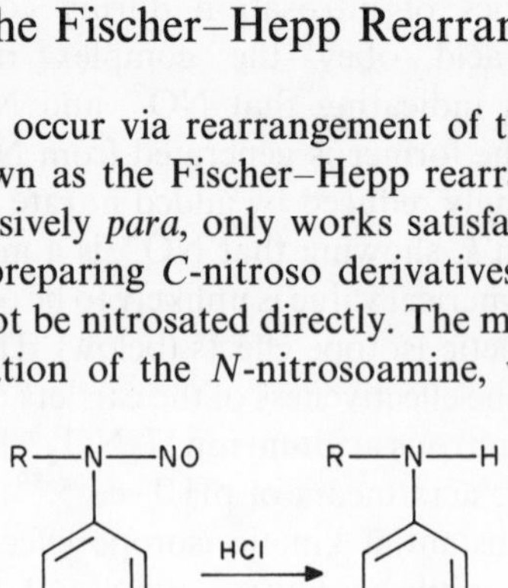

(7.14)

either intramolecularly to give the *p*-nitrosoaromatic, or nitrosonium ion and free amine are formed;[46] however, the ion is captured by other nucleophiles present before it can nitrosate the amine.[47]

7.4 NITRATION

Nitration, which has been described in many major reviews,[48–52] has been the most studied of all electrophilic aromatic substitutions, and was probably the first to be observed. This focus of attention in both preparative and kinetic studies is due to the following:

(1) The reaction is very easy to carry out, owing to the high reactivity of the electrophile; very unreactive aromatics may be readily nitrated.

(2) The nitro products have higher boiling and melting points with wider differences between isomers, than for virtually any other derivatives. Consequently, separation into pure isomers is relatively easy, and this was a vital factor in early quantitative studies.

(3) Yields are generally good or excellent, partly owing to the electron-withdrawing nature of the nitro group; mononitration products are therefore readily obtained.

(4) The products are readily converted (via reduction and diazotization) into a wide range of other derivatives.

(5) Many of the products are commercially important, e.g. explosives.

Despite this intensive scrutiny, it is paradoxical that some aspects of the mechanism are imperfectly understood, in particular the nature of the electrophile or its mode of attack under certain conditions.

7.4.1 Conditions of Nitration

Nitration can be brought about using a wide variety of reagents:

(i) Nitric acid, either aqueous or anhydrous.
(ii) Nitric acid with added inorganic acids, such as sulphuric, perchloric, or trifluoromethanesulphonic acid.
(iii) Nitric acid in organic solvents such as nitromethane or acetic acid.
(iv) Nitric acid in acetic anhydride.
(v) Acyl nitrates in organic solvents.
(vi) Dinitrogen tetroxide or dinitrogen pentoxide.
(vii) Nitronium salts.
(viii) Nitroalkanes, alkyl nitrates, and metal nitrates.
(ix) Nitrous acid, giving nitrosation which is followed by oxidation.
(x) Nitrocyclohexadienones.

(xi) Pernitric acid and nitryl halides.
(xii) Protonated methyl nitrate (gas-phase nitration).

Although there are marked differences in these conditions, which cause wide variations in the ease of nitration of a given compound, the mechanism of nitration under most of them is believed to be essentially the same. The attacking entity is thought to be the nitronium ion, NO_2^+ (nitryl cation), which in some cases may be abstracted from a species NO_2X by the aromatic substrate. One mechanistic problem therefore is to discover how much association exists between NO_2^+ and X^- in the rate-determining step of the reaction. The mechanism of nitration by the nitronium ion may be considered in terms of eqs 7.15–7.17; in some cases there is evidence that an additional step may be involved.

$$HNO_3 \underset{k_{-1}}{\overset{k_1}{\rightleftharpoons}} NO_2^+ \quad (7.15)$$

$$ArH + NO_2^+ \underset{k_{-2}}{\overset{k_2}{\rightleftharpoons}} ArNO_2H^+ \quad (7.16)$$

$$ArNO_2H^+ \xrightarrow{k_3} ArNO_2 + H^+ \quad (7.17)$$

(i) Nitration by nitric acid

Concentrated nitric acid contains mainly molecular nitric acid. This is not the nitrating species, since the rate of nitration is considerably altered by the addition of sulphuric acid, or of nitrate ions (neither of which should alter significantly the concentration of nitric acid). Cryoscopic,[53] conductimetric,[54] and Raman spectroscopic[55] studies indicate that nitronium ion is present (to the extent of ca 3%) through self-protonation according to eq. 7.18. Lines in the Raman spectrum at 1400 and 1050 cm^{-1} were assigned to NO_2^+ and NO_3^-, respectively, and their concentrations are equal, which rules out a significant presence of the nitric acidium ion formed according to eq. 7.19. Similar lines are also observed in other nitrating systems, namely dinitrogen pentoxide in nitric acid (1400, 1050),[56] nitric acid in perchloric or selenic acids (1400),[57] and crystalline nitronium perchlorate (1400),[58] which is completely ionized into nitronium ions and perchlorate ions. The single line for the nitronium ion shows that it is linear, i.e. $O{=}N^+{=}O$. Kinetic studies in the presence of stronger mineral acids [Section (ii), below] also show nitration by $H_2NO_3^+$ to be unimportant.

$$2HNO_3 \rightleftharpoons NO_2^+ + NO_3^- + H_2O \quad (7.18)$$

$$2HNO_3 \rightleftharpoons H_2NO_3^+ + NO_3^- \quad (7.19)$$

Nitration of unreactive aromatics in concentrated nitric acid follows the rate law, rate $= k_1[\text{ArH}]$, because formation of nitronium ion is not rate-determining, and its precursor nitric acid is present in large excess.

Addition of water to nitric acid reduces the rate of nitration of unreactive aromatics, as might be expected in view of eq. 7.18, but the kinetic order is unchanged. However, nitration of reactive aromatics shows mixed first- and zeroth-order kinetics.[59] The implication is that nitric acidium ion is present in these weaker media, and that this then heterolyses to give nitronium ion in a step which is slow relative to the subsequent nitration; the observed kinetics then follow. There is thus competition between water and the aromatic for nitronium ion and it has been shown that zeroth-order kinetics results when the aromatic is ca 20 000 times more reactive than water towards nitronium ion. Confirmation that the rate-determining step is formation of the nitronium ion is provided by the rate of ^{18}O exchange between nitric acid and water being the same as the initial rates of the zeroth-order nitrations.[60]

The presence of nitrous acid causes ten–twentyfold rate reductions for a hundredfold increase in its concentration. Nitrous acid is present mainly as dinitrogen tetroxide, which is ionized according to eq. 7.10, so the rate retardation follows from the suppression of eq. 7.18 by nitrate ion.[59] The anticatalytic effect of nitrous acid is greater in more dilute nitric acid solutions, because the water reduces the self-protonation of nitric acid according to eq. 7.18. The concentration of nitrate ions is therefore lower, making the nitration rate more sensitive to other sources of these ions.

(ii) Nitration by nitric acid in strong mineral acids

Addition on strong mineral acids such as sulphuric acid to nitric acid produces a higher concentration of nitronium ion, through protonation to the nitric acidium ion followed by rapid dissociation. This is represented by the overall equilibrium in eq. 7.20, supported by an *i*-factor of 3.82 in cryoscopic measurements[61] (The difference from the expected value of 4 is attributed to incomplete protonation of water molecules.) Moreover, the Raman spectrum showed lines at 1400 cm^{-1} (NO_2^+) and 1050 cm^{-1}, the latter here arising from HSO_4^-.[62]

$$HNO_3 + 2H_2SO_4 \rightleftharpoons NO_2^+ + H_3O^+ + 2HSO_4^- \qquad (7.20)$$

Because the rate of formation of nitronium ion is more rapid in this medium, nitration gives essentially second-order kinetics: rate $= k_2[\text{ArH}][HNO_3]$,[63,64] and shows a rate maximum at around 90 wt-% sulphuric acid,[63,65–67] probably owing to a combination of causes: protonation[68] (most of the compounds contain the group X = O), hydrogen bonding,[69] and a decrease in the activity coefficient ratio, $f(NO_2^+)/f^{\ddagger}$[59,70] (the latter being indicated by a rate maximum for $PhNMe_3^+$, which can neither protonate nor hydrogen bond[70,71]).

Addition of water to solutions of nitric acid in 90 wt-% H_2SO_4 produces reductions in rate which parallel the decrease in nitronium ion concentration.[63]

Nitration is observed even when nitronium ion is no longer detectable, but support for nitronium ion being the nitrating species is provided by the rates paralleling the H_R, rather than the H_0 acidity scale.[63,64] Half-lives for the nitronium ion have been calculated to be 10^{-4}, 10^{-6}, 5×10^{-8}, and 10^{-9} s in 80, 68.3, and 60.4 wt-% H_2SO_4, and water, respectively.[72,73]

The ease of formation of nitronium ion from nitric acid depends on the extent of protonation of nitric acid and hence the strength of the catalysing acid. It follows that the effectiveness of catalysing acids is, for a given wt-%, perchloric > sulphuric > phosphoric;[74,75] trifluoromethansulphonic acid, which is much stronger than any of these, is an extremely powerful catalyst for nitration, and converts the nitric acid completely to nitronium ions.[76]

Nitration in strong mixed acid media exhibits a problem that also shows up under other nitrating conditions, namely that a limiting rate (known as the encounter rate) is reached (Section 2.10). At the encounter limit, no compound, regardless of its intrinsic reactivity towards electrophilic substitution, nitrates significantly faster than the encounter rate. Another consequence of this limiting nitration rate is that the *Additivity Principle* breaks down. The phenomenon is well demonstrated by the following relative rates for nitration in 68.3 wt-% H_2SO_4 at 25 °C: benzene, 1.0; toluene, 17; *o*-xylene, 38; *m*-xylene, 38; *p*-xylene, 38; mesitylene, 36; anisole, 13; *o*-methylanisole, 22; *p*-methylanisole, 21; biphenyl, 16; naphthalene, 28; 2-methylnaphthalene, 28; 1-methoxynaphthalene, 35; phenol, 24.[77] Similar rate spreads have been obtained for nitration in 63.2 wt-% H_2SO_4,[77] 61.05 wt-% $HClO_4$,[77] and 85.7 wt-% H_3PO_4;[75] a limiting rate is also obtained in nitration in 88.0 wt-% $MeSO_3H$.[78] The encounter limit also causes nitration of 1,2,3,4- and 1,2,4,5-tetramethylbenzene to take place only 41- and 20-fold faster, respectively, than their nitro derivatives.[79] The encounter rate is a function of viscosity, showing that diffusion control is a contributory factor. Thus relative mesitylene:benzene nitration rates decrease along a series of perchloric and sulphuric acid media of increasing viscosities,[77] and the greater the viscosity of a given medium, the lower is the intrinsic reactivity of an aromatic that shows onset of an encounter nitration rate.[75] The observed encounter nitration rate is also close to the calculated rate coefficient for encounter between two molecules in these viscous media,[77] but this may be fortuitous.

The limiting rate found under encounter conditions implies that reaction occurs at almost every collision, in which case there ought not to be any positional selectivity within a given aromatic. This is not the case, however, and this paradox has led to the proposal that the nitronium ion and the aromatic form a discreet 'encounter pair', the rate of formation of which is rate-controlling.[80] This subsequently forms a Wheland intermediate, and only in the latter is the product isomer distribution determined; the formation of the encounter pair is thought to occur by diffusion together of the components, not by pre-association.[72] The *ortho*:*para* ratios for nitration of various aromatics in phosphoric acid are lower than those obtained in sulphuric acid.

This is apparently not due to differences in size of the solvated nitronium ion, since chlorobenzene, the least reactive compound examined and unlikely to be reacting at the encounter rate, gave similar ratios under both conditions. It was therefore proposed that viscosity can affect rates of translation (from encounter pair to Wheland intermediate, but in a manner difficult to envisage), so that less hindered sites in highly reactive substrates are relatively favoured in the more viscous solvent.[81]

The concept of the 'encounter pair,' which is almost unique to nitration and under this condition, leads to a number of difficulties, including the evaluation of its structure. Two possibilities are π-complexes and, more probably, radical–radical cation pairs [see also Section 7.4.2(ii)], but there is no compelling evidence as yet that either are involved.[82] Another factor that may contribute is the change in shape of the nitro group from linear in the ion to bent in the product. Calculations have also indicated that in the reaction with ethene, structures with nitronium ion bridged across adjacent carbon atoms are more stable than that with it attached to one carbon atom;[83] such structures could be involved in the encounter pairs.

Another difficulty that arises in nitration in these media concerns oxygen-containing substituents. Nitration of anisole in aqueous perchloric acid gives *ortho*:*para* ratios that vary from 1.58 in 50.6 wt-% acid to 0.80 in 72.7 wt-% acid.[74] Similar decreases accompanied nitration of anisole in 54–82 wt-% sulphuric acid (1.8–0.7),[84] and of phenol in 58–80 wt-% sulphuric acid (2.4–0.9).[85] This has been attributed to increased hydrogen bonding with increasing acidity; hydrogen bonding also causes decrease in the ratios with increasing acidity in the nitration of methyl *N*-phenylcarbamate, α-chloroacetanilide, acetanilide, dihydroquinolone, and indolin-2-one.[86]

Finally, the rate of nitration of 4-nitrotoluene above 90 wt-% sulphuric acid has been found to increase considerably faster than the increase in nitronium ion concentration, determined by ^{14}N n.m.r.[87] Moreover, the values of k_1, k_{-1}, and k_2[ArH] (eqs. 7.15 and 7.16) for nitration of anisole were determined, and although k_{-1} was 10-fold less than k_2[ArH], the expected zeroth-order kinetics were not obtained. The encounter rate was also calculated to be 100-fold greater than k_2[ArH], so the normally observed rate limit is anomalous. Rates of nitration of deactivated aromatics by nitric acid containing various amounts of N_2O_5 also increase more rapidly than the nitronium ion concentration.[88] These results suggest that current views on the nitration mechanism in these media may need modifying. More work may be needed to establish with certainty the homogeneous nature of nitration in these mixed acids. It is the writer's experience based on hydrogen exchange carried out in sulphuric acid media, that solutions can appear to be perfectly homogeneous when the kinetics show that they are not. Nitration would then occur at the interface between the acid and the dispersed microdroplets (giving the usual positional selectivity), diffusion from within the droplets to their surfaces producing the limit on the nitration rate. Nitration in the presence of a small amount of co-solvent or with ultrasonic agitation might help to evaluate this possibility.

(iii) Nitration by nitric acid in organic solvents

These mixtures improve the solubility of aromatics and ensure homogeneous nitration; solvents most commonly used are trifluoroacetic acid, acetic acid, nitromethane, tetramethylene sulphone (sulpholane), and carbon tetrachloride. The nitronium has not been detected spectroscopically in these media, but its presence is inferred from kinetic studies.

Trifluoroacetic acid (TFA) is both a good solvent and a strong acid, so permitting high substrate concentrations and fast reaction rates; however, it is expensive and thus not suitable for large-scale nitration unless recovery of the acid is contemplated. Nitration of toluene by nitric acid–TFA[89] (or by sodium nitrate–TFA)[90] gives both an isomer distribution and reactivity relative to benzene typical of nitration of the nitronium ion. A detailed study of nitration of a range of aromatics showed that the relative reactivity of toluene to benzene was the same by both competition and kinetic methods, but both mesitylene and *p*-xylene were more reactive when measured by the former method.[91] This was attributed to their ability to nitrate under competition conditions (when they are present in high concentration) with ion-paired nitronium ions.

In this medium, durene (in contrast to mesitylene and *p*-xylene) is nitrated through the intervention of radical cations (which are predicted to be involved only with reactive aromatics). Attack of nitronium ions occurs at a position bearing a methyl group (*ipso* attack) to give the usual Wheland intermediate (**4**). This then undergoes homolytic C—N cleavage in a step having a low energy barrier to give the radical cation–radical pair (**5**), followed by migration of the nitro group to the adjacent unoccupied site giving **6** and thence **7**.[92]

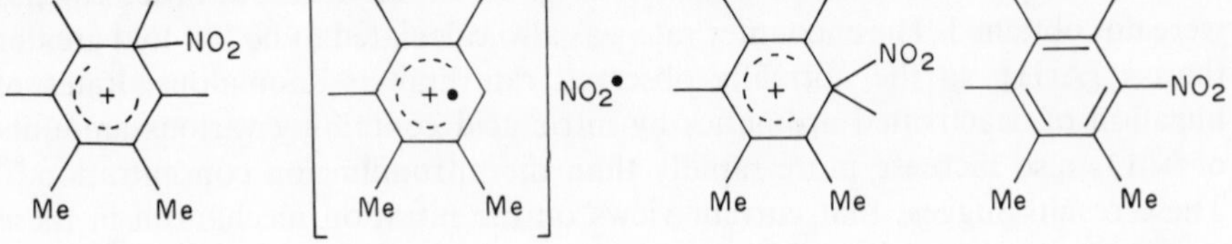

Me Me NO_2 Me Me (**4**) Me Me Me Me $NO_2^{\bullet}$ (**5**) Me Me NO_2 Me Me (**6**) Me Me NO_2 Me Me (**7**)

The kinetics of nitrations using excess nitric acid in nitromethane, sulpholane, and acetic acid are zeroth order in aromatic if this is fairly reactive but first order if it is unreactive.[59,93,94] These results thus parallel those above for nitration in aqueous nitric acid, i.e. the rate-determining step in the zeroth-order nitration is heterolysis of the nitric acidium ion (eq. 7.21). This is confirmed by the kinetic effect of the addition of large amounts of water (for nitration in acetic acid). The water competes with the aromatic for the nitronium ion and a change to kinetics first order in aromatic results. Similarly, the zeroth-order rate coefficients decrease with decreasing concentration of nitric acid.[59,93,94]

$$H_2NO_3^+ \rightleftharpoons NO_2^+ + H_2O \tag{7.21}$$

Some additional confirmation comes from nitration in acetic acid and in nitromethane under high pressure.[95] It was argued that the two- and fivefold increases in the zeroth- and first-order rates, respectively, produced by a pressure of 2000 atm, were consistent with the volume decreases expected in the respective rate-limiting steps.

Nitrations in nitromethane show a greater tendency towards zeroth-order kinetics, and thus require larger additions of water than does nitration in acetic acid, in order to cause a change to first-order kinetics; nitration in sulpholane shows intermediate behaviour.

Nitration in carbon tetrachloride can show anomalous results[96] owing to incursion of heterogeneity arising from the lower polarity of the co-solvent.[97] If this is avoided then the usual decrease in kinetic order with increase in reactivity of the aromatic is found. However, the zeroth-order rate coefficients increased according to the fifth power of the nitric acid concentration, and the reaction has a negative activation energy. Both features may be attributed to the formation of aggregates of nitric acid molecules from which the nitronium ions are produced, and it may be significant that negative activation energies are obtained in various molecular brominations carried out in carbon tetrachloride (Section 9.3.2).

(iv) Nitration by nitric acid in acetic anhydride

Nitric acid in acetic anhydride provides a potent nitrating mixture which is also a good solvent; reaction occurs ca 10^4 times faster than for an equivalent concentration of nitric acid in nitromethane or sulpholane. The acid should always be added to the anhydride (not the reverse) and the temperature kept below ca 65 °C, otherwise explosions may occur.

Nitric acid and acetic anhydride combine rapidly at room temperature to give acetyl nitrate (eq. 7.22) provided acetic anhydride is in excess, and dinitrogen pentoxide (eq. 7.23) if nitric acid is in excess.[98] The dinitrogen pentoxide is either covalent or, at higher nitric acid concentrations, ionized to give nitronium ions (eq. 7.24) which may be detected spectroscopically.[99] These reactions are fairly rapid at room temperature, but slow at −10 °C, so if the reagents are mixed at low temperature (preferred for safety reasons) they must either be allowed to warm to room temperature or allowed to stand for 24 h, otherwise nitration will be very slow.[100,101] A key function of the acetic anhydride is therefore to remove water from the system. Since both eqs 7.22 and 7.23 are acid catalysed, the use of stronger nitric acid causes the nitrating species to be formed more rapidly and the difference in potency of reagents mixed at room temperature and low temperature diminishes.[101]

$$Ac_2O + HNO_3 \rightleftharpoons AcONO_2 + HOAc \tag{7.22}$$

$$Ac_2O + 2HNO_3 \rightleftharpoons N_2O_5 + 2HOAc \tag{7.23}$$

$$N_2O_5 \rightleftharpoons NO_2^+ + NO_3^- \tag{7.24}$$

An additional nitrating species present in these media is protonated acetyl nitrate (**8**)[100] (which may also considered as nitronium ion solvated by acetic acid, **9**).[102] There are thus may conceivable nitrating agents present and to these we must add the possibility of nitration via nitrosation [see Section 7.4.1(ix)].

$CH_3C^+(OH)—O—NO_2$

(8)

Me—C(—O—H)(=O : $^+$N(=O)→O)

(9)

Nitration of benzene with nitric acid (0.4–2.0 M) in acetic anhydride is first order in benzene and second order in nitric acid and retarded by added nitrate ion.[103,106] These results are therefore consistent with nitration by the nitronium ion formed via eqs 7.23 and 7.24. Earlier work had indicated that nitration of toluene and mesitylene was third order in nitric acid,[104,105] but more recent work has shown that if purified acetic anhydride is used then the order in acid is only two.[106] A zeroth-order nitration also appeared to be obtainable with xylenes, mesitylene, and anisole,[107] but again later work has shown that this is a medium effect arising from the relatively high concentrations of aromatic used in these studies.[108] Crucial to the rejection of the zeroth-order mechanism was the observation that the apparent zeroth-order rates were different for each aromatic, which cannot apply if formation of the nitrating species was rate-determining. The appearance of the pseudo-zeroth-order nitration may account for the toluene:benzene reactivity ratios obtained under competition conditions (38) being lower than that obtained under kinetic conditions (50);[103] side-reactions may also lower the former value.[109]

A number of difficulties prevent clear identification of the nitrating species in this medium. First is the retarding effect of added nitrate ion,[103,104,106,110] which suggests that a cation must be involved, nitronium ion or protonated acetyl nitrate being strong possibilities. The nitronium ion is favoured by the calculated nitration rates for toluene in acetic anhydride and acetic acid being the same,[106] and by the similarity in the isomer distribution for nitration of toluene under these and other conditions where only the nitronium ion is

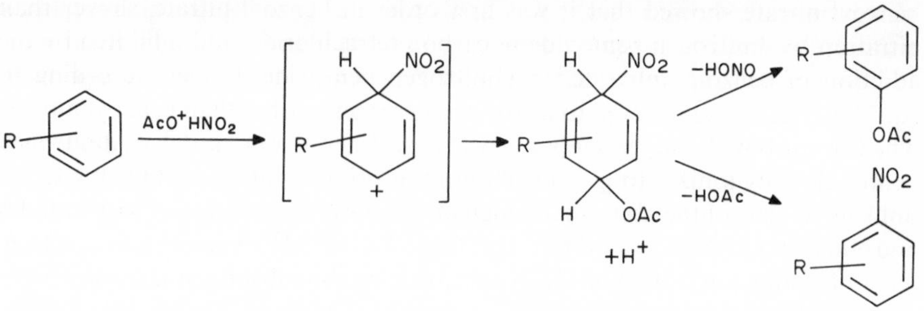

Scheme 7.1. Nitration and acetoxylation via addition of protonated acetyl nitrate.

likely (Table 7.1). Protonated acetyl nitrate should have a larger steric requirement. On the other hand, ethers, anilides, and other aromatics give much greater *ortho*:*para* ratios for nitration in acetic anhydride, and similar to those obtained in nitration with acyl nitrates where the reagent is considered to be either the acyl nitrate itself or dinitrogen pentoxide (Tables 7.3 and 7.12). Further, the rate ratios obtained with nitric acid–acetic anhydride are significantly greater than those obtained under other conditions and are, moreover, very dependent on the quality of the acetic anhydride. Thus the anisole:mesitylene:*m*-xylene:toluene:benzene reactivity ratios are: 650:650:220:34:1 (purified anhydride)[106]; 1500:5000:870:50:1 (unpurified anhydride)[105,106];—:355:136:23:1 (acetic acid).[106] The reason for this is not clear, but the conductivity of purified acetic anhydride alters on standing; acetic anhydride is known to readily undergo thermal elimination to give ketene and acetic acid[111] and this may take place to a significant extent at room temperature.

Nitration in acetic anhydride is strongly catalysed by added sulphuric acid.[112] A probable cause is conversion of acetyl nitrate into protonated acetyl nitrate, whence the good leaving group acetic acid provides the nitronium ion.

Acetyl nitrate adds to alkenes to give *cis*-nitroacetates.[100] A similar addition process also occurs with aromatics, subsequent loss of nitrous acid giving acetoxylation (see also Chapter 8), and of acetic acid giving nitration (Scheme 7.1). The common process for both reactions results in the ratio of nitration to acetoxylation (first observed with *o*- and *m*-xylene) remaining constant for a 330-fold change in overall reactivity.[112] The much greater stability of aromatics compared with alkenes means that a more electrophilic reagent is required for the initial step, i.e. protonated acetyl nitrate, and consequently acetoxylation is greatly accelerated by added sulphuric acid.[112] This process is described in greater detail in Section 7.4.2(ii).

(v) Nitration by acyl nitrates in organic solvents

Nitration can be carried out by acyl nitrates in organic solvents, usually carbon tetrachloride. Kinetic studies of the nitration of benzene by benzoyl nitrate showed that it was first order in benzoyl nitrate, slower than nitration by dinitrogen pentoxide in carbon tetrachloride, and inhibited by the addition of benzoic anhyride.[113] Dinitrogen pentoxide, formed according to eq. 7.25, is therefore believed to be the active reagent rather than benzoyl nitrate, which accords with the greater polarity of the former. It is not known if dinitrogen pentoxide fully ionizes to give nitronium ion prior to nitration, but this seems unlikely in view of the abnormally high *ortho*:*para* ratios obtained with anilides and ethers in nitration by acyl nitrates (Table 7.12).[114,115] Nucleophilic displacement of nitrate ion from dinitrogen pentoxide by the substituent is the probable cause of this anomally.

$$2PhCOONO_2 \rightleftharpoons (PhCO)_2O + N_2O_5 \qquad (7.25)$$

Acyl nitrates may be produced *in situ* from acyl chlorides and silver nitrate.[102,116] Use of aroyl chlorides produced no change in isomer ratios, so providing further evidence that the aroyl nitrate itself is not the nitrating reagent.[116]

(vi) Nitration by dinitrogen tetroxide and dinitrogen pentoxide

Dinitrogen tetroxide in non-polar solvents (e.g. CCl_4) will nitrate aromatics,[117] giving first the nitrosoaromatic, which is subsequently oxidized to the nitro derivative. This may involve attack by covalent N_2O_4, although the anomalous substitution pattern observed with fluoranthene [different from the (predicted) order found in hydrogen exchange (Section 3.1.2.12) and nitration by NO_2 in $CHCl_2$ or HNO_3–Ac_2O] could indicate a radical mechanism.[118] In sulphuric acid, dinitrogen tetroxide is ionized into nitrosonium ions and nitronium ions (eq. 7.26), and since the latter are the more powerful electrophiles, nitration proceeds through these.[119]

$$N_2O_4 + 3H_2SO_4 \rightleftharpoons NO_2^+ + NO^+ + H_3O^+ + 3HSO_4^- \qquad (7.26)$$

Protonation of N_2O_4 under less drastic conditions (by passing NO_2 into trifluoroacetic acid) also yields nitronium ion and nitrous acid (eq. 7.27), formation of the former being indicated by the rate–orientation patterns obtained with anisole, toluene, benzene, and chlorobenzene.[120]

$$N_2O_4 + CF_3CO_2H \rightleftharpoons NO_2^+ + HNO_2 + CF_3CO_2^- \qquad (7.27)$$

Nitration by dinitrogen pentoxide in relatively non-polar solvents, e.g. CCl_4, $CHCl_3$, $MeNO_2$, and MeCN, follows second-order kinetics, rate $= k_2[ArH][N_2O_5]$.[121] This is consistent with either covalent N_2O_5, the ion pair $[NO_2^+NO_3^-]$, or free nitronium ions formed by dissociation of this ion pair, being the electrophile. These latter possibilities have been disproved by the fact that added nitrate ions *increase* the rate, as do other anions, and the rate is not much affected by an increase in solvent polarity.

The reaction is autocatalysed since the nitric acid byproduct protonates N_2O_5 to give NO_2^+, and added nitric acid also catalyses the reaction, the order in nitric acid being > 2. The catalysis involves protonation analogous to that shown to occur when N_2O_5 is dissolved in sulphuric acid (eq. 7.28),[121] and the high order in added nitric acid follows from this.

$$N_2O_5 + 3H_2SO_4 \rightleftharpoons 2NO_2^+ + H_3O^+ + 3HSO_4^- \qquad (7.28)$$

Under the catalysed conditions, the isomer distribution obtained in nitration of chlorobenzene (27% *ortho*, 73% *para*) is similar to that (33% *ortho*, 67% *para*) obtained in nitration by HNO_3–H_2SO_4, indicating the nitronium ion to be the electrophile. However, N_2O_5 in CCl_4 gives only 43% *para* product, further indicating that here a different electrophile, covalent N_2O_5, is involved.[121] Likewise, nitration of fluorobenzene by N_2O_5 in CCl_4 gives a higher *ortho*:*para* ratio (0.39) than HNO_3–H_2O_4 (0.14).[102]

(vii) Nitration by nitronium salts

Olah and co-workers were responsible for generating interest in nitration by nitronium salts such as $NO_2^+BF_4^-$, $NO_2^+PF_6^-$, $NO_2^+ClO_4^-$, $NO_2^+AsF_6^-$, and $NO_2^+HS_2O_7^-$; many studies have been carried out using solutions of these salts in nitromethane, tetramethylene sulphone, and other solvents.[79,122–138] The former two reagents are commercially available and are therefore those most commonly used. The non-aqueous conditions are especially advantageous for nitrating substrates susceptible to hydrolysis.

Nitration of nitrobenzene by nitronium tetrafluoroborate in sulphuric acid, methanesulphonic acid, and acetonitrile was found to be first order in both nitronium salt and aromatic.[124] In the first two solvents, rate coefficients were similar for nitration by nitric acid and by the nitronium salt, indicating a common nitrating entity which must be the nitronium ion. With acetonitrile the rate coefficients were lower, consistent with nitration by ion pairs $[NO_2^+X^-]$. Evidence supporting the presence of ion pairs consists of the following:

(1) Increasing the size of the anion in a series of nitronium salts causes a decrease in the proportion of *ortho* substitution in nitration of alkylbenzenes.[125]

(2) Raman spectra and cryoscopic measurements gave no indication of the separate existence of the nitronium ion.[125]

(3) Nitronium trifluoromethanesulphonate, $NO_2^+CF_3SO_3^-$, is a very powerful nitrating species.[139] However, if all of the nitronium salts were fully ionized, the ease of nitration should be independent of the stability of the counter ion.

(4) Nitration of aromatics containing either lone pairs or π-clouds by nitronium tetrafluoroborate gives enhanced *ortho*:*para* ratios (as do nitrations by dinitrogen pentoxide or acetyl nitrate), indicating that nucleophilic displacement of the counter ion by the substituent is occurring.[129,130]

(5) For a range of nitrotoluenes, the intrinsic nitration rates were in every case larger when using nitric acid–sulphuric acid than when using nitronium hexafluorophosphate in nitromethane.[79] This is consistent with extra energy being required to displace the counter ion from the nitronium ion under the latter conditions.

Olah *et al.*[123] found that nitration with nitronium tetrafluoroborate gave very low substrate selectivities (determined by the competition method), combined with normal positional selectivities, and this was subsequently shown to be due to mixing control arising from the very high reactivity of the reagent.[127,131] Reaction tends to be complete before the reagents have time to mix properly, and true competitive conditions are not established. In particular, nitration of 1,2-diphenylethane by nitronium tetrafluoroborate gave a ratio of dinitro to mononitro products that was much greater than statistical. In this molecule the two benzene rings are close-coupled yet effectively electronically isolated from each other, i.e. a nitro substituent in one ring has little effect on the reactivity of the other. The product ratio showed that some molecules had a greater chance of reacting than had others, i.e. mixing was slow compared with nitration.[131] Mixing control also accounts for the predominant formation of 3,6-dinitrodurene in nitration of durene with nitronium hexafluorophosphate in nitromethane.[134]

Nitronium salts have poor solubility in most organic solvents, and consequently under heterogeneous conditions nitration will be much slower, and mixing control should not be a problem. This could explain the reported change in the toluene:benzene reactivity ratio for reaction with nitronium hexafluorophosphate from the typically abnormal values of ca 2 in nitromethane and sulpholane to normal values of ca 28 in ethyl acetate and dioxane.[135] Alternatively, it may be due to the formation of (less reactive) adducts between the nitronium ion and the solvents.[136] Similar adducts are formed with pyridines, the sigma lone pair from the pyridine nitrogen being donated to that in the nitronium ion, giving e.g. 1-nitropyridinium tetrafluoroborate.[137] These reagents are consequently much less reactive, and give toluene:benzene rate ratios of ca 40.

(viii) Nitration by nitroalkanes, alkyl nitrates, and metal nitrates

In principle, any compound XNO_2 in which X is a good anionic leaving group should be capable of nitration, the ease of this depending on the stability of X^-.

It follows that nitroalkanes, RNO_2, are poor nitrating agents unless the alkyl group contains strongly electron-withdrawing substituents. This condition is satisfied in tetranitromethane, which has therefore been used in a number of studies, though mainly with very reactive aromatics such as phenols,[140] *N*,*N*-dialkylanilines,[141] and azulene.[142]

Alkyl nitrates, $RONO_2$, have greater electron withdrawal from the nitro group than in nitroalkanes, but even so they are unable to nitrate aromatics unless a protic or Lewis acid is present to coordinate with the oxygen of the RO group, thereby increasing the electron withdrawal.[51] Since R is more electron releasing than H, it would be expected that ionization of $RHNO_3^+$ to ROH and NO_2^+ would be less complete than in the case of the nitric acidium ion (eq. 7.21). Consequently, nitration of toluene gives *ortho*:*para* ratios that decrease according to the size of the group R; for a given alkyl nitrate the ratio was substantially lower (0.64) when using polyphosphoric acid as catalyst than when using sulphuric acid (1.42).[143] This indicates that the acid anion is associated with the electrophilic complex.

In BF_3-catalysed nitration with methyl nitrate, a 'normal' toluene:benzene rate ratio of 25.5 was obtained, showing that mixing control is not a problem in this system.[144]

Nitration with metal nitrates permits the use of higher temperatures and thus nitration will take place without a catalyst, e.g. benzene may be nitrated by heating at ca 300 °C with copper(II) or silver nitrate. The order of effectiveness of metal nitrates is Ag > K > Na > Pb > Ba; Lewis acid catalysts increase the reaction rate in the usual order, viz. $AlCl_3 > FeCl_3 > BF_3 > SiCl_4$.[145] Metal nitrates have limited solubility in organic solvents, but silver nitrate is soluble in acetonitrile, and the BF_3-catalysed reaction in this medium gives a toluene:benzene rate ratio of 24.5, typical of many other nitrations.[146]

Metal nitrates [especially copper(II) nitrate] in acetic anhydride have frequently been used for nitration especially for compounds that are either acid sensitive, or react uncontrollably fast with other nitrating reagents (as for example does thiophene). Zinc nitrate is the most effective nitrating agent under these conditions.[147] Copper(II) nitrate on K10 montmorillonite clay in the presence of excess acetic anhydride in hexane or dichloromethane gives very high *para*:*ortho* ratios with halogenobenzenes, e.g. 35 for fluorobenzene.[148] This was attributed to polarizability of the halogens, but this is untenable since it is well established from the charge distribution in the Wheland intermediate (Section 3.1.1.2) that under conditions of maximum polarizability the ratio would be just below 1.0; a steric effect is the most probable explanation of the results.

Nitration with titanium(IV) nitrate shows a typical electrophilic substitution pattern, but free nitronium ion is probably not involved.[149]

(ix) Nitration via nitrosation

Nitrous acid has both a catalytic and an anticatalytic effect on nitration, the former applying dilute (ca 6 M) nitric acid, the latter in more concentrated solutions.

(a) *The anticatalytic effect.* Nitrous acid retards both zeroth- and first-order nitrations by nitric acid alone, in acetic acid, or in nitromethane, without changing the kinetic order. At low nitrous acid concentrations, and in the absence of water, the retardation is proportional to $[HNO_2]^{0.5}$, and at higher concentrations or in the presence of water it is proportional to $[HNO_2]^{1.5}$.[59] These dependences were attributed to lowering of the nitronium ion concentration by nitrate ion and nitrite ion, respectively.[59] More recently it has been shown that both dependences can arise solely from the presence of nitrate ion [from ionization of dinitrogen tetroxide (eq. 7.10)] causing reversal of eq. 7.18.[150]

(b) *The catalytic effect.* The more common effect is *C*-nitrosation followed by oxidation (eqs 7.29 and 7.30) observed with aromatic amines and phenols which may be nitrated by dilute (< 5 M) solutions of nitric acid in water or acetic acid, provided nitrous acid is present. Since nitrous acid is continually reformed its concentration remains constant throughout the reaction.[151] However, oxidative side-reactions sometimes occur and lead to an increase in nitrous acid concentrations so that autocatalysis is observed. These side-reactions can be suppressed by using low temperatures and aromatic substrates bearing deactivating substituents, whence the reaction can be separated into two kinetic terms:[151] rate $= k_0(1 + a[HNO_2]^{0.5})^{-1}$ and rate $= k_2[ArH][HNO_2]$. The former is characteristic of nitration by nitronium ion in the presence of low concentrations of nitrous acid [see (a), above], whereas the latter describes nitration according to eqs 7.29 and 7.30. The dual mechanism is reflected in *ortho*:*para* ratios that vary widely according to the nitrous acid concentration.[151,152]

$$ArH + HNO_2 \xrightleftharpoons{\text{slow}} ArNO + H_2O \qquad (7.29)$$

$$ArNO + HNO_3 \xrightarrow{\text{fast}} ArNO_2 + HNO_2 \qquad (7.30)$$

Cation radicals appear to be involved in some nitrous acid-catalysed nitrations. For example, nitration of aromatic amines in 85–100 wt-% H_2SO_4 involves the anilinium ion, and is greatly accelerated by the addition of nitrous acid; this is not due to *C*-nitrosation, which is very slow. Isotopic labelling showed that $> 80\%$ of the *p*-NO_2 group in *p*-nitro-*N*,*N*-dimethylaniline comes from the stoichiometric nitric acid, so nitric acid-catalysed *C*-nitrosation cannot be involved. The mechanism appears to involve reaction of nitrosonium ion with the aromatic to give the $PhNMe_2{}^{+\cdot}NO^{\cdot}$ radical pair. This may either be oxidized by nitric acid to give the $PhNMe_2{}^{+\cdot}NO_2{}^{\cdot}$ radical pair, followed by combination of the radicals to give *p*-nitro-*N*,*N*-dimethylaniline, or the aromatic radical cations may combine, so accounting for the formation of *N*,*N*,*N′*,*N′*-tetramethylbenzidine.[153,154]

Ridd and co-workers have used Chemically Induced Dynamic Nuclear Polarization (CIDNP) to detect the presence of these associated radical pairs. Certain states of the radical pairs are able to recombine faster than others, so escape of radicals (from the solvent cage) to give reaction will be more probable from radical pairs with a slower recombination rate. Consequently, the reaction products will have preferred spin states which can be detected by ^{15}N n.m.r. This method has detected cation radical pairs in nitrous acid-catalysed nitration in trifluoroacetic acid of mesitylene and 4-nitrophenol;[155] in this latter reaction, which gives 2,4-dinitrophenol, some nitrodenitration also occurred, as shown by migration of the original 4-nitro group to the 2-position. The kinetic form, rate $= k[\text{ArOH}][\text{HNO}_2]$ and rate $= k[\text{ArOH}]$, observed at low and high nitrous acid concentrations, respectively, in the catalysed nitration of 4-nitrophenol in aqueous solution,[156] has also been shown to be consistent with a $\text{PhO}^{+\cdot}\text{NO}^{\cdot}$ radical pair.[157] However, an alternative proposal is that the reaction proceeds via initial formation of phenyl nitrite, PhONO,[156,158] which dissociates into phenoxyl radical and nitric oxide; the latter is oxidized to NO_2 and this then carries out substitution.[158a]

Nitrous acid-catalysed nitration of naphthalene gives rise to a term *second order* in naphthalene, and this becomes more significant at higher naphthalene concentrations. This unique result has been attributed to the formation of the cation radical of the π-dimer of naphthalene (eq. 7.31), which reacts with NO_2 radical (formed as in eq. 7.32) in the product-forming step (eq. 7.33). Since stabilization of radical cations by π-dimers occurs widely, it follows that paths involving electron transfer in nitration will be more significant as the concentration of the aromatic is increased.[158b]

$$\text{ArH} + \text{ArH} + \text{NO}^+ \rightleftharpoons (\text{ArH})_2{}^{+\cdot} + \text{NO}^{\cdot} \tag{7.31}$$

$$\text{NO}^{\cdot} + \text{NO}_2{}^+ \rightleftharpoons \text{NO}^+ + \text{NO}_2{}^{\cdot} \tag{7.32}$$

$$(\text{ArH})_2{}^{+\cdot} + \text{NO}_2{}^{\cdot} \longrightarrow \text{ArNO}_2 + \text{ArH} + \text{H}^+ \tag{7.33}$$

(x) Nitration with nitrocyclohexadienones

Bromocyclohexadienones have been used for brominating aromatics (Section 9.3.9), and the comparable reagents, nitrocyclohexadienones, e.g. **10**, have been introduced recently for mononitration in good yield of highly activated substrates under mild conditions.[159] For example, 1-hydroxynaphthalene gives 2- and 4-nitro derivatives in a ratio of 0.8–1.5, depending on the conditions,

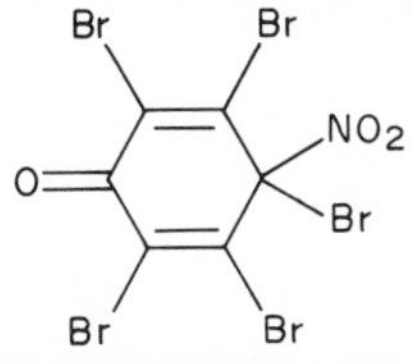

Br Br NO2 O Br Br Br

(10)

whereas nitric acid produces much disubstitution with oxidation as the main reaction. The efficacy of the reagent stems from it regaining aromaticity after delivering the nitronium ion.

(xi) Nitration with pernitric acid and nitryl halides

Nitration has been accomplished with pernitric acid, HNO_4, but the expected hydroxylation side-reaction also occurs.[160] Nitryl halides, NO_2X, will nitrate, the ease depending on the electronegativity of X. The efficacy is increased by the presence of Lewis acids (LA), giving either NO_2^+ or XNO.O:LA as the electrophiles, depending on the affinity of the Lewis acid for coordination to oxygen[161] [cf. acylation with acyl halides, Section 6.7.1.(ii)].

(xii) Gas-phase nitration

In contrast to earlier attempts at gas-phase nitration,[162] the reaction of protonated methyl nitrate [solvated nitronium ion, $MeO^+(H)NO_2$] with aromatics gives partial rate factors typical of an electrophilic substitution (see Section 7.4.3) and a ρ factor of ca -3.9 (σ^+ values appropriate to the gas phase were not used). A limiting nitration rate appears to exist; anisole is less reactive than expected, probably owing to incursion of a proton-transfer mechanism.[163a,b]

Nitration of a range of alkylaromatics with protonated alkyl nitrate gave closely similar substrate:benzene rate ratios (ca 8), indicating further that nitration at the encounter rate was occurring. This was confirmed by the intramolecular selectivity for substitution in each ring of benzylmesitylene (**11**) being > 20, whereas the intermolecular selectivity for substitution in the corresponding 1,2,3,5-tetramethylbenzene and toluene was only 1.5.[163c]

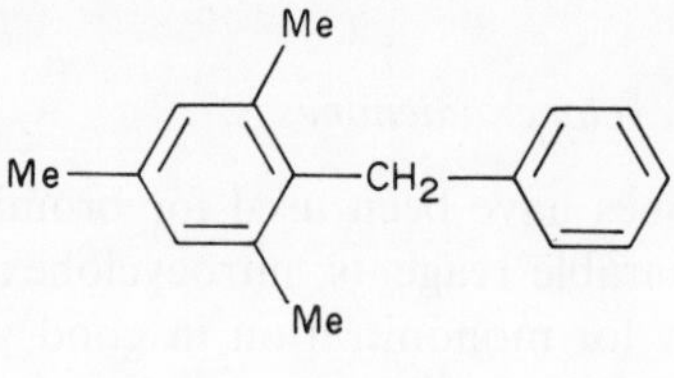

(11)

7.4.2 Mechanism of Nitration

(i) Isotope effects

Except when the formation of the nitronium ion provides the rate-determining step in nitration [Section 7.4.1(i) and (iii)], the kinetics are usually second order, the rate being proportional to the concentrations of the nitrating agent and the aromatic compound.[63,64,66,71,164] The transition state of the rate-determining step therefore contains both of these species, but the kinetics give no information as to whether addition of the nitrating agent followed by loss of a proton is a two-step or a concerted process. The isolation of an adduct of trifluoromethylbenzene and the nitronium ion (Section 2.1.1) provides evidence for the former path in this case, but more general evidence comes from kinetic isotope-effect measurements. There is no isotope effect in the nitration under a range of conditions of benzene and a number of its derivatives of widely differing reactivities.[71,165,166] In these cases the C—H bond is not significantly weakened in the rate-determining step. This is evidence against the synchronous mechanism and indicates that the rate-determining step consists of addition of the nitronium ion to the aromatic nucleus. When such addition is rate-determining, it is the formation of a σ- rather than a π-complex which is the critical step, and nitrations may therefore be represented as in eq. 7.34.

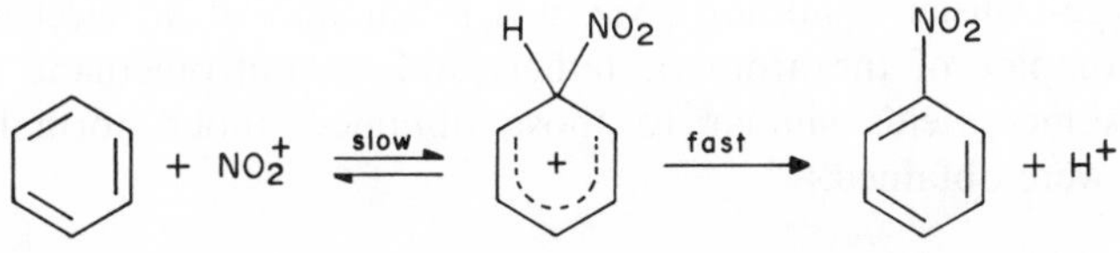

$$C_6H_6 + NO_2^+ \underset{}{\overset{\text{slow}}{\rightleftharpoons}} [C_6H_6NO_2]^+ \xrightarrow{\text{fast}} C_6H_5NO_2 + H^+ \qquad (7.34)$$

It follows from the general rate equation (eq. 2.9, Chapter 2) that k_2 is only kinetically significant if $k_2[B] \ll k_{-1}$. The latter applies in nitration of very sterically hindered aromatics **12** and **13**. For **12** the values of k_H/k_D are, for various Z, 1.0 (H), 2.3 (F), 3.0 (NO_2), and 3.7 (Me),[167] and for **13** the values are 1.445 (Z = I) and 1.15 (Z = Br).[168] Large values (2.25–6.1) have been obtained for 9-nitration of anthracene by $NO_2^+BF_4^-$ in sulpholane, acetonitrile, or nitromethane.[169] Not only is the 9-position very sterically hindered, but the intermediate **14** is very stable since it contains two benzenoid rings; the energy required for return of the intermediate to starting materials is thus small, making k_{-1} particularly rapid, hence the isotope effect follows.

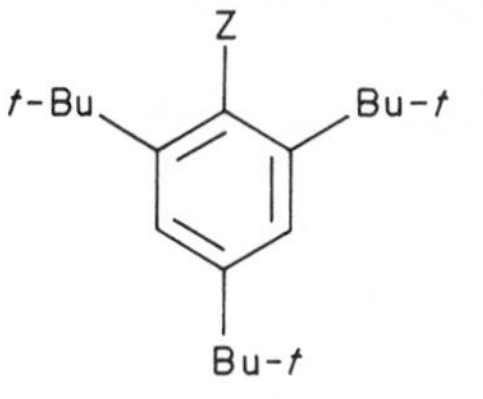

(**12**) Z=H, F, NO_2, Me

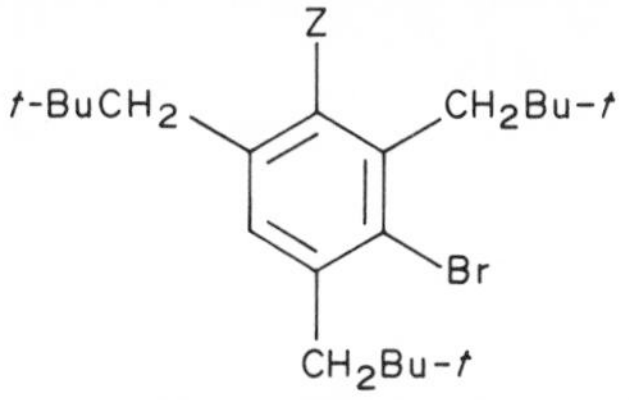

(**13**) Z=Br, I

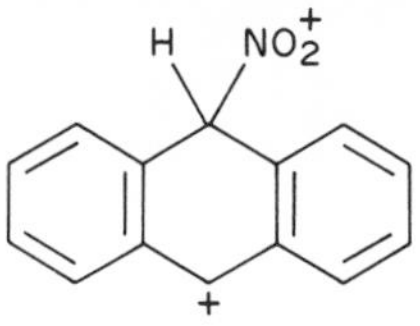

(**14**)

(ii) Electron-transfer mechanism

The proposal that the transfer of electrons from the aromatic to the nitronium ion takes place singly rather than as a pair (electron-transfer mechanism,

Section 2.1.29[170]) has received some support. Two main mechanisms may exist: the first, which is a special case involving the formation of an aromatic radical cation through reaction of nitrosonium ion with the aromatic [as described in Section 7.4.1(ix)] has good experimental evidence. Evidence for the comparable mechanism involving the nitronium ion is harder to obtain. Conclusions from earlier experiments using electrochemically generated radical cations[171] were subsequently shown to be invalid as a result of concurrent nitration by an alternative acid-catalysed mechanism involving N_2O_4.[172] Eberson and Radner[173] have drawn attention to the difference between the energies of the transition states for the electron-transfer and normal process on the one hand, and between $ArH^{+\bullet} + NO_2$ and $ArH + NO_2$ on the other. These do not parallel each other, and the differences make it possible to observe the presence of radical cations only in nitration of very reactive aromatics. Recent work has shown that under charge-transfer conditions, viz. u.v. irradiation of an electron donor–acceptor complex of the aromatic (ether) and tetranitromethane in various solvents, isomer yields similar to those obtained under normal nitration conditions were obtained.[174]

(iii) Ipso *substitution*

In many nitrations, Wheland intermediates are formed by *ipso* substitution, which usually takes place at the site of highest electron density in the ring (although recent theoretical calculations contradict this[175]). These intermediates can provide routes to some nitro isomers additional to that involving direct nitration at a given site.

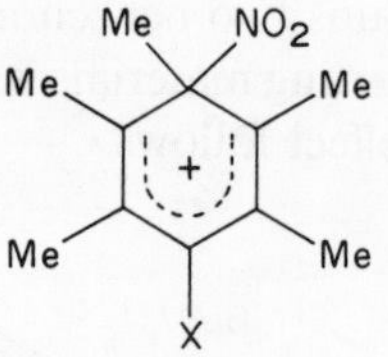

(15) **(16)**

X=Me, F, Cl, Br

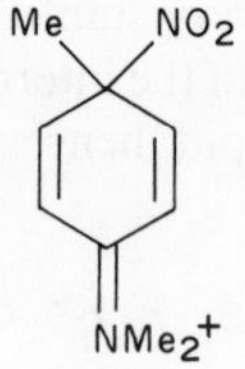

The Wheland intermediates are in some cases stable enough to be isolated as salts, e.g. **15**,[176] **16** and derivatives.[154,177] These intermediates can then react in a number of different ways:

(1) By loss of the substituent at the site attacked by the nitronium ion. This reaction has long been known and is described fully in Chapter 10. In a few cases displaced methyl groups resubstitute elsewhere in the same ring.[178,179]

(2) By reaction with nucleophiles, e.g. acetate, fluoride, nitrate, water, and aryl rings; the latter reaction is described in Section 6.2.

Many adducts have been described[180–184] and some typical examples are **17**–**28**.

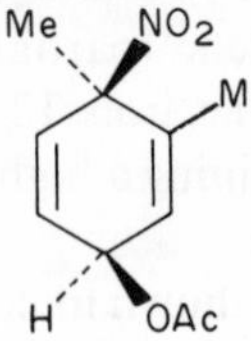

(**17**) *cis* and *trans* (**18**) 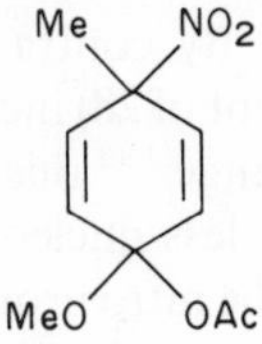 (**19**) 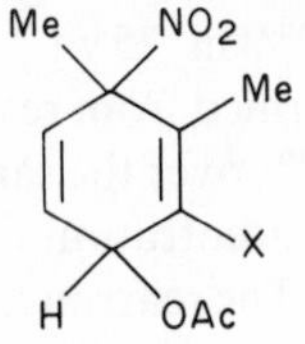 (**20**) X = CN, COR

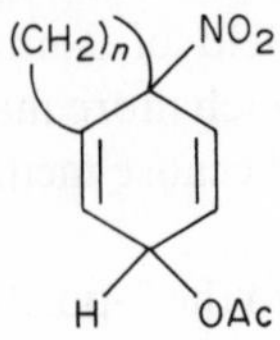

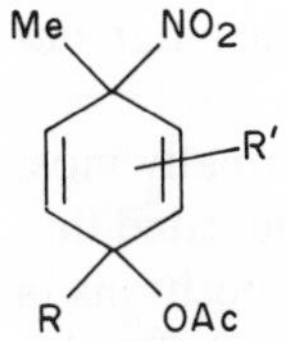

(**21**) R′ = H, Me; R = H, Me, *i*-Pr, *t*-Bu (**22**) R = *t*-Bu, hal, OMe, NHAc (**23**) 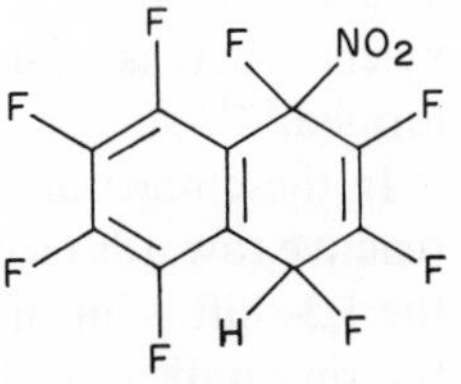 (**24**)

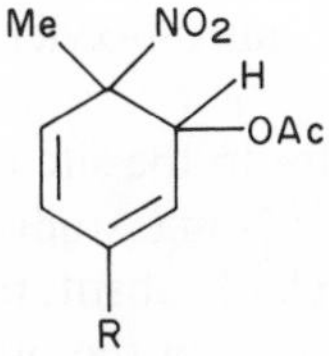

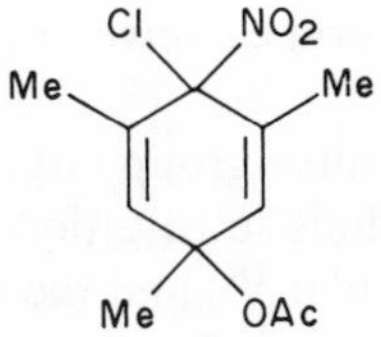

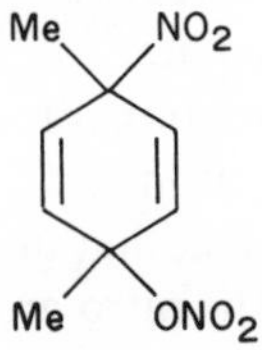

(**25**) 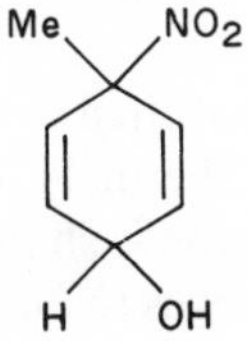 (**26**) 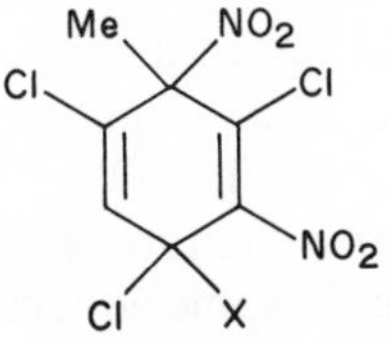 (**27**) $X = HSO_4^-$ (?) (**28**)

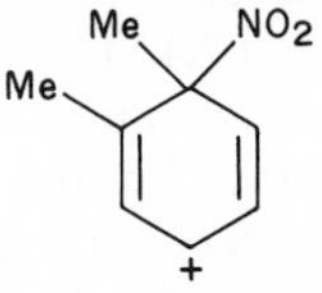

The adducts may then lose nitrous acid from the nitroacetate to give the acetoxy derivatives (Section 8.3), and from adducts such as **26** to give phenols that in turn may undergo further nitration, so accounting for the formation of nitrophenols in many nitrations.[185,186] Phenols may also be formed from dienones which arise from decomposition of the adducts.[89,179,181,187–189]

(3) By rearrangement of the nitro group to either the *ortho* or *meta* position. This is the mechanism by which the isomer yields in nitration may differ substantially according to the conditions. Whether migration occurs in the Wheland intermediate, or whether the intermediate is nucleophilically substituted, depends on the nucleophilicity of the solvent. This is shown in the

nitration of *o*-xylene by HNO_3 in aqueous H_2SO_4, where *ipso* substitution proceeds via the intermediate **28**. The yield of 4-nitro-*o*-xylene increases regularly with increasing sulphuric acid concentration, e.g. from 23% in 54 wt-% H_2SO_4 to 42% in 75 wt-% H_2SO_4. By contrast, the yield of 3-nitro-*o*-xylene (partially formed from rearrangement of **28**) increases much more markedly, i.e. from 12 to 58% over the same acid range;[185] this follows because media containing a higher concentration of acid are less nucleophilic.

The rearrangement of the nitro group in **28** to the 3-position was shown to be a 1,2- rather than a 1,3-shift by labelling experiments, which also showed that the 1,2-migration to the adjacent site bearing the methyl group occurs 50 times faster than to the adjacent site bearing hydrogen;[190] these relative migration rates are to be expected from symmetry considerations. The 1,2-shift was also shown by acid-catalysed solvolysis of **17**; 1,3-migration can give either 3- or 4-nitro-*o*-xylene whereas 1,2-migration can give only 3-nitro-*o*-xylene; only the latter was formed.[191]

In these reactions, the nitro group migrates to the site that is normally most reactive towards electrophilic substitution.[192] N.m.r. studies have indicated that the 1,2-shift is intramolecular,[187] and the lack of substitution at all positions in the ring indicates the absence of radical cations in the process; they have also apparently been ruled out by other studies.[190] Nevertheless, recent CIDNP studies of *ipso* nitration of 1,2,4,5-tetramethylbenzene followed by migration of the nitro group to the vacant 3-position (assumed to be a 1,2-shift although in this molecule it could equally be a 1,3-shift) indicates that radical cations are produced in homolysis of the C—N bond in the intermediate [see Section 7.4.1.(iii)] but not in the initial formation of the intermediate.[92]

1,3-Migration can also occur as demonstrated by the formation of 2,3-dimethyl-5-nitrobenzonitrile **(29)** from thermal decomposition of **20** (X = CN).[182] The thermal nature of this reaction is shown by the fact that under solvolytic conditions the same adduct rearranges to **30**.[183] Here the nitro group migrates in both a 1,2-intramolecular and intermolecular fashion since it can cross-nitrate into 4-fluorophenol, with 2,3-dimethylbenzonitrile being also formed. The thermal rearrangement of the diene **20** (X = CN) probably proceeds via a tautomeric 1,3-shift to give **31**, since this will then very readily undergo *cis* β-elimination of acetic acid; rearrangement prior to, and to facilitate, β-thermal

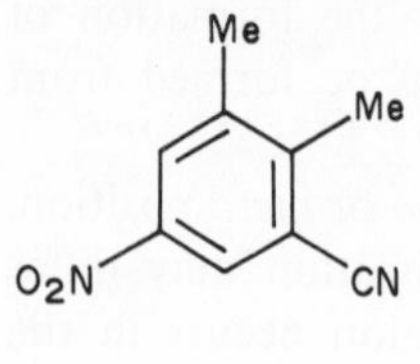

(29) 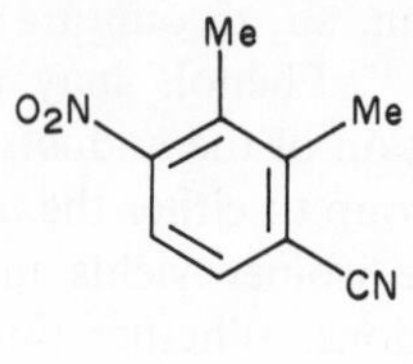 **(30)** 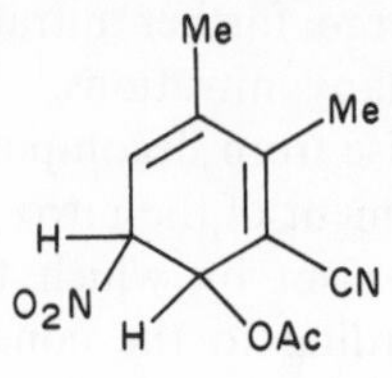 **(31)** 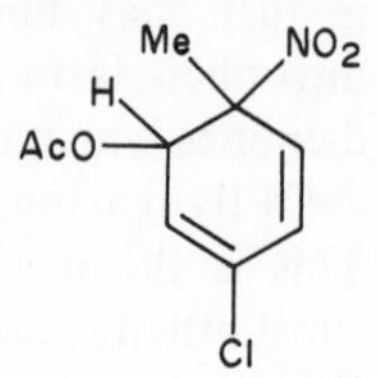 **(32)**

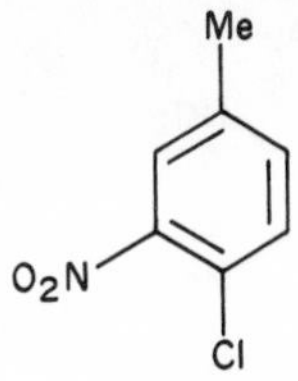

(33) 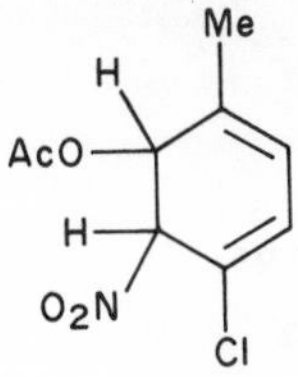 (34) 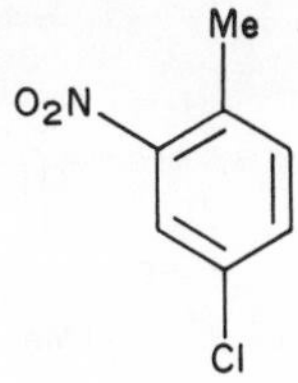 (35)

elimination is a well documented process.[193] Likewise, the thermal formation of 4-chloro-3-nitrotoluene (**33**) from **32** may proceed via initial rearrangement to **34**; solvolysis of **32**, by contrast, gives 4-chloro-2-nitrotoluene (**35**), again accompanied by cross-nitration. Formation of the 8-nitro derivative **37** by solvolysis of **36** also indicated a 1,3-nitro shift;[184] the deuterium label showed that the nitro group migrated rather than the alkyl chain. Thermal 1,3-migrations also occur in formation of 2,4-dinitrophenol from **38**[187] and in nitration of pentamethylbenzene by nitric acid in HSO_3F to give nitropentamethylbenzene. Here the initially formed Wheland intermediate is **39**,[188] where the nitro group again attaches to the site of highest electron density.

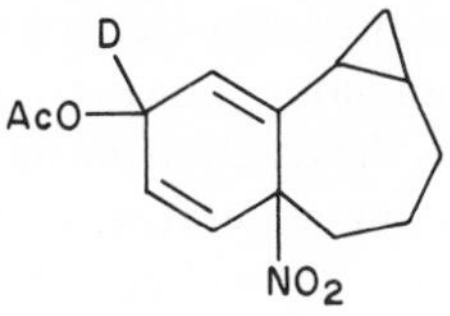

(36) 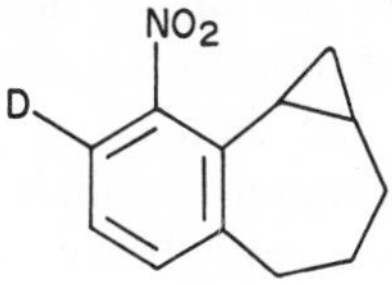 (37)

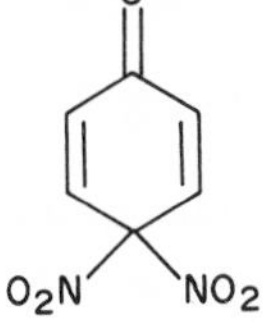

(38) 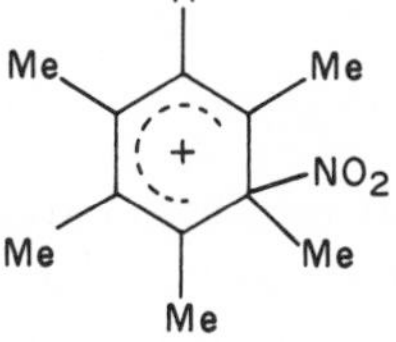 (39)

Each of the *N,N*-dimethylanilines **40** undergo *ipso* substitution by the nitrous acid-catalysed mechanism in 60–70 wt-% H_2SO_4, but in 76–83 wt-% H_2SO_4 it is the conjugate acid that reacts, the extent of *ipso* substitution at the 4-position then varying from 0% ($R^1, R^2 = H$) to 84% ($R^1, R^2 = Me$).[194] The isotope effects for the subsequent 1,3-rearrangement of **40** ($R^1 = H$, $R^2 = Me$) to the *o*-nitroamine **41** varies from 1.0 in 25 wt-% H_2SO_4 to 4.8 in 75.9 wt-% H_2SO_4. Thus, either the 1,3-

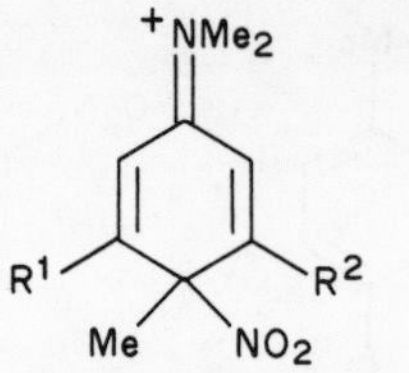

(40) **(41)**

$R^1 = R^2 = H, Me$

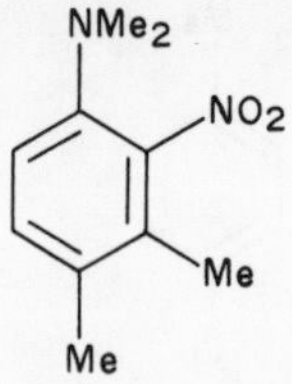

rearrangement or the subsequent proton loss can, depending on acidity, be rate-determining; the overall rearrangement rate (which occurs ca 500 times faster than in the corresponding cyclohexadienone) is not very acidity-dependent, however.[195]

7.4.3 Substituent Effects

Many substituent data are available for nitration, particularly for deactivated compounds where the high reactivity of the electrophile renders accessible information that is difficult to obtain in many other electrophilic substitutions. Interpretation of the data is not unambiguous, however. In particular, the existence of an encounter-rate limit renders the reaction unsuitable for determining the reactivities of aromatics of greater than moderate reactivity. Also, data for some compounds, e.g. alkylbenzenes, indicate the electrophile to be the same under all conditions, whereas other data, e.g. for ethers, indicate this not to be the case. Thirdly, problems of mixing control (with either very reactive nitronium ion sources[123,127,131] or in non-polar solvents[196]) can give erroneous relative reactivities.

(i) Simple alkyl substituents

The most frequently measured parameter in nitration is the toluene:benzene rate ratio. Wide variations in value have been obtained but most of these variations are due to mixing control and/or solubility effects. Mixing control[131] accounts for the low ratio of 1.7 obtained with nitronium salts in tetramethylene sulphone,[125] and may also account for the ratio of 17 obtained[125] with nitric acid in tetramethylene sulphone. With nitryl chloride and a range of Lewis acid catalysts, ratios of 11.2–39.3 were obtained, but these became a constant 27.7 $\pm$ 0.9 in the presence of nitromethane as solvent.[197] The former variation was attributed to the electrophile being a donor–acceptor complex in the non-polar hydrocarbon solvent, but mixing control is likely to have been a contributory factor. It clearly accounts for the drop in the ratio from 32.6–46.4 obtained in

Table 7.1 Partial rate factors for nitration of simple alkylbenzenes

Toluene			Ethylbenzene			Isopropylbenzene			*tert*-Butylbenzene			Reagent	Temperature/°C	Ref.
o	*m*	*p*	*o*	*m*	*p*	*o*	*m*	*p*	*o*	*m*	*p*			
42	2.5	58							5.5	4.0	75	HNO_3–10% aq. HOAc	45	198
43.7[a]	1.9[a]	46.8[a]							3.6[a]	3.6[a]	54.8	HNO_3–Ac_2O	25	199
47	3.0	62										HNO_3–Ac_2O	0	200
50.3	1.84	55.4							4.6	4.6	67.9	HNO_3–Ac_2O	25	201
49.7	1.3	60	31.4	2.3	69.5	14.8	2.4	71.6	4.5	3.0	75.5	HNO_3–Ac_2O	0	202
170	5.6	185				43	6.5	253				HNO_3–Ac_2O–CH_2Cl_2	−25	203
42.4	1.9	62.6										$AcONO_2$	25	204
37	2.8	47										HNO_3–$MeNO_2$	30	200
38.9	1.2[a]	45.8							5.5	3.7	71.6	HNO_3–$MeNO_2$	25	199
48.7	2.5	56.1	32.1	1.6	67.1							HNO_3–$MeNO_2$	25	125
51.7	2.2	60.1										HNO_3–CF_3CO_2H	25	205
41[b]	1.3[b]	60[b]										N_2O_4–CF_3CO_2H	25	120
9.0	1.1	10.4	7.9	0.7	16.5	5.6	0.9	23.0	4.3	2.0	37.8	$MeO^+(H)NO_2$	37.5	163a,b

[a]Calculated from the data in ref. 199.
[b]Varied according to the rate of additon of N_2O_4.

nitration by N-nitropyridinium and N-nitroquinolinium salts in acetonitrile to ca 14 in nitromethane.[137] High ratios are obtained using $NO_2^+PF_6^-$ in sulpholane in the presence of two molar equivalents of ethers, attributed to nitration by nitroxonium and nitrosulphonium ions, e.g. 65.7 in the presence of dimethyl sulphide.[197]

Partial rate factors for nitration of alkylbenzenes under conditions where these complications are not thought to be a significant problem are given in Table 7.1. Variations in values under different (solution) conditions at similar temperatures are small, indicating a common electrophile under all conditions [though data for other substituents contradict this (see Tables 7.3 and 7.12)]. Differences that exist are due largely to difficulties in determining accurate relative rates by the competition method, evident from the values obtained by different workers using the same conditions; the f_m values for toluene < 2.5 are particularly suspect (see also ref. 202). The high and variable ratios (34–50) obtained under some conditions using nitric acid in acetic anhydride have also been noted above [Section 7.4.1(iv)]. Under other conditions, the average relative rates are 27 (0 °C), 25.5 (25 °C), and 24 (45 °C), and the average isomer distributions in nitrations carried out in solution are as shown in Table 7.2.

The *p-t*-Bu > *p*-Me activation order applies under all conditions in Table 7.1, this being the order of both inductive and hyperconjugative electron release. It is obtained in nitration owing to the high reactivity of the nitronium ion, such that in the transition state a relatively small amount of charge is delocalized into the aromatic ring, making steric hindrance to solvation unimportant.[206] It then follows that the highest *p-t*-Bu:*p*-Me rate ratio should be observed in the gas phase, and this is indeed the case. Olah *et al.*[146] have reported the *p*-Me > *p-t*-Bu activation order for nitration by $AgNO_3$–BF_3 in MeCN, and this follows from the very large solvating counter ion ($AgOBF_3^-$) involved here [see Section 1.4(iii)].

The data show that steric hindrance to nitration is substantial, although it is less than in most other electrophilic substitutions. The *ortho*:*para* ratio for toluene may be reduced to as low as 0.49 by using neopentyl nitrate in polyphosphoric acid,[143] probably owing to association between the nitronium ion and the acid anion. Steric hindrance is, as expected, slightly less for nitrations

Table 7.2 Average isomer distributions (%) in nitration of simple alkylbenzenes

Compound	*ortho*	*meta*	*para*
Toluene	59	3.0	38
Ethylbenzene	49	2.5	43.5
Isopropylbenzene	35	4.5	60
tert-Butylbenzene	11	8.0	81

carried out under high pressure. Thus the *ortho*, *meta*, and *para* partial rate factors for nitration of *tert*-butylbenzene by HNO_3–H_2SO_4 in HOAc at 45 °C change from 4.4, 3.4, and 75 at 1 kg cm^{-2} to 4.1, 3.0, and 59.7, respectively, at 2000 kg cm^{-2}; these results also indicate that the ρ factor decreases with increasing pressure.[207]

(ii) Bulky alkyl substituents

Isomer ratios ($\frac{1}{2}o:p$) for a series of alkylbenzenes, $Ph(CH_2)_nMe$ ($n = 1$–16), in nitration by HNO_3–Ac_2O showed no significant differences, but the ratios (ca 0.42) were higher than those obtained using HNO_3–H_2SO_4–HOAc.[208] This contrasts with another study of bulkier compounds that gave virtually identical ratios under both conditions,[209] and yet another that showed the former condition to give lower ratios.[210] Moreover, the reagent formed by mixing the components at room temperature and then cooling to −40 °C, the reaction temperature, gave substantially lower ratios for isopropyl- and *tert*-butylbenzenes than with the reagent mixed at −40 °C.[210] It seems probable that either the electrophile is not the same under each condition, or some *ipso* substitution followed by rearrangement is occurring; this conclusion is reinforced by results for cyclopropylbenzene, below.

The ratios for nitration with HNO_3–Ac_2O–$MeNO_2$ were also fairly constant for compounds $PhCH_2R$ (R = Me, Et, Pr, *i*-Pr, *t*-Bu, cyclohexyl), although significantly smaller for R = cyclobutyl.[211] However, for compounds $PhCHRR^1$ and $PhCRR^1R^2$ (R, R^1, R^2 variously Me, Et, *i*-Pr, and *t*-Bu) the ratios decrease regularly with increasing bulk.[209] In general, bulkier substituents tended to be the more electron releasing as they are more polarizable (i.e. C–C hyperconjugation in them is greater).

(iii) Cycloalkyl substituents

Nitration of cyclopropylbenzene gives enhanced but variable *ortho*:*para* ratios (Table 7.3). The highest ratios are evidently obtained at low temperatures but it is

Table 7.3 Isomer ratios ($\frac{1}{2}o:p$) in nitration of cyclopropylbenzene

Ratio	Reagent	Temperature/°C	Ref.
1.25	HNO_3–Ac_2O	10–20	212
1.00	HNO_3–Ac_2O, prepared at −40 °C	−40	210
2.0–2.35	HNO_3–Ac_2O, prepared at room temp.	−40	210, 213
2.39	$AcONO_2$–CH_2Cl_2	−25	203
1.64	$MeO^+(H)NO_2$	37.5	163a,b
1.41	HNO_3–Ac_2O–$MeNO_2$	25	209
1.05	HNO_3–H_2SO_4	40	210

Table 7.4 Partial rate factors for nitration of cycloalkylbenzenes, PhR

	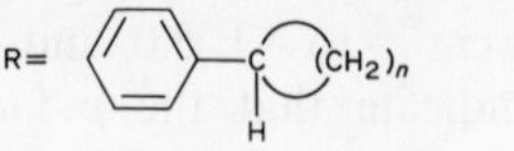 R = C_6H_5–CH(CH$_2$)$_n$ (H)			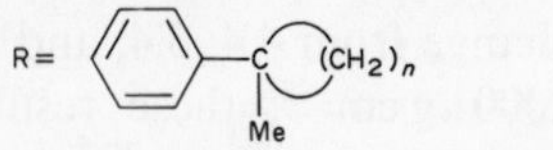 R = C_6H_5–C(Me)(CH$_2$)$_n$		
n	f_o	f_m	f_p	f_o	f_m	f_p
3	>221	>0.9	>157	115	2.4	159
4	35	2.7	82	5.8	3.3	102
5	26	3.5	100	4.66	7.2	94
6	17	4.4	89			
7	16	4.7	105			

not clear if this enhancement is due to involvement of a different electrophile, or to temperature-variable *ipso* substitution (followed by rearrangement); *ipso* substitution has been shown to occur in nitration of some arylcyclopropanes by HNO_3–Ac_2O.[214]

Partial rate factors for nitration of cycloalkylbenzenes by HNO_3-Ac_2O-$MeNO_2$ are shown in Table 7.4.[209] Notable features are the following:

(1) The *para*-cycloalkyl substituents activate in the order cyclopropyl > cyclopentyl > cyclohexyl ≈ cyclobutyl, paralleling that found in hydrogen exchange and solvolysis [see Section 3.1.2.1.(viii) for interpretation].

(2) The strong activation by the *para*-cyclopropyl group compared with other alkyl substituents, and which arises through strain-assisted C–C hyperconjugation [see Section 3.1.2.1(vii)] is conformationally dependent. This is demon-

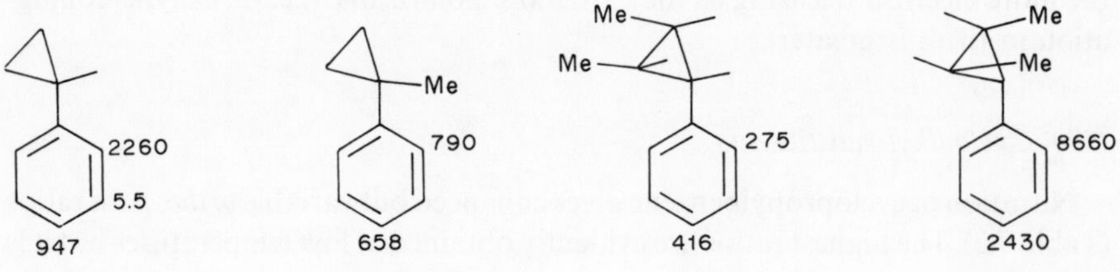

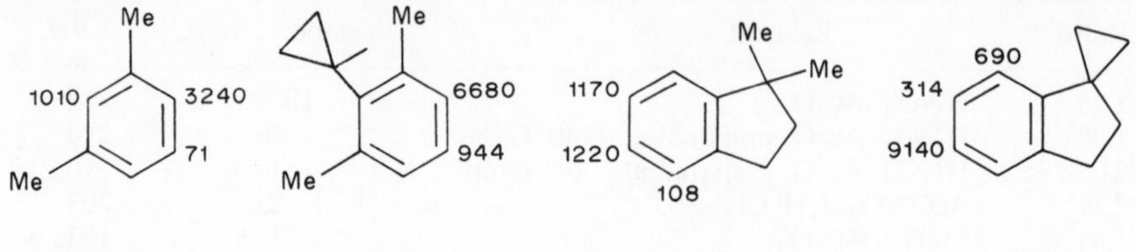

Scheme 7.2. Partial rate factors for nitration with HNO_3–Ac_2O–CH_2Cl_2 at −25 °C.

strated more clearly by the partial rate factors in Scheme 7.2, obtained using HNO_3–Ac_2O–CH_2Cl_2 at $-25\,°C$, a condition under which both the rate data shown appeared to be unaffected by encounter control, and acetoxylation was only minor.[203]

Activation is maximal in the bisected conformation where the electrons of the C—C bond are coplanar with the p-orbitals of the aryl ring. In 1-methylcyclopropylbenzene (**43**), steric hindrance between the methyl group and the *ortho*-hydrogen of the aryl ring prevents this conformation being achieved, and so the activation is less than in **42**. A comparable interaction accounts for the low reactivity of **44**, whilst in **45** steric interactions are *reduced* on adopting the bisected conformation, so the reactivity here is greater. Steric hindrance prevents adoption of the bisected conformation in **47**, so that relative to **46** activation at the position *para* to cyclopropyl is only 13.3-fold. By contrast, comparison of **49** (in which the bisected conformation is enforced) and **48** shows *p*-cyclopropyl to activate 7.5-fold relative to *p*-isopropyl, whereas there is only a 3.7-fold reactivity difference between **42** and isopropylbenzene (Table 7.1).

The relative activations by cyclopropyl and 1-methylcyclopropyl will also depend on the conformation of the cyclopropyl group under given conditions, and superimposed on this will be the effect of C–C vs. C–H hyperconjugation at the 1-position; the latter for example probably accounts for the 1-methylcyclopropyl substituent being *more* electron-releasing than cyclopropyl in hydrogen exchange (Section 3.1.2.1).

(3) The *ortho* position of cyclopropylbenzene is considerably more reactive than the *para* position, as shown in Table 7.3. Moreover, the partial rate factors in Scheme 7.2 show that the *ortho*:*para* ratio is dependent on conformation, being larger the more activating the substituent is, viz. 0.66 (**44**), 1.20 (**43**), 2.39 (**42**), and 3.56 (**45**). This provides very compelling evidence for initial coordination of the electrophile with the π-cloud of the substituent[215] (as proposed for ethers[115] and biphenyl[101]), since this should be maximal when the cyclopropyl group is in the bisected conformation.

There have been a number of studies of the nitration of other cyclopropyl-substituted aromatics,[216] of *trans*-2-methylcyclohexylbenzene (which is much more sterically hindered than the *cis* isomer), and of 1-adamantylbenzene (which

Table 7.5 Partial rate factors for nitration of 4-X-1-phenylbicyclo[2.2.2]octanes

X	f_o	f_m	f_p	X	f_o	f_m	f_p
H[a]	10.9	6.87	123	Cl	2.92	2.76	37.8
OMe	4.84	4.12	63.5	F	3.16	2.82	40.6
CO_2Me	4.91	4.01	61.5	CN	2.02	1.96	27.6
Br	2.46	2.52	57.1	NO_2	1.81	1.69	23.5

[a]Alkyl groups gave similar factors.

is much more hindered than the 2-isomer).[209] Partial rate factors for nitration of 4-substituted-1-phenylbicyclo[2.2.2]octanes by HNO_3–Ac_2O ar 25 °C (Table 7.5)[201] indicated the involvement of π-inductive effects in addition to the σ-inductive effects expected in this system; the partial rate factors also show bicyclo[2.2.2]octyl to be more electron releasing than *tert*-butyl (cf. hydrogen exchange, Chapter 3, Tables 3.2 and 3.3).

(iv) Polyalkyl substituents

There have been various nitration studies of aromatics containing more than one alkyl substituent.[79,105,144,146,217,218] However, the onset of encounter control of the reaction rates under some conditions, and the occurrence of *ipso* substitution followed by rearrangement, render quantitative interpretation of the data unsafe. Typical anomalies are the high 3-:4-positional reactivity ratio in *o*-xylene[218] and the low reactivities of *m*-xylene and mesitylene compared with their isomers.[144,146] Nitration with methyl nitrate–boron trifluoride has been recommended for selective mononitration of polyalkylbenzenes.[144]

(v) Cyclic alkyl substituents

Nitration of various benzocyclenes, e.g. **50** and **51**, has been carried out,[219–222] but the results are complicated by *ipso* substitution followed by rearrangement. The most notable feature of the results is that the $\alpha:\beta$ reactivity ratio diminishes with increasing strain in the side-chain (note that the result for indane in ref. 221 is in error[222]), the interpretation of this being given in Section 3.1.2.3. Thus, for example, benzcyclobutene (**50**, $n = 2$) and **51** and derivatives give almost exclusive β-substitution. Noteworthy also is the fact that the α-reactivity decrease is accompanied by β-reactivity increase as n diminishes; this is predicted only by the interpretation based on bond-strain effects, but not that[221] which attributes the low α-reactivity to increased s-character of certain orbitals of the bridgehead atoms.

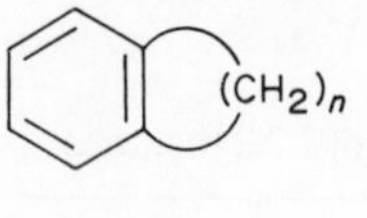

$(CH_2)_n$

(50) $n = 2$–4

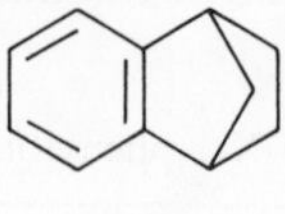

(51)

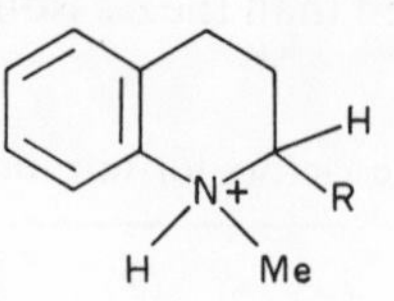

(52) R=H, Me, Et, *i*-Pr, *t*-Bu

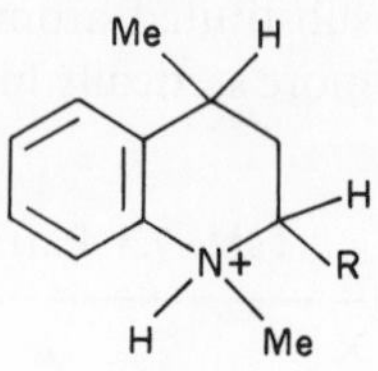

(53) R=H, Me, Et, *t*-Bu

Nitration of 2-alkyl-tetrahydroquinolinium ions (**52**, **53**) by HNO_3–H_2SO_4

showed *no* 5-substitution, which was likewise attributed to increase in strain in the N-containing ring on going to the transition state for 5-substitution. The overall 6- and 7-substitution rate was only marginally affected by the nature of the alkyl group, but the 7-:6-rate ratio was substantially larger when R was bulky.[223] It seems most probable that conjugative electron release to the 6-position (which is itself minor) is reduced when R is bulky owing to the nitrogen lone pair being twisted more out of plane of the aromatic ring.

(vi) Substituted alkyl groups

Partial rate factors for nitration of some compounds $PhCH_2X$ at 25 °C (Table 7.6) show that as X becomes more electron withdrawing the effect of CH_2X changes from activating to deactivating, yet the orientation remains *ortho,para*. This latter combination parallels the behaviour of, for example, the halogen substituents, and provides important evidence confirming that the origin of the electron-supplying effect of the methyl group is mainly conjugative (hyperconjugation, Section 1.4.3).[204] Thus for groups such as CH_2CN the overall electronic effect is $-I$, $+M$. The trend towards *meta* orientation with increasing electron withdrawal may also be found in $PhCH_2OSO_2X$ compounds, as X is made more electron withdrawing.[228]

In Table 7.6 the decrease in the f_o values parallels that in the f_p values, except in the case of the phosphorus-containing and β,β,β-trichloroethyl substituents for which steric hindrance is likely to be significant.

Table 7.6 Partial rate factors for nitration of compounds $PhCH_2X$ at 25 °C

X	f_o	f_m	f_p	Reagent	Ref.
H	42.4	1.9	62.6	$AcONO_2$	204
Ph	11		36	HNO_3–Ac_2O	224a
	30.5[a]		56.4[a]	HNO_3–Ac_2O	224b
$4\text{-}NO_2C_6H_4$	14.7		34.0	HNO_3–Ac_2O	224b
OMe	10.0	1.3	16.3	$AcONO_2$	204
$PO(OEt)_2$	2.3		10.9	$AcONO_2$	224
CO_2Et	6.3	1.5	7.55	$AcONO_2$	204
CO_2H	1.9		2.7	HNO_3–CF_3CO_2H	225
Cl	0.72	0.30	2.24	$AcONO_2$	204
CN	0.25	0.21	1.15	$AcONO_2$	204
$P^+(OH)_3$	0.15		0.90	HNO_3–CF_3CO_2H	225
CH_2CCl_3	0.10	0.27	0.73	HNO_3–CF_3CO_2H	226
SO_2Et	0.24	0.15	0.58	HNO_3–Ac_2O	227
NO_2	0.08	0.20	0.17	$AcONO_2$	204

[a] At 0 °C. For compounds $Ph(CH_2)_nPh$, the partial rate factors f_o and f_p are 41.6, 61.0 ($n = 2$), 44.7, 60.7 ($n = 3$), 46.5, 60.8 ($n = 4$),[224a] showing the diminishing inductive effect of phenyl as the chain length increases.

Table 7.7 Isomer distributions (%) and partial rate factors for α-halotoluenes, PhX

X	*o*	*m*	*p*	$10^5 f_o$	$10^5 f_m$	$10^5 f_p$	Conditions	Ref.
CH_3	56.1	2.5	41.4				$AcONO_2$, 25 °C	204
CH_2Cl	33.6	13.9	52.5				$AcONO_2$, 25 °C	204
$CHCl_2$	23.3	33.8	42.9				HNO_3–H_2SO_4, 25 °C	229
CCl_3	6.8	64.5	28.7				HNO_3–H_2SO_4, 25 °C	229
	16.8	62.5	20.7				HNO_3–H_2SO_4, 25 °C	233
	13	68	19	3.5	18	10	$NO_2^+BF_4^-$–sulpholane, 25°C	226
CF_3	6.0	91.0	3.0		6.7[a]		HNO_3–H_2SO_4, 0 °C	230
	6.7	91.0	2.3	0.5	6.8	0.345	$NO_2^+BF_4^-$–sulpholane, 25 °C	226
					1100		$MeO^+(H)NO_2$, 37.5 °C	163a,b

[a]Ref 231.

The substitution of a second and then a third electron-withdrawing group into methyl changes the orientation from predominantly *ortho*, *para* to *meta*, as shown in Table 7.7. The result for trifluoromethylbenzene compared with trichloromethylbenzene follows from the greater electron withdrawal by CF_3. The value of the ratio $\log f_p/\log f_m$ for CF_3 is greater than for other comparable electron-withdrawing groups such as CCl_3, NO_2, and NMe_3^+,[226] so indicating additional conjugative deactivation of the *para* position through C–F (negative) hyperconjugation (see Section 1.4.3); this contributes 17% to the overall free energy of activation. The $C(CF_3)_3$ substituent also produces *meta* orientation in nitration.[232]

Partial rate factors for nitration of triptycene at the α- and β-positions are 1.6 and 67, respectively, and a nitro group in one ring reduces the reactivity of the others 5.5-fold.[234] The β:α reactivity ratio is higher than in hydrogen exchange (Section 3.1.2.4), indicating that the lower demand for conjugative stabilization of the transition state in nitration renders more important the $-I$ effect of the bridgehead carbon atom arising through strain (cf. Section 3.1.2.3). Nitration of compound **54** takes place preferentially in the coplanar central rings rather than in the lateral ones, even though electronic effects in each should be identical.[235] This strongly points to transannular stabilization of the transition state for

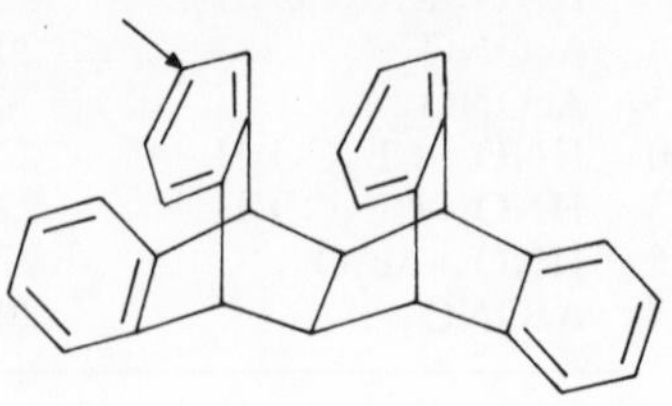

(**54**)

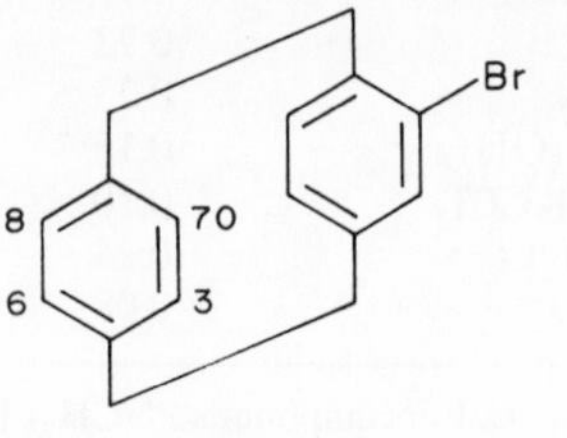

Br
8
70
6
3

(**55**)

substitution in the coplanar rings, in the manner found for cyclophanes (Section 6.7.3). Nitration of 4-bromo[2.2]paracyclophane (**55**) gave the positional yields shown. The high yield pseudo *ipso* to the most basic site in the bromine-containing ring was attributed to this acting as an internal base in the product-forming step of the reaction.[236]

(vii) Substituted alkenyl and alkynyl groups

These groups combine $-I$ and $+M$ effects and thus, like the halogens, are *ortho,para*-orientating (Table 7.8). They show a tendency towards *meta* orientation as electron withdrawal in the side-chain increases; this tendency is also produced by protonation which occurs under more acidic conditions. The $CH{=}CHNMe_3{}^+$ and $CH{=}CHSO_2Cl$ substituents are also reported to be almost entirely *ortho,para*-orientating.[240] The deactivating effect of a *para*-nitro substituent in azobenzene (39-fold) is less than in stilbene (115-fold), which may reflect the differences in transmission abilities of the corresponding double bonds; the effect is smaller in 2,4-dinitrostilbene (7.75-fold) in accordance with the *Reactivity–Selectivity Principle.* The high reactivity at the *ortho* position of azobenzene suggests that initial coordination of the electrophile at nitrogen occurs.

(viii) Halogen substituents

Partial rate factors for nitration of halogenobenzenes are shown in Table 7.9. Notable features are as follows:

(1) Deactivation of the *meta* positions is in the order $F = Cl > Br > I$ which is approximately that of the $-I$ effects of the halogens. This shows that the demand for resonance stabilization of the transition states in nitration is small, because in other reactions, e.g. hydrogen exchange (Section 3.1.2.5), secondary relay of resonance to the *meta* positions makes fluorine relatively less deactivating. In 4-halogenoanisoles, partial rate factors for nitration at the 2-position (i.e. *meta* to halogen) are 0.119 (I), 0.077 (Br), and 0.069 (Cl);[244] the order is the same as that above, but the values are much smaller, which may be largely due to nitration of anisole, the reference, being encounter controlled.

(2) The order of reactivity for the *para* positions is $F = I > Cl > Br$, largely accountable in terms of a combination of inductive electron withdrawal and conjugative electron release. The latter is not large, however, so fluorine does not activate, in contrast to its behaviour in some other reactions. The high reactivity of iodobenzene relative to the other halogens seems anomalous.

(3) The *ortho*:*para* ratio decreases in the order $I > Br > Cl > F$, which follows from the relative strengths of the $-I$ effects of the halogens, and indicates that steric hindrance to nitration is small. However, the very marked difference in he ratio between iodine and bromine, coupled with the well established nitro- (or

Table 7.8 Isomer distributions (%) and partial rate factors for unsaturated compounds PhX

X	o	m	p	f_o	f_m	f_p	Reagent	Ref.
$CH{=}CHCO_2H$				3.4		2.8	$AcONO_2$–Ac_2O	225
				1.0		1.8	HNO_3–CF_3CO_2H	225
$CH{=}CHPO(OH)_2$				6.4		5.8	$AcONO_2$–Ac_2O	225
				0.9		0.6	HNO_3–CF_3CO_2H	225
$CH{=}CHNO_2$	31	2	67				HNO_3	237
	34	8	58	0.0031	0.00073	0.0105	HNO_3–H_2SO_4	238
$C(NO_2){=}CH(4\text{-}C_6H_4NO_2)$	32	21	48				HNO_3	237
$CH{=}CHPh$	42		58	73.1		201	HNO_3–Ac_2O	224a
$CH{=}CH(4\text{-}C_6H_4NO_2)$	39.7		60.3	0.57		1.74	HNO_3–Ac_2O	224a
$CH{=}CH[2,4\text{-}C_6H_3(NO_2)_2]$	27.5		72.5	0.013		0.068	HNO_3–Ac_2O	224a
$CH{=}CH[2,4,6\text{-}C_6H_2(NO_2)_3]$	20.3		79.7	0.0012		0.0088	HNO_3–Ac_2O	224a
$N{=}NPh$	79.3		20.7	15.1		7.9	HNO_3–Ac_2O	224a
$N{=}N(4\text{-}C_6H_4NO_2)$	58		42	0.141		0.204	HNO_3–Ac_2O	224a
$C{\equiv}CCO_2H$	27	8	65				HNO_3	239
$C{\equiv}CCO_2Et$	36	6	58				HNO_3	239

Table 7.9 Partial rate factors for nitration of halogenobenzenes

Fluorobenzene			Chlorobenzene			Bromobenzene			Iodobenzene					
o	*m*	*p*	*o*	*m*	*p*	*o*	*m*	*p*	*o*	*m*	*p*	Reagent	$T/^{\circ}C$	Ref.
0.04	—	0.77										HNO_3-Ac_2O	25	202
0.056[a]	0.0009[a]	0.79[a]										HNO_3	0	241
0.24	—	1.3	0.32	0.082	0.69							$MeO^+(H)NO_2$	37.5	163a, b
			0.029[a]	0.0009[a]	0.137[a]	0.033[a]	0.0011[a]	0.112[a]	0.252	0.012	0.78	$AcONO_2-MeNO_2$	25	242
			0.05	—	0.2							$N_2O_4-CF_3CO_2H$	25	120
									0.149	0.007	0.46	$AcONO_2-Ac_2O$	25	242

[a]Calculated using the PhX:PhH ratios of 0.15 (F), 0.033 (Cl), 0.030 (Br), and 0.18 (I) obtained with acetyl nitrate at 18 °C.[243]

Table 7.10 $\frac{1}{2}$ *ortho:para* ratios in the nitration of fluorobenzene and chlorobenzene

Fluorobenzene	Chlorobenzene	Condition	Ref.
0.19	0.66	N_2O_5–CCl_4	102
0.07	0.27	HNO_3–H_2SO_4	102
—	0.20	$AcONO_2$–H_2SO_4	243
—	0.19	$AcONO_2$–MeCN	102
0.03	0.14	$AcONO_2$–Ac_2O	102
—	0.11	$AcONO_2$–CCl_4	102
—	0.15	$NO_2^+BF_4^-$–sulpholane	123

nitroso-)deiodination (Section 10.61), strongly suggests that some *ortho* nitration of iodobenzene occurs via *ipso* substitution followed by rearrangement. It is significant that Bird and Ingold[243] were unable to obtain consistent values for the relative rates of nitration of iodobenzene and benzene, the former sometimes reacting faster than the latter, whilst Roberts *et al.*[242] obtained substantially different values for this ratio using different solvents, namely 0.13 ($MeNO_2$) and 0.22 (Ac_2O). Iodobenzene is also abnormally reactive in nitration by nitric acid–aqueous sulphuric acid,[66f,245] the relative reactivities in 67.1 wt-% acid being 1.0 (H), 0.133 (F), 0.068 (Cl), 0.056 (Br), and 0.245 (I). All of these anomalies would follow from *ipso* substitution.

The *ortho:para* ratio for chlorobenzene and fluorobenzene also varies substantially according to the reagent (Table 7.10). This implies that under some conditions the substituent nucleophilically displaces the counter ion and coordinates with the electrophile, which subsequently migrates to the *ortho* position; the electrophile must therefore vary with conditions.

(ix) Polyhalogen substituents

The effects of multiple chlorine substituents has been determined[66f] and it was found that additivity breaks down badly (Table 7.11). Contributing factors may

Table 7.11 Ratios of observed to calculated reactivities in nitration of chloroaromatics

No.	Substituents	f_{obs}/f_{calc}	Position	No.	Substituents	f_{obs}/f_{calc}
1	1,2-Cl_2	4.25	3	6	1,3,5-Cl_3	1.0
2	1,2-Cl_2	0.9	4	7	1,2,3,4-Cl_4	9.0
3	1,3-Cl_2	0.18	2	8	1,2,3,5-Cl_4	4.0
4	1,3-Cl_2	0.57	4	9	1,2,4,5-Cl_4	13.8
5	1,4-Cl_2	7.3	2	10	1,2,3,4,5-Cl_5	244

be steric hindrance (e.g. entry no. 3), *ipso* substitution (e.g. entry no. 1), reduced $-I$ effects due to C—Cl bond elongation, and enhanced $+M$ conjugative electron release in the very unreactive systems, these latter being especially significant in the more chlorinated compounds.[246] Parallel effects are observed in gas-phase elimination of 1-arylethyl acetates, which takes place via partial carbocation formation adjacent to the aromatic ring and thus provides a gas-phase measure of electrophilic aromatic reactivity.[246]

(x) *Amines, anilides, and ethers*

Isomer ratios and partial rate factors obtained in the nitration of these compounds under various conditions are given in Table 7.12. For the anilides and ethers the ratios are very dependent on the conditions [and show the same pattern of ratio–reagent dependence found for cyclopropylbenzene (Table 7.3)]. Explanations for these results are twofold; neither is uniquely satisfactory.

(1) It is evident that the ratios are smaller the more acidic the conditions (and this is true also for nitration of phenols and cresols),[254] and under a given condition are smaller for acetanilide than for anisole. This suggested protonation as the cause, but this has been ruled out for both molecules, since if they were protonated then substantial *meta* substitution should occur.[249] The alternative explanation is that the substituent is hydrogen bonded so that it acquires either sufficient positive charge, or possibly bulk, to render attack at the *ortho* position less favourable.

(2) Whilst the evidence indicates that an acidity-dependent interaction such as hydrogen bonding must occur, this cannot be a full explanation because under many conditions the ratio is higher than statistical, higher than in all other reactions, and cannot be accounted for theoretically. This has led to proposals of specific interactions between the electrophile whereby the heteroatom displaces the counterion.

Halvarson and Melander[248] suggested the involvement of structure **56** with subsequent six-centre rearrangement. It follows that increased electron density on the side-chain heteroatom should lead to a greater *ortho*:*para* ratio and *vice versa*. This would account, for example, for the increase in the ratio with *increasing* size of the substituent in nitration of a range of ethers PhOR by HNO_3–Ac_2O[208] (in this work the ratios also decreased markedly with increasing temperature). It would also account for the low ratios obtained with trifluoroanisole.

Nucleophilic displacement followed by nitronium ion migration was proposed to account for the high ratio obtained in nitration of anisole by nitronium tetrafluoroborate.[129]

A further variation was proposed by Knowles and Norman,[204] whereby the reagent was assumed to be a mixture of dinitrogen pentoxide, giving rise to specific *ortho* interaction involving a six-membered cyclic transition state in **57**

Table 7.12 Isomer distributions (%) and partial rate factors in nitration of amines, anilides, and ethers, PhX

X	*ortho*	*para*	$\frac{1}{2}o:p$	f_o	f_p	Conditions	Ref.
OMe	37.5	62.5	0.26			N_2O_4–CF_3CO_2H, 25 °C	120
	31	67	0.23			HNO_3–H_2SO_4, 45 °C	247
	40	58	0.34			HNO_3, 45 °C	247
	41	59	0.35			$MeO^+(H)NO_2$, 37.5 °C	163a, b
	44	55	0.40			HNO_3–HOAc, 65 °C	247
	69	31	1.1			$NO_2^+BF_4^-$–sulpholane	129
	71	28	1.27			HNO_3–Ac_2O, 10 °C	248
	60	39	0.76			HNO_3–Ac_2O, 45 °C	208
	75	25	1.5			$BzONO_2$–MeCN, 0 °C	248
CH_2OMe	29	53	0.27			HNO_3–H_2SO_4, 25 °C	249
	39	49	0.40			HNO_3, 25 °C	115
	51	42	0.61			HNO_3–Ac_2O, 25 °C	204
$(CH_2)_2OMe$	32	59	0.27			HNO_3–H_2SO_4, 25 °C	115
	39	49	0.40			HNO_3, 25 °C	115
	41	56	0.37			HNO_3–$MeNO_2$, 25 °C	115
	62	34	0.91			HNO_3–Ac_2O, 25 °C	115
	66	30	1.1			$AcONO_2$–MeCN, 0 °C	115
	69	28	1.23			N_2O_5–MeCN, 0 °C	115
$(CH_2)_3OMe$	44	52	0.42			HNO_3, 0 °C	115
	43	53	0.41			$AcONO_2$–MeCN, 0 °C	115
OCF_3	10	90	0.05			$NO_2^+BF_4^-$–$MeNO_2$	134
OPh	51	49	0.52	120	230	HNO_3–Ac_2O, 0 °C	224a
$O(2\text{-}C_6H_4NO_2)$	31.5	68.5	0.23	0.44	1.9	HNO_3–Ac_2O, 0 °C	224a
$O(3\text{-}C_6H_4NO_2)$	38	62	0.31	1.3	4.35	HNO_3–Ac_2O, 0 °C	224a

$O(4\text{-}C_6H_4NO_2)$	24.1	75.9	0.16	0.49	3.1	HNO_3–Ac_2O, 0 °C	224a
$O[2,4\text{-}C_6H_3(NO_2)_2]$	26	74	0.18	0.011	0.063	HNO_3–Ac_2O, 0 °C	224a
$O[2,4,6\text{-}C_6H_2(NO_2)_3]$	22	78	0.14	0.00069	0.0049	HNO_3–Ac_2O, 0 °C	224a
SPh	37.4	62.6	0.30	900	3010	HNO_3–Ac_2O, 0 °C	224a
$S(2\text{-}C_6H_4NO_2)$	34	66	0.26	2.1	8.3	HNO_3–Ac_2O, 0 °C	224a
$S(3\text{-}C_6H_4NO_2)$	40.2	59.8	0.34	5.7	17	HNO_3–Ac_2O, 0 °C	224a
$S(4\text{-}C_6H_4NO_2)$	35	65	0.27	3.3	12	HNO_3–Ac_2O, 0 °C	224a
$S[2,4\text{-}C_6H_3(NO_2)_2]$	22.1	77.9	0.14	0.034	0.24	HNO_3–Ac_2O, 0 °C	224a
$S[2,4,6\text{-}C_6H_2(NO_2)_2]$	17.4	82.6	0.11	0.0014	0.013	HNO_3–Ac_2O, 0 °C	224a
NHPh	76	24	1.58	8.4×10^5	5.3×10^5	HNO_3–Ac_2O, 0 °C	224b
$NH(2\text{-}C_6H_4NO_2)$	72.5	27.5	1.32	2600	2000	HNO_3–Ac_2O, 0 °C	224a
$NH(3\text{-}C_6H_4NO_2)$	61.5	38.5	0.80	2800	3500	HNO_3–Ac_2O, 0 °C	224a
$NH(4\text{-}C_6H_4NO_2)$	66.1	33.9	0.97	2700	2700	HNO_3–Ac_2O, 0 °C	224a
$NH[2,4\text{-}C_6H_3(NO_2)_2]$	33.3	66.7	0.25	3.7	14.8	HNO_3–Ac_2O, 0 °C	224a
$NH[2,4,6\text{-}C_6H_2(NO_2)_3]$	40	60	0.33	0.044	0.133	HNO_3–Ac_2O, 0 °C	224a
$NHCOCH_3$	19	79	0.12			HNO_3–H_2SO_4, 20 °C	250
	24	77	0.15			90% HNO_3, – 10 °C	229
	44.5	55.5	0.40			HNO_3–72.7% $HClO_4$, 25 °C	74
	59.4	37.6	0.79			HNO_3–50.8% $HClO_4$, 25 °C	74
	77	23	1.70			HNO_3–Ac_2O, 25 °C	251
	—	—	1.80			$NO_2^+BF_4^-$–MeCN, – 20 °C	252
$NHCOCF_3$	30	70	0.21			HNO_3–80% H_2SO_4, 25 °C	251
	34	66	0.26			HNO_3–Ac_2O, 25 °C	251
$NHCO_2CH_3$	—	—	1.89			HNO_3–Ac_2O, 25 °C	251
	—	—	0.82			HNO_3–H_2SO_4, 25 °C	251
$NHSO_2CH_3$	50	50	0.50			HNO_3–80% H_2SO_4, 25 °C	251
	60	40	0.75			HNO_3–Ac_2O	253

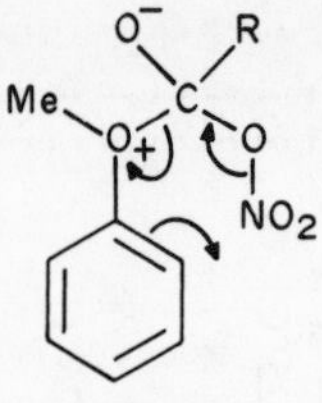

(56) 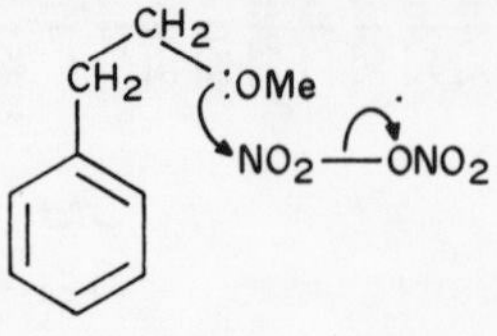 (57) 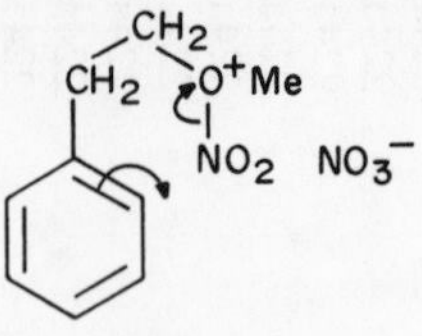 (58)

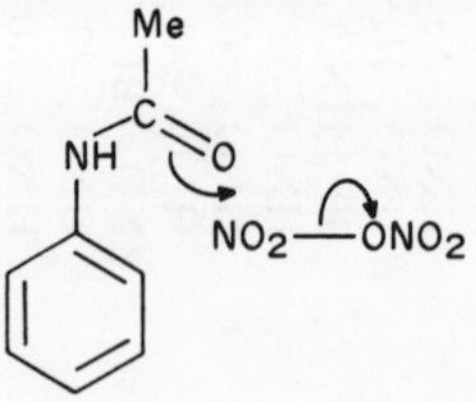

(59) 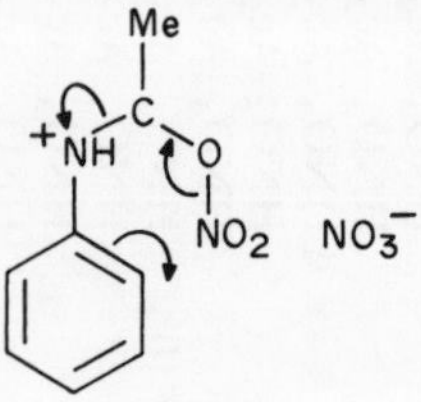 (60)

and **58**, and nitronium ion (produced by heterolysis) which substituted at each position in the normal way. The specific *ortho* nitration would parallel the decreasing ease of formation of the necessary cyclic transition states, i.e. $6 > 5 > 7$, so accounting for the smaller effect in nitration of benzyl methyl ether, and the absence of any effect with methyl 3-phenylpropyl ether (Table 7.12).

Coordination of the electrophile with the substituent was also proposed to account for the high ratios under certain conditions obtained with acetanilide;[255] 1-acetylaminonaphthalene also gives a high 2-:4-isomer ratio (2.3) in nitration by nitric acid–acetic anhydride.[256] The low ratios for both α,α,α-trifluoroacetanilide and *N*-(methylsulphonyl)aniline and their invariance with conditions strongly supports this view. However, it is now clear that this cannot involve coordination with the carbonyl group, as in **59** and **60**;[115] nitration of dihydro-2-

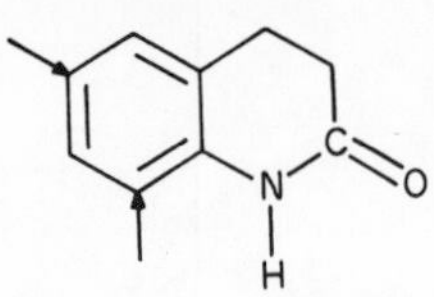

(61) 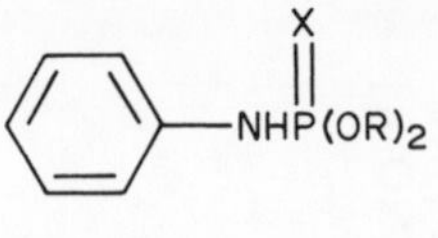 (62) (63) (64)

X=O,S;
R=Me,Et

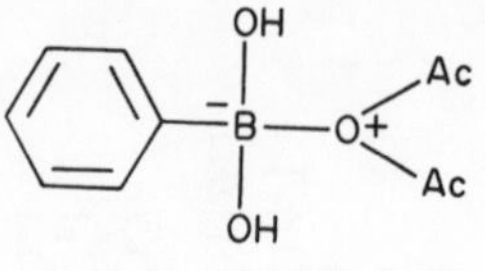

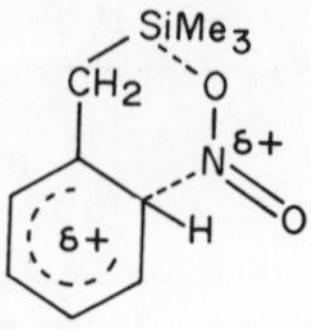

quinolone (**61**), in which such coordination could not aid 8-substitution, also gives a much higher 8-:6-substitution ratio.[253] Coordination with nitrogen is thus more probable and the fact that methyl-*N*-phenylcarbamate, in which the electron density on nitrogen is higher than in acetanilide, gives a slightly higher ratio (Table 7.12) is consistent with this.

(3) Although *ipso* substitution has not been shown to occur with these compounds, it would seem to be a feasible alternative explanation of these results.[257]

Nitration of *N*-arylphosphoramidates (**61**, X = O) in protic media[258] gives very high *meta* yields in contrast to acetanilide and *N*-(methylsulphonyl)aniline (Table 7.12), indicating that here the protonated form is involved; by contrast, *N*-phosphorthioamidates (**62**, X = S) gave almost entirely the *para* product (93%) (consistent with nitration of the free base, and steric hindrance arising from the bulky substituent). With nitronium tetrafluoroborate the amidate gave mainly *ortho*, *para* nitration with a high *ortho*:*para* ratio typical of this reagent.

(4) Partial rate factors for diphenyl ether are less than for diphenyl sulphide, indicating that the former is hydrogen bonded. This is also shown by the fact that a 4-nitro substituent reduces the reactivity of the ether only 74-fold compared with 251-fold for the thioether. For diphenylamine the corresponding factor is 196-fold, but this may not be reliable since the rate for the parent compound is comparable to the encounter rate. The results for all three series of compounds are self-consistent, however, in showing the substituent deactivation order 2-nitro- > 4-nitro- > 3-nitrophenyl. The effects of the nitro substituents are not additive, and are in accordance with the *Reactivity–Selectivity Principle.*

(xi) Substituents containing boron and silicon

These elements are electropositive with respect to carbon so each has a $+I$ effect. Boron has only six electrons in its outer orbital, whilst silicon has unfilled d-orbitals, so each can accept a pair of π-electrons from the aromatic ring, resulting in a $-M$ effect; this $+I$, $-M$ combination is unique, and results in *meta* orientation with activation.

Nitration of $PhSiMe_3$ gave partial rate factors of $f_o = 2.95$, $f_m = 2.2$, and isomer ratios of 22:73:5 (HNO_3–H_2SO_4) and 63:23:14 (HNO_3–Ac_2O).[259] The high *ortho*:*para* ratio is typical of $-M$ substituents (Section 11.2.4) and the higher ratio for nitration in acetic anhydride (with accompanying decrease in *meta* substitution) was attributed to formation of the complex **63**, with a consequent greater $+I$ effect.

Nitration of $PhSiMe_3$ gave partial rate factors of $f_o = 2.95$, $f_m = 2.2$, and $f_p = 1.25$.[260] Nitration of $Ph(CH_2)_nSiMe$, compounds gave approximate values of f_o, f_p of 91.8, 32.4 ($n = 1$), 13.5, 32.0 ($n = 2$), 13.4, 27.6 ($n = 3$), and 28.0, 30.3 ($n = 4$).[261] The very high value of f_o for $n = 1$ was attributed to coordination of the electrophile with the side-chain, thereby involving **64** as a transition state. The minimum in reactivity for $n = 3$ confirms the behaviour in hydrogen exchange (Section 3.1.2.4) and protiodesilylation [Section 4.7.1(ii)].

The effects of carborane substituents have been determined in nitration. In the icosohedral carboranyl group $B_{10}H_{10}C_2H$, the (hexacoordinate) carbons may be in either a 1,2-, 1,7- or 1,12-relationship, these being pseudo-*ortho*, *meta*, and *para*, respectively. The phenyl group may be attached either to carbon (as in 1-phenyl-*o*-carborane and 1-phenyl-*m*-carborane), or to boron (as in 3-phenyl-*o*-carborane and 2-phenyl-*m*-carborane). Nitration of the first three compounds (HNO_3–CCl_4) is said to occur mainly *para*,[262] and the latter gives an *ortho*:*meta*:*para* substitution ratio of 4:1:7.[263] 1-Phenyl-*m*-carborane is similar in reactivity to chlorobenzene, with that of 1-phenyl-*o*-carborane being somewhat lower. In the $B_8H_8C_2H$ carboranyl substituents, the (pentacoordinate) carbons may be in a 1,2-, 1,6-, or 1,10-relationship. Nitration of the phenyl derivative of the 1,10-isomer goes into the *para* position.[264]

(xii) Positive poles

The effects of positive poles have been studied more comprehensively in nitration than in any other reaction. Aromatics containing mercury, thallium, bismuth, tin, lead, iodine, nitrogen, phosphorus, arsenic, antimony, oxygen, sulphur, selenium, and iodine bonded to the ring as positive centres show mainly *meta* substitution.[265] The nuclear positions are all strongly deactivated, and the deactivation and tendency towards *meta* orientation increase, the more electronegative the element. The *meta* orientation is ascribed to the *para* intermediate **65** being of higher energy than the *meta* intermediate **66**, because of the unfavourable juxtaposition of positive charges in the former.[266]

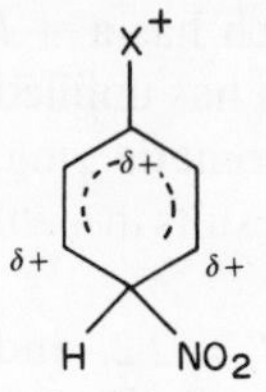

(65) 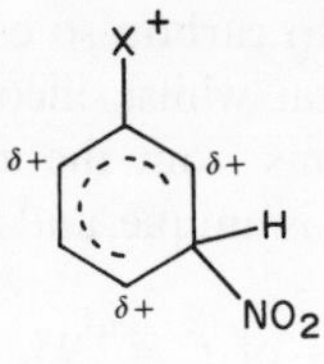 **(66)** 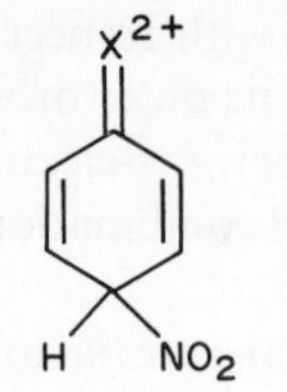 **(67)** 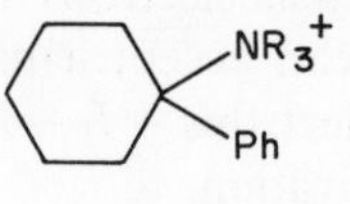 **(68)** **(69)**

(68) **a**, R=H; **b**, R=Me

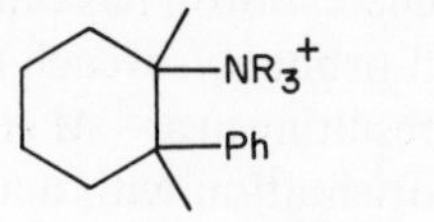

(69) **a**, R=H (*cis*); **b**, R=H (*trans*); **c**, R=Me (*cis*); **d**, R=Me (*trans*)

The deactivation by positive poles was formerly attributed to the $-I$ effect of the pole, but more recent evidence indicates that the field effect may be partially responsible. Moreover, substantial *para* substitution may occur; these features are discussed below. The isomer yields also show considerable acidity dependence; for example, the $\frac{1}{2}m:p$ ratio for nitration of $PhNH_3^+$ changes from 0.31 in 82.0 wt-% H_2SO_4 to 0.88 in 100 wt-% H_2SO_4. Hydrogen bonding of the pole has been suggested as a possible cause.[267,268] The change in the ratio for a given change in acid concentration decreases along the series $PhNH_3^+ > PhNH_2Me^+ > PhNHMe_2^+$, which could be due to increasing steric hindrance either to hydrogen bonding or to solvation in this highly solvating medium. It follows that in quantitative comparisons of the effects of poles, both the medium and the groups attached to the pole atom should be kept constant. Partial rate factors are given in Table 7.13.

Notable features of these results are as follows:

(1) The *meta*:*para* rate ratios for the poles are much higher than would be expected, and much higher than obtained, for example, with the nitro substituent. This has been attributed partially to the field effect (which would produce relatively less deactivation at the *para* position).[280] More important is likely to be conjugative electron release to the *para* position from the pole (either from the lone pair giving rise to the intermediate **67** or from N–R hyperconjugation) and thus for example the NH_3^+ substituent has a substantial σ_R° value of -0.18.[281,282] Conjugative electron release from the oxygen pole in triphenyloxonium ion, Ph_3O^+, appears to be very strong since it is reported to give almost 100% *para* nitration, whereas the triphenylsulphonium ion Ph_3S^+ gives *meta* nitration;[283] this is the opposite order to that which would be expected on the basis of electronegativities.

(2) The relative reactivities of the Group V poles, N:P:As:Sb, are 1.0:5.3:41.5:1957 at the *meta* positions and 1.0:0.87:14.1:913 at the *para* positions. The reactivity of the phosphorus pole at the *para* position appears to be low, and has been ascribed to $-M$ electron withdrawal into the empty d-orbitals.[268] The anomaly is, however, very readily explained in terms of the substantial $+M$ effect of the nitrogen pole noted above causing the *para* reactivity of NMe_3^+ to be abnormally high (cf. the high *para* reactivity of fluorine relative to the other halogens, Section 3.1.2.5). The relative reactivities of the Group VI poles, S:Se, are by contrast the same (1.0:23) at both positions.

(3) The reactivity decreases on changing from NH_3^+ to NMe_3^+, and the *meta*:*para* ratio decreases. For a given acid concentration the *meta* reactivity decreases 40-fold, whilst the *para* reactivity decreases 195-fold.[268] This is consistent with steric hindrance to solvation which will be very important in these highly solvating media, and will adversely affect *para* substitution in $PhNMe_3^+$.

(4) Interposing methylene groups between the aryl ring and the poles severely attenuates the deactivation as expected, and at a given acidity the reactivities relative to benzene ($=1$) for $Ph(CH_2)_nNMe_3^+$ have been determined as 3.39×10^{-8}, 7.94×10^{-5}, 0.224, and 3.16 for $n=0$, 1, 2, and 3, respectively.[274] For the less deactivating phosphorus-containing poles the values for $Ph(CH_2)_nP(OH)_3^+$ are 1.1×10^{-5}, 0.18, and 3.2 for $n=0$, 1 and 2, respec-

Table 7.13 Partial rate factors for nitration of positive poles, PhX

X	H_2SO_4/wt-%	f_o	f_m	f_p	Ref.
NH_3^+	82.0	19×10^{-8}	138×10^{-8}	451×10^{-8}	267
	98.0	4.3×10^{-8}	173×10^{-8}	213×10^{-8}	267–269
NH_2Me^+	98.0	—	57×10^{-8}	49×10^{-8}	268
$NHMe_2^+$	98.0	—	123×10^{-9}	71×10^{-9}	268
NMe_3^+	98.7	—	46.7×10^{-9}	11.5×10^{-9}	270
PMe_3^+	98.7	—	24.7×10^{-8}	1.0×10^{-8}	270
$AsMe_3^+$	98.7	—	194×10^{-8}	$<16.2 \times 10^{-8}$	270
$SbMe_3^+$	75.9	8.4×10^{-6}	91.4×10^{-6}	10.5×10^{-6}	270
SMe_2^+	98.1	4.4×10^{-10}	1.1×10^{-8}	14.7×10^{-10}	271
	98.7	2.6×10^{-10}	11.2×10^{-9}	94.4×10^{-11}	272
$SeMe_2^+$	98.1	71.9×10^{-10}	25.3×10^{-8}	33.7×10^{-9}	271
$CH_2NH_3^+$	80.0	1.4×10^{-3}	3.1×10^{-3}	4.7×10^{-3}	273
$(CH_2)_2NH_3^+$	73.0	8.5×10^{-2}	5.4×10^{-2}	0.50	273
$(CH_2)_3NH_3^+$	73.0	0.75[c]	0.19[c]	6.4[c]	273
$CH_2NMe_3^+$	80.0	1.6×10^{-6}	6.9×10^{-5}	1.6×10^{-5}	273
$(CH_2)_2NMe_3^+$	68.3	—	4.7×10^{-3}	—	274
	73.0	2.3×10^{-2}	6.2×10^{-2}	0.60	273
$(CH_2)_3NMe_3^+$	68.3	—	1.6×10^{-2}	—	274
	73.0	0.59	0.21	6.9	273

$CH_2PMe_3^+$	a	2.6×10^{-3}	3.9×10^{-3}	26.8×10^{-3}	275
$CH_2AsMe_3^+$	a	6.7×10^{-3}	2.6×10^{-3}	57.7×10^{-3}	275
$CH_2SMe_2^+$	77.3	38.4×10^{-5}	93.8×10^{-5}	21.0×10^{-4}	271
$(CH_2)_2SMe_2^+$	69.0	6.0×10^{-2}	6.6×10^{-2}	0.95	273
$CH_2SeMe_2^+$	75.5	56.4×10^{-5}	3.5×10^{-4}	41.5×12^{-4}	271
$P(OH)_3^+$	72.3	—	6.0×10^{-6}	27×10^{-6}	276
$P(OH)Me_2^+$	84.7[b]	1.6×10^{-8}	14.8×10^{-8}	—	277
$As(OH)_3^+$	83.5	—	1.7×10^{-8}	0.7×10^{-8}	276
$SMeOH^+$	98.4	4.5×10^{-10}	12.4×10^{-9}	17.8×10^{-10}	271
$PhSOH^+$	97.8	11.5×10^{-10}	27.4×10^{-9}	85.3×10^{-10}	271
2-Pyridinium	84.5[b]	1.9×10^{-6}	8.5×10^{-6}	2.2×10^{-5}	278, 279
4-Pyridinium	77.5[b]	1.3×10^{-4}	2.1×10^{-4}	5.9×10^{-4}	278
4-CH_2-pyridinium	69.5[b]	6.1×10^{-2}	2.6×10^{-2}	0.70	278
2-Pyridinium-*N*-oxide	78.0[b]	1.8×10^{-5}	47×10^{-5}	22×10^{-5}	279
1-$C_6H_{10}NH_3^+$ (**68a**)	72.4	—	3.5×10^{-3}	—	280
2-$C_6H_{10}NMe_3^+$ (**68b**)	75.0	—	1.8×10^{-4}	—	280
cis-2-$C_6H_{10}NH_3^+$ (**69a**)	72.9	—	2.6×10^{-2}	—	280
trans-2-$C_6H_4NH_3^+$ (**67b**)	72.7	—	2.3×10^{-2}	—	280
cis-2-$C_6H_4NMe_3^+$ (**69c**)	73.0	—	4.1×10^{-3}	—	280
trans-2-$C_6H_4NMe_3^+$ (**69d**)	72.9	—	6.6×10^{-3}	—	280

[a]HNO_3–$MeNO_2$.
[b]Average of acid range used.
[c]Values given in ref. 273 are in error.

tively.[276] The results for the nitrogen poles indicated that a field effect operates, although competitive nitrations in media of different dielectric suggested the converse.[274]

(5) Further work with the ions **68** and **69** showed a reactivity difference of 23-fold (*cis*) and 37-fold (*trans*) between **69c, d** and **68b** compared with a much bigger factor (ca 550-fold) between $PhCH_2NMe_3^+$ and $Ph(CH_2)_2NMe_3^+$. Likewise, the difference is 7.4-fold (*cis*) and 6·6-fold (*trans*) between **69a, b** and **68a** compared with 17-fold between $PhCH_2NH_3^+$ and $Ph(CH_2)_2NH_3^+$. These conformational dependences imply a field component of the deactivation, although the marked differences in effect between the different poles shows that some other major factors also operate.

(6) To provide information regarding the operation of a direct field effect, Ridd and coworkers[284] studied the rates of nitration of compounds **70** and **71**. If a field effect operates then **71** should be less reactive than **70** for a given value of n (except when $n = 1$, since here the distance between the pole and the ring is the same in both systems). The data (Table 7.14) show that in each case compounds **71** are the least reactive. However, the fact that this true even for $n = 1$ indicates that a large part of the difference arises from steric hindrance to solvation of one side of the aromatic ring, which is known markedly to affect substitution rates; an additional factor may be the ease of transmission of the field effect through the solvent-free cavity. The solvation effect is confirmed by the increasing reactivities for increasing values of m (with n constant) for the poles (**71**, $X = SMe^+$). On the other hand, the difference in reactivities between **70** and **71** for $n = 1$, $m = 8$

Table 7.14 Partial rate factors for nitration of cyclic and open-chain positive poles

$X(CH_2)_n$–C_6H_4–$(CH_2)_nX$ **(70)**

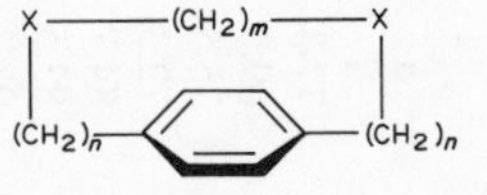

X—$(CH_2)_m$—X bridging $(CH_2)_n$–C_6H_4–$(CH_2)_n$ **(71)**

X	n	f	X	n	m	f
NH_3^+	1	2.95×10^{-6}	NH_2^+	1	8	2.80×10^{-9}
	2	5.40×10^{-2}		2	4	4.90×10^{-4}
	3	2.80		—	—	—
NMe_3^+	1	7.02×10^{-11}	NMe_2^+	1	8	4.43×10^{-11}
	—	—		2	4	3.40×10^{-4}
	3	2.10		3	2	1.20×10^{-3}
SMe_2^+	1	2.79×10^{-9}	SMe^+	1	6	1.76×10^{-10}
	—	—		1	8	4.03×10^{-10}
	—	—		1	10	1.01×10^{-9}
	2	1.46×10^{-2}		2	4	1.76×10^{-5}
				2	6	2.85×10^{-4}
				2	8	4.23×10^{-4}

depends markedly on the pole, being 1050, 1.6, and 7 for NH_3^+, NMe_3^+, and SMe_2^+, respectively, so an additional factor must operate.

(7) Nitration of the dimethylphenyloxosulphonium ion, $PhSOMe_2^+$, in 98 wt-% H_2SO_4 goes 100% *meta* and takes place 500 times slower than nitration of $PhSMe_2^+$.[285] The substituent is therefore extremely deactivating ($f_m \approx 2 \times 10^{-11}$), giving $\sigma_m \approx 1.6$ (cf. 1.38 given in ref. 285).

(8) The $P(OH)(CH_2Cl)_2^+$ substituent is less deactivating than $P(OH)Me_2^+$,[277] which may be due to diminished protonation in the former.

(xiii) Dipolar substituents

Each group here is of the $-I, -M$ type, a combination which produces strong deactivation of the *ortho* and *para* positions, and hence *meta* orientation. High *ortho*:*para* ratios are obtained, which vary according to the *meta*:*para* ratios (Table 7.15), showing them to have an electronic origin,[286] described in Section 11.2.4. Partial rate factors and the positional selectivities (Table 7.15) show significant medium dependence due to hydrogen bonding (protonation may also be involved)[295] which produces an increase in the amount of *meta* product and a corresponding increase in the *ortho*:*para* ratio.

The abnormal isomer distribution in nitration of *tert*-butyl phenyl ketone is attributed to twisting of the carbonyl group out of the plane of the ring, due to the bulk of the *tert*-butyl group, and this attenuates the $-M$ effect.[292]

(xiv) Polysubstituted aromatics

Substituent effects are not additive. In the case of benzene containing two or more electron-supplying substituents, additivity breaks down because the encounter limit is reached. When an electron-supplying and an electron-withdrawing group are present, the reactivity is greater than predicted by the effects of the individual substituents. This arises because the demand for resonance stabilization of the transition states is increased by the electron-withdrawing substituent, so necessitating greater electron release from the electron-supplying substituent (polarizability effect). The magnitude of this effect is position-dependent, as shown by the ratios of observed: calculated partial rate factors for nitration in sulphuric acid (**72–74**).[297]

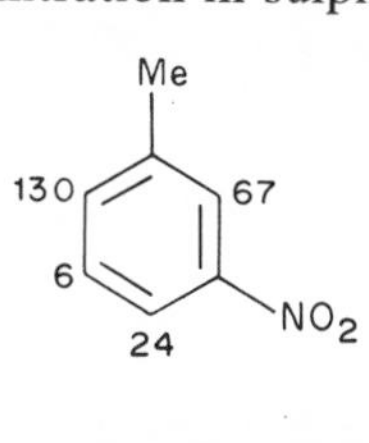

Me; 130; 67; 6; 24; NO_2

(72)

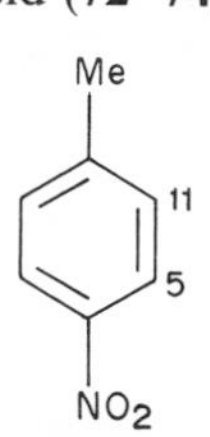

Me; 11; 5; NO_2

(73)

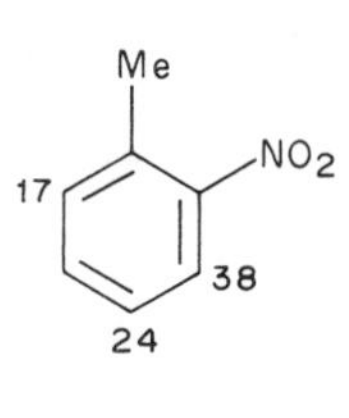

Me; NO_2; 17; 38; 24

(74)

Table 7.15 Isomer distributions (%) and partial rate factors for aromatics (PhX) containing dipolar substituents

X	*o*	*m*	*p*	f_o	f_m	f_p	Reagent	*T*/°C	Ref.
NO_2	6.4	93.2	0.3	—	—	—	HNO_3	0	241
	6.1	91.8	2.1	1.08×10^{-8}	16.2×10^{-8}	7.26×10^{-9}	HNO_3–H_2SO_4	25	287
	10.5	84.7	4.8	—	—	—	HNO_3–CF_3CO_2H	75	288
CN	13.8	85.0	1.5	—	—	—	HNO_3–oleum	0	289
	16.8	80.8	1.95	—	—	—	HNO_3	0	289
	25	70	4.4	—	—	—	HNO_3–CF_3CO_2H	75	288
CO_2H	18.5	80.2	1.3	—	—	—	HNO_3	0	241
CHO	—	90.8	—	—	—	—	HNO_3–oleum	−8	290
	—	72	—	—	—	—	HNO_3	−8	290
	32	63	4.3	—	—	—	HNO_3–CF_3CO_2H	75	288
COMe	—	90	—	—	—	—	HNO_3–oleum	−8	290
	19.5	78.5	2.0	1.02×10^{-7}	2.77×10^{-7}	1.55×10^{-8}	HNO_3–98.1% H_2SO_4	25	291
	26.4	71.6	2.0				HNO_3–80.3% H_2SO_4	25	291
COBu-*t*	30	44	26	—	—	—	HNO_3	25	292
$CONH_2$		70	⩽3	—	—	—	HNO_3	−15	293
CO_2Et	28.3	68.4	3.3	—	—	—	HNO_3	30	241
	24.1	72.0	4.0	2.6×10^{-3}	7.9×10^{-3}	0.9×10^{-3}	$AcONO_2$	18	294
SOMe	3.6	82.7	13.7	—	—	—	HNO_3–99.8% H_2SO_4	25	295
	3.1	52.9	44.0	—	—	—	HNO_3–83.5% H_2SO_4	25	295
SO_2Me	5.1	94.0	<2	2.0×10^{-9}	3.2×10^{-8}	—	HNO_3–83.9% H_2SO_4	25	296
SO_2^-	34.1	56.3	9.6	2.3×10^{-6}	3.8×10^{-6}	1.3×10^{-6}	HNO_3–83.9% H_2SO_4	25	296
SO_2Et	8.1	88.6	3.3	0.9×10^{-3}	9.3×10^{-3}	0.7×10^{-3}	HNO_3–Ac_2O	25	275

In **73** the factor *ortho* to methyl is greater than that *meta* to it because the conjugative effect of methyl is relayed to the former. A comparable factor is obtained at the 5-position of **72**, but the factors are substantially larger at the other positions because these are all deactivated much more by the conjugative ($-M$) effect of nitro. This in turn produces much greater electron release at these positions of **72** which are *ortho,para* to methyl. The factors in **74** would be comparable to those in **73** were it not for the intervention of a second factor, namely the forcing of the nitro group out of the plane of the aromatic ring thereby reducing deactivation by its $-M$ effect, so increasing the overall reactivity. This factor also accounts for the higher reactivity of *o*-chloronitrobenzene relative to nitrobenzene[63]). The relative reactivities of **72**–**74** are therefore 1.0:1.1:2.2. Similar results were obtained for nitration by nitronium salts, which also showed that in 1,3-dinitrobenzenes, methyl activated the *ortho* and *para* positions 468- and 2705-fold, respectively. This high *para*:*ortho* ratio was attributed to the extreme conjugative electron release from methyl,[132] which is also evident in the relative reactivities of the ions **75**. [298] In 82 and 98 wt-% H_2SO_4 the methyl compound (R′ = Me) is 2600 and 3500 times as reactive, respectively, as the unsubstituted compound, the corresponding *tert*-butyl values (R′ = *t*-Bu) being 288 and 207 (cf. the activation by *ortho*-alkyls given in Table 7.1). The effect of increased conjugative electron release from substituents under conditions of high electron demand is also evident in the relative rates of nitration of $4\text{-}NO_2C_6H_4X$ and C_6H_5X compounds, where $X = $ Me, F, Cl, Br.[299]

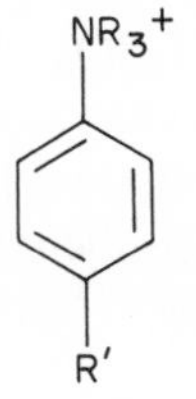

NR_3^+ ... R′ **(75)** NR_3^+ ... OMe **(76)** NR_3^+ ... Me Me **(77)**

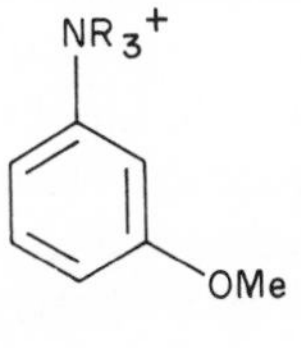

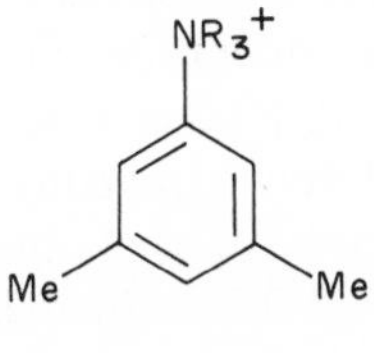

In the nitration of methyl- and methoxy-substituted anilinium ions, e.g. **76** and **77**, considerably greater substitution occurs *ortho* to the pole than in the unsubstituted ions.[300] In the latter the differential *ortho,para* reactivity was ascribed to hyperconjugative release of electrons from the NR_3^+ group, but there is less demand for this in the substituted compounds, hence the *ortho*:*para* ratio becomes reduced.

The departure from additivity due to the reduced deactivation by $-M$ groups being twisted out of the plane of the aromatic ring will in particular reduce the deactivation of positions *ortho* and *para* to the group. A typical example of the result of this may be seen in nitration of 2-nitro-1,4-dialkylbenzenes which give much more substitution *ortho*:*para* to nitro than in nitrobenzene itself.[301]

The departure from additivity arising from the presence of a number of deactivating groups is exemplified by the results for the polyhaloaromatics, given in Section 7.4.3. (ix).

(xv) Biphenyl and derivatives

Partial rate factors for nitration of biphenyl and its mononitro derivatives are given in Table 7.16. The results for biphenyl parallel those in hydrogen exchange (Section 3.1.2.10). In the nitrobiphenyls the second nitro group enters the unsubstituted ring, and the deactivation of one ring by a nitro group is least when the group is in the 3-position, showing that its deactivating influence from the 2- and 4-positions is partly conjugative. The 4-nitro group reduces the reactivity of the 4′-position by a factor of only about 30, whereas in nitrobenzene the 4-position is deactivated 10^7-fold greater (Table 7.15). In linear free-energy terms this represents a ca fourfold reduction in transmission of substituent effects between the rings, as found in other reactions (see Section 3.1.2.10).

When an activating substituent is present, nitration occurs in both rings if the substituent is in the 4-position, and mainly in the substituted ring if the substituent is in the 2- or 3-position.[303] The proportions of substitution in each ring can vary markedly according to the acidity of the medium for substituents that can either protonate or hydrogen bond, e.g. NHAc.[304]

Biphenyl gives *ortho*:*para* ratios that depend markedly on the medium, the ratios obtained with nitric acid–acetic anhydride (e.g. 2.13)[302] being higher than those obtained under other conditons (e.g. 0.6 in HNO_3–H_2SO_4),[305] higher than theoretical, or any other reaction including hydrogen exchange (which always gives results closely paralleling theoretical prediction). Similar variations in ratio with conditions have been obtained with *p*-terphenyl,[306] 4-phenylpyimidine,[307] 2,2′-bithienyl,[308] 2,3′-bithienyl,[308] *trans*-2-styrylthiophene,[309] and 1-methyl-4-phenylpyrazole.[310]

For biphenyl, earlier work indicated that the low ratios were due to heterogeneous nitration,[311] but this was subsequently disproved;[101] in the absence of nitrosation the ratio of 0.6 is raised to 1.4.[43] High ratios are also obtained with N_2O_5–MeCN (2.8, −20 °C), and $NO_2^+BF_4^-$–sulpholane (2.0, 10 °C), and increase with decreasing temperature as does that obtained with

Table 7.16 Partial rate factors for nitration of biphenyl and nitrobiphenyls

Compound	f_2	f_3	f_4	Conditions	Ref.
Biphenyl	1.8	0.18	5.04	$MeO^+(H)NO_2$, 37.5 °C	163a, b
Biphenyl	41	<0.6	38	HNO_3–Ac_2O, 0 °C	302
2′-Nitrobiphenyl	0.28	0.03	1.25	HNO_3–Ac_2O, 0 °C	302
3′-Nitrobiphenyl	1.4	0	3.4	HNO_3–Ac_2O, 0 °C	302
4′-Nitrobiphenyl	0.35	<0.01	1.3	HNO_3–Ac_2O, 0 °C	302

HNO_3–Ac_2O (2.8, –40 °C);[130] this pattern with the latter reagent is also found with *p*-terphenyl.[306] The high ratios are obtained under the same conditions that give high ratios in the nitration of anisole, etc., indicating that one ring nucleophilically displaces the counter ion from the nitrating species, subsequent migration favouring *ortho* substitution (**78**).[101] Alternatively, *ipso* substitution may occur,[254] though a report of a ratio of 3.3 in nitration by HNO_3–45% H_2SO_4 is said to discount this since reaction of the intermediate with the solvent should be very rapid,[312] as shown by results for nitration of *o*-xylene [Section 7.4.2(iii)(3)]. Nitrations in these weak sulphuric acid media are also complicated by heterogeneity and dinitration,[43] and yields can be considerably less than quantitative.[312] The cause of these anomalously high ratios in nitration of biphenyl has yet to be fully resolved.

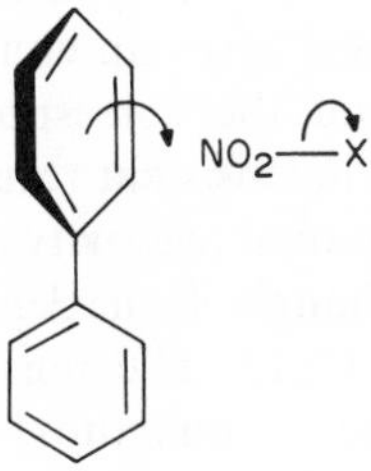

NO2—X

18.6 30 22.8

163 125 Ph Ph

(**78**) 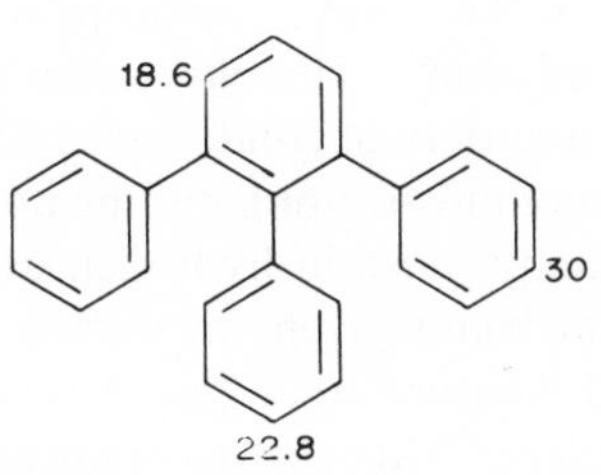 (**79**) 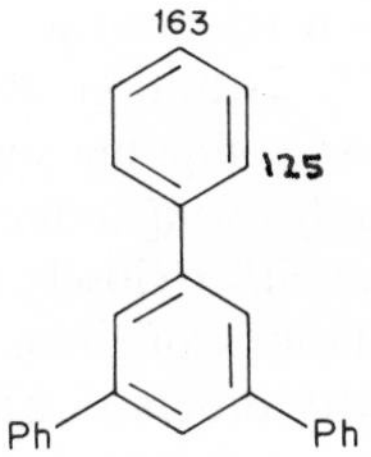 (**80**)

The positional reactivity order for nitration of *p*-terphenyl is 4 > 2 > 2′ (≫ 3), as expected;[306] likewise, quaterphenyl nitrates in the 4-position.[313] Isomer yields in the nitration of pentafluorobiphenyl by fuming nitric acid–sulphuric acid, viz. *ortho* (25), *meta* (18), *para* (57%), are unaffected by nitrosation;[314] fuming nitric acid gives less *meta* derivative.[315] Only acetoxylation occurs if nitric acid–acetic anhydride is used.[314] Published partial rate factors for nitration of fluorene, viz. f_2 2040, f_3 60, f_4 944, are greater than for biphenyl (as in hydrogen exchange, Section 3.1.2.10), but do not correspond with the given rate data.[224b] Nitration of some substituted fluorenes takes place in the expected positions.[316] Partial rate factors for nitration of 1,2,3-triphenylbenzene by fuming HNO_3–Ac_2O (**79**)[317] show the effect of twisting of the adjacent phenyl groups further out of coplanarity than in biphenyl. Those for 1,3,5-triphenylbenzene (**80**)[318] are anomalous since the *meta* phenyl groups in one ring should decrease the reactivity in the other (cf. Tables 7.16 and 3.15); pentaphenylbenzene nitrates in the central ring (50%).[318] The nitration positions in *o*-, *m*-, and *p*-terphenyl[318] and biphenylene[319] are as predicted by hydrogen exchange (Table 3.15).

(xvi) Polycyclic aromatics

Partial rate factors for nitration of some polycyclics by HNO_3–Ac_2O at 0 °C were determined as follows (positions in parentheses): naphthalene, 470(1), 50(2);

phenanthrene, 490(9), 360(1), 300(3), 92(2), 79(4); triphenylene, 600(1,2); chrysene, 3500 (6); pyrene, 17 000 (1); perylene, 77 000 (3); benzo[*a*]pyrene, 108 000 (6); coronene, 1150 (1); anthanthrene, 156 000 (6).[224b] The accuracy of some of these values is doubtful (that for triphenylene, for example, is clearly wrong, cf. Table 3.17 and ref. 320), and some may be affected by encounter control; redetermination of them might be appropriate. For example, nitration of perylene with dilute nitric acid in dioxane has been shown to give 3-nitro- *and* 1-nitroperylene in 56 and 24% yields, respectively.[321] Radical-cation mechanisms are not significant for compounds of reactivity at least as high as that of pyrene.[320] Partial rate factors for fluoranthene (calculated from the data in ref. 322) are 330 (1), 1365 (3), 564 (7) and 846 (8).

Nitration of naphthalene has been studied in depth, and the α:β reactivity ratio was shown to vary widely with the conditions,[323] probably owing to nitration via nitrosation since the presence of nitrous acid gives an enhanced ratio.[324] Nitration of 1- and 2-methylnaphthalenes and the corresponding methoxy compounds under a variety of conditions gives inconsistent results[323] (probably owing to incursion of nitrosation), but the positional reactivity orders parallel either closely or exactly those in hydrogen exchange (Table 3.22), the explanations of these orders being given in Section 3.1.2.13. The positional reactivity orders in 2,5- and 2,6-dimethyl-, 2,6-, 2,7-, and 2,3-dimethoxy-, and 6-methoxy-2-methylnaphthalene[325] are each as predicted from the known effects of the individual substituents. Partial rate factors have been determined for nitration of 1,2-, 1,3-, 1,4-, 1,5-, 1,8-, 2,3-, 2,6-, and 2,7-dimethylnaphthalenes by HNO_3–Ac_2O at 0 °C.[326] There are some inconsistencies within the data, attributable to steric hindrance and possibly encounter control, but the positional reactivity orders are for the most part as predicted by the data for hydrogen exchange (Table 3.22); the results confirm the fairly strong 2,6-conjugative interaction and the very poor 2,3-interaction (cf. Section 3.1.2.13). Each of 1- and 2-acetylaminonaphthalenes nitrate at the expected position, the ratio of 2-:4-substitution in the former being relatively unaffected by 5- or 7-nitro groups, though decreased by an 8-nitro group evidently owing to steric hindrance and by a 6-nitro group because of the strong conjugation between the 2- and 6-positions.[327] Nitration of 1- and 2-nitronaphthalenes takes place at each of the α-positions of the unsubstituted rings.[328] Dealkylation followed by realkylation occurs in nitration of 1,8-di-*tert*-butylnaphthalene so the main product is 4-nitro-1,3,6,8-tetra-*tert*-butylnaphthalene.[329] With $NO_2^+BF_4^-$, heptafluoro-1*H*-naphthalene nitrates at the 1-position.[330]

Nitration of 1,8-naphthylene disulphide (**81**) in the 2- and 4-positions[331] and 5-nitration of benzo[*c*]fluorene (**82**)[332] are both expected results. Benz[*a*]-anthracene nitrates in the 5-position,[333] as do benzo[*c*]phenanthrene,[334] and hexahelicene,[335] each result paralleling that in hydrogen exchange (Table 3.17). For hexahelicene a second isomer was assumed to be the 8-nitro compound, but it is more probably the 7-isomer (cf. Table 3.17). Nitration of 3-nitro-

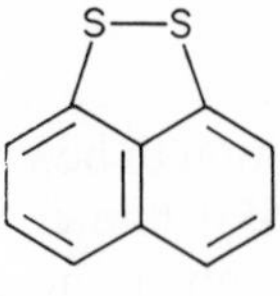

(81) 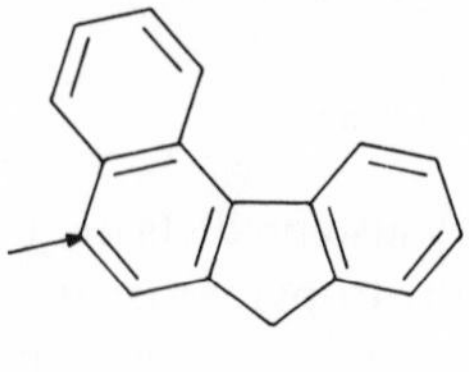 **(82)** 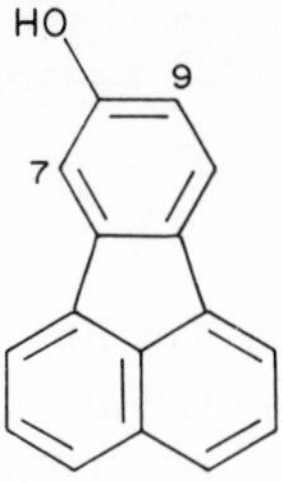 **(83)** 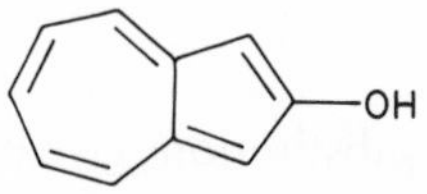 **(84)**

fluoranthene gives 3,9-dinitrofluoranthene,[336] consistent with the intrinsic reactivity of fluoranthene (Table 3.17) and conjugative deactivation of the 8-position by nitro. The unidentified isomer from nitration of 2-nitrofluoranthene should be the 2,9-dinitro derivative. A 2-acetylamino substituent directed nitration into the 3-position,[336] and 3-methylfluoranthene gives mainly 8- together with some 2- and 4-nitration.[337] All these are expected, the lack of 9-substitution of the 2-acetamido compound following from the poor inter-ring conjugation in fluoranthene due to strain [Section 3.1.2.13(iii)]. Anomalous, however, is the 70% of 7-nitration of 8-hydroxyfluoranthene (**83**) with HNO_3–HOAc (although 9-nitration occurred in the presence of acetone).[338] The high intrinsic reactivity of the 9-position (Table 3.17) coupled with the high 8,9-bond order arising from bond fixation (to avoid placing double bonds in the five-membered ring) should produce exclusive 9-nitration. Partial rate factors for nitration of acenaphthene ($f_2 = 16000$, $f_4 = 22200$)[326] are much greater than in naphthalene (as in hydrogen exchange), but this is an electronic effect rather than due to relief of steric hindrance (cf. ref. 326).

(xvii) Miscellaneous aromatics

Nitration of 2-hydroxyazulene (**84**) gives a mixture of 1-nitroazulen-2-one and 6-nitro-2-hydroxyazulene, both 1- and 6-positions being conjugated with hydroxy.[339] Nitration of porphyrin goes first into the *meso* α-position and then, it is thought, into the β-position rather than the γ-position.[340] Both of the latter are conjugated with the α-position but, as in benzene, conjugation between more remote sites is greater, so α-nitro deactivates the γ-position more strongly. Metal derivatives of octaalkylporphyrins undergo nitration more readily than the parent compound.[341] Nitration of octaethylporphyrin with $Zn(NO_3)_2 \cdot 6H_2O$ in Ac_2O yields mono-, di-, tri-, and tetranitro derivatives, with here a preference for α,γ-substitution.[342] Nitration of triphenylphosphonium cyclopentadienylide (**85**) gives the 2-derivative,[343] showing the relative unimportance of steric hindrance in five-membered rings. Nitration of diazocyclopentadiene (**86**) gives the 2- and 3-isomers in a 2:1 ratio.[344] Nitration of azupyrene (structure **66**, Chapter 5) is anomalous is giving a 1-:4-substitution ratio of 0.03 whereas other

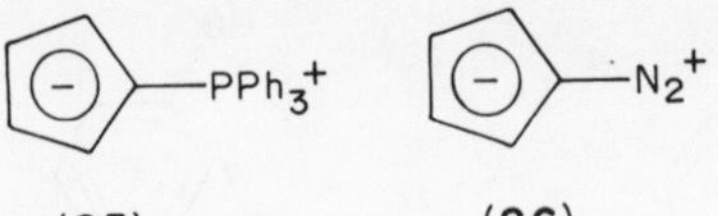

(85) (86)

substitutions give ratios of ca 10;[345] theory also predicts the 1-position to be the most reactive. Hexamethyl-*trans*-15,16-dihydropyrene (structure **65**, Chapter 5) nitrates in both 2- and 4-positions, whereas acylation takes place only in the 2-position;[346] this may reflect differences in steric requirements of the two reactions.

7.5 NITRAMINE REARRANGEMENT

Some *N*-nitramines rearrange in acidic media to give *o*- (mainly) and *p*-nitroanilines (eq. 7.35). The isomer distribution is entirely different from that obtained by separate nitration of the amines,[347] and the presence of $K^{15}NO_3$ does not result in ^{15}N enrichment in the nitro products.[348] This, coupled with the exclusive formation of *ortho* and *para* products, led to the proposal of an intramolecular 'cartwheel' mechanism, involving migration of the nitro group in a six-centre process.[348] However, this required improbable initial rearrangement of the bonding between the amino and nitro groups.

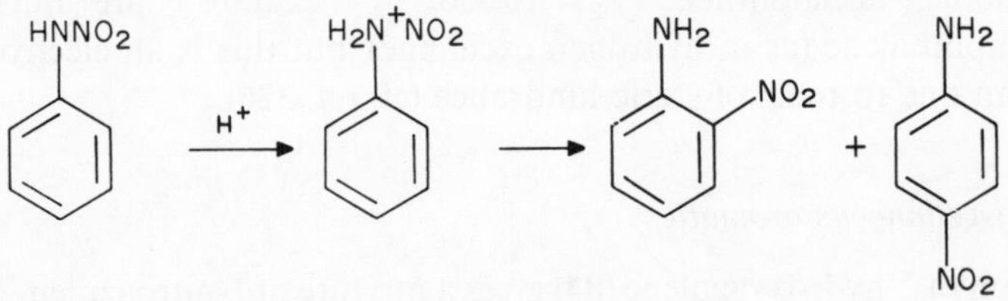

HNNO2 —(H+)→ H2N+NO2 → NH2 / NO2 (ortho) + NH2 / NO2 (para) (7.35)

In a more recent study of concurrent rearrangement of *N*-methyl-*N*-nitroaniline and 4-fluoro-*N*-methyl-*N*-nitro[^{15}N]aniline, *N*-methyl-4-nitroaniline containing excess ^{15}N was obtained, but this could be prevented by the addition of hydroquinone.[349] Thus the mechanism appears partially to involve an intermolecular radical process in this case; rate-determining fission to a radical–radical cation pair within a solvent cage was proposed. Intramolecular migration of the NO_2 radical within the cage (1,3-shifts) then gives the reaction products; if some radicals diffuse out of and back into the cage, then a small amount of crossing of the ^{15}N label can occur. The intermediacy of radical cations has been confirmed by CIDNP studies,[350] and is consistent with the ρ-factor of -3.7 obtained in a Hammett correlation with σ^+ values.[351] If the *ortho* position in a molecule is blocked, then exclusive substitution at the *para* position occurs, as in rearrangement *N*-methyl-*N*-nitro-9-aminoanthracene to 10-nitroanthrone.[352]

'Normal' nitration of amines may in some cases proceed via *N*-nitration followed by heterolysis of the N—N bond and ring nitration by the nitronium ion. The extent to which heterolysis rather than homolysis occurs appears to be dependent on the reactivity of the amine.[353]

7.6 PHOSPHONATION

At 600 °C methyl but-2-enylphosphonate decomposes to butadiene and monomeric metaphosphate, $MeOPO_2$. The latter will react at −60 °C with *N*,*N*-diethylaniline to give 4-*N*,*N*-diethylaminobenzenephosphonate (eq. 7.36).[354]

$$MeOPO_2 + C_6H_5NEt_2 \longrightarrow MeO(O)(O^-)P{-}C_6H_4{-}NHEt_2^+ \quad (7.36)$$

REFERENCES

1. R. A. Abramovitch, S. R. Challand, and E. F. V. Scriven, *J. Org. Chem.*, 1972, **37**, 2705.
2. P. Kovacic, R. L. Russell, and R. P. Bennett, *J. Am. Chem. Soc.*, 1964, **86**, 1588.
3. H. Takeuchi, K. Takano, and K. Koyama, *J. Chem. Soc., Chem. Commun.*, 1982, 1254; H. Takeuchi and K. Takano, *J. Chem. Soc., Chem. Commun.*, 1983, 447.
4. K. Nakamura, A. Ohno, and S. Oka, *Synthesis*, 1974, 882.
5. H. Takeuchi and E. Masturbara, *J. Chem. Soc., Perkin Trans. 1*, 1984, 981.
6. H. Bock and K.-L. Kampa, *Angew. Chem., Int. Ed. Engl.*, 1965, **4**, 783.
7. G. Sosnovsky and D. J. Rawlinson, *Adv. Free-Radical Chem.*, 1972, **4**, 203; F. Minisci, *Synthesis*, 1973, 1; *Top Curr. Chem.*, 1976, **62**, 1.
8. P. Kovacic, R. M. Lange, J. L. Foote, C. T. Goralski, J. J. Hiller, and J. A. Levisky, *J. Am. Chem. Soc.*, 1964, **86**, 1650; P. Kovacic and J. A. Levisky, *J. Am. Chem. Soc.*, 1966, **88**, 1000; J. W. Strand and P. Kovacic, *J. Am. Chem. Soc.*, 1973, **95**, 2977; P. Kovacic and A. K. Harrison, *J. Org. Chem.*, 1967, **32**, 207; P. Kovacic, K. W. Field, P. D. Roskos, and F. U. Scalzi, *J. Org. Chem.*, 1967, **32**, 585; P. Kovacic and R. J. Hopper, *Tetrahedron*, 1967, **23**, 3965, 3977.
9. P. Kovacic and J. L. Foote, *J. Am. Chem. Soc.*, 1961, **83**, 743.
10. K. Shudo, T. Ohta, and T. Okamoto, *J. Am. Chem. Soc.*, 1981, **103**, 645.
11. F. W. Wassmundt and S. J. Padegimas, *J. Am. Chem. Soc.*, 1967, **89**, 7131; J. March and J. S. Engenito, *J. Org. Chem.*, 1981, **46**, 4304.
12. Y. Kikugawa and M. Kawasi, *J. Am. Chem. Soc.*, 1984, **106**, 5729.
13. P. Kovacic, R. P. Bennett, and J. L. Foote, *J. Am. Chem. Soc.*, 1962, **84**, 759.
14. H. Zollinger, *Azo and Diazo Chemistry*, Interscience, New York, 1961, pp. 210–265; R. Taylor, *Comprehensive Chemical Kinetics*, Vol. 13, Elsevier, Amsterdam, 1972,

pp. 50–54; I. Szele and H. Zollinger, *Top. Curr. Chem.*, 1983, **112**, 1; K. H. Saunders and R. L. M. Allen, *Aromatic Diazo Compounds*, Arnold, London, 1985.
15. J. B. Conant and W. D. Peterson, *J. Am. Chem. Soc.*, 1930, **52**, 1220.
16. C. R. Hauser and D. S. Breslow, *J. Am. Chem. Soc.*, 1941, **63**, 418.
17. R. Wistar and P. D. Bartlett, *J. Am. Chem. Soc.*, 1941, **63**, 413.
18. L. P. Hammett, *Physical Organic Chemistry*, McGaw-Hilll, New York, 1940, p. 314.
19. K. H. Meyer and H. Tochtermann, *Chem. Ber.*, 1921, **54**, 2283.
20. V. Baranek and M. Vecera, *Collect. Czech. Chem. Commun.*, 1970, **35**, 3402.
21. I. Dobas, J. Panchartek, V. Sterba, and M. Vecera, *Collect. Czech. Chem. Commun.*, 1970, **35**, 1288.
22. H. Kropacova, J. Panchartek, V. Sterba, and K. Valter, *Collect. Czech. Chem. Commun.*, 1970, **35**, 3287.
23. (a) H. Zollinger, *Helv. Chim. Acta*, 1953, **36**, 1730; (b) J. Kavalek, J. Panchartek, and V. Sterba, *Collect. Czech. Chem. Commun.*, 1970, **35**, 3470.
24. V. Machacek, J. Panchartek, V. Sterba, and M. Vecera, *Collect. Czech. Chem. Commun.*, 1970, **35**, 844.
25. O. Machackova, V. Sterba, and K. Valter, *Collect. Czech. Chem. Commun.*, 1972, **37**, 1851.
26. V. Sterba and K. Valter, *Collect. Czech. Chem. Commun.*, 1972, **37**, 1327.
27. I. Dobas, V. Sterba, and M. Vecera, *Chem. Ind.* (*London*), 1968, 1814.
28. N. N. Bubnov, K. A. Bilevitch, L. A. Poljakova, and O. Yu. Okhlobytsin, *J. Chem. Soc., Chem. Commun.*, 1972, 1058.
29. S. Koller and H. Zollinger, *Helv. Chim. Acta*, 1970, **53**, 78; J. R. Penton and H. Zollinger, *Helv. Chim. Acta*, 1971, **54**, 573.
30. H. Zollinger, *Helv. Chim. Acta*, 1955, **38**, 1597, 1617, 1623; H. F. Hodson, O. A. Stamm, and H. Zollinger, *Helv. Chim. Acta*, 1958, **41**, 1816.
31. R. Ernst, O. A. Stamm, and H. Zollinger, *Helv. Chim. Acta*, 1958, **41**, 2274.
32. S. B. Hanna, C. Jermini, and H. Zollinger, *Tetrahedron Lett.*, 1969, 4415.
33. H. Zollinger, in ref. 14, pp. 182–187.
34. (a) R. P. Kelly, J. R. Penton, and H. Zollinger, *Helv. Chim. Acta*, 1982, **65**, 122; Y. Ogata, Y. Nakagawa, and M. Inaishi, *Bull. Chem. Soc. Jpn.*, 1981, **54**, 2853. (b) T. K. Miller, J. T. Sharp, H. R. Sood, and E. Stefaniuk, *J. Chem. Soc., Perkin Trans. 2*, 1884, 823.
35. D. L. H. Williams, *Nitrosation*, Cambridge University Press, Cambridge, 1988.
36. B. C. Challis, R. J. Higgins, and A. J. Lawson, *J. Chem. Soc., Perkin Trans. 2*, 1973, 1831.
37. E. L. Blackhall, E. D. Hughes, and C. K. Ingold, *J. Chem. Soc.*, 1952, 28.
38. J. H. Ridd, *Q. Rev. Chem. Soc.*, 1961, **15**, 418; B. C. Challis and J. H. Ridd, *J. Chem. Soc.*, 1962, 5197, 5208; B. C. Challis, L. F. Larkesworthy, and J. H. Ridd, *J. Chem. Soc.*, 1962, 5203.
39. B. C. Challis and R. J. Higgins, *J. Chem. Soc., Perkin Trans. 2*, 1973, 1597.
40. B. C. Challis, R. J. Higgins, and A. J. Lawson, *J. Chem. Soc., Chem. Commun.*, 1970, 1223; L. R. Dix and R. B. Moodie, *J. Chem. Soc., Perkin Trans. 2*, 1986, 1097.
41. V. I. Minkin, R. M. Minyaev, I. A. Yudilevich, and M. E. Kletskii, *J. Org. Chem. USSR*, 1985, **21**, 842.
42. B. C. Challis and R. J. Higgins, *J. Chem. Soc., Perkin Trans.* 2, 1972, 2365.
43. R. Taylor, *Tetrahedron Lett.*, 1972, 1755.
44. A. Castro, E. Iglesias, J. R. Leis, M. Mosquera, and M. E. Pena, *Bull. Soc. Chim. Fr.*, 1987, 83.
45. D. L. H. Williams, in *The Chemistry of Functional Groups, Supplement F*, Ed. S. Patai, Wiley, New York, 1982, pp. 127–153.

46. E. Yu. Belyaev and T. I. Nikulicheva, *Org. React. USSR*, 1971, **7**, 165; T. D. B. Morgan, D. L. H. Williams, and J. A. Wilson, *J. Chem. Soc., Perkin Trans. 2*, 1973, 473; D. L. H. Williams and J. A. Wilson, *J. Chem. Soc., Perkin Trans.2*, 1974, 13; I. D. Biggs and D. L. H. Williams, *J. Chem. Soc., Perkin Trans.2*, 1975, 107; 1976, 601; D. L. H. Williams, *Int. J. Chem. Kinet.*, 1975, **7**, 215; *Tetrahedron*, 1975, **31**, 1343; *J. Chem. Soc., Perkin Trans. 2*, 1982, 801.
47. T. I. Aslapovskaya, E. Yu. Belyaev, V. P. Kumarev, and B. A. Porai-Koshits, *Org. React. USSR*, 1968, **5**, 189; T. D. B. Morgan and D. L. H. Williams, *J. Chem. Soc., Perkin Trans. 2*, 1972, 74.
48. C. K. Ingold, *Structure and Mechanism in Organic Chemistry*, 2nd edn., Bell, London, 1969, Ch. 6.
49. P. B. D. de la Mare and J. H. Ridd, *Aromatic Substitution, Nitration and Halogenation*, Butterworths, London, 1959, pp. 57–77.
50. R. Taylor, *Comprehensive Chemical Kinetics*, Vol. 13, Elsevier, Amsterdam, 1972, pp. 10–47.
51. (a) J. G. Hoggett, R. B. Moodie, J. R. Penton, and K. Schofield, *Nitration and Aromatic Reactivity*, Cambridge University Press, Cambridge, 1971; (b) K. Schofield, *Aromatic Nitration*, Cambridge University Press, Cambridge, 1980.
52. G. A. Olah and S. J. Kuhn, *Friedel–Crafts and Related Reactions*, Wiley, New York, 1964, Ch. 44.
53. R. J. Gillespie, E. D. Hughes, and C. K. Ingold, *J. Chem. Soc.*, 1950, 2552.
54. W. H. Lee and D. J. Millen, *J. Chem. Soc.*, 1956, 4463.
55. C. K. Ingold, and D. J. Millen, *J. Chem. Soc.*, 1950, 2612.
56. B. Susz and E. Briner, *Helv. Chim. Acta*, 1935, **18**, 378.
57. C. K. Ingold, D. J. Millen, and H. G. Poole, *Nature* (*London*), 1946, **158**, 480; *J. Chem. Soc.*, 1950, 2576.
58. D. J. Millen *J. Chem. Soc.*, 1950, 2606.
59. C. A. Bunton and E. A. Halevi, *J. Chem. Soc.*, 1952, 4917.
60. E. D. Hughes, C. K. Ingold, and R. I. Reed, *J. Chem. Soc.*, 1950, 2400; ref. 49, p. 73.
61. R. J. Gillespie, *J. Chem. Soc.*, 1950, 2493.
62. L. Medard, *C. R. Acad. Sci.*, 1934, **199**, 1615; J. Chédin, *C. R. Acad. Sci.*, 1935, **200**, 1397; 1936, **202**, 220; *Ann. Chim.* (*Paris*), 1937, **8**, 243; *Mem. Serv. Chim. État.*, 1944, **31**, 113.
63. F. H. Westheimer and M. S. Kharasch, *J. Am. Chem. Soc.*, 1946, **68**, 1871.
64. A. M. Lowen, M. A. Murray, and G. Williams, *J. Chem. Soc.*, 1950, 3318; N. C. Deno and R. Stein, *J. Am. Chem. Soc.*, 1956, **78**, 578.
65. H. Martinsen, *Z. Phys. Chem.*, 1905, **50**, 385; 1907, **59**, 605.
66. (a) G. M. Bennett, J. C. D. Brand, D. M. James, T. G. Saunders, and G. Williams, *J. Chem. Soc.*, 1947, 474; (b) T. G. Bonner, F. Bowyer, and G. Williams, *J. Chem. Soc.*, 1953, 2650; (c) B. Surfleet and P. A. H. Wyatt, *J. Chem. Soc.*, 1965, 6524; (d) M. A. Akand and P. A. H. Wyatt, *J. Chem. Soc. B*, 1967, 1326; (e) R. B. Moodie, J. R. Penton, and K. Schofield, *J. Chem. Soc. B*, 1969, 578; (f) R. G. Coombes, D. H. G. Crout, J. G. Hoggett, R. B. Moodie, and K. Schofield, *J. Chem. Soc. B*, 1970, 347; (g) M. I. Vinnik, Zh. E. Grabovaskaya, and L. N. Arzamaskova, *Russ. J. Phys. Chem.*, 1967, **41**, 580; N. C. Marziano, M. Sampoli, F. Pinna, and A. Passerini, *J. Chem. Soc., Perkin Trans. 2*, 1984, 1163.
67. R. B. Moodie, K. Schofield and T. Yoshida, *J. Chem. Soc., Perkin Trans. 2*, 1975, 788.
68. Ref. 49, pp. 63–64.
69. R. J. Gillespie and D. J. Millen, *Q. Rev. Chem. Soc.*, 1948, **2**, 277.
70. R. J. Gillespie and D. G. Norton, *J. Chem. Soc.*, 1953, 971.
71. T. G. Bonner, F. Bowyer, and G. Williams, *J. Chem. Soc.*, 1952, 3274.

72. M. R. Draper and J. H. Ridd, *J. Chem. Soc., Perkin Trans. 2*, 1981, 94.
73. R. B. Moodie, K. Schofield, and P. G. Taylor, *J. Chem. Soc., Perkin Trans. 2*, 1979, 133.
74. R. B. Moodie, K. Schofield, and P. N. Thomas, *J. Chem. Soc., Perkin Trans. 2*, 1978, 318.
75. H. W. Gibbs, L. Main, R. B. Moodie, and K. Schofield, *J. Chem. Soc., Perkin Trans. 2*, 1981, 848.
76. C. L. Coon, W. G. Blucher, and M. E. Hill, *J. Org. Chem.*, 1973, **38**, 4243.
77. R. G. Coombes, R. B. Moodie, and K. Schofield, *J. Chem. Soc. B*, 1968, 800.
78. J. W. Barnett, R. B. Moodie, K. Schofield, P. G. Taylor, and J. B. Weston, *J. Chem. Soc., Perkin Trans. 2*, 1979, 747.
79. A. K. Manglik, R. B. Moodie, K. Schofield, E. Dedouglu, A. Dutly, and P. Rys, *J. Chem. Soc., Perkin Trans. 2*, 1981, 1358.
80. J. W. Barnett, R. B. Moodie, K. Schofield, and J. B. Weston, *J. Chem. Soc., Perkin Trans. 2*, 1975, 648.
81. R. B. Moodie, K. Schofield, and A. D. Wait, *J. Chem. Soc., Perkin Trans. 2*, 1984, 921.
82. Ref. 51b, pp. 105–111.
83. F. Bernardi and W. J. Herre, *J. Am. Chem. Soc.*, 1973, **95**, 3079.
84. J.W. Barnett, R. B. Moodie, K. Schofield, J. B. Weston, R. G. Coombes, J. G. Golding, and G. D. Tobin, *J. Chem. Soc., Perkin Trans. 2*, 1977, 248.
85. R. G. Coombes, J. G. Golding, and P. Hadjigeorgiou, *J. Chem. Soc., Perkin Trans. 2*, 1979, 1451.
86. R. B. Moodie, P. N. Thomas, and K. Schofield, *J. Chem. Soc., Perkin Trans. 2*, 1977, 1693.
87. D. S. Ross, K. F. Kuhlmann, and R. Malhotra, *J. Am. Chem. Soc.*, 1983, **105**, 4299.
88. R. B. Moodie and R. J. Stephens, *J. Chem. Soc., Perkin Trans. 2*, 1987, 1059.
89. H. C. Brown and R. A. Wirkkala, *J. Am. Chem. Soc.*, 1966, **88**, 1447.
90. V. A. Spitzer and R. Stewart, *J. Org. Chem.*, 1974, **39**, 3936.
91. R. B. Moodie, K. Schofield and G. D. Tobin, *J. Chem. Soc., Perkin Trans. 2*, 1977, 1688.
92. A. H. Clemens, J. H. Ridd, and J. P. B. Sandall, *J. Chem. Soc., Perkin Trans. 2*, 1985, 1227.
93. G. A. Benford and C. K. Ingold, *J. Chem. Soc.*, 1938, 929.
94. J. C. Hoggett, R. B. Moodie, and K. Schofield, *J. Chem. Soc. B*, 1969, 1.
95. D. W. Coilett and S. D. Hamann, *Trans. Faraday Soc.*, 1961, **57**, 2231.
96. T. G. Bonner, R. A. Hancock, F. R. Rolle, and G. Yousif, *J. Chem. Soc. B*, 1970, 314.
97. R. G. Coombes, *J. Chem. Soc. B*, 1969, 1256.
98. R. Vandoni and P. Viala, *Mém. Serv. Chim. Etat.*, 1945, **32**, 80.
99. O. Mantsch, N. Bodor, and F. Hodorsan, *Rev. Roum. Chim.*, 1968, **13**, 1435.
100. F. G. Bordwell and E. W. Garbisch, *J. Am. Chem. Soc.*, 1960, **82**, 3588.
101. R. Taylor, *J. Chem. Soc. B*, 1966, 727.
102. A. K. Sparks, *J. Org. Chem.*, 1966, **31**, 2299.
103. M. A. Paul, *J. Am. Chem. Soc.*, 1958, **80**, 5329.
104. S. R. Hartshorn, R. B. Moodie, and K. Schofield, *J. Chem. Soc. B*, 1971, 1256.
105. S. R. Hartshorn, R. B. Moodie, K. Schofield, and M. J. Thompson, *J. Chem. Soc. B*, 1971, 2447.
106. N. C. Marziano, R. Passerini, J. H. Rees, and J. H. Ridd, *J. Chem. Soc., Perkin Trans. 2*, 1977, 1361.
107. A. Fischer, A. J. Read, and J. Vaughan, *J. Chem. Soc.*, 1964, 3691; J. G. Hoggett, R. B. Moodie, and K. Schofield, *J. Chem. Soc., Chem. Commun.*, 1969, 605; S. R. Hartshorn,

J. G. Hoggett, R. B. Moodie, K. Schofield, and M. J. Thompson, *J. Chem. Soc. B*, 1971, 2461.
108. N. C. Marziano, J. H. Rees, and J. H. Ridd, *J. Chem. Soc., Perkin Trans. 2*, 1974, 600.
109. G. W. Gray and D. Lewis, *J. Chem. Soc.*, 1961, 5156.
110. S. C. Narang and M. J. Thompson, *Aust. J. Chem.*, 1975, **28**, 385.
111. R. Taylor, *J. Chem. Soc., Perkin Trans. 2* 1983, 89.
112. A. Fischer, A. J. Read, and J. Vaughan, *J. Chem. Soc.*, 1964, 3691.
113. V. Gold, E. D. Hughes, and C. K. Ingold, *J. Chem. Soc.*, 1950, 2467.
114. Ref. 49, p. 76.
115. R. O. C. Norman and G. K. Radda, *J. Chem. Soc.*, 1961, 3030.
116. M. E. Kurz, L. T. A. Yang, E. P. Zahora, and R. C. Adams, *J. Org. Chem.*, 1973, **38**, 2271.
117. T. G. Bonner, R. A. Hancock, G. Yousif, and F. R. Rolle, *J. Chem. Soc. B*, 1969, 1237.
118. G. Squadrito, D. F. Church, and W. A. Pryor, *J. Am. Chem. Soc.*, 1987, **109**, 6535.
119. L. A. Pinck, *J. Am. Chem. Soc.*, 1927, **49**, 2536; A. I. Titov and A. N. Banyshrikova, *Zh. Obshch. Khim.*, 1937, **6**, 1800; 1937, **7**, 667.
120. R. O. C. Norman, W. J. E. Parr, and C. B. Thomas, *J. Chem. Soc., Perkin Trans. 1*, 1974, 369.
121. V. Gold, E. D. Hughes, C. K. Ingold, and G. H. Williams, *J. Chem. Soc.*, 1950, 2452.
122. G. A. Olah and S. J. Kuhn, *J. Am. Chem. Soc.*, 1961, **83**, 4564.
123. G. A. Olah, S. J. Kuhn, and S. H. Flood, *J. Am. Chem. Soc.*, 1961, **83**, 4571, 4581.
124. G. A. Olah and S. J. Kuhn, *J. Am. Chem. Soc.*, 1962, **84**, 3684.
125. G. A. Olah, S. J. Kuhn, S. H. Flood, and J. C. Evans, *J. Am. Chem. Soc.*, 1962, **84**, 3687.
126. L. L. Ciaccio and R. A. Marcus, *J. Am. Chem. Soc.*, 1962, **84**, 1838.
127 W. S. Tolgyesi, *Can. J. Chem.*, 1965, **43**, 343.
128. G. A. Olah and N. A. Overchuk, *Can. J. Chem.*, 1965, **43**, 3273.
129. P. Kovacic and J. J. Hiller, *J. Org. Chem.*, 1965, **30**, 2871.
130. R. Taylor, *Tetrahedron Lett.*, 1966, 6093.
131. P. F. Christy, J. H. Ridd, and N. D. Stears, *J. Chem. Soc. B*, 1970, 797; J. H. Ridd, *Acc. Chem. Res.*, 1971, **4**, 248; A. Gustaminza and J. H. Ridd, *J. Chem. Soc., Perkin Trans. 2*, 1972, 813.
132. G. A. Olah and H. C. Lin, *J. Am. Chem. Soc.*, 1974, **96**, 549.
133. E. Hunziker, P. C. Myhre, J. R. Penton, and H. Zollinger, *Helv. Chem. Acta*, 1975, **58**, 230.
134. S. B. Hanna, E. Hunziker, T. Saito, and H. Zollinger, *Helv. Chim. Acta*, 1969, **52**, 1537; E. Hunziker, J. R. Penton, and H. Zollinger, *Helv. Chim. Acta*, 1971, **54**, 2043.
135. S. V. Naidenov, Yu. V. Gak, and E. L. Golod, *J. Org. Chem. USSR*, 1982, **18**, 1731.
136. G. A. Olah, *Industrial and Laboratory Nitrations*, ACS Symposium Series, No. 22, American Chemical Society, Washington, DC, 1976.
137. C. A. Cupas and R. L. Pearson, *J. Am. Chem. Soc.*, 1968, **90**, 4742.
138. G. A. Olah, T. Yamamoto, T. Hashimoto, J. G. Shin, N. Trivedi, B. P. Singh, M. Piteau, and J. A. Olah, *J. Am. Chem. Soc.*, 1987, **109**, 3708.
139. C. L. Coon, W. G. Blucher, and M. E. Hill, *J. Org. Chem.*, 1973, **38**, 4203; M. Schmeiser, P. Sartori, B. Lippsmeier, *Z. Naturforsch., Teil B*, 1973, **28**, 573; L. M. Yagulpolskii, I. I. Maletina, V. V. Orda, *J. Org. Chem. USSR*, 1974, **10**, 2240; S. K. Yarbo and R. E. Noftle, *J. Fluorine Chem.*, 1975, **6**, 187; F. Effenberger and J. Geke, *Synthesis*, 1975, 40.
140. S. Allsop. F. Allsop, and J. Kenner, *J. Chem. Soc.*, 1923, 2314.
141. E. Schmidt and H. Fischer, *Chem. Ber.*, 1920, **53**, 1529; E. Schmidt, H. Fischer, and R. Schumacher, *Chem. Ber.*, 1921, **54**, 1414.

142. A. G. Anderson, R. Scotoni, E. J. Cowles, and C. G. Fritz, *J. Org. Chem.*, 1957, **22**, 1193; K. Hafner and K.-L. Moritz, *Justus Liebigs Ann. Chem.*, 1962, **656**, 40.
143. S. M. Tsang, A. P. Paul, and M. P. DiGiaimo, *J. Org. Chem.*, 1964, **29**, 3387.
144. G. A. Olah and H. C. Lin, *J. Am. Chem. Soc.*, 1974, **96**, 2892.
145. A. V. Topchiev, *Nitration of Hydrocarbons and Other Organic Compounds*, Pergamon Press, London, 1959.
146. G. A. Olah, A. P. Fung, S. C. Narang, and J. A. Olah, *J. Org. Chem.*, 1981, **46**, 3533.
147. K. Fukunaga and M. Kimyra, *Nippon Kagaku Kaishi*, 1973, 1306.
148. P. Laszlo and P. Pennetreau, *J. Org. Chem.*, 1987, **52**, 2407.
149. D. W. Amos, D. A. Baines, and G. W. Flewett, *Tetrahedron Lett.*, 1973, 3191; R. G. Coombes and L. W. Russell, *J. Chem. Soc., Perkin Trans. 2*, 1974, 830.
150. Ref. 51b, p. 78.
151. C. A. Bunton, E. D. Hughes, C. K. Ingold, D. I. H. Jacobs, M. H. Jones, J. G. Minkoff, and R. I. Reed, *J. Chem. Soc.*, 1950, 2628.
152. J. Glazer, E. D. Hughes, C. K. Ingold, A. T. James, G. T. Jones, and. E. Roberts, *J. Chem. Soc.*, 1950, 2657.
153. J. C. Giffney, D. J. Mills, and J. H. Ridd, *J. Chem. Soc., Chem. Commun.*, 1976, 19; J. C. Giffney and J. H. Ridd, *J. Chem. Soc., Perkin Trans. 2*, 1979, 618.
154. F. Al-Omran, K. Fujiwara, J. C. Giffney, and J. H. Ridd, *J. Chem. Soc., Perkin Trans. 2*, 1918, 518.
155. A. H. Clemens, J. H. Ridd, and J. P. B. Sandall, *J. Chem. Soc., Perkin Trans. 2*, 1984, 1659, 1667.
156. A. P. Gosney and M. I. Page, *J. Chem. Soc., Perkin Trans. 2*, 1980, 1783.
157. M. Ali and J. H. Ridd, *J. Chem. Soc., Perkin Trans. 2*, 1986, 327.
158. (a) U. Al-Obaidi and R. B. Moodie, *J. Chem. Soc., Perkin Trans. 2*, 1985, 467; (b) J. R. Leis, M. E. Pẽna, J. H. Ridd, *J. Chem. Soc., Chem. Commun.*, 1988, 670.
159. M. Lemaire, A. Guy, J. Roussel, and J.-P. Guette, *Tetrahedron*, 1987, **43**, 835.
160. N. A. Vysotskaya and A. E. Brodsky, *Chem. Abstr.*, 1967, **67**, 21232.
161. Ref. 52, pp. 1400–1407.
162. R. C. Dumbar, J. Shen, and G. A. Olah, *J. Am. Chem. Soc.*, 1972, **94**, 6862; J. D. Morrison, K. Stanning, and J. M. Tedder, *J. Chem. Soc., Perkin Trans. 2*, 1981, 967.
163. (a) M. Attina, F. Cacace, *J. Am. Chem. Soc.*, 1986, **108**, 318; (b) M. Attina, F. Cacace, and M. Yanez, *J. Am. Chem. Soc.*, 1987, **109**, 5092; (c) M. Attina, F. Cacace, and G. de Petris, *Angew. Chem., Int. Ed. Engl.*, 1987, **26**, 1176.
164. K. Lauer and R. Oda, *J. Prakt. Chem.*, 1936, **144**, 176; R. Oda, and V. Veda, *Bull. Inst. Chem. Res. Jpn.*, 1941, **20**, 335.
165. L. Melander, *Nature* (*London*), 1949, **163**, 599; *Ark. Kemi*, 1951, **2**, 211.
166. W. M. Lauer and W. E. Noland, *J. Am. Chem. Soc.*, 1953, **75**, 3689.
167. P. C. Myhre, M. Beng, and L. L. James, *J. Am. Chem. Soc.*, 1968, **90**, 2205.
168. K. Olsson, *Acta Chem. Scand.*, 1972, **26**, 3555.
169. H. Cerfontain and A. Telder, *Recl. Trav. Chim. Pays-Bas*, 1967, **86**, 371; G. A. Olah, S. C. Narang, R. Malhotra, and J. A. Olah, *J. Am. Chem. Soc.*, 1979, **101**, 1805.
170. J. Kenner, *Nature* (*London*), 1945, **156**, 369.
171. C. L. Perrin, *J. Am. Chem. Soc.*, 1977, **99**, 5516.
172. L. Eberson, L. Jönsson, and F. Radner, *Acta Chem. Scand., Ser B*, 1978, **32**, 749.
173. L. Eberson and F. Radner, *Acta Chem. Scand. Ser. B*, 1986, **40**, 71; *Acc. Chem. Res.*, 1987, **20**, 53.
174. S. Sankararama, W. A. Haney, and J. K. Kochi, *J. Am. Chem. Soc.*, 1987, **109**, 5235.
175. J. Feng, X. Zheng, and M. C. Zerner, *J. Org. Chem.*, 1986, **51**, 4531.
176. G. A. Olah, H. C. Lin, and Y. K. Mo, *J. Am. Chem. Soc.*, 1972, **94**, 3667; G. A. Olah, H. C. Lin, and D. A. Forsyth, *J. Am. Chem. Soc.*, 1974, **96**, 6908.

177. K. Fujiwara, J. C. Giffney, and J. H. Ridd, *J. Chem. Soc., Chem. Commun.*, 1977, 301; P. Helsby and J. H. Ridd, *J. Chem. Soc., Perkin Trans. 2*, 1983, 311.
178. H. Suzuki and K. Nakamura, *J. Chem. Soc., Chem. Commun.*, 1972, 340; H. Suzuki, *Synthesis*, 1977, 217; A. N. Detsina, V. I. Mamatyuk, and V. A. Koptyug, *Bull. Acad. Sci. USSR*, 1973, 2122; *J. Org. Chem. USSR*, 1977, **13**, 122.
179. A. N. Detsina and V. A. Koptyug, *J. Org. Chem. USSR*, 1972, **8**, 2263.
180. D. J. Blackstock, A. Fischer, K. E. Richards, and G. J. Wright, *J. Chem. Soc., Chem. Commun.*, 1970, 641; D. J. Blackstock, J. R. Cretney, A. Fischer, M. R. Hartshorn, K. E. Richards, J. Vaughan, and G. J. Wright, *Tetrahedron Lett.*, 1970, 2793; A. Fischer, C. C. Greig, A. L. Wilkinson, and D. R. A. Leonard, *Can. J. Chem.*, 1972, **50**, 2211; A. Fischer and D. R. A. Leonard, *Can. J. Chem.*,1972, **50**, 3367; 1976, **54**, 1795; A. Fischer and A. L. Wilkinson, *Can. J. Chem.*, 1972, **50**, 3988; A. Fischer and J. N. Ramsay, *J. Chem. Soc., Perkin Trans. 2*, 1973, 237; *J. Am. Chem. Soc.*, 1974, **96**, 1614; *Can. J. Chem.*, 1974, **52**, 3960; A. Fischer and C. C. Greig, *Can. J. Chem.*, 1978, **56**, 1348; A. Fischer, C. C. Greig, and R. Röderer, *Can. J. Chem.*, 1975, **53**, 1570; A. Fischer and R. Röderer, *Can. J. Chem.*, 1976, **54**, 423, 3978; A. Fischer and K. C. Teo, *Can. J. Chem.*, 1978, **56**, 258, 1758; A. Fischer and S. S. Seyan, *Can. J. Chem.*, 1978, **56**, 1348; A. Fischer, G. N. Henderson, and R. J. Thompson, *Aust. J. Chem.*, 1978, **31**, 1241; R. C. Hahn and D. L. Strack, *J. Am. Chem. Soc.*, 1974, **96**, 4335; A. H. Clemens, M. P. Hartshorn, K. E. Richards, and G. J. Wright, *Aust. J. Chem.*, 1977, **30**, 103; R. C. Hahn, H. Shosenji, and D. L. Strack, in *Industrial and Laboratory Nitrations*, Eds L. F. Albright and C. Hanson, ACS Symposium Series, No. 22, American Chemical Society, Washington, DC, 1976, p. 95; O. I. Osina and V. D. Steingarts, *J. Org. Chem. USSR*, 1974, **10**, 335; 1976, **12**, 1482; A. Fischer, D. L. Fykes, and G. N. Henderson, *J. Chem. Soc., Chem. Commun.*, 1980, 513; R. B. Moodie, M. A. Payne, and K. Schofield, *J. Chem. Soc., Chem. Commun.*, 1985, 1457; A. Fischer, D. L. Fykes, G. N. Henderson, and S. R. Mahasay, *Can. J. Chem.*, 1986, **64**, 1764.
181. D. J. Blackstock, M. P. Hartshorn, A. J. Lewis, K. E. Richards, J. Vaughan, and G. J. Wright, *J. Chem. Soc. B*, 1971, 1212.
182. A. Fischer and C. C. Greig, *Can. J. Chem.*, 1974, **52**, 1231.
183. C. Bloomfield, R. B. Moodie, and K. Schofield, *J. Chem. Soc., Perkin Trans. 2*, 1983, 1003.
184. R. C. Hahn and M. B. Groen, *J. Am. Chem. Soc.*, 1973, **95**, 6128.
185. J. W. Barnett, R. B. Moodie, K. Schofield, and J. B. Weston, *J. Chem. Soc., Perkin Trans. 2*, 1975, 648.
186. P. C. Myhre, *J. Am. Chem. Soc.*, 1972, **94**, 7921; T. Banwell, C. S. Morse, P. C. Myhre, and A. Vollmar, *J. Am. Chem. Soc.*, 1977, **99**, 3042.
187. V. I. Mamatyuk, B. G. Derendyaev, A. N. Detsina, and V. A. Koptyug, *J. Org. Chem. USSR*, 1974, **10**, 2506.
188. A. N. Detsina, V. I. Mamatyuk, and V. A. Koptyug, *J. Org. Chem. USSR*, 1977, **13**, 122.
189. H. J. Lewis and R. Robinson, *J. Chem. Soc.*, 1934, 1523; A. J. M. Reuvers, F. F. van Leeuwen, and A. Sinnema, *J. Chem. Soc., Chem. Commun.*, 1972, 828; B. A. Collins, K. E. Richards, and G. J. Wright, *J. Chem. Soc., Chem. Commun.*, 1972, 1216; A. H. Clemens, M. P. Hartshorn, K. E. Richards, and G. J. Wright, *Aust. J. Chem.*, 1977, **30**, 113; C. E. Barnes and P. C. Myhre, *J. Am. Chem. Soc.*, 1978, **100**, 973; M. P. Hartshorn, K. E. Richards, R. S. Thompson, and J. Vaughan, *Aust. J. Chem.*, 1981, **34**, 1345; A. M. Chittenden *et al.*, *Aust. J. Chem.*, 1982, **35**, 2229; M. P. Hartshorn, H. T. Ing, K. E. Richards, R. S. Thompson, and J. Vaughan, *Aust. J. Chem.*, 1982, **35**, 221; M. J. Gray *et al.*, *Aust. J. Chem.*, 1982, **35**, 1237; M. P. Hartshorn, H. T. Ing, K. E. Richards, K. H. Sutton, and J. Vaughan, *Aust. J. Chem.*, 1982, **35**, 1635; C.

Bloomfield, A. K. Manglik, R. B. Moodie, K. Schofield, and G. D. Tobin, *J. Chem. Soc., Perkin Trans. 2*, 1983, 75; G. C. Cross, A. Fischer, G. N. Henderson, and T. A. Smyth, *Can J. Chem.*,1984, **62**, 1446.
190. C. E. Barnes, and P. C. Myhre, *J. Am. Chem. Soc.*, 1978, **100**, 975.
191. P. C. Myhre, *J. Am. Chem. Soc.*, 1972, **94**, 7921.
192. R. Taylor, *Specialist Periodical Report on Aromatic and Heteroaromatic Chemistry*, Chemical Society, London, 1974, Vol. 3, p. 246.
193. R. Taylor, *J. Chem. Res. (S)*, 1978, 267.
194. F. Al-Omran and J. H. Ridd, *J. Chem. Soc., Perkin Trans. 2*, , 1983, 1185.
195. P. Helsby and J. H. Ridd, *J. Chem. Soc., Perkin Trans. 2*, 1983, 1191.
196. A. Hirose, K. Matsui, and S. Sekiguchi, *Bull. Chem. Soc. Jpn.*, 1972, **45**, 2955.
197. H. C. Lin, *PhD Thesis*, Case Western Reserve University, 1972.
198. H. Cohn, E. D. Hughes, M. H. Jones, and M. G. Peeling, *Nature (London)*, 1952, **169**, 291.
199. L. M. Stock, *J. Org. Chem.*, 1961, **26**, 4120.
200. C. K. Ingold, A. Lapworth, E. Rothstein, and D. Ward, *J. Chem. Soc.*, 1931, 1959.
201. S. Sotheeswaran and K. J. Toyne, *J. Chem. Soc., Perkin Trans. 2*, 1977, 2042.
202. J. R. Knowles, R. O. C. Norman, and G. K. Radda, *J. Chem. Soc.*, 1960, 4885.
203. L. M. Stock and P. E. Young, *J. Am. Chem. Soc.*, 1972, **94**, 4247.
204. J. R. Knowles and R. O. C. Norman, *J. Chem. Soc.*, 1961, 2938.
205. H. C. Brown and R. A. Wirkkala, *J. Am. Chem. Soc.*, 1966, **88**, 1447.
206. R. Taylor, *J. Chem. Res. (S)*, 1985, 318.
207. T. Asano, R. Goto, and A. Sera, *Bull. Chem. Soc. Jpn.*, 1967, **40**, 2208.
208. L. Brandt and R. Verbesselt, *Tetrahedron*, 1972, **28**, 89.
209. J. M. A. Baas and B. M. Wepster, *Recl. Trav. Chim. Pays-Bas*, 1972, **91**, 285, 517, 831.
210. R. Ketcham, R. Cavestri, and D. Jambotkar, *J. Org. Chem.*, 1963, **28**, 2139.
211. J. M. A. Baas and B. M. Wepster, *Recl. Trav. Chim. Pays-Bas*, 1971, **90**, 1081.
212. R. C. Hahn, T. F. Corbin, and H. Shechter, *J. Am. Chem. Soc.*, , 1968, **90**, 3404.
213. Yu. S. Shabarov, V. K. Potapov, and R. Ya. Levina, *J. Gen. Chem. USSR*, 1964, **34**, 3171.
214. S. S. Mochalov, N. B. Matveeva, I. P. Stepanova, and Yu. S. Shabarov, *J. Org. Chem. USSR*, 1976, **12**, 1514.
215. W. Kurtz, P. Fischer, and F. Effenberger, *Chem. Ber.*, 1973, **106**, 525.
216. Yu. S. Shabarov, S. S. Mochalov, and O. M. Khryashchevskaya, *J. Org. Chem. USSR*, 1970, **6**, 2446; Yu. S. Shabarov and S. N. Burenko, *J. Org. Chem. USSR*, 1971, **7**, 2737; Yu. S. Shabarov and S. S. Mochalev, *J. Org. Chem. USSR*, 1973, **9**, 54, 728, 2061; Yu. S. Shabarov, S. S. Mochalov, and S. A. Ermishkina, *Proc. Acad. Sci. USSR*, 1973, **211**, 663; Yu. S. Shabarov and S. N. Burenko, *J. Org. Chem. USSR*, 1973, **9**, 1791; Yu. S. Shabarov, S. S. Mochalov, and I. N. Kuzmina, *J. Org. Chem. USSR*, 1974, **10**, 759; O. M. Nefedov and R. N. Shafrom, *J. Org. Chem. USSR*, 1974, **10**, 481; Yu. S. Shabarov, S. S. Mochalov, N. B. Matveeva, *J. Org. Chem. USSR*, 1975, **11**, 565; S. A. Ermishkina, S. S. Mochalov, and Yu. S. Shabarov, *J. Org. Chem. USSR*, 1975, **11**, 369; S. S. Mochalov, V. D. Novokreshchennykh, and Yu. S. Shabarov, *J. Org. Chem. USSR*, 1976, **12**, 1019.
217. B. Van de Graaf and B. M. Wepster, *Recl. Trav. Chim. Pays.-Bas*, 1966, **85**, 619.
218. A. Fischer, J. Vaughan, and G. J. Wright, *J. Chem. Soc. B*, 1967, 368; R. G. Coombes and L. W. Russell, *J. Chem. Soc. B.*, 1971, 2443.
219. J. B. F. Lloyd and P. A. Ongley, *Tetrahedron*, 1964, **20**, 2185.
220. J. Vaughan, G. J. Welch, and G. J. Wright, *Tetrahedron*, 1965, **21**, 1665.
221. H. Tanida and R. Muneyki, *J. Am. Chem. Soc.*, 1965, **87**, 4794.
222. M. W. Galley and R. C. Hahn, *J. Am. Chem. Soc.*, 1974, **96**, 4337; *J. Org. Chem.*, 1976, **41**, 2006.

223. J. H. P. Utley and T. A. Vaughan, *J. Chem. Soc., Perkin Trans. 2*, , 1972, 2343.
224. (a) G. P. Sharnin and I. F. Falyakhov, *J. Org. Chem. USSR*, 1973, **9**, 751; (b) M. J. S. Dewar and D. S. Urch, *J. Chem. Soc.*, 1958, 3079.
225. T. A. Modro, W. F. Reynolds, and E. Skorupowa, *J. Chem. Soc., Perkin Trans. 2*, 1977, 1479.
226. G. Grynkiewicz and J. H. Ridd, *J. Chem. Soc. B*, 1971, 716.
227. F. L. Riley and E. Rothstein, *J. Chem. Soc.*, 1964, 3860.
228. C. K. Ingold, E. H. Ingold, and F. R. Shaw, *J. Chem. Soc., Perkin Trans. 2*, 1927, 813.
229. A. F. Holleman, *Chem. Rev.*, 1925, **1**, 187.
230. R. J. Albers and E. C. Kooyman, *Recl. Trav. Chim. Pays-Bas*, 1964, **83**, 930.
231. R. G. Coombes, R. B. Moodie, and K. Schofield, *J. Chem. Soc. B*, 1969, 52.
232. L. M. Yagulpolskii, N. V. Kondratenko, N. I. Delyagina, B. L. Dyalkin, and I. L. Knunyants, *J. Org. Chem. USSR*, 1973, **9**, 669.
233. L. M. Yagulpolskii, N. G. Pavlenko, S. N. Soludushenkov, and Ya. A. Fialkov, *Ukr. Khim. Zh.*, 1966, **32**, 849.
234. J. H. Rees, *J. Chem. Soc., Perkin Trans. 2*, 1975, 945.
235. S. J. Cristol and D. C. Lewis, *J. Am. Chem. Soc.*, 1967, **89**, 1476.
236. H. J. Reich and D. J. Cram, *J. Am. Chem. Soc.*, 1969, **91**, 3505.
237. J. W. Baker and I. S. Wilson, *J. Chem. Soc.*, 1927, 842.
238. R. B. Moodie, K. Schofield, P. G. Taylor, and P. J. Baillie, *J. Chem. Soc., Perkin Trans. 2*, 1981, 842.
239. J. W. Baker, K. E. Cooper, and C. K. Ingold, *J. Chem. Soc.*, 1928, 426.
240. F. G. Bordwell and K. Rohde, *J. Am. Chem. Soc.*, 1948, **70**, 1191; W. E. Truce and J. A. Simms, *J. Org. Chem.*, 1957, **22**, 762.
241. A. F. Holleman, *Recl. Trav. Chim. Pays-Bas*, 1906, **24**, 140; *Chem. Rev.*, 1925, **1**, 187.
242. J. D. Roberts, J. K. Sanford, F. L. J. Sixma, H. Cerfontain, and R. Zagt, *J. Am. Chem. Soc.*, 1954, **76**, 4525.
243. M. L. Bird and C. K. Ingold, *J. Chem. Soc.*, 1938, 918.
244. C. L. Perrin and G. A. Skinner, *J. Am. Chem. Soc.*, 1971, **93**, 3389.
245. N. C. Marziano, A. Zingales, and V. Ferlito, *J. Org. Chem.*, 1977, **42**, 2511.
246. E. Glyde and R. Taylor, *J. Chem. Soc., Perkin Trans. 2*, 1977, 1541.
247. P. H. Griffiths, W. A. Walkey, and H. B. Watson, *J. Chem. Soc.*, 1934, 631.
248. K. Halvarson and L. Melander, *Ark. Kemi*, 1957, **11**, 77.
249. J. R. Knowles and R. O. C. Norman, *J. Chem. Soc.*, 1961, 3888.
250. F. Arnall and T. Lewis, *J. Soc. Chem. Ind.*, 1929, **48**, 159T.
251. S. R. Hartshorn, R. B. Moodie, and K. Schofield, *J. Chem. Soc. B*, 1971, 2454.
252. B. M. Lynch, C. M. Chen, and Y.-Y. Wingfield, *Can. J. Chem.*, 1969, **46**, 1141.
253. R. B. Moodie, P. N. Thomas, and K. Schofield, *J. Chem. Soc., Perkin Trans. 2*, 1977, 1693.
254. R. G. Coombes, J. G. Golding, and P. Hadjigeorgiou, *J. Chem. Soc., Perkin Trans. 2*, 1979, 1451.
255. Ref. 49, p. 76.
256. S. R. Hartshorn and K. Schofield, *J. Chem. Soc., Perkin Trans. 2*, 1972, 1652.
257. R. Taylor, *Specialist Periodical Report on Aromatic and Heteroaromatic Chemistry*, Chemical Society, London, 1974, Vol. 2, p. 244.
258. G. W. Buchanan and S. H. Preusser, *J. Org. Chem.*, 1982, **47**, 5029.
259. D. R. Harvey and R. O. C. Norman, *J. Chem. Soc.*, 1962, 3822.
260. J. L. Speier, *J. Am. Chem. Soc.*, 1953, **75**, 2930.
261. E. A. Chernyshev, H. E. Dolgaya, and A. D. Petrov, *Bull. Akad. Sci. USSR*, 1960, 1323.
262. L. I. Zakharkin and V. N. Kalinin, *Proc. Acad. Sci. USSR*, 1965, 904; L. I.

Zakharkin, V. I. Stanko, and A. I. Klimova, *J. Gen. Chem. USSR*, 1965, **35**, 393; V. I. Stanko and A. V. Bobrov, *J. Gen. Chem. USSR*, 1965, **35**, 1994.
263. L. I. Zakharkin and V. N. Kalinin, *J. Gen. Chem. USSR*, 1973, **43**, 853.
264. L. I. Zakharkin, V. N. Kalinin, E. G. Rys, and B. A. Kvasov, *Bull. Akad. Sci. USSR*, 1972, 458.
265. D. Vorländer and E. Siebert, *Chem. Ber.*, 1919, **52**, 283; D. Vorländer, *Chem. Ber.*, 1925, **58**, 1893; F. R. Goss, C. K. Ingold, and I. S. Wilson, *J. Chem. Soc.*, 1926, 2440; F. R. Goss, W. Hanhart, and C. K. Ingold, *J. Chem. Soc.*, 1927, 250; C. K. Ingold and I. S. Wilson, *J. Chem. Soc.*, 1927, 810; C. K. Ingold, F. R. Shaw, and I. S. Wilson, *J. Chem. Soc.*, 1928, 1280; J. W. Baker and W. G. Moffitt, *J. Chem. Soc.*, 1930, 1722; F. Challenger and E. Rothstein, *J. Chem. Soc.*, 1934, 1258; R. B. Sandin, F. T. McClure, and F. Irwin, *J. Am. Chem. Soc.*, 1939, **61**, 3061.
266. J. D. Roberts, F. T. McClure, and J. J. Drysdale, *J. Am. Chem. Soc.*, 1951, **73**, 2181.
267. S. R. Hartshorn and J. H. Ridd, *J. Chem. Soc. J. Chem. Soc. B*, 1968, 1063, 1068.
268. M. Brickman, J. H. P. Utley, and J. H. Ridd, *J. Chem. Soc.*, 1965, 6851.
269. M. Brickman and J. H. Ridd, *J. Chem. Soc.*, 1965, 6845.
270. A. Gastaminza, T. A. Modro, J. H. Ridd, and J. H. P. Utley, *J. Chem. Soc. B*, 1968, 534.
271. H. M. Gilow, M. de Shazo, and M. van Cleave, *J. Org. Chem.*, 1971, **36**, 1745.
272. N. C. Marziano, E. Maccarone, and R. C. Passerini, *J. Chem. Soc. B*, 1971, 745; *Tetrahedron Lett.*, 1972, 17; N. C. Marziano, E. Maccarone, G. M. Cimino, and R. C. Passerini, *J. Chem. Soc., Perkin Trans. 2*, 1974, 1098.
273. R. Danieli, R. Ricci, H. M. Gilow, and J. H. Ridd, *J. Chem. Soc., Perkin Trans. 2*, 1974, 1477.
274. T. A. Modro and J. H. Ridd, *J. Chem. Soc. B*, 1968, 528.
275. F. L. Riley and E. Rothstein, *J. Chem. Soc.*, 1964, 3860, 3872.
276. T. A. Modro and A. Piekos, *Tetrahedron*, 1972, **28**, 3867; 1973, **29**, 2561.
277. E. Malinski, A. Piekos, and T. A. Modro, *Can. J. Chem.*, 1975, **53**, 1468.
278. F. de Sarlo and J. H. Ridd, *J. Chem. Soc. B*, 1971, 712.
279. A. R. Katritzky and M. Kingsland, *J. Chem. Soc. B*, 1968, 862.
280. C. K. Ingold, *Structure and Mechanism in Organic Chemistry*, 2nd edn., Bell, London, 1969, p. 301.
281. A. R. Katritzky and R. D. Topsom, *Angew. Chem., Int. Ed. Engl.*, 1970, **9**, 87.
282. J. H. Rees, J. H. Ridd, and A. Ricci, *J. Am. Chem. Soc.*, 1976, 294.
283. A. N. Nesmeyanov, T. P. Tolstaya, L. S. Isaeva, and A. V. Grib, *Dokl. Akad. Nauk SSSR*, 1960, **133**, 602.
284. R. Danieli, A. Ricci, and J. H. Ridd, *J. Chem. Soc., Perkin Trans. 2*, 1972, 1547; 1976, 290.
285. K. Kamiyama, H. Minato, and M. Kobayashi, *Bull. Chem. Soc. Jpn.*, 1973, **46**, 2255.
286. Ref. 49, p. 82.
287. J. G. Tillett, *J. Chem. Soc.*, 1962, 5142; A. D. Mésure and J. G. Tillett, *J. Chem. Soc. B*, 1966, 669.
288. B. Östman and I. Lingren, *Acta Chem. Scand.*, 1970, **24**, 1105.
289. J. P. Wibaut and R. van Strik, *Recl. Trav. Chim. Pays-Bas*, 1958, **77**, 317.
290. J. W. Baker and W. G. Moffitt, *J. Chem. Soc.*, 1931, 314.
291. R. B. Moodie, J. R. Penton, and K. Schofield, *J. Chem. Soc. B*, 1969, 578.
292. S. D. Barker, R. K. Norris, and D. Randles, *Aust. J. Chem.*, 1981, **34**, 1875.
293. K. E. Cooper and C. K. Ingold, *J. Chem. Soc.*, 1927, 836.
294. C. K. Ingold and M. S. Smith, *J. Chem. Soc.*, 1938, 905.
295. N. C. Marziano, E. Maccarone, and R. C. Passerini, *J. Chem. Soc. B*, 1971, 745.
296. R. B. Moodie, K. Schofield, and T. Yoshida, *J. Chem. Soc., Perkin Trans. 2*, 1975, 788.

297. J. G. Tillett, *J. Chem. Soc.*, 1962, 5142.
298. J. H. P. Utley and T. A. Vaughan, *J. Chem. Soc. B*, 1968, 196.
299. J. C. D. Brand and R. C. Paton, *J. Chem. Soc.*, 1952, 281; R. G. Coombes, D. H. G. Crout, J. G. Hoggett, R. B. Moodie, and K. Schofield, *J. Chem. Soc. B*, 1970, 347.
300. R. S. Cook, R. Phillips, and J. H. Ridd, *J. Chem. Soc., Perkin Trans. 2*, 1974, 1166.
301. C. D. Johnson and M. J. Northcott, *J. Org. Chem.*, 1967, **32**, 2029.
302. O. Simamura and Y. Mizuno, *Bull. Chem. Soc. Jpn.*, 1957, **30**, 190; *J. Chem. Soc.*, 1958, 3875.
303. T. van Hove, *Bull. Acad. Belg. Cl. Sci.*, 1922, **8**, 505; H. G. Dennett and E. E. Turner, *J. Chem. Soc.*, 1926, 476; R. J. W. Le Févre and E. E. Turner, *J. Chem. Soc.*, 1926, 2041; 1930, 1158; R. J. W. Le Févre, D. D. Moir, and E. E. Turner, *J. Chem. Soc.*, 1927, 2330; H. A. Scarborough and W. A. Waters, *J. Chem. Soc.*, 1927, 1133; W. Blackey and H. A. Scarborough, *J. Chem. Soc.*, 1927, 3000; H. C. Gull and E. E. Turner, *J. Chem. Soc.*, 1929, 491; W. S. M. Grieve and D. H. Hey, *J. Chem. Soc.*, 1932, 1888, 2245; 1933, 968; D. H. Hey, *J. Chem. Soc.*, 1932, 2637; D. H. Hey and E. R. B. Jackson, *J. Chem. Soc.*, 1934, 645; E. E. J. Marler and E. E. Turner, *J. Chem. Soc.*, 1931, 1359; F. R. Shaw and E. E. Turner, *J. Chem. Soc.*, 1932, 285; L. Mascarelli, D. Gatti, and B. Longo, *Gazz. Chim. Ital.*, 1933, **63**, 654; F. H. Case, *J. Am. Chem. Soc.*, 1942, **64**, 1848; F. H. Case and R. U. Schock, *J. Am. Chem. Soc.*, 1943, **65**, 2086.
304. F. Bell, *J. Chem. Soc.*, 1928, 2770; H. A. Scarborough and W. A. Waters, *J. Chem. Soc.*, 1927, 89.
305. R. L. Jenkins, R. McCullough, and C. F. Booth, *Ind. Eng. Chem.*, 1930, **22**, 1.
306. G. W. Gray and D. Lewis, *J. Chem. Soc.*, 1961, 5156.
307. B. M. Lynch and L. Poon, *Can. J. Chem.*, 1967, **45**, 1431.
308. C. Dell'Erba, G. Garbarino, and G. Guanti, *J. Heterocycl. Chem.*, 1971, **8**, 849; 1974, **11**, 1017.
309. A. Arcoria, E. Maccarone, and G. A. Tomaselli, *J. Heterocycl. Chem.*, 1973, **10**, 153.
310. P. Cohen-Fernandes and C. L. Halbraken, *Recl. Trav. Chim. Pays-Bas*, 1972, **91**, 1185.
311. C. J. Billing and R. O. C. Norman, *J. Chem. Soc.*, 1961, 3885.
312. R. G. Coombes and J. G. Golding, *Tetrahedron Lett.*, 1976, 77.
313. M. L. Scheinbaum, *J. Chem. Soc., Chem. Commun.*, 1969, 1235.
314. R. Taylor, *J. Chem. Soc., Perkin Trans. 2*, 1973, 253.
315. J. M. Birchall, R. N. Haszeldine, and H. Woodfine, *J. Chem. Soc., Perkin Trans. 1*, 1973, 1121.
316. E. K. Weisburger and J. E. Weisburger, *J. Org. Chem.*, 1959, **24**, 1511; L. H. Klemm, E. Huber, and C. E. Klopfenstein, *J. Org. Chem.*, 1964, **29**, 1960; E. O. Arene and D. A. H. Taylor, *J. Chem. Soc. J. Chem. Soc. C*, 1966, 481.
317. D. Buza and W. Polaczkowa, *Tetrahedron*, 1965, **12**, 3415.
318. G. P. Sharnin, I. E. Moisek, E. E. Gryazin, and I. F. Falyakhov, *J. Org. Chem. USSR*, 1967, **3**, 1792.
319. W. Baker, J. W. Barton, and J. F. W. McOmie, *J. Chem. Soc.*, 1958, 2666.
320. F. Radner, *Acta Chem. Scand., Ser. B*, 1983, **37**, 65; L. Eberson and F. Radner, *Acta Chem. Scand., Ser. B*, 1980, **34**, 739; 1984, **38**, 861; 1985, **39**, 343; 1986, **40**, 71.
321. J. J. Looker, *J. Org. Chem.*, 1972, **37**, 3379.
322. A. Streitwieser and R. C. Fahey, *J. Org. Chem.*, 1962, **27**, 2352.
323. P. G. E. Alcorn and P. R. Wells, *Aust. J. Chem.*, 1965, **18**, 1377, 1391.
324. D. S. Ross, K. D., Moran, and R. Malhotra, *J. Org. Chem.*, 1983, **48**, 2118.
325. C. W. J. Chang, R. E. Moore, and P. J. Scheuer, *J. Chem. Soc. C*, 1967, 840; J.-C. Richer and Y. Pépi, *Can. J. Chem.*, 1965, **43**, 3443; J.-C. Richer and A. Rossi, *Can. J. Chim.*, 1967, **47**, 3935; D. T. Clark and D. J. Fairweather, *Tetrahedron*, 1969, **25**, 5525.

326. A. Davies and K. D. Warren, *J. Chem. Soc. B*, 1969, 873.
327. H. H. Hodgson and H. S. Turner, *J. Chem. Soc.*, 1843, 391; E. R. Ward and P. R. Wells, *J. Chem. Soc.*, 1961, 4859; H. H. Hodgson and J. Walker, *J. Chem. Soc.*, 1933, 1205; A. V. Topchiev, *Nitration of Hydrocarbons and other Organic Compounds*, Pergamon Press, London, 1969.
328. H. H. Hodgson and J. Walker, *J. Chem. Soc.*, 1933, 1346; E. R. Ward and J. G. Hawkins, *J. Chem. Soc.*, 1954, 2975.
329. K. J. Falci, R. W. Franck, and E. Soykan, *J. Org. Chem.*, 1975, **40**, 2547.
330. O. I. Osina and V. D. Shteingarts, *J. Org. Chem. USSR*, 1974, **10**, 329.
331. B. I. Stepanov, V. Ya. Rodionov, and S. E. Voinova, *J. Org. Chem. USSR*, 1977, **13**, 768.
332. R. Bolton, *J. Chem. Soc. (S)*, 1977, 149.
333. G. M. Badger and J. W. Cook, *J. Chem. Soc.*, 1940, 409.
334. M. S. Newman and A. I. Kosak, *J. Org. Chem.*, 1949, **14**, 375.
335. P. M. op den Brouw and W. H. Laarhaven, *Recl. Trav. Chim. Pays.-Bas*, 1978, **97**, 265.
336. E. H. Charlesworth and C. V. Lithown, *Can. J. Chem.*, 1969, **47**, 1595.
337. H. F. Andrew, N. Campbell, and N. H. Wilson, *J. Chem. Soc., Perkin Trans. 1*, 1972, 755.
338. M. I. Shenbar and V. I. Kirichenko, *J. Org. Chem. USSR*, 1971, **7**, 1544.
339. T. Nozoe, T. Asao, and M. Oda, *Bull. Chem. Soc. Jpn.*, 1974, **47**, 681.
340. J. E. Drach and F. R. Longo, *J. Org. Chem.*, 1974, **39**, 3282.
341. R. Grigg, G. Shelton, A. Sweeney, and A. W. Johnson, *J. Chem. Soc., Perkin Trans. 1*, 1972, 1789.
342. E. Watanabe, S. Nishimura, H. Ogoshi, and Z. Yoshida, *Tetrahedron*, 1975, **31**, 1385.
343. Z. Yoshida, S. Yoneda, Y. Murata, and H. Hashimoto, *Tetrahedron Lett.*, 1971, 1523.
344. D. J. Cram and R. D. Partos, *J. Am. Chem. Soc.*, 1963, **85**, 1273.
345. A. G. Anderson and E. D. Daugs, *J. Org. Chem.*, 1987, **52**, 4391.
346. H. B. Renfroe, J. A. Gurney, and L. A. R. Hall, *J. Org. Chem.*, 1972, **37**, 3045.
347 E. D. Hughes and G. T. Jones, *J. Chem. Soc.*, 1950, 2678.
348. S. Brownstein, C. A. Bunton, and E. D. Hughes, *Chem. Ind. (London)*, 1956, 981.
349. W. N. White and J. T. Golden, *J. Org. Chem.*, 1970, **35**, 2759.
350. J. H. Ridd and J. P. B. Sandell, *J. Chem. Soc., Chem. Commun.*, 1982, 261.
351. W. N. White and J. R. Klink, *J. Org. Chem.*, 1970, **35**, 965.
352. D. V. Banthorpe and J. G. Winter, *J. Chem. Soc., Perkin Trans. 2*, 1972, 1259.
353. J. H. Ridd and E. V. Scriven, *J. Chem. Soc., Chem. Commun.*, 1972, 64.
354. A. C. Satterthwait and F. H. Westheimer, *J. Am. Chem. Soc.*, 1978, **100**, 3197.

CHAPTER 8

Oxygen, sulphur, and selenium electrophiles

8.1 HYDROXYLATION

Although there is no clear evidence for the existence of OH^+ as a discrete entity, a hydroxyl group can be transferred from certain peroxy compounds to the aromatic ring without its covalent pair, and thus meets the criterion of an electrophilic reagent. Since the hydroxy group is strongly activating towards electrophilic substitution, further reaction usually occurs so yields are generally low. Moreover, oxidation frequently occurs, giving rise to quinones, and it is significant that quinone formation takes place at sites normally the most reactive towards electrophilic substitution.

8.1.1 Hydroxylation with Acidified Hydrogen Peroxide

Support for the proposal that OH^+ should be present in acidified solutions of hydrogen peroxide[1] was first provided by the hydroxylation of mesitylenes with H_2O_2 in HOAc and H_2SO_4.[2] It is possible that the actual mechanism involves displacement on protonated hydrogen peroxide by the reactive aromatic compounds (eq. 8.1). The reactivities towards oxidation of a range of aromatic ethers under these conditions parallels the reactivities of the ethers to other electrophilic substitutions;[3] this work showed a common feature in that methoxy groups, but not alkyl or halogen substituents, are often displaced from the aromatic ring. More recently, HF[4] and superacids[5] have been used, the advantage of the latter being that the phenolic products are protonated and so do not undergo further substitution; in this work toluene and other alkylbenzenes gave anomalously high *ortho*:*para* ratios, a common feature of hydroxylations (see below).

$$\mathrm{ArH} + \mathrm{O(H)}\text{—}\mathrm{OH_2}^+ \longrightarrow \mathrm{Ar}^+\!\!<^{\mathrm{H}}_{\mathrm{OH}} + \mathrm{H_2O} \longrightarrow \mathrm{ArOH} + \mathrm{H_3O^+} \qquad (8.1)$$

A variation of the above method uses a Lewis acid in place of the protic acid.[6,7] Here the advantage is that the acid coordinates to the oxygen of the product, again retarding subsequent reaction. With $AlCl_3$ as catalyst the toluene:benzene rate ratio was 19.1.[7]

8.1.2 Hydroxylation with Peracids

Polycyclic aromatic compounds are oxidized by perbenzoic acid at the nuclear position most susceptible to electrophilic attack,[8–10] which led to the suggestion[10] that perbenzoic acid is an electrophilic reagent. This was confirmed by the finding that the ease of oxidation of methoxy-substituted benzenes by perbenzoic acid increased with the number of methoxy substituents.[11] At least 2 mol of peracid are consumed for each mole of aromatic oxidized, and the kinetics are second order; the mechanism shown in eq. 8.2 was proposed.[11]

MeO, OMe, OMe —($PhCO_3H$, slow)→ OH, MeO, OMe, OMe —($PhCO_3H$, fast)→ H, O^+, MeO, OMe, HO, OMe

—($-H^+$, $-MeOH$)→ O, MeO, OMe, O (8.2)

Polarization of a peracid as in eq. 8.3 will be aided by electron withdrawal in R. Consequently, trifluoroperacetic acid (in CH_2Cl_2) is more effective than other peracids in aromatic oxidations.[12] Monohydroxylation of anisole and diphenyl ether gave, in each case, a substantially larger yield of the *ortho* than the *para* product (PhOMe, 27 and 7%; PhOPh, 35 and 12%, respectively). These high *ortho*:*para* ratios were shown to be due to the subsequent preferential oxidation of the *p*-hydroxy compounds.[13] TFPA is normally prepared *in situ* by treating TFA with H_2O_2, so acidified hydrogen peroxide could in principle be the electrophilic reagent, but this was shown not to be the case.[14]

$$HO{-}O_2R \longrightarrow HO^+ + {}^-O_2CR \qquad (8.3)$$

Partial rate factors (o, m, p) for hydroxylation of some aromatics PhR with

TFPA at 25 °C are R = OMe, 1170, —, 830; Me, 27.5, 0.8, 13.7; and F, 0.14, —, 1.34.[15] The reactivity sequence and the *ortho*, *para*-directing effects are typical of an electrophilic substitution. The anomalously high *ortho*:*para* ratios for toluene and anisole may arise as described above; the result for *meta* substitution in toluene is thought to be due to experimental error. Intermediates (see Section 2.1.2) are evidently formed under these conditions since hydroxylation of 2,6-dimethylphenol gave a Diels–Alder dimer of 6-hydroxy-2,6-dimethyl-2,4-cyclohexadienone (**1**) in addition to 2,6-dimethylbenzoquinone.

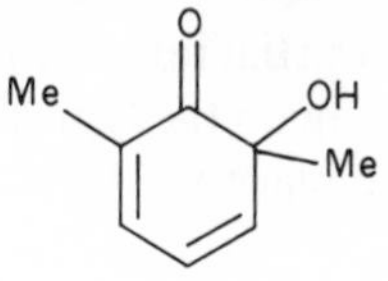

O
Me OH
Me

(**1**)

During hydroxylation of polymethylbenzenes with TFPA, methyl groups migrate[8] (also found using H_2O_2 and superacids)[5] and this has been attributed to *ipso* hydroxylation followed by a 1,2-shift of methyl.[8] A similar 1,2-shift of hydrogen also evidently occurs in the hydroxylation of acetanilide labelled with either deuterium or tritium at the *para* position, since the *p*-hydroxylated product contained ca 8% of the isotopic label.[16]

Electron withdrawal in the trifluoroacetate ion can be greatly increased by Lewis acids (which also coordinate with the hydroxy product), consequently TFPA–BF_3 is a very good hydroxylating species. For example, mesitol can be obtained from mesitylene in 88% yield.[17]

Peroxymonophosphoric acid, H_3PO_5, also provides a stable leaving group, and hydroxylation with this takes place ca 100 times more readily than with perbenzoic acid.[18] Likewise, hydroxylation accompanies nitration with pernitric acid, HNO_4.[19]

8.1.3 Miscellaneous Hydroxylating Reagents

tert-Butyl hydroperoxide–$AlCl_3$ hydroxylates alkylbenzenes with relative rates for *m*-xylene, *p*-xylene, *o*-xylene, toluene, and benzene of 137, 63.5, 62, 14 and 1, respectively, and the *ortho*, *meta*, and *para* yields for toluene were 56, 8, and 36%, respectively.[20] Reaction probably occurs by initial formation of the *tert*-butyl aryl ether, which would be instantaneously decomposed under the acidic conditions.

Hydroxylation may be obtained through the use of dialkyl peroxydicarbonates [$(ROCO_2)_2$] with Lewis acids, the reaction giving initially the aryl alkyl carbonate which is rapidly hydrolysed to the phenol if the conditions are sufficiently acidic (as is the case when $AlCl_3$ is used).[21,22] The reaction gives good yields and substantially lower *ortho*:*para* ratios than other methods. The extent to which the mechanism is either electrophilic or free-radical seems to depend on the catalyst; with $AlCl_3$ it is electrophilic, with $CuCl_2$ it is free radical, and with $FeCl_3$ both may be involved. With diisopropyl peroxydicarbonate and $AlCl_3$ the toluene:benzene rate ratio is 9, and the *ortho*, *meta*, and *para* isomer distribution is 34, 11, and 55%, respectively, but with $AlBr_3$ the rate ratio is increased to 17, and the halogenation side reaction becomes more significant.[22]

Hydroxylation has also been achieved with hypofluorous acid, HOF, which gives an anisole:benzene rate ratio of 120 (cf. 530 for hydroxylation with trifluoroperacetic acid, TFPA). Partial rate factors (*ortho*, *para*) were anisole 255, 175 and toluene 27, 14, showing again the typical high *ortho*, *para* reactivity. The reaction was shown not to take place via HF-catalysed reaction with H_2O_2.[23]

8.2 AROXYLATION

Electrophilic aroxylation to give ethers can be achieved by the thermal decomposition of arylpyridinium tetrafluoroborates at 180 °C. This produces the ion ArO^+ (**2**) which may directly substitute the aromatic giving the ether, or delocalization of the charge to the ring *ortho* carbon may occur to give the aryl cation (**3**) and hence arylation (see also Section 6.2) (eq. 8.4).[24] Aroxylation is

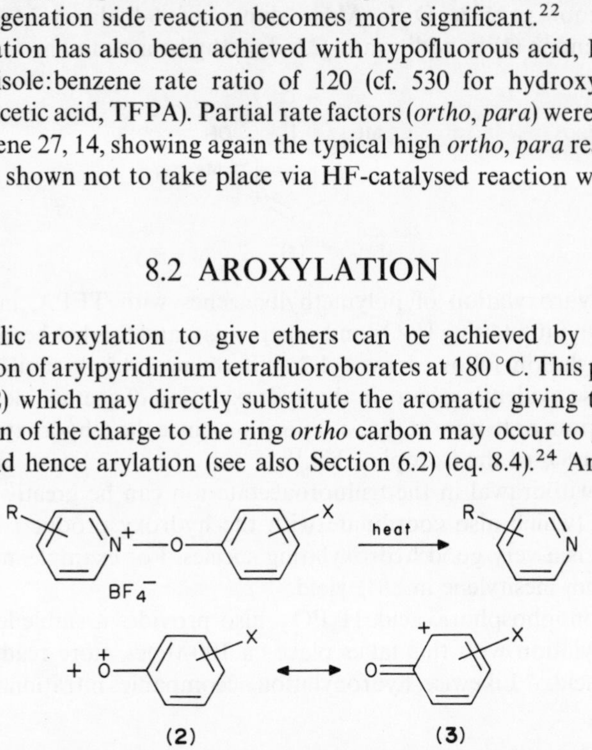

(8.4)

favoured over arylation by greater electron withdrawal by X (due, it is thought, to greater destabilization of the ion **3** than of **2** because X is nearer to the charge in **3**). The aroxylation:arylation rate ratio is also increased by greater nucleophilicity in the aromatic; this is consistent with attack by the aromatic occurring here sooner and thus before much delocalization can take place.

The *ortho*:*para* ratio for reaction with anisole ($R = p\text{-}NO_2$) is 0.41, but much larger (3.0) for arylation.

8.3 BENZOOXYLATION AND ACETOXYLATION

These reactions, represented by eq. 8.5, give rise to esters; benzooxylation is described incorrectly in the literature as benzoyloxylation.

$$\mathrm{ArH + RCOOX \xrightarrow{H^+} ArOCOR + HX} \tag{8.5}$$

The TFA-catalysed decomposition of benzoyl peroxide (eq. 8.6) gives rise to both electrophilic phenylation (giving biphenyls) and benzooxylation (giving benzoates).[25] The ratio of the two reactions also depends on the nucleophilicity of the aromatic (cf. Section 8.2), the yield of ester relative to biphenyl varying from 14 for reaction with anisole to 0.2 for reaction with nitrobenzene. The first-formed electrophile is probably $PhCO_2^+$, which may or may not have time to extrude CO_2 before reaction with the aromatic. The isomer distributions (*ortho, meta, para*) for benzoyloxylation are anisole 29, 0, 71; toluene 51, 9, 40; chlorobenzene 45, 5, 50; and nitrobenzene <1, 88, 12. It is notable that the distribution for anisole is identical to that obtained in aroxylation.

$$\mathrm{PhC(=O)OOC(=O)Ph \xrightarrow{H^+} Ph\overset{+}{C}(OH)OOC(=O)Ph \longrightarrow PhCO_2H + PhCO_2^+ (\longrightarrow Ph^+ + CO_2)} \tag{8.6}$$

In contrast to benzooxylation, acetoxylation seems only to have been investigated in the context of a side-reaction accompanying nitration by HNO_3 in acetic anhydride [Section 7.4.1(iv)],[26,27] and halogenation in acetic acid [Section 9.2.2(i)].[28] The reaction could in principle by expected to occur with acetyl peroxide and a suitable acid.

The ratio of acetoxylation to nitration is constant for a given aromatic, over a wide (330-fold) reactivity range, but decreases with increasing reactivity of the aromatic. Consequently, acetoxylation has a larger ρ factor than nitration, and a value of -18 has been estimated.[26] The reaction is strongly accelerated by H_2SO_4, so that protonated acetyl nitrate is thought to be the electrophile. The intermediate cation is attacked by acetate to give a diene, many examples of which have been isolated (see structures **15**–**21** in Chapter 7),[27] which then either loses acetic acid to give the nitro product or nitrous acid to give the acetate (see Scheme 7.1). An addition–elimination mechanism is thus involved.

8.4 SULPHONOXYLATION

Reaction of arylsulphonyl peroxides with aromatics gives aryl sulphonates (eq. 8.7).[29] A radical mechanism may be involved unless electron-withdrawing groups are present in the aryl group of the peroxide. The reaction then shows no e.s.r. signal,[30] and hydrogen is not abstracted from the alkyl groups of alkylbenzenes.[31] The reaction is first order in peroxide, is not acid-catalysed, does not give a kinetic isotope effect, and in methylene chloride is first order in all aromatics studied (in ethyl acetate however, a reaction zeroth order in aromatic

Table 8.1 Partial rate factors for reaction of $(RC_6H_4SO_2O)_2$ with XAr

X	R = *o*-NO_2			R = *m*-NO_2			R = *m*-CF_3		
	o	*m*	*p*	*o*	*m*	*p*	*o*	*m*	*p*
OMe				2600		32 000			
Me	14.1	1.1	49.5	18.5	2.0	75.5			
Et	10.5	1.5	44.8	17.3	2.5	76.2	14.8	2.6	52
i-Pr	6.0	2.4	36.8	10.6	2.9	63.2	10.3	2.8	46.1
t-Bu	1.0	3.4	25.6	2.8	7.0	57.1			
Br				0.29[a]	0.042[a]	2.1[a]			
CO_2Me				0.0059	0.017	0.0043			
NO_2				0.0017	0.0046	0.0015			
CH_2OH							9.2	3.1	37.8
COMe							0.15	0.22	0.15

[a]For R = *p*-NO_2 these values become 0.37, 0.068, and 2.1, respectively.

was detected in the reaction with benzene).[32]

$$ArH + Ar'SO_2OOSO_2Ar' \longrightarrow ArOSO_2Ar' + Ar'SO_3H \qquad (8.7)$$

The presence of the strongly electron-withdrawing groups in the aryl ring of the peroxide, combined with the greater electronegativity of the SO_2 moiety compared with the CO moiety, accounts for the reaction taking place without a catalyst, in contrast to benzooxylation. Reaction may be envisaged as occurring by nucleophilic displacement of one OSO_2Ar' group by attack of the aromatic ring upon the other, this being aided by electron withdrawal in Ar′.

Partial rate factors have been obtained for the reaction of the *m*- and *p*-nitro- and *m*-trifluoromethyl-substituted peroxides (Table 8.1).[31–33] These show clearly the electrophilic nature of the reaction (confirmed by other workers),[34] which is strongly sterically hindered as expected. Moreover, greater electron withdrawal in the aryl ring of the peroxide creates a more reactive electrophile which is correspondingly less selective. This was also confirmed by the reactivities relative to benzene of *p*-xylene and mesitylene, which were 200 and 1350, and 340 and 2420 for reaction with the *p*- and *m*-nitro compounds, respectively.[31,32] The ρ factor for the reaction with the *m*-nitro peroxide is -4.4, so the electrophile is very reactive.

8.5 SULPHENYLATION

This is the sulphur equivalent of aroxylation and gives thioethers (and derivatives).[35,36] The reaction, represented by eq. 8.8, has also been referred to as aryl- or alkylthiolation, sulphenation, and sulphuration. The reaction may carried out successfully with R = Ar′ if iron is present as a catalyst, and substitution in the positions normally most reactive towards electrophilic substitution suggests that the reaction is of this type.[35] The arylsulphenyl chlorides also disproportionate to give diaryl disulphides, Ar'SSAr', and chlorine which chlorinates some of the aromatic ArH, so a mixture of products is obtained.[35,37]

$$ArH + RSCl \longrightarrow ArSR + HCl \qquad (8.8)$$

Symmetrical diaryl sulphides may be obtained by the reaction of aromatic ArH with sulphur halides (eq. 8.9).[35]

$$2ArH + S_xCl_2 \xrightarrow{\text{Fe or FeCl}_3} ArSAr + 2HCl + (S) \qquad (8.9)$$
$$(x = 1 \text{ or } 2)$$

The method given in eq. 8.8 is less satisfactory with MeSCl because it is unstable. However, the MeS group has been introduced successfully (50–85% yields) by the use of methyl methylthiosulphonate, $MeSSO_2Me$, and likewise PhS by the use of $PhSSO_2Me$. The activating effect of MeS and consequent polysubstitution prevents higher yields being obtained.[38] This appears to account for the variety of polysubstituted products obtained in the sulphenylation of benzene with polyphenylene sulphide–$AlCl_3$; the products do not arise simply from fragmentation of the PPS because the yield is 1.6–1.8 times the amount of this reagent that is used.[39]

Reaction of acetylsulphenyl chloride, MeCO·SCl, with aromatics gives the product ArSS·COMe.[40] The mechanism proposed for the reaction included a termolecular process, but this must be considered improbable.

8.6 THIOCYANATION

Thiocyanation with thiocyanogen (eq. 8.10, X = SCN) takes place with very reactive aromatics such as phenols and amines,[41] azulene,[42] and azupyrene.[43] Less reactive aromatics, e.g. anilides and ethers,[44] can be thiocyanated by using the more polar thiocyanogen chloride (eq. 8.10, X = Cl).[44]

$$ArH + XSCN \longrightarrow ArSCN + HX \qquad (8.10)$$

8.7 SULPHINYLATION

Sulphinylation (eq. 8.11) produces sulphoxides, and has been studied for R = p-MeC_6H_4.[45] Relative reactivities have been determined as follows (benzene = 1): toluene, 420; *o*-xylene, 7600; *m*-xylene, 59 000; *p*-xylene, 970; and mesitylene, 250 000. The reaction is severely sterically hindered, as shown by 99.2% of substitution in *o*-xylene taking place at the 4-position. Electron withdrawal in the aryl ring gave increased toluene:benzene rate ratios,[45] showing that free cations are not produced, and this is due to the lower electron withdrawal by the SO group compared with the SO_2 group (see sulphonylation). Reaction therefore takes place by nucleophilic displacement by the aromatic on the sulphinyl chloride [see also acylation, Section 6.7.1.(iii)]. The mechanistic interpretation of these results in the original paper is incorrect (cf. ref. 46).

$$ArH + RSOCl \xrightarrow{AlCl_3} ArSOR + HCl \tag{8.11}$$

In attempted sulphinylation of anisole with *p*-toluenesulphinyl chloride using either $ZnCl_2$ or BF_3–Et_2O as catalysts, the sulphoxide product was reduced to the diaryl sulphide, the reduction being apparently carried out by excess sulphinyl halide–catalyst complex.[47] This sulphenylation does not occur if $AlCl_3$, $SnCl_4$, or $SbCl_5$ is used as the catalyst.

Trimethylsulphonium tetrafluoroborate, $Me_3{}^{\delta+}BF_4{}^-$ reacts with anisole to give *p*-$MeOC_6H_4S^+Me_2$; the electrophile here may be the cation radical $Me_2S^{+\bullet}$.[48]

8.8 SULPHONYLATION

Sulphonylation (eq. 8.12) produces sulphones, and the mechanism of the reaction differs according to the conditions under which it is carried out. It is also possible to sulphonylate using sulphones, but these are much less reactive than sulphonyl chlorides.[49]

$$ArH + RSO_2Cl \xrightarrow{\text{Lewis acid}} ArSO_2R + HCl \tag{8.12}$$

The relative efficiencies of catalysts (in methanesulphonylation) are $AlCl_3 > FeCl_3 > SbCl_5$,[50] so $AlCl_3$ is the catalyst normally employed. This coordinates with both the sulphonyl chloride starting material and especially the sulphone product, which is therefore isolated as a 1:1 addition complex[51] [cf. acylation, Section 6.7.1(ii)].

8.8.1 Mechanism of Sulphonylation

Studies have been carried using solutions of either excess sulphonyl chloride or some other solvent. The mechanism of the reaction resembles acylation, in that reactive aromatics substitute by attacking the undissociated electrophilic reagent whereas unreactive ones require prior dissociation.

(i) Sulphonylation in excess sulphonyl halide[51–53]

Benzenesulphonylation in excess benzenesulphonyl chloride as solvent gave second-order kinetics, rate = $k_2[AlCl_3][PhMe]$, for reaction with toluene, and an order, said to be three-halves, rate = $[AlCl_3]^{\frac{1}{2}}[PhCl]$, for reaction with chlorobenzene,[52] but the evidence for the latter is less convincing; the concentration of $AlCl_3$ is equivalent to the concentration of the 1:1 addition complex $PhSO_2Cl{\cdot}AlCl_3$. The reaction with toluene was retarded by the presence of a less polar solvent, cyclohexane, indicating that the addition complex must be ionized prior to reaction (eq. 8.13). The rate-determining step is then attack of the aromatic on the ionized entity, $PhSO_2^+AlCl_4^-$ (eq. 8.14).

$$PhSO_2Cl{\cdot}AlCl_3 \rightleftharpoons PhSO_2^+AlCl_4^- \qquad (8.13)$$

$$ArH + PhSO_2^+AlCl_4^- \rightleftharpoons ArSO_2Ph{\cdot}AlCl_3 + HCl \qquad (8.14)$$

$$PhSO_2^+AlCl_4^- \rightleftharpoons PhSO_2^+ + AlCl_4^- \qquad (8.15)$$

The kinetics for substitution in the less reactive chlorobenzene were said to be consistent with the necessity for predissociation of the ionized complex (eq. 8.15). However, the kinetic argument given in support of this[52] is incorrect, and also contradicts the view of the same workers that *acylation* by a similarly ionized complex would give second-order kinetics [see Section 6.7.1.(iii)]. While it is possible that predissociation occurs, the fractional order in $AlCl_3$ (and hence the complex) could, for example, be related to an anticipated lower tendency for complexing of the sulphone product bearing an electron-withdrawing group in the aromatic ring.

The relative effects of substituents in the aryl ring of the sulphonyl halide are[53] 4-OMe 3.0, 4-I 0.7, 4-Br(Cl) 0.5, and 3-NO_2 0.06 (similar effects were observed in the reaction with naphthalene),[54a] and this is consistent with attack of the aromatic on the polarized complex [cf. acylation, Section 6.7.1(ii)]. The kinetic isotope effect, $k_H{:}k_D$, of 0.86[54b,c] which indicated the last step of the reaction to be non-rate contributing may be in error; a more recent determination (for *p*-toluenesulphonylation) gave a value of 1.9–2.5.[55]

Sulphonylation with alkylsulphonyl chlorides gives concurrent alkylation, and this is more significant the greater the stability of the alkyl cation.[49,56] Chlorosulphonylbenzoyl chloride, $ClO_2SC_6H_4COCl$, gives acylation rather than sulphonylation,[57] although this does not necessarily show that the former is the faster reaction, since the electrophile is this acylation has a much stronger electron-withdrawing group present than does the electrophile in sulphonylation.

(ii) Sulphonylation in the presence of solvents

With the sulphonyl halide no longer in excess, this now appears directly in the rate equations. For benzenesulphonylation in nitromethane, the greater polarity of the medium could be expected to facilitate ionization of the halide–catalyst complex (eq. 8.15). With very reactive aromatics, e.g. mesitylene, the kinetics of benzenesulphonylation are zeroth order in aromatic and second order overall, rate = k_2[$AlCl_3$][$PhSO_2Cl$], whereas for compounds of the reactivity of benzene or less, the kinetics are third order, rate = k_3[ArH][$AlCl_3$][$PhSO_2Cl$].[58] In the former case, ionization of the complex (eq. 8.13) is the rate-determining step, whereas for reaction of less reactive aromatics, the ionization is fast compared with subsequent reaction of the electrophile with the aromatic (eq. 8.14), which is then rate determining.[58] The kinetic isotope effect, k_H:k_D, was reported as ca 1.0 for benzenesulphonylation in the solvents $PhNO_2$, $MeNO_2$, and $CFCl_3$.[55,59,60] However, a more recent determination for *p*-toluenesulphonylation gave values of 1.5 ($MeNO_2$) and 3.3 (CH_2Cl_2).[61]

For *p*-toluenesulphonylation in dichloromethane, the kinetics were also third order, but i.r. showed no evidence of free cations, and substantial kinetic isotope effects (2.0–2.8) were also obtained. Reaction appears here to involve rate-determining attack of the aromatic on the halide–catalyst complex, with proton loss from the intermediate being partially rate determining.[62]

8.8.2 Substituent Effects

These show a number of inconsistencies which suggest that the mechanism is particularly dependent on the conditions, and mixing control (Section 2.10) may also be involved.

For benzenesulphonylation of toluene in nitrobenzene, the toluene:benzene rate ratio is 8.0, and partial rate factors (*ortho*, *meta*, *para*) are 6.8, 2.1, and 30.2, respectively; alkylbenzenes gave the Baker–Nathan activation order,[58] as expected under the highly solvating conditions (Section 1.4.3). These data suggest, unexpectedly, that the electrophile is more reactive than the nitronium ion, and it is possible that mixing control is involved. For example *p*-methoxybenzenesulphonylation gave a toluene:benzene rate ratio of 82, whereas *p*-nitrobenzenesulphonylation gave a ratio of only 2.8,[63] and mixing control is clearly implicated in the latter reaction. (Similarly, sulphonation by Et_2NSO_2Cl gave a ratio of 42, whereas with $EtSO_2Cl$ a ratio of only 3.8 was obtained.[63])

p-Bromobenzenesulphonylation (no added solvent) gave a toluene:benzene rate ratio of 41.3,[53] indicating that a less reactive electrophile is involved than in benzenesulphonylation. However, if the electrophile was the free cation, it should be more reactive, so nucleophilic displacement by the aromatic upon the electrophile complex may occur here. Methanesulphonylation gives a *para*:*meta* ratio of 2.4[50] (cf. 7.2 for benzenesulphonylation), showing the electrophile to be less selective and therefore more reactive, which is not consistent with both reagents substituting by the same mechanism, since the positive charge on sulphur should be greatest in benzenesulphonylation. The lack of *ortho* substitution of halobenzenes in methanesulphonylation[50] is presumably due to the lack of demand for resonance stabilization of the transition state under these conditions.

Partial rate factors (f_o, f_p) for *p*-toluenesulphonylation of toluene in various solvents are —, 52.5 (CH_2Cl_2), 12.6, 31.4 ($MeNO_2$), and 16.0, 38.3 ($PhNO_2$), and under the former conditions, $f_p^{Cl} = 0.224$, $f_p^{Br} = 0.164$.[61]

8.9 SULPHONATION

Sulphonation is an important electrophilic substitution because it is easy to carry out, is very sterically hindered, and the sulphonic acid products are uniquely water soluble. The latter property is particularly valuable for dyestuffs, many of which are therefore sulphonic acids. The bulk of the sulphonic acid group makes it an excellent blocking group, forcing subsequent electrophilic substitutions to take place at remote sites, the sulphonic acid being subsequently removed by acid (protiodesulphonation, Section 4.12). This latter demonstrates another feature of the reaction, namely that it is reversible.

The severe steric hindrance to sulphonation gives rise to two other aspects. First, the reaction products tend to rearrange to the thermodynamically most stable isomers,[64–67] i.e. those in which steric hindrance is less. The best known example of this is the rearrangement, on heating, of naphthalene-1-sulphonic acid to the 2-sulphonic acid. This may occur alongside normal substitution so that whereas sulphonation of naphthalene with concentrated H_2SO_4 at 80 °C gives 96% of the 1-sulphonic acid, at 165 °C 85% of the 2-sulphonic acid is obtained.[68] The substitution pattern in sulphonation tends to be atypical for an electrophilic substitution, not only as a result of having this thermodynamic rather than kinetic control of substitution, but also because of both the severe steric hindrance, and the reversibility of the reaction; the strong acid conditions may also cause migration of alkyl groups during substitution.

Sulphonation may be accomplished by using either concentrated aqueous sulphuric acid, fuming sulphuric acid, SO_3 in aprotic solvents, or halosulphonic acids. Disubstitution can occur and other byproducts, produced to varying extents when free SO_3 is present, are sulphonic anhydrides (from intermolecular dehydration of monosulphonic acids, or intramolecular dehydration of *ortho*-disulphonic acids) and diaryl sulphones [from electrophilic substitution by the sulphonic acid (or sulphonyl chloride that can be produced when chlorosulphonic acid is used)]; the sulphones may also be further sulphonated.

8.9.1 Mechanism of Sulphonation

Elucidation of the mechanism of sulphonation is complicated by the large number of species that may be present under the reaction conditions, especially in sulphuric acid; the topic has been reviewed.[69,70] The bulk of the electrophile means that in the S_E2 mechanism for substitution (eq. 2.4), k_{-1} becomes large relative to k_2, and a kinetic isotope effect results (see Section 2.1.3). Kinetic isotope effects have been measured both intermolecularly and intramolecularly, i.e., in the latter case the relative amount of substitution at two equivalent sites, one of which is deuteriated, in a symmetrical molecule gives the isotope effect. As in other electrophilic substitutions, the isotope effect increases with increasing hindrance to substitution, as a result of the increase in k_{-1}. The *ortho*:*para* ratios for sulphonation vary markedly with conditions, indicating that the electrophile differs accordingly.[67]

(i) *Sulphonation by sulphur trioxide in aprotic solvents*

Sulphur trioxide forms complexes with $MeNO_2$, $PhNO_2$, and 1,4-dioxane, but not with halogenated solvents. Sulphonation in each of these solvents produces the pyrosulphonic acid $ArSO_2OSO_2OH$ due to complexing of SO_3 with the sulphonic acid, and the problem is that it is not possible to ascertain at what stage this occurs.

For sulphonation in halogenated solvents, the reaction is first order in both aromatic and SO_3,[71] consistent with the mechanism given by eq. 8.16. Since sulphonic acids and SO_3 react to give crystalline solids,[72] it is feasible that complexing with the SO_3 occurs after formation of the sulphonic acid. However, Bosscher and Cerfontain have proposed that the second molecule of SO_3 is attached in a fast step prior to transfer of the proton from the aromatic ring to the sulphonate anion.[71] The proton transfer step is not rate-limiting except when the reaction site is significantly hindered sterically. Thus values of $k_H:k_D$ (at ca $-35\,°C$) are 1.23 (benzene), 1.1 (1,4-dichlorobenzene),[60,71] 2.0 (naphthalene), and 3.1 (1,2,4,5-tetramethylbenzene);[73] with SO_3 in dichloroethane, naphthalene gave a value of 1.23 at $-5\,°C$ (although this was thought to be due to a secondary effect since octadeuterionaphthalene was used).[74]

$$ArH + SO_3 \longrightarrow Ar^{+}\!\!<^{H}_{SO_3^-} \longrightarrow ArSO_3H \qquad (8.16)$$

Hinshelwood and coworkers[75] found sulphonation in $PhNO_2$ to be second order in SO_3, and this order has been found also when using $MeNO_2$ and 1,4-dioxane as solvents.[76] This is consistent with one molecule of SO_3 assisting polarization of the other,[75] this behaviour being common for electrophilic substitutions involving formally neutral electrophiles; the second molecule of SO_3 cannot assist removal of the proton in a rate-determining step (cf. ref. 75) since a large kinetic isotope effect would then be observed.

By contrast, Cerfontain[70] proposed the mechanism given by eqs 8.17–8.19, in which 8.18 is rate limiting and 8.17 is fast.[70]

$$ArH + SO_3 \overset{\text{fast}}{\rightleftharpoons} Ar^{+}\begin{matrix} H \\ SO_3^{-} \end{matrix} \quad (8.17)$$

$$Ar^{+}\begin{matrix} H \\ SO_3^{-} \end{matrix} + SO_3 \overset{\text{slow}}{\rightleftharpoons} Ar^{+}\begin{matrix} H \\ S_2O_6^{-} \end{matrix} \quad (8.18)$$

$$Ar^{+}\begin{matrix} H \\ S_2O_6^{-} \end{matrix} \longrightarrow ArS_2O_6H \quad (8.19)$$

The last step of the reaction is relatively fast but becomes rate-limiting when the reaction site is sterically hindered. Values of the kinetic isotope effect, $k_H:k_D$, that have been obtained in $MeNO_2$ at 0 °C are 1.1 (benzene),[60] 1.2 (mesitylene),[77] 5.6 (1,2,4,5-tetramethylbenzene),[73] and 1.8 (naphthalene);[73] benzene also gave a value of 1.35 in $PhNO_2$ at 25 °C.[60] In dioxane, values obtained were 3.8 (1,6-methano[10]annulene, 12 °C) and 7 (anthracene, 40 °C).[73]

(ii) Sulphonation by sulphuric acid

The rate of sulphonation by sulphuric acid increases as the concentration increases, but there is a dramatic rate increase in the region of 100 wt-% acid.[78,79] This suggests that the mechanism in sulphuric acid containing free water is different from that containing free SO_3. Sulphonation in sulphuric acid is accompanied by protiodesulphonation, which has a higher activation energy at any given acid concentration. The difference in activation energies increases with increasing acidity, and consequently the rates of desulphonation increase less readily with increasing acid concentration than do rates of sulphonation.[80] In aqueous sulphuric acid no byproducts are produced.

There have been a large number of studies of the mechanism of sulphonation in sulphuric acid media,[79–85] and a variety of conclusions reached concerning the mechanism. Unravelling the mechanism is a formidable kinetic task because the reaction products are difficult to isolate, desulphonation may be occurring, the formation of water during the kinetic runs causes the rate coefficients to decrease with time, and there are a large number of species present. Any equilibrium which produces a water molecule then requires a further molecule of sulphuric acid to ionize the water via eq. 8.20.

$$H_2O + H_2SO_4 \rightleftharpoons H_3O^+ + HSO_4^- \quad (8.20)$$

Hence equilibria leading to possible sulphonating species are eqs 8.21–8.24.

$$2H_2SO_4 \rightleftharpoons H_3SO_4^+ + HSO_4^- \quad (8.21)$$

$$2H_2SO_4 \rightleftharpoons SO_3 + H_3O^+ + HSO_4^- \quad (8.22)$$

$$3H_2SO_4 \rightleftharpoons HSO_3^+ + H_3O^+ + 2HSO_4^- \quad (8.23)$$

$$4H_2SO_4 \rightleftharpoons S_2O_6 + 2H_3O^+ + 2HSO_4^- \quad (8.24)$$

$$SO_3 + H_2SO_4 \rightleftharpoons H_2S_2O_7 \quad (8.25)$$

Many of the conclusions of earlier work were shown subsequently to be deficient to various extents.[69] The following is the current view; the nature of the electrophile cannot be precisely defined under some conditions since it may be represented in different ways which are nevertheless kinetically equivalent:

(a) The electrophile is predominantly $H_3SO_4^+$ in <80 wt-% acid, changes to predominantly $H_2S_2O_7$ in >85 wt-% acid,[83–85] and in oleum it is SO_3.[82] Thus with increasing acidity the electrophile progresses from SO_3 solvated by H_3O^+, then by H_2SO_4, through to the poorly or unsolvated species. Consistent with this interpretation is the fact that the concentration of H_2SO_4 becomes greater than that of H_3O^+ in ca 91 wt-% acid,[86] and the rate of hydrogen exchange (dependent on the proton concentration) increases less with increasing acidity than does the rate of sulphonation.[87] Moreover, the very much higher reactivity of unsolvated SO_3 accounts for the dramatic increase in sulphonation rate at 100 wt-% acid, noted above.

These studies necessitated the use of a range of aromatics of decreasing reactivity with increasing acid strength; the acid strength at which reaction occurs predominantly by one electrophile changes to reaction via another depends on the reactivity of the aromatic.

(b) Since the electrophile changes with acidity, changes in isomer ratios with acidity could be expected. This is found, and $SO_3(H_3O^+)$ *appears* to be bulkier than $SO_3(H_2SO_4)$ since the *ortho*:*para* ratios for toluene[88] and ethylbenzene,[89] and the 3:4 ratio for *o*-xylene,[90] each decrease with decreasing acidity. The changes in the ratios do not give a direct measure of the relative sizes of the electrophiles, however, since $SO_3(H_3O^+)$ is the less reactive and hence more selective electrophile, as shown by the fact that the α:β reactivity ratio for naphthalene increases with decreasing acidity.[91] It is also possible, although this has not been considered previously, that the decreasing proportion of *ortho* substitution with decreasing acidity is not due to involvement of a larger electrophile, but rather to protiodesulphonation (which is sterically accelerated and therefore faster at *ortho* positions) being more important at lower acidities (Section 4.12).

(c) Measurements of kinetic isotope effects in these media is complicated in principle by the occurrence of acid-catalysed hydrogen exchange. However, at high acidities this is less significant and values of k_H:k_D of 1.6–2.1, independent of acidity, were obtained for sulphonation of aryltrimethylammonium ions and nitrobenzene at 25 °C;[82] for sulphonation of chlorobenzene, values of 1.25 (95 wt-% acid) to 2.5 (97.5 wt-% acid) have been reported, and this is consistent with the fact that the concentration of the base HSO_4^- becomes very low at around these acidities.[83]

The mechanism is therefore believed to vary with conditions in a manner unique for an electrophilic substitution, as follows.

In <85 wt-% acid the mechanism consists of eqs 8.26–8.28:

$$ArH + H_3SO_4^+ \underset{}{\overset{slow}{\rightleftharpoons}} Ar^+\!\!<^{H}_{SO_3^-} + H_3O^+ \quad (8.26)$$

$$Ar^+\!\!<^{H}_{SO_3^-} + HSO_4^- \overset{fast}{\rightleftharpoons} ArSO_3^- + H_2SO_4 \quad (8.27)$$

$$ArSO_3^- + H_3O^+ \overset{fast}{\rightleftharpoons} ArSO_3H + H_2O \quad (8.28)$$

In 90–97 wt-% acid the electrophile becomes H_2SO_4–solvated SO_3, so the first step of the mechanism is replaced by eq. 8.29:

$$ArH + H_2S_2O_7 \overset{slow}{\rightleftharpoons} Ar^+\!\!<^{H}_{SO_3^-} + H_2SO_4 \quad (8.29)$$

In 97–100 wt-% acid the observed isotope effect indicates that removal of the proton from the intermediate is rate limiting, hence reaction of the electrophile with the aromatic (eq. 8.29), is now fast, and eq. 8.27 is slow.

In oleum, the electrophile is thought to be unsolvated SO_3 which, being more reactive than the solvated species, reacts with the aromatic in a relatively fast step (eq. 8.30), followed by eqs 8.27 and 8.28, one of which must be a slow step.

$$ArH + SO_3 \overset{fast}{\rightleftharpoons} Ar^+\!\!<^{H}_{SO_3^-} \quad (8.30)$$

(iii) Sulphonation by halosulphonic acids

Fluorosulphonic acid, HSO_3F, readily dissociates into HF and SO_3,[92] so is not a sulphonating reagent of choice. It is a more powerful sulphonating species than concentrated sulphuric acid[93] and gives sulphonic acids, sulphones, and sulphonyl fluorides,[94] but there have been no mechanistic studies.

Chlorosulphonic acid similarly gives sulphonic acids, sulphones, and sulphonyl chlorides, the latter being formed only in the presence of an excess of chlorosulphonic acid.[95] An earlier mechanism proposed for formation of the sulphonyl chloride (eqs 8.31 and 8.32)[96] is incorrect since the reaction given by eq. 8.32 has been shown to be slower than the observed overall rate of formation of the sulphonyl chloride.[97] Chlorosulphonation of benzene or toluene in 1,2-dichloroethane is first order in aromatic and third order in chlorosulphonic acid, the latter order probably being due to the formation of the ion pairs as the electrophile, as shown in eq. 8.33; in this work toluene was five times as reactive as benzene.

$$ArH + HSO_3Cl \rightleftharpoons ArSO_3H + HCl \quad (8.31)$$

$$ArSO_3H + HSO_3Cl \rightleftharpoons ArSO_2Cl + H_2SO_4 \quad (8.32)$$

$$3ClSO_3H \rightleftharpoons SO_2Cl^+ + 2SO_3Cl^- + H_3O^+ \quad (8.33)$$

A study of the sulphonation of toluene, benzene, chlorobenzene, and *o*-xylene by $ClSO_3H$ in CH_2Cl_2 indicated the electrophile to be ClS_2O_6H, whereas in $MeNO_2$ as solvent it appeared to be $MeNO_2 \cdot SO_3^+$. Under the latter conditions the ρ factor for the reaction is ca -10.[98] Sulphonation of benzene with chlorosulphonic acid gave kinetic isotope effects, $k_H:k_D$, of 1.55–1.70.[61]

Sulphonation also occurs with sulphuric acid in acetic acid or TFA.[99] In mixtures of Ac_2O with H_2SO_4, oleum, or $ClSO_3H$, or of SO_3 or oleum in AcOH, the reagent apparently is acetylsulphuric acid, $AcOSO_2OH$.[100]

8.9.2 Substituent Effects

Partial rate factors for sulphonation vary widely according to the conditions, e.g. the relative reactivity of toluene to benzene decreases from ca 100 in 79.0 wt-% acid to 30 in 97.0 wt-% acid.[101,102] This accords with the belief that the electrophile, and the mechanism for sulphonation vary with conditions. It is important, though, to be certain that both substrates are completely soluble in the reaction medium, which is a poor solvent, especially at low acidities.

(i) Alkyl substituents

Partial rate factors for sulphonation of alkylbenzenes are given in Table 8.2; values of f_p^{Me} have also been determined as 301, 150, and 140 in 81.9, 85.5, and 86.3 wt-% H_2SO_4, respectively.[73]

The Baker–Nathan activation order is obtained due to steric hindrance to solvation in the highly solvating medium (see Section 1.4.3), and this presumably

Table 8.2 Partial rate factors for sulphonation of alkylbenzenes, PhR

R	f_o	f_m	f_p	Conditions	T/°C	Ref.
Me	57	11	2100	$ClSO_3H$ in $MeNO_2$	25	98
	68	12	2400	$ClSO_3H$ in $MeNO_2$	0	98
	114	13	3100	$ClSO_3H$ in $MeNO_2$	−25	98
	6.0	10.4	182	SO_3 in SO_2	−10	104
	69	6.7	490	77.8 wt-% H_2SO_4	25	105
	63.4	5.7	258	82.3 wt-% H_2SO_4	25	101
	51	4.1	174	84.3 wt-% H_2SO_4	25	105
	37	3.0	84	89.1 wt-% H_2SO_4	25	89,103
Et	16		82	89.1 wt-% H_2SO_4	25	89,103
i-Pr	0.8		28	89.1 wt-% H_2SO_4	25	89,103
t-Bu	0	3.0	53	86.3 wt-% H_2SO_4	25	103
	0	1.5	18	89.1 wt-% H_2SO_4	25	89,103
t-BuCH$_2$	1.2	5.7	82	89.9 wt-% H_2SO_4	25	106

also accounts for the lower activation by the neopentyl substituent relative to *tert*-butyl, since hydrogen exchange (Section 3.1.2.1) shows that the electron release by these substituents is in the reverse direction. The relative proportions of substitution *ortho* to these substituents also confirms the greater steric requirement of the *tert*-butyl substituent.

The proportion of *ortho* substitution decreases with increasing acidity due to increasing steric requirements of the electrophiles as noted above. For *tert*-butylbenzene, the *para*:*meta* substitution ratio, which decreases on going to stronger aqueous H_2SO_4, increases above 100 wt-% acid, reinforcing the proposal that there is a change in mechanism at this acidity.[103] The large steric requirement for sulphonation is also confirmed by the percentages of *ortho* substitution for toluene, which are 6% (using $ClSO_3H$, FSO_3H, $H_2S_2O_7$, PhS_2O_6H, and MeS_2O_6H) and ca 13% (using SO_3 in $MeNO_2$).[107] The latter decreases with increasing temperature, as was also found using SO_3 or $ClSO_3H$ in various solvents.[108]

In sulphonation of $PhNMe_3^+$ in oleum, a 4-Me substituent increased the reactivity 406-fold and 240-fold in 100.9 and 102.6 wt-% H_2SO_4, respectively.[82] Here methyl is *ortho* to the reaction site, so these are extremely large f_o values, especially in a reaction that is very sterically hindered, and wholly different from those obtained in the aqueous acid. In sulphonation of nitrobenzene, the activating effect of a 4-Me substituent was only 28-fold owing to protonation of the nitro group in the methyl derivative, but not in nitrobenzene itself.[82] Correction for this gives an activation of ca 210-fold, again very large for a f_o value. For sulphonation of nitrobenzene by SO_3 in $PhNO_2$ at 40 °C, a 4-Me substituent also gave a large (121-fold) activation, whilst 4-OMe (i.e. *ortho* methoxy) activated 8.0×10^5-fold.[82]

(ii) Polyalkyl and cyclialkyl substituents

The relative rates of sulphonation of some polymethylbenzenes (Table 8.3) again show the decreasing rate spread with increasing acidity. This variation is also shown by the isomer yields given in Scheme 8.1; the high steric hindrance causes marked departures between calculated and observed partial rate factors for sulphonation in 90.1 wt-% acid, especially at the 2-position of *m*-xylene (40-fold), and at the highly buttressed 4-position of 1,2,3-trimethylbenzene (20-fold).[109] The 5-position of 1,2,4-trimethylbenzene was also 17.5-fold less reactive than predicted, attributed to an 'earlier transition state' than that for reaction of benzene and toluene upon which the calculations were based. However, hydrogen exchange results (Section 3.1.2.2) show that even for a reaction of much higher ρ factor this produces only a minor discrepancy, and it is significant that in sulphonation the reactivity of the buttressed 6-position was almost exactly that predicted, suggesting that an experimental artifact accounts for the anomalously low reactivity of the 5-position. Partial rate factors for

Table 8.3 Relative rates for sulphonation of polymethylbenzenes, ArR, by H_2SO_4 of different concentrations

R	77.8 wt-%[90,101]	81.5 wt-%[79,83]	84.3 wt-%[90,101]
H	1.0	1.0	1.0
Me	106	47	25
1,4-Me_2	265	210	180
1,2-Me_2	387	236	—
1,3-Me_2	1660	840	620
1,2,4-Me_3	—	808	530
1,2,3-Me_3	—	1200	660
1,3,5-Me_3	—	2750	1020

sulphonation of *o*-xylene by $ClSO_3H$ in $MeNO_2$ are 230 and 11 800 at the 3- and 4-positions, respectively,[98] indicating the electrophile to be much less reactive and more hindered under this condition.

90.1 wt-% H_2SO_4:

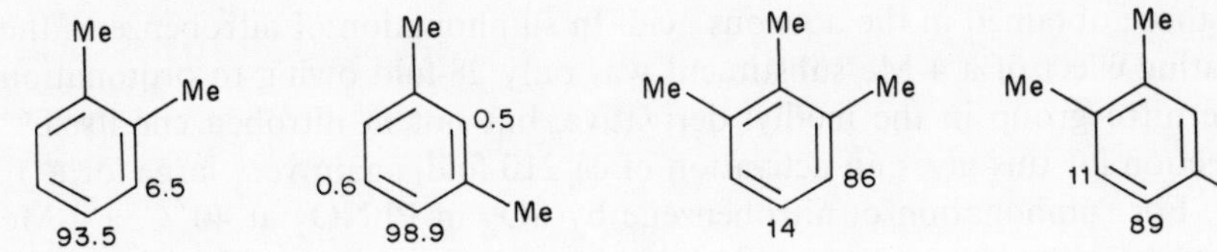

98.4 wt-% H_2SO_4:

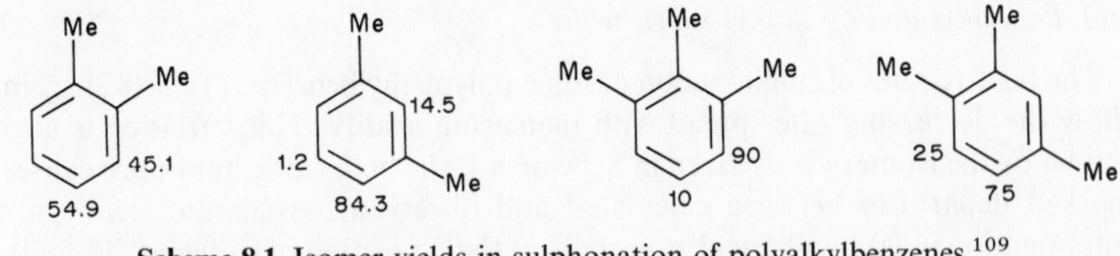

Scheme 8.1 Isomer yields in sulphonation of polyalkylbenzenes.[109]

Sulphonation of polyethylbenzenes showed similar features to the above, and in addition the sulphonic acids derived from 1,2,3,5- and 1,2,4,5-tetraethylbenzenes were unstable owing to the extreme steric hindrance. Consequently, in 98.4 wt-% H_2SO_4 they underwent rearrangement to give, in each case, 2,3,4,5-tetraethylbenzenesulphonic acid.[110] It is notable here that the steric hindrance to the formation of the sulphonic acids parallels that found for the methylbenzenes, as indicated by the observed:calculated reactivity ratios noted above. Sulphonation of polyisopropylbenzenes under the same conditions gives even more hindered products so that here those derived from the trialkylbenzenes are unstable and rearrange (accompanied by some dealkylation), whereas those derived from the tetralkylbenzenes dealkylate.[111] Thus, 2,3,5,6-tetraisopropylbenzenesulphonic acid gave 2,4,5-triisopropyl- and 3,5-diisopropylbenzenesulphonic acids, each of these being formed also from 2,4,6-triisopropylbenzenesulphonic acid.

t-Bu, Bu-t → (80%) t-Bu, HO_3S, Bu-t ← (6%) t-Bu, t-Bu → (67%) t-Bu, SO_3H

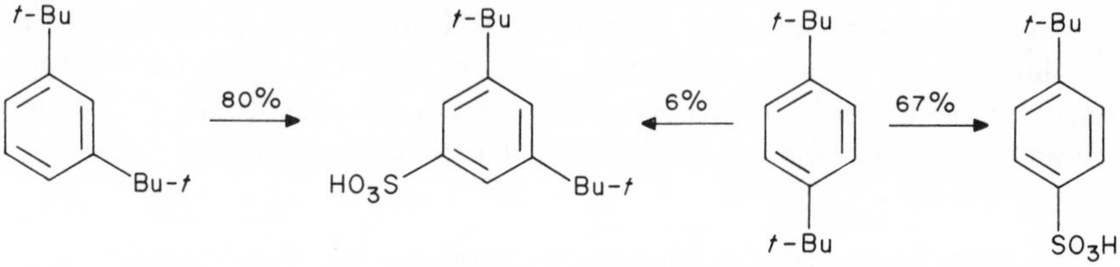

(8.34)

Sulphonation of poly-*tert*-butylbenzenes shows additional features due to the extreme difficulty of substitution *ortho* to the *tert*-butyl group. Thus, sulphonation of 1,3- and 1,4-di-*tert*-butylbenzenes by 103 wt-% acid (eq. 8.34) gives rise to dealkylation, alkyl group migration, and substitution *meta* to the alkyl groups.[112] The use of oleum for this study produced substantial amounts of sulphonic anhydrides (both intermolecular and intramolecular) as byproducts.

Sulphonation of benzcycloalkenes by sulphuric acid[113] contrasts with other electrophilic substitutions in that indane does not exhibit abnormally low reactivity of the position α to the side-chain, compared with *o*-dialkylbenzenes and tetralin (cf. Section 3.1.2.3). Thus for *o*-xylene, indane, tetralin and benzsuberane the $f_\alpha : f_\beta$ values were 0.43, 0.67, 0.86, and 0.26, respectively. This is due here to the importance of the reduced steric requirement of the α-position in indane, a view reinforced by the results obtained using SO_3 in $MeNO_2$, conditions giving much greater steric hindrance, the corresponding ratios being 0.07, 0.18, 0.13, and 0.03. Thus the bulkiest substituents give much lower ratios, whilst indane gives the highest ratio in the group. Sulphonation of benzcyclobutene by SO_3–dioxane in $CFCl_3$ or by SO_3–$MeNO_2$ in $MeNO_2$ gives an *ipso*:3-:4-substitution ratio of 25:5:70, so that here the effect of strain outweighs any other consideration. The *ipso* substitution gives rise to the sultone of 2-phenylethylbenzene-*o*-sulphonic acid (**4**, R = H).[114]

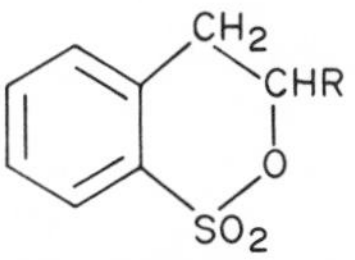

CH_2, CHR, O, SO_2

(**4**) 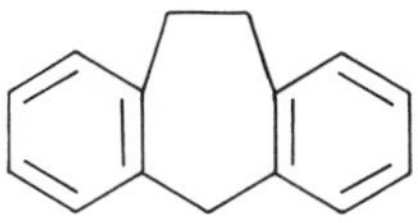 (**5**)

(iii) Substituted alkyl groups

Values of $f_o:f_p$[115] show that steric hindrance to substituents increases along the series $Me < CH_2Me < (CH_2)_2R < CH_2CHR_2 \ll CH_2Ph < CH_2Bu$-*t* and cyclobutyl < cyclopentyl < cyclohexyl ≈ $CHMe_2 \ll CHMePh < CHPh_2$, as expected. For compounds $Ph(CH_2)_nPh$, the amount of *ortho* substitution increases, as expected, with increasing *n*.[116] Sulphonation of 9,10-dihydroanthracene, 10,11-dihydro-5*H*-benzo[*a,d*]cycloheptene (**5**) and fluorene gave in each case the disulphonic acid, with one SO_3H group in each benzenoid ring. For fluorene the monosulphonation pattern paralleled that for hydrogen exchange (Section 3.1.2.10), except that the amount of 4-substitution was reduced through steric hindrance. Sulphonation of triptycene confirmed the low $\alpha:\beta$

Table 8.4 Partial rate factors for sulphonation of compounds $Ph(CH_2)_nX$

X	*n*	f_o	f_m	f_p
SO_3H	0	$<0.04 \times 10^{-8}$	4.0×10^{-8}	0.2×10^{-8}
	1	0.19×10^{-4}	2.3×10^{-4}	4.6×10^{-4}
	2	0.07	0.03	1.9
	3	0.6	0.32	15
	4	6.0		49
	5	12.3		64
	6	14.1		86
OSO_3H	2	<0.1		3.5
	3	1.5		21
	4	4.3		46
	5	12		75
	6	12		93
	10	24		130
NO_2	0	$\leqslant 4 \times 10^{-13}$	9×10^{-11}	3×10^{-12}
	2	0.03	0.07	1.0
	3	0.7	0.8	20
NH_3^+	0	7×10^{-11}	2×10^{-9}	4×10^{-9}
	1	1.2×10^{-5}	1.3×10^{-4}	1.8×10^{-4}
	2	0.017	0.027	0.39
	3	0.6	0.5	13.4
	4	2.4	1.8	62
NMe_3^+	0	$<4 \times 10^{-12}$	16×10^{-11}	6×10^{-11}
	1	5×10^{-8}	10×10^{-7}	4×10^{-7}
	2	$\sim 5 \times 10^{-3}$	0.018	0.18
	3	~0.2	0.7	8.9
	4	~0.6	1.8	52

reactivity ratio found in nitration and hydrogen exchange (Section 3.1.2.4).[116]

In 90–98 wt-% H_2SO_4 ($H_2S_2O_7$), sulphonic acids $Ph(CH_2)_nSO_3H$ are present in the non-ionized form, and partial rate factors (Table 8.4)[117] increase regularly with increasing n, as expected; the $f_o:f_p$ values also increase in this direction. Under the same conditions the corresponding alkanols, $(PhCH_2)_nOH$, are present as the hydrogensulphates and these gave very similar partial rate factors to the sulphonic acids (Table 8.4).[117] The partial rate factors were necessarily obtained under a range of acidities, and consequently the relative values of one compound to another may have a significant margin of error. For the alkanol $Ph(CH_2)_3OH$ in 98.4 wt-% acid the *ortho* derivative (obtained in 12% relative yield) undergoes a further series of reactions, involving the intermediate formation of alkenes and carbocations, to give, ultimately, the sultone (**4**, R = Me).[118]

Partial rate factors for nitro compounds, $Ph(CH_2)_nNO_2$, and amines, $Ph(CH_2)_nNH_3^+$, and their *N*-methyl derivatives also show a regular increase with increasing *n*.[119] As in nitration [Section 7.4.3.(xii)], the methylated poles are less reactive than the unmethylated poles and this may be attributed to steric hindrance to solvation in the highly solvating media; in addition, the point charge in the unmethylated poles may be more easily dispersed by solvation. For $n = 0$, the nitro compound was more deactivating than for either of the corresponding poles, whereas for compounds with $n = 2$ or 3 the converse is true. This fact, coupled with the abnormally low reactivity of nitrobenzene, and the fact that its reactivity was determined in 102–108 wt-% acid, shows that nitrobenzene reacts here as the hydrogen bonded, or possibly even the protonated species.

(iv) Halogen substituents

Sulphonation of halogenobenzenes requires fairly severe conditions, and consequently anhydrides are among the products obtained. The *para*-sulphonic acids are the main products due to steric hindrance. The strongly acidic conditions cause iodo compounds to protiodeiodinate, and this is followed by iodination of the starting material. Thus, for example, with 96 wt-% H_2SO_4 at 120 °C iodobenzene gives 54% *p*-iodobenzenesulphonic acid and 32% *p*-diiodobenzene.[120] Similarly, sulphonation of iodonaphthalene gives tri- and tetraiodonaphthalenes amongst the products,[121] and iodotoluenes also deiodinate.[122]

Partial rate factors for sulphonation are given in Table 8.5. A comprehensive study of sulphonation of halogenotoluenes by either 98 wt-% H_2SO_4 at 25 °C or SO_3 in $MeNO_2$ at 0 °C gave results which were largely independent of the reagent and in agreement with the predicted electronic and steric effects of the substituents.[122] Compounds as unreactive as pentafluorobenzene may be sulphonated.[123]

Table 8.5 Partial rate factors for sulphonation of halogenobenzenes, PhX

X	*ortho*	*meta*	*para*	Conditions	Ref.
F	0.013	<0.003	2.8	88–96 wt-% H_2SO_4, 25 °C	123
Cl	0.0016	0.010	0.39	88–96 wt-% H_2SO_4, 25 °C	123
	0.009[a]		~0.30	SO_3 in $PhNO_2$, 40 °C	75c
	0.32[a]	0.30[a]	~4.1	100 wt-% H_2SO_4 in $PhNO_2$, 40 °C	124
			0.14	$ClSO_3H$ in $MeNO_2$, 25 °C	98
Br			~0.20	91 wt-% H_2SO_4, 12 °C	85
			~0.26	SO_3 in $PhNO_2$, 40 °C	75c
			3.7	100 wt-% H_2SO_4 in $PhNO_2$, 40 °C	124

[a]Calculated assuming that the additivity principle applies.

(*v*) *Sulphonic acids*

There have been a number of studies of the sulphonation of sulphonic acids, additional to those described in Section (iii) above. The results are generally in line with prediction. For example, the reactivity order for polymethylbenzenesulphonic acids is 2,3,4-Me_3- > 2,4-Me_2- > 2,3-Me_2- > 2,4,5-Me_3- > 2,4,6-Me_3- > 2-Me- > 4-Me-benzenesulphonic acid, the results showing the product of electronic and steric (especially buttressing) effects.[125] The isomer distributions are, as usual, dependent on the sulphuric acid concentration but approach constant values below 104 wt-% and above 115 wt-% acid; in this latter region the substrate is believed to be $ArSO_3H_2^+$. By assuming additivity, partial rate factors may be calculated for the effect of the methyl substituent in ≤ 104 wt-% H_2SO_4 as follows:[126] f_o, 250, 160, and 67; f_m, 11, 15, 56, 200; f_p, 620, 8200.[125] Various factors cause the values to be too variable to permit any meaningful conclusions. For example, the *ortho* value of 67 arises through buttressing, whereas the high *para* value of 8200 (derived from 2-methylbenzenesulphonic acid) is due to the SO_3H group being twisted out of the plane of the aromatic ring compared with benzene sulphonic acid itself; the very high *meta* values are clearly anomalous.

Sulphonation of either 2- or 4-substituted benzenesulphonic acids with 115 wt-% H_2SO_4 gave only the corresponding 2,4-disulphonic acid,[126] whereas the 3-substituted acids sulphonated at the 4-, 5-, and 6-positions in positional yields commensurate with the electronic effects of the substituents (F, Cl, Br, Me, *t*-Bu, OMe, and NH_3^+).[127] The use of oleum in this work also resulted in the formation of 4-substituted benzene-1,2-disulphonic anhydrides.

(vi) Biphenyl and derivatives

Sulphonation of biphenyl by H_2SO_4 in the acid range 81.5–90.6 wt-% acid was said to involve $H_3SO_4^+$ as electrophile up to an acid concentration of 86 wt-% but to change thereafter.[128] [The evidence for his latter is, however, weak, since within the quoted experimental error a plot of log (k/s^{-1}) vs log $a(H_3SO_4^+)$ is linear throughout the whole acid range.] In 86 wt-% acid f_p^{Ph} was calculated as 600 ± 300, giving $\rho = -10 \pm 1.2$.

With H_2SO_4 in $MeNO_2$ at 25 °C, biphenyl is sulphonated 1600, 240, and 380 times more readily than the 2-, 3-, and 4-biphenylsulphonic acids respectively.[129] The enhanced former value arises because biphenyl is twisted out of coplanarity by the bulky *o*-SO_3H group. A more detailed study of sulphonation of biphenylsulphonic acids and other biphenyl derivatives in 86.8–95.0 wt-% H_2SO_4 (in which the electrophile was said to be $H_2S_2O_7$, but see above) gave the partial rate factors in Scheme 8.2; the sulphonic acid group is thought to react as the anion under these conditions.[130] The reactivity of the 4-sulphonic acid relative to the 4-nitro compound parallels that found in desilylation (Table 4.2), which suggests that the SO_3H group reacts as the same species (whatever that may be) in both reactions. However, in contrast to desilylation, the CO_2H substituent is more deactivating than NO_2 in sulphonation, suggesting that here it is protonated (cf. ref. 130). There have been a number of other studies of the effects of substituents in sulphonation of biphenyl, all of which lead to the expected results.[131]

0.26 ≤0.025
0.48
SO_3^-
<0.034 0.007
13.4
SO_3^-
0.055 0.005
9.1
SO_3^-
<0.006
0.6
NO_2
<0.001
0.1
COMe
<0.004
0.38
CO_2H

Scheme 8.2. Partial rate factors for sulphonation.

The reactivities of the 4-positions is commensurate with the electron-withdrawing effects of the substituents, coupled with the twisting of the *ortho*-substituted acid out of coplanarity. However, the results for the *meta* positions are anomalous and suggest experimental error; in particular, the high reactivity of the 3′-position of the 2-sulphonic acid may refer to substitution at the 4-

position (which would not be conjugatively deactivated by the SO_3H group, itself twisted out of coplanarity with the benzene ring).

The order of stability of biphenyldisulphonic acids towards isomerization in 75 wt-% H_2SO_4 at 140 or 180 °C is 2,3′ < 2,2′ < 2,4′ ≪ 4,4′ < 3,3′.[132] The instability of the 2-isomers follows from steric hindrance, whilst the greater reactivity of the 3′- compared with the 2′- and 4′-isomers reflects the ease of electrophilic substitution, the process taking place via initial protiodesulphonation. The products were either the 4,4′- or the 3,4′-disulphonic acids.

(vii) Hydroxy, methoxy, and amino substituents

Phenols sulphonate either in the ring, or at oxygen. The latter occurs readily if there are electron-withdrawing substituents in the ring, although it is less common in *ortho*-substituted compounds owing to steric hindrance. Ring sulphonation with H_2SO_4 gives *ortho* and *para* products, and the *ortho*:*para* ratio decreases with increasing acid concentration and temperature, owing to thermodynamic control of the reaction products.[133] Long reaction times and elevated temperatures lead, for the same reason, to as much as 38% of *m*-hydroxybenzenesulphonic acid.[134a] *Ortho* and *para* partial rate factors for sulphonation by 77.8 wt-% H_2SO_4 ($H_3SO_4^+$) at 25 °C are 3400 and 7300 for phenol and 2000 and 7100 for anisole, respectively.[134b] These are relatively low and indicate that the substituents are hydrogen bonded with the solvent. The values of $\log f_o$:$\log f_p$ (0.85, 0.91) are close to the theoretically predicted value (Section 11.2.4), the lower value for anisole being consistent with the result of steric hindrance.

The effects of substituents in sulphonation of anisoles and phenols are very small,[134b,135] consistent with the effect of differential hydrogen bonding; greater electron supply increases the bonding, which attenuates the normal rate increase which would otherwise be expected.

In sulphonation of 2,6-disubstituted phenols with SO_3 in $MeNO_2$, the first-formed phenyl hydrogensulphate is converted to the sulphonic aicd via *O*-desulphonation followed by *C*-sulphonation if the phenol is in excess, but the sequence is reversed if SO_3 is in excess. The ratio of 3- to 4-substitution (which varied as expected according to the electronic and steric effects of the substituents) also varied according to conditions, probably as a result of variable electron release by oxygen, this being conformationally dependent.[136]

Sulphonation of 4-hydroxyazobenzene takes place on an *N*-protonated species, but electron release from OH is sufficiently powerful to produce 4′-substitution; steric hindrance cannot account for the lack of 3-substitution (cf. ref. 137).

Amines undergo both *C*- and *N*-sulphonation and the thermodynamic stability of the products is likely to be more important for sulphonation of phenols in view of the greater steric hindrance. Thus mainly *meta, para* products are obtained, with a high *meta*:*para* ratio;[138,139] more *ortho* and less *meta* product may be obtained in the sulphonation of aniline (as the anilinium sulphate) by using weaker acids, even though higher temperatures are then necessary.[140] Studies are complicated by complex formation between SO_3 and amines, and these complexes are thought to be intermediates on the reaction pathway for ring sulphonation.[138] With sulphuric acid, amines also give amine hydrogen-sulphates, which on heating give the *p*-aminobenzenesulphonic acid (or the *ortho* compound if the *para* position is blocked).[141] A pathway for amine sulphonation could therefore be that shown in eq. 8.35 involving a sulphamic acid intermediate.[142]

$$PhNH_3{}^+HSO_4{}^- \xrightarrow{-H_2O} PhNHSO_3H \longrightarrow p\text{-}NH_3{}^+C_6H_4SO_3{}^- \quad (8.35)$$

This is supported by the finding that *N*-alkylsulphamic acids will sulphonate anisole, aniline, and *N*,*N*-dimethylaniline.[143] However, a study of the sulphonation of sulphamic acids showed the presence of *o*- and *p*-sulphophenylsulphamic acids as intermediates, and the rate of sulphonation and the isomer distribution of the products depended on the H_2SO_4 concentration.[144] It seems, therefore, that the reaction proceeds via initial *C*-sulphonation and subsequent *N*-desulphonation.

(viii) Polycyclics

Partial rate factors for sulphonation of the 1- and 2-positions of naphthalene decrease from 1020 and 180 in 79 wt-% H_2SO_4 to 540 and 110, respectively, in 83.4 wt-% acid[145] (note that the values were miscalculated in the original paper). The ratio $\log f_1:\log f_2$ (ca 1.33) is smaller than that found for reactions in which steric hindrance is either small or non-existent[146] (Section 11.2.4), confirming that there is steric hindrance to 1-substitution. In stronger sulphuric acid, in which steric hindrance is considered to be less significant, the proportion of 1-substitution actually *decreases*, which is surprising.[145]

Sulphonation of 1-methylnaphthalene by 90.1 wt-% H_2SO_4 at 25 °C gives the positional reactivity order $4 > 2 > 5$, which is in agreement with the predictions of hydrogen exchange [Section 3.1.2.13. (i)];[109b] 1-alkylnaphthalenes generally give mainly 4-sulphonation.[147] Further substitution using stronger acids goes mainly in sites conjugated with the methyl group, but avoids sterically hindered positions, so the final products are the 2,4,7-, 2,4,6-, and 2,5,7-trisulphonic acids in 76, 16, and 8% yields, respectively.[109b]. Sulphonation with SO_3 in $MeNO_2$ at 12 °C gives > 95% of 4-substitution,[109b] indicating much greater steric hindrance with this reagent.

The reported sulphonation of 2-methylnaphthalene with 93 wt-% H_2SO_4[148] is anomalous. The positional order was said to be $7 > 8 > 6 > 3 = 4 > 1 > 5$ at 0 °C, but this cannot be correct insofar as sulphonation of the β-positions is concerned, since the 6-position should be much more strongly activated by the 2-methyl group than should the 7-position (cf. hydrogen exchange, Table 3.22). Indeed, earlier reports of the sulphonation of 2-methylnaphthalene with 93–96 wt-% acid showed that 70–84% of substitution occurs in the 6-position.[149] The low reactivity of the 1-position relative to the 8- and 4-positions is also unexpected and indicates severe steric hindrance here.

The principal position of sulphonation of dimethylnaphthalenes[150] is that which is predicted on the basis of the electronic effects of the methyl groups determined in hydrogen exchange [Section 3.1.2.13.(vi)], except where steric hindrance becomes of overriding importance. Thus the 2,6-Me_2, 2,3-Me_2, and 1,8-Me_2 compounds substitute in the 3-, 5-, and 4-positions, respectively, rather than the corresponding 1-, 1-, and 2-positions predicted.

The greater conjugative activation by OH relative to Me, coupled with the lower steric hindrance, means that compared with the corresponding methylnaphthalenes, 1-naphthol gives proportionally more 2-sulphonation,[151] whereas 2-naphthol gives mainly 1- rather than 6-sulphonation;[152] concentrated H_2SO_4 was used in each study. The effects of the amino substituent are only explicable in terms of substantial reaction occurring on the protonated amine. Thus sulphonation of 1-naphthylamine takes place mainly in the 5-position together with some in the 4-position,[153] whereas sulphonation of 2-naphthylamine goes mainly in the 5- and 8-positions.[154]

Sulphonation of naphthalenesulphonic acids (believed to react as the sulphonate anions) gives the partial rate factors shown in Scheme 8.3.[155] For the 1-sulphonic acid, electronic and steric effects prevent substitution at the 2-, 3-, 4-, and 8-positions. Substitution occurs at the 5-position because the 1,5-conjugative interaction is very weak (cf. Table 3.23), and conjugative deactivation of the 7(*β*)-position makes this less reactive than the 6(*β*)-position. For the 2-sulphonic acid, conjugative and steric effects account for the absence of 1-, 3-, and 6-substitution, and conjugative deactivation of the 8(*α*)-position causes it to be less reactive than the 5(*α*)-position. The same order of positional reactivities is observed for sulphonation of the nitro[156] and carboxyl[157] derivatives, each of which has the same electronic effect as SO_3^-.

Sulphonation of acenaphthene is anomalous. With concentrated H_2SO_4 at 100 °C 5-substitution is obtained,[158] although a more recent report (in which the yields were calculated from the disubstitution pattern because of solubility difficulties) gave the 3-position as the most reactive,[159] and this is also true for

SO_3^-

0.0021

0.0067

0.0122

0.032 SO_3^-

0.0133 0.007

Scheme 8.3 Partial rate factors for sulphonation of naphthalenesulphonic acids.

sulphonation with $ClSO_3H$ in $PhNO_2$ at 0 °C.[158] Since the 5-position is intrinsically more reactive than the 3-position (Table 3.22), the results imply that there is appreciable steric hindrance to 5-substitution under some conditions. It is also significant that sulphonation of the 3-sulphonic acid gave the yields in (**6**),[159] for here the 6-position (≡ 5-position) is more reactive than the 8-position (≡ 3-position), yet neither is conjugated with the SO_3H group; sulphonation of the 5-sulphonic acid (**7**) took place in the 3- and 8-positions,[159] as expected.

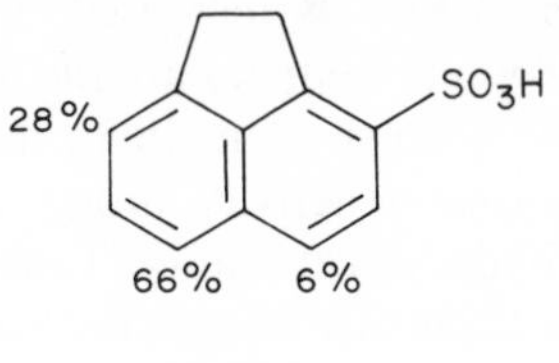

28% SO_3H 66% 6%

(**6**)

72% 28% SO_3H

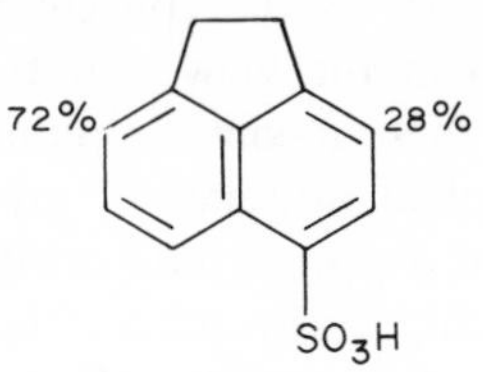

(**7**)

Sulphonation of anthracene with $ClSO_3H$–HOAc gives equal amounts of 1- and 2-anthracenesulphonic acid, the 1:2-product ratio decreasing with increasing temperature;[160] the 2-sulphonic acid has been isolated in sulphonation by 67% H_2SO_4 at 130 °C.[161] The absence of any substitution at the 9-position is due to steric hindrance which causes the reverse reaction, protiodesulphonation, to be very rapid in these protic media[162] (an alternative explanation, that 9-sulphonation does not take place because this position is protonated, the cation produced then directing into the 1- and 2-positions,[163] is untenable).[164]

Under aprotic conditions, i.e. using SO_3–dioxane, the 1-, 2-, and 9-sulphonic acids are obtained in 26, 8, and 66% yields, respectively,[164] which reinforces the above conclusion; use of SO_3–pyridine/isoparaffin gives substantially less of the 9-isomer[165].

The behaviour of 9-alkylanthracenes towards sulphonation is unpredictable.[166] Steric hindrance appears to make side-chain sulphonation and subsequent reactions more attractive in most cases, although 9-neopentylanthracene did sulphonate in the 4- and 10-positions, presumably because here there is also steric hindrance to side-chain sulphonation. In some cases a sultone (**8**) is formed between the 10-position and the alkyl group in the 9-position.[167] Each of 1- and 2-methyl-, 1,2-, 1,3-, and 2,3-dimethyl-, and 2,3,6,7-tetramethylanthracenes sulphonate in the expected positions after allowance is made

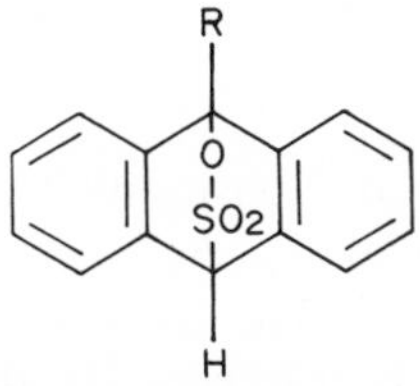

R O SO_2 H

(**8**)

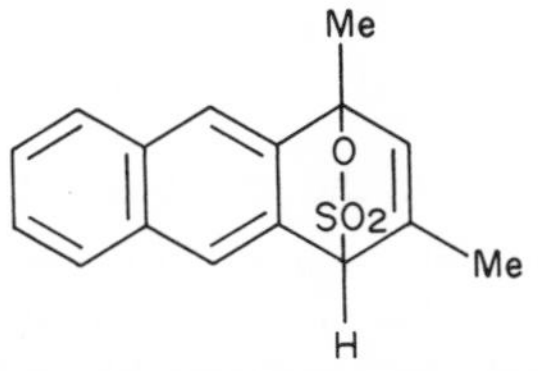

Me O SO_2 Me H

(**9**)

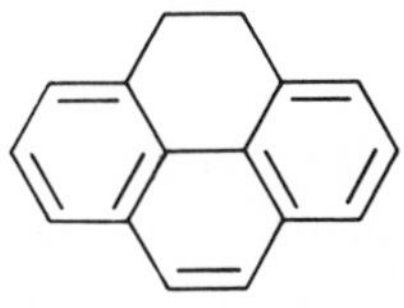

(**10**)

for steric hindrance; 1,3-dimethylanthracene also gave some of the sultone **9**,[166] understandable, like the formation of **8**, in terms of substantial retention of aromaticity in the product.

Early studies of sulphonation of phenanthrene showed that the positional reactivity order was[168] $3 > 2 > 9 > 1$, which differs from the theoretical order (Table 3.20) because of steric hindrance (which also accounts for the complete absence of the 4-isomer). A more recent study showed that the order changes to $9 > 1 > 3 > 2$ as the concentration of H_2SO_4 is increased,[169] and this is consistent with the view that the electrophile changes to the less hindered $H_2S_2O_7$ in these media. Sulphonation of a range of methyl-, dimethyl-, and tetramethylphenanthrenes[170] gives positional reactivity orders entirely consistent with those predicted from data for hydrogen exchange (Tables 3.23–3.25), and the bond orders in phenanthrene, after allowance is made for steric hindrance (especially at the 4-position). The tetramethyl compounds disulphonated, as did 4,5-ethanophenanthrene (**10**), the products here being the 1,6- and 1,8-disulphonic acids in 30 and 70% yields, respectively.[170]

Sulphonations of pyrene,[171] perylene,[172] chrysene,[173] and fluoranthene[174] follow the predictions of hydrogen exchange (Table 3.17). Further substitution in perylene goes in the 9- and then the 10-position (both are conjugatively deactivated by the 3-SO_3H substituent, but the interaction is greater at the 10-position). Conjugative activation of the 6- and 8-positions accounts for the observed sulphonation of 1-methylpyrene;[175] some 3-sulphonation probably also occurs.

Sulphonations of biphenylene (in which both sites are free of steric hindrance) goes exclusively in the 2-position, predicted by the hydrogen exchange data (Table 3.17).[176] Interestingly, disulphonation gives the 2,6- and 2,7-disulphonic acids in the proportions 65:35.[176] It might be supposed (see ref. 176) that the 7-position ought to be very strongly conjugatively deactivated by the 2-SO_3H group relative to the 6-position, but structure **11** shows why this is not the case. Conjugative deactivation places a double bond in the four-membered ring, thereby increasing strain considerably, and this is unfavourable.

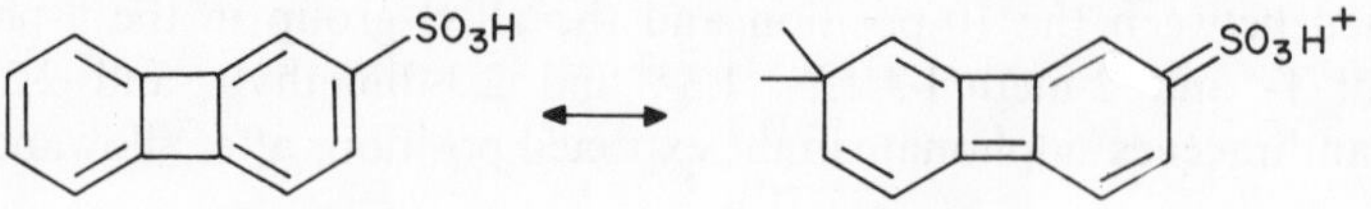

SO3H ⟷ SO3H+

(**11**)

Triphenylene sulphonates in the predicted (and unhindered) 2-position[176,177] (cf. hydrogen exchange, Table 3.17). Further substitution gives the 2,6-, 2,7-, and 2,11-disulphonic acids in yields of 59, 1, and 40%, respectively. The low yield of the 2,7-isomer follows from the 7-position being conjugatively deactivated by the 2-SO_3H group (**12**). The 11-position is in principle also conjugated with the 2-

position, but to achieve this involves complete loss of aromaticity giving structure **13**, which is therefore unimportant.

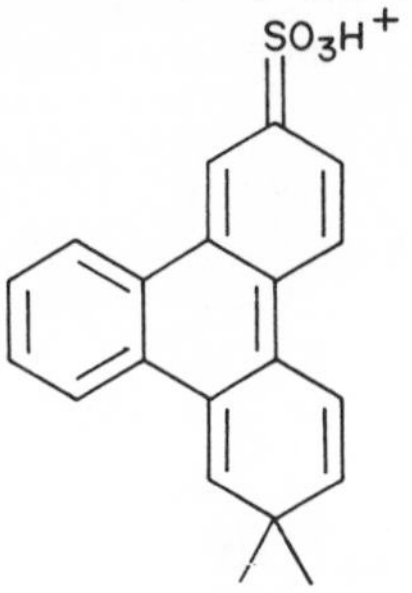

(12)

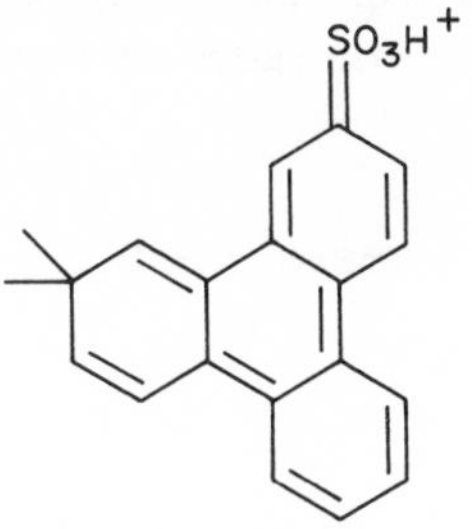

(13)

(ix) Annulenes and ferrocene

Sulphonation of 1,6-methano[10]annulene and its 11-fluoro- and 11,11-difluoro derivatives[178] and of azulene[179] takes place in the positions predicted by hydrogen exchange (cf. Section 3.1.2.16). Ferrocene has also been sulphonated, and disulphonation gives the 1,1′-disulphonic acid,[180] i.e. the electron-withdrawing SO_3H group directs into the unsubstituted ring.

8.10 SELENOCYANATION

This reaction is the selenium equivalent of thiocyanation, but since selenocyanogen is less stable than thiocyanogen, the electrophilic reagent has to be generated *in situ*. The first study used triselenodicyanide prepared from lead selenocyanate and bromine, and this reacted with aniline and *N,N*-dimethylaniline in the *para* position (eq. 8.36).[181]

$$C_6H_5NH_2 + Se_3(CN)_2 \longrightarrow 4\text{-}CNSeC_6H_4NH_2 + 2Se + HCN \qquad (8.36)$$

Selenocyanogen may be prepared *in situ* from potassium selenocyanate and bromine, and this method was used to selenocyanate 2,6-di-*tert*-butylphenol in the 4-position.[182]

8.11 SELENYLATION

Reaction of arylselenium chlorides with either phenols[183] or anilines[184] produces diarylselenides (eq. 8.37).

$$ArSeCl + C_6H_5OH \longrightarrow 4\text{-}ArSeC_6H_4OH + HCl \qquad (8.37)$$

With electron-rich aromatics, phenylselenodimethylsulphonium tetrafluoroborate, $PhSeS^+Me_2\,BF_4^-$ at 0 °C gives ca 50% yields of the *para*-substituted diarylselenide (eq. 8.38).[185] In keeping with the electrophilic nature of the reaction, substitution goes *ortho*, *para* to NMe_2 which activates more than Me or OMe.

$$Me_2N\text{–}C_6H_5 + PhSeS^+Me_2\ BF_4^- \longrightarrow Me_2N\text{–}C_6H_4\text{–}SePh + Me_2S^+\ BF_4^- \quad (8.38)$$

8.12 SELENONATION

Selenonation is more difficult than sulphonation because selenium compounds are more strongly oxidizing than the corresponding sulphur compounds. Treatment of aromatic with selenium trioxide in SO_2 produces the *para*-substituted arylselenonic acids (eq. 8.39), together with diarylselenones and selenic acids.[186] Selenonation may also be brought about by using selenic acid in acetic anhydride, and the reaction under these conditions (probably involving acetylselenic acid as electrophile) has similar selectivity to sulphonation with H_2SO_4 in Ac_2O.[187]

$$SeO_3 + PhX \longrightarrow 4\text{-}XC_6H_4SeO_3H \quad (8.39)$$
$$(X = H, Me, Cl, Br)$$

REFERENCES

1. M. G. Evans and N. Uri, *Trans. Faraday Soc.*, 1949, **45**, 224.
2. D. H. Derbyshire and W. A. Waters, *Nature* (*London*), 1950, **165**, 401.
3. H. Davidge, A. G. Davies, J. Kenyon, and R. F. Mason, *J. Chem. Soc.*, 1958, 4569.
4. J. A. Vesley and L. Schmerling, *J. Org. Chem.*, 1970, **35**, 4028.
5. G. A. Olah and R. Ohnishi, *J. Org. Chem.*, 1978, **43**, 865; G. A. Olah, A. P. Fung, and T. Keumi, *J. Org. Chem.*, 1981, **46**, 4305.
6. J. D. McClure and P. H. Williams, *J. Org. Chem.*, 1962, **27**, 627.
7. M. E. Kurz and G. J. Johnson, *J. Org. Chem.*, 1971, **36**, 3184.
8. J. Böeseken and M. L. von Königsfeldt, *Recl. Trav. Chim. Pays-Bas*, 1935, **54**, 313; J. Böeseken and C. F. Metz, *Recl. Trav. Chim. Pays-Bas*, 1935, **54**, 345.
9. H. Fernholtz, *Angew. Chem.*, 1948, **60A**, 62, *Chem. Ber.*, 1951, **84**, 110.
10. I. M. Roitt and W. A. Waters, *J. Chem. Soc.*, 1949, 3060.
11. S. L. Friess, A. H. Soloway, B. K. Morse, and W. C. Ingersoll, *J. Am. Chem. Soc.*, 1952, **74**, 1305.
12. R. D. Chambers, P. Goggin, and W. K. R. Musgrave, *J. Chem. Soc.*, 1959, 1804.
13. J. C. McClure and P. H. Williams, *J. Org. Chem.*, 1962, **27**, 627.
14. J. D. McClure, *J. Org. Chem.*, 1963, **28**, 69.

15. A. J. Davidson and R. O. C. Norman, *J. Chem. Soc.*, 1964, 5404.
16. D. Jerina, J. Daly, W. Landis, B. Witkop, and S. Udenfriend, *J. Am. Chem. Soc.*, 1967, **89**, 3347.
17. C. A. Buehler and H. Hart, *J. Am. Chem. Soc.*, 1963, **85**, 2177; H. Hart, *Acc. Chem. Res.*, 1971, **4**, 337.
18. Y. Ogata, Y. Sawaki, K. Tomizawa, and T. Ohno, *Tetrahedron*, 1981, **37**, 1485.
19. N. A. Vysotskaya and A. E. Brodsky, *Chem. Abstr.*, 1967, **67**, 21232.
20. S. Hashimoto and W. Kuike, *Bull. Chem. Soc. Jpn.*, 1970, **43**, 293.
21. G. A. Razuvaev, N. A. Kartashova, and L. S. Boguslavskaya, *J. Gen. Chem. USSR*, 1964, **34**, 2108.
22. P. Kovacic and M. E. Kurz, *J. Org. Chem.*, 1966, **31**, 2011, 2459; *J. Chem. Soc., Chem. Commun.*, 1966, 321; *J. Am. Chem. Soc.*, 1965, **87**, 4811; 1966, **88**, 2008; 1967, **89**, 4960; P. Kovacic and S. T. Morneweck, *J. Am. Chem. Soc.*, 1965, **87**, 1566.
23. E. H. Appleman, R. Bonnett, and B. Mateen, *Tetrahedron*, 1977, **33**, 2119.
24. R. A. Abramovitch, M. N. Inbasekaran, and S. Kato, *J. Am. Chem. Soc.*, 1973, **95**, 5428; R. A. Abramovitch, G. Alvernhe, R. Barnik, N. L. Dassanayake, M. N. Inbasekaran, and S. Kato, *J. Am. Chem. Soc.*, 1981, **103**, 4558.
25. N. Kamigata, H. Minato, and M. Kobayashi, *Bull. Chem. Soc. Jpn.*, 1974, **47**, 894.
26. A. Fischer, J. Vaughan, and G. J. Wright, *J. Chem. Soc. B*, 1967, 368.
27. Refs 112, 220, and 180–184 of Ch. 7.
28. P. C. Myhre, G. S. Owen, and L. L. James, *J. Am. Chem. Soc.*, 1968, **90**, 2115.
29. L. W. Crovatt and R. L. McKee, *J. Org. Chem.*, 1959, **24**, 2031.
30. R. L. Dannley and G. E. Corbett, *J. Org. Chem.*, 1966, **31**, 153.
31. R. L. Dannley, J. E. Gagen, and O. J. Stewart, *J. Org. Chem.*, 1970, **35**, 3076.
32. R. L. Dannley, J. E. Gagen, and K. Zak, *J. Org. Chem.*, 1973, **38**, 1.
33. R. L. Dannley and P. K. Tornstrom, *J. Org. Chem.*, 1975, **40**, 2278; R. L. Dannley and R. V. Hoffman, *J. Org. Chem.*, 1975, **40**, 2426.
34. E. M. Levi, P. Kovacic, and J. F. Gormish, *Tetrahedron*, 1970, **26**, 4537.
35. T. Fujisawa, T. Kobori, N. Ohtsuka, and G. Tsuchihashi, *Tetrahedron Lett.*, 1968, 4533, 5071.
36. N. Karasch, S. J. Potempa, and H. L. Wehrmeister, *Chem. Rev.*, 1946, **39**, 269; C. M. Buess and N. Karasch, *J. Am. Chem. Soc.*, 1950, **72**, 3529; H. Brintzinger, H. Schmahl, and H. Witte, *Chem. Ber.*, 1952, **85**, 338; H. Brintzinger and M. Langheck, *Chem. Ber.*, 1953, **86**, 557; N. Kharasch and R. Swidler, *J. Org. Chem.*, 1954, **19**, 1704; R. D. Scheutz and W. L. Fredericks, *J. Org. Chem.*, 1962, **27**, 1301; B. S. Farah and E. E. Gilbert, *J. Org. Chem.*, 1963, **28**, 2807; R. T. Wragg, *J. Chem. Soc.*, 1964, 5482.
37. D. W. Grant, D. R. Hogg, and J. L. Wardell, *J. Chem. Res. (S)*, 1987, 392.
38. J. K. Boscher, E. W. A. Kraak, and H. Kloosterziel, *J. Chem. Soc., Chem. Commun.*, 1971, 1365.
39. V. A. Sergeev, V. I. Nedelkin, V. U. Novikov, and T. V. Grechukina, *Bull. Acad. Sci. USSR*, 1983, **32**, 1756.
40. T. Fujisawa and N. Kobayashi, *J. Org. Chem.*, 1971, **36**, 3546.
41. J. L. Wood, *Org. React.*, 1946, **3**, 240; R. Q. Brewster and W. Schroeder, *Org. Synth.*, 1943, Coll. Vol. 2, 574.
42. A. G. Anderson and R. N. McDonald, *J. Am. Chem. Soc.*, 1959, **81**, 5669.
43. A. G. Anderson and E. D. Daugs, *J. Org. Chem.*, 1987, **52**, 4391.
44. R. G. R. Bacon and R. G. Guy, *J. Chem. Soc.*, 1960, 318.
45. G. A. Olah and J. Nishimura, *J. Org. Chem.*, 1974, **39**, 1203.
46. R. Taylor, *Specialist Periodical Report on Aromatic and Heteroaromatic Chemistry*, Chemical Society, London, 1975, Vol. 3, p. 253.
47. T. Fujisawa, M. Kakatani, and N. Kobayashi, *Bull. Chem. Soc. Jpn.*, 1973, **46**, 3615.

48. Y. L. Chow and K. Iwon, *J. Chem. Soc., Perkin Trans. 2*, 1980, 931.
49. G. A. Olah, J. Nishimura, and Y. Yamada, *J. Org. Chem.*, 1974, **39**, 2430.
50. W. E. Truce and C. W. Vrieson, *J. Am. Chem. Soc.*, 1953, **75**, 5053.
51. S. C. J. Olivier, *Recl. Trav. Chim. Pays-Bas*, 1915, **35**, 166.
52. F. R. Jensen and H. C. Brown, *J. Am. Chem. Soc.*, 1958, **80**, 4042.
53. S. C. J. Olivier, *Recl. Trav. Chim. Pays-Bas*, 1914, **33**, 91, 244; 1915, **35**, 109.
54. (a) Y. Yoshi, A. Ito, T. Hirashima, S. Shinkai, and O. Minabe, *J. Chem. Soc., Perkin Trans. 2*, 1988, 777; (b) H. Cerfontain and A. Telder, *Recl. Trav. Chim. Pays-Bas*, 1965, **84**, 1613; (c) H. Cerfontain, H. J. Hofman, and A. Telder, *Recl. Trav. Chim. Pays-Bas*, 1964, **83**, 493.
55. M. Kobayashi, K. Honda, and A. Yamaguchi, *Tetrahedron Lett.*, 1968, 487.
56. E. G. Willard and H. Cerfontain, *Recl. Trav. Chim. Pays-Bas*, 1973, **92**, 739.
57. E. C. Dart and G. Holt, *J. Chem. Soc., Perkin Trans. 1*, 1974, 1403.
58. F. R. Jensen and H. C. Brown, *J. Am. Chem. Soc.*, 1958, **80**, 4038.
59. F. R. Jensen and H. C. Brown, *J. Am. Chem. Soc.*, 1958, **80**, 4046.
60. J. K. Bosscher and H. Cerfontain, *J. Chem. Soc. B*, 1968, 1524.
61. M. P. van Albada, and H. Cerfontain, *Recl. Trav. Chim. Pays-Bas*, 1972, **91**, 499.
62. M. Kobayashi, H. Minato, and Y. Kohara, *Bull. Chem. Soc. Jpn.*, 1970, **43**, 234, 520.
63. G. A. Olah, S. Kobayashi, and J. Nishimura, *J. Am. Chem. Soc.*, 1973, **95**, 564.
64. A. F. Holleman, *Chem. Rev.*, 1925, **1**, 187; F. J. Stubbs, C. D. Williams, and C. N. Hinshelwood, *J. Chem. Soc.*, 1948, 1065; R. J. Gillespie, *J. Chem. Soc.*, 1950, 2493, 2516.
65. O. Jacobsen, *Chem. Ber.*, 1886, **19**, 1209; 1887, **20**, 896.
66. M. Kilpatrick and M. W. Meyer, *J. Phys. Chem.*, 1961, **65**, 1312.
67. A. A. Spryskov and O. K. Kachurin, *Zh. Obshch. Khim.*, 1958, **28**, 2213.
68. C. M. Suter and A. W. Weston, *Org. React.*, 1946, **3**, 141.
69. R. Taylor, *Comprehensive Chemical Kinetics*, Elsevier, Amsterdam, 1972, Vol. 13, pp. 56–77.
70. H. Cerfontain, *Mechanistic Aspects in Aromatic Sulphonation and Desulphonation*, Interscience, New York, 1968.
71. J. K. Bosscher and H. Cerfontain, *Recl. Trav. Chim. Pays-Bas*, 1968, **57**, 873; *Tetrahedron*, 1968, **24**, 6543.
72. L. Leierson, R. W. Bost, and R. LeBaron, *Ind. Eng. Chem.*, 1948, **40**, 508.
73. K. Lammertsma and H. Cerfontain, *J. Chem. Soc., Perkin Trans. 2*, 1980, 28.
74. B. V. Passat and N. V. Korotchenkova, *J. Org. Chem. USSR*, 1971, **7**, 2075.
75. (a) D. R. Vicary and C. N. Hinshelwood, *J. Chem. Soc.*, 1939, 1372; (b) K. D. Wadsworth and C. N. Hinshelwood, *J. Chem. Soc.*, 1944, 469. (c) E. Dresel and C. N. Hinshelwood, *J. Chem. Soc.*, 1944, 649.
76. H. Cerfontain and A. Koeberg-Telder, *Recl. Trav. Chim. Pays-Bas*, 1970, **89**, 569.
77. H. Cerfontain, A. Koeberg-Telder, C. Ris, and Z. R. H. Schaasberg-Nienhuis, *J. Chem. Soc., Perkin Trans. 2*, 1975, 970.
78. H. Cerfontain, *Recl. Trav. Chim. Pays-Bas*, 1961, **80**, 296; 1965, **84**, 551.
79. M. Kilpatrick and M. W. Meyer, *J. Phys. Chem.*, 1961, **65**, 530.
80. J. Pinnow, *Z. Elektrochem.*, 1915, **21**, 380; 1917, **23**, 243.
81. H. Martinson, *Z. Phys. Chem.*, 1908, **62**, 713; I. S. Ioffe, *Zh. Obshch. Khim.*, 1933, **3**, 437; R. Lanz, *Bull. Soc. Chim. Fr.*, 1935, **2**, 2092; K. Lauer and R. Oda, *J. Prakt. Chem.*, 1935, **142**, 258; 1935, **144**, 32; *Chem. Ber.*, 1937, **70**, 333; K. Lauer and K. Irie, *J. Prakt. Chem.*, 1936, **145**, 281; K. Lauer and Y. Hirata, *J. Prakt. Chem.*, 1936, **145**, 287; K. Lauer, *Chem. Ber.*, 1937, **70**, 1707; W. A. Cowdrey and D. S. Davies, *J. Chem. Soc.*, 1949, 1871; V. Gold and D. P. N. Satchell, *J. Chem. Soc.*, 1956, 1635; T. F. Young and

G. E. Walrafen, *Trans. Faraday Soc.*, 1961, **57**, 34; A. W. Kaandorp, H. Cerfontain, and F. L. J. Sixma, *Recl. Trav. Chim. Pays-Bas*, 1962, **81**, 969.
82. J. C. D. Brand, *J. Chem. Soc.*, 1950, 1004; J. C. D. Brand and W. C. Horning, *J. Chem. Soc.*, 1952, 3922; J. C. D. Brand, A. W. P. Jarvie and W. C. Horning, *J. Chem. Soc.*, 1959, 3844.
83. C. W. F. Kort and H. Cerfontain, *Recl. Trav. Chim. Pays-Bas*, 1967, **86**, 865.
84. O. I. Kachurin, A. A. Spryskov, and E. V. Kovalenko, *Chem. Abstr.*, 1963, **59**, 7337, 13781; O. I. Kachurin and A. A. Spryskov, *Chem. Abstr.*, 1964, **61**, 1723.
85. M. Kilpatrick, M. W. Meyer, and M. L. Kilpatrick, *J. Phys. Chem.*, 1960, **64**, 1433.
86. P. A. H. Wyatt, *Faraday Discuss. Chem. Soc.*, 1957, **24**, 162; *Trans. Faraday Soc.*, 1960, **56**, 490.
87. C. Eaborn and R. Taylor, *J. Chem. Soc.*, 1960, 3301.
88. H. Cerfontain, F. L. J. Sixma, and L. Vollbracht, *Recl. Trav. Chim. Pays-Bas*, 1963, **82**, 659.
89. H. de Vries and H. Cerfontain, *Recl. Trav. Chim. Pays-Bas*, 1967, **86**, 873.
90. A. J. Prinsen and H. Cerfontain, *Recl. Trav. Chim. Pays-Bas*, 1969, **88**, 833.
91. H. Cerfontain and A. Telder, *Recl. Trav. Chim. Pays-Bas*, 1967, **86**, 527.
92. R. J. Gillespie, *Acc. Chem. Res.*, 1968, **1**, 202.
93. U. Svanholm and V. D. Parker, *J. Chem. Soc., Perkin Trans. 2*, 1972, 962.
94. W. Steinfopf, *J. Prakt. Chem.*, 1927, **117**, 1.
95. L. Harding, *J. Chem. Soc.*, 1921, 1261; L. I. Levina, S. N. Patrakova, and D. A. Patruskev, *J. Gen Chem. USSR*, 1958, **28**, 2464.
96. B. Y. Yasnitskii, *Zh. Obshch. Khim.*, 1953, **23**, 107, 1953.
97. S. M. Chizhlik and B. V. Passat, *J. Org. Chem. USSR*, 1975, **11**, 1637.
98. M. P. van Albada and H. Cerfontain, *J. Chem. Soc., Perkin Trans. 2*, 1977, 1548.
99. C. Eaborn and R. Taylor, *J. Chem. Soc., Perkin Trans. 2*, 1961, 247.
100. A. Casadevall, A. Commeyras, P. Paillous, and H. Collet, *Bull. Soc. Chim. Fr.*, 1970, 719; M. Schmidt and K. E. Pichl, *Z. Anorg. Chem.*, 1961, **335**, 244.
101. A. W. Kaandrop, H. Cerfontain, and F. L. J. Sixma, *Recl. Trav. Chim. Pays-Bas*, 1963, **82**, 113, 565.
102. Ref. 69, p. 72.
103. J. M. Arends and H. Cerfontain, *Recl. Trav. Chim. Pays-Bas*, 1966, **85**, 93.
104. J. J. Duvall, *Diss. Abstr.*, 1964, **24**, 4993.
105. H. Cerfontain, A. W. Kaandorp, and L. Vollbracht, *Recl. Trav. Chim. Pays-Bas*, 1963, **82**, 923.
106. C. Ris, Z. R. H. Schaasberg-Nienhuis, and H. Cerfontain, *Tetrahedron*, 1973, **29**, 3165.
107. M. P. van Albada, H. Cerfontain, and A. Koeberg-Telder, *Recl. Trav. Chim. Pays-Bas*, 1972, **91**, 33.
108. A. A. Spryskov and B. G. Gnedin, *J. Gen. Chem. USSR*, 1963, **33**, 1069.
109. (a) H. Cerfontain, A. Koeberg-Telder, C. Ris, and Z. R. H. Schaasberg-Nienhuis, *J. Chem. Soc., Perkin Trans. 2*, 1975, 970. (b) K. Lammertsma, C. J. Verlaan, and H. Cerfontain, *J. Chem. Soc., Perkin Trans. 2*, 1978, 719.
110. A. Koeberg-Telder and H. Cerfontain, *J. Chem. Soc., Perkin Trans. 2*, 1977, 717.
111. H. Cerfontain, A. Koeberg-Telder, and C. Ris, *J. Chem. Soc., Perkin Trans. 2*, 1977, 720.
112. C. Ris and H. Cerfontain, *J. Chem. Soc., Perkin Trans. 2*, 1975, 1438.
113. H. Cerfontain, Z. R. H. Nienhuis, and W. A. Zwart Voorspuy, *J. Chem. Soc., Perkin Trans. 2*, 1972, 2087; H. Cerfontain, A. Koeberg-Telder, and E. van Kuipers, *J. Chem. Soc., Perkin Trans. 2*, 1972, 2091.
114. J. B. F. Loyd and P. A. Ongley, *Tetrahedron*, 1965, **21**, 245; A. Koeberg-Telder and H. Cerfontain, *J. Chem. Soc., Perkin Trans. 2*, 1974, 1206.

115. H. Cerfontain and Z. R. H. Schaasberg-Nienhuis, *J. Chem. Soc., Perkin Trans. 2*, 1974, 536.
116. Z. R. H. Schaasberg-Nienhuis, H. Cerfontain, and T. A. Kortekaas, *J. Chem. Soc., Perkin Trans. 2*, 1979, 844.
117. H. Cerfontain and Z. R. H. Schaasberg-Nienhuis, *J. Chem. Soc., Perkin Trans. 2*, 1976, 1776, 1780.
118. A. Koeberg-Telder, F. van de Griendt, and H. Cerfontain, *J. Chem. Soc., Perkin Trans. 2*, 1980, 358.
119. R. Bregman and H. Cerfontain, *J. Chem. Soc., Perkin Trans. 2*, 1980, 33.
120. G. S. Neumann, *Justus Liebigs Ann. Chem.*, 1887, **241**, 33; J. Troeger and F. Hurdelbrink, *J. Prakt. Chem.*, 1902, **65**, 82.
121. H. Suzuki and N. Yamamoto, *Bull. Chem. Soc. Jpn.*, 1972, **45**, 289.
122. H. Cerfontain, A. Koeberg-Telder, K. Laali, H. J. A. Lamprechts, and P. de Wit, *Recl. Trav. Chim. Pays-Bas*, 1982, **101**, 390.
123. C. W. Kort and H. Cerfontain, *Recl. Trav. Chim. Pays-Bas*, 1967, **86**, 865; 1969, **88**, 860, 1298.
124. F. J. Stubbs, C. D. Williams, and C. N. Hinshelwood, *J. Chem. Soc.*, 1948, 1065.
125. A. Koeberg-Telder and H. Cerfontain, *J. Chem. Soc., Perkin Trans. 2*, 1973, 633.
126. H. Cerfontain, A. Koeberg-Telder, and W. A. Zwart Voorspuy, *Can. J. Chem.*, 1972, **50**, 1574.
127. A. Koeberg-Telder, C. Ris, and H. Cerfontain, *J. Chem. Soc., Perkin Trans. 2*, 1974, 98; A. Koeberg-Telder, H. J. A. Lamprechts, and H. Cerfontain, *J. Chem. Soc., Perkin Trans. 2*, 1985, 1241.
128. T. A. Kortekaas and H. Cerfontain, *J. Chem. Soc., Perkin Trans. 2*, 1977, 1560.
129. A. P. Zaraiski and O. I. Kachurin, *J. Org. Chem. USSR*, 1973, **9**, 1017.
130. T. A. Kortekaas, H. Cerfontain, and J. M. Gall, *J. Chem. Soc., Perkin Trans. 2*, 1978, 445.
131. Ref. 70, Tables 4.7 and 4.8; T. A. Kortekaas and H. Cerfontain, *J. Chem. Soc., Perkin Trans. 2*, 1979, 224.
132. T. A. Kortekaas and H. Cerfontain, *J. Chem. Soc., Perkin Trans. 2*, 1978, 742.
133. F. Olsen and J. C. Goldstein, *Ind. Eng. Chem.*, 1924, **16**, 66; B. H. Chase and E. McKeown, *J. Chem. Soc.*, 1963, 50; Y. Muramoto, *Chem. Abstr.*, 1956, **50**, 9946.
134. (a) B. I. Karavaev and A. A. Spryskov, *J. Gen. Chem. USSR*, 1963, **33**, 1840; (b) H. Cerfontain, Z. R. H. Schaasberg-Nienhuis, R. G. Coombes, P. Hadjigeorgiou, and G. P. Tucker, *J. Chem. Soc., Perkin Trans. 2*, 1985, 659.
135. A. F. Campbell, *J. Chem. Soc.*, 1922, 847.
136. H. Cerfontain, A. Koeberg-Telder, H. J. A. Lamprechts, and P. de Wit, *J. Org. Chem.*, 1984, **49**, 4917.
137. W. M. J. Strachan and E. Buncel, *Can. J. Chem.*, 1969, **47**, 4011.
138. E. R. Alexander, *J. Am. Chem. Soc.*, 1946, **68**, 969.
139. R. Gnehm and T. Scheutz, *J. Prakt. Chem.*, 1901, **63**, 405; I. S. Uppal and K. Venkataraman, *J. Soc. Chem. Ind.*, 1938, **57**, 410; G. V. Shriolkar, I. S. Venkataraman, *J. Indian Chem. Soc.*, 1940, **17**, 443; A. N. Kurakin, *Zh. Obshch. Khim.*, 1948, **18**, 2089.
140. P. K. Maarsen and H. Cerfontain, *J. Chem. Soc., Perkin Trans. 2*, 1977, 1008.
141. W. Huber, *Helv. Chim. Acta*, 1932, **15**, 1372.
142. C. M. Suter, *The Organic Chemistry of Sulphur*, Wiley, New York, 1948, p. 247.
143. F. L. Scott, J. A. Barry, and W. J. Spillane, *J. Chem. Soc., Perkin Trans. 1*, 1972, 2663.
144. P. K. Maarsen and H. Cerfontain, *J. Chem. Soc., Perkin Trans. 2*, 1977, 921.
145. H. Cerfontain and A. Telder, *Recl. Trav. Chim. Pays-Bas*, 1967, **86**, 527.
146. R. Taylor and G. G. Smith, *Tetrahedron*, 1963, **19**, 937.

147. K. Elbs and B. Christ, *J. Prakt. Chem.*, 1923, **106**, 17; W. E. Bachman and L. H. Klemm, *J. Am. Chem. Soc.*, 1950, **72**, 4911; L. H. Klemm and S. S. Rawluis, *J. Org. Chem.*, 1952, **17**, 613.
148. P. H. Gore and A. S. Siddiquei, *J. Chem. Soc., Perkin Trans. 1*, 1972, 2344.
149. R. N. Shreve and J. H. Lux, *Ind. Eng. Chem.*, 1943, **35**, 306; J. Reichel, A. Balint, A. Demian, and W. Schmidt, *Chem. Abstr.*, 1965, **62**, 11746.
150. K. Lammertsma and H. Cerfontain, *J. Chem. Soc., Perkin Trans. 2*, 1979, 673.
151. M. Conrad and W. Fischer, *Justus Liebigs Ann. Chem.*, 1893, **273**, 102; P. Friedlander and R. Taussig, *Chem. Ber.*, 1897, **30**, 1456.
152. I. I. Vorontsov and P. P. Sokolova, *Chem. Abstr.*, 1937, **31**, 1794; I. I. Vorontsov, *Zh. Prikl. Khim.*, 1948, **21**, 1002; N. N. Woroshtow, *Chem. Ber.*, 1929, **62**, 57; H. Iida and M. Ohkawa, *J. Chem. Soc. Jpn.*, 1955, **58**, 995.
153. E. Schmidt and B. Schaal, *Chem. Ber.*, 1874, **7**, 1367; O. N. Witt, *Chem. Ber.*, 1886, **19**, 578.
154. C. Butler and F. A. Royle, *J. Chem. Soc.*, 1923, 1649; A. Corbellini, *Chem. Abstr.*, 1928, **22**, 1972; N. N. Vorontsov, *Chem. Abstr.*, 1932, **26**, 4599; H. Iida and M. Ohkawa, *Chem. Abstr.*, 1957, **51**, 2680.
155. P. de Wit and H. Cerfontain, *Can. J. Chem.*, 1983, **61**, 1453.
156. A. A. Spryskov and N. A. Ovsyankina, *Zh. Obshch. Khim.*, 1946, **16**, 1057; A. A. Danish, M. Silverman, and W. A. Tajima, *J. Am. Chem. Soc.*, 1954, **76**, 6144.
157. F. A. Royle and J. A. Schedler, *J. Chem. Soc.*, 1923, 1641.
158. K. Dziewonski and T. Stolyhwo, *Chem. Ber.*, 1924, **57**, 151; G. T. Morgan and V. E. Yarsley, *J. Soc. Chem. Ind.*, 1925, **44**, 513.
159. H. Cerfontain and Z. R. H. Schaasberg-Nienhuis, *J. Chem. Soc., Perkin Trans. 2*, 1974, 989.
160. M. Battegay and P. Brandt, *Bull. Soc. Chim. Fr.*, 1922, **31**, 910; 1923, **33**, 1667; J. O. Morley, *J. Chem. Soc., Perkin Trans. 2*, 1976, 1560.
161. *Ger. Pat.* 72226, 73961, 76280 (1893).
162. P. H. Gore, *J. Org. Chem.*, 1957, **22**, 135.
163. Ref. 70, p. 73.
164. H. Cerfontain, A. Koeberg-Telder, C. Ris, and C. Schenk, *J. Chem. Soc., Perkin Trans. 2*, 1975, 966.
165. J. O. Morley, *J. Chem. Soc., Perkin Trans. 2*, 1976, 1554.
166. F. van de Griendt, C. P. Visser, and H. Cerfontain, *J. Chem. Soc., Perkin Trans. 2*, 1980, 911.
167. F. van de Griendt and H. Cerfontain, *J. Chem. Soc., Perkin Trans. 2*, 1980, 13, 19, 23.
168. L. F. Fieser, *J. Am. Chem. Soc.*, 1929, **51**, 2460, 2471; A. Werner, *Justus Liebigs Ann. Chem.*, 1902, **321**, 248; I. S. Yoffe, *Chem. Abstr.*, 1941, **35**, 4009; S. M. G. Solomon and D. J. Hennesey, *J. Org. Chem.*, 1957, **22**, 1649.
169. O. I. Kachurin, E. S. Fedorchuk, and V. Ya. Vasilenko, *J. Org. Chem. USSR*, 1973, **9**, 1956, 1961.
170. H. Cerfontain, A. Koeberg-Telder, K. Laali, and H. J. A. Lamprechts, *J. Org. Chem.*, 1982, **47**, 4069.
171. H. Vollman, H. Becker, M. Corell, and H. Streeck, *Justus Liebigs Ann. Chem.*, 1937, **531**, 1; E. Tietze and O. Bayer, *Justus Liebigs Ann. Chem.*, 1939, **540**, 189; Y. Abe and Y. Nagai, *Chem. Abstr.*, 1962, **57**, 8520.
172. C. Marschalk, *Bull. Soc. Chim. Fr.*, 1927, **41**, 74.
173. A. Schmelzer, *US Pat.*, 2 032 505 (1936).
174. T. Holbro and N. Campbell, *J. Chem. Soc.*, 1957, 2652; N. Campbell and N. H. Kier, *J. Chem. Soc.*, 1955, 1233.

175. H. Cerfontain, K. Laali, and H. J. A. Lamprechts, *Recl. Trav. Chim. Pays-Bas*, 1983, **102**, 210.
176. H. Cerfontain, K. Laali, and H. J. A. Lamprechts, *Recl. Trav. Chim. Pays-Bas*, 1982, **101**, 313.
177. A. Schmelzer, *Ger. Pat.*, 654 283 (1934).
178. H. Cerfontain, H. Goossens, A. Koeberg-Telder, C. Kruk, and H. J. A. Lamprechts, *J. Org. Chem.*, 1984, **49**, 3097.
179. W. Schroth and K. Achtelik, *Z. Chem.*, 1963, **3**, 426; W. Triebs and W. Schroth, *Justus Liebigs Ann. Chem.*, 1954, **586**, 202.
180. V. Weinmayr, *J. Am. Chem. Soc.*, 1955, **77**, 3009; G. R. Knox and P. L. Pauson, *J. Chem. Soc.*, 1958, 692.
181. F. Challenger, A. T. Peters, and J. Halevy, *J. Chem. Soc.*, 1926, 1648.
182. E. Müller, H. B. Stegman, and R. Scheffler, *Justus Liebigs Ann. Chem.*, 1962, **657**, 5.
183. L. R. M. Pitombo, *Chem. Ber.*, 1959, **92**, 745; N. Marziano and R. Passerini, *Gazz. Chim. Ital.*, 1964, **94**, 1137.
184. O. Behagel and H. Siebert, *Chem. Ber.*, 1933, **66**, 708.
185. P. G. Gassman, A. Miura, and T. Miura, *J. Org. Chem.*, 1982, **47**, 951.
186. K. Distal, Z. Zak, and M. Cernik, *Chem. Ber.*, 1971, **104**, 2044.
187. C. Ris and H. Cerfontain, *J. Chem. Soc., Perkin Trans. 2*, 1973, 2129.

CHAPTER 9

Electrophilic Halogenation

Halogenation is customarily brought about in one of three ways: (1) with molecular halogen, e.g. Br_2; (2) with molecular halogen in the presence of a Lewis acid catalyst such as a metallic halide or iodine; or (3) with halogen that is more positively charged than in (2), such as Cl^+, the solvated species $Cl—OH_2^+$, or $Cl^{\delta+}—OCl^{\delta-}$.

The reactivity of the reagent increases from (1) to (3), and from I to F. A molecular halogen has a non-polar bond and is only weakly electrophilic; it requires a fairly strong donation of electrons from the aromatic nucleus in order that the reactants may surmount the activation energy barrier. In (2), a Lewis acid catalyst polarizes the inter-halogen bond as in structure (**1**), thereby rendering one of the two halogen atoms more electrophilic. In (3), this polarization is considerably more powerful: in the extreme case a free halogen cation may be present through complete polarization of the bond.

$$Br—Br\ AlBr_3 \longrightarrow \underset{(\mathbf{1})}{Br^{\delta+}\cdots Br\cdots AlBr_3{}^{\delta-}}$$

The electrophilicity of a halogen atom X should therefore be increased as the electron-withdrawing ability of the atom to which it is bonded is increased, i.e. the order of reactivities should be $XOH < XOX < XOAc < XX < XX$ (catalysed) $< XOR_2{}^+ < X^+$; this order corresponds closely to that observed, but there are some exceptions that are noted later.

Because there is a gradation in electrophilicity, it is not satisfactory to categorize halogenation as either 'positive' or 'molecular' in the traditional manner. The subsequent presentation emphasizes conditions of halogenation, and describes the nature of the electrophile under each condition insofar as this is known.

9.1 FLUORINATION

The fluorine—fluorine bond is particularly weak, so that reaction of elementary fluorine with aromatics tends to involve very rapid free-radical chain

Table 9.1 Partial rate factors for fluorination of PhX by F_2–N_2 in $CFCl_3$ at $-78\,°C$

X	*ortho*	*meta*	*para*	X	*ortho*	*meta*	*para*
Me	8.5	1.55	8.2	OMe	123	8.1	76.1
CN	0.018	0.041	0.017	F	0.26	0.16	1.56
NO_2	0.005	0.041	0.011	Cl	0.19	0.77	0.44
CF_3	0.014	0.058	0.036	Br	0.08	0.61	0.43

processes, with consequent explosions. Three techniques have therefore been devised to overcome this difficulty. The first used CF_3OF (which would be polarized as $F^{\delta+}$----$OCF_3{}^{\delta-}$) and with 2-acetylaminonaphthalene a 50% yield of the 1-fluoro derivative was obtained.[1] The 1-position in this compound is the most activated towards electrophilic substitution.

The second method uses solutions of fluorine in an inert medium. With MeCN at low temperature an electrophilic substitution pattern was obtained for nitro- and fluorobenzenes, although toluene gave a very high *ortho*:*para* ratio,[2] indicating that radical processes were involved. Use of fluorine in nitrogen, TFA, or TFA–BF_3 at low temperature gave typical electrophilic substitution patterns for nitrobenzene, trifluoromethylbenzene, benzoic acid, and toluene, although again the latter gave a high *ortho*:*para* ratio indicative of the involvement of radicals.[3] Re-examination of this method using very dilute (>1 mol-%) solutions of fluorine in nitrogen or argon at low temperature in various fluorocarbons as solvents, and allowing the reaction to proceed to less than 0.01%, gave the partial rate factors shown in Table 9.1. These gave a good correlation with σ^+ values with $\rho = -2.45$;[4] this is equivalent to a value of -1.60 at 25 °C, so the reagent is very unselective, indeed the least selective of all electrophiles. Since there should then be little demand for conjugative electron release from the substituents, the reactivity at the *ortho* positions of the halogenobenzenes indicates a significant steric effect (cf. the discussion of hydrogen exchange of these compounds, Section 3.1.2.5). The high reactivity of the *ortho* position of anisole parallels its behaviour in nitration under some conditions, where coordination of the electrophile with the substituent is believed to be responsible.

The third method uses xenon difluoride in CCl_4 or CH_2Cl_2 as the fluorinating reagent and gives reasonable yields with typical electrophilic substitution patterns. The reaction is catalysed by HF (which presumably polarizes the xenon—fluorine bond) and may involve a radical cation as the electrophile since biphenyls and polyphenyls are among the reaction products.[5]

9.2 CHLORINATION

A wide variety of chlorination methods are available. Because many chloroaromatics are important fine chemicals and pharmaceuticals, recent developments have concentrated on ways of chlorinating specific sites in aromatics (especially phenols and their ethers) in order to minimize isomer separation procedures and to avoid producing byproducts with difficult disposal problems.

9.2.1 Chlorination by Hypochlorous Acid, its Esters and Acetyl Derivative

(i) Hypochlorous acid

Chlorination by hypochlorous acid, HOCl, is catalysed by mineral acids,[6] by Cl^-, and by ClO^-, the last two due to the formation of Cl_2 (eq. 9.1)[7] and Cl_2O,[8] respectively. In the absence of these nucleophiles the kinetics for reactive aromatics such as anisole, phenol, and *p*-dimethoxybenzene followed eq. 9.2 at low aromatic concentrations (< 0.1 M) and eq. 9.3 at higher concentrations.[5]

$$HOCl + H^+ + Cl^- \longrightarrow Cl_2 + H_2O \tag{9.1}$$

$$\text{Rate} = k_1[HOCl] + k_2[HOCl][H^+] \tag{9.2}$$

$$\text{Rate} = k_1[HOCl] + k_2[HOCl][H^+] + k_3[HOCl][H^+][ArH] \tag{9.3}$$

The three terms in eq. 9.3 represent fission of HOCl to give Cl^+, of H_2OCl^+ to give Cl^+, and attack of the aromatic on H_2OCl^+, hence the last term becomes significant at higher aromatic concentrations. It also follows that as the concentration of the aromatic is decreased, the point at which the rate becomes independent of this concentration should be lower the more reactive is the aromatic, i.e. the most reactive aromatics remain effective in competing with OH^- and H_2O for Cl^+ down to a lower concentration. This was observed, the concentrations at which a significant decrease in rate occurred being 0.05 M for methyl *p*-tolyl ether, 0.005 M for anisole, and 0.0015 M for phenol; for the most reactive compounds, e.g. methyl *m*-tolyl ether, eq. 9.3 applied at all concentrations.[9]

The alternative interpretation of the second term in eq. 9.3, i.e. that it represents a slow proton transfer to HOCl, has been ruled out on two grounds: first, the rate of this proton transfer is much higher than the first-order chlorination rate,[10] and second, the third term should disappear when the reactivity of the aromatic is very high since then the reaction responsible for the second term would be dominant.[9]

Support for Cl^+ being an electrophile comes from measurements of solvent isotope effects, the rate of chlorination in H_2O being only half that in D_2O, consistent with the higher acid dissociation constants of protium- than deuterium-containing substrates, appropriate to formation of Cl^+ being rate-determining; if the formation of H_2OCl^+ was rate-determining the rate would have been slower in D_2O.[11] This kinetic evidence, however, is difficult to reconcile with thermodynamic calculations of the equilibrium constants for the formation of either Cl^+ or H_2OCl^+.[10] These lead to concentrations for the two species which are so low that neither could conceivably be the attacking entity. The calculations for H_2OCl^+ could be in error,[12] but Cl^+ appears to be ruled out as an electrophile; $ClAgCl^+$ has been suggested as an alternative positive chlorinating species in acidified HOCl[12] (silver perchlorate is added to minimize the concentrations of Cl^- and ClO^-).

Reinvestigation of the mechanism using anisole as substrate has indicated a different kinetic form (eq. 9.4).[13] Here there are terms bimolecular in HOCl and these have been interpreted in terms of rate-determining formation of chlorine monoxide, the anhydride of hypochlorous acid (eq. 9.5); the third term represents rate-determining attack of aromatic upon preformed H_2OCl^+, as before. The acceptance of this alternative mechanism stands or falls on its ability to explain why chlorination under these conditions is much less sterically hindered than chlorination with Cl_2 (see Section 9.2.2).

$$\text{Rate} = k_2[\text{HOCl}]^2 + k_3[\text{HOCl}]^2[\text{H}^+] + k'_3[\text{HOCl}][\text{H}^+][\text{ArH}] \quad (9.4)$$

$$2\text{HOCl} \rightleftharpoons \text{Cl}_2\text{O} + \text{H}_2\text{O} \quad (9.5)$$

(ii) Esters of hypochlorous acid

tert-Butyl hypochlorite is the most stable ester of HOCl and can be distilled. It may be prepared by reacting chlorine with *tert*-butyl alcohol in the presence of calcium carbonate to remove HCl. The selectivity of an acidified solution of *tert*-butyl hypochlorite is similar to that of an acidified solution of hypochlorous acid, indicating that the electrophile is the same under both conditions.[14] By contrast, when the reagent is used under non-acidic conditions the selectivity resembles that of molecular chlorine.[14]

A study of the chlorination of anisole and phenol with *tert*-butyl hypochlorite showed that the *ortho*:*para* ratio for the former (0.65) was independent of acidity, whereas for phenol it varied from 0.43 at pH 4.0 to 4.3 at pH 10. Phenoxide ion also gives a high ratio, and the increase in ratio with lower acidity was attributed to an increased amount of reaction via this ion.[15]

Recent developments in chlorination with *t*-BuOCl include the use in the presence of zeolite X, which gives very high *para* selectivity with toluene (97%) and halogenobenzenes (e.g. 92% with PhCl).[16] It has also been used (as also have dichloramine-T and *N*,*N*-dichlorourethane) in the presence of silica, the advantage here being that no HCl byproduct is produced.[17]

(iii) The acetyl derivative of hypochlorous acid

The acetyl derivative of HOCl is more commonly known as chlorine acetate, ClOAc. This may be prepared from reaction of mercury(II) acetate with chlorine in acetic acid, but is also present in solutions of HOCl in acetic acid (eq. 9.6). Since it is readily hydrolysed by water, its concentration decreases the more aqueous the acetic acid medium becomes. The rate of chlorination of toluene under non-catalytic conditions (governed by eq. 9.7) falls rapidly on adding water to acetic acid, and then passes through a maximum in 50% aqueous HOAc. The fall-off in rate is attributable to reduction in the concentration of chlorine acetate as the medium becomes more aqueous, and the subsequent rise in rate to improved solvation (cf. hydrogen exchange, Section 3.1).

$$HOCl + HOAc \rightleftharpoons ClOAc + H_2O \qquad (9.6)$$

$$Rate = k_2[ArH][HOCl] \qquad (9.7)$$

$$Rate = k_2[ArH][HOCl] + k_3[ArH][HOCl]f[H^+] \qquad (9.8)$$

A similar rate vs medium composition profile is observed under catalysed conditions (although there are complications arising from experimental discrepancies).[18-20] Here the rate is governed by eq. 9.7 at low acid concentrations and by eq. 9.8 at high concentrations,[18] and it is probable that under these conditions both ClOAc and $ClOAcH^+$ are electrophiles.

Separate experiments showed that ClOAc is more reactive than Cl_2,[18] which contrasts with prediction based on the electron affinity of X in ClX. This led to the proposal that the transition state for chlorination by chlorine acetate involves a six-centre cyclic transition state (**2**) in which Cl—O bond-breaking is aided by intramolecular hydrogen bonding.[18]

ArH + MeCO2Cl ⟶ +Ar(H······O)(Cl⁻---O)C—Me ⟶ +Ar(H)(Cl) + ⁻O(⁻O)C+—Me

(2)

A notable feature of chlorination by ClOAc, as in the case of ClOH, is the apparent absence of steric hindrance, and indeed both reagents give very similar $\log f_o : \log f_p$ ratios.

9.2.2 Chlorination by Molecular Chlorine

Chlorination may be achieved with molecular chlorine either in the form of the gas, or by reagents which produce it.

(i) Chlorine

Chlorination may be readily achieved using chlorine in chlorinated hydrocarbons, nitro compounds, acetonitrile, and carboxylic acids. Of these, studies using acetic acid as solvent have been the most common.

Chlorination of aromatic ethers and anilides using Cl_2 in HOAc follows second-order kinetics (eq. 9.9), and the reaction rate is only slightly increased by the addition of HCl.[21] Since the reaction is neither strongly catalysed by acids nor strongly retarded by chloride ion or acetate ion,[22] chlorination by either Cl^+, $AcOHCl^+$, or AcOCl cannot occur, and molecular chlorine must be the reactive species.[23] The kinetic form has been amply confirmed by other workers, and applies also to chlorination in other solvents.[24–43] Rates of chlorination increase with increasing polarity of the solvent in the order 1,2-$C_2H_4Cl_2 \ll Ac_2O \approx HOAc \approx MeCN < PhNO_2 < MeNO_2 \ll$ TFA.[31,32] In the least polar solvents, chlorination is catalysed by HCl, ICl_2, $ZnCl_2$, and TFA, leading to kinetics first order in catalyst. This indicates either that the catalyst interacts with chlorine in a preliminary fast step to give a polarized complex which is the electrophile, or that the Cl—Cl bond is broken by the catalyst M, as in eqs 9.9–9.11, this mechanism having previously been proposed for bromination.[44]

$$ArH + Cl_2 \rightleftharpoons ArHCl_2 \tag{9.9}$$

$$ArHCl_2 + M \xrightarrow{\text{slow}} ArHCl^+ + MCl^- \tag{9.10}$$

$$ArHCl^+ \longrightarrow ArCl + H^+ \tag{9.11}$$

Determinations of the kinetic isotope effects, $k_H:k_D$, in molecular chlorination in polar solvents gave values of 0.92 for 3-bromo-1,2,4,5-tetramethylbenzene[45] and 0.85 for naphthalene.[46] These show that breaking of the C—H bond is non-rate-determining so that the normal mechanism of electrophilic substitution (eq. 2.4), in which the second step is fast, applies. The small inverse effect obtained is attributed to changes in hybridization from sp^2 to sp^3 at the carbon undergoing substitution, on passing through the rate-determining transition state for the reaction.

An addition–elimination mechanism can also apply,[47] however, particularly with those molecules for which loss of resonance energy on going to the addition transition state is less severe. This is most readily shown in chlorination of phenanthrene in HOAc which yields 9-chlorophenanthrene (34%), *cis*- and *trans*-9,10-dichloro-9,10-dihydrophenanthrene (48%), and *cis*- and *trans*-9-acetoxy-10-chloro-9,10-dihydrophenanthrenes. The latter adducts can eliminate either HOAc or HCl, yielding the chloro and acetoxy derivatives respectively (Scheme 9.1), in a ratio of ca 3:1.[47] Adducts are also obtained in chlorination of biphenyl (which gives 1-phenyl-3,4,5,6-tetrachlorocyclohexene) and fluorene,[47] naphthalene (which gives 21% of tetrachlorotetralins and 13% of acetoxy chlorides),[46,48] 1- and 2-methylnaphthalene,[49] and 3,4-dimethylphenol, its methyl ether and acetate (Scheme 9.2).[50] Scheme 9.2 shows that 1,3-migration of chlorine occurs [cf. nitration, Section 7.4.2. (iii)]. Addition products have also been proposed as intermediates in the chlorination of diphenyl ether in the absence of a solvent,[51] and have been isolated in chlorination of anisole, 4-chloroanisole (but not the 2- or 3-chloro isomers),[52] and of 1-substituted 2-naphthols.[53]

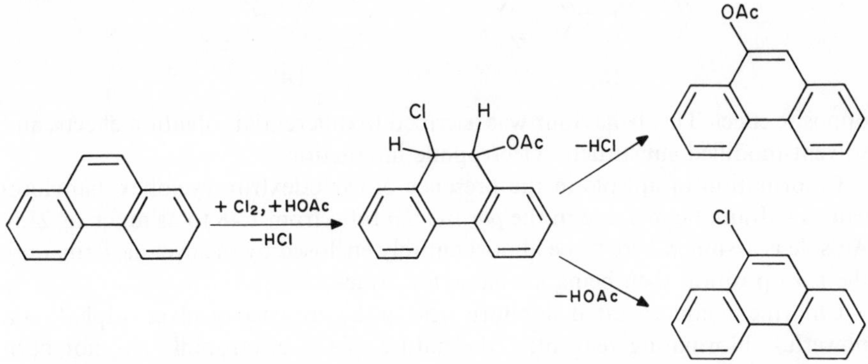

Scheme 9.1. Addition–elimination in reaction of phenanthrene with Cl_2–HOAc.

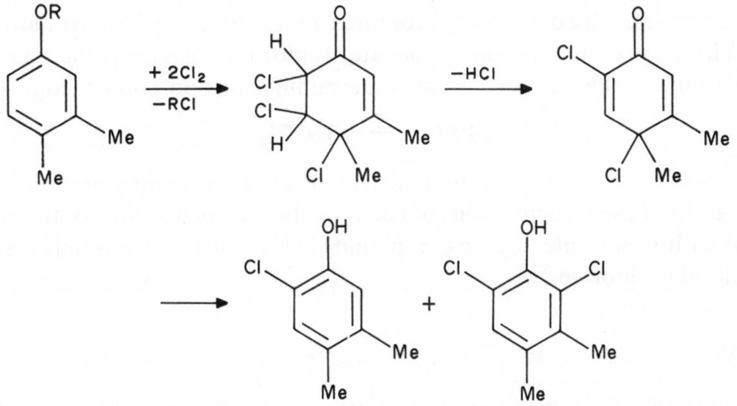

Scheme 9.2. Addition–elimination in chlorination of 3,4-dimethylphenol and derivatives.

The formation of *cis* adducts in phenanthrene is little affected by changes in the nature of the solvent, whereas the reaction rate is altered considerably. This latter rules out the possibility that a non-polar concerted transition state such as **3** is involved, and points to the involvement of a charged intermediate such as **4**.[47]

The effects of solvents on the *ortho*:*para* ratio in the chlorination of anisole by chlorine have been examined. Solvents of high dielectric show a decrease in the ratio with decreasing dielectric, whereas solvents of low dielectric show the

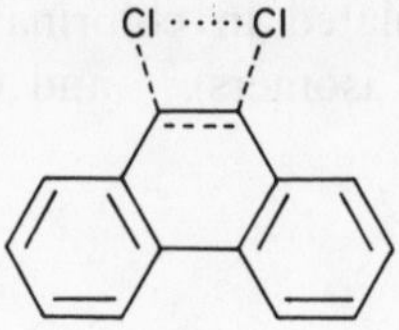

(3) 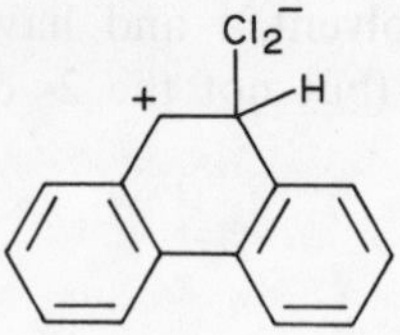 (4)

opposite effect. This behaviour was ascribed to differential solvation effects, and solvent-modified substituent–electrophile interactions.[54]

Chlorination of anisole in the presence of cyclodextrin (cyclohexylamylose) causes a dramatic increase in the *para*:*ortho* ratio from 1.48 to as much as 21.6. Anisole is assumed here to be almost entirely enclosed by the cyclodextrin, only the *para* position then being available for attack.[55]

Chlorine in concentrated sulphuric acid in the presence of silver sulphate is a powerful chlorinating reagent;[56] the nature of the electrophile has not been investigated but it is probably chlorine bisulphate, $ClOSO_2OH$.

(ii) Iodobenzene dichloride

Iodobenzene dichloride dissociates to molecular chlorine according to eq. 9.12.[31a,c] The dissociation is promoted by catalysts such as trifluoroacetic acid. With very reactive aromatics the rate of chlorination is faster than the rate of dissociation, so that eq. 9.12 is rate-determining. In non-polar media there is

$$PhICl_2 \rightleftharpoons PhI + Cl_2 \qquad (9.12)$$

evidence that reaction with particularly reactive aromatics occurs without dissociation of the reagent, indicated also by the stereochemistry of the adducts formed with unsaturated cyclic compounds being different from that produced by molecular chlorine.[57]

(iii) N-Chloroamines and N-chloroamides: the Orton rearrangement

In non-polar solvents and particularly in the presence of initiators, *N*-chloroamides can chlorinate by a free-radical mechanism. However, in polar solvents and in the presence of HCl, chlorine is produced, eq. 9.13.[21,24] Since HCl is a byproduct of the chlorination, the reaction is continuous once initiated, and the stationary concentration of chlorine can be maintained at any desired level according to the amount of HCl added initially. If other acids are used with acetic acid as solvent, then ClOAc becomes the chlorinating reagent.

$$R_2NCl + HCl \rightleftharpoons R_2NH + Cl_2 \qquad (9.13)$$

The amine R_2NH must be unreactive so that it is not preferentially chlorinated. This does happen in the Orton rearrangement of *N*-chloroanilides (eq. 9.14), which is thought to be intermolecular because the *ortho*:*para* ratio is the same as in chlorination of acetanilide under the same conditions. However, in non-polar solvents an intramolecular process may be involved since treatment of aromatic amines with *N*-chlorosuccinimide in benzene gives relatively large amounts of *ortho* substituents; similar results are obtained with *N*-chloro-*N*-methylaniline in CCl_4.[58] A *nucleophilic* substitution may also be involved, as in the rearrangement of the *N*-chloroamines, $4\text{-}RC_6H_4N(Cl)CMe_3$, in ethanol buffered with HOAc–NaOAc. Here the reaction is strongly accelerated by the *para* substituents ($\rho = -6.35$), indicating rate-determining formation of a nitrenium cation (eq. 9.15), the liberated Cl^- then substituting at the position *ortho* to the cation.[59]

$$C_6H_5N(Ac)Cl \xrightarrow{HCl} o\text{- and } p\text{-}ClC_6H_4NHAc \qquad (9.14)$$

$$4\text{-}RC_6H_4N(Cl)CMe_3 \xrightarrow{-Cl^-} RC_6H_4N^+CMe_3 \qquad (9.15)$$

Use of bulky *N*-chloroamines (such as *N*-chloropiperidine) in the presence of TFA gives with anisole almost exclusive *para* substitution and high overall yields ($>90\%$).[60] Similar results are obtained using aqueous H_2SO_4 as acid, although the *para*:*ortho* ratio increases as the medium is made more aqueous. The acid catalysis and the absence of any effect of chloride ion indicated that the electrophile is R_2HN^+Cl (eq. 9.16),[60,61] and this was confirmed by identical results being obtained with *N*-chloroammonium ions, e.g. *N*-chlorotriethylammonium chloride, Et_3N^+Cl.

$$R_2NCl + H^+ \rightleftharpoons R_2HN^+Cl \qquad (9.16)$$

The preference for *para* substitution does not seem to be a steric effect since 3-methylanisole also substitutes 100% in the 4-position; radical-cations may be involved.[60]

(*iv*) *Chlorine in the presence of a catalyst*

Heterolysis of the Cl—Cl bond may be aided through the use of a Lewis acid catalyst, which therefore produces a substantial increase in chlorination rate and a substantial decrease in selectivity. For example, the rates of chlorination of some alkylaromatics, relative to that of benzene ($=1$), viz. 2445 (toluene), 14 200 (*p*-xylene), 247 000 (*m*-xylene), and ca 5×10^6 (mesitylene), are at least an order of magnitude smaller in the presence of a catalyst.[62,63]

The $SnCl_4$-catalysed chlorination of aromatics in the absence of a solvent follows eq. 9.17,[62] so the formation of the electrophile appears here to be rate-determining.

$$\text{Rate} = k_2[Cl_2][SnCl_4] \qquad (9.17)$$

$$\text{Rate} = k_2[Cl_2][ArH] + [Cl_2][ArH][ZnCl_2] \qquad (9.18)$$

Zinc chloride-catalysed chlorination of alkylbenzenes in HOAc is governed by eq. 9.18,[31a] and a similar rate expression (although with greater emphasis on the third-order term) applies in the $AlCl_3$-catalysed chlorination in nitro solvents.[64] In the latter work, rates were measured by a direct method, and gave k(toluene:benzene) rate ratios at 0 °C of 247 (in $PhNO_2$) and 215 (in $MeNO_2$). These are much larger than the values obtained (along with those for other alkylbenzenes) by Olah *et al.*,[63] who used the competition method. It was concluded that the competition method is unsuitable for determining large reactivity differences, and that the previously proposed π-complex mechanism for chlorination (and bromination) under these conditions[63] is invalid.

The increased bulk of the electrophile, probably $Cl^+MX_nCl^-$ produces a reduction in the *ortho*:*para* ratio.[65] Solvation of the electrophile produces reduced reactivity (and hence increased selectivity) and a lower *ortho*:*para* ratio,[66] as expected.

Silica enhances chlorination in CCl_4, evidently producing a more polar environment, fairly high *para*:*ortho* ratios being obtained.[67]

9.2.3 Chlorination with Sulphuryl Chloride

The kinetics of chlorination of anisoles by sulphuryl chloride, SO_2Cl_2, in chlorobenzene follow eq. 9.19. The lack of rate retardation by chloride ion shows that pre-equilibria to form Cl^+ or SO_2Cl^+ may be ruled out. Similarly, the invariance of rate with added SO_2 shows that a pre-equilibrium to form Cl_2 does not occur. The electrophile is believed to be $Cl^{\delta+}\cdots SO_2Cl^{\delta-}$, and a cyclic transition state (**5**) was considered probable.[68] The polar nature of the reaction is shown by chlorination taking place 1000 times faster in $PhNO_2$ than in PhCl (although here some dissociation to give Cl_2 does occur); the kinetic isotope effect, $k_H:k_D$, is 1.0 so loss of the proton occurs in a fast step.[68]

$$\text{Rate} = k_2[\text{ArH}][\text{SO}_2\text{Cl}_2] \qquad (9.19)$$

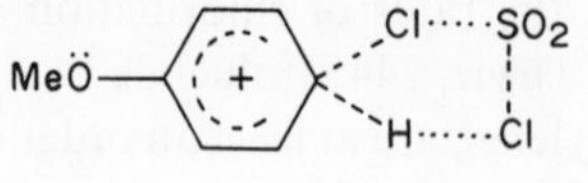

MeÖ Cl····SO2 H······Cl

(**5**)

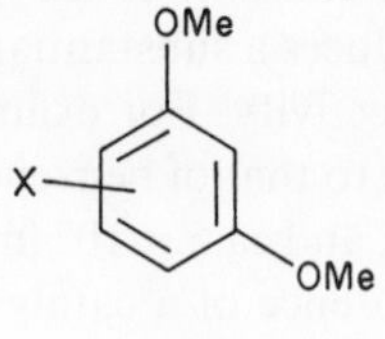

OMe X OMe

(**6**)

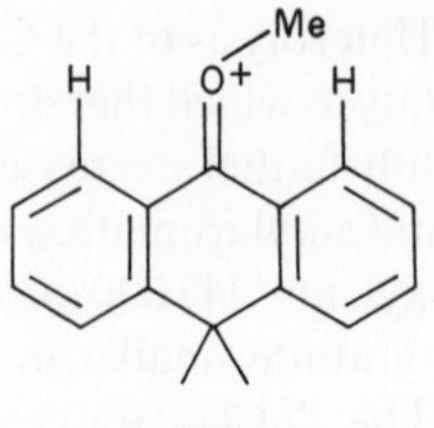

Me O+ H H

(**7**)

The reaction (in $PhNO_2$ as solvent) gave a Hammett correlation with $\rho = -7.2$ at 25 °C,[69] whilst substituents in 1,3-dimethoxybenzene (**6**) gave a ρ value of -4.0 for the reaction carried out in PhCl.[69] The effects of substituents at the

9-position of anthracene on rates of chlorination at the 10-position (in PhCl) have been measured.[70] Analysis of these data shows them to correlate with σ^+ values with $\rho = -2.1$, the lower value here being commensurate with the high reactivity of the substrate. Interestingly, the methoxy and nitro substituents activate less and deactivate less, respectively, than the correlation predicts, which probably reflects the inability of them to become coplanar with the aromatic ring as for example in 7, the mesomeric interactions thus being impeded. (This contrasts with protiodesilylation of substituted thiophenes, where the nitro substituent is exceptionally *deactivating* because it is able to become *more* coplanar with the five-membered ring than in benzene.[71])

Chlorination with SO_2Cl_2 in the presence of Ph_2S and $AlCl_3$ gives increased *para* selectivity in the reaction with phenols and ethers, attributed to a bulk effect.[72]

9.2.4 Chlorination with Metal Halides

Various metal chlorides appear to act as electrophilic chlorinating species. The kinetics of antimony pentachloride chlorination follow eq. 9.20, interpreted in terms of slow decomposition of the intermediate ArH–$SbCl_5$ complex, assisted in part by a second molecule of $SbCl_5$.[73]

$$\text{Rate} = k_2[\text{ArH}][\text{SbCl}_5] + [\text{ArH}][\text{SbCl}_5]^2 \tag{9.20}$$

Chlorination by $TiCl_4$ in peroxyacids gives a very high *ortho*:*para* ratio, believed to be due to formation of HOCl (eq. 9.21).[74]

$$\text{HOOH} + \text{TiCl}_4 \longrightarrow \text{HOTiCl}_3 + \text{HOCl} \tag{9.21}$$

An industrially useful reagent is $CuCl_2$, which gives $>90\%$ yields in the chlorination of phenol and its 2-methyl and 3,5-dimethyl derivatives.[75] By carrying out the reaction in the presence of HCl and H_2O_2 it is possible to regenerate the $CuCl_2$ continuously, and no inorganic byproducts are produced.[76a] Supported on alumina, $CuCl_2$ becomes a much more effective chlorinating reagent and gives very high *para* selectivity with anisole (*para*:*ortho* ratio = 32.3).[76b] In chlorination with $CuCl_2$ in the presence of peroxodisulphate, the aromatic may react as a radical cation.[77]

Chlorination by thallium(III) chloride tetrahydrate in CCl_4 gave a toluene:benzene reactivity ratio of 26 with an *ortho*:*para* ratio of 1.1. These values became 43 and 1.8 in the presence of Cl_2, and to up to 133 and 2.3 with Cl_2 and other thallium salts. The lower values under the former condition indicated that free chlorine was not involved and that the electrophile was bulky, $TlCl_2^+$ being proposed.[78]

9.2.5 Miscellaneous Chlorination Methods

The reaction of activated aromatics with *S*-halodimethylsulphonium halides, e.g. $Me_2SCl^+Cl^-$ (prepared from Me_2S and Cl_2) in dichloromethane, gives $>97\%$ *para* selectivity and high yields (74–94%).[79]

Anisole is chlorinated almost exclusively in the *para* position, by 2,3,4,5,6,6-hexachloro-2,4-dienone (**8**) or by 2,3,4,4,5,6-hexachloro-2,5-dienone (**9**), which may reasonably be attributed to steric hindrance. With phenol, however, **8** gives high *ortho* selectivity, especially in non-polar solvents, the probable cause being hydrogen bonding between the carbonyl group and the phenolic hydrogen, so facilitating chlorine transfer from the CCl_2 group (**10**).[80] Chlorination accompanies sulphonation of anthracene by chlorosulphuric acid, $ClSO_3H$.[81]

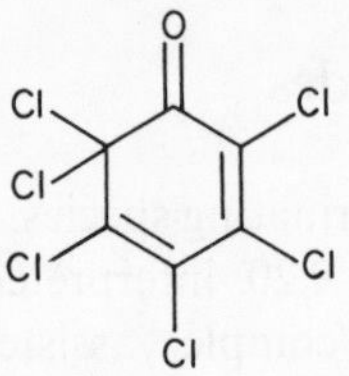

(**8**) 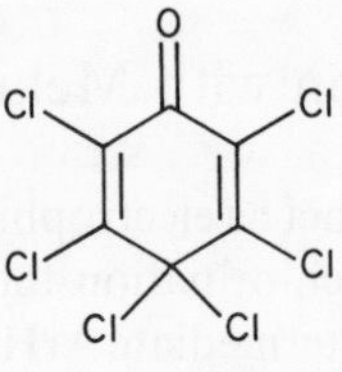 (**9**) 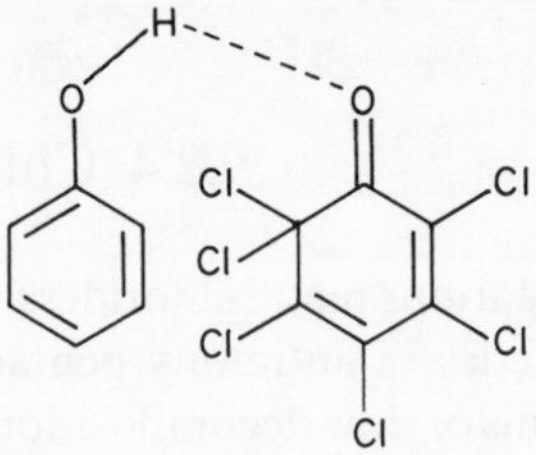 (**10**)

9.3 BROMINATION

9.3.1 Bromination by Hypobromous Acid and its Acetyl Derivative

(i) Hypobromous acid

Bromination by HOBr shows kinetic features similar to those for chlorination with HOCl (Section 9.2.1). Thus, addition of mineral acid produces a large increase in the bromination rate in water or aqueous dioxan,[6,82] (the rate is slower in 50% aqueous dioxan than in water)[83] and the kinetics following eq. 9.22. These were interpreted as involving bromination by either Br^+ or, more probably, by H_2OBr^+, the former reagent being deemed improbable on grounds similar to those advanced against Cl^+ as the electrophile in chlorination by H^+–HOCl. In bromination of sodium 4-anisate by HOBr in phosphate buffers (pH 7–8) at 25 °C, hypobromous acid was shown to be a 2000 times less effective brominating reagent than molecular bromine.[84] Another report gave the relative reactivities of the electrophiles H_2OBr^+:Br_2:HOBr as ca 10^6:6700:1.[85]

$$\text{Rate} = k_3[\text{ArH}][\text{HOCl}](f[\text{H}^+]) \qquad (9.22)$$

$$\mathrm{HOBr} \underset{\mathrm{ArH}}{\overset{\mathrm{H^+}}{\rightleftharpoons}} \mathrm{H_2OBr^+} \xrightarrow{\mathrm{ArH}} \mathrm{ArH.H_2OBr^+}; \quad \mathrm{HOBr} \xrightarrow{\mathrm{ArH}} \mathrm{ArH.HOBr} \xrightarrow{\mathrm{H^+}} \mathrm{ArH.H_2OBr^+} \underset{}{\overset{\text{slow}}{\rightleftharpoons}} \mathrm{Ar^+(H)(Br)} \xrightarrow{\text{fast}} \mathrm{ArBr}$$

Scheme 9.3. Pathway for bromination by hypobromous acid in aqueous perchloric acid.

More recent work on bromination (with HOBr–aqueous $HClO_4$) with compounds ranging in reactivity from mesitylene to 2,6-dimethylpyridinium ion showed the rates to parallel those for nitration, whilst the kinetic form of eq. 9.22 holds over the whole range. This indicated that there must be a common brominating species, yet the concentration of Br^+ at the lowest acidities would be too low to give the observed rates, even if reaction occurred at every encounter. It was proposed that the effective electrophile is H_2OBr^+, but that this is formed by protonation of a complex of the aromatic compound and HOBr in a pre-equilibrium step (Scheme 9.3),[86] the existence of a protropic pre-equilibrium being indicated by a solvent isotope effect, $k(H_2O)/k(D_2O)$, of 2.2. For highly reactive aromatics in feebly acidic media the lower pathway is followed, whereas for less reactive aromatics in more acidic media the upper pathway becomes more important.

Bromination with potassium bromate–H_2SO_4 is a particularly useful method for brominating unreactive aromatics in high yield, and 65 wt-% H_2SO_4 is the optimum acid strength; above 70% acid the reaction becomes uncontrollably rapid.[87] Hypobromous acid is believed to be formed via protonation and loss of oxygen, giving H_2OBr^+ as electrophile (or $AcOHBr^+$ if the reaction is carried out in aqueous HOAc).[88]

Measurement of kinetic isotope effects, k_H:k_D, for positive bromination show very clearly that as steric strain in the intermediate becomes greater, then the intermediate becomes partitioned between product formation and reversion to the starting materials, leading to an isotope effect (Section 2.1.3). Thus, whereas bromination of benzene by HOBr–$HClO_4$ gives no isotope effect,[89] compounds **11** give small effects of 1.05 (X = H) and 1.16 (X = Me),[90] compounds **12** give large effects of 1.27, 1.37, and 1.49 for X = Cl, Br, and I, respectively,[91] whilst the very hindered **13** gives a very large value of 10.[92]

(11) X=H, Me — 1-X-2,6,4-triethyl-3-methylbenzene (Et, Et, Me, Et substituents)

(12) X=Cl, Br, I — 1-X-2,4,6-tris(CH_2Bu-t)benzene

(13) 1,3,5-tri-t-Bu-benzene

(ii) The acetyl derivative of hypobromous acid

Bromination of biphenyl by HOBr in 75% aqueous HOAc indicated that part of the reaction was governed by eq. 9.23, so that bromine acetate is an electrophile under these conditions. The relative rate of bromination of biphenyl to that of benzene was 1270 (at 0 °C) compared with 26.9 if mineral acid was added, whence the electrophile becomes H_2OBr^+ or $HOAcBr^+$. The log f_o:log f_p value also changed from 0.75 in the absence of mineral acid to 0.88 in the presence, confirming a change in the nature of the electrophile with conditions.[93]

$$\text{Rate} = k_2[\text{ArH}][\text{BrOAc}] \qquad (9.23)$$

Bromination by bromine acetate is faster than that by bromine, and a cyclic hydrogen-bonded transition state within which the O—Br bond is broken may be responsible [cf. chlorine acetate, Section 9.2.1.(iii)].

Bromine acetate is believed to be the electrophile in bromination with Br_2 and peroxyacetic acid in HOAc.[94]

Bromination of 1,3,5-tri-*tert*-butylbenzene with preformed bromine acetate (or with Br_2–Ag^+–HOAc) gives acetoxylation in addition to bromination (**14**–**17**), and this becomes the major reaction if sodium acetate is present; acetoxylation is also observed with other reactive polyalkylbenzenes.[95] Compounds **14** and **16** are formed via bromination and bromodebutylation, respectively, whilst **15** and **17** are formed from bromoacetoxy adducts via loss of *t*-BuBr and HBr, respectively.

Me₃C CMe₃ CMe₃ → Br Me₃C CMe₃ CMe₃ (**14**) + Me₃C CMe₃ OAc (**15**) + Me₃C CMe₃ Br (**16**) + OAc Me₃C CMe₃ CMe₃ (**17**)

9.3.2 Bromination by Molecular Bromine

The ease of handling bromine is the probable reason for the majority of studies of bromination with molecular bromine having been carried out with bromine itself rather than a precursor (cf. molecular chlorination).

(i) Bromine

Bromination with bromine has been carried out using a wide range of solvents, e.g. acetic acid,[22,29,44,96–106] trifluoroacetic acid,[107] superacids,[108–110] nitromethane,[111] chloroform,[112], carbon tetrachloride,[113] dimethylformamide,[114] sulphur dioxide,[115], water,[116–118] and perchloric and sulphuric acids.[119] The reaction is always first order in aromatic, and at least first order in Br_2, but higher orders in Br_2 are commonly observed. For example, the full expression for bromination in HOAc is given by eq. 9.24.[44,120]

$$\text{Rate} = k_2[\text{ArH}][\text{Br}_2] + k_3[\text{ArH}][\text{Br}_2]^2 + k_4[\text{ArH}][\text{Br}_2]^3 \qquad (9.24)$$

The second-order term represents unassisted heterolysis of **18** to **19** (eq. 9.25), while the higher order terms may represent the assistance by one or more molecules of bromine, in breaking the Br—Br bond in the intermediate, as shown in eq. 9.26. Alternatively, they may represent polarization of one molecule of Br_2 by another to give, e.g., $Br^{\delta+} \cdots {Br_3}^{\delta-}$ prior to attack on the aromatic. It is noteworthy that the existence of the intermediates in bromination has been provided by the isolation of **20** and related compounds,[112] the stability of which derives from the enormous conjugative electron release by the *o*- and *p*-NMe_2 groups.

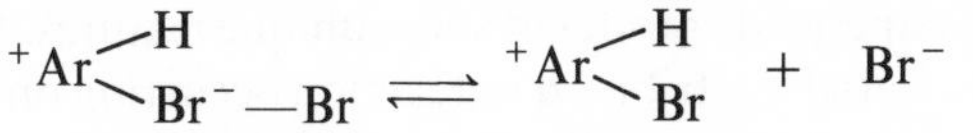

$$^{+}\text{Ar}\begin{matrix}\text{H}\\ \text{Br}^{-}\text{—Br}\end{matrix} \rightleftharpoons {}^{+}\text{Ar}\begin{matrix}\text{H}\\ \text{Br}\end{matrix} + \text{Br}^{-} \qquad (9.25)$$

(18) **(19)**

$$\text{ArHBr}_2 + \text{Br}_2 \longrightarrow {}^{+}\text{ArHBr} + {\text{Br}_3}^{-} \qquad (9.26)$$

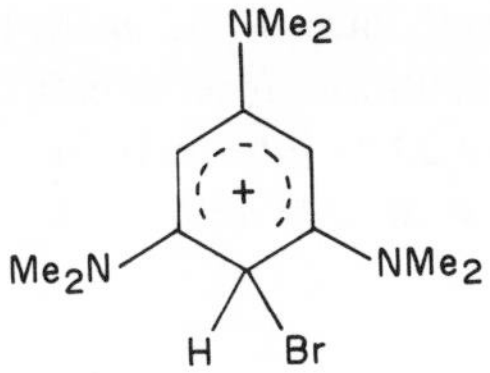

NMe2

Me2N NMe2

H Br

(20)

The reaction order is reduced, and the rate usually increased, by the addition of water, salts, or acids.[32,100–103] The higher polarity of the medium produced in this way is evidently sufficient to permit heterolysis (eq. 9.25) to occur unassisted. Data obtained in HOAc media show that higher order terms are less significant in bromination of more reactive aromatics, since for these substitution is more readily achieved by the weaker electrophiles. The lower electrophilicity of Br_2 compared with Cl_2 accounts for the lack of high-order processes in chlorination, and an additional factor is that the ion ${Cl_3}^-$ is less stable than ${Br_3}^-$ and so is less readily formed.[121]

Since the ion IBr_2^- is more stable than Br_3^-, it follows that bromination rates are increased by the addition of iodine; iodine bromide is formed according to eq. 9.27, in which the equilibrium lies well to the right,[122] and this replaces, and is more efficient than, a molecule of Br_2 in eq. 9.26.[103,123–126]

$$I_2 + Br_2 \rightleftharpoons 2IBr \tag{9.27}$$

Hydrogen bromide is a byproduct of bromination with bromine and interferes with the kinetics by reversing eq. 9.24, by removing Br_2 as Br_3^-, a very unreactive electrophile (Section 9.3.3), and it may also protonate the aromatic, thereby reducing its reactivity. One method of overcoming this is to brominate in the presence of an amine, chosen so as to give an insoluble amine hydrobromide. This combination (at $-70\,°C$) has been found to be very useful for the specific *ortho*-bromination of phenols; the process here is thought to involve formation of phenyl hypobromite, PhOBr, which then rearranges.[127]

Pyridine catalyses bromination, this being assumed originally to be due to the polarization of bromine through formation of pyridinium bromide, $pyBr^{\delta+} \cdots Br^{\delta-}$. However, it has been found that pyridinium salts also accelerate bromination in HOAc and $CHCl_3$ to a comparable extent, and no more effectively than other ammonium salts. It was concluded that the acceleration is due to a salt, i.e. polarity, effect.[128]

Bromination in superacids produces some unique features. Phenols give *meta*-bromo derivatives, initially attributed to reaction occurring on $PhOH_2^+$.[108] This was criticized[109] on the basis that Ph_3O^+ is *ortho, para* directing in nitration[129] and it was proposed that a phenol, e.g. 4-methylphenol, is initially brominated in the 2-position giving **21** as intermediate. Since under the highly acidic conditions this cannot lose a proton from oxygen as readily as bromine can migrate, rearrangement as in **21** to give **22** was postulated. More recently it has been noted that phenol itself cannot be *meta*-brominated and therefore a substituent is required to facilitate *ipso* substitution, which, in the case of 4-bromophenol, must involve **23** as intermediate; rearrangement as before then gives **22**;[110a] rearrangements of this kind can also take place in sulphuric acid.[110b]

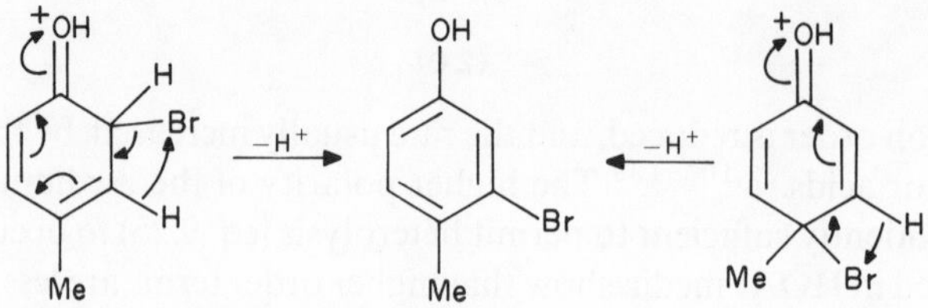

(21) **(22)** **(23)**

Bromination by Br_2 in liquid SO_2 is extremely selective,[130a] more so even than Br_2 in TFA, for which ρ has been determined as -16.7.[107b] It is noteworthy that whereas the latter poorly solvating medium gives the 'inductive' order of activation by *p*-alkyl groups, the 'Baker–Nathan' order is obtained in SO_2. Here solvation is better, so that steric hindrance to solvation becomes very important and masks the electron-releasing order of the alkyl groups (see Section 1.4.3). Bromination under very selective conditions is an excellent method of obtaining maximum yields of a given isomer.[130b]

(ii) Bromine in the presence of a catalyst

Iodine is a more effective catalyst than either $AlCl_3$, $SbCl_3$, PCl_3, PCl_5, or $SnCl_4$ in catalysing bromination by Br_2.[131] The rate maximum occurs at different I_2:Br_2 ratios for different aromatics, which reflects the variable degree of catalysis;[126,132,133] dimers of IBr are evidently needed for the less reactive compounds.[134–136]

The relative effectiveness of metal halide catalysts is $MgBr_2 < ZnBr_2 \approx HgBr_2 < CdBr_2 < BeBr_2$, whilst the halides of Group II metals have very little effect.[137] More detailed studies of the effect of $ZnCl_2$ indicated that it catalyses only that part of the reaction that is first order in Br_2, excess bromine being apparently a more effective catalyst than $ZnCl_2$ for the higher order components.[100] Relative rates of $FeCl_3$-catalysed bromination of a range of aromatics in $MeNO_2$ are very dependent on the method of mixing the reagents. The widest rate spread, with minimal dibromination, was obtained by dilution of the bromine with the solvent prior to adding to the other reagents (Table 9.2). Dibromination was also absent if $FeCl_3$ was added to the bromine solution prior to addition to the aromatic–halide–solvent mixture, but in this case the selectivity was greatly reduced. This was attributed to the presence of a much more reactive electrophile that reacted with the aromatic via rate-determining formation of a π-complex.[138] However, mixing control (Section 2.10) is the probable cause since reduction in the concentrations of Br_2 and $FeCl_3$ caused a 10-fold increase in selectivity. Moreover, the selectivity in the $AlCl_3$-catalysed bromination of toluene and benzene showed great variation with substrate concentration (measured by the competition method) and at the maximum was only half that determined by direct rate measurements.[62]

Dioxan accelerates bromination through the formation of dioxan dibromide, in which the Br—Br bond is polarized; the kinetics follow eq. 9.28.[139]

$$\text{Rate} = k_2[\text{ArH}][\text{dioxan dibromide}]^2 \qquad (9.28)$$

Table 9.2 Relative rate of $ZnCl_2$-catalysed reaction of Br_2 with ArX at 25 °C

X	H	Me	1,2-Me_2	1,3-Me_2	1,4-Me_2	1,3,5-Me_3	F	Cl	Br
$k_{rel.}$	1	7.4	34.3	534	31.5	> 1000	0.27	0.12	0.10

Bromination of 4-nitrobiphenyl on a silica surface occurs rapidly and gives a *para*:*ortho* ratio of 4. The reaction must occur between the aromatic and bromine adsorbed on the silica, since the rate decreases the more of the surface that has the aromatic adsorbed on it.[140]

Bromination of some phenolic compounds in the presence of $TiCl_4$ gives substitution mainly *ortho* to the OH group (cf. chlorination, Section 9.2.4), whereas for the corresponding methyl ethers substitution occurs mainly *para*.[141] It has been suggested that a complex is formed between $TiCl_4$ and the OH group, but this does not explain the orientation satisfactorily.[142]

Bromination by Br_2 in the presence of thallium triacetate goes mainly in the *para* position,[143] and this is an electronic rather than a steric effect;[144] the selectivity, however, is less than that of Br_2 in TFA, which is both easier and safer to use. In general, Tl^{I} salts give rise to a much more reactive electrophile than do Tl^{III} salts.[145]

9.3.3 Bromination by Tribromide Ion

Surprisingly, the tribromide ion, Br_3^-, present in media containing Br_2, is an electrophile. Its expected lower reactivity than the latter species is confirmed by the decrease in rate produced by the addition of Br^-.[101,115] The tribromide ion is a significant brominating agent only for the most reactive aromatics, and even in the bromination of PhO^- and naphthalene by Br_2, only 1–3% and 0.3%, respectively, of reaction occurred via Br_3^-.[102a,115]

9.3.4 Bromination by Bromine Phosphate

The rate of bromination of sodium 4-anisate in phosphate buffers increased along with the concentration of phosphate buffer, whereas varying the concentration of the acid component of the buffer produced no rate change. The rate change was therefore attributed to eq. 9.29 giving bromine phosphate as a more reactive electrophile. This would be consistent with the second dissociation constant of phosphoric acid being much greater than the dissociation constant of water.[84]

$$HOBr + H_2SO_4^- \rightleftharpoons BrHPO_4^- + H_2O \qquad (9.29)$$

9.3.5 Bromination by Bromine Sulphate

Bromine sulphate may be a brominating species since the bromination of benzoic acid by HOBr was catalysed much more effectively by H_2SO_4 than by

$HClO_4$ of the same acidity. It is considerably less reactive than H_2OBr^+, so that its participation becomes noticeable only with the more reactive aromatics.[88]

9.3.6 Bromination by α-Bromoketones and Aluminium Halides

Both α,α-dibromo- and α,α,α-tribromoketones, $ArCOCHBr_2$ and $ArCOCBr_3$, in the presence of aluminium halides will brominate aromatics, and the reaction rate is increased, as expected, by electron withdrawal in the aryl group.[146] Electron withdrawal by X in $ArCOCXBr_2$ also increases the rate as expected. The aluminium chloride evidently coordinates to the carbonyl oxygen and aids polarization of the C—Br bond.

9.3.7 Bromination by *N*-Bromoamides

Because the N—Br bond is less polar than the N—Cl bond, *N*-bromoamides are more inclined to react by a free-radical mechanism than the corresponding *N*-chloroamides. In the presence of carboxylic acids (and thus also a polar environment), *N*-bromoanilides rearrange to the corresponding *o*- and *p*-bromoacetanilides. However, if the reaction is carried out in the presence of anisole (which becomes brominated in the *ortho* and *para* positions), the *ortho*:*para* ratio is dependent on both the catalysing acid and the nature of R (eq. 9.30). This shows that free acyl hypobromite is not produced, and the mechanism is believed to involve three steps: (1) protonation of the substrate, (2) heterolytic fission of the N—Br bond, and (3) rearrangement of the resulting ion pair.[147]

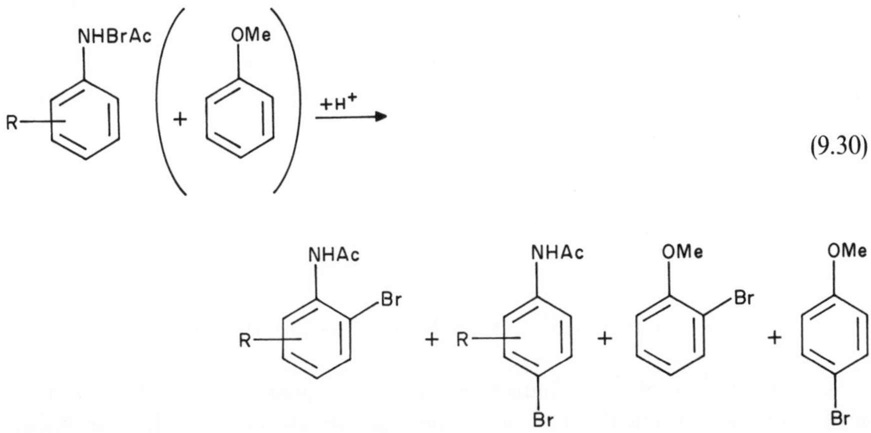

NHBrAc OMe

R– + $\xrightarrow{+H^+}$ (9.30)

NHAc Br NHAc OMe Br OMe

R– + R– + +

Br Br

9.3.8 Bromination with Interhalogens

Both bromine chloride, BrCl, and bromine fluoride, BrF, are more polar than molecular bromine and so are more effective brominating species. Thus BrF (produced quantitatively by passing a 10% solution of F_2 in N_2 into Br_2 in $CHCl_3$ at $-75\,^{\circ}C$), electrophilically brominates in proton-donating solvents at low temperature without a catalyst. Anisole brominated quantitatively, giving a *para*:*ortho* ratio of 4, toluene gave a 90% yield of equal amounts of both isomers, and *tert*-butylbenzene gave a 90% yield of the *para* isomer only. Even 1,3-dinitrobenzene and ethyl benzoate could be brominated in the *meta* positions in ca 95% yields.[148]

Bromination with BrCl is also rapid,[84] and this reagent may be formed in bromination by Br_2 and $SbCl_5$ in CCl_4, which gives very high *para* selectivities (up to 99% for halogenobenzenes), and a high substrate selectivity of 257 for toluene; lower selectivities are obtained using 1,2-dichloroethane as solvent, and this is useful for brominating unreactive aromatics.[149]

9.3.9 Bromination with Bromocyclohexadienones

Just as nitration may be carried out with cyclohexadienone derivatives, so too may these be used for bromination.[150] In this case 2,4,4,6-tetrabromocyclohexa-2,5-dienone (**24**) was used, and the brominating ability derives from the fact that aromaticity is gained by loss of bromine. However, the lower electron withdrawal compared with the dienone used for nitration (structure **10**, Chapter 7) means that homolytic C—Br cleavage is more likely here. Some indication that non-polar reactions interfere are provided by the decrease in the *ortho*:*para* ratio as the medium is made more polar; *N*-bromosuccinimide behaved similarly[151] (free-radical reactions give high *ortho*:*para* ratios).

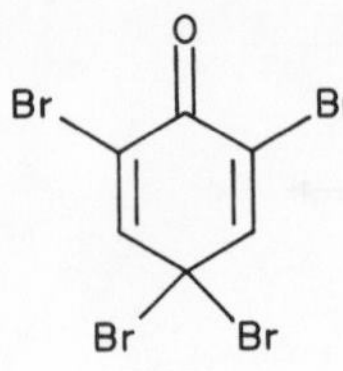

O
Br Br
Br Br

(24)

9.3.10 Isotope Effects in Bromination by Bromine

Since bromination by bromine has a 'late' transition state, it could be anticipated that kinetic isotope effects might be observed in suitable cases. Thus, whereas there is no effect in the $ZnCl_2$-catalysed bromination of pentamethylbenzene,[152] positive effects have been found under many other conditions. As in the case of bromination by H_2OBr^+, the isotope effect increases with increasing steric hindrance and decreasing reactivity in the aromatic. Those shown in Scheme 9.4 are for bromination by Br_2–HOAc [use of $MeNO_2$ as solvent gave a larger value (2.7) for **28**]; a larger value (3.6) was also obtained for the very hindered 1,3,5-tri-*tert*-butylbenzene for bromination by Br_2–Ag^+–HOAc–dioxan.[153] Scheme 9.4 also shows values for bromination by Br_2 in DMF.[154]

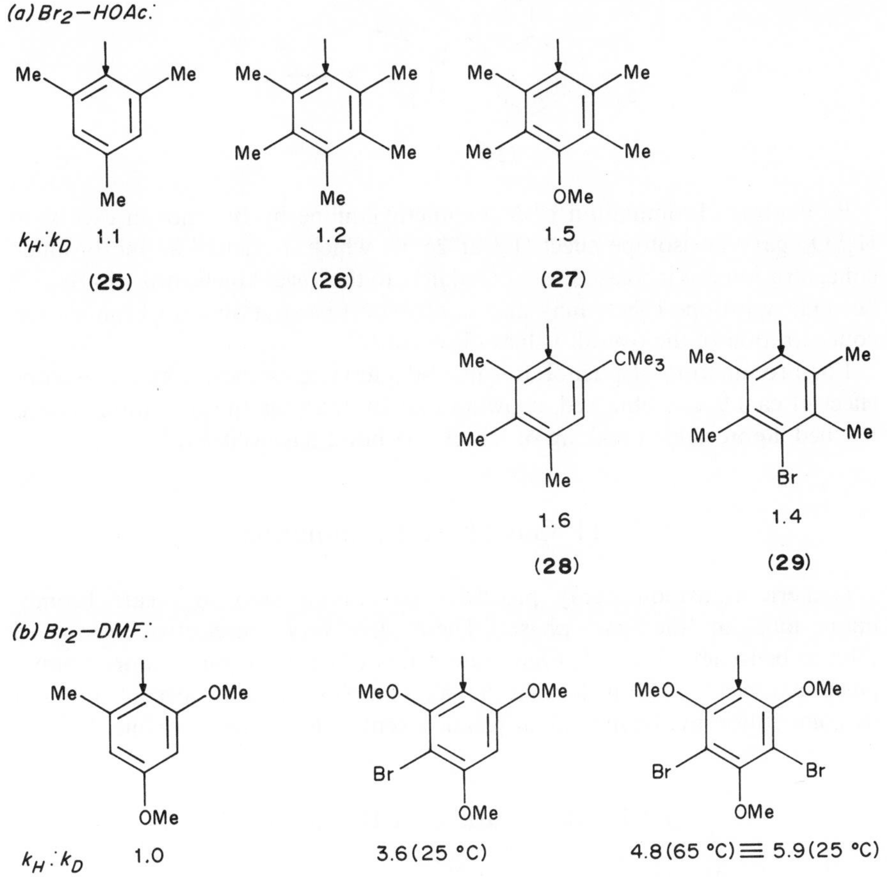

(a) Br_2–HOAc:

	(25)	(26)	(27)	(28)	(29)
$k_H:k_D$	1.1	1.2	1.5	1.6	1.4

(b) Br_2–DMF:

$k_H:k_D$	1.0	3.6 (25 °C)	4.8 (65 °C) ≡ 5.9 (25 °C)

Scheme 9.4. Kinetic isotope effects for bromination with bromine.

Increasing the concentration of Br^- should, by reversing eq. 9.25 or making the reaction occur via the less reactive Br_3^-, cause the isotope effect to increase.[155,156] This was found in the bromination of 4-methoxybenzenesulphonic acid,[157] and the disodium salt of 2-naphthol-6,8-disulphonic acid (a compound which gives an isotope effect through the high stability of the intermediate **30**).[155] Moreover, the value of $k_H:k_D$ diminished as the concentration of Br_2 was increased; this can be explained in terms of the additional Br_2 molecules catalysing Br—Br bond breaking, thereby lowering the isotope effect. Fast formation of the intermediate **30** was also indicated by the bromination rate being almost identical with that for bromination by HOBr (which gave an isotope effect of 2.1).[102]

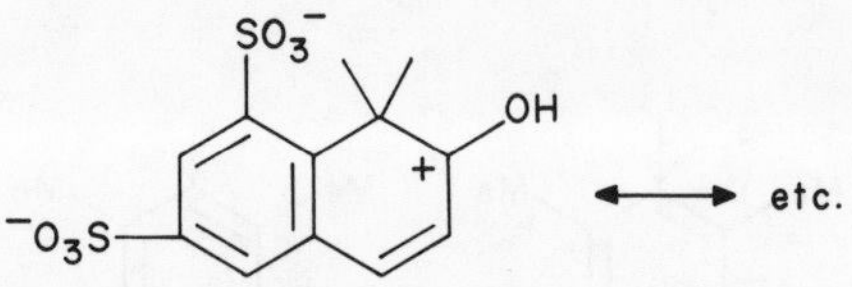

(30)

By contrast, bromination of *N*,*N*-dimethylaniline by Br_2 (not in excess) in H_2SO_4 gave an isotope effect (1.8 at 25 °C) which *decreased* as the bromide concentration was increased, attributed here to the lower kinetic order in Br_2.[158] Secondary isotope effects may also contribute to variation with bromide ion concentration of the overall values observed.[159]

For bromination of phenol and its *O*-deuteriated isomer, a kinetic isotope effect of ca 1.9 was obtained, showing that by the time the transition state is reached, appreciable breaking of the O—H bond has occurred.[160]

9.3.11 Gas-phase Bromination

Isomeric transition decay processes have been used to create bromonium ions in the gas phase. These are very unselective, giving a toluene:benzene:halogenobenzene rate ratios of ca 2:1:1, but a more normal positional selectivity. It is thought that a π-complex is initially formed, followed by competition by the individual reaction centres for the electrophile.[161,162]

9.3.12 Base-catalysed Bromination

Like a number of other electrophilic substitutions, bromination is able to undergo the base-catalysed mechanism under certain conditions. Thus, whereas 1,3,5-trimethoxybenzene reacts normally with *tert*-butyl hypobromite in hexamethylphosphoramide (and the reaction is retarded by potassium *tert*-butoxide), 1,3,5-tribromobenzene does not react with *tert*-butyl hypobromite unless KOBu-*t* is present. The reason is that the latter aromatic is reacting via the base-catalysed mechanism (eqs 9.31a and 9.31b).[163]

$$\text{1,3,5-Br}_3\text{C}_6\text{H}_3 \xrightarrow{\text{KOBu-}t} \text{2,4,6-Br}_3\text{C}_6\text{H}_2^-\ \text{K}^+ + t\text{-BuOH} \tag{9.31a}$$

$$\text{2,4,6-Br}_3\text{C}_6\text{H}_2^-\ \text{K}^+ \xrightarrow{t\text{-BuOBr}} \text{1,2,3,5-Br}_4\text{C}_6\text{H}_2 + t\text{-BuOK} \tag{9.31b}$$

9.4 IODINATION

9.4.1 Iodination with Iodine

Iodination with iodine has led to some contradictory kinetics and conclusions. The following account gives the currently accepted position.

Iodination of [2,4,6-$^2H_3$]phenol gives a kinetic isotope effect, $k_H:k_D$, of 4.0,[164] and for 2-iodination of 4-nitrophenol the isotope effect increased from 2.3 to 5.5 with increasing [I^-];[165] similar results have been obtained for 4-iodination of 2,6-dibromophenol (which gives a larger isotope effect than for bromination).[166] This shows that there is a rapid formation of the intermediate followed by rate-determining loss of the proton (eqs 9.32 and 9.33); increasing the concentration of I^- increases the rate of reversal of eq. 9.32, $k_2:k_{-1}$ decreases and the isotope effect is increased (Section 2.1.3). This effect of [I^-] on the isotope effect, coupled with the dependence of rate on $[I^-]^{-1}$ rules out the alternative pre-equilibrium step (eq. 9.34) involving a positive iodinating species, proposed earlier to account for the apparent dependence of rate on $[I^{-1}]^{-2}$ in the iodination of phenol and amines.[167] The latter work gave the relative reactivities of $PhO^-:PhNH_2:PhOH$ as $9.2 \times 10^9:3.7 \times 10^5:1$. Phenols are 5-fold more reactive than their methyl ethers,[168] and in this work it was shown that the order in I_2 may be either first or second, depending on whether the concentrations of H^+ and I^- are high or low, respectively; this may account for discrepancies in some of the earlier work noted above.

$$I_2 + ArH \overset{\text{fast}}{\rightleftharpoons} ArHI^+ + I^- \tag{9.32}$$

$$ArHI^+ \overset{\text{slow}}{\rightleftharpoons} ArI + H^+ \tag{9.33}$$

$$H_2OI^+ + ArH \overset{\text{fast}}{\rightleftharpoons} ArHI^+ + H_2O \tag{9.34}$$

Since phenol is a very reactive aromatic, it might not be expected to show an isotope effect, this being the case in many other reactions. However, iodination takes place on the phenoxide ion and so gives rise to a particularly stable intermediate, e.g. **31**. In many other cases where isotope effects are observed, the intermediate is probably a moderately stable species.

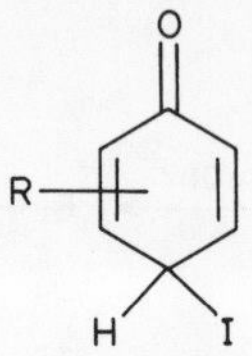

(31)

Isotope effects have also been observed in iodination with I_2–aqueous KI of aromatic amines, aminocarboxylic acids, and aminosulphonic acids, but not with aminesulphonates.[169] The kinetic form of these reactions is given by eq. 9.35, where B is a basic component of the buffer solution. The first term is observed both in reactions which give a positive isotope effect and in those which do not, showing that the second molecule of aromatic base cannot be assisting removal of a proton from the intermediate (otherwise the magnitude of the isotope effect would be a function of base concentration, Section 2.1.3). The second molecule of amine may therefore be involved in the iodinating entity, e.g. $ArNR_2I^+$, and the same may be true of the base B in the second term.[169] A similar interpretation has been given to account for the invariance in the isotope effect with base concentration in iodination of glyoxaline,[170] and for the correlation of the accelerating effect of bases with their nucleophilicities rather than their basicities.[171]

$$\text{Rate} = k[\text{ArNR}_2]^2[\text{`iodine'}] + k_3[\text{ArNR}_2][\text{`iodine'}][\text{B}] \qquad (9.35)$$

Just as kinetic isotope effects become increased, the greater the steric hindrance in the aromatic [see Sections 9.3.1(i) and 9.3.10], so too they may be increased by greater hindrance in the base. Thus, in iodination of azulene, the $k_H:k_D$ values increased from 2.0 to 6.5 as the base was changed from pyridine to 2,4,6-trimethylpyridine.[172]

9.4.2 Iodination with Iodine and an Oxidizing Agent

Iodine in the presence of silver sulphate–sulphuric acid is a useful reagent for unreactive aromatics; the effective electrophile may be I^+ or a solvated derivative.[173] A mixture of iodine–$NaClO_4$–HOAc–dioxane has also been used, and with 1,3,5-trineopentylbenzene this gave a kinetic isotope effect $k_H:k_D$, of 2.63, whereas bromination and chlorination in DMF gave values of 1.1 and 1.01, respectively, which demonstrates nicely the result of an increase in steric hindrance with increasing size of halogen.[174]

The rate dependence on concentration of N_2O_4 and acid in iodination by iodine in a mixture of nitric and acetic acids indicated that the electrophile is $[INO_2H]^+$.[175] A positive iodinating species, I_3^+, was suggested as the electrophile in iodination by I_2–KIO_3–H_2SO_4; this gave a ρ factor of -6.4, similar to that for positive bromination, and kinetic isotope effects of 2.1 and 3.4 for benzoic acid and nitrobenzene, respectively.[176] A positive iodinating species is probably also involved in iodination with I_2–HIO_3 in HOAc–H_2SO_4,[177] with I_2–peracetic acid (protonated acetyl hypoiodite was considered to be formed as in eqs 9.36 and 9.37[178]), and with iodine in HNO_3–H_2SO_4, partial rate factors of 28.5 (4-Me), 12.2 (2-Me), 2.05 (3-Me), 0.33 (4-Cl), and 0.039 (3-CO_2H), and a ρ factor of -4.5 being obtained.[179a]

$$I_2 + MeCO_3H + H_2O \xrightleftharpoons{\text{slow}} 2HOI + MeCO_2H \tag{9.36}$$

$$HOI + MeCO_2H + H^+ \xrightleftharpoons{\text{fast}} H_2O + (MeCO_2HI)^+$$

$$\xrightarrow{\text{ArH, fast}} ArI + MeCO_2H + H^+ \tag{9.37}$$

A highly reactive positive iodinating species is produced by the anodic oxidation of I_2 in TFA, and this will even iodinate nitrobenzene in the *meta* position in high yield, the extent of which depends on the solvent. The mechanism does not involve anodic oxidation of the aromatic to a cation radical and subsequent reaction with iodine, because this would give other isomers.[179b]

9.4.3 Iodination with Iodine Acetate and Iodine Trifluoroacetate

These reagents, CH_3CO_2I and CF_3CO_2I, are the acetyl and trifluoroacetyl derivatives, respectively, of hypoiodous acid. The former may be responsible for the large increase in rate with increase in the concentration of the buffer acid in the iodination of phenol at constant pH.[167b] Iodine acetate may also be present in the iodination of pentamethylbenzene by I_2–$Hg(OAc)_2$, since $Hg(OAc)_2$ speeds up the reaction, consistent with it disturbing eq. 9.38;[180] rate data indicated iodine acetate to be less reactive than bromine acetate. This is confirmed by a study of iodination by iodine trifluoroacetate, which showed the reactivity order to be $Br_2 \approx ICl \approx CH_3CO_2H \ll CF_3CO_2H < CH_3CO_2Br \ll CF_3CO_2Br$, the relative values being ca $1:10^4:10^6:10^{10}$.[181]

$$I_2 + Hg(OAc)_2 \rightleftharpoons IOAc + HgIOAc \tag{9.38}$$

9.4.4 Iodination by Iodine Monochloride

Iodination by ICl in HOAc gives third-order kinetics, i.e. first order in aromatic and second order in ICl, and these tend towards second order as the temperature is raised.[182] This resembles molecular bromination, suggesting that ICl is the reagent; the third-order kinetics then result from the second molecule of ICl assisting breakage of the I—Cl bond (eq. 9.39), whilst at higher temperature this heterolysis can occur unassisted.

$$\mathrm{Ar^{+}(H)(I{-}Cl\cdots I{-}Cl)} \longrightarrow \mathrm{Ar^{+}(H)(I)} + \mathrm{ICl_2^-} \qquad (9.39)$$

The kinetics for iodination by ICl in the presence of $ZnCl_2$ are first order in aromatic, ICl, and $ZnCl_2$, so here the Lewis acid must assist heterolysis of the I—Cl bond.[182] In TFA the order is two overall because the higher polarity of the solvent permits unassisted heterolysis of the I—Cl bond, whereas in CCl_4 the order in ICl is three, evidently because *two* molecules of ICl are required to promote the heterolysis.[182]

Iodination by ICl in aqueous HCl is believed to involve a positive iodinating species.[183]

Iodination by ICl (or I_2) catalysed by $SbCl_5$ in CCl_4 gives substantial (ca 105) toluene:benzene selectivity, together with an *ortho*:*para* ratio of 1.4.[149]

9.5 ASTATINATION

Oxidation of astatine with acidified dichromate gives an electrophile believed to be the univalent astatine cation, which reacted with benzene to give up to 49% of astatobenzene. Halogenobenzenes gave the isomeric yields shown in Table 9.3, which confirmed the electrophilic nature of the reaction.[184]

9.6 SUBSTITUENT EFFECTS IN HALOGENATION

9.6.1 Alkyl, Cyclic Alkyl, and Substituted Alkyl Groups

The electrophiles in halogenation are of widely differing types, and these differences show up in their reactivities and selectivities, exemplified by the data for halogenation of alkylbenzenes at 25 °C under a variety of conditions (Table 9.4). Notable features of these data are:

(1) Halogenation by entities believed to be positively charged gives a relatively small spread of rates (selectivity) consistent with their greater predicted reactivity

Table 9.3 Yields in astatination of halogenobenzenes, PhX

X	*ortho*	*meta*	*para*
F	7		93
Cl	15.5	1.6	83
Br	30	2	78

Table 9.4 Partial rate factors for halogenation of alkylbenzenes, PhR

R	f_o	f_m	f_p	Log f_o:log f_p	Conditions	No.[a]	Ref.
Me	134	4	82	1.11	HOCl, H^+	1	6
	198	4	132	1.08	HOCl, 75% HOAc, H^+	2	185
	306		237	1.05	ClOAc, aq. HOAc	3	18
	534		552	0.995	Cl_2, aq. HOAc	4	32
	520	4.2	560	0.99	Cl_2, HCO_2H	5	186a
	780	5.5	750	1.01	Cl_2, TFA	6	186a
	617	5	820	0.96	Cl_2, HOAc	7	34
	1830	9.1	6250	0.86	Cl_2, MeCN	8	32
	2430	8.3	9500	0.89	Cl_2, $MeNO_2$	9	32
	61	2.0	48	1.06	HOBr, H^+, aq. dioxane	10	186b
	600	5.5	2420	0.82	Br_2, aq. HOAc	11	187
	1360	10	12700	0.76	Br_2, TFA	12	107
	4340		42400	0.785	Br_2, 93.3% aq. TFA	13	188
	3700	12	57000	0.75	Br_2, liq. SO_2	14	189
	144	6.6	230	0.91	ICl–$ZnCl_2$, HOAc	15	182
Et	450		840	0.91	Cl_2, HOAc	16	37
	465		1800	0.82	Br_2, aq. HOAc	17	37
	1460		19500	0.74	Br_2, TFA	18	107
	2600		53500	0.72	Br_2, liq. SO_2	19	189
i-Pr	218		650	0.83	Cl_2, HOAc	20	37
	180		1200	0.73	Br_2, aq. HOAc	21	37
	1090		25000	0.69	Br_2, TFA	22	107
	475		35500	0.59	Br_2, liq. SO_2	23	189
t-Bu	13.6	2.6	38.5	0.715	HOBr, H^+, aq. dioxane	24	190
			158		ClOAc, aq. HOAc		18
	108		341	0.80	Cl_2, aq. HOAc	25	32
	56.6	6.0	401	0.67	Cl_2, HOAc	26	32
	114	9.4	500	0.76	Cl_2, HCO_2H	27	186a
	5.2	7.3	805	0.25	Br_2, aq. HOAc	28	191
	330	33	1100	0.83	Cl_2, TFA	29	186a
	345		3140	0.725	Cl_2, MeCN	30	32
		34	19300		Br_2, TFA		107
	10		22500	0.23	Br_2, liq. SO_2	31	189
		119	59100		Br_2, 93.3% aq. TFA		191

[a]These refer to the points in Fig. 9.1.

compared with that of molecular reagents. Molecular bromination gives a greater spread of rates than molecular chlorination as expected, since the kinetic evidence shows it to involve a less reactive electrophile (Section 9.3.2).

(2) In each condition under which the reactivity of all four alkylbenzenes has been measured, there is a regular decrease in the $\log f_o$:$\log f_p$ ratio as the alkyl group is made larger, confirming the existence of steric hindrance. The decrease in the ratio with increasing size is also greater for bromination than for chlorination, as expected.

(3) The partial rate factors for molecular chlorination of toluene are strongly dependent on the nature of the medium. The $\log f_o$:$\log f_p$ ratio decreases as the partial rate factors increase, and in the graphical representation in Fig. 9.1 the points for these and other chlorinations, and also brominations, lie on a smooth curve. These results indicate that the decrease in the *ortho*:*para* ratio which accompanies the increase in the selectivity of the reagent as measured by f_p^{Me} is largely an electronic rather than a steric effect; similar behaviour is observed in hydrogen exchange (Table 3.2). The points which lie below the curve of Fig. 9.1 refer mainly to halogenations of the other alkylbenzenes, especially of *tert*-butylbenzene, indicating here the superimposition of steric hindrance to *ortho* substitution on the electronic effects which are otherwise responsible for determining the *ortho*:*para* ratio.

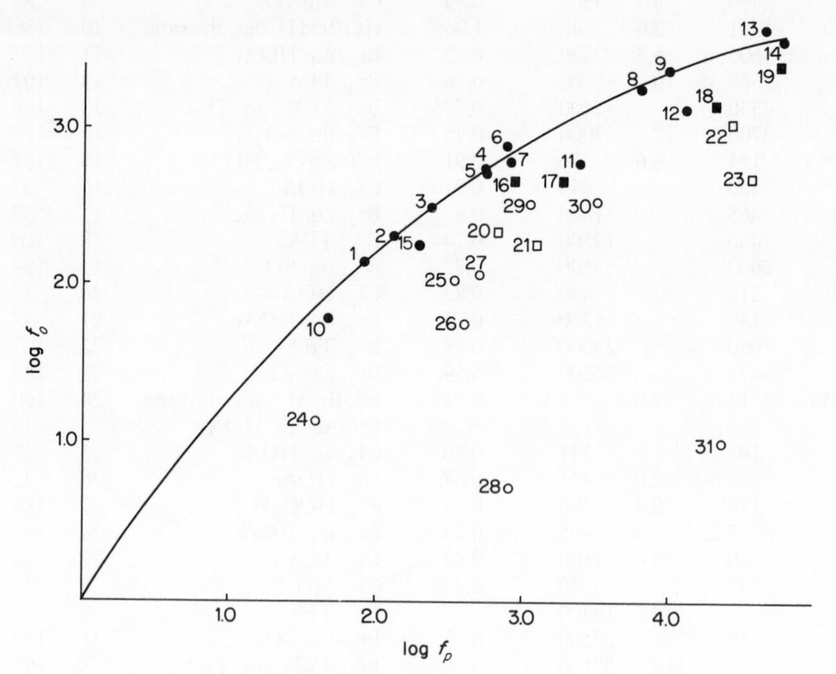

Figure 9.1. Plot of $\log f_o$ against $\log f_p$. ●, Toluene, ■, ethylbenzene; □, isopropylbenzene; ○, *tert*-butylbenzene.

(4) Activation by the *para*-alkyl groups follows the Baker–Nathan order in all good solvating media, owing to steric hindrance to solvation (Section 1.4.3). As in hydrogen exchange, the inductive order is followed in the poorly solvating trifluoroacetic acid. As water is added to TFA, solvation is improved and so the $\log f_p^{t\text{-Bu}}:\log f_p^{\text{Me}}$ ratio for bromination decreases, being 1.044 (TFA), 1.031 (93.3% aqueous TFA), and 1.004 (78.3% aqueous TFA).[191] Similar results are obtained for chlorination in TFA,[186a] whereas for chlorination in the better solvating HOAc, addition of water produces only a very small increase in the ratio (1.077 to 1.081).[186a] The partial rate factors for bromination also pass through a maximum in 93.3% aqueous acid,[188,191] which probably reflects better differential solvation of the ground and transition states for the alkylbenzenes relative to benzene itself. It is noteworthy that hydrogen exchange rates also pass through a maximum in aqueous TFA of similar concentration (Section 3.1) owing to a balance between increased solvation and decreased reactivity of the electrophile.

The relative rates of bromination by Br_2 in HOAc of some alkylbenzenes, PhR, have also been determined as follows: R = $MeCH_2$, 100; $MeCH_2CH_2$, 80; Me_2CHCH_2, 53; Me_3CCH_2, 30.[192] The decrease in reactivity with increasing size of the alkyl group is now known to be due to steric hindrance to solvation.[193]

Bromination of cyclopropylbenzene by Br_2–CCl_4 at −20 °C gives mainly *para* substitution, but ring opening occurs at higher temperatures;[194] the dichlorocyclopropyl group also directs bromination in the *para* position.[195]

Indane brominates mainly in the β-position, whereas tetralin brominates mainly in the α-position,[196] the explanation being that given for hydrogen exchange of these compounds (Section 3.1.2.3); similar results are obtained for chlorination under various conditions.[185] Compounds **32** brominate in the α-position for $n = 4$ and in the β-position for $n = 2$, 3, 5, and 7; this general pattern is also true for compounds **33** (X = O, NAc)[197] and is attributable to increase in strain in the bond common to both rings on going to the transition state for α-substitution (see Section 3.1.2.3).

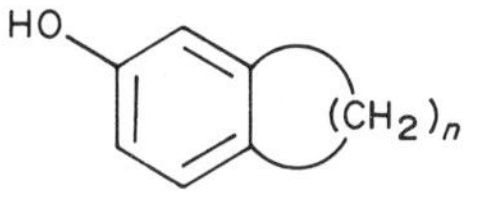

HO $(CH_2)_n$ HO $(CH_2)_n$ X X

(**32**) (**33**) (**34**)

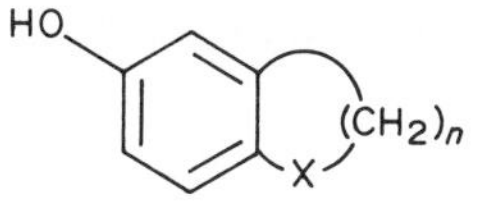

X=O, NAc

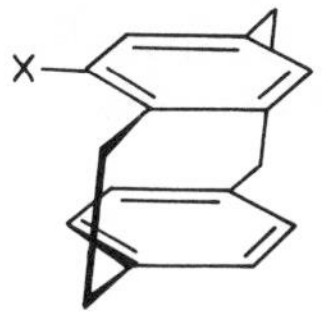

Partial rate factors for the chlorination of diphenylmethane in 75% aqueous HOAc at 25 °C are $f_o = 30.7$, $f_p = 37.6$, i.e. less than in toluene as expected, with the reactivity of the *ortho* position being attenuated most. In 98% HOAc the values are slightly smaller, and smaller also in the presence of perchloric acid.[198]

Bromination of substituted [2,2]paracyclophanes (**34**, X = CO_2Me, COMe, CO_2H, NO_2) gave unexpected behaviour, with bromine entering only the pseudo *gem* position. This was attributed to the oxygen of the substituent acting as an internal base for removal of the proton from the adjacent ring in the transition state. For X = CN and Br this is less feasible on either steric or electronic grounds, and consequently neither compound gave any *gem* substitution. The substitution patterns were also anomalous, however, e.g. the bromo compound gave 5% *p*-, and 16%, 26%, and 6% pseudo *o*-, *m*-, and *p*-dibromo derivatives, respectively. Here again, proton removal from the intermediate is thought to take place via transfer to the adjacent ring (e.g. Scheme 9.5).[199]

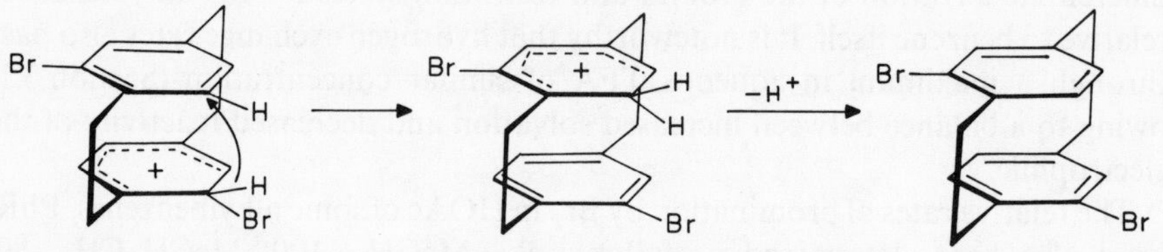

Scheme 9.5. Mechanism for anomalous bromination pattern in bromo-[2,2]-paracyclophane.

9.6.2 Amines, Anilides, Ethers, and Esters

Partial rate factors for halogenation of these compounds are shown in Table 9.5.

The *para* activation orders NHAc > NMeAc and OH > OMe may be due to (1) a greater importance of X–H than X–Me hyperconjugation [this seems unlikely since it is now well established that C–Me is greater than C–H hyperconjugation (Section 1.4.3)]; (2) steric hindrance to solvation; (3) steric inhibition of attainment of coplanarity between the substituent p-orbitals and those of the aromatic ring in the transition state (**35**). Although the last effect is clearly significant, as shown by the data for the O-n-Pr and O-n-Bu substituents, and more especially by data below, the general results for the OR substituents suggest that it is not of overriding importance.

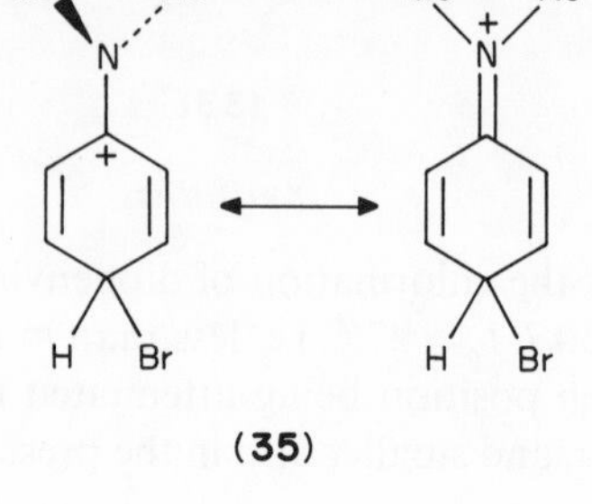

(**35**)

Table 9.5 Partial rate factors for halogenation of compounds PhX at 25 °C

X	*o*	*m*	*p*	Conditions	Ref.
NMe_2			3×10^{19}	Br_2, aq. HOAc	29, 200
			8×10^{13}	Br_2, H_2O	201
NMeAc			1.4×10^{6}	Br_2, H_2O	29, 200
NHAc			1.3×10^{9}	Br_2, H_2O	29, 200
	6.2×10^{5}	0.37	2.5×10^{6}[a]	Cl_2, HOAc	202
OAc	6.5	0.071	87	Cl_2, HOAc	202
OCOPh	5.5	0.059	52	Cl_2, HOAc	203
OH			3.7×10^{12}	Br_2, aq. HOAc	29, 200
OMe	7.9×10^{4}		5.6×10^{5}	$HOBr/H^+$	186b
	6.1×10^{6}		4.6×10^{7}	Cl_2, HOAc	21, 35
	8.7×10^{7}	2.0–3.75	1.1×10^{10}	Br_2, aq. HOAc	29, 200, 204
	6.6×10^{7}		3.5×10^{9}	Br_2, H_2O	201
OEt	1.3×10^{8}	4.7	2.8×10^{10}	Br_2, H_2O	204
O-n-Pr			3.74×10^{10}	Br_2, H_2O	204
O-*i*-Pr			5.85×10^{10}	Br_2, H_2O	204
O-n-Bu			2.76×10^{10}	Br_2, H_2O	204
OPh			3.33×10^{7}	Br_2, H_2O	204
SPh			3.42×10^{5}	Cl_2, $MeNO_2$	205

[a]For the 2-Me and 2,6-Me_2 derivatives the corresponding values are 6.2×10^5 and 1.2×10^3.

Table 9.6 Relative rates of bromination of phenols and anisoles

Compound	*k*(rel.)	Compound	*k*(rel.)
Anisole	1	Phenol	90
2-Methylanisole	6.3	2-Methylphenol	450
2,6-Dimethylanisole	0.4	2,6-Dimethylphenol	550

The effect of inhibition of conjugation has been confirmed by results for the bromination of phenols and anisoles (Table 9.6).[206] Introduction of a second methyl group into an *ortho* position in phenol produces a far smaller increase in bromination rate than does the introduction of the first methyl group; for anisole the second methyl group actually decreases the reaction rate. The results are consistent with the view that two *ortho* substituents, by preventing the O—H or O—Me bond from being coplanar with the aromatic ring, also prevent coplanarity between the substituent and aryl p-orbitals.

The same situation is found in the chlorination of acetanilide and its *ortho*-substituted derivatives (Table 9.5). Thus, whereas one *o*-methyl group causes a fourfold decrease in the rate, the second methyl group causes a further 500-fold decrease (see also ref. 207). Likewise, the rates of bromination of *N*,*N*-dimethyl- and *N*,*N*-diethylaniline in strong aqueous acid were decreased by *o*-methyl substituents whereas *m*- or *p*-methyl substituents produced a marked rate increase.[116]

Two factors are evidently involved overall as shown by the following. Whereas *o*-methylation of acetanilide reduces the rate of *p*-chlorination by a factor of 20 (after allowance for the effect of the methyl substituent *meta* to the reaction site), *N*-methylation reduces it 1000-fold. Since the effects of hindrance to conjugation should be comparable in the two molecules, the marked rate reduction by the *N*-methyl substituent must be due to some other cause.[208] The most probable explanation is steric hindrance to solvation, for although hindrance on the aromatic rings will be similar in each molecule, a key difference is that most of the charge in the transition state resides on nitrogen, and it is in the *N*-methylated compound that this is most hindered.

A further possibility is that a portion of the reaction takes place on phenoxide ion in the case of phenol and amide ion in the case of anilines and anilides. The importance of the former has been demonstrated in chlorination of phenol and anisole by *tert*-butyl hypochlorite.[15] Anisole gave an *ortho*:*para* ratio that was independent of acidity, whereas that for phenol was 0.43 at pH 4.0 and 4.3 at pH 10.0; the increase in ratio with decreasing acidity for phenol is due to reaction taking place on the phenoxide ion.

Bromination of thioanisole by Br_2 in HOAc is ca 80 times slower than bromination of anisole. However, thioanisole is reported to be six times *more* reactive if 95% aqueous HOAc is used;[209] the reason for this is unclear, but hydrogen bonding of the ethers could be responsible.

Catalysed chlorination of $PhOCF_3$ gives 23% *ortho*, 6% *meta* and 71% *para* substitution.[210] The greater extent of *meta* substitution compared with that found with anisole is attributable to the greater $-I$ effect of OCF_3 compared with OMe; bromination gave less *ortho* substitution as a result of increased steric hindrance. By contrast, neither chlorination nor bromination gave any *ortho* substitution with 2,2-dichlorocyclopropoxybenzene.[211]

9.6.3 Deactivating Substituents

The chlorination of a number of compounds containing deactivating substituents with $HOCl–H^+$ (Table 9.7) has thrown light on the reason for the high *ortho*:*para* ratios produced by these substituents in various reactions, and nitration in particular;[212] also included in Table 9.7 are some more recent results (in parentheses) for $FeCl_3$-catalysed chlorination.

The trend in *ortho*:*para* ratios is very similar to that in nitration (Table 7.15); the ratio increases as the electron-attracting power of the substituent is increased. Since the attacking species in the two reactions are stereochemically different, the high ratios in nitration cannot be ascribed to a steric interaction between substituent and reagent. Further, the trend in *meta*:*para* ratios in chlorination is

Table 9.7 Isomer distribution (%) for chlorination of compounds PhX containing deactivating substituents

X	*o*	*m*	*p*	$\frac{1}{2}o:p$	$\frac{1}{2}m:p$	Ref.
(SiF_3)	27.7	71.7	0.6	27	70	214
NO_2	17.6	80.9	1.5	5.9	27.0	212
CN	23.2	73.9	2.9	4.0	12.7	212
$(SiCl_3)$	28	69	3	4.7	11.5	215
CHO	30.7	63.5	5.8	2.6	5.5	212
CF_3	15.7	80.2	4.1	1.9	7.9	212
$(SiCl_2Ph)$	31	57	12	2.4	1.3	216
(CH_2SiF_3)	49.5	5.0	45.5	0.54	0.06	217
Br	39.7	3.4	56.9	0.35	0.03	212
Cl	36.4	1.3	62.3	0.29	0.01	212

similar to that in the *ortho*:*para* ratios, suggesting that the latter has mainly an electronic rather than a steric basis; a similar conclusion has been reached from the nitration data.[213] The electronic factors which underlie these high *ortho*:*para* ratios are discussed fully in Section 11.2.4.

Partial rate factors f_m, f_p for chlorination of halogenobenzenes by Cl_2 in HOAc are F, 5.6×10^{-3}, 3.93; Cl, 2.3×10^{-3}, 0.046; and Br, 3.2×10^{-3}, 0.31.[37] The relative reactivities of the *meta* position show that resonance effects are transmitted to these positions. Partial rate factors for SOPh substituent in chlorination by Cl_2 in $MeNO_2$ are 6.94, 11.1 and 241 for the *ortho*, *meta* and *para* positions, respectively.[205] This surprising result contrasts with protiodesilylation, where the SOMe substituent is strongly deactivating (Table 4.2). It appears, then, that the SOR substituent is either unexpectedly polarizable or is protonated under the conditions used in desilylation.

Partial rate factors for bromination by HOBr–H_2SO_4 of positive poles PhX are shown in Table 9.8.[186b,218]

The arsenic pole is more reactive than the nitrogen pole and is a better *meta* director, whereas the nitrogen pole is relatively reactive at the *para* position; the sulphur poles are less electron withdrawing than the nitrogen poles. These results

Table 9.8 Partial rate factors for bromination of positive poles PhX

X	*o*	*m*	*p*
NMe_3^+	$< 16 \times 10^{-8}$	515×10^{-8}	515×10^{-8}
$CH_2NMe_3^+$	126×10^{-5}	210×10^{-5}	168×10^{-5}
$(CH_2)_2NMe_3^+$	0.68	0.087	1.53
$CH_2SMe_2^+$	437×10^{-5}	374×10^{-5}	659×10^{-5}
$(CH_2)_2SMe_2^+$	1.25	0.11	1.61
$AsMe_3^+$	123×10^{-7}	111×10^{-6}	$< 74 \times 10^{-7}$

parallel those in nitration [Section 7.4.3.(xii)], and the same interpretations given there apply here.

Bromination of cyclophanes incorporating positive poles gives results similar to those obtained in nitration, and again the same explanation applies. Thus compounds **70** in Table 7.14, viz. $n = 1$, $X = NMe_3^+$, $f = 6.25 \times 10^{-7}$; $n = 1$, $X = SMe_2^+$, $f = 7.52 \times 10^{-6}$; $n = 2$, $X = SMe_2^+$, $f = 0.41$, are each more reactive than the corresponding compounds **71** in Table 7.14, viz. $n = 1$, $m = 8$, $X = NMe_2^+$, $f = 1.31 \times 10^{-7}$; $n = 1$, $m = 6$, $X = SMe^+$, $f = 3.86 \times 10^{-7}$; $n = 2$, $m = 6$, $X = SMe^+$, $f = 0.015$.[219]

9.6.4 Substituents Containing Boron and Silicon

Halogenation of the icosohedral boranes $B_{10}H_{10}C_2H_2$ [Section 7.4.3.(xi)] takes place only at boron with an ease that shows the relative reactivities to be (pseudo) *ortho*- > *meta*- > *para*-carborane; the last isomer monohalogenates only, whereas the other isomers undergo polyhalogenation. In the chlorination of 1-phenyl-*o*-carborane, chlorine enters the phenyl ring first, whereas in bromination only the second bromine atom enters the phenyl ring. This suggests that the *o*-carborane has rather similar reactivity to that of benzene, differences in polarizability accounting for the preferred site for initial substitution. In 3-phenyl-*o*-carborane and 1-phenyl-*m*-carborane, the phenyl rings are preferentially brominated.[220]

Partial rate factors have been determined for the I_2- or $FeCl_3$-catalysed chlorination of a range of Si-containing substituents.[214,217,221] Results obtained under the latter condition, which exhibits lower selectivity, are given in Table 9.9. Chlorination of the more reactive compounds was accompanied by substantial chlorodesilylation, which arises because the corresponding SiR_3 substituents are better leaving groups. The results show the expected decreasing reactivity with

Table 9.9 Partial rate factors for chlorination of Si-containing compounds, PhX

X	*o*	*m*	*p*	X	*o*	*m*	*p*
$SiMe_3$	8.38	5.33	8.70				
$SiMe_2F$	0.79	0.89	0.42	$SiMe_2Cl$	0.97	1.15	0.50
$SiMeF_2$	0.105	0.17	0.03	$SiMeCl_2$	0.157	0.215	0.06
SiF_3	0.0054	0.014	0.0002	$SiCl_3$[a]	0.0142	0.0216	0.004
CH_2SiMe_3	1840	0	7130				
CH_2SiMe_2F	204	0	1470				
CH_2SiMeF_2	17.3	0	46.4				
CH_2SiF_3	1.48	0.15	2.72				

[a]Values of 0.0106, 0.0261, and 0.00226 have also been obtained.[214]

increasing halogen substitution, and F deactivates more than Cl. The large loss of reactivity with increasing halogen substitution of the groups CH_2SiR_3 is far greater than could arise from the inductive effect of the halogen, and is due to decreased C–Si hyperconjugation, which is the primary cause of the high reactivity of these groups (Section 3.1.2.4).

9.6.5 Polysubstituted Benzenes

There have been many investigations into the additivity of methyl substituent effects in halogenation.[22a,31a,33–35,182a,187,201] Some results, given in Table 9.10, show that the *Additivity Principle* holds remarkably well, deviations for chlorination and bromination being no greater than for the hydrogen-exchange reaction (Table 3.4), in which there is no steric hindrance. The deviations are ascribable to decreasing selectivity with decreasing reactivity of the aromatic substrate, i.e. for the more reactive compounds the transition state is shifted towards the ground state. An alternative view is that the transition state position is not significantly altered, but rather that the substituents are able to respond variably to demands for resonance stabilization of the transition state; this is amplified by other data below. The larger deviations in iodination are due to steric hindrance.

For bromination by bromine in water, the spread of rates is greater than under the conditions given in Table 9.10, but again additivity holds very well, the ρ factor for the methyl substituent effects in the polymethylbenzenes being -10.7 compared with -11.6 for the reaction of monosubstituted benzenes under these

Table 9.10 Observed and calculated rates of halogenation of polymethylbenzenes[34,182a,187]

Positions of methyl groups	Chlorination		Bromination		Iodination	
	Obs.	Calc.	Obs.	Calc.	Obs.	Calc.
None	1	1	1	1	0.007	—
1-	344	344	605	605	1	1
1,4-	2.1×10^3	2.0×10^3	2.5×10^3	2.2×10^3	7.15	7.15
1,2-	2.1×10^3	2.4×10^3	5.3×10^3	5.5×10^3	9.3	9.3
1,3-	1.85×10^5	2.3×10^5	5.1×10^5	5.4×10^5	164	164
1,2,4-	—	—	1.5×10^6	1.7×10^6	540	640
1,2,3-	—	—	1.7×10^6	2.7×10^6	—	—
1,3,5-	—	—	1.9×10^8	4.4×10^8	1.3×10^4	2.7×10^4
1,2,4,5-	—	—	2.8×10^6	3.6×10^6	164	339
1,2,3,4-	—	—	1.1×10^7	1.5×10^7	—	—
1,2,3,5-	—	—	4.2×10^8	1.6×10^9	1.1×10^4	1.2×10^5
1,2,3,4,5-	—	—	8.1×10^8	4.4×10^9	5.3×10^3	3.9×10^5

conditions.[201] Substituted anisoles and *N*,*N*-dimethylanilines showed much greater departure from additivity, the corresponding ρ factors being -6.5 for *ortho*-substituted anisoles, -5.6 for *para*-substituted anisoles, and -2.2 for *para*-substituted *N*,*N*-dimethylanilines.[201] Although these departures are in the direction expected for operation of the *Reactivity–Selectivity Principle*, i.e. the substituent effects becomes smaller the more reactive the compound, it could be that for the *N*,*N*-dimethylamino compounds at least, encounter control of bromination is occurring.[222] This view is supported by the anomalously low *para* partial rate factor for *N*,*N*-dimethylaniline (Table 9.5). Similar large departures from additivity have been observed in chlorination *ortho* to the NHAc group of compounds (**36**, R = NHAc) by Cl_2–HOAc; the ρ factor is thus only -4.7 instead of -10, which applies to chlorination of monosubstituted benzenes.

Many substituent effects have been determined by application of the *Additivity Principle*. The effects of *ortho* substituents OR in bromination of compounds **37** by HOBr–HOAc were R = Me, 1; Et, 2; *i*-Pr, 3–4; n-Bu, 2.7; *i*-Bu, 2.7; n-Am, 2.3; *i*-Am, 2.6; n-Oct, 2.2; CH_2CO_2H, 0.08; $(CH_2)_2CO_2H$, 0.5; $CHMeCO_2H$, 0.1; $(CH_2)_2Br$, 0.16; $(CH_2)_3Br$, 0.56.[223] These were virtually independent of the nature of either X (H, Cl, Br, CO_2Me) or Y (Cl, Br, CO_2Me, CO_2Et, NO_2),[223] and show the usual decreasing inductive effect with distance from the aromatic ring.

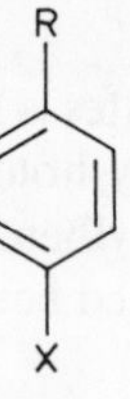

(**36**) 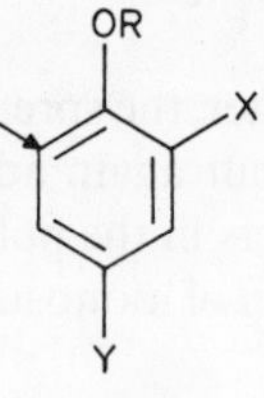 (**37**) 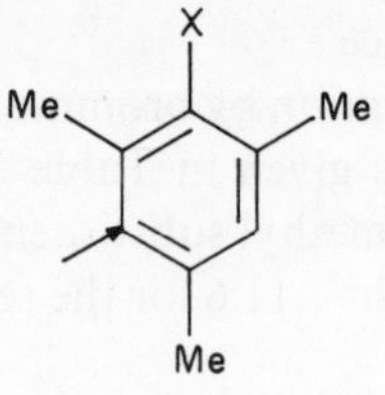 (**38**)

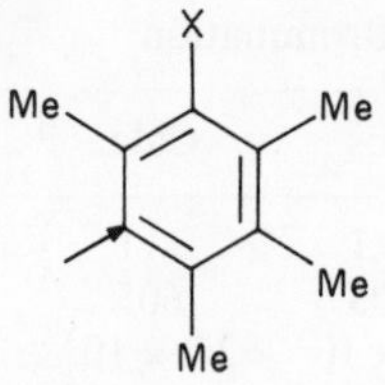

(**39**) 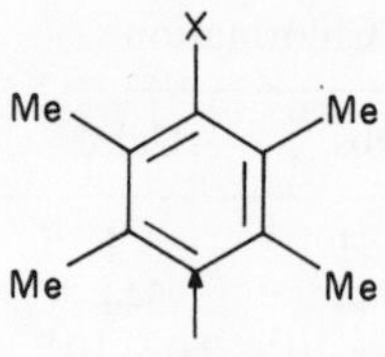 (**40**)

Partial rate factors for *meta* substituents have been determined in bromination by Br_2–$MeNO_2$ of compounds **38** and **39**,[111,224] and relative rates for bromination of compounds **36** (R = OMe) by HOBr–HOAc[223] (Table 9.11). They confirm in each case that *m*-F is less deactivating than the other halogen substituents (cf. Section 9.6.3), owing to secondary relay of conjugative effects to the *meta* position. This secondary relay is also apparent in the data for the OH,

Table 9.11 Partial rate factors and relative rates for bromination of PhX

X	Mesitylenes (**38**) (*meta* position)	Isodurenes (**39**) (*meta* position)	Anisoles (**36**)[a] (*meta* position)	Durenes (**40**) (*para* position)
H	1	1	—	1
F	8.6×10^{-4}	1.2×10^{-3}	1	4.62
Cl	4.9×10^{-4}	6.4×10^{-4}	0.63	0.145
Br	4.3×10^{-4}	6.3×10^{-4}	0.47	0.062
I	1.1×10^{-3}	3.3×10^{-3}	—	0.080
CN	8.7×10^{-7}	—	—	3.1×10^{-6}
CO_2Me	—	—	0.10	—
CO_2Et	—	—	0.11	—
NO_2	—	—	0.0015	—
OH	14.2	—	—	—
OMe	0.19	—	—	1.64×10^5
SMe	0.025	—	—	9.7×10^2

[a]Relative rates, R = OMe.

OMe, and SMe substituents, since in the absence of a relayed resonance effect the relative reactivities would be governed by the inductive effect and therefore lie in the order SMe > OMe > OH. The spread of rates is also less in the more reactive compounds **39**, consistent with the transition state for the more reactive compounds being displaced towards the reactants.

Compounds **40**[111,224] show that halogens activate the *para* position in the order F > H > Cl > I > Br and, with the exception of the iodo substituent, for which the *para* reactivity cannot be measured because of the formation of iodobenzene dichloride [Section 9.2.2.(ii)], this order is also followed in molecular chlorination (Section 9.6.3) (the bromination of 1-halonaphthalenes gives the slightly different order, F > H > I > Cl > Br[225]).

The stronger activating effect of OMe than of SMe parallels that in hydrogen exchange (Section 3.1.2.7), but the activating effect of *p*-OMe is far less in durene (1.64×10^5) than in anisole (1.1×10^{10}), owing to steric inhibition of resonance.[224] Steric inhibition of resonance also causes *p*-NO_2 to deactivate less in halogenation of durene than predicted, although in mesitylene and isodurene *m*-NO_2 deactivated *more* than predicted (especially in bromination),[226] showing that in these polysubstituted compounds steric hindrance is also a contributory factor in producing the overall reactivity. Deviations from additivity of substituent effects in bromination of methoxyaromatic ketones[227] are also attributable to steric inhibition of coplanarity between the acyl group and the aromatic ring. Interestingly, this feature causes a high ratio of di- to monosubstitution in the bromination of 3-acetylanisole; bromine enters the 4-position and causes the acetyl group to become less coplanar with the aromatic ring, which is thus deactivated less and further bromination is facilitated.[228]

The data in Table 9.11 show that the cyano group deactivates the *meta* position more than the *para* position, whereas the converse is normally true. This anomaly can be attributed to the fact that in the mesitylenes (**38**), the cyano group is able to withdraw electrons conjugatively from the methyl groups more effectively than it can in compounds **40**. Moreover, the methyl groups in **40** that suffer this conjugative electron withdrawal are themselves not conjugated with the reaction site and therefore have a relatively small effect on the overall reactivity of the molecule.

The effect of steric hindrance to coplanarity is also manifest in rate data for the chlorination of compounds **41**–**44**.[206] For **41** and **42**, the agreement between the calculated and observed rates is reasonable considering the fact that chlorination of them is accompanied by some *ipso* substitution involving addition–elimination pathways. The 178-fold attenuation of the reactivity of compound **44** is evidently due to steric inhibition of conjugation between the acetoxy group and the aromatic ring, which actually causes the acetoxy group to deactivate here. The 19-fold attenuation of the reactivity of compound **43**, where reaction occurs *meta* to the acetoxy group, shows that there must be substantial relay of the conjugative effects to the *meta* positions (cf. Section 1.4.2).

In the bromination of *meta*- and *para*-disubstituted benzenes containing substituents Me, OMe, and $(CH_2)_nX$, where $n = 0$–2 and $X = NMe_3^+$ and SMe_2^+, the aromatic was less reactive than predicted if both groups were strongly electron withdrawing.[229] If one substituent was strongly deactivating whilst the other was activating (e.g. Me or OMe), or if both substituents were moderately deactivating [i.e. $(CH_2)_2X$], then the observed reactivity was greater than predicted. The reasons for these deviations are described in Section 11.1.5.

Estimates of *meta* partial rate factors for deactivating substituents have been determined using the *Additivity Principle* (Table 9.12).[25b,29,83,230,231] There are wide variations in values for a given substituent for the same reaction but using different substrates, and this serves to emphasize the deficiency in the method.

	(**41**)	(**42**)	(**43**)	(**44**)
$10^2 k_2/\mathrm{l\,mol^{-1}\,s^{-1}}$ (obs.)	0.0055	0.67	0.078	62
$10^2 k_2/\mathrm{l\,mol^{-1}\,s^{-1}}$ (calc.)	0.0038	1.1	1.5	11 000

Scheme 9.6. Observed and calculated rates of chlorination of aryl acetates.

Table 9.12 Estimates of *meta* partial rate factors based on the *Additivity Principle*

CO_2H	CO_2Et	COPh	NO_2	Cl	CN	NMe_3^+	Reaction
7.8×10^{-4}	1.5×10^{-3}	2.3×10^{-3}	3.6×10^{-4}	0.0026			Cl_2–HOAc
			7.4×10^{-5}	0.031[a]			Cl_2–HOAc
			5.4×10^{-6}				Cl_2–HOAc
			9.5×10^{-8}		8.7×10^{-7}		Br_2–$MeNO_2$
2.25×10^{-2}			4.8×10^{-5}			4.8×10^{-5}	HOBr–H^+

[a]Under these conditions $f_o = 0.21$ and $f_p = 0.44$.

Comparison of values for *m*-Cl under a variety of conditions[231c] shows that the deactivation is consistently smaller the more reactive the system, as noted in many instances above.

9.6.6 Biphenyl and its Derivatives

Representative data for the halogenation of biphenyl are given in Table 9.13.[38,83,93,102b,232] The *ortho*:*para* ratio decreases with increasing selectivity of the reagent (and this is true also in chlorination under various conditions).[233] This is partly due to increasing steric hindrance as the bond between the ring and the electrophile becomes shorter in reactions with later transition states,[233] but is largely a mathematical consequence of the selectivity effect, as the $\log f_o$:$\log f_p$ values demonstrate.[234] As in chlorination of toluene (Table 9.4), reaction of biphenyl with ClOAc gives anomalously high *ortho* reactivity, with $f_o = 20.3$, $f_p = 21.3$.[235] Steric hindrance is obviously severe in iodination of biphenyl by acetyl hypoiodite, the corresponding values being 0 and 100; this reagent also gives low *ortho* reactivities for toluene, diphenylmethane, and 1,2-diphenylethane.[236]

Table 9.13 Partial rate factors for halogenation of biphenyl

	Ph-benzene	Ph-benzene	Ph-benzene	Ph-benzene	Ph-benzene
ortho	10.7	21.5	249	~400	130
meta	0.28				
para	15.6	37.5	780	~3000	3940
	HOBr–H^+– aq. dioxane	HOBr–H^+– HOAc	Cl_2–HOAc	BrOAc	Br_2–50% aq. HOAc
ortho:*para* ratio	0.685	0.573	0.319	0.133	0.033
$\log f_o$:$\log f_p$	0.86	0.85	0.83	0.75	0.59

Table 9.14 Partial rate factors for chlorination of alkylbiphenyls by Cl_2 in HOAc

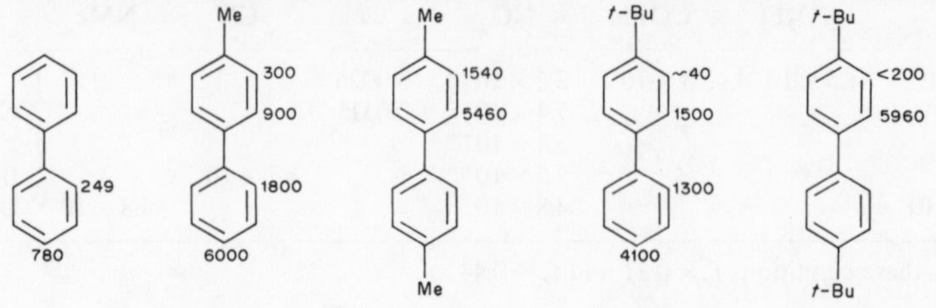

(45) Me; 300; 900; 1800; 6000 — biphenyl: 249; 780

(46) Me; 1540; 5460; Me

(47) *t*-Bu; ~40; 1500; 1300; 4100

(48) *t*-Bu; <200; 5960; *t*-Bu

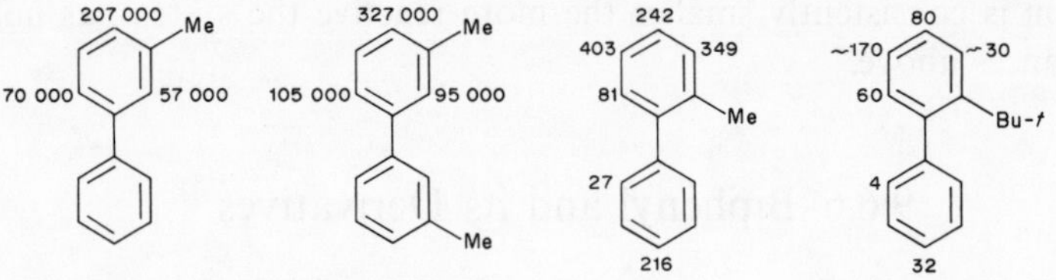

(49) 207 000; Me; 70 000; 57 000

(50) 327 000; Me; 105 000; 95 000; Me

(51) 242; 403; 349; 81; Me; 27; 216

(52) 80; ~170; ~30; 60; Bu-*t*; 4; 32

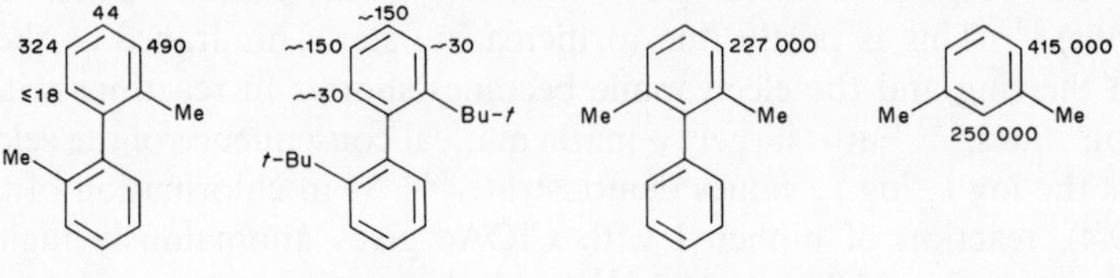

(53) 44; 324; 490; ≤18; Me; Me

(54) ~150; ~150; ~30; ~30; Bu-*t*; *t*-Bu

(55) 227 000; Me; Me

(56) 415 000; Me; Me; 250 000

An extensive study of the molecular chlorination of alkyl-substituted biphenyls has yielded the results given in Table 9.14.[237] The following deductions can be made.

(1) Comparison of the values between either of the pairs toluene and **45, 55** and **56**, or between toluene and **51** (3- or 5-position), yields remarkably consistent values for f_m^{Ph} of 0.49, 0.55, 0.57, and 0.49 (cf. 0.7 obtained from chlorination of acetanilide and its 4-phenyl derivative[35]).

(2) Comparison of the values for the 2′-position in **45** and the 2-position in **46** shows *m*-Me to activate by a factor of 3, which may be compared with a value of 5 obtained by the direct method (Table 9.4). Similar treatment of the data for compounds **47** and **48** shows *m*-*t*-Bu to activate 4-fold.

(3) Comparison of the values for compound **45** and biphenyl shows that the 4-Me substituent increases the reactivity of the 2′- and 4′-positions by a factor of ca 7.5, as compared with the factor of 820 in toluene. The ratios of the logarithms

of these activating effects (3.3-fold) gives, in linear free enegy terms, a measure of the reduced transmission of substituent effects between the phenyl rings. The value is in good agreement with those obtained in a number of other reactions [Section 3.1.2.10(5)].

(4) Whereas the ratios of the reactivities of the 4-positions in toluene and *tert*-butylbenzene is 2.04 (Table 9.4), the ratios of the reactivities at the 4′-positions in **45** and **47** is only 1.46. This demonstrates further the importance of steric hindrance to solvation in governing the relative reactivities of compounds containing these alkyl groups (Section 1.4.3). In the biphenyls, a smaller fraction of the transition state charge is delocalized on to the carbon bearing the alkyl groups than is the case in benzene. Consequently, steric hindrance to solvation is less important in the biphenyls and attenuation of the reactivity of the *tert*-butyl compound becomes less.

(5) Compounds **51**–**53** are much less reactive than predicted by the *Additivity Principle*, and they are much less reactive than biphenyl itself, despite the presence of the activating alkyl groups. The substituents prevent the two nuclei from approaching the coplanar arrangement necessary for maximum resonance interaction between them; the positive charge which resides in the system in the transition state is less able to be delocalized over both rings. Calculations based on results for biphenyl, **51**, **53**, and toluene (Table 9.4) show that the activating effect of the phenyl substituent is reduced 15-fold by one *o*-Me group and 100-fold by two *o*-Me groups.

The connection between the stereochemistry and the reactivity of biphenyl is further illustrated by the data in Table 9.15.[238] As the bridging side-chain brings the aromatic rings closer to the coplanar configuration, so the conjugation between the rings, and hence the reactivity, are increased. These results parallel those in hydrogen exchange (Table 3.14), as do the changes in relative positional reactivities between open and closed structures, owing to strain in the central ring [Sections 3.1.2.10(11) and 11.2].

Table 9.15 Partial rate factors for chlorination by Cl_2 in HOAc

Bridge	Partial rate factors
CH_2	300 000[a]; 39 000
$(CH_2)_2$	500; 26600; 3900; 1150
$(CH_2)_3$	200; 1300; 340; 220
$(CH_2)_2$–CHMe	113; 1150
CH_2–$(CHMe)_2$–CH_2	840; 235
N–Ac (open)	6200; 24 200
N–Ac (bridged)	4300; 8600; 8600; 122 000

[a]For bromination by Br_2–aq. HOAc, $f_2 = 1.03 \times 10^7$.[239]

Pentafluorobiphenyl was reported to give 2′- and 4′-bromo derivatives in a 1:10 ratio in bromination by Br_2–$AlBr_3$.[240] However, results for hydrogen exchange predicted that significant *meta* substitution should be obtained, and the *ortho*:*para* ratio should be higher.[241] Use of high-resolution g.l.c. showed that the 4′-bromo derivative was in fact a 3′- and 4′-bromo mixture. Moreover, bromination under less sterically demanding conditions (Br_2–HOAc–I_2) gave *ortho*, *meta*, and *para* derivatives in 12, 23, and 65% yields, respectively, these proportions being close to those predicted from hydrogen exchange.[241]

The relative reactivities of terphenyls reported for bromination (and also alkylation), viz. *ortho* > *meta* > *para*,[242] is anomalous, cannot be explained by any known substituents effects, and arises from the reaction mixtures being heterogeneous;[243] the observed reactivity order is that of decreasing solubility.

9.6.7 Naphthalene and Polycyclic Compounds

Partial rate factors for halogenation of polycyclics (Table 9.16) follow the generally observed reactivity pattern (cf. hydrogen exchange, Table 3.17), and the data correlate reasonably well with reactivity parameters.[239]

Bromination of 2-methylnaphthalene gives the positional reactivity order 1 > 8 > 4 > 5 > 6 > 3 > 7.[244] This order is that observed in hydrogen exchange and nitration, except that the relative reactivities of the 5- and 6-positions are reversed. This is not unexpected, because although the 2-methyl group strongly activates the 6-position, in a reaction of high ρ factor the higher intrinsic reactivity of the 5(α)-position will become relatively more significant. Bromination of dimethylnaphthalenes generally follows prediction, except for the 1,8-dimethyl isomer, where steric crowding becomes very significant.[244]

Table 9.16 Partial rate factors for halogenation of polycyclics

Compound	Position	Reagent		
		Br_2–aq. HOAc	Cl_2–HOAc	$ClAgCl^+$
Naphthalene	1	1.84×10^5	4.9×10^4	33 000
	2	1860	—	170
Phenanthrene	9	2.23×10^6	9.7×10^5	—
Fluoranthene	3	6.90×10^6	—	—
Triphenylene	1	—	4600	660
	2	—	1700	120
Chrysene	6	3.75×10^7	—	—
Pyrene	1	4.26×10^{10}	—	—
1,2-Benzanthracene	7	1.46×10^{11}	—	—
Anthracene	9	2.36×10^{12}	—	—
Acenaphthene	5	1.65×10^{11}	—	—

Differences between the correlations of rate vs $1/[Br^-]$ in bromination of 1,5-dimethylnaphthalene (at the 4-position) and 2,3-dimethylnaphthalene (at the 1-position) indicated that the mechanism differs for each, with C—H bond breaking being partially rate-determining for the former compound, substitution in which must be more hindered than in the latter.[245] Buttressing by the adjacent methyl groups in the latter compound must therefore be unusually small, and this may be attributed to the exceptionally long 2,3-bond in naphthalene.

Bromination of 1,8-bridged naphthalenes (**57**) follows the reactivity order $n = 2 > 3 > 4 >$ (1,8-dimethylnaphthalene),[246] as found in hydrogen exchange (Table 3.22) and for the reasons given in Section 3.1.2.13.(vi). Bromination of acenaphthene gives much higher 5:3 product ratios in chlorination (2.2–3.3) compared with bromination (29–48),[247] indicating that substitution at the 3-position is very sterically hindered. Bromination of 1*H*-cyclobuta-[*de*]naphthalene (**57**, $n = 1$) parallels results for acetylation[248] (see Section 6.7.3 for interpretation of these results).

$(CH_2)_n$

(**57**)

Bromination of hexahelicene gives mainly the 5-bromo derivative, together with what was assumed to be the 8-isomer.[249] However, hydrogen exchange results (Table 3.14)[250] indicate that the 7-position is the next most reactive after the 5-position. As in nitration, bromination of benzo[*c*]fluorene occurs at the 5-position.[251] As in nitration and hydrogen exchange (Table 3.15), halogenation of biphenylene goes mainly in the 2-position[252] (see Section 3.1.2.10 for an explanation of this result).

REFERENCES

1. D. H. R. Barton, A. K. Ganguly, R. H. Hesse, S. N. Loo, and M. M. Pechet, *J. Chem. Soc., Chem. Commun.*, 1968, 806.
2. V. Grakauskas, *J. Org. Chem.*, 1970, **35**, 723.
3. N. B. Kazmina, L. S. German, I. D. Rubin, and I. L. Knunyants, *Proc. Acad. Sci. USSR*, 1970, **194**, 1329.
4. F. Cacace, P. Giacomello, and A. P. Wolff, *J. Am. Chem. Soc.*, 1980, **102**, 3511.
5. M. J. Shaw, H. H. Hyman, and R. Filler, *J. Am. Chem. Soc.*, 1970, **92**, 6498; *J. Org. Chem.*, 1971, **36**, 2917.
6. D. H. Derbyshire and W. A. Waters, *J. Chem. Soc.*, 1951, 73; P. B. D. de la Mare, J. T. Harvey, M. Hassan, and S. Varma, *J. Chem. Soc.*, 1958, 2756.
7. F. G. Soper and G. F. Smith, *J. Chem. Soc.*, 1926, 1582.
8. E. A. Shilov, N. P. Kanyaev, and A. P. Otmennikova, *Zh. Fiz. Khim.*, 1936, **8**, 909.

9. (a) P. B. D. de la Mare, A. D. Ketley, and C. A. Vernon, *J. Chem. Soc.*, 1954 1290; (b) P. B. D. de la Mare and J. H. Ridd, *Aromatic Substitution*, Butterworths, London, 1959, Ch. 9.
10. R. P. Bell and E. Gelles, *J. Chem. Soc.*, 1951, 2734.
11. C. G. Swain and A. D. Ketley, *J. Am. Chem. Soc.*, 1955, **77**, 3410.
12. J. Arotsky and M. C. R. Symons, *Q. Rev. Chem. Soc.*, 1962, **16**, 282; P. B. D. de la Mare and L. Main, *J. Chem. Soc. B*, 1971, 90.
13. C. G. Swain and D. R. Crist, *J. Am. Chem. Soc.*, 1972, **94**, 3195.
14. D. R. Harvey and R. O. C. Norman, *J. Chem. Soc.*, 1961, 3604.
15. Y. Ogata, M. Kimura, Y. Kondo, and H. Katoh, *J. Chem. Soc., Perkin Trans. 2*, 1984, 451.
16. K. Smith, M. Butters, and B. Nay, *Synthesis*, 1985, 1157.
17. K. Smith, M. Butters, W. E. Paget, and B. Nay, *Synthesis*, 1985, 1155.
18. P. B. D. de la Mare, I. C. Hilton, and S. Varma, *J. Chem. Soc.*, 1960, 4044.
19. G. Stanley and J. Shorter, *J. Chem. Soc.*, 1958, 246, 256; P. B. D. de la Mare, I. C. Hilton, and C. A. Vernon, *J. Chem. Soc.*, 1960, 4039.
20. R. Taylor, *Comprehensive Chemical Kinetics*, Elsevier, Amsterdam, 1972, Vol. 13, pp. 89–91.
21. K. J. P. Orton and A. E. Bradfield, *J. Chem. Soc.*, 1927, 986.
22. (a) P. B. D. de la Mare and P. W. Robertson, *J. Chem. Soc.*, 1943, 279; (b) P. W. Robertson, R. M. Dixon, W. G. M. Goodwin, I. R. McDonald, and J. F. Scaife, *J. Chem. Soc.*, 1949, 294.
23. P. W. Robertson, *J. Chem. Soc.*, 1949, 294.
24. K. J. P. Orton and H. King, *J. Chem. Soc.*, 1911, 1369; K. J. P. Orton, F. G. Soper, and G. Williams, *J. Chem. Soc.*, 1928, 998.
25. (a) A. E. Bradfield and B. Jones, *J. Chem. Soc.*, 1928, 1006, 3073; 1931, 2903; (b) *Trans. Faraday Soc.*, 1941, **37**, 726; (c) A. E. Bradfield, W. O. Jones, and F. Spencer, *J. Chem. Soc.*, 1931, 2907.
26. R. E. Roberts and F. G. Soper, *Proc. R. Soc. London, Ser. A*, 1933, **140**, 71.
27. B. Jones, *J. Chem. Soc.*, 1934, 210; 1935, 1831, 1835; 1936, 1231, 1854; 1938, 1414; 1941, 267, 358, 445; 1942, 418, 676; 1943, 430, 445.
28. K. Lauer and R. Oda, *Chem. Ber.*, 1936, **69**, 1061.
29. P. W. Robertson, P. B. D. de la Mare, and B. E. Swedlund, *J. Chem. Soc.*, 1953, 782.
30. M. J. S. Dewar and T. Mole, *J. Chem. Soc.*, 1957, 342.
31. (a) L. J. Andrews and R. M. Keefer, *J. Am. Chem. Soc.*, 1957, **79**, 4348, 5169; (b) 1959, **81**, 1063; (c) 1960, **82**, 5823; (d) 1962, **84**, 3635.
32. L. M. Stock and A. Himoe, *J. Am. Chem. Soc.*, 1961, **83**, 1937, 4605.
33. E. Baciocchi and G. Illuminati, *Chem. Ind.* (*London*), 1958, 917; *J. Am. Chem. Soc.*, 1964, **86**, 2677; E. Baciocchi and L. Mandolini, *J. Chem. Soc. B*, 1967, 1361.
34. H. C. Brown and L. M. Stock, *J. Am. Chem. Soc.*, 1957, **79**, 5175, 5615.
35. P. B. D. de la Mare and M. Hassan, *J. Chem. Soc.*, 1958, 1519.
36. S. F. Mason, *J. Chem. Soc.*, 1959, 1233.
37. L. M. Stock and F. W. Baker, *J. Am. Chem. Soc.*, 1962, **84**, 1660.
38. P. B. D. de la Mare, D. M. Hall, M. M. Harris, M. Hassan, E. A. Johnson, and N. V. Klassen, *J. Chem. Soc.*, 1962, 3784.
39. P. B. D. de la Mare and E. A. Johnson, *J. Chem. Soc.*, 1963, 4076.
40. P. B. D. de la Mare, E. A. Johnson, and J. S. Lomas, *J. Chem. Soc.*, 1964, 5317.
41. O. M. H. el Dusouqui and M. Hassan, *J. Chem. Soc. B*, 1966, 374.
42. G. Marino, *Tetrahedron*, 1965, **21**, 843.
43. P. B. D. de la Mare, O. M. H. el Dusouqui, and E. A. Johnson, *J. Chem. Soc. B*, 1966, 521.

44. P. W. Robertson, P. B. D. de la Mare, and W. T. G. Johnson, *J. Chem. Soc.*, 1943, 276.
45. E. Baciocchi, G. Illuminati, and G. Sleiter, *Tetrahedron Lett.*, 1960, (23), 30.
46. P. B. D. de la Mare and J. S. Lomas, *Recl. Trav. Chim. Pays-Bas*, 1967, **86**, 1082.
47. G. H. Beaven, P. B. D. de la Mare, M. Hassan, E. A. Johnson, and N. V. Klassen, *J. Chem. Soc.*, 1961, 2749; 1962, 988; P. B. D. de la Mare, N. V. Klassen, and R. Koenigsberger, *J. Chem. Soc.*, 1961, 5285; P. B. D. de la Mare *et al.*, *J. Chem. Soc. B*, 1969, 717; P. B. D. de la Mare, *Acc. Chem. Res.*, 1974, **7**, 369.
48. P. B. D. de la Mare and R. Koenigsberger, *J. Chem. Soc.*, 1964, 5327; P. B. D. de la Mare, M. D. Johnson, J. S. Lomas, and V. Sanchez del Olmo, *J. Chem. Soc. B*, 1966, 827.
49. G. Cum, P. B. D. de la Mare, J. S. Lomas, and M. D. Johnson, *J. Chem. Soc. B*, 1967, 244; G. Cum, P. B. D. de la Mare, and M. D. Johnson, *J. Chem. Soc. C*, 1967, 1590.
50. P. B. D. de la Mare and B. N. B. Hannan, *J. Chem. Soc., Chem. Commun.*, 1971, 1324.
51. W. D. Watson and H. E. Hennis, *J. Org. Chem.*, 1979, **44**, 1155.
52. W. D. Watson and J. P. Heeschen, *Tetrahedron Lett.*, 1974, 695.
53. D. J. Calvert, P. B. D. de la Mare, and H. Suzuki, *J. Chem. Soc., Perkin Trans. 2*, 1983, 255.
54. K. Seguchi, T. Asano, A. Sera, and R. Goto, *Bull. Chem. Soc. Jpn.*, 1970, **43**, 3318.
55. R. Breslow and P. Campbell, *J. Am. Chem. Soc.*, 1969, **91**, 3085.
56. J. H. Gorvin, *Chem. Ind. (London)*, 1951, 910.
57. C. J. Berg and E. S. Wallis, *J. Biol. Chem.*, 1946, **162**, 683; D. H. R. Barton and R. Miller, *J. Am. Chem. Soc.*, 1950, **72**, 370; S. J. Cristol, F. R. Stermitz, and P. S. Ramey, *J. Am. Chem. Soc.*, 1956, **78**, 4939.
58. R. S. Neale, R. G. Shepers, and M. R. Walsh, *J. Org. Chem.*, 1964, **29**, 3390; P. Haberfield and D. Paul, *J. Am. Chem. Soc.*, 1965, **87**, 5502.
59. P. G. Gassman and G. A. Campbell, *J. Am. Chem. Soc.*, 1971, **93**, 2567.
60. J. R. Lindsay-Smith, L. C. McKeer, and J. M. Taylor, *J. Chem. Soc., Perkin Trans. 2*, 1987, 1533; 1988, 385.
61. M. D. Carr and B. D. England, *Proc. Chem. Soc.*, 1958, 350.
62. L. Le Page and J. C. Jungers, *Bull. Soc. Chim. Fr.*, 1960, 525.
63. G. A. Olah, S. J. Kuhn, and B. A. Hardie, *J. Am. Chem. Soc.*, 1964, **86**, 1055.
64. S. Y. Caille and R. Corriu, *J. Chem. Soc., Chem. Commun.*, 1967, 1251; *Tetrahedron*, 1969, **25**, 2005.
65. P. Kovacic and A. K. Sparks, *J. Am. Chem. Soc.*, 1960, **82**, 5740; *J. Org. Chem.*, 1961, **26**, 1310.
66. J. Dolansky, J. Vcelak, and V. Chvalovsky, *Collect. Czech. Chem. Commun.*, 1973, **38**, 3823.
67. C. Yaroslavsky, *Tetrahedron Lett.*, 1974, 3395.
68. R. Bolton and P. B. D. de la Mare, *J. Chem. Soc. B*, 1967, 1044; P. B. D. de la Mare and H. Suzuki, *J. Chem. Soc. C*, 1967, 1586; R. Bolton, *J. Chem. Soc. B*, 1970, 1770.
69. R. Bolton, *J. Chem. Soc. B*, 1968, 712, 714.
70. R. Bolton, D. B. Hibbert, and S. Parand, *J. Chem. Soc., Perkin Trans. 2*, 1986, 981.
71. R. Taylor, *Chemistry of Heterocyclic Compounds*, Ed. S. Gronowitz, Vol. 44, Part 2, Wiley, New York, 1986, p. 26.
72. W. D. Watson, *Tetrahedron Lett.*, 1976, 2591; *J. Org. Chem.*, 1985, **50**, 2145.
73. R. Corriu and C. Coste, *Tetrahedron*, 1969, **25**, 4949.
74. G. K. Chip and J. S. Crossert, *Can. J. Chem.*, 1972, **50**, 1233.
75. H. P. Crocker and R. Walser, *J. Chem. Soc. C*, 1970, 1982.
76. (a) M. Sugitona *et al.*, *Jpn. Pat.*, 56–131535, 1981; (b) M. Kodomari, S. Takahashi, and S. Yoshitomi, *Chem. Lett.*, 1987, 1901.

77. A. Ledwith and P. J. Russell, *J. Am. Chem. Soc.*, 1975, **97**, 1503.
78. S. Uemura, K. Sohma, and M. Okano, *Bull. Chem. Soc. Jpn.*, 1972, **45**, 860.
79. G. A. Olah, L. Ohannesian, and M. Arvanaghi, *Synthesis*, 1986, 868.
80. A. Guy, M. Lemaire, and J.-P. Guette, *Tetrahedron*, 1982, 2339.
81. J. O. Morley, *J. Chem. Soc., Perkin Trans. 2*, 1976, 1554.
82. W. J. Wilson and F. G. Soper, *J. Chem. Soc.*, 1949, 3376.
83. P. B. D. de la Mare and I. C. Hilton, *J. Chem. Soc.*, 1962, 997.
84. D. H. Derbyshire and W. A. Waters, *J. Chem. Soc.*, 1950, 564.
85. F. M. Vainshtein and E. A. Shilov, *Proc. Acad. Sci. USSR*, 1960, **133**, 821.
86. H. M. Gilow and J. H. Ridd, *J. Chem. Soc., Perkin Trans. 2*, 1973, 1321.
87. J. J. Harrison, J. P. Pellegrini, and C. M. Selwitz, *J. Org. Chem.*, 1981, **46**, 2169.
88. Y. Furuya, A. Morita, and I. Urasaki, *Bull. Chem. Soc. Jpn.*, 1968, **41**, 997.
89. P. B. D. de la Mare, T. M. Dunn, and J. T. Harvey, *J. Chem. Soc.*, 1957, 923.
90. A. Nilsson and K. Olsson, *Acta Chem. Scand.*, 1969, **23**, 7.
91. J. Marton, *Acta Chem. Scand.*, 1969, **23**, 3321.
92. P. C. Myhre, *Acta Chem. Scand.*, 1960, **14**, 219.
93. P. B. D. de la Mare and J. L. Maxwell, *J. Chem. Soc.*, 1962, 4829.
94. Y. Ogata, Y. Furuya, and K. Okano, *Bull. Chem. Soc. Jpn.*, 1964, **37**, 960.
95. P. C. Myhre, G. S. Owen, and L. L. James, *J. Am. Chem. Soc.*, 1968, **90**, 2115.
96. A. E. Bradfield, G. I. Davies, and E. Long, *J. Chem. Soc.*, 1949, 1389.
97. S. J. Branch and B. Jones, *J. Chem. Soc.*, 1954, 2317.
98. P. W. Robertson, *J. Chem. Soc.*, 1954, 1267.
99. A. E. Bradfield, B. Jones, and K. J. P. Orton, *J. Chem. Soc.*, 1929, 2810.
100. R. M. Keefer, A. Ottenberg, and L. J. Andrews, *J. Am. Chem. Soc.*, 1956, **78**, 255.
101. S. F. Mason, *J. Chem. Soc.*, 1958, 4329.
102. (a) E. Berliner and M. C. Beckett, *J. Am. Chem. Soc.*, 1957, **79**, 1425; (b) E. Berliner and J. C. Powers, *J. Am. Chem. Soc.*, 1961, **83**, 905; (c) U.-J. P. Zimmerman and E. Berliner, *J. Am. Chem. Soc.*, 1962, **84**, 3953; (d) E. Berliner and B. J. Landry, *J. Org. Chem.*, 1962, **27**, 1083.
103. L. M. Yeddanapalli and N. S. Gnanapragasam, *J. Chem. Soc.*, 1956, 4934; *J. Indian Chem. Soc.*, 1959, **36**, 745.
104. J. Rajaram and J. C. Kuriacose, *Aust. J. Chem.*, 1968, **21**, 3069.
105. E. Berliner, D. M. Falcione, and J. L. Riemenschneider, *J. Org. Chem.*, 1965, **30**, 1812.
106. I. K. Lewis, R. D. Topsom, J. Vaughan, and G. J. Wright, *J. Org. Chem.*, 1968, **33**, 1497.
107. (a) H. C. Brown and R. A. Wirkkala, *J. Am. Chem. Soc.*, 1966, **88**, 1447. (b) P. Alcais, *J. Chim. Phys.*, 1966, **103**, 1443.
108. J.-C. Jacquesy, M.-P. Jouannetaud, and S. Makini, *J. Chem. Soc., Chem. Commun.*, 1980, 110.
109. J. M. Brittain, P. B. D. de la Mare, and P. A. Newman, *Tetrahedron Lett.*, 1980, 4111.
110. (a) A. Fischer and G. N. Henderson, *Can. J. Chem.*, 1983, **61**, 1045; (b) J. M. Brittain, P. B. D. de la Mare, and P. D. McIntyre, *J. Chem. Soc., Perkin Trans. 2*, 1979, 933.
111. G. Illuminati and G. Marino, *J. Am. Chem. Soc.*, 1956, **78**, 4975.
112. P. Menzel and F. Effenberger, *Angew. Chem., Int. Ed. Engl.*, 1972, **11**, 922.
113. R. M. Keefer, J. H. Blake, and L. J. Andrews, *J. Am. Chem. Soc.*, 1954, **76**, 3062.
114. E. Helgstrand, *Acta Chem. Scand.*, 1965, **19**, 1583.
115. P. Castellonese and P. Villa, *Helv. Chim. Acta*, 1984, **67**, 2097.
116. R. P. Bell and E. N. Ramsden, *J. Chem. Soc.*, 1958, 161; R. P. Bell and T. Spencer, *J. Chem. Soc.*, 1959, 1156; R. P. Bell and D. J. Rawlinson, *J. Chem. Soc.*, 1961. 63.
117. J. E. Dubois, P. Alcaio, and G. Barbier, *C. Z. Acad. Sci.*, 1962, **254**, 3000.
118. E. Berliner and F. Gaskin, *J. Org. Chem.*, 1967, **32**, 1660.
119. R. P. Bell and P. De Maria, *J. Chem. Soc. B*, 1969, 1057.

120. N. H. Briggs, P. B. D. de la Mare, and D. Hall, *J. Chem. Soc., Perkin Trans. 2*, 1977, 106.
121. Ref. 9b, p. 124.
122. D. M. Yost, T. F. Anderson, and F. Skoog, *J. Am. Chem. Soc.*, 1933, **55**, 552.
123. L. Bruner, *Z. Phys. Chem.*, 1902, **41**, 514.
124. P. W. Robertson, J. E. Allan, K. N. Haldane, and M. G. Simmers, *J. Chem. Soc.*, 1949, 933.
125. T. Tsurata, K. Sasaki, and J. Furukawa, *J. Am. Chem. Soc.*, 1952, **74**, 5995; 1954, **76**, 994.
126. J. H. Blake and R. M. Keefer, *J. Am. Chem. Soc.*, 1955, **77**, 3707.
127. D. E. Pearson, R. D. Wysong, and C. V. Breder, *J. Org. Chem.*, 1967, **32**, 2359.
128. G. E. Dunn and B. J. Blackburn, *Can. J. Chem.*, 1974, **52**, 2552.
129. A. N. Nesmeyanov, T. P. Tolstaya, L. S. Isaeva, and A. V. Grib, *Dokl. Akad. Nauk SSSR*, 1960, **133**, 602.
130. (a) J. P. Causelier, *Bull. Soc. Chim. Fr.*, 1971, 1785; 1972, 762; (b) H. V. Ansell and R. Taylor, *J. Chem. Soc. B*, 1968, 526.
131. C. C. Price, *J. Am. Chem. Soc.*, 1936, **58**, 2101; *Chem. Rev.*, 1941, **29**, 37; C. C. Price and C. E. Arntzen, *J. Am. Chem. Soc.*, 1938, **60**, 2835.
132. P. W. Robertson, J. E. Allan, K. N. Haldane, and M. G. Simmers, *J. Chem. Soc.*, 1949, 933.
133. N. S. Gnanapragasam, N. V. Rao, and L. M. Yeddanapalli, *J. Indian Chem. Soc.*, 1959, **36**, 777.
134. T. Tsurata, K. Sasaki, and J. Furukawa, *J. Am. Chem. Soc.*, 1952, **74**, 5995; 1954, **76**, 994.
135. R. Josephson, R. M. Keefer, and L. J. Andrews, *J. Am. Chem. Soc.*, 1961, **83**, 2128.
136. J. Rajaram and J. C. Kuriacose, *Aust. J. Chem.*, 1969, **22**, 1193.
137. R. Pajeau, *C. R. Acad. Sci.*, 1938, **207**, 1420; *Bull. Soc. Chim. Fr.*, 1939, **6**, 1187.
138. G. A. Olah, S. J. Kuhn, S. H. Flood, and B. A. Hardie, *J. Am. Chem. Soc.*, 1964, **86**, 1039, 1044.
139. V. S. Karpinski and V. D. Lyashenko, *J. Gen. Chem. USSR*, 1960, **30**, 164; 1962, **32**, 3922; 1963, **33**, 599.
140. M. J. Rosen and J. Gandler, *J. Phys. Chem.*, 1971, **75**, 887.
141. T. M. Cresp, M. V. Sargent, J. A. Elix, and D. P. H. Murphy, *J. Chem. Soc., Perkin Trans. 1*, 1973, 340.
142. R. Taylor, *Specialist Periodical Report on Aromatic and Heteroaromatic Chemistry*, Vol. 2, Chemical Society, London, 1974, Vol. 2, p. 239.
143. A. McKillop, D. Bromley, and E. C. Taylor, *J. Org. Chem.*, 1972, **37**, 88.
144. K. L. Erickson and H. W. Barowsky, *J. Chem. Soc., Chem. Commun.*, 1971, 1596.
145. S. Uemura, K. Sohma, M. Okano, and K. Ichikawa, *Bull. Chem. Soc. Jpn.*, 1971, **44**, 2490.
146. V. F. Traven, V. A. Smrchek, and B. I. Stepanov, *J. Org. Chem. USSR*, 1972, **8**, 1810; 1973, **9**, 585.
147. J. M. W. Scott and J. G. Martin, *Can. J. Chem.*, 1965, **43**, 732; P. D. Golding, S. Reddy, J. M. W. Scott, V. A. White, and J. G. Winter, *Can. J. Chem.*, 1981, **59**, 839.
148. S. Rozen and M. Brand, *J. Chem. Soc., Chem. Commun.*, 1987, 752.
149. S. Uemura, A. Onoe, and M. Okano, *Bull. Chem. Soc. Jpn.*, 1974, **47**, 147.
150. G. Hallas and J. D. Hepworth, *Educ. Chem.*, 1974, **11**, 25.
151. V. Calo, L. Lopez, G. Pesce, F. Ciminale, and P. E. Todesco, *J. Chem. Soc., Perkin Trans. 2*, 1974, 1189.
152. R. Josephson, R. M. Keefer, and L. J. Andrews, *J. Am. Chem. Soc.*, 1961, **83**, 3562.
153. E. Baciocchi, G. Illuminati, G. Sleiter, and F. Stegel, *J. Am. Chem. Soc.*, 1967, **89**, 125.
154. E. Helgstrand, *Acta Chem. Scand.*, 1965, **19**, 1583.

155. M. Christen and H. Zollinger, *Helv. Chim. Acta*, 1962, **45**, 2057, 2066.
156. E. Berliner and K. Scheuller, *Chem. Ind.* (*London*), 1960, 1444.
157. B. T. Baliga and A. N. Bourns, *Can. J. Chem.*, 1966, **44**, 379.
158. P. G. Farrell and S. F. Mason, *Nature* (*London*), 1959, **183**, 250.
159. A. Ehrlich and E. Berliner, *J. Org. Chem.*, 1972, **37**, 4186.
160. P. B. D. de la Mare and O. M. H. El Dusouqui, *J. Chem. Soc. B*, 1967, 251.
161. F. Cacace and G. Stöcklin, *J. Am. Chem. Soc.*, 1972, **94**, 2519.
162. E. J. Knust, A. Halpern, and G. Stöcklin, *J. Am. Chem. Soc.*, 1974, **96**, 3733.
163. M. H. Mach and J. F. Bunnett, *J. Am. Chem. Soc.*, 1974, **96**, 936.
164. E. Grovenstein and D. C. Kilby, *J. Am. Chem. Soc.*, 1957, 79, 2972.
165. E. Grovenstein and N. S. Aprahamian, *J. Am. Chem. Soc.*, 1962, **84**, 212.
166. E. Grovenstein *et al.*, *J. Am. Chem. Soc.*, 1973, **95**, 4261.
167. (a) F. G. Soper and G. F. Smith, *J. Chem. Soc.*, 1927, 2757; (b) B. S. Painter and F. G. Soper, *J. Chem. Soc.*, 1947, 342; (c) E. Berliner, *J. Am. Chem. Soc.*, 1950, **72**, 4003; 1951, **73**, 4307.
168. V. Machacek, V. Sterba, and K. Valter, *Collect. Czech. Chem. Commun.*, 1972, **37**, 3073.
169. E. Shilov and F. Weinstein, *Nature* (*London*), 1958, **182**, 1300; *Dokl. Akad. Nauk SSSR*, 1958, **123**, 93.
170. A. Grimison and J. H. Ridd, *J. Chem. Soc.*, 1959, 3013.
171. L. Schutte and E. Havinga, *Tetrahedron*, 1970, **26**, 2297.
172. E. Grovenstein and F. C. Schmalstieg, *J. Am. Chem. Soc.*, 1967, **89**, 5084.
173. D. H. Derbyshire and W. A. Waters, *J. Chem. Soc.*, 1950, 3694.
174. J. Marton, *Acta Chem. Scand.*, 1969, **23**, 3329.
175. A. R. Butler and A. P. Sanderson, *J. Chem. Soc., Perkin Trans. 2*, 1974, 1214.
176. J. Arotsky, A. C. Darby, and J. B. A. Hamilton, *J. Chem. Soc., Perkin Trans. 2*, 1973, 595.
177. H. Suzuki, *Bull. Chem. Soc. Jpn.*, 1971, **44**, 2871.
178. Y. Ogata and K. Nakajima, *Tetrahedron*, 1964, **20**, 43; Y. Ogata and I. Urasaki, *J. Chem. Soc. C*, 1970, 1689.
179. (a) A. M. Sedov and A. N. Novikov, *J. Org. Chem. USSR*, 1971, **7**, 524; (b) R. Lines and V. D. Parker, *Acta Chem. Scand., Ser. B*, 1980, **34**, 47.
180. E. M. Chen, R. M. Keefer, and L. J. Andrews, *J. Am. Chem. Soc.*, 1967, **89**, 428.
181. J. R. Barnett, L. J. Andrews, and R. M. Keefer, *J. Am. Chem. Soc.*, 1972, **94**, 6129.
182. (a) L. J. Andrews and R. M. Keefer, *J. Am. Chem. Soc.*, 1956, **78**, 5623; (b) 1957, **79**, 1412.
183. E. Berliner, *J. Am. Chem. Soc.*, 1956, **78**, 3632.
184. L. Vasharosh, Yu. V. Norseev, and V. A. Khalkin, *Proc. Acad. Sci. USSR*, 1982, **266**, 297.
185. O. M. H. El Dusouqui, A. R. H. El Nadi, M. Hassan, and G. Yousif, *J. Chem. Soc., Perkin Trans. 2*, 1976, 357, 359.
186. (a) L. M. Stock and A. Himoe, *J. Am. Chem. Soc.*, 1969, **91**, 1452; (b) R. Danieli, A. Ricci, H. M. Gilow, and J. H. Ridd, *J. Chem. Soc., Perkin Trans. 2*, 1974, 1477.
187. L. M. Stock and H. C. Brown, *J. Am. Chem. Soc.*, 1957, **79**, 1421.
188. W. M. Schubert and D. F. Gurka, *J. Am. Chem. Soc.*, 1969, **91**, 1443.
189. J. P. Causelier, *Bull. Soc. Chim. Fr.*, 1971, 1785; 1972, 762.
190. P. B. D de la Mare and J. T. Harvey, *J. Chem. Soc.*, 1956, 36.
191. L. M. Stock and M. R. Wasielewski, *J. Org. Chem.*, 1971, **36**, 1002.
192. E. Berliner and F. Berliner, *J. Am. Chem. Soc.*, 1949, **71**, 1195.
193. W. M. Archer, M. A. Hossaini, and R. Taylor, *J. Chem. Soc., Perkin Trans. 2*, 1982, 181; M. A. Hossaini and R. Taylor, *J. Chem. Soc., Perkin Trans. 2*, 1982, 187.
194. R. T. Lahonde, P. B. Ferrara and A. D. Debboli, *J. Org. Chem.*, 1972, **37**, 1094.

195. O. M. Nefedov and R. N. Shafran, *J. Org. Chem. USSR*, 1974, **10**, 481.
196. J. Vaughan, G. J. Welch and G. J. Wright, *Tetrahedron*, 1965, **21**, 1665.
197. J. L. G. Nilsson, H. Selander, H. Sievertsson, I. Skänberg, and K.-G. Svensson, *Acta Chem. Scand.*, 1971, **25**, 94; K.-G. Svensson, H. Selander, M. Karlsson, and J. L. G. Nilsson, *Tetrahedron*, 1973, **29**, 1115.
198. M. Hassan and G. Yousif, *J. Chem. Soc. B*, 1968, 459; 1969, 591.
199. H. J. Reich and D. J. Cram, *J. Am. Chem. Soc.*, 1968, **90**, 1365; 1969, **91**, 3505.
200. H. C. Brown and L. M. Stock, *J. Am. Chem. Soc.*, 1960, 1942.
201. J. E. Dubois, J. J. Aaron, P. Alcais, J. P. Doucet, F. Rothenberg, and R. Uzan, *J. Am. Chem. Soc.*, 1972, **94**, 6823.
202. P. B. D. de la Mare, N. S. Isaacs, and M. J. McGlane, *J. Chem. Soc., Perkin Trans. 2*, 1976, 784.
203. J. M. Brittain, P. B. D. de la Mare, and J. M. Smith, *J. Chem. Soc., Perkin Trans. 2*, 1981, 1629.
204. K. V. Seshadri and R. Ganesan, *Tetrahedron*, 1972, **28**, 3827.
205. A. C. Boicelli, R. Danieli, A. Mangini, A. Ricci, and G. Pirazzini, *J. Chem. Soc., Perkin Trans. 2*, 1974, 1343.
206. G. Baddeley, N. H. P. Smith and M. A. Vickars, *J. Chem. Soc.*, 1956, 2455; P. B. D. de la Mare, *Tetrahedron*, 1959, **5**, 107.
207. P. B. D. de la Mare, B. N. B. Hannan, and N. S. Isaacs, *J. Chem. Soc., Perkin Trans. 2*, 1976, 1389.
208. Ref. 9b, p. 145.
209. S. Ahmed and J. L. Wardell, *Tetrahedron Lett.*, 1971, 3089.
210. G. A. Olah *et al.*, *J. Am. Chem. Soc.*, 1987, **109**, 3708.
211. A. A. Retinski, V. P. Tolmasova, G. P. Kotomanova, and S. M. Shostakovskii, *Bull. Acad. Sci. USSR*, 1968, 1756.
212. R. O. C. Norman and G. K. Radda, *J. Chem. Soc.*, 1961, 3610.
213. Ref. 9b, p. 82.
214. B. Lepeska, V. Bazant, and V. Chvalovsky, *J. Organomet. Chem.*, 1970, **23**, 41.
215. P. P. Alikhanov, G. V. Motsarev, and K. I. Sakodynskii, *Proc. Acad. Sci. USSR*, 1978, **238**, 82.
216. G. V. Motsarev, V. T. Inshakova, V. R. Rozenberg, and V. I. Kolbasov, *J. Gen. Chem. USSR*, 1977, **47**, 2351.
217. J. Vcelak and V. Chvalovsky, *J. Organomet. Chem.*, 1970, **23**, 47.
218. A. Gustaminza, J. H. Ridd, and F. Roy, *J. Chem. Soc. B*, 1969, 684.
219. R. Danieli, A. Ricci, and J. H. Ridd, *J. Chem. Soc., Perkin Trans. 2*, 1976, 290.
220. H. D. Smith, T. A. Knowles and H. Schroeder, *Inorg. Chem.*, 1965, **4**, 107; L. I. Zakharin and V. N. Kalinin *et al.*, *J. Gen. Chem. USSR*, 1966, **36**, 1703; 1967, **37**, 889; 1970, **40**, 115; *Bull. Acad. Sci. USSR*, 1966, 549, 1946, 1882; 1968, 1683, 2532.
221. B. Lepeska and V. Chvalovsky, *Collect. Czech. Chem. Commun.*, 1969, **34**, 3553.
222. V. A. Koptyug, N. F. Salakhutdinov, and A. N. Detsina, *J. Org. Chem. USSR*, 1984, **20**, 1143.
223. S. Branch and B. Jones, *J. Chem. Soc.*, 1955, 2921.
224. G. Illuminati, *J. Am. Chem. Soc.*, 1958, **80**, 4941, 4945.
225. P. B. D. de la Mare and P. W. Robertson, *J. Chem. Soc.*, 1948, 100.
226. E. Baciocchi and G. Illuminati, *J. Am. Chem. Soc.*, 1964, **86**, 2677.
227. J. Aaron, J. Dubois, F. Krausz, and R. Martin, *J. Org. Chem.*, 1973, **38**, 300.
228. T. J. Broxton, L. W. Deady, J. D. McCormack, L. C. Kam, and S. H. Toh, *J. Chem. Soc., Perkin Trans. 1*, 1974, 1769.
229. R. Danieli, A. Ricci, H. M. Gilow, and J. H. Ridd, *J. Chem. Soc., Perkin Trans. 2*, 1974, 1477.
230. Ref. 9b, p. 146.

231. (a) O. M. H. el Dusouqui and M. Hassan, *J. Chem. Soc. B*, 1966, 374; (b) O. M. H. el Dusouqui, M. Hassan, and B. Ibrahim, *J. Chem. Soc. B*, 1969, 589; (c) 1970, 926.
232. P. B. D. de la Mare and M. Hassan, *J. Chem. Soc.*, 1957, 3004.
233. H. Weingarten, *J. Org. Chem.*, 1962, 4347.
234. R. Taylor and G. G. Smith, *Tetrahedron*, 1963, **19**, 937.
235. M. Hassan and S. A. Osman, *J. Chem. Soc.*, 1965, 2194.
236. Y. Ogata, I. Urasaki, and T. Ishibashi, *J. Chem. Soc., Perkin Trans. 1*, 1972, 180.
237. P. B. D. de la Mare and E. A. Johnson, *J. Chem. Soc.*, 1963, 4076.
238. P. B. D. de la Mare, E. A. Johnson, and J. S. Lomas, *J. Chem. Soc.*, 1964, 5317; P. B. D. de la Mare, O. M. H. el Dusouqui, and E. A. Johnson, *J. Chem. Soc. B*, 1966, 521.
239. L. Altschuler and E. Berliner, *J. Am. Chem. Soc.*, 1966, **88**, 5837; G. W. Burton, P. B. D. de la Mare, L. Main, and B. N. B. Hannan, *J. Chem. Soc., Perkin Trans. 2*, 1972, 265.
240. P. J. N. Brown, M. T. Chaudhry, and R. Stephens, *J. Chem. Soc. C*, 1969, 2747.
241. R. Taylor, *J. Chem. Soc., Perkin Trans. 2*, 1973, 253.
242. S. V. Zakharova and E. P. Kaplan, *Bull. Acad. Sci. USSR*, 1971, 2695.
243. Y. F. El-din Shafig and R. Taylor, *J. Chem. Soc., Perkin Trans. 2*, 1978, 1263.
244. J. B. Kim, C. Chen, J. K. Krieger, K. R. Judd, C. C. Simpson, and E. Berliner, *J. Am. Chem. Soc.*, 1970, **92**, 910.
245. E. Berliner, J. B. Kim, and M. Link, *J. Org. Chem.*, 1968, **33**, 1160.
246. K. Lewis, R. D. Topsom, J. Vaughan, and G. J. Wright, *J. Org. Chem.*, 1968, **33**, 1497.
247. L. I. Denisova, N. A. Morozova, V. A. Plakhov, and A. I. Tochilkin, *J. Org. Chem. USSR*, 1966, **2**, 276.
248. F. E. Friedli and H. Schechter, *J. Org. Chem.*, 1985, **50**, 5710.
249. P. M. op den Brouw and W. H. Laarhoven, *Recl. Trav. Chim. Pays-Bas*, 1978, **97**, 265.
250. W. J. Archer, Y. F. El-din Shafig, and R. Taylor, *J. Chem. Soc., Perkin Trans. 2*, 1981, 675.
251. R. Bolton, *J. Chem. Res. (S)*, 1977, 149.
252. W. Baker, J. W. Barton, and J. F. W. McOmie, *J. Chem. Soc.*, 1958, 2666.

CHAPTER 10

The Replacement of a Substituent X by a Substituent Y

Reactions in which a substituent X is replaced by an electrophilic reagent Y comprise the largest class of aromatic substitutions. Thus Y may be any one of the electrophiles described in Chapters 5–9 and X may be any one of the groups whose replacement by a proton was described in Chapter 4. This gives a theoretical total of around 700 different reactions, but only a small proportion of the possible combinations are as yet known. Some reactions feature amongst the earliest recorded electrophilic substitutions; they have recently been given the description '*ipso*' substitutions (for a defintion of *ipso* partial rate factors, see Chapter 4).

Mechanistic investigations and quantitative results for substituent effects in these reactions are relatively few, but for some of the processes accurate partial rate factors are available because measurements can be made on individual compounds which can be obtained highly pure (cf. Chapter 4). Care is necessary to ensure that deprotonation reactions are not occurring concurrently and giving erroneous rate data. In some reactions it has been shown that overall replacement of X by Y is either accompanied by, or occurs wholly by, protolytic cleavage of X, followed by substitution of Y into the parent aromatic.

Since the proton-replacement reactions in Chapters 5–9 are subject to steric hindrance, and the protonolyses in Chapter 4 are subject to steric acceleration, the present reactions are subject to a combination of these effects, some of which are discussed in detail later. A unique effect is encountered in cleavages by nitronium ion (which is linear). This electrophile differs from most in that bonding takes place at its centre, rather than at one 'end', and it also changes shape during substitution. In nitrodeprotonation this does not cause substantial steric hindrance, consistent with the expectation that in the transition state the oxygens will already be bending away from the substituent (and interactions can be minimized further by rotation about the C—N bond). However, in *ipso* substitutions the oxygens will be interacting substantially with the substituent if this is bulky (**1**), e.g. CMe_3, $SiMe_3$, I, and these interactions cannot be minimized by rotation. This is not the case for the corresponding nitroso cleavages (**2**), and this explanation[1] accounts for nitroso cleavages of bulky groups occurring much faster than the corresponding nitro cleavages (see Sections 10.23, 10.28, 10.36, 10.60).

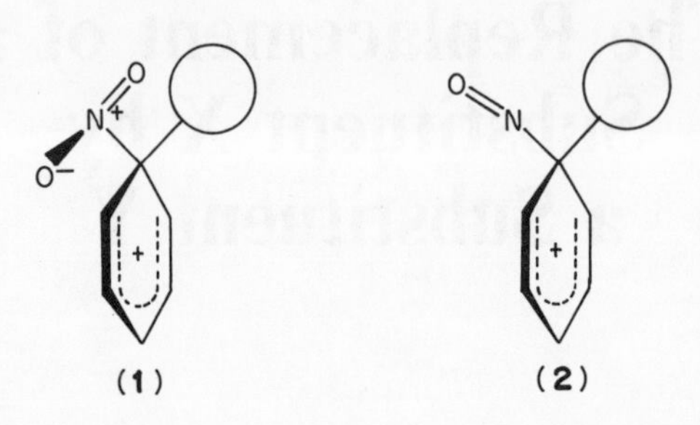

Some of the reactions are useful for introducing groups into the aromatic ring to give derivatives which are otherwise very difficult to obtain. This is because the C—X bond tends to be weaker than the C—H bond, so reaction conditions milder than those needed for the direct substitutions may be employed, thereby minimizing side-reactions. Moreover, it is possible to 'activate' positions, e.g. *ortho* to bulky groups, which would be difficult to substitute by the direct methods.

The sequencing of the reactions is as follows: both X and Y are taken in order across the Periodic Table (considering the element that is attached to the aromatic ring), but the leaving group takes sequence priority. Thus deboronations precede desilylations, and acyldesilylation precedes halodesilylation.

10.1 SILYLDELITHIATION

The reactions of aryllithium compounds with a variety of electrophiles such as water and halogens are widely known, but consideration of these processes is beyond the scope of this book, especially as they have not been investigated in the context of electrophilic substitutions. However, replacement of lithium by a trialkylsilyl group (eq. 10.1), using ethyldimethylsilyl compounds with X = Cl, OEt, and H, has been so examined.[2] For X = Cl, the groups *m*-Me, *p*-Me, and *m*-CF_3 in the aromatic ring have a negligible effect on the rate of reaction, but for X = OEt or H a small spread of rates occurs: *p*-Me > *m*-Me ≈ H > *m*-CF_3. This is the order expected for an electrophilic substitution, and the small range of rates indicates that there is little change in electron density in the aromatic ring in passage from the reactants to the transition state; this is consistent with the high polarity of the Ar—Li bond compared with Ar—H.

$$ArLi + R_3SiX \longrightarrow ArSiR_3 + LiX \qquad (10.1)$$

If the reacting silicon compound contains more than one group X, further substitution can occur, its extent depending largely on the steric influences in the groups Ar and R.[3]

The reaction (and also stannyldelithiation) has been used to prepare the 3-derivatives of benzocyclobutene. Electrophilic substitution at the 3-position is extremely difficult [Sections 2.6 and 4.7.1(ii)], but lithiation (Section 5.1) occurs there readily, and the Ar—Li bond can then be cleaved by Me_3SiCl or Me_3SnCl compounds.[4]

10.2 HALODEAURATION

The group $AuP(Ph)_3$ is cleaved from ferrocene by halogens to give the haloferrocene (eq. 10.2); bis-ferrocenyl is a byproduct of the reaction.[5]

$$FerrAuP(Ph)_3 + Hal \longrightarrow FerrHal + HalAuP(Ph)_3 \qquad (10.2)$$

10.3 NITROSODEMAGNESIATION

Many of the reactions of aryl Grignard reagents with electrophiles are electrophilic aromatic substitutions but, like those of the aryllithium reagents, are outside the scope of this book. A typical example is the reaction with NOCl, giving the nitrosoaromatic (eq. 10.3)[6] compounds which are otherwise difficult to obtain; other similar nitroso cleavages are described below.

$$ArMgCl + NOCl \longrightarrow ArNO + MgCl_2 \qquad (10.3)$$

10.4 MERCURIDEMERCURIATION

Just as hydrogen on an aromatic ring can undergo exchange, so too can other groups. The best documented of these reactions is mercuridemercuriation in which diaryl- and alkylarylmercury compounds react with mercury(II) halides in various solvents, including non-polar ones (eqs. 10.4 and 10.5). The reactions are second order, unaffected by the addition of water, are inhibited by iodide ion (in the reaction with HgI_2), and, in the similar reactions that take place with alkylmercurials, occur with retention of configuration in the alkyl group so that the reaction probably involves front-side attack.[7–11] Activation entropies for the reaction are consistent with a fairly cyclic and therefore symmetrical transition state structure which has been formulated, for example, as **3**; this may

Table 10.1 Partial rate factors for mercuridemercuriation

Reaction	Substituent							
	4-OMe	4-Me	4-Ph	H	4-F	4-Cl	4-Br	$4\text{-}NO_2$
$Ar_2Hg + HgI_2$, 25 °C	36.3	6.65	1.14	1	0.21	0.047		
$ArHgBr + HgBr_2$, 20 °C	411	9.5		1			0.214	0.009

alternatively be represented by **4**, in which the full arrow indicates greater electron transfer than the broken arrow.

$$Ar_2Hg + HgX_2 \rightleftharpoons 2ArHgX \tag{10.4}$$

$$ArHgR + HgX_2 \rightleftharpoons ArHgX + RHgX \tag{10.5}$$

Investigations into the extent of symmetry show this to vary according to the nature of the various groups involved. Thus, whereas reaction of dialkylmercurials with $^{203}HgCl_2$ gives almost equal distribution of label in the products,[9] in the corresponding reaction with PhHgEt most of the label ended up in the phenylmercury(II) chloride, showing that most of the cleavage of the organomercurial compound occurs between Ar and Hg.[11] Similar exchange reactions between Ph_2Hg and PhHgR[12] and between ArHgBr and $HgBr_2$[13] have confirmed the above general mechanistic picture.

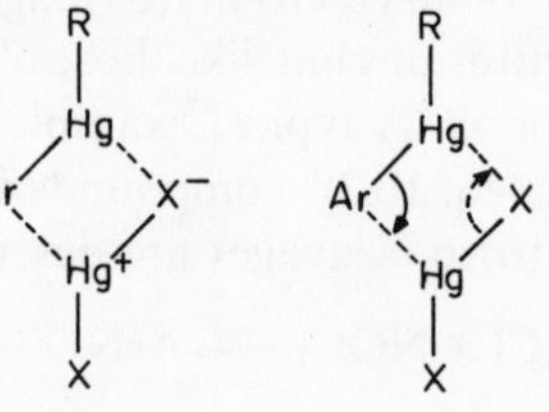

(**3**) (**4**)

Partial rate factors for mercuridemercuriation (Table 10.1) demonstrate the electrophilic nature of the reaction.[7,13] Surprisingly, the data for reactions of the diarylmercurials[7] correlated with σ values rather than σ^+ values, giving in consequence an anomalously high ρ of -5.9. For reaction of the arylmercury(II) bromides a satisfactory correlation with σ^+ values was obtained, with $\rho = -3.4$.

10.5 PLUMBYLDEMERCURIATION

The reaction of lead tetrakistrifluoroacetate in TFA with 4-fluorophenylmercury(II) trifluoroacetate gives 4-fluorophenyllead(IV) tristrifluoroacetate (eq. 10.6). The alternative mechanism, involving protolytic cleavage followed by plumbylation, was shown not to occur.[14]

$$Pb(OCOCF_3)_4 + ArHgOCOCF_3 \xrightarrow{TFA} ArPb(OCOCF_3)_3 + Hg(OCOCF_3)_2 \qquad (10.6)$$

10.6 NITROSODEMERCURIATION

Nitrosoaromatics are produced by the reaction of NOCl with either diarylmercurials or arylmercury(II) chloride (eq. 10.7).[15] The mercury(II) salt-catalysed nitration of toluene is thought to proceed via mercuriation followed by nitrosodemercuriation (yields were greatly reduced in the presence of urea) and then oxidation.[16] Nitrosodemercuriation may also be brought about by reaction with sodium nitrite–TFA or with N_2O_3.[17,18]

$$ArHgCl + NOCl \longrightarrow ArNO + HgCl_2 \qquad (10.7)$$

10.7 HALODEMERCURIATION

Both bromo- and iododemercuriation have been reported. The former reaction has been used as a valuable method for preparing polyhalogenoaromatics, particularly those containing deactivating substituents.[19] Thus, for example, treatment of nitrobenzene with fused mercury(II) trifluoroacetate results in substitution of five $HgOCOCF_3$ groups in the ring. Subsequent treatment with Br_3^- gives pentabromonitrobenzene.

Cleavage of arylmercury(II) bromides with iodine (eq. 10.8) produces the iodoaromatics; the rate data correlated with σ^+ values with $\rho = -2.87$.[20] The electrophilic nature of the reaction was also shown by the reaction of Ph_2Hg with iodine being 860 times faster than that of $(C_6F_5)_2Hg$ (in dioxane at 19.8 °C); in aprotic dipolar HMPA, however, the rate differences become much smaller, possibly owing to incursion of the S_E1 mechanism for compounds containing strongly electron-withdrawing groups (which react much more readily by this mechanism).[21]

$$ArHgBr + I_2 \longrightarrow ArI + HgBrI \qquad (10.8)$$

10.8 SULPHODEMERCURIATION

Arylsulphonic acids are obtained by the reaction of arylmercury(II) chlorides with >100% sulphuric (eq. 10.9). If, however, weaker acid is used, the product distribution becomes the same as that obtained with the unmercuriated aromatic,[22] showing that here protiodemercuriation occurs, followed by sulphonation.

$$ArHgCl + SO_3(H_2O) \longrightarrow ArSO_3H + HgClOH \qquad (10.9)$$

10.9 PLUMBYLDETHALLIATION

This reaction closely resembles reaction 10.5 and has been shown to occur between 4-halogenophenylthallium(III) bistrifluoroacetate and lead tetrakistrifluoroacetate (eq. 10.10).[14]

$$Pb(OCOCF_3)_4 + ArTl(OCOCF_3)_2 \xrightarrow{TFA} ArPb(OCOCF_3)_3 + Tl(OCOCF_3)_3 \qquad (10.10)$$

10.10 NITROSODETHALLIATION

This provides another route to nitroso compounds. Treatment of arylthallium bistrifluoroacetates in HCl (this gives the intermediate arylthallium dichlorides) with NOCl gives the nitroso compound in good yield; the four-centre transition state **5** may be involved.[23] Sodium nitrite in TFA may also be used as the electrophilic reagent [with $ArTl(OAc)ClO_4$ compounds] and here the intermediate nitroso compounds are oxidized to the nitroaromatics; N_2O_3 has also been used as the electrophile.[17,18]

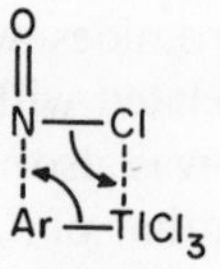

O
N—Cl
Ar—TlCl$_3$

(5)

10.11 NITRODETHALLIATION

Reaction of N_2O_4 with 4-methylphenylthallium bistrifluoroacetate gives a 95% yield of 4-nitrotoluene, and the method has been recommended for selective *para* nitration.[24] It is possible that this reaction takes place via nitrosodethalliation followed by oxidation, although no nitroso intermediates could be isolated under these conditions.[18]

Although the $Tl(OCOCF_3)_2$ substituent appears to be electron withdrawing, so withdrawing electrons from the *ipso* position, it is this very position to which the electrophile becomes preferentially attached. This high reactivity of the *ipso* position implies, therefore, that C–Tl hyperconjugation, like C–Hg hyperconjugation,[25] is very strong (see Chapter 4, Introduction, for the reasoning behind this assumption).

10.12 HYDROXYDETHALLIATION

The formation of 1,4-quinones by treatment of arylthallium bistrifluoroacetates with 90% hydrogen peroxide in TFA is believed to proceed via hydroxydethalliation, and the method may be a simple one for synthesizing a range of quinones.[26]

10.13 IODODETHALLIATION

It has long been known that treatment of arylthallium dichlorides (or bromides) with KI leads to rapid formation of the aryl iodide, via intermediate formation of the corresponding arylthallium diiodide.[27,28] However, this does not seem to involve electrophilic substitution. By contrast, such a substitution does appear to be involved in the formation (in some cases quantitatively) of aryl iodides from arylthallium bistrifluoroacetates (eq. 10.11).[29]

$$ArTl(OCOCF_3)_2 + I_2 \xrightarrow{TFA} ArI + TlI(OCOCF_3)_2 \qquad (10.11)$$

10.14 MERCURIDEBORONATION

In 1882 Michaelis and Becker[30] showed that arylboronic acids react readily with mercury(II) chloride in aqueous solution give arylmercury(II) chlorides and boric acid (eq. 10.12).

$$ArB(OH)_2 + HgCl_2 \xrightarrow{H_2O} ArHgCl + B(OH)_3 + HCl \qquad (10.12)$$

The related reaction between benzeneboronic acid and phenylmercury(II) perchlorate in aqueous ethanol is first order in each reagent.[31] Possible electrophiles are $PhHgClO_4$, $PhHg^+$, $PhHgOH_2{}^+$, and PhHgOH, while the substrate may be $PhB(OH)_2$ or $PhB(OH)_3{}^-$. From the variation in reaction rate with pH and with the addition of phosphoric acid, acetic acid, and dihydrogenphosphate ion in buffered solutions, it has been concluded that the relative species are $PhHg^+$ and $PhB(OH)_3{}^-$ or kinetically equivalent pairs [e.g. $PhHgOH_2{}^+$ and $PhB(OH)_3{}^-$]. Possible transitions states are 6–8, but the effects of substituents (which might permit a distinction to be made) have not been determined.

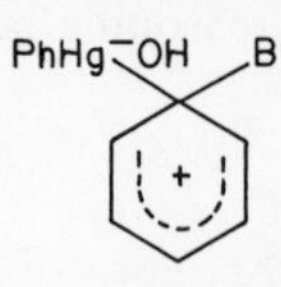

PhHg—OH B(OH)$_2$ (6) PhHg OH B(OH)$_2$ (7) PhHg B$^-$(OH)$_3$ (8)

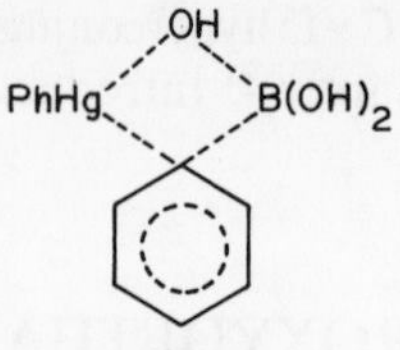

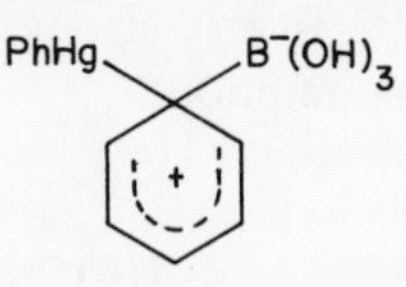

10.15 THALLIODEBORONATION

The parallel between the reactions of mercury and thallium is further emphasized by thalliodeboronation, which occurs between thallium(III) bromide or chloride and phenylboronic acid.[27]

10.16 NITRODEBORONATION

Nitration of phenylboronic acid takes place mainly *meta* to the $B(OH)_2$ substituent, but some nitrobenzene is also produced, showing that nitrodeboronation also occurs (eq. 10.13).[32,33]

$$PhB(OH)_2 + HNO_3 \longrightarrow PhNO_2 + B(OH)_3 \tag{10.13}$$

10.17 HYDROXYDEBORONATION

Treatment of an arylboronic acid with hydrogen peroxide results in replacement of the boronic acid group by hydroxyl (eq. 10.14).[32]

$$ArB(OH)_3 + H_2O_2 \longrightarrow ArOH + B(OH)_3 \tag{10.14}$$

The mechanism is usually complex and four different reaction paths may be involved, depending on the conditions.[34] Each requires stoichiometric concentrations of peroxide and boronic acid, and three are characterized by (1) a pH independence (in the pH range 1–3), (2) a pH dependence, and (3) a pH dependence coupled with a second-order dependence on boronic acid concentration (one molecule of this acid acting as a catalyst since the reaction stoichiometry does not alter). Each of these reactions is thought to involve the formation of the entity **9**, which rearranges to **10**, the latter then being hydrolysed in a fast step. Substituent effects are small and random for these three paths.

$$\underset{\textbf{(9)}}{Ph{-}\overset{\displaystyle OH}{\underset{\displaystyle O^+}{\overset{|}{\underset{|}{B^-}}}}{-}OH} \longrightarrow \underset{\textbf{(10)}}{Ph{-}O{-}B(OH)_2} \xrightarrow[\text{fast}]{H_2O} PhOH + B(OH)_3 \tag{10.15}$$

The fourth reaction occurs in strong acids, but the mechanism is uncertain. Since acidified solutions of hydrogen peroxide provide an electrophilic hydroxylating entity (Section 8.1.1), nuclear hydroxylation may occur to give the intermediate **11**, from which $B(OH)_2$ is eliminated. Consistent with this suggestion are the typical electrophilic substituent effects, although here $\log k_{rel.}$ values are correlated more satisfactorily with σ than with σ^+ values, giving a very small ρ factor for the reaction of ca -1.2.[35]

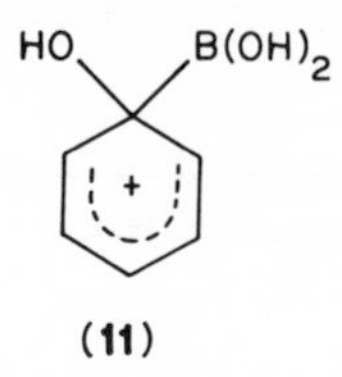

(**11**)

10.18 HALODEBORONATION

The replacement of the boronic acid groups in arylboronic acids by bromine or iodine (eq. 10.16) is first order in both halogen and acid. The rate of the reaction with bromine in acetic acid is increased by the addition of water, and the rate is linearly related to the molecular halogen concentration, indicating that molecular halogen is the attacking species. Further, a plot of log (rate) against pH is linear and of unit gradient, and since the logarithm of the concentration of the boronate anion is related to pH in this way, **12** appears to be the reactive substrate and eq. 10.17 the mechanism.[36]

$$\mathrm{ArB(OH)_2 + Hal_2 + H_2O \longrightarrow ArHal + B(OH)_3 + HHal} \qquad (10.16)$$

$$\mathrm{ArB(OH)_2 + H_2O \xrightarrow{fast} (H^+) + \underset{(\mathbf{12})}{ArB(OH)_3^-} + Br_2 \xrightarrow{slow} ArBr + B(OH)_3 + Br^-} \qquad (10.17)$$

The effects of substituents in the reaction (Table 10.2) confirm that the reaction is a typical electrophilic substitution. The data of Kuivila and coworkers[36] were found to correlate with the Yukawa–Tsuno equation (Section 11.1.3), with $\rho = -3.84$ and $r = 2.29$.[37] This contradicted the assumption that the resonance interaction in the transition state of an electrophilic substitution should parallel the amount of charge developed (as measured by the ρ factor).[38] However, re-examination of the reaction[39] revealed that the data for the 4-MeO compound, upon which the correlations were largely based, could not have been measured since the half-life is only 0.01 s! Moreover, the 4-Ph compound was ten times less soluble than the concentration stated to have been used for the kinetic studies in the original work, and for the 3-Me compound concurrent bromodeprotonation

Table 10.2 Partial rate factors for bromodeboronation in 80% aqueous HOAc at 25 °C

Substituent	$k_{rel.}$ (ref. 37)[a]	$k_{rel.}$ (ref. 39)	Substituent	$k_{rel.}$
4-OMe	1.45×10^6	1.1×10^4	4-Br	0.413
4-Me	78.7	51.5	3-I	0.072
4-Ph	21.7	13.0	3-Br	0.044
3-Me	3.33	3.00[b]	3-CO_2Et	0.044
4-F	2.81	2.56	3-F	0.039
H	1.0	1.0	3-Cl	0.035
4-Cl	0.54		4-CO_2Et	0.010
4-I	0.497		3-NO_2	0.003

[a]Data for activating substituents are in error; see text.
[b]Corrected for concurrent bromodeprotonation.

was appreciable. Revised kinetic data given in Table 10.2 show that an almost linear correlation with σ^+ values is obtained, i.e. $r \approx 1.0$.[39]

10.19 ALKYLDEALKYLATION

The $ZnCl_2$-catalysed reaction of 4-nitrobenzyl chloride with either hexamethylbenzene or 2,3,4,5,6-pentamethyldiphenylmethane results in replacement of Me or CH_2Ph, respectively, by the 4-nitrobenzyl group.[40]

The BF_3-catalysed cyclization of 4-(1-naphthyl)butanol (**13**) to 1,2,3,4-tetrahydrophenanthrene (**14**) takes place by two routes.[41] Normal substitution occurs at the 2-position, but because the 1-position is much more reactive towards electrophilic substitution, 16% of *ipso* substitution occurs, and this is then followed by rearrangement; the extent of *ipso* substitution increases to 71% if a methoxy group is in the *para* position in the naphthalene ring.

CH_2OH (**13**) → (**14**) + [spiro compound] —rearranges→ **14**

10.20 ACYLDEALKYLATION

There have been many reports of this reaction (eq. 10.18), which takes place

$$ArR + R'COCl \xrightarrow{AlCl_3} ArCOR' + RCl \tag{10.18}$$

in high yield. It shows common *ipso* substitution features[42] in that secondary alkyl groups are the most readily displaced, and this reflects the high steric requirement of the acylating species (Section 6.7.3). Displacement of *tert*-butyl groups occurs only when acyldeprotonation is sterically unfavourable, e.g. in the formation of 4-acetyl-*tert*-butylbenzene (71%) from 1,4-di-*tert*-butylbenzene.[43] The ethyl group has been successfully displaced from hexaethylbenzene by a wide variety of acylating reagents.[44]

Alkyl group migration can accompany the reaction, as expected [Section 6.1.1.(ii)].

10.21 DIAZODEALKYLATION

α-Hydroxyalkyl groups *para* to *N,N*-dialkylamino groups can be replaced by diazo groups (eq. 10.19), a method useful for preparing aldehydes and ketones.[45] This dealkylation is facilitated by the high stability of the α-hydroxyalkyl carbocation. The 4-$Me_2NC_6H_4CH_2$ substituent also provides a highly stable leaving group and is thus readily cleaved in this reaction.[46]

$$4\text{-}Me_2NC_6H_4CHROH + ArN_2^+ \longrightarrow 4\text{-}Me_2NC_6H_4NNAr + RCHO \tag{10.19}$$

10.22 NITROSODEALKYLATION

This closely parallels the above reaction in that the nitroso group can replace either PhCH(OH)– or 4-$Me_2NC_6H_4CH_2$– if these are *para* to NMe_2 in the benzene ring.[47]

10.23 NITRODEALKYLATION

This is well known reaction which has been reviewed.[48] Its electrophilic nature is indicated by the ready occurrence at sites most activated towards electrophilic substitution. The more highly branched alkyl groups are most readily replaced, reflecting the stability of the leaving carbocation, but replacement of methyl groups is known,[49,50] and takes place 2–3 times faster than nitration *meta* to a methyl group, the *ipso* partial rate factor being 4.7.[50] Replacement of *tert*-butyl groups takes place *less* readily than replacement of isopropyl groups, and this has been attributed to steric hindrance to approach of the electrophile to the *ipso* position in the *tert*-butyl compounds.[51]

A typical example is the nitration of 1,4-diisopropylbenzene, which gives mainly 4-nitroisopropylbenzene (**15**) together with some 2-nitro-1,4-diisopropylbenzene (**16**).[51] Likewise, 1,4-di-*tert*-butylbenzene gives 4-nitro-*tert*-butylbenzene and the rate of this is increased by activating substituents at the 2-position.[52] In favourable cases nitrodealkylation can be the exclusive reaction, as for example in nitration of 1,2,4,5-tetraisopropylbenzene to give **17**; the alternative nitrodeprotonation at the 3-position is sterically hindered.[53] Another example is the exclusive formation of 5-nitro-1,3-diisopropylbenzene in nitration of 1,3,5-triisopropylbenzene.[51] However, the corresponding *tert*-butyl compound does not undergo *ipso* substitution.[54a]

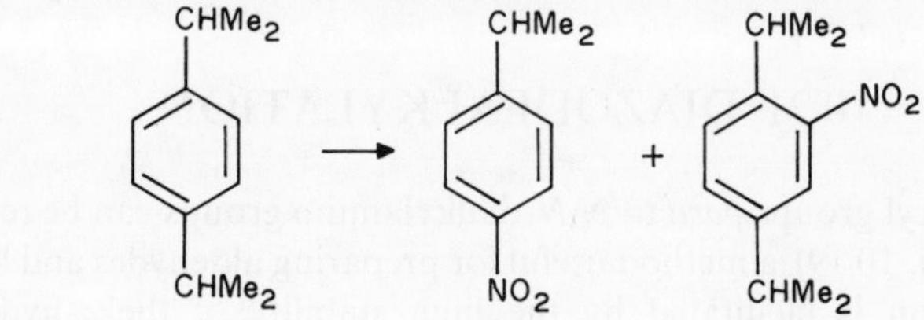

CHMe2 / CHMe2 → CHMe2 / NO2 **(15)** + CHMe2 / NO2 / CHMe2 **(16)**

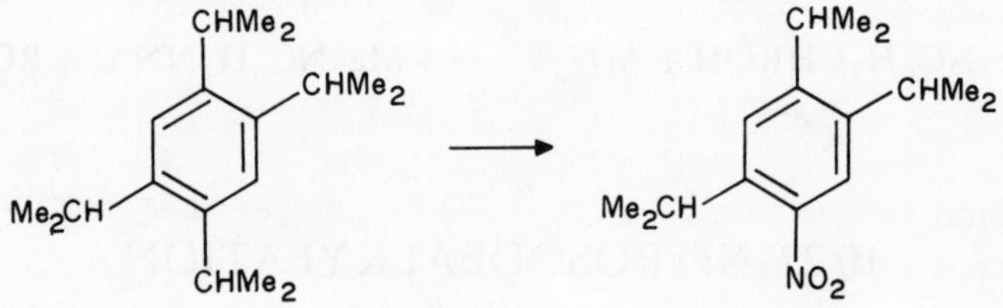

CHMe2 / CHMe2 / Me2CH / CHMe2 → CHMe2 / CHMe2 / Me2CH / NO2 **(17)**

The steric acceleration of the reaction is nicely demonstrated by nitration of *o*- and *p*-isopropyltoluenes. Whereas the *para* isomer gives only 10%, the *ortho* isomer gives 25–43% of nitrodeisopropylation, yet each position should be activated to a similar extent.[49,51] Steric acceleration provides the explanation for the nitrode-*tert*-butylation at both the 1- and 4-positions of 2-nitro-1,4-di-*tert*-butylbenzene, these positions being respectively *ortho* and *meta* to the nitro group.[54b] Similar reasoning accounts for the nitrode-*tert*-butylation at the 2-positions of 2,4,6-tri-*tert*-butylbromo- and -nitrobenzenes.[54a,55]

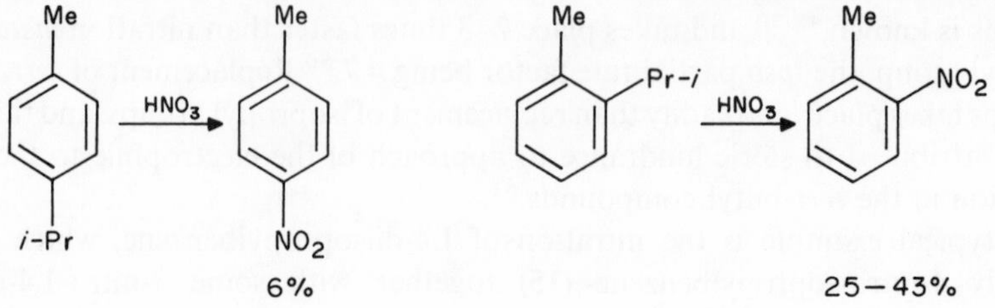

Me / *i*-Pr $\xrightarrow{HNO_3}$ Me / NO_2 6%

Me / Pr-*i* $\xrightarrow{HNO_3}$ Me / NO_2 25–43%

The ease of removal of isopropyl groups is such that it will take place even at deactivated positions, if other positions in the molecule are not particularly reactive. Thus 20% of the nitration of 4-chloroisopropylbenzene occurs at the carbon bearing the isopropyl group.[49] However, strong activation is needed for poorer leaving groups and this is both available in, and accounts for, the nitrodealkylation of cyclotriveratrylene (**18**)[56] and galbulin (**19**).[57a] From nitration of polyalkylbenzenes, the ρ factor for nitrodemethylation has been estimated to be between -8 and -12.[57b]

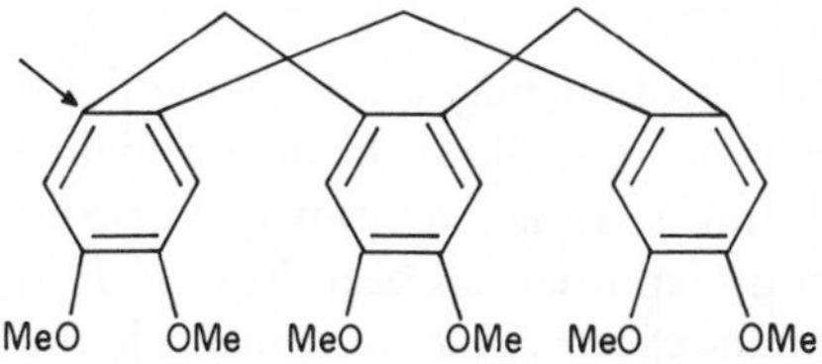

MeO OMe MeO OMe MeO OMe

(**18**)

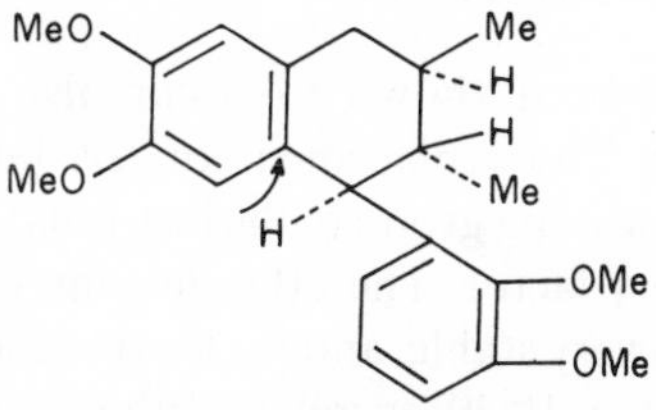

MeO MeO Me H H Me H OMe OMe

(**19**)

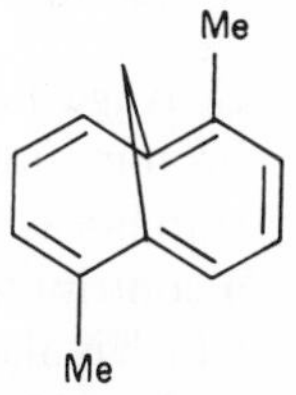

Me Me

(**20**)

10.24 SULPHODEALKYLATION

Steric hindrance to sulphonation is so severe that sulphodealkylation occurs under circumstances that would not favour other *ipso* substitutions. For example, 2,7-dimethyl-1,6-methano[10]annulene (**20**) undergoes sulphodemethylation at the 7-position in addition to sulphonation at the 5-position.[58] The sites activated towards sulphodeprotonation are all very sterically hindered, so cleavage of the methyl group readily occurs, and this is favoured by the high intrinsic reactivity of the annulene.[59] Likewise, sulphonation of 1,4-di-*tert*-butylbenzene gives 67% of 4-*tert*-butylbenzenesulphonic acid, 1,3,5-tri-*tert*-butylbenzene gives mainly the 3,5-di-*tert*-butylbenzenesulphonic acid,[60] and 1,2,4,5-tetraisopropylbenzene gives 2,4,5-triisopropylbenzenesulphonic acid;[53] various *tert*-butylated phenols also undergo sulphodebutylation in accord with expectation.[61]

Some of these reactions may proceed partly or wholly via protiodebutylation followed by sulphonation, but this cannot be the case when aprotic reagents are used. An example of the latter is the sulphodebutylation of 1,4-di-*tert*-butylbenzene by SO_3.[62]

10.25 HALODEALKYLATION

Halode-*tert*-butylation accompanies the chlorination and bromination of *tert*-butylbenzene and derivatives under various conditions;[63–72] in one study[66] the

Table 10.3 *ipso* factors for halogenode-*tert*-butylation of *t*-BuAr

f	Reaction	Ar	Position	Ref.
1.0	Cl_2, aq. HOAc	Benzene	1	63
0.8	Cl_2, aq. HOAc	Biphenyl	4	69[a]
0.27	Br_2, HOAc	2,6-Di-*tert*-butylphenol	4	74
1.4	$HOBr–H^+$, aq. dioxane	Benzene	1	64[b]
1.45	$HOBr–H^+$, aq. dioxane	1,3-Di-*tert*-butylbenzene	5	71

[a]Calculated from data in ref. 69.
[b]Calculated from data in ref. 64.

cleavage has been shown to occur also via protiodebutylation followed by bromination. The replacement of *tert*-butyl groups follows from the higher stability as a leaving group of the *tert*-butyl cation; steric acceleration may also be a contributory factor. The other leaving group reported has been $Ar'CHOH^+$, which is also very stable, and for this reaction the effects of the substituents in the aryl ring Ar′ (eq. 10.20) correlate with σ^+ values with $\rho = -1.24$.[73] This confirms that the group leaves as a cation, and the small ρ factor is due to the highly stabilizing effect upon this ion of the α-OH group.

$$4\text{-MeOC}_6\text{H}_4\text{CHOHAr}' + \text{Br}_2 \longrightarrow 4\text{-MeOC}_6\text{H}_4\text{Br} + \text{ArCHO} + \text{HBr} \quad (10.20)$$

The susceptibility of 1,3,5-tri-*tert*-butylbenzene to *ipso* substitution has already been noted (Section 10.25), and in bromination ($Br_2–CCl_4$) 71% of the bromo product is 3,5-dibromo-*tert*-butylbenzene.[72] (If bromination is carried out in acetic acid, acetoxyde-*tert*-butylation is an accompanying reaction.)[71]

Some *ipso* factors are available for halogenodebutylation (Table 10.3). Since breaking of the C—C bond should be easier than breaking of a C—H bond, and the electron density at the *ipso* carbon should be higher than in the corresponding unsubstituted molecule, factors significantly greater than 1.0 could be expected. The observed values (and the differences under various halogenating conditions) therefore reinforce the conclusion (above), that *ipso* substitution of *tert*-butyl groups is sterically hindered.

Partial rate factors have also been determined for substituent effects in chlorode-*tert*-butylation as follows: 4-*t*-Bu, 456; 4-Me, 730; 4-Ph, 615; and 4-(4′-C_6H_4), 2085. The values are very similar to those obtained for chlorination under the same conditions.[69]

10.26 DIAZODEACYLATION

The groups COMe and COH can be displaced from the *para* position of *N*,*N*-dimethylaniline by aryldiazonium ions.[46]

10.27 NITROSODEACYLATION AND NITROSODECARBOXYLATION

In the same compounds as in the preceding reaction, the COMe and COH can be displaced by nitrous acid.[47]

$$ArCO_2H + HONO \longrightarrow ArNO + H_2O + CO_2 \qquad (10.21)$$

Kinetic studies have been carried out on the nitrosodecarboxylation of 3,5-dibromo-4-hydroxybenzoic acid (eq. 10.21).[75] This takes place 13 times faster than the corresponding nitrosodeprotonation. This high *ipso* substitution factor may reflect the greater ease of C—C compared with C—H bond breaking, and the fact that the latter is partially rate-determining in nitrosation.[76] Since a C—H bond is not being broken in the reaction, no base catalysis is observed.

10.28 NITRODEACYLATION AND NITRODECARBOXYLATION

The electrophilic nature of this well documented reaction[48] is shown by its occurrence only at positions strongly activated towards electrophilic substi-

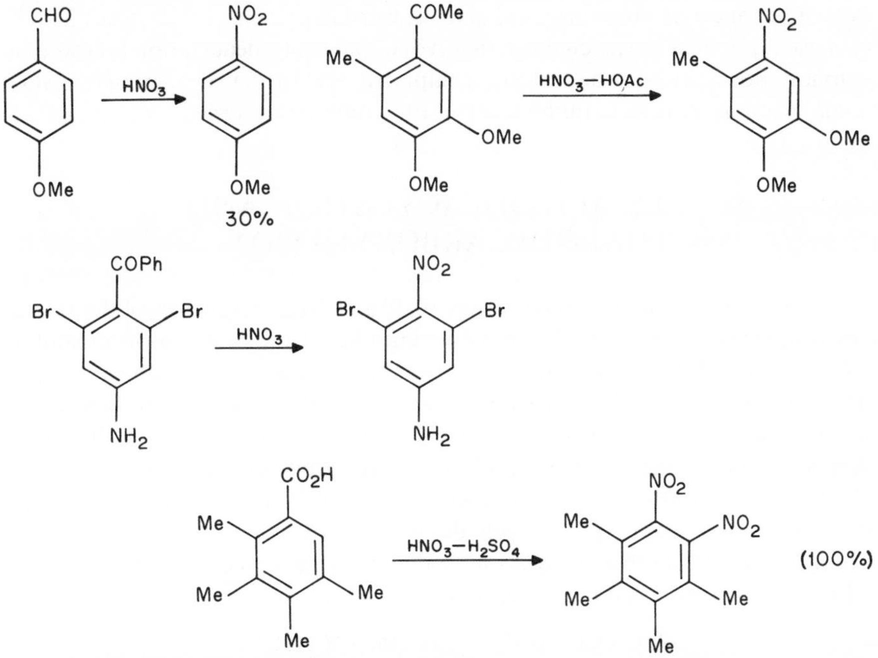

CHO, OMe —HNO_3→ NO_2, OMe 30%

COMe, Me, OMe, OMe —HNO_3—HOAc→ NO_2, Me, OMe, OMe

COPh, Br, Br, NH_2 —HNO_3→ NO_2, Br, Br, NH_2

CO_2H, Me, Me, Me, Me —HNO_3—H_2SO_4→ NO_2, NO_2, Me, Me, Me, Me (100%)

Scheme 10.1. Nitrodeacylation and nitrodecarboxylation reactions.

tution. The reactions are slower than nitrodeprotonation, which may again reflect steric hindrance to *ipso* substitution by the nitronium ion, and the fact that the electron density at the *ipso* carbon should be lower than in the parent aromatic. This latter constraint should also apply to nitrosodecarboxylation which nevertheless gives a high *ipso* factor, so this is a further demonstration that nitroso cleavages are much faster than nitro cleavages (see Introduction).

Some examples of the reaction are given in Scheme 10.1.[77–79] Also notable is the fact that 3,4,5-trimethoxybenzoic acid undergoes nitrodecarboxylation, whereas 2,3,4-trimethoxybenzoic acid does not, even though the *ipso* position is more strongly activated in the latter.[80] Nitrodeprotonation is more sterically hindered in the former compound than in the latter, and this evidently suffices to favour replacement of the carboxyl group. 3,4,5-Trimethoxyacetylbenzene also undergoes nitrodeacetylation (70%),[81] as does 3,4,5,6-tetraalkyl-2-acetylphenol.[82]

10.29 SULPHODEACYLATION

A clear example of sulphodeacylation is the reaction of 1,1′-diacetylferrocene with SO_3 in dichloroethane, to give ferrocene-1,1′-disulphonic acid.[83] The reactions of dimesityl ketone, 2,4-diacetylmesitylene, and 2,4,6,2′,6′-pentamethylbenzophenone with sulphuric acid give 2,4,6-trimethylbenzenesulphonic acid in each case, but here the mechanism involves initial protiodeacylation followed by sulphonation.[84] It is probable that in sulphuric acid media containing free SO_3, some reaction by direct sulphodeacylation would be observed.

10.30 HALODEACYLATION AND HALODECARBOXYLATION

These reactions are described by eqs 10.22 and 10.23. Examples of the former reaction are the formation of 2- and 4-bromophenol and 2- and 4-bromoaniline by treatment of 2- and 4-hydroxy- and -aminobenzaldehydes with bromine.[85] The reaction of either 2,4,6-trimethoxybenzaldehyde or 2,4,6-trimethoxyacetophenone with sulphuryl chloride gave chlorodeacylation, together with chlorodeprotonation at the two free sites.[86] In the reaction with the aldehyde, it was shown that chlorodeformylation must precede chlorodeprotonation, since 3,5-dichloro-2,4,6-trimethoxybenzaldehyde would not undergo chlorodeformylation. Moreover, chlorobenzoylation could not be made to occur, indicating that $PhCO^+$ is a poorer leaving group than $MeCO^+$.

$$\mathrm{ArCR'O + Hal_2 \longrightarrow ArHal + R'Hal + CO} \qquad (10.22)$$

$$\mathrm{ArCO_2H + Hal_2 \longrightarrow ArHal + HHal + CO_2} \qquad (10.23)$$

Examples which illustrate the electrophilic nature of halodecarboxylation are the formation of 2,4,6-tribromophenol from 2- or 4-hydroxybenzoic acid and bromine,[87] and the formation of 2,4,6-triiodophenol from 4-hydroxybenzoic acid and iodine.[88] Similarly, anthranilic acid and 4-aminobenzoic acid undergo halodecarboxylation.[89] Surprisingly, however, 2,6-dihydroxy-4-methylbenzoic acid does not undergo halodecarboxylation with molecular bromine or chlorine.[90]

The formation of tribromophenol from 2- and 4-hydroxybenzoic acid occurs via the corresponding 3,5-dibromohydroxy acids[91] (curiously, this sequence of halodeprotonation and halodecarboxylation is the exact opposite of that noted above for halodeacylation). The subsequent bromodecarboxylation of these intermediates in aqueous HOAc is first order in both the acid and bromine, but the pseudo second-order rate coefficients obtained are inversely proportional to the product, $[Br^-][H^+]^2$. Further, 2,6-dibromophenol is brominated 700 times faster than 3,5-dibromo-4-hydroxybenzoic acid, and 2,4-dibromophenol is brominated 100 times faster than 3,5-dibromo-2-hydroxybenzoic acid. Hence (1) the reaction cannot involve rapid decarboxylation followed by rate-determining bromination, since these rate differences would not then occur, and (2) it cannot involve rate-determining protiodecarboxylation followed by rapid bromination, since the rates would then be independent of the bromine concentration.

The following mechanism was proposed on the basis of the observed kinetic dependences. Rate-determining bromination of the substrate (eq. 10.24) gives the intermediate **21**. (An alternative to eq. 10.24 and kinetically indistinguishable from it involves bromination of the phenate ion formed in a rapid pre-equilibrium.)

$$\mathrm{ArOH} + \mathrm{Br_2} \xrightarrow{\text{slow}} \mathbf{21} + \mathrm{H^+} + \mathrm{Br^-} \qquad (10.24)$$

The intermediate **21** ionizes in a fast step (eq. 10.25), and the resulting carboxylate anion then eliminates carbon dioxide (eq. 10.26).

The kinetic dependences are, however, also consistent with an alternative mechanism in which the species H_2OBr^+ and $^-O{-}Ar{-}CO_2^-$ are formed in fast pre-equilibria. Being of opposite charge, these would react rapidly together to give the intermediate **22**, which would then lose CO_2 in the rate-determining step (eq. 10.26).

The first mechanism has been shown to be correct. The reaction exhibits a carbon-13 isotope effect (in the CO_2 product) which varies from 1.002 in the absence of added bromide ion to 1.045 in the presence of added bromide ion. The second mechanism would have an isotope effect if eq. 10.26 were rate-determining, but its value should be independent of $[Br^-]$. According to the first mechanism, the intermediate **21** is partitioned between product formation (eqs 10.25 and 10.26), and reversion to starting materials by the reverse of eq. 10.24, and the latter course is favoured by added bromide ion. A carbon-13 isotope effect should occur if eq. 10.26 is partially rate-determining, and should be

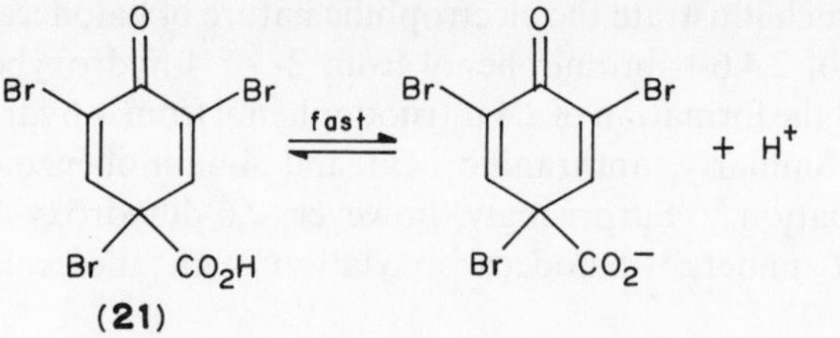

fast

+ H^+ (10.25)

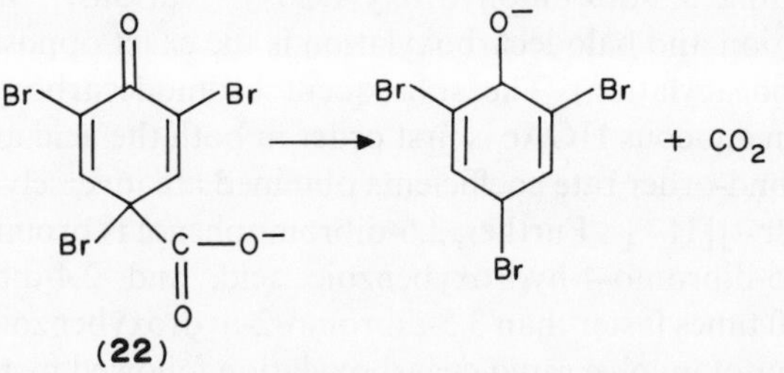

(**21**)

+ CO_2 (10.26)

(**22**)

increased, as is observed, by the addition of bromide ion, which facilitates the reversion of **21** to the reactants.

A notable feature is the fact that whereas bromodecarboxylation takes place 100–700 times *slower* than bromodeprotonation, nitrosodecarboxylation takes place *faster* than nitrosodeprotonation. This may reflect greater steric hindrance in bromodecarboxylation or, alternatively, differences in reacting species (phenol vs phenate ion) in the two brominations.

10.31 MERCURIDESILYLATION

Cleavage of trialkylsilyl groups from the aromatic ring by mercury(II) acetate in (aqueous) acetic acid (eq. 10.27)[92–95] is a convenient method of preparing aromatic mercury compounds.

$$ArSiR_3 + Hg(OAc)_2 \longrightarrow ArHgOAc + R_3SiOAc \qquad (10.27)$$

The mechanism is uncertain, the reaction being first order in each reagent only in 80% aqueous HOAc (and then only if equal quantities of reagents are taken); in less aqueous media the reaction order becomes three or more.[96] The reaction is thought to involve non-ionized mercury(II) acetate and acetoxymercury(II) ions, $AcOHg^+$, as electrophiles, the reactivity difference between them being 15-fold for reaction with 4-methylphenyltrimethylsilane, compared with twofold for mercuriation of anisole.[97] This result is unexpected since the selectivity should be similar for substrates of similar reactivity, as is the case here. The participation of the cyclic structure **23** in the transition state for reaction with the neutral electrophile[98] may be a contributory factor.

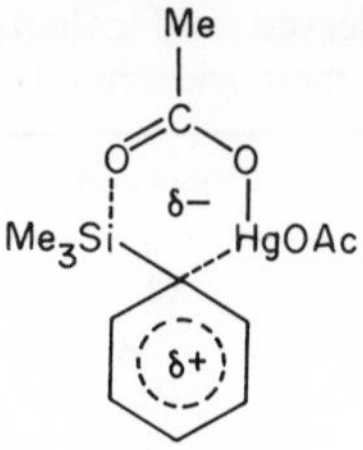

(23)

The relative rates of mercuridesilylation of 4-$MeOC_6H_4SiR_3$ at 25 °C were $R_3 = Me_3$, 190; Me_2CH_2Cl, 130; Me_2Ph, 69; $MePh_2$, 6.7; Me_2NHEt, 2.9; (4-$MeC_6H_4)_3$, 2.5; (3-$MeC_6H_4)_3$, 1.55: Ph_3, 1.0; (4-$ClC_6H_4)_3$, 0.5; and (3-$ClC_6H_4)_3$, 0.15.[99] For substituted aryl substituents the results are similar to those observed in protiodesilylation [Section 4.7.1.(i)], both sets of data arising from the relative stabilities of the leaving groups. However, between the Me_3 and Ph_3 compounds the rate difference is over 3 times greater than in protiodesilylation, showing that, as expected, steric hindrance to *ipso* substitution is much greater in mercuridesilylation.

Substitution effects have been obtained for the reaction carried out in either aqueous or glacial acetic acid (Table 10.4).[94,95]

There are two features of interest here:

(1) Activation by both *m*- and *p*-alkyl groups is in the inductive order. This is because there is little charge developed in the aromatic ring in the transition state, and consequently steric hindrance to solvation of this charge (which causes the reverse order, see Section 1.4.3) is not significant here.

(2) The *o*-Ph substituent activates slightly less than *p*-Ph. In mercuriation, the phenyl substituent exerts strong steric hindrance to *ortho* substitution (Section 5.3.2), whereas in protiodesilylation the *o*-Ph substituent leads to steric acceleration [Section 4.7.(ii)]. Mercuridesilylation should be subject to both

Table 10.4 Partial rate factors for mercuridesilylation

Substituent	*ortho*	*meta*	*para*	Conditions
Me	10.8	1.99	11.5	HOAc
	11.0	2.5	17.5	Aq. HOAc
Et	—	—	11.5	HOAc
i-Pr	—	3.96	12.0	HOAc
t-Bu	—	5.50	14.0	HOAc
Ph	—	0.68	2.73	HOAc
	2.5	0.58	3.3	Aq. HOAc
3:4-Benzo	2.9[a]			Aq. HOAc

[a]The 2-position of naphthalene.

Table 10.5 Observed and calculated partial rate factors for mercuridesilylation in HOAc

Substituent	Observed	Calculated
2-Me	11.3	—
3-Me	2.6	—
4-Me	10.7	—
2,3-Me_2	43.0	29.4
3,4-Me_2	27.2	26.9
3,5-Me_2	3.55	6.7
2,5-Me_2	24.3	26.9
2,4-Me_2	~160	121
2,6-Me_2	Very fast	128

steric hindrance and steric acceleration, and it is notable[100] that the value of $\log f_o^{Ph}:\log f_p^{Ph}$ of 0.77 is approximately the mean of the values obtained from mercuriation and protiodesilylation (0.80). Since all three reactions have closely similar selectivities and probably therefore have similar configuration at the transition state, this result suggests that steric hindrance and steric acceleration are closely balanced in this particular example of mercuridesilylation.

The *Additivity Principle* holds satisfactorily (Table 10.5),[92] except for the 2,3- and 2,6-dimethyl derivatives, which are more reactive than expected. Both discrepancies may be attributed to steric acceleration; for the 2,3-compound steric crowding in the ground state arises from a buttressing effect;[101] the low reactivity of the 3,5-dimethyl compound is anomalous.

Mercuridesilylation (and also halodesilyl- and -destannylation) has been used to make the corresponding 3-substituted benzcyclobutenes, which are otherwise difficult to obtain.[102]

10.32 THALLIODESILYLATION

The similarity between reactions of thallium and mercury is shown by the thalliodesilylation of arylsilanes by thallium(III) trifluoroacetate in TFA to give arylthallium bistrifluoroacetates.[103] The yields (ca 90%) are not as high as in mercuridesilylation because the trifluoroacetic acid causes concurrent protiodesilylation.

10.33 ALKYLDESILYLATION

The $AlCl_3$-catalysed reaction between phenyltrimethylsilane and alkyl halides results in alkyldesilylation, e.g. benzyl bromide gives diphenylmethane (67%)

(eq. 10.28).[104]

$$PhSiMe_3 + PhCH_2Br \xrightarrow{AlCl_3} PhCH_2Ph + SiMe_3Br \qquad (10.28)$$

10.34 ACYLDESILYLATION

The $AlCl_3$-catalysed reaction of arylsilicon compounds with acid chlorides gives aromatic ketones (eq. 10.29). If water is used instead of the chlorides, then the aromatic hydrocarbon is produced.

$$ArSiCl_3 + RCOCl \xrightarrow{AlCl_3} ArCOR + SiCl_4 \qquad (10.29)$$

It was believed originally that arylaluminium compounds (eq. 10.30) are intermediates in these reactions, and that these subsequently react as in eqs 10.31 and 10.32 to give the observed products.[105] This was disproved by the observation that only catalytic quantities of $AlCl_3$ are required; if eq. 10.32 applied the $AlCl_3$ would be consumed.[106] Moreover, if water is rigidly excluded, then no aromatic hydrocarbon is produced, showing that no arylaluminium intermediate is formed.[107] In the presence of water, the aryl hydrocarbon arises from protiodesilylation of the aryl silicon compound by the very strong acid $HAlCl_4$ (formed from $AlCl_3$, and HCl which is produced by hydrolysis of $AlCl_3$).

$$ArSiCl_3 + AlCl_3 \longrightarrow ArAlCl_2 + SiCl_4 \qquad (10.30)$$

$$ArAlCl_2 + RCOCl \longrightarrow ArCOR + AlCl_3 \qquad (10.31)$$

$$ArAlCl_2 + H_2O \longrightarrow ArH + HOAlCl_2 \qquad (10.32)$$

The most commonly observed reaction is acetyldesilylation (eq. 10.29, R = Me). It has been shown that acetyldesilylation does not take place by protiodesilylation followed by acetylation, or indeed by acetylation followed by protiodesilylation, since trichloro-*m*-tolylsilane gave only *m*-methylacetophenone; if reaction occurred through either of these paths, other isomers would be obtained.[107] A four-centre cyclic transition state (**24**) was proposed for the reaction.

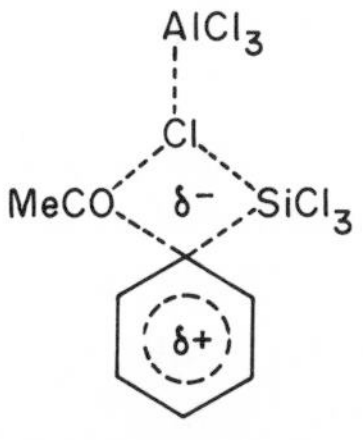

AlCl3

Cl

MeCO δ− SiCl3

δ+

(**24**)

Acetyldesilylation of $PhSiMe_3$ takes place 3600 times faster than acetylation of a position in benzene. This *ipso* factor is similar to that which applies to protiodesilylation (10^4).[104]

10.35 DIAZODESILYLATION

Triphenyl(4-*N*,*N*-dimethylaminophenyl)silane undergoes diazodesilylation on treatment with 4-nitrophenyldiazonium ion (eq. 10.33), but the reaction takes place less readily than diazonium coupling of *N*,*N*-dimethylaniline.[108] This suggests that steric hindrance to *ipso* substitution is substantial here.

$$4\text{-}Me_2NC_6H_4SiPh_3 + 4\text{-}O_2NC_6H_4N_2{}^+X^- \longrightarrow 4\text{-}Me_2C_6H_4N{=}NC_6H_4NO_2\text{-}4 + SiPh_3X \quad (10.33)$$

10.36 NITROSODESILYLATION AND NITRODESILYLATION

Many of the processes that are described as nitrodesilylations (eq. 10.34) take place by nitrosodesilylation followed by oxidation. This appears to arise because, as noted in the Introduction to Chapter 10, there is high steric hindrance to nitrodesilylation, which parallels that in nitrode-*tert*-butylation (Section 10.23). Theoretical calculations indicate that the nitrosonium ion may interact with the aromatic π-cloud (in a way that the nitronium ion can not) before moving to form the σ-complex.[109] This would create minimal steric interactions, as previously suggested by Taylor.[1]

$$ArSiR_3 + HNO_3 \longrightarrow ArNO_2 + HOSiR_3 \quad (10.34)$$

Nitrodesilylation[110] can be complicated by associated protiodesilylation in strongly acidic media, although the former is the faster reaction since, 1,4-bis(trimethylsilyl)benzene reacts with HNO_3 in Ac_2O to give 80% of trimethyl-4-nitrophenylsilane.[111] Protiodesilylation can be avoided by use of copper nitrate in acetic anhydride, and under these conditions nitrodesilylation takes place less readily than nitrodeprotonation,[112] the *ipso* factor being 0.66.[113]

The importance of nitrosodesilylation was shown in the reaction with HNO_3 in Ac_2O.[114] A solution prepared and used at 15 °C gave a 'nitro' desilylation:nitration ratio of 0.64. This was reduced to 0.14 by the addition of urea, increased to 4.8 by the addition of nitrous fumes, and increased to 90 by preheating the reagent briefly at 100 °C (**Care!**). Moreover, a solution of nitrous fumes in Ac_2O gave only nitrosodesilylation, and under these conditions a 4-Me substituent activates ca 50 times.

Sodium nitrite (or N_2O_3) in TFA also produces nitrosodesilylation, but with

ArH nitrodeprotonation occurs. Cleavage of organo-mercury, -thallium, -tin, -lead, or -bismuth groups also gives nitrosoaromatics.[17] Presumably the severer conditions required for nitrodeprotonation facilitate the subsequent oxidation.

10.37 SULPHODESILYLATION

Aryltrimethylsilanes and aryltriethylgermanes are cleaved by SO_3 in CCl_4 to give compounds of the type $ArSO_2OMR_3$.[115] The electrophilic nature of the reactions is indicated by reaction of 1,4-bis(triethylgermyl)benzene with SO_3, which leads to the replacement of only one of the Et_3Ge substituents, since the introduction of one sulphonic ester group deactivates the product to further substitution.

The products, $ArSO_2OMR_3$, are readily hydrolysed to the corresponding sulphonic acids, and this gives the reactions preparative value for it is possible to prepare (via the Grignard reagent) a specific arylsilicon (or arylgermanium) compound which may then be converted into the sulphonic acid; preparation of certain sulphonic acids (e.g. *m*-tolylsulphonic acid) is otherwise difficult.

Trimethylsilylchlorosulphonate, Me_3SiSO_3Cl, has also been used as a sulphonating reagent in this reaction.[116]

10.38 HALODESILYLATION

Aryl—silicon bonds are cleaved by halogen (eq. 10.35), and the reaction shows many similarities to halogenodeprotonation. Thus the reaction is first order in arylsilane, second order in bromine for bromination in CCl_4,[117] mixed first and second order for bromination by bromine in acetic acid,[94,95,118–120] first order for both bromination by BrCl[120] and chlorination by Cl_2,[119,120] and second order for iodination by ICl or IBr.[120] Reactions with Br_2, ICl, and IBr are complicated by trihalide formation. For example, in bromination by Br_2 the order in bromine decreases as a kinetic run proceeds[94,95] (and to a similar extent to that observed in molecular bromodeprotonation), owing to reaction of bromine with Br^- (formed in the reaction), to give the unreactive tribromide ion.

$$ArSiR_3 + Hal_2 \longrightarrow ArHal + HalSiR_3 \tag{10.35}$$

For bromination by Br_2, the order in bromine increases with both increasing concentration and decreasing reactivity of the aromatic. The former effect may be attributed to decreasing polarity of the reaction medium, whereas the latter is due to the need to increase the reactivity of the electrophile through polarization by additional molecules of bromine (see Section 9.3.2). However, for a given aromatic the kinetic order is lower in bromodesilylation than for bromodeprotonation, whilst the reaction takes place faster by a factor of $2 \times 10.^8$ This large *ipso* factor reflects the weak electrophile involved, so that bromodesilylation is greatly facilitated by the weakness of the C—Si bond. The rate difference between chlorodesilylation and chlorodeprotonation can be estimated from literature data[119] to be ca 10^3–10^4, and the smaller factor here accords with the more reactive electrophile involved in molecular chlorination. Likewise, bromodesilylation of trimethylphenylsilane occurs only 2–3 times slower than chlorodesilylation, whereas a much greater reactivity difference applies in halodeprotonation. Iododesilylation by ICl takes place about eight times faster than chlorodesilylation, whereas iododeprotonation takes place about 200 times less readily than chlorodeprotonation;[121] the reason for this difference is unclear.

Bromodesilylation of methyl-1-naphthyl-(4-methoxyphenyl)silane by Br_2 in benzene of carbon tetrachloride takes place with inversion of configuration at silicon. This shows that the organosilyl group is not detached from the aromatic ring as a free siliconium ion, but that (from the fact of *inversion* of configuration) nucleophilic attack at silicon assists cleavage of the aryl—silicon bond. The reaction therefore cannot proceed through a four-centre transition state (**25**), since this would require retention of configuration.[122,123] The results are, however, consistent with a six-centre transition state (**26**), which could follow from the high kinetic order in bromine that applies in non-polar solvents.

(**25**) (**26**)

The relative rates of bromodesilylation of $4\text{-}MeOC_6H_4SiR_3$ compounds by Br_2 in 98.5% aqueous HOAc at 25 °C are as follows: $R_3 = Me_3$, 1100; Me_2Ph, 355; $MePh_2$, 15; $(PhCH_2)_3$, 30; Me_2CH_2Cl, 65; $(i\text{-}Pr)_3$, ca 5; $(EtO)_3$, 8.5; $(4\text{-}MeOC_6H_4)_3$, 89; $(3\text{-}MeC_6H_4)_3$, 1.5; Ph_3, 1.0; $(4\text{-}ClC_6H_4)_3$, 0.12; $(3\text{-}ClC_6H_4)_3$, 0.052; and $(2\text{-}MeC_6H_4)_3$, 0.023.[94,95,99,123] These show the same general features as in both protiodesilylation or mercuridesilylation, but a larger rate spread [consistent with the larger ρ factor for the reaction (see below)]. In particular,

Table 10.6 Partial rate factors for bromodesilylation of $ArSiMe_3$

Substituent in Ar	*f*	Substituent in Ar	*f*	Substituent in Ar	*f*
2:3-Benzo	195	3:4-Benzo	11.5	4-F	0.68
2-Me	81.5	$3\text{-}CH_2SiMe_3$	8.5	3-Ph	0.41
4-Me	48.8	$4\text{-}SiMe_3$	3.05	4-Cl	0.092
4-Et	45.4	3-Me	2.9	4-I	0.088
4-*i*-Pr	32.5	2-Ph	1.81	4-Br	0.071
4-*t*-Bu	29.2	H	1.0	3-Cl	0.003
4-Ph	12.5				

steric hindrance to bromodesilylation is evident as shown by the relative rates of the trimethyl- and triisopropylsilanes, and the deactivation by a 2-Me substituent in the phenyl rings. For compounds $PhSiX_3$ (X = F, Cl), the rates of bromodesilylation are slower than for bromodeprotonation.[124]

Partial rate factors for bromodesilylation in 98.5% aqueous HOAc at 25 °C (Table 10.6) correlate very satisfactorily with the Yukawa–Tsuno equation (Section 11.1.3), with $\rho = -6.8$ and $r = 0.79$, confirming that the transition state occurs earlier along the reaction coordinate than that for bromodeprotonation.

The effect of steric acceleration in the reaction is evident from two sets of data. First, 2-Me activates more strongly than 4-Me. Second, the ratio of the logarithms of the reactivities of the 1- and 2-positions in naphthalene is 2.16, and this is greater than the value in a number of analogous reactions.

The ratio $\log f_1 : \log f_2$ for naphthalene is almost exactly the mean 2.14 of the values obtained in protiodesilylation (2.72) and molecular bromination (1.56), thus demonstrating a precise balance between steric hindrance and steric acceleration. This is not the case for substitution in biphenyl, for whereas protiodesilylation at the 2-position is sterically accelerated, bromodesilylation is hindered *more* than is bromodeprotonation, the respective $\log f_o^{Ph} : \log f_p^{Ph}$ values being 1.74, 0.45, and 0.235. An explanation consistent with the facts is the following. Biphenyl is non-planar in solution and becomes more coplanar in the transition state, especially in reactions with a substantial demand for resonance stabilization. This restricts access to one face of the phenyl ring at the 2-position. The leaving group in protiodesilylation and the entering group in bromodeprotonation are able to use the least hindered pathway. By contrast, in bromodesilylation one or other of the bulky groups must use the hindered pathway, so steric hindrance will be more severe than in the component reactions.

Bromodesilylation has been used to evaluate the transmission of electronic effects between the 2- and 7-positions in 1,6-methano[10]annulene. Bromodesilylation at the 2-position took place with increasing ease for a series of 7-substituents in the order $CHO < CO_2Me < Br < SiMe_3 < Me <$ O-*t*-Bu, and this is the expected order for electrophilic substitution.[125]

10.39 MERCURIDESTANNYLATION

The cleavage of arylstannanes by mercury(II) salts in methanol (eq. 10.36) is first order in each reagent, and the efficacy of the salts is $Hg(OAc)_2 > HgCl_2 > HgI_2 > HgI_3^-$.[126] The transition state is believed to be of low polarity, and this is confirmed by the small partial rate factors for the reaction of $ArSn(cyclohexyl)_3$ compounds with $Hg(OAc)_2$ in THF at 20 °C (Table 10.7).[127] These are smaller than for protiodestannylation (Section 4.9) and correlate poorly with σ values, giving $\rho \approx -3.0$.

$$PhSnMe_3 + HgX_2 \longrightarrow PhHgX + Et_3SnX \qquad (10.36)$$

Table 10.7 Partial rate factors for mercuridestannylation of XC_6H_4Sn-(cyclohexyl)$_3$

X	4-MeO	4-Me	3-Me	H	3-MeO	4-Cl	3-Cl
f	8.30	1.76	1.21	1.0	0.72	0.25	0.029

10.40 ACYLDESTANNYLATION

This reaction takes place more readily than acyldesilylation,[104,128] in accordance with the relative strengths of the C—Si and C—Sn bonds.

10.41 NITROSODESTANNYLATION

Cleavage of arylstannanes by nitrosyl chloride in dichloromethane at low temperature gives nitrosoaromatics (eq. 10.37), and is the optimum method of

$$ArSnMe_3 + NOCl \longrightarrow ArNO + Me_3SnCl \tag{10.37}$$

making the latter in view of the very mild reaction conditions used. Yields varied from 25 to 80%, being higher when electron-supplying groups are present in the aromatic ring, confirming that the reaction is an electrophilic substitution.[129] N_2O_3 in TFA has also been used as the electrophile.[17]

10.42 HALODESTANNYLATION

Aryl—tin bonds are cleaved by bromine and iodine (e.g. eq. 10.38). In carbon tetrachloride, the reaction with iodine is first order in aromatic and second order in iodine,[130] whereas in methanol it is only first order in iodine, the rate here being dependent on the ionic strength of the medium.[131]

$$ArSnR_3 + I_2 \longrightarrow ArI + ISnR_3 \tag{10.38}$$

With CCl_4 as solvent, the relative rates as R in SnR_3 is varied are as follows: R = C_6H_{11}, 5.4; Et, 5.1; Me, 1.0; Ph, 0.018. The greater reactivity of the cyclohexyl compound compared with the methyl compound is opposite to that obtained in protiodesilylation, degermylation, and destannylation. The order in the latter reactions was attributed to steric hindrance to solvation at the reaction site. Carbon tetrachloride is poorly solvating, and consequently steric hindrance to solvation will be unimportant in iododestannylation and the reversal of the order then follows. A second consequence of this poor solvation is noted below.

Partial rate factors have been determined for iododetricyclohexylstannylation in CCl_4[130] and for iododetrimethylstannylation in MeOH[131] (Table 10.8).

Table 10.8 Partial rate factors for iododestannylation of $ArSnR_3$

Substituent	$R = C_6H_{11}$	Me	Substituent	$R = C_6H_{11}$	Me
4-OMe	69	64	3-OMe	2.2	—
4-$Sn(C_6H_{11})_3$	20	—	4-$SiMe_3$	—	1.01
4-*t*-Bu	13.9	—	H	1.0	1.0
4-*i*-Pr	12.1	—	2-Ph	0.34	—
4-Et	10.1	—	4-F	0.22	—
4-Me	7.5	4.95	4-Cl	0.10	—
3-Me	4.2	1.5	4-Br	0.08	0.24
4-Ph	2.9	—	3-Cl	0.039	—
4-$SnMe_3$	—	1.71	4-CO_2H	0.0145	—

Whilst the spread of rates is consistent with an electrophilic substitution, no satisfactory correlation exists between the data obtained in CCl_4 and the Yukawa–Tsuno equation. The discrepancies are not attributable to the occurrence of iododeprotonation as a side-reaction, for it has been shown to be absent even for the 3-OMe compound, which would be the most susceptible to it. The imprecision of the correlation suggests a significant proportion of π-complex bonding in the transition state of the rate-determining step, and this is consistent with (1) the small spread of rates which indicates that the transition state should not be far displaced from the π-complex, and (2) the fact that aromatic compounds form π-complexes with halogens in CCl_4.[132] The mechanism under these conditions has therefore been proposed as shown in eq. 10.39, where the appropriate π-complex is **27**, in which iodine is associated with one bond in the aromatic ring.[130] Since the stability of the π-complex will depend on the electron density in this double bond and this is similarly situated with respect to both *meta* and *para* substituents, the enhanced reactivities shown by *meta* substituents and the reduced reactivities shown by *para* substituents (compared with those which would be required for a successful Yukawa–Tsuno correlation) are understandable.

The poor solvation by CCl_4 leads to the highest $\log f_p^{t\text{-Bu}} : \log f_p^{Me}$ ratio (1.31) obtained for an electrophilic substitution, and it is as high as that obtained in the gas phase;[133] the reverse order is obtained only in highly solvating media or if the electrophile has a large counterion (Section 1.4.3).

$$\xrightarrow{I_2} \longrightarrow + R_3SnI + I_2 \qquad (10.39)$$

(27)

Table 10.9 Partial rate factors for iododestannylation of $ArSnR_3$ in methanol

	Ar			
R	1-Naphthyl	2-Naphthyl	9-Phenanthrenyl	3-Pyrenyl
Me	2.52	2.33	1.46	—
n-Bu	3.72	2.40	—	12.3
i-Pr	4.06	1.81	—	—

For iodostannylation in methanol, the data plot satisfactorily using the Yukawa–Tsuno equation with $\rho = -2.96$ and $r = 0.65$, so demonstrating the difference between this reaction and that carried out in CCl_4. The greater activation by 4-$SnMe_3$ than by 4-$SiMe_3$ is notable (see also ref. 130).

Another difference from the results obtained in CCl_4 is the decrease in rates as R is made larger in $PhSnR_3$, being 1.0, 0.24, and 0.036 for R = Me, n-Bu, and *i*-Pr, respectively.[131] Under these good solvating conditions, steric hindrance to solvation of the transition state becomes important. As R is made larger, so steric acceleration from crowded positions should become more important. This is seen to be so from the partial rate factors obtained for polycyclics (Table 10.9). The log f_1:log f_2 ratios for naphthalene are 1.09, 1.50, and 2.36 for R = Me, n-Bu, and *i*-Pr, respectively.

Bromodestannylation of aryltrimethylstannes by bromine in methanol gives a smaller ρ factor (-2.58) than for iododestannylation.[134] The *ipso* factor here is 2×10^{12}, and thus considerably larger than in bromodesilylation, as expected in view of the relative strengths of the C—Si and C—Sn bonds.

10.43 IODODEPLUMBYLATION

Iododeplumbylation in methanol (eq. 10.40) is first order in plumbane and in iodine. The reaction occurs much more readily than iododestannylation under the same conditions,[135] so paralleling the relative reactivities of the stannanes and plumbanes towards protiodesilylation.

$$PhPbMe_3 + I_2 \longrightarrow PhI + IPbMe_3 \qquad (10.40)$$

10.44 DIAZONIUM EXCHANGE

Arylazo groups are able to displace other arylazo groups from a variety of aromatics (e.g. eq. 10.41).[46,136] The ability of a given group to carry out this displacement parallels its reactivity in diazodeprotonation,[136] showing that an electrophilic substitution is involved. A group can displace another only if it is more electrophilic than the leaving group.[136]

$$4\text{-}Me_2NC_6H_4NNPh + ArN_2X \longrightarrow 4\text{-}Me_2NC_6H_4NNAr + PhN_2X \quad (10.41)$$

10.45 NITROSODEARYLAZONIATION

This reaction occurs, e.g., on treatment of 4-dimethylaminoazobenzene with nitrous acid, which gives nitrosobenzene (eq. 10.42).[47]

$$4\text{-}Me_2NC_6H_4N{=}NPh + HONO \longrightarrow 4\text{-}Me_2NC_6H_4NO + PhN_2OH \quad (10.42)$$

10.46 NITRODEARYLAZONIATION

Nitration of 1-phenylazo-2-hydroxynaphthalene (**28**) gives 1,6-dinitro-2-hydroxynaphthalene (**29**).[137] The reaction is facilitated not only by the substantial intrinsic reactivity of the 1-position of naphthalene, but also by the very high activation by the OH group across the high-order 1,2-bond; the second nitro group enters the 6-position, which is the next most reactive site in 2-hydroxynaphthalene. A 1-(4-nitrophenyl)azo group could not be cleaved in the same way.[138] The reaction may proceed via nitrosodearylazoniation.

N=NPh OH NO₂ OH O₂N

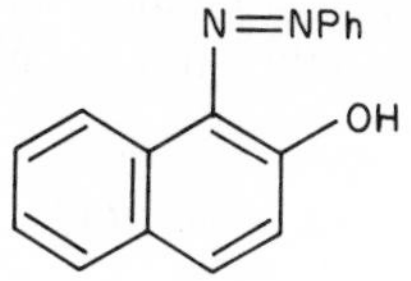

(**28**) 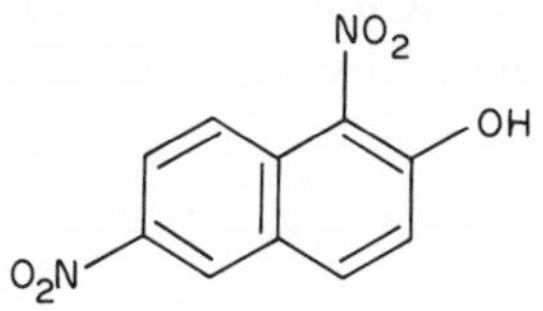 (**29**)

10.47 BROMODEARYLAZONIATION

The reaction of bromine in acetic acid with 1-phenylazo-2-hydroxynaphthalene gives rise to the corresponding 1,1-dibromoketone derivative, and clearly bromodearylazoniation must be involved as the first step.[138]

10.48 DIAZODENITRATION

This reaction, which may also be described as arylazodenitration, has been found to take place between 1-nitro-2-hydroxynaphthalene and 4-nitrophenyldiazonium ions.[138] It is the reverse of reaction 10.47, and is favoured by the high reactivity of the electrophile and the substrate.

10.49 NITRODENITRATION (NITRO EXCHANGE)

Examples of nitro groups attacking *ipso* positions bearing nitro groups have been observed.[139] Exchange of the nitro groups has been shown by labelling experiments in nitrous acid-catalysed nitration of 4-nitrophenol to 2,4-dinitrophenol. Part of the reaction (involving a radical pair) proceeds via *ipso* substitution at the 4-position, followed by 1,3-migration (believed to be intramolecular) of one of the nitro groups to the 2-position (eq. 10.43).[140]

$\xrightarrow{HNO_3-TFA}$ (10.43)

10.50 NITRODEPHOSPHONATION

Nitration of 2- and 4-methoxyphenyl phosphonic acid by nitric acid in acetic anhydride gives the corresponding 2- and 4-nitroanisoles in 3 and 5% yields, respectively. The reaction (eq. 10.44) is indicated to be an electrophilic substitution since benzenephosphonic acid itself gave only a trace of nitrobenzene.[141]

$$ArPO_3H_2 + HNO_3 \longrightarrow ArNO_2 + H_3PO_4 \qquad (10.44)$$

10.51 NITRODETHIOPHENYLATION

Nitrodethiophenylation accompanies nitration of diphenyl sulphides. Thus 4-acetylaminodiphenyl sulphide gives 4-nitroacetanilide; the 4-*N*,*N*-dimethylamino derivative similarly gives 4-nitro-*N*,*N*-dimethylaniline.[142]

10.52 DIAZODESULPHONATION

Reaction of naphthalene-1- or -2-sulphonic acid with 4-chlorophenyldiazonium ions gives replacement of the sulphonic acid group by the diazo group (eq. 10.45) in yields which can be quantitative, depending on the solvent.[143] Similar reactions have also been observed with reactive heteroaromatics.[144]

$$ArSO_3H + 4\text{-}ClC_6H_4N_2X \longrightarrow 4\text{-}ClC_6H_4N_2Ar + HSO_3X \qquad (10.45)$$

10.53 NITRODESULPHONATION

Nitration of aromatic sulphonic acids can give nitrodesulphonation when the sulphonic acid group is attached to a carbon activated towards electrophilic substitution.[48,79a,145,146] The examples shown are illustrative.

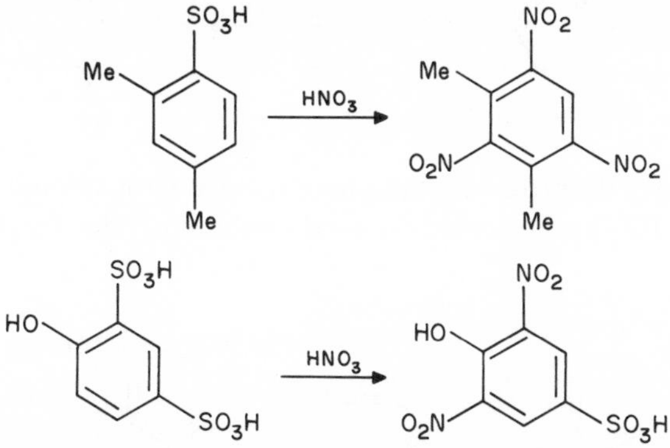

Kinetic studies have not been reported, and it is not known whether the reaction also takes place via protiodesulphonation followed by nitration. Some indication that two processes are involved is provided by the variation of the nitration:nitrodesulphonation ratio with acid concentration, in reaction of 4-hydroxybenzenesulphonic acid.[146] In general, nitration occurs more readily than, and therefore precedes, nitrodesulphonation.

10.54 HALODESULPHONATION

There are numerous examples of the replacement of the sulphonic acid group in an arylsulphonic acid, by chlorine, bromine, or iodine.[147] Reaction occurs when the sulphonic acid group is *ortho* or *para* to alkyl, OH, OR, or NH_2, showing it to be an electrophilic substitution. Bromodesulphonation is faster for an *ortho*- than for a *para*-substituted arylsulphonate,[148] showing the reaction to be sterically accelerated since the halogen is smaller than the sulphonic acid group it replaces. Typical reactions are shown in eqs 10.46,[149] 10.47,[150,151] and 10.48.[148]

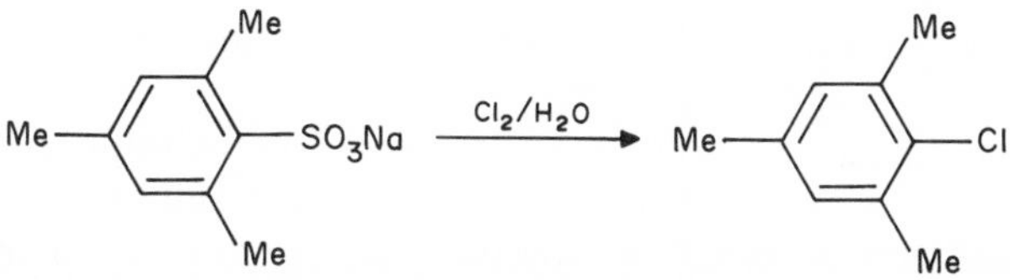

(10.46)

$$MeO-C_6H_4-SO_3Na \xrightarrow{Br_2/H_2O} MeO-C_6H_4-Br \qquad (10.47)$$

$$H_2N-C_6H_2Br_2-SO_3H \xrightarrow{I_2/AcOH} H_2N-C_6H_2Br_2-I \qquad (10.48)$$

Investigation of the bromodesulphonation of aromatic sulphonate salts by Br_2 in aqueous $HClO_4$ has indicated the mechanism illustrated by eqs 10.49–10.51.

$$HO-C_6H_2Br_2-SO_3^- \underset{}{\overset{fast}{\rightleftharpoons}} {}^-O-C_6H_2Br_2-SO_3^- + H^+ \qquad (10.49)$$

$${}^-O-C_6H_2Br_2-SO_3^- + Br_2 \overset{fast}{\rightleftharpoons} \textbf{30} + Br^- \qquad (10.50)$$

$$\textbf{30} \xrightarrow{slow} {}^-O-C_6H_2Br_3 + SO_3 \qquad (10.51)$$

Evidence for this as follows.

(1) The reaction is first order in bromine and in sulphonate.

(2) Addition of acid decreases the reaction rate.

(3) When dilute aqueous solutions of bromine and the sulphonate are mixed, the bromine colour and the u.v. spectrum of the sulphonate immediately disappear, and are replaced by spectra characteristic of compounds with structures similar to **30**. Immediate addition of iodide ion gives an iodine titre

equal to the amount of bromine added, and this is consistent with eq. 10.50 being fast and reversible. Likewise, addition of NaBr reduces the reaction rate due to the removal of bromine as tribromide ion, and therefore a decrease in the concentration of **30**.

(4) The spectrum ascribed to **30** is slowly replaced by that of tribromophenol, and the addition of iodide ion after various intervals gives an iodine titre which decreases with time. Bromine is therefore consumed in a slow irreversible step (eq. 10.51).

(5) For bromodesulphonation of compounds which give intermediates much less stable than **30**, eq. 10.50 becomes rate-determining.

The mechanism has also been shown to be stepwise rather than synchronous, since sulphur kinetic isotope effects for cleavage of the sulphonic acid group from sodium 4-methoxybenzenesulphonate depended on bromide ion concentration, arising from the reversal of eq. 10.50.[152]

10.55 HALODETELLURIATION

Reaction of iodine or bromine with aryl- or diaryltellurium(IV) compounds results in cleavage of the C—Te bond by halogen, e.g. eq. 10.52 ($X = Cl$, OAc, or $OCOCF_3$). The reaction occurs readily with $R = MeO$ or Me, but poorly with $R = H$ or Br, indicating an electrophilic substitution.[153].

$$4\text{-}RC_6H_4TeX_3 + I_2 \longrightarrow 4\text{-}RC_6H_4I + ITeX_3 \qquad (10.52)$$

10.56 LITHIODEBROMINATION

Reaction of n-butyllithium with substituted bromobenzenes in hexane gave replacement of bromine by lithium, the substituent effects giving a Hammett correlation, with $\rho \approx 2.0$.[154]. The transition state is thus *negatively* charged, and this is a clear example of an *ipso* substitution taking place via the S_E1 mechanism (cf. Section 5.1), although some synchronous character may be involved.

10.57 ALKYLDEBROMINATION

In the preparation of 1-(4-bromophenyl)adamantane by the $AlCl_3$-catalysed reaction of adamantyl bromide with bromobenzene at low temperature, 1,4-diadamantylbenzene was produced as a byproduct, as a result of alkyldebromination (eq. 10.53).[155]

$$PhBr + 2AdBr \xrightarrow{AlCl_3} 4\text{-}AdC_6H_4Ad + HBr + Br_2 \qquad (10.53)$$

Methyldehalogenation (which shows the leaving group order Br > Cl ≫ F) has been observed in the gas phase, but the mechanism differs from that in solution. Attack by CH_5^+ gives HX, an arenium ion, and CH_4, the last two species then combining to give the normal intermediate.[156]

10.58 ACYLDEHALOGENATION

Acetylation of 4-bromo-*m*-xylene is accompanied by some acetyldebromination; the displaced Br^+ reacts further with starting material to give 4,6-dibromo-*m*-xylene.[157] The reaction may occur via protiodebromination followed by acetylation. Benzoyldechlorination, observed in $AlCl_3$-catalysed benzoylation of 1,4-dichlorobenzene, arises mainly from protiodechlorination followed by benzoylation, but true benzoyldechlorination also occurs. The latter gives rise to abnormal products, e.g. 3,4-dichlorobenzophenone, attributed to 1,3-migration of chloronium ion from the *ipso* intermediate.[158]

Reaction of 1,3-dibromoazulene with DMF–$POCl_3$ results in replacement of one or both of the bromines by CHO groups.[159]

10.59 DIAZODEHALOGENATION

Rates of diazodehalogenation of 1-halo-2-hydroxynaphthalene-6-sulphonic acids (relative to the non-halogenated compound) by 4-chlorodiazonium ions are 0.0070 (Cl), 0.0089 (Br), and 0.149 (I).[160] These relative rates are almost exactly the same as those found for nitro-(or nitroso-)dehalogenation of 4-haloanisoles (Section 10.60), implying a common feature for both reactions. These relative rates may therefore be either (1) a measure of the leaving group abilities (which requires that the second step of the reaction is rate-determining), or (2) a measure of the ease of attachment of the electrophile to the carbon bearing the halogens. Cleavage of bromine was claimed to be rate-determining, but cleavage of the other carbon—halogen or carbon—hydrogen bonds was not; this curious result may reflect experimental difficulties.

10.60 NITROSO- OR NITRODEHALOGENATION

Halogens in aromatics in positions activated towards electrophilic substitution can be replaced by nitro groups. The relative rates of cleavage of the carbon—halogen bonds by nitric acid in 4-haloanisoles follows the order 0.18 (I), 0.079 (Br), 0.061 (Cl), and 1.0 (H).[161] This order [quantitatively very similar to that found for diazodehalogenation (Section 10.59)] suggests that nitrodechlorination should be relatively rare, which is the case. Nitration of pentachlorobenzene gives rise to 2% of the hexachloro derivative,[162] which may arise from *ipso* substitution followed by intermolecular migration of chloronium ion (cf. benzoyldechlorination, Section 10.58). Nitrodechlorination accompanied by a 1,3-shift of chloronium ion accounts for the results shown in eq. 10.54.[163] Nitrodechlorination of 3,4,6-trichloro-*o*-xylene gives rise to 4,6-dichloro-3,5-dinitro-*o*-xylene.[164] Nitrodebromination is more common than nitrodechlorination, and shows behaviour similar to that shown in eq. 10.54.[49,77,81,165,166]

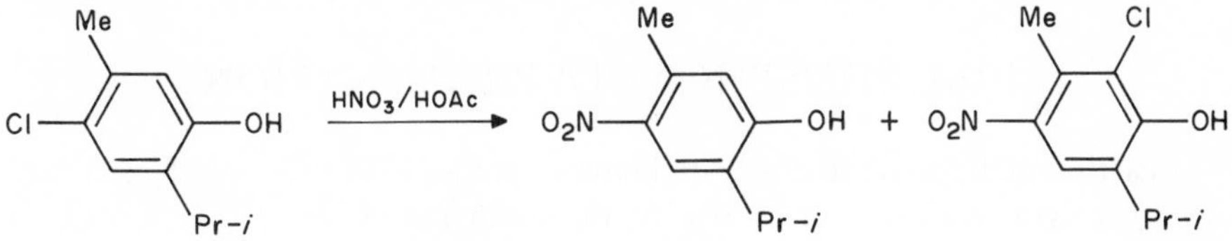

(10.54)

The above reactions may in fact take place via nitrosodehalogenation followed by oxidation. This pathway will certainly be more probable the weaker is the carbon—halogen bond and the more reactive is the aromatic. The parallel between the rates of dehalogenation of the 4-haloanisoles and rates of diazodehalogenation noted above suggests that nitrosodehalogenation is involved, since the aryldiazonium and nitrosonium ions generally show very similar behaviour in electrophilic substitution. However, it is significant that neither 2-bromo- or -chloroanisoles nor 2-bromo-*m*-xylene can be nitrated at the *ipso* position, whereas this was readily observed with 1,4-dibromobenzene and 4-bromotoluene.[166,167] This is consistent with steric hindrance to *ipso* substitutions involving the nitro group, evident in other reactions (see, e.g., Sections 10.23 and 10.36).

For cleavage of iodine the nitroso pathway has been established. Thus, reaction of 4-iodoanisole with nitric acid gives 4-nitroanisole, with migration (probably intramolecularly) of the displaced iodine to the 2-position giving 2-iodo-4-nitroanisole as a further product.[168] Intermolecular migration is apparently also involved since 2,4-diiodoanisole is also obtained (eq. 10.55).[168]

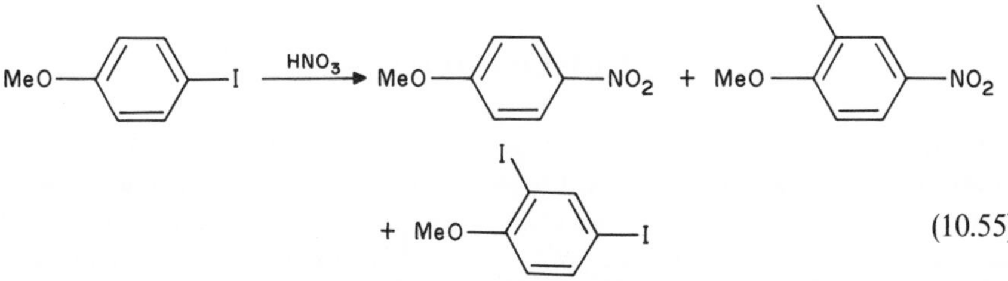

(10.55)

Kinetic isotope effects for a range of iodoaromatics studied showed that nitrodeiodination does not take place via slow protiodeiodination followed by nitrodeprotonation. The absence of any effect of urea on the nitrodeiodination:nitrodeprotonation ratio led to the belief that nitrosodeiodination was not involved, except for the most reactive compounds.[169,170] However, urea is not vary effective as a nitrous trap and, moreover, the rate of deiodination of some compounds at least is increased on addition of nitrite.[170,171] Other reports of nitrodeiodination confirmed the above general details.[172]

10.61 PHOSPHONODEBROMINATION

Reaction of diphenyl ethers or diphenyl sulphides with PCl_3–$AlCl_3$, followed by hydrolytic work-up, gives rise to phosphinylation (eq. 10.56, X = O, S). However, reaction of the corresponding compounds with Z = 4-Br is accompanied by phosphonodebromination to give the phosphonic acids **31**. An electrophilic mechanism is indicated by the failure of the 3-bromo compounds to undergo the same reaction. If reaction is carried out on 4,4′-dimethyldiphenyl sulphide, then the carbon—sulphur bond is also cleaved to give 4-methylphenylphosphonic acid (**32**).[173]

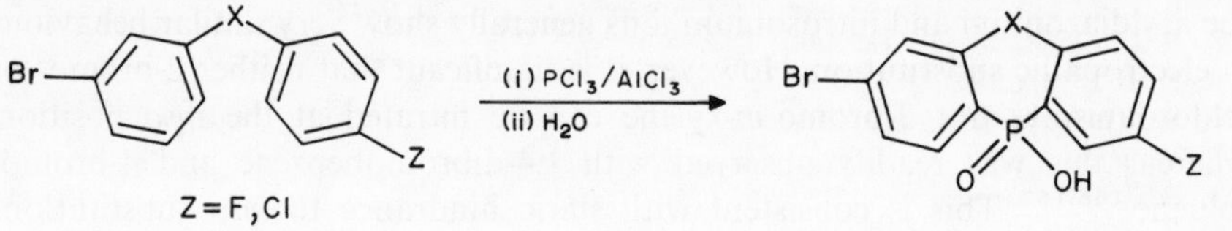

Br–C6H4–X–C6H4–Z (Z = F, Cl) —(i) $PCl_3/AlCl_3$; (ii) H_2O→ cyclic phosphinic acid (Br, X, Z; P(=O)OH)

(10.56)

Br–C6H4–X–C6H4–P(=O)(OH)2 **(31)**   Me–C6H4–P(OH2)(=O) **(32)**

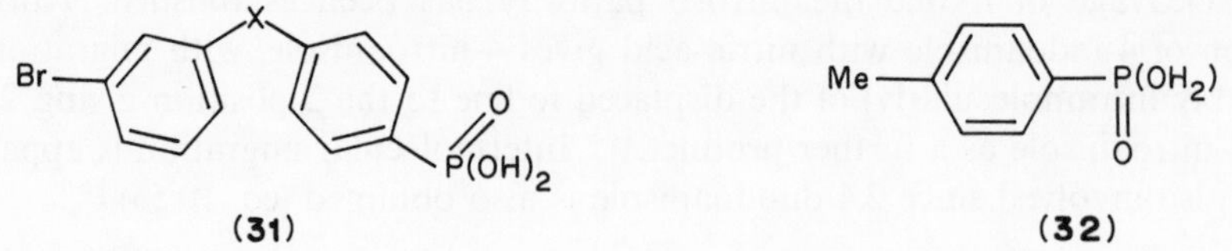

10.62 SULPHODEIODINATION

Both 2- and 4-iodotoluenes undergo sulphodeiodination, but this is more important for the *para* isomer, because steric acceleration (which would normally favour *ortho* substitution) is outweighed by increased steric hindrance due to the bulk of the entering electrophile. With SO_3 in $MeNO_2$ the reaction takes place directly, but with 98 wt-% H_2SO_4 part of the overall reaction at least occurs via protiodeiodination, followed by sulphonation.[174]

10.63 HALOGEN EXCHANGE

In the gas phase, both Br^+ and I^+ are able to displace halogen from halobenzenes, the ease of replacement being Br > Cl > F. Parallel with this replacement was an increase in the amount of *ortho* product formed, which suggests that much of the latter is derived through a 1,2-shift from an *ipso* intermediate.[175]

REFERENCES

1. R. Taylor, *Specialist Periodical Report on Aromatic and Heteroaromatic Chemistry*, Chemical Society, London, 1976, Vol. 4, p. 259.
2. C. Eaborn and D. R. M. Walton, *J. Chem. Soc.*, 1963, 5626.
3. C. Eaborn, *Organosilicon Compounds*, Butterworths, London, 1960, pp. 19–25.
4. C. Eaborn, A. A. Najam, and D. R. M. Walton, *J. Chem. Soc., Chem. Commun.*, 1972, 840.
5. E. G. Perevalova, D. A. Lemenovskii, K. I. Grandberg, and A. N. Nesmeyanov, *Proc. Acad. Sci. USSR*, 1971, **199**, 643.
6. B. Oddo, *Gazz. Chim. Ital.*, 1909, **39**, I, 659.
7. R. E. Dessey and R. K. Lee, *J. Am. Chem. Soc.*, 1960, **82**, 689.
8. R. E. Dessey, Y. K. Lee, and J.-Y. Kim, *J. Am. Chem. Soc.*, 1961, **83**, 1163.
9. H. B. Charman and C. K. Ingold, *J. Chem. Soc.*, 1959, 2523, 2530.
10. S. Winstein, T. G. Traylor, and C. S. Garner, *J. Am. Chem. Soc.*, 1955, **77**, 3741; S. Winstein and T. G. Traylor, *J. Am. Chem. Soc.*, 1956, **78**, 2597.
11. A. N. Nesmeyanov and O. A. Reutov, *Proc. Acad. Sci. USSR*, 1962, **144**, 405; *Tetrahedron*, 1964, **20**, 2803.
12. F. R. Fensen and J. Miller, *J. Am. Chem. Soc.*, 1964, **86**, 4735; I. P. Beletskaya, G. A. Artamkina, and O. A. Reutov, *Proc. Acad. Sci. USSR*, 1966, **166**, 242.
13. I. I. Zakharycheva, I. P. Beletskaya, and O. A. Reutov, *J. Org. Chem. USSR*, 1969, **5**, 2023, 2028.
14. J. R. Kalman, J. T. Pinhey, and S. Sternhell, *Tetrahedron Lett.*, 1972, 5369.
15. L. I. Smith and F. L. Taylor, *J. Am. Chem. Soc.*, 1935, **57**, 2460.
16. L. M. Stock and T. L. Wright, *J. Org. Chem.*, 1977, **42**, 2875.
17. S. Uemura, A. Toshimitsu, and M. Okano, *J. Chem. Soc., Perkin Trans. 1*, 1978, 1076.
18. S. Uemura, A. Toshimitsu, and M. Okano, *Bull. Chem. Soc. Jpn.*, 1976, **49**, 2582.
19. G. B. Deacon and G. J. Farquharson, *Aust. J. Chem.*, 1976, **29**, 627.
20. O. Itoh, H. Taniguchi, A. Kawabe, and K. Ichikawa, *Kogyo Kagaku Zasshi*, 1966, **69**, 913.
21. I. P. Beletskaya, L. V. Savinykh, and O. A. Reutov, *J. Organomet. Chem.*, 1971, **26**, 13.
22. K. Lauer, *J. Prakt. Chem.*, 1933, **138**, 81.
23. E. C. Taylor, R. H. Danforth, and A. McKillop, *J. Chem. Soc., Perkin Trans. 2*, 1973, **38**, 2088.
24. B. Davies and C. B. Thomas, *J. Chem. Soc., Perkin Trans. 1*, 1975, 65.
25. W. Hanstein and T. G. Traylor, *Tetrahedron Lett.*, 1967, 4451.
26. G. K. Chip and J. S. Grossert, *J. Chem. Soc., Perkin Trans. 1*, 1972, 1629.
27. F. Challenger and B. Parker, *J. Chem. Soc.*, 1931, 1462.
28. A. McKillop *et al.*, *J. Am. Chem. Soc.*, 1971, **93**, 4841.

29. N. Ishikawa and A. Sekiya, *Bull. Chem. Soc. Jpn.*, 1974, **47**, 1680.
30. A. Michaelis and P. Becker, *Chem. Ber.*, 1882, **15**, 180.
31. H. G. Kuivila and T. C. Muller, *J. Am. Chem. Soc.*, 1962, **84**, 377.
32. A. D. Ainley and F. C. Challenger, *J. Chem. Soc.*, 1930, 2171.
33. D. R. Harvey, and R. O. C. Norman, *J. Chem. Soc.*, 1962, 3822.
34. H. G. Kuivila, *J. Am. Chem. Soc.*, 1954, **76**, 870; 1955, **77**, 4014; H. G. Kuivila and R. A. Wiles, *J. Am. Chem. Soc.*, 1955, **77**, 4830; H. G. Kuivila and A. G. Armour, *J. Am. Chem. Soc.*, 1957, **79**, 5659.
35. H. C. Brown and Y. Okamoto, *J. Am. Chem. Soc.*, 1958, **80**, 4979.
36. H. G. Kuivila and E. K. Easterbrook, *J. Am. Chem. Soc.*, 1951, **73**, 4629; H. G. Kuivila and A. R. Hendrickson, *J. Am. Chem. Soc.*, 1952, **74**, 5068; H. G. Kuivila and E. J. Soboczenski, *J. Am. Chem. Soc.*, 1954, **76**, 2675; H. G. Kuivila and R. M. Williams, *J. Am. Chem. Soc.*, 1954, **76**, 2679; H. G. Kuivila and L. E. Benjamin, *J. Am. Chem. Soc.*, 1955, **77**, 4834; H. G. Kuivila, L. E. Benjamin, C. J. Murphy, A. D. Price, and J. H. Polevy, *J. Org. Chem.*, 1962, **27**, 825.
37. Y. Yukawa and Y. Tsuno, *Bull. Chem. Soc. Jpn.*, 1959, **27**, 825.
38. R. W. Bott and C. Eaborn, *J. Chem. Soc.*, 1963, 2139.
39. C. A. Holder and R. Taylor, unpublished results.
40. Yu. V. Pozdnyakovich, G. B. Kondratova, and S. M. Shein, *J. Org. Chem. USSR*, 1982, **18**, 2313.
41. A. H. Jackson, P. V. R. Shannon, and P. W. Taylor, *J. Chem. Soc., Perkin Trans. 2*, 1981, 286.
42. D. V. Nightingale, H. B. Hucker, and O. L. Wright, *J. Org. Chem.*, 1953, **18**, 244.
43. G. F. Hennon and S. F. de. C. McCleese, *J. Am. Chem. Soc.*, 1942, **64**, 242.
44. H. Hopff and A. K. Wick, *Helv. Chim. Acta*, 1960, **43**, 1473.
45. M. Stiles and A. J. Sisti, *J. Org. Chem.*, 1960, **25**, 1691; A. J. Sisti, J. Burgmaster, and M. Fudim, *J. Org. Chem.*, 1962, **27**, 279; A. J. Sisti, J. Sawinski, and R. Stout, *J. Chem. Eng. Data*, 1964, **9**, 108.
46. M. Colonna, L. Greci, and M. Poloni, *J. Chem. Soc., Perkin Trans. 2*, 1982, 455.
47. M. Colonna, L. Greci, and M. Poloni, *J. Chem. Soc., Perkin Trans. 2*, 1984, 165.
48. D. V. Nightingale, *Chem. Rev.*, 1947, **40**, 117.
49. A. K. Manglik, R. B. Moodie, K. Schofield, G. D. Tobin, R. G. Coombes, and P. Hadjigeorgiou, *J. Chem. Soc., Perkin Trans. 2*, 1980, 1606.
50. A. Fischer and G. J. Wright, *Aust. J. Chem.*, 1974, **27**, 217.
51. G. A. Olah and S. J. Kuhn, *J. Am. Chem. Soc.*, 1964, **86**, 1067.
52. K. Bantel and H. Musso, *Chem. Ber.*, 1969, **102**, 696.
53. A. Newton, *J. Am. Chem. Soc.*, 1943, **65**, 2434, 2439.
54. (a) P. C. Myhre, M. Beug, and L. L. James, *J. Am. Chem. Soc.*, 1968, **90**, 2105; (b) A. J. Hoefnagel, J. H. A. Nonnink, A. van Veen, P. E. Verkade, and B. M. Wepster, *Recl. Trav. Chim. Pays-Bas*, 1969, **88**, 386.
55. (a) P. C. Myhre and M. Beug, *J. Am. Chem. Soc.*, 1966, **88**, 1568, 1569; (b) P. C. Myhre, M. Beug, K. S. Brown, and B. Östman, *J. Am. Chem. Soc.*, 1971, **93**, 3555.
56. T. Sato, T. Akina, S. Akabori, H. Ochi, and K. Hata, *Tetrahedron Lett.*, 1969, 1767.
57. (a) J. B. McAlpine and N. V. Riggs, *Aust. J. Chem.*, 1975, **28**, 831; (b) A. H. Clemens, M. P. Hartshorn, K. E. Richards, and G. J. Wright, *Aust. J. Chem.*, 1977, **30**, 103, 113.
58. K. Lammertsma and H. Cerfontain, *J. Am. Chem. Soc.*, 1978, **100**, 8244.
59. R. Taylor, *J. Chem. Soc., Perkin Trans. 2*, 1975, 1287.
60. C. Ris and H. Cerfontain, *J. Chem. Soc., Perkin Trans. 2*, 1975, 1438.
61. H. J. A. Lamprechts, J. Mul, and H. Cerfontain, *J. Chem. Soc., Perkin Trans. 2*, 1985, 677.
62. E. E. Gilbert, *Chem. Rev.*, 1962, **62**, 549.

63. P. B. D. de la Mare, J. T. Harvey, M. Hassan, and S. Varma, *J. Chem. Soc.*, 1957, 131.
64. P. B. D. de la Mare and J. T. Harvey, *J. Chem. Soc.*, 1957, 131.
65. R. Taylor, *PhD Thesis*, University of London, 1959.
66. J. M. A. Baas and B. M. Wepster, *Recl. Trav. Chim. Pays-Bas*, 1966, **85**, 457.
67. Yu. V. Pozdnyakovich, V. V. Borodovitsyn, and S. M. Shein, *J. Org. Chem. USSR*, 1981, **17**, 2343.
68. J. M. Brittain, P. B. D. de la Mare, P. A. Newman, and W. S. Chin, *J. Chem. Soc., Perkin Trans. 2*, 1982, 1193.
69. P. B. D. de la Mare and E. A. Johnson, *J. Chem. Soc.*, 1963, 4076.
70. J. R. Lindsay-Smith, L. C. McKeer, and J. M. Taylor, *J. Chem. Soc., Perkin Trans. 2*, 1988, 385.
71. P. C. Myhre, G. S. Owen, and L. L. James, *J. Am. Chem. Soc.*, 1968, **90**, 2115.
72. P. D. Bartlett, M. Roha, and M. Stiles, *J. Am. Chem. Soc.*, 1954, **76**, 2349.
73. E. M. Arnett and G. B. Klingensmith, *J. Am. Chem. Soc.*, 1965, **87**, 1023, 1032, 1038.
74. E. Baciocchi and G. Illuminati, *J. Am. Chem. Soc.*, 1967, **89**, 4017.
75. K. M. Ibne-Rasa, *J. Am. Chem. Soc.*, 1962, **84**, 4962.
76. R. Taylor, *Comprehensive Chemical Kinetics*, Elsevier Amsterdam, 1972, Vol. 13, 372.
77. K. Auwers and A. Kockritz, *Justus Liebigs Ann. Chem.*, 1907, **352**, 320.
78. V. J. Harding and C. Weizmann, *J. Chem. Soc.*, 1910, **97**, 1126.
79. (a) M. P. de Lange, *Recl. Trav. Chim. Pays-Bas*, 1926, **45**, 19; (b) L. Elion, *Recl. Trav. Chim. Pays-Bas*, 1924, **42**, 867; (c) L. I. Smith and S. A. Harris, *J. Am. Chem. Soc.*, 1935, **57**, 1289.
80. V. J. Harding, *J. Chem. Soc.*, 1911, **99**, 1585.
81. R. Royer, P. Demerseman, and S. Risse, *Bull. Soc. Chim. Fr.*, 1974, 1691.
82. G. M. Benedikt and L. Traynor, *Tetrahedron Lett.*, 1987, 763.
83. A. N. Nesmeyanov and B. N. Strunin, *Proc. Acad. Sci. USSR*, 1961, **137**, 275.
84. J. A. Farooqui and P. H. Gore, *Tetrahedron Lett.*, 1977, 2983; J. A. Farooqui, P. H. Gore, E. F. Saad, D. N. Waters, and G. F. Moxon, *J. Chem. Soc., Perkin Trans. 2*, 1980, 835; P. H. Gore, E. F. Saad, D. N. Waters, and G. F. Moxon, *Int. J. Chem. Kinet.*, 1982, **14**, 55.
85. A. W. Francis and A. J. Hill, *J. Am. Chem. Soc.*, 1935, **57**, 1289.
86. J. Strating, L. Thijs, and B. Zwanenburg, *Recl. Trav. Chim. Pays-Bas*, 1966, **85**, 291.
87. M. A. Cahours, *Ann. Chim. Phys.*, 1845, **13**, 87; R. Benedikt, *Justus Liebigs Ann. Chem.*, 1879, **199**, 127; E. J. Smith, *Chem. Ber.*, 1978, **11**, 1225; E. Lellmann and R. Grothman, *Chem. Ber.*, 1884, **17**, 2724; L. H. Farinholt, A. P. Stuart, and D. Twiss, *J. Am. Chem. Soc.*, 1940, **62**, 123.
88. P. Welensky, *Justus Liebigs Ann. Chem.*, 1874, **174**, 99.
89. F. Ullman and E. Kopetschni, *Chem. Ber.*, 1911, **44**, 425.
90. H. J. Rylance, *J. Chem. Soc.*, 1963, 5579.
91. E. Grovenstein and U. V. Henderson, *J. Am. Chem. Soc.*, 1956, **78**, 569; E. Grovenstein and G. A. Ropp, *J. Am. Chem. Soc.*, 1956, **78**, 2560.
92. R. A. Benkeser, R. A. Hickner, and D. I. Hoke, *J. Am. Chem. Soc.*, 1958, **80**, 5294.
93. R. A. Benkeser, T. V. Listen, and G. Stanton, *Tetrahedron Lett.*, 1960, (15), 1.
94. F. B. Deans, C. Eaborn, and D. E. Webster, *J. Chem. Soc.*, 1959, 3031.
95. C. Eaborn, Z. Lasocki, and D. E. Webster, *J. Chem. Soc.*, 1959, 3034.
96. C. Eaborn and D. E. Webster, personal communication.
97. J. R. Chipperfield, G. D. France, and D. E. Webster, *J. Chem. Soc., Perkin Trans. 2*, 1972, 405.
98. L. M. Stock and H. C. Brown, *Advances in Physical Organic Chemistry*, Ed. V. Gold, Academic Press, New York, 1963, Vol. 1, p. 44.
99. R. C. Moore, *PhD Thesis*, University of Leicester, 1961.

100. R. Taylor and G. G. Smith, *Tetrahedron*, 1963, **19**, 937.
101. M. Rieger and F. Westheimer, *J. Am. Chem. Soc.*, 1950, **72**, 28.
102. C. Eaborn, A. A. Najam, and D. R. M. Walton, *J. Chem. Soc., Perkin Trans. 1*, 1972, 2481.
103. H. C. Bell, J. R. Kalman, J. T. Pinhey, and S. Sternhell, *Tetrahedron Lett.*, 1974, 3391.
104. K. Dey, C. Eaborn, and D. R. M. Walton, *Organomet. Chem. Rev.*, 1970, **1**, 151.
105. W. E. Evison and F. S. Kipping, *J. Chem. Soc.*, 1931, 2774; Z. M. Manulkin, *Zh. Obshch. Khim.*, 1948, **18**, 299; B. N. Dolgov and O. K. Panina, *Zh. Obshch. Khim.*, 1948, **18**, 1293; A. Ya. Yakubovich and G. V. Motsarev, 1953, **23**, 771, 1059; G. A. Russell, *J. Am. Chem. Soc.*, 1959, **81**, 4815, 4825.
106. E. G. Rochow and W. F. Gilliam, *J. Am. Chem. Soc.*, 1945, **67**, 1772.
107. J. D. Austin, C. Eaborn, and J. D. Smith, *J. Chem. Soc.*, 1963, 4744.
108. S. V. Sunthankar and H. Gilman, *J. Org. Chem.*, 1950, **15**, 1200.
109. V. I. Minkin, R. M. Minyaev, I. A. Yudilevich, and M. E. Kletskii, *J. Org. Chem. USSR*, 1985, **21**, 842.
110. F. S. Kipping, *J. Chem. Soc.*, 1907, 209; N. W. Cusa and F. S. Kipping, *J. Chem. Soc.*, 1932, 2205.
111. F. B. Deans and C. Eaborn, *J. Chem. Soc.*, 1957, 498.
112. R. A. Benkeser and H. Landesman, *J. Am. Chem. Soc.*, 1954, **76**, 904.
113. J. L. Speier, *J. Am. Chem. Soc.*, 1953, **75**, 2930.
114. C. Eaborn, Z. S. Salih, and D. R. M. Walton, *J. Chem. Soc., Perkin Trans. 2*, 1972, 172.
115. R. W. Bott, C. Eaborn, and T. Hashimoto, *J. Chem. Soc.*, 1963, 3906.
116. P. Bourgeois, R. Calas, E. Jousseaume, and J. Gerval, *J. Organomet. Chem.*, 1975, **84**, 165.
117. R. A. Benkeser and A. Torkelson, *J. Am. Chem. Soc.*, 1954, **76**, 1253.
118. C. Eaborn and D. E. Webster, *J. Chem. Soc.*, 1957, 4449.
119. C. Eaborn and D. E. Webster, *J. Chem. Soc.*, 1960, 179.
120. J. R. Chipperfield, R. Z. Fernandez, and D. E. Webster, *J. Chem. Soc., Perkin Trans. 2*, 1979, 117.
121. L. M. Stock and A. R. Spector, *J. Org. Chem.*, 1963, **28**, 3272.
122. C. Eaborn and O. W. Steward, *J. Chem. Soc.*, 1965, 521; L. H. Sommer, K. W. Michael, and W. D. Korte, *J. Am. Chem. Soc.*, 1967, **89**, 868.
123. C. Eaborn and D. E. Webster, *J. Chem. Soc.*, 1960, 179.
124. J. Vcelak and V. Chvalovsky, *Collect. Czech. Chem. Commun.*, 1973, **38**, 1055.
125. K. Takahashi, K. Ohnishi, and K. Takase, *Chem. Lett.*, 1985, 1079.
126. M. H. Abraham and M. R. Sedaghat-Herati, *J. Chem. Soc., Perkin Trans. 2*, 1978, 729.
127. H. Hashimoto and Y. Morimoto, *J. Organomet. Chem.*, 1967, **8**, 271.
128. J. R. Pratt, F. H. Pinkerton, and S. F. Thames, *J. Organomet. Chem.*, 1972, **38**, 29.
129. E. H. Bartlett, C. Eaborn, and D. R. M. Walton, *J. Chem. Soc. C*, 1970, 1717.
130. R. W. Bott, C. Eaborn, and J. A. Waters, *J. Chem. Soc.*, 1963, 681.
131. O. Buckman, M. Grosjean, and J. Nasielski, *Helv. Chim. Acta*, 1964, **47**, 1679, 1688, 2037.
132. L. J. Andrews and R. M. Keefer, *J. Am. Chem. Soc.*, 1952, **78**, 4500; 1955, **77**, 2164.
133. E. Glyde and R. Taylor, *J. Chem. Soc., Perkin Trans. 2*, 1977, 678.
134. P. Alcais and J. Nazielski, *J. Chim. Phys.*, 1969, **66**, 95.
135. A. Delhaye, J. Nazielski, and M. Planchon, *Bull. Soc. Chim. Belg.*, 1960, **69**, 134.
136. S. F. Filippuichev and N. A. Chekalin, *Chem. Abstr.*, 1935, **29**, 5087.
137. R. Huisgen, in *Azo and Diazo Chemistry*, Ed. H. Zollinger, Interscience, London, 1961, p. 241.

138. N. J. Bunce, *J. Chem. Soc., Perkin Trans. 1*, 1974, 870.
139. R. B. Moodie, M. A. Payne, and K. Schofield, *J. Chem. Soc., Chem. Commun.*, 1983, 233; *J. Chem. Soc., Perkin Trans. 2*, 1985, 1457.
140. A. H. Clemens, J. H. Ridd, and J. P. B. Sandall, *J. Chem. Soc., Chem. Commun.*, 1983, 343.
141. T. A. Modro and A. Piekos, *Phosphorus*, 1974, **3**, 195.
142. A. Campagnini, M. Santagati, N. Marziano, and P. Passerini, *Ann. Chim. (Rome)*, 1970, **60**, 527, 537.
143. P. B. Fischer and H. Zollinger, *Helv. Chim. Acta*, 1972, **55**, 2146.
144. Y. K. Yurev and N. K. Sadovaya, *J. Gen. Chem. USSR*, 1964, **34**, 2210; A. P. Terentev and M. A. Schadchina, *Dokl. Akad. Nauk SSSR*, 1947, **55**, 231.
145. H. E. Armstrong, *Chem. Ber.*, 1874, **7**, 404; J. Post, *Chem. Ber.*, 1874, **7**, 1322; H. Lampricht, *Chem. Ber.*, 1885, **18**, 2172; W. J. Karslake and W. J. Morgan, *J. Am. Chem. Soc.*, 1908, **30**, 828; F. Olsen, *Ind. Eng. Chem.*, 1924, **16**, 66; K. Lésniak and T. Urbański, *Nitro Compounds, Tetrahedron*, 1964, **20**, Suppl. 1, 61.
146. R. King, *J. Chem. Soc.*, 1921, 2105.
147. C. M. Suter, *Organic Chemistry of Sulphur*, Wiley, New York, 1945, pp. 394–412; H. Cerfontain, *Mechanistic Aspects in Aromatic Sulphonation and Desulphonation*, Interscience, London, 1968, pp. 196–197.
148. J. J. Sudburough and J. V. Lakhumalani, *J. Chem. Soc.*, 1917, 41.
149. R. L. Datta and J. C. Bhoumik, *J. Am. Chem. Soc.*, 1921, **43**, 303.
150. L. G. Cannell, *J. Am. Chem. Soc.*, 1957, **79**, 2927, 2932.
151. A. N. Meldrum and M. S. Shah, *J. Chem. Soc.*, 1923, 1982.
152. B. T. Boliga and A. N. Bourns, *Can. J. Chem.*, 1966, **44**, 363.
153. S. Uemura, S. I. Fukuzawa, M. Wakasugu, and M. Okano, *J. Organomet. Chem.*, 1981, **214**, 319.
154. H. R. Rogers and J. Houk, *J. Am. Chem. Soc.*, 1982, **104**, 522.
155. W. J. Archer, M. A. Hossaini, and R. Taylor, *J. Chem. Soc., Perkin Trans. 2*, 1982, 181.
156. M. Speranza and F. Cacace, *J. Am. Chem. Soc.*, 1977, **99**, 3051.
157. T. A. Elwood, W. R. Flack, K. J. Inman, and P. W. Rabideau, *Tetrahedron*, 1974, **30**, 535.
158. M. Godfrey, P. Goodman, and P. H. Gore, *Tetrahedron*, 1976, **32**, 841.
159. Yu. N. Pornshev and E. M. Tereschenko, *J. Org. Chem. USSR*, 1975, **11**, 655.
160. P. B. Fischer and H. Zollinger, *Helv. Chim. Acta*, 1972, **55**, 2139.
161. C. L. Perrin and G. A. Skinner, *J. Am. Chem. Soc.*, 1971, **93**, 3389.
162. T. G. Jackson, J. E. Norris, and R. C. Legendre, *J. Org. Chem.*, 1971, **36**, 3638.
163. P. W. Robertson and H. V. A. Briscoe, *J. Chem. Soc.*, 1912, **101**, 1964.
164. L. E. Hinckel, *J. Chem. Soc.*, 1920, **117**, 1301.
165. D. L. Fox and E. E. Turner, *J. Chem. Soc.*, 1930, 1115; R. G. Coombes, D. H. G. Crout, J. G. Hoggett, R. B. Moodie, and K. Schofield, *J. Chem. Soc. B*, 1970, 347.
166. R. B. Moodie, K. Schofield, and J. B. Weston, *J. Chem. Soc., Perkin Trans. 2*, 1976, 1089.
167. C. Bloomfield, A. K. Manglik, R. B. Moodie, K. Schofield, and G. D. Tobin, *J. Chem. Soc., Perkin Trans. 2*, 1983, 75.
168. A. R. Butler and A. P. Sanderson, *J. Chem. Soc., Perkin Trans. 2*, 1972, 989.
169. K. Olsson and P. Martinsen, *Acta Chem. Scand.*, 1972, **26**, 3549; K. Olsson, *Acta Chem. Scand., Ser. B*, 1974, **28**, 322.
170. A. R. Butler and A. P. Sanderson, *J. Chem. Soc., Perkin Trans. 2*, 1974, 1784.
171. K. Olsson, *Acta Chem. Scand., Ser. B*, 1975, **29**, 405; K. Olsson and I. Johansson, *Acta Chem. Scand., Ser. B*, 1978, **32**, 297.

172. A. Zweig, K. B. Hoffman, and G. W. Nachtigall, *J. Org. Chem.*, 1977, **42**, 4049; I. Johansson, *Acta Chem. Scand. Ser. B*, 1981, **35**, 723.
173. I. Granoth, A. Kalir, Z. Pelah, and E. D. Bergmann, *J. Chem. Soc., Chem. Commun.*, 1969, 260; I. Granoth, Y. Segall, and A. Kalir, *J. Chem. Soc., Perkin Trans. 1*, 1973, 1972.
174. H. Cerfontain, A. Koeberg-Telder, K. Laali, H. J. A. Lamprechts, and P. de Wit, *Recl. Trav. Chim. Pays-Bas*, 1982, **101**, 390.
175. E. J. Kunst, A. Halpern, and G. Stöcklin, *J. Am. Chem. Soc.*, 1974, **96**, 3733.

CHAPTER 11

Quantitative Evaluation of Electronic and Steric Effects in Electrophilic Aromatic Substitution

11.1 QUANTITATIVE TREATMENTS OF STRUCTURE–REACTIVITY RELATIONSHIPS

In Section 1.4.4, the application of the Hammett equation (eq. 1.7) to reactions in the side-chains of substituted aromatics was described. It was shown that the equation in this form is inapplicable to electrophilic aromatic substitutions because in the passage from the initial to the transition state of these reactions, changes occur in the conjugative interaction between the substituent and the reaction centre which do not normally have counterparts in the side-chain reactions. In fact, there are certain side-chain reactions where these changes do occur: two examples are the S_N1 solvolysis of benzyl halides[1] and the pyrolysis of 1-arylethyl acetates.[2] In the former reaction, for example, the transition state has considerable carbocation character, and is stabilized by a substituent of the $+M$ type (such as OMe) in the *para* position, as depicted by structure **1**.

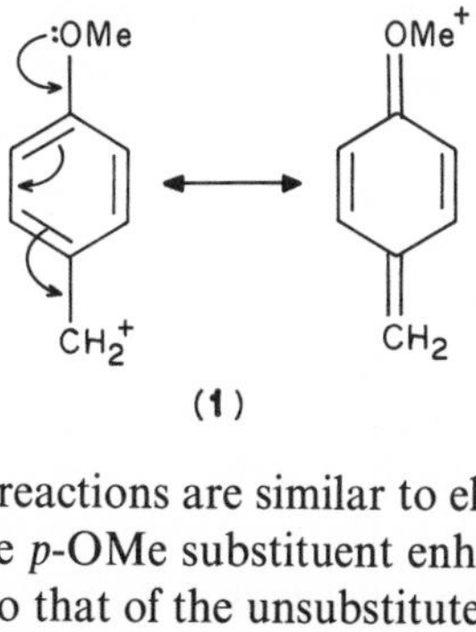

:OMe OMe$^+$ CH_2^+ CH_2

(1)

:OMe OMe$^+$ E^+ H E H

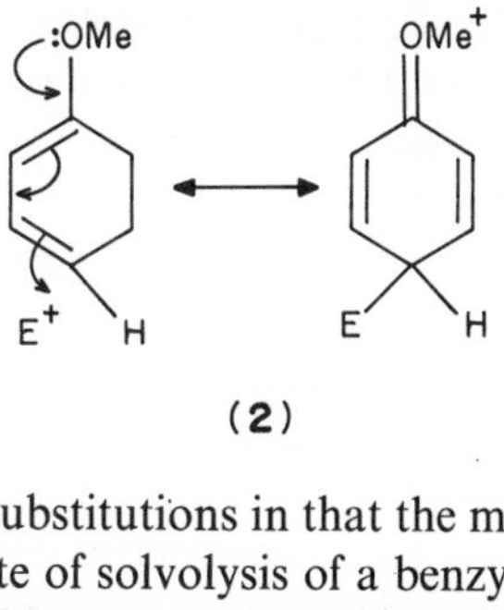

(2)

These reactions are similar to electrophilic substitutions in that the means by which the *p*-OMe substituent enhances the rate of solvolysis of a benzyl halide relative to that of the unsubstituted compound is analogous to that by which it enhances the rate of nuclear substitution by an electrophile (**2**). It follows, therefore, that it should in principle be possible to derive a set of sigma constants which are applicable to all reactions of this type.[3] This is true (to a first approximation, see below), and the specific set of constants are known as *electrophilic substituent constants*, designated σ^+ because they apply to reactions which have electron-deficient (positive) reaction centres, with which the substituents can conjugate.

Ideally, a standard set of σ^+ values would be determined from rate data in an electrophilic substitution. In practice, the solvolysis of 1-aryl-1-methylethyl chlorides (*tert*-cumyl chlorides) by aqueous acetone at 25 °C was employed (eq. 11.1),[4] because relatively few accurate rate data for electrophilic substitution were available. This reaction also had the advantage of a smaller ρ factor (-4.54) than most electrophilic substitutions, so that a wider range of rates could be measured under a given condition without recourse to overlap techniques. The ρ factor was scaled (using data for *meta* substituents) to be directly comparable to that (1.0) which applies to the ionization of benzoic acids. The σ^+ values so derived were found to correlate reactivity data for electrophilic substitutions by means of the modified Hammett equation (eq. 11.2) (known as the Hammett–Brown equation) with much greater precision than σ values previously used.[5] (Because the rho factors were scaled to be equivalent, use of the designation ρ^+ rather than ρ is fairly uncommon).

$$X\text{-}C_6H_4\text{-}C(Me)_2\text{-}Cl \xrightarrow{H_2O} X\text{-}C_6H_4\text{-}C(Me)_2\text{-}OH + HCl \qquad (11.1)$$

$$\log k/k_0 = \rho^+\sigma^+ \qquad (11.2)$$

The solvolysis reaction does, however, have some disadvantages. First, it is impossible to isolate and purify the compounds containing electron-supplying substituents in the aromatic ring because they undergo β-elimination of HCl on heating. Kinetic studies had therefore to be carried out on solutions of the alcohol precursors in HCl. Secondly, aqueous acetone is an excellent solvating medium, a result of which is that steric hindrance to solvation is very severe under these conditions. This means that not only do bulky alkyl groups activate in the Baker–Nathan order, so that the solvolysis rates for all bulky *p*-alkyl groups are depressed, but the solvolysis rates for all *m*-alkyl substituents are also depressed [see Chapter 1, Section 1.4.3 and structures **52a,b** and **54a**–**c**)]. It turns out that the σ^+ values for *m*-alkyl groups determined from this reaction do not correlate any electrophilic aromatic substitution satisfactorily. This has generally not been recognized, because points for these substituents in Hammett-type plots fall close to the origin, a mathematical consequence of this being to obscure what are in fact large deviations. Errors of 50% in a σ^+ value of 0.07 become 'unnoticed,' whereas a similar percentage error in a value of 0.7 would be regarded as extremely serious.

A more suitable reaction is the pyrolysis of 1-arylethyl acetates (eq. 11.3).[2] The transition state for this reaction has a partial carbocation at the α-side-chain

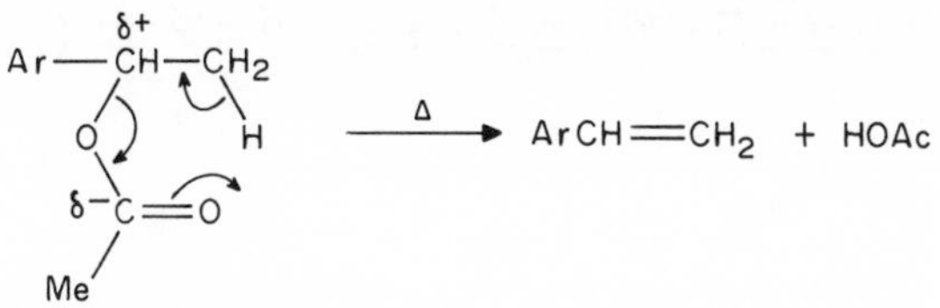

$$\mathrm{Ar{-}\overset{\delta+}{C}H{-}CH_2(H){-}O{-}\overset{\delta-}{C}(Me){=}O} \xrightarrow{\Delta} \mathrm{ArCH{=}CH_2 + HOAc} \qquad (11.3)$$

carbon, and has three advantages over the solvolysis reaction:

(1) There is no solvent so solvation effects are entirely absent, and the results obtained are entirely unambigous.

(2) The compounds can be prepared in a pure state.

(3) The ρ factor for the reaction, -0.66 at 600 K, means that a wider range of reactivities can be measured under identical conditions than for any other reaction with a cerbocationic transition state. For certain applications, other esters may be used, the ρ factors being closely similar.[6–8]

The pyrolysis gives an excellent correlation against σ^+ values determined from the solvolysis,[2] *except* for those substituents (bulky *p*-alkyl,[2,9,10] *m*-alkyl,[7] *m*-phenyl,[2] *m*- and *p*-trimethylsilyl[11]) the solvolysis data for which are adversely affected by steric hindrance to solvation (Section 11.1.4). The pyrolysis data for the *m*-CF_3[12] and *m*-NO_2[8] substituents also do not correlate very well with the solvolysis results, but in fact predict more accurately their effects in *all* electrophilic substitution and related reactions than do the solvolysis data. The pyrolysis has been very successfully employed for the determination of σ^+ values for heteroaromatics.[13]

Other reactions and spectroscopic methods have been used to determine σ^+ values; in particular, the hydrogen exchange (detritiation) reaction has proved very suitable because of both the freedom from steric hindrance and the relative unimportance of steric hindrance to solvation in trifluoroacetic acid, the exchange medium. A compilation of σ^+ values for substituents and, where available, the corresponding σ values are given in Table 11.1. Because the data have been obtained by various workers using different methods, the data from

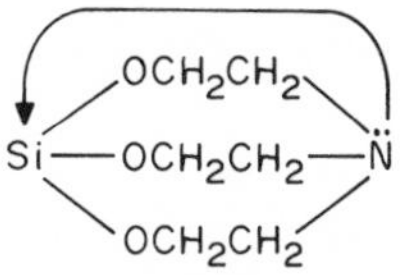

$\mathrm{Si(OCH_2CH_2)_3N}$ **(3)** $\qquad$ X–C$_6$H$_4$– **(4)** $\qquad$ 2-naphthyl **(5)**

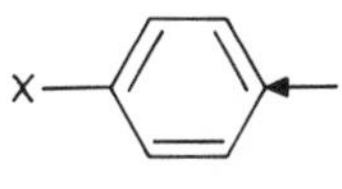

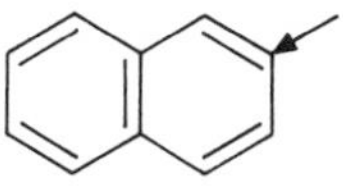

one set of experiments may not be precisely related to those obtained from another, especially where resonance interactions in the processes differ

Table 11.1 Substituent constants

Substituent	σ_m	σ_m^+	σ_p	σ_p^+	Ref.
CH_2^-	—	—	—	−11.2	14
$CH^-CO_2^-$	—	—	—	−8.7	14
NH^-	—	—	—	−7.6	14
CH^-CN	—	−1.1	—	−5.4	14
CH^-CO_2Et	—	—	—	−5.0	14
O^-	−0.71	−1.5	−0.81	−2.5	14, 15
$CH_2CO_2^-$	−0.1	—	−0.02	−0.6	14, 15
CO_2^-	−0.1	−0.03	0.0	−0.02, −0.45	5, 14, 15
NMe_2	−0.15	—	−0.63	−1.74	15, 16
NH_2	0.0	−0.16	−0.57	−1.47	15, 17
NHPh	—	—	−0.45	−1.4	15, 18
NHCOMe	0.12	0.08	−0.09	−0.62	15, 19
NHCOPh	0.02	—	−0.19	−0.6	15, 18
$N{=}N^+{=}N^-$	0.37	—	0.08	−0.54	20
(3,4-CH_2CH_2O)	—	—	—	−0.98[a]	16
(3,4-OCH_2O)	—	—	—	−0.675[b]	16
OMe	0.11	0.05	−0.28	−0.78	5, 15
OEt	0.1	—	−0.24	−0.81	15, 16, 21
O-n-Pr	—	—	—	−0.83	21
O-*i*-Pr	—	—	—	−0.85	16, 21
O-n-Bu	—	—	—	−0.81	21
OCH_2Ph	—	—	−0.41	−0.65	15, 22
OH	0.13	—	−0.38	−0.59[c]	23
OPh	0.25	—	−0.32	−0.53	15, 24a
OCOMe	0.39	0.335	0.31	−0.12	15, 19
OCOPh	0.21	0.23	—	−0.11	15, 24b, 25
$OGe(OPh)Me_2$	—	—	—	−0.86	26
SMe	0.15	0.16	0.0	−0.55[d]	5, 15, 16, 27
SPh	0.17	—	0.18	−0.455[d]	15, 27
CH_2HgCH_2Ph	—	—	—	−1.12	22

$CH_2CHMeHgCHMeCH_2Ph$	—	—	—	−0.25	22
$CH_2B(OH)_3^-$	—	—	—	−1.11	22
CH_2silatrane[e]	—	—	—	−0.48	28
$CH_2Si(OMe)_3$	—	—	—	−0.25	28
$CH_2SiMe(OMe)_2$	0.04	—	0.01	−0.40	15, 28
CH_2SiMe_2OMe	0.04	—	−0.02	−0.45	15, 28
$C(SiMe_3)_3$	—	—	−0.27	−0.52 (−0.79)	29, (30)
$CH(SiMe_3)_2$	—	—	−0.33	−0.62 (−0.76)	29, (30)
$CH_2SiMe_2SiMe_3$	—	—	—	(−0.72)	(30)
SiH_3	—	—	—	−0.27	28
CH_2SiMe_3	−0.16	−0.16[f]	−0.21	−0.56 (−0.66)	15, 29, (30)
CH_2SiEt_3	—	—	—	−0.57	28
CH_2Si-n-Pr_3	—	—	—	−0.58	28
CH_2SiPh_3	—	—	—	−0.38 (−0.4)	31, (32)
CH_2GeMe_3	—	—	—	−0.63	31
CH_2GeEt_3	—	—	—	−0.67	28
CH_2GePh_3	—	—	—	−0.51 (−0.6)	31, (32)
CH_2SnMe_3	—	—	—	−0.81 (−0.93)	31, (22)
CH_2SnPh_3	—	—	—	−0.73 (−0.75)	31, (32)
CH_2PbMe_3	—	—	—	−1.03	31
CH_2PbPh_3	—	—	—	−0.90 (−1.08)	31, (22)
$(CH_2)_2SiMe_3$	—	—	—	−0.33	29
$(CH_2)_3SiMe_3$	—	—	—	−0.31	29
$(CH_2)_4SiMe_3$	—	—	—	−0.32	29
$(CH_2)_2PbPh_3$	—	—	—	−0.22	22
$C_{10}H_9Fe$ (ferrocenyl)	—	—	—	−0.7	33
2-C_4H_3O (2-furanyl)	0.06	—	0.02	−0.39	34
2-C_4H_3S (2-thiophenyl)	0.09	—	0.05	−0.33	34
CH_2SiCl_3	—	—	—	−0.145	35
$(CH_2)_2SiCl_3$	—	—	—	−0.155	35
$(CH_2)_3SiCl_3$	—	—	—	−0.21	35
$(CH_2)_4SiCl_3$	—	—	—	−0.255	35
1-Homocubyl	—	—	—	−0.75	36
1-Bicyclo[3.1.0]hexyl	—	—	0.15	−0.71	15, 36

(*continued*)

Table 11.1 (*continued*)

Substituent	σ_m	σ_m^+	σ_p	σ_p^+	Ref.
7-Bicyclo[4.1.0]heptyl	—	—	—	−0.68	22
1-Bicyclo[4.1.0]heptyl	—	—	0.01	−0.60	15, 36
1-Bicyclo[2.1.1]hexyl	—	—	—	−0.50	36
1-Bicyclo[2.2.1]heptyl	—	—	—	−0.40	36
1-Bicyclo[2.2.2]octyl	—	—	−0.25	-0.44^{g}	15, 37
1-Adamantyl	—	—	−0.24	-0.425^{g}	15, 37
2-*exo*-Norbornyl	—	—	—	-0.42^{g}	37
2-*endo*-Norbornyl	—	—	—	-0.39^{g}	37
Cyclopropyl	−0.07	—	−0.21	−0.47	15, 38
1-Methylcyclopropyl	—	—	—	−0.525	38
Cyclobutyl	−0.13	—	−0.13	−0.345	15, 39
2-Butyl	−0.08	—	−0.19	−0.325	15, 39
Cyclopentyl	−0.15	—	−0.13	−0.35	15, 39
3-Pentyl	—	—	—	−0.325	39
Cyclohexyl	−0.15	—	−0.13	−0.34, -0.38^{g}	15, 39, 40
3-Hexyl	—	—	—	−0.325	39
Neopentyl	−0.08	—	−0.10	-0.35^{g}	15, 37
t-Bu	−0.10	-0.19^{g}	−0.20	−0.365	15, 40
i-Pr	−0.08	-0.135^{g}	−0.15	−0.330	15, 25
Et	−0.07	-0.11^{g}	−0.15	−0.315	15, 25
Me	−0.07	$-0.10^{g,h}$	−0.17	−0.31	5, 7, 15
CH_2Ph	−0.08	−0.07	−0.09	−0.23	15, 27
$(CH_2)_2Ph$	−0.03	−0.085	−0.12	−0.265	15, 41
$CHPh_2$	−0.03	−0.04	−0.04	−0.185	15, 41
CPh_3	−0.01	−0.07	0.02	−0.21	15, 41
$CH_2CH{=}CMe_2$	—	—	—	−0.26	42
$CH_2CH{=}CH_2$	—	−0.025	—	−0.22	42
$CMe_2CH{=}CH_2$	—	—	—	−0.21	42
$(CH_2)_2CH{=}CH_2$	—	—	—	−0.26	42
$CH{=}CHPh$	0.03	—	−0.07	−1.0	43

$CH{=}C(CN)_2$	—	—	—	~0.55–0.8	22
$C{\equiv}CPh$	0.14	—	0.16	−0.03	43
$C{\equiv}CH$	0.20	—	0.23	0.18	44
Silatrane[e]	—	—	—	−0.35	28
3,7,10-Me_3-silatrane[e]	—	—	—	−0.40	28
$SiMe(SiMe_3)_2$	—	—	—	−0.81[i]	30
$(SiMe_2)_2SiMe_3$	—	—	—	−0.77[i]	30
$SiMe_2SiMe_3$	—	—	—	−0.62[i]	30
$SiMe(CH_2)_3$	—	—	—	−0.20[i]	30
$SiMe_3$	−0.04	−0.16[g]	−0.07	−0.09[g,j]	11, 15
$Ge(SMe)_3$	—	—	—	−0.27	26
$GePh(C_6H_{13})_2$	—	—	0.09	0.0	26
$GePh_2C_6H_{13}$	—	—	0.10	0.02	26
$Ge(CH{=}CHC_4H_9)_3$	—	—	0.10	0.02	26
$GePh_2CH_2CH{=}CHCH_3$	—	—	0.11	0.04	26
$GePh_2CH_2OMe$	—	—	0.11	0.04	26
$GePh_2H$	—	—	0.11	0.04	26
GePhHBr	—	—	0.14	0.10	26
$GePhCl(CH_2)_4COMe$	—	—	0.14	0.10	26
$GeHCl(CH_2)_5Me$	—	—	0.15	0.11	26
$GePh_3$	0.05	—	0.08, 0.15	0.11	15, 26
$GePh_2(CH_2)_4COMe$	—	—	0.16	0.13	26
GePhHCl	—	—	0.16	0.13	26
$SnMe_3$	0.0	—	0.0	−0.12	45
CH_2CO_2R	—	—	—	−0.16	18
CH_2CO_2H	—	—	−0.07	−0.02	15, 46
CH_2COMe	—	—	—	−0.01	46
CH_2CN	0.16	—	0.18	0.12	15, 46
CH_2OMe	0.02	—	0.02	−0.05	15, 46
CH_2OEt	0.02	—	0.02	0.00	15, 46
CH_2OH	0.01	—	0.01	0.01	15, 46
CH_2Cl	0.11	—	0.12	−0.01	15, 18
CH_2Br	0.12	—	0.14	−0.06	15, 46
CH_2F	0.10	0.105	0.10	—	15, 47

(*continued*)

Table 11.1 (*continued*)

Substituent	σ_m	σ_m^+	σ_p	σ_p^+	Ref.
2,2-Dichlorocyclopropyl	—	—	—	−0.02	48
CH_2SO_3H	—	0.64	—	0.59	49
$(CH_2)_2SO_3H$	—	0.28	—	−0.05	49
$(CH_2)_3SO_3H$	—	0.09	—	−0.21	49
$(CH_2)_4SO_3H$	—	—	—	−0.30	49
$(CH_2)_5SO_3H$	—	—	—	−0.32	49
$(CH_2)_6SO_3H$	—	—	—	−0.34	49
SOPh	~0.51	—	0.46	−0.19[d]	5, 50
Ph	0.06	0.0[g]	0.02	−0.21[k]	15, 40, 51
C_6H_4Me-4	—	—	—	−0.32	27
C_6H_4Cl-4	—	—	—	−0.19	27
C_6H_4Br-4	—	—	—	−0.18	27
C_6H_4Cl-3	—	—	—	−0.15	27
C_6H_4F-2	—	—	—	−0.095	52
C_6H_4F-3	—	—	—	−0.075	52
C_6H_4F-4	—	—	—	−0.165	52
C_6F_5	−0.12	0.285	−0.03	0.225	52, 53
H	0.0	0.0	0.0	0.0	
D	−0.001	—	−0.001	−0.001	15, 54
OCF_3	0.365	—	0.335	0.07	5, 55
OSO_2Me	0.39	—	0.36	0.15	22
$OSO_2C_6H_4Me$-4	—	—	—	~0.1	22
N=NPh	0.29	—	0.33	0.17[l]	15, 56
$N{=}NC_6H_4OMe$-4	—	—	—	0.03[l]	56
$N{=}NC_6H_4Me$-4	—	—	—	0.13[l]	56
$N{=}NC_6H_4F$-4	—	—	—	0.18[l]	56
$N{=}NC_6H_4Br$-4	—	—	—	0.21[l]	56
$N{=}NC_6H_4Br$-3	—	—	—	0.24[l]	56
$N{=}NC_6H_2Me_2$-2,6	—	—	0.31	0.18[l]	56
N=N-*t*-Bu	0.24	—	0.28	0.15[l]	56
N=N(O)Ph	0.24	—	0.27	0.06	57

F	0.335	0.35	0.06	−0.075	5, 15
Cl	0.375	0.40	0.225	0.115	5, 15
Br	0.39	0.405	0.23	0.15	5, 15
I	0.35	0.36	0.21	0.135	5, 15
$SiCl_3$	0.48	—	0.56	0.355	35, 58
CO_2H	0.35	0.32	0.44	0.42	5, 15
CO_2Me	0.36	0.37	0.46	0.49	5, 15
CO_2Et	0.36	0.365	0.46	0.48	5, 15
CF_3	0.43	0.565	0.54	0.61	5, 12, 15
CN	0.61	0.56	0.675	0.66	5, 15
NO_2	0.72	0.73	0.79	0.79	5, 8, 15
SMe_2^+	1.00	1.38	0.90	—	15, 59
NMe_3^+	1.01	(0.67)[m]	0.88	0.41	5, 15
$CH_2NHMe_2^+$	0.40	—	0.43	0.50	15, 46
$CH_2NMe_3^+$	0.40	—	0.44	0.50	15, 46
N_2^+	1.65	—	1.93	1.88	60
$OMe.AlCl_3$	—	—	—	0.1	61
$CN.AlCl_3$	—	—	—	1.2	61
$CO_2Me.AlCl_3$	—	—	—	1.2	61
$COPh.AlCl_3$	—	—	—	1.3	61
$COMe.BCl_3(AlCl_3, SbCl_5)$	—	—	—	1.6 (1.5,0.9)	61
$COH.AlCl_3(SbCl_5)$	—	—	—	1.9 (1.0)	61

[a] Refers to the 5-position of coumarin.
[b] Refers to the 5-position of benzo-1,3-dioxole.
[c] This value obtained by physical measurements under non-polar conditions. For reactions in polar media much higher values are obtained, e.g. −1.0 in protiodesilylation, probably owing to partial ionization of the substituent. For some other values, see ref. 22.
[d] No single value is satisfactory here because of the high polarizability of the substituent; see text.
[e] The structure of silatrane is **3**.
[f] Deduced from data for protiodesilylation, Section 4.7.
[g] Value applicable to solvent-free conditions. Smaller values will be required the more solvating are the reaction conditions (Section 11.1.4).
[h] Values up to −0.13 apply in solvent-free reactions of high electron demand (Section 11.1.4).
[i] These were obtained under conditions that gave exhalted values for substituents of the C—Si type (see above).
[j] Because of the high polarizability of the substituent, values up to −0.24 have been recorded for this substituent.[28,30]
[k] Either higher values, due to increased electron demand enforcing coplanarity, or lower values, due to steric hindrance to solvation, can apply.
[l] Widely varying values have been obtained, probably owing to the high polarizability of the substituent and hydrogen bonding (see text).
[m] See text.

significantly. This point is dealt with fully in Section 11.1.4. In view of this and possible errors in the literature data, it is not meaningful to quote the parameters to better than 0.005 unit, and values have been rounded off accordingly. Where a choice of values exists, the one given is that which correlates most satisfactorily a range of reactivity data. The substituents are arranged in order of approximately decreasing electron supply, those having similar characteristics being grouped together.

The Hammett equation can also be extended to describe reactivities of polycyclic and related molecules. A sigma constant (say) $\sigma^+_{p\text{-X}}$ is generally understood to describe the effect of the substituent X at the *para* position in PhX. However, it is also a measure of the reactivity of the *para* position of PhX (**4**) relative to that of a position in benzene. Consequently, sigma values may also be used to describe the reactivity of a position in a polycyclic (or heteroaromatic) molecule relative to that in a position in benzene. Thus $\sigma^+_{2\text{-naph}}$ is a measure of the reactivity of the 2-position in reactivity of the 2-position in naphthalene (**5**) relative to that of a position in benzene. In the case of naphthalene, the 'substituent' is actually 3:4-benzo (—CH═CH—CH═CH—), but for larger polycyclics this notation becomes inconvenient and is not therefore used. Table 11.2 gives σ^+ values, designated $\sigma_{\text{Ar}}{}^+$, for positions in polycyclic, etc., molecules, calculated as described in ref. 27.

These σ^+ values correlate the rates of electrophilic substitutions with far greater precision than that obtained using σ values. Some important features are as follows:

(1) The largest differences between the σ and σ^+ values for particular substituents occur for *para* substituents of the $+M$ type, the latter values being uniformly more negative than the former, and this is to be expected from the earlier discussion. Consider, for example, the *p*-F substituent. Its σ value is positive, indicating that its $-I$ effect normally outweighs its $+M$ effect. In an electrophilic substitution, however, the lone pair on fluorine can conjugate with the positive charge in the transition state in a manner that is not possible in reactions for which σ values apply. The $+M$ effect then outweighs the $-I$ effect and so a negative sigma value is required to describe the overall electron-releasing effect.

(2) It would be expected that σ^+ values and σ values for electron-withdrawing substituents should be approximately the same. This is not always the case, however, and among possible reasons for this may be inaccuracy in the derived data. This was certainly found to be the case for the *m*-nitro substituent, for which the revised value of σ^+ of 0.73 gives a more satisfactory correlation of all electrophilic substitution data and related reactions[8] than the original value of -0.674,[5] and is also closer to the corresponding σ value.

(3) Another discrepancy is found with the CF_3 substituent. For this the σ^+ values for both *meta* and *para* positions are each considerably more positive than the corresponding σ values, and it is possible that solvent interactions are in some

Table 11.2 Values of $\sigma_{Ar}{}^{+}$ applicable to electrophilic aromatic substitution

Aromatic compound	σ^{+} value (position)	Ref.
1,2-Dimethylbenzene	−0.37 (4)	62
Benzcyclopentane	−0.41 (5)	63
Benzcyclohexane	−0.41 (6)	63
Benzcycloheptane	−0.40 (7)	64
p-Terphenyl	−0.28 (4)	65
m-Terphenyl	−0.24 (4)	65
o-Terphenyl	−0.22 (4)	65
1,3,5-Triphenylbenzene	−0.22 (4)	65
Biphenylene	−0.23 (1), −0.47 (2)	66
Triphenylene	−0.32 (1), −0.245 (2)	67
9,10-Dihydroanthracene	−0.30 (1), −0.32 (2)	62
Anthracene	−0.445 (1), −0.35 (2), −0.81 (9)	67
Triptycene	−0.19 (1), −0.29 (2)	62
Fluorene	−0.15 (1), −0.48 (2), −0.24 (3), −0.425 (4)	68
9-Methylfluorene	−0.48 (2)	69
9,9-Dimethylfluorene	−0.49 (2)	69
2-Methylfluorene	−0.55 (7)	70
2-Chlorofluorene	−0.415 (7)	70
2-Bromofluorene	−0.40 (7)	70
9,10-Dihydrophenanthrene	−0.39 (2)	68
Phenanthrene	−0.34 (1), −0.25 (2), −0.39 (3), −0.33 (4), −0.365 (9)	68
9-Fluorophenanthrene	−0.24 (1), −0.435 (10)	71
9-Chlorophenanthrene	−0.28 (10)	71
9-Bromophenanthrene	−0.26 (10)	71
9-Iodophenanthrene	−0.265 (10)	71
1-Methylphenanthrene	−0.44 (9)	72
2-Methylphenanthrene	−0.42 (9)	72
3-Methylphenanthrene	−0.51 (9)	72
4-Methylphenanthrene	−0.425 (9)	72
5-Methylphenanthren	−0.41 (9)	72
6-Methylphenanthrene	−0.40 (9)	72

(continued)

Table 11.2 (*continued*)

Aromatic compound	σ^+ value (position)	Ref.
7-Methylphenanthrene	−0.425 (9)	72
8-Methylphenanthrene	−0.415 (9)	72
9-Methylphenanthrene	−0.62 (10)	71
2,7-Dimethylphenanthrene	−0.525 (9)	73
3,6-Dimethylphenanthrene	−0.47 (9)	73
4,5-Dimethylphenanthrene	−0.545 (9)	73
2,4,5,7-Tetramethylphenanthrene	−0.615 (9)	73
3,4,5,6-Tetramethylphenanthrene	−0.635 (9)	73
Cyclopenta[*def*]phenanthrene	−0.505 (1), −0.43 (2), −0.475 (3), −0.44 (9)	74
Naphthalene	−0.35 (1), −0.25 (2)	75
1-Methylnaphthalene	−0.525 (2), −0.305 (3), −0.57 (4), −0.39 (5)	
	−0.275 (6), −0.305 (7), −0.465 (8),	75
2-Methylnaphthalene	−0.635 (1), −0.185 (3), −0.40 (4), −0.365 (5)	
	−0.395 (6), −0.27 (7), −0.405 (8)	75
1-Fluoronaphthalene	−0.22 (2), 0.07 (3), −0.425 (4), −0.195(5), −0.205 (8)	75
2-Fluoronaphthalene	−0.375 (1), 0.02 (3)	75
1-Chloronaphthalene	−0.12 (2), −0.02 (3), −0.285 (4), −0.175 (5)	
	−0.02 (6), −0.055 (7), −0.195 (8)	75
2-Chloronaphthalene	−0.115 (1), −0.095 (3), −0.05 (4), −0.18 (5)	
	−0.165 (6), −0.075 (7), −0.25 (8)	75
1-Bromonaphthalene	−0.095 (2), −0.02 (3), −0.25 (4), −0.165 (5)	
	−0.035 (7), −0.20 (8)	75
2-Bromonaphthalene	−0.26 (1), −0.075 (3)	75
1-Iodonaphthalene	−0.125 (2), −0.055 (3), −0.265 (4), −0.215 (8)	75
2-Iodonaphthalene	−0.295 (1), −0.10 (3)	75
1-Phenylnaphthalene	−0.485 (4)	75
1,8-Dimethylnaphthalene	−0.575 (2), −0.375 (3), −0.61 (4)	76
2,3-Dimethylnaphthalene	−0.685 (1), −0.50 (5), −0.44 (6)	76
2,6-Dimethylnaphthalene	−0.685 (1), −0.445 (3), −0.515 (4)	76
2,7-Dimethylnaphthalene	−0.75 (1), −0.505 (3), −0.445 (4)	76
1,5-Dimethylnaphthalene	−0.515 (2), −0.40 (3), −0.59 (4)	76
1,4-Dimethylnaphthalene	−0.535 (2), −0.38 (7), −0.44 (8)	76

1,3-Dimethylnaphthalene	− 0.625 (2), − 0.805 (4), − 0.475 (6)	76
1,3-$(CH_2)_{10}$-naphthalene	− 0.81 (4)	77
1,3-$(CH_2)_8$-naphthalene	− 0.85 (4)	77
1,3-$(CH_2)_7$-naphthalene	− 0.855 (4)	77
Acenaphthane	− 0.71 (2), − 0.40 (3), − 0.76 (4)	78
Perinaphthane	− 0.715 (2), − 0.36 (3), − 0.68 (4)	78
Fluoranthene	− 0.27 (1), − 0.21 (2), − 0.45 (3), − 0.275 (7), − 0.415 (8)	79
3-Bromofluoranthene	− 0.145 (2), − 0.34 (4), 0.025 (5), − 0.365 (8)	79
8-Bromofluoranthene	− 0.175 (2), − 0.41 (3), − 0.34 (4), − 0.165 (5)	79
Benzo[*c*]phenanthrene	− 0.365 (1), − 0.35 (2), − 0.30 (3), − 0.38 (4), − 0.45 (5), − 0.39 (6)	80
Pentahelicene	− 0.395 (1), − 0.385 (2), − 0.34 (3), − 0.435 (4), − 0.485 (5), − 0.44 (6), − 0.46 (7)	81
Hexahelicene	− 0.46 (1), − 0.415 (2), − 0.36 (3), − 0.45 (4), − 0.505 (5), − 0.435 (6), − 0.495 (7), − 0.49 (8)	82
Benz[*a*]anthracene	− 0.415 (5), − 0.75 (7)	83
7-Methylbenz[*a*]anthracene	− 0.505 (5)	83
12-Methylbenz[*a*]anthracene	− 0.495 (5), − 0.945 (7)	83
Benzo[*a*]naphtho[1,2-*h*]anthracene	− 0.41 (1), − 0.28 (3), − 0.35 (4), − 0.46 (5)	82
Chrysene	− 0.34 (1), − 0.26 (2), − 0.285 (3), − 0.325 (4), − 0.395 (5), − 0.465 (6)	84
Pyrene	− 0.68 (1), − 0.22 (2), − 0.36 (5)	85
Perylene	− 0.705 (1)	85
Coronene	− 0.44 (1)	67
crown-Hexa-*o*-phenylene	− 0.205 (4)	86
helical-Hexa-*o*-phenylene	− 0.34 (4)	86
crown-Tetra-*o*-phenylene	− 0.14 (4)	86
Azulene	− 1.6	87
1,6-Methano[10]annulene	− 0.80 (2)	88
11,11-Difluoro-1,6-methano[10]annulene	− 0.41 (2), − 0.25 (3)	88
1,6:8,13-Propane-1,3-diylidene[14]annulene	− 0.965	89
Ferrocene	− 1.3 (1)	90
o-Carborane	0.31 (1), 0.10 (3)	91
m-Carborane	0.23 (1)	91

way responsible. However, what should be noted in particular here is that the earlier value for σ^+ for *m*-CF_3 of 0.52[5] (and still used by many workers) is considerably in error. The revised value of 0.565 (determined in the gas phase) correlates with better precision not only the data for *all* electrophilic substitutions, but also data for other reactions[12] (see also, for example, the data for addition of TFA to styrenes, and gas-phase basicities of styrenes[92]).

(4) The σ^+ value for *m*-NMe_3^+ (0.36), determined from the solvolysis,[5] incorrectly predicts the effect of this group in the two reactions, positive bromination and protiodegermylation, for which it has been accurately measured. These reactions both predicted a value of 0.67; the value of 0.36 (still given in fairly recent reviews) should not be used. The failure of the solvolysis here is due to the abnormally high entropy of activation found for *m*-$Me_3N^+C_6H_4CMe_2Cl$.[93] The fact that the σ^+ value is here less positive than the σ value arises because of the substantial electron release from the N—Me bond which occurs under conditions of high electron demand (N–C hyperconjugation), and is transmitted to the *meta* position by secondary relay. This electron release also accounts for the substantial differences in the sigma values at the *para* position.

(5) The Hammett–Brown treatment can be criticized on the grounds that it assumes that the resonance interaction between a substituent and the reaction centre in the transition state is independent of the reagent. This assumption would be correct if the transition state was either identical with the Wheland intermediate or had the same structure, irrespective of the reagent, but if this was so it would be difficult to understand why different reagents have different selectivities, i.e. give different ρ factors. There is evidence (discussed below) that the extent to which the aromatic sextet is deformed (and hence the conjugation between a substituent and the reaction centre is altered) at the transition state is partly determined by the nature of the reagent.

In practice, therefore, the use of σ^+ values leads to a number of incorrect predictions of reactivities, especially for highly polarizable substituents. Sulphur-containing substituents in particular show anomalous behaviour, possibly because of the opportunities for combining $-I$, $+M$, and $-M$ effects.[24a] The $SiMe_3$ substituent also shows wide variations in its effect, although a substantial part of the variation here is due to the incursion of steric hindrance to solvation, the second cause of unsatisfactory generality of the σ^+ values (Section 11.1.4).

The difficulties associated with substituents of this type are highlighted by σ^+ values for the Ph—N=N— group, which may be determined from a number of reactions. These are -0.4 (chlorination, AcOCl, aqueous AcOH, 25 °C); -0.187 (nitration, HNO_3–Ac_2O, 0 °C); -0.1 (chlorination, Cl_2, aqueous AcOH, 25 °C); -0.04 (chlorination, Cl_2, AcOH, 25 °C); and 0.09 (bromination, HOBr–H^+, aqueous dioxane, 25 °C).[56]

The variable polarizability effect may also be illustrated by comparing the ratios $\log f^X/\log f^Y$, which should depend only on the nature of the substituents X and Y, but be independent of the ρ values of the reactions studied. The data in

Table 11.3 Variation in the values of $\log f_p^{Me}/\log f_p^{Ph}$ and $\log f_p^{Me}/\log f_{3:4}^{benzo}$

Reaction	f_p^{Me}	$\frac{\log f_p^{Me}}{\log f_p^{Ph}}$	$\frac{\log f_p^{Me}}{\log f_{3:4}^{benzo}}$
Protiodestannylation	5.6	3.01	
Protiodegermylation	14.0	2.66	4.54
Mercuridesilylation (aq. HOAc)	17.5	2.40	2.69
Protiodesilylation	18.0	2.81	3.96
Mercuriation	23.2	1.69	1.05
Bromodesilylation	48.8	1.54	
Positive bromination	59.0	1.48	
Nitration (HNO_3–Ac_2O, 0 °C)	60.0	1.18	
Detritiation (TFA–aq. $HClO_4$)	313	1.45	1.39
Detritiation (TFA)	450	1.20	1.22
Benzoylation	626	1.17	
Acetylation	749	1.20	
Molecular chlorination	820	1.01	
Molecular bromination (85% aq. HOAc)	2420	0.98	1.02

Table 11.3 show that this is untrue, for example, for the cases when X = *p*-Me, and Y = *p*-Ph and 3:4-benzo (2-position of naphthalene), respectively. In each case the ratio decreases fairly regularly as f_p^{Me} (a measure of the reactivity of the reagent) increases, whereas constant values for the ratio of 1.48 and 1.24, respectively, are predicted from the σ^+ values. These results show that the phenyl and benzo substituents are each more polarizable than methyl.

Another example of the polarizability effect shows up in the reactivity produced by the *p*-F substituent, this being particularly noticeable because of the proximity in behaviour to the *p*-H 'substituent'. Thus, whereas the σ^+ value predicts that *p*-F should activate in all reactions, in fact it deactivates in protiodesilylation, protiodegermylation, protiodestannylation, ethylation, nitration, mercuridemercuriation, bromodesilylation, and iododestannylation.

The predicted order for the other *para*-halogen substituents is H > Cl > I > Br. This order is obtained in bromination of polymethylhalobenzenes, hydrogen exchange, bromodeboronation and bromodesilylation, but the orders H > Cl > Br > I and H > Cl > Br ≈ I apply in protiodesilylation and protiodegermylation, respectively. Similarly, the order for the *meta*-halogen substituents is not constant, as reference to the preceding chapters shows. In consequence, the effects of the halogens in either *meta* or *para* positions cannot be adequately described in terms of single substituent constants.

The deficiencies arising from polarizability have led to an elaboration of the Hammett equation, described in Section 11.1.3.

11.1.1 Selectivity

The value of the ratio σ_p^+/σ_m^+ should be constant under ideal conditions, and it is then a simple mathematical consequence of the Hammett equation that f_p/f_m will increase as the ρ factor (and hence f_p) increases. This has been referred to (especially in relation to the methyl substituent effect for which most data exist) as the *Selectivity Effect* (this term also describes the variable discrimination between different aromatics implied by different ρ factors).

A further mathematical consequence is that if $\log f_o/\log f_p$ for a given substituent is constant, indicating a constant steric effect (or absence of one), then f_o/f_p will decrease with decreasing f_p and ρ.[2] It is important therefore to discuss steric effects wherever possible in terms *only* of $\log f_o/\log f_p$ ratios rather than *ortho*:*para* ratios.

11.1.2 Extensions of the Hammett Equation

It is appropriate here to summarize both the various sigma parameters that are in use, and modifications of the Hammett equation, since the plethora of these can cause confusion and engender a lack of confidence amongst those less familiar with the underlying principles of linear free-energy correlations.

As established above, the σ^+ parameter is necessary for correlating reactions where there is the possibility of resonance interaction between an electron-donating substituent and an electron-deficient reaction centre. For comparison purposes one needs ideally a parameter that describes the situation in which there is no corresponding resonance interaction. It would be convenient if the parameter σ did this, since there is a vast range of σ values available. Unfortunately, there are secondary conjugative interactions, whereby the negative charge on oxygen lowers the acidity of the benzoic acids by inhibiting ionization of the O—H bond (eq. 11.4). It was hoped, therefore, that a set of parameters, free from secondary resonance interactions, could be obtained from measurements of the ionization constants for phenylacetic acids (**6**) or hydrolysis of the corresponding esters; the new parameters were designated σ^0.[94]

MeÖ–C₆H₄–C(=O)–O—H ⟷ MeO⁺=C₆H₄=C(–O⁻)–O—H (11.4)

$ArCH_2CO_2H$

(**6**)

Unfortunately, it transpired that the methylene group is able to transmit a small amount of resonance. Consequently, a further set of resonance-free parameters, designated σ^n were devised,[95] the method being essentially to select the minimum value from all those that were obtained for a given substituent in reactions that were considered to have low resonance interactions. Thus σ^n values are more 'resonance-free' than σ^0 values,[6b,96] but unfortunately the compilation is smaller than that for σ^0. In summary, the series of sigma values that describe substituent effects with an increasing resonance component is $\sigma^n < \sigma^0 < \sigma < \sigma^+$.

Attempts have also been made to separate substituent effects into inductive and resonance components. There have been two methods of doing this. One involves dissecting the sigma parameter into an inductive component, σ_I (determined from ionization constants for substituted acetic acids XCH_2CO_2H), and a resonance component, σ_R, as noted in Section 1.4.4, eqs 1.8 and 1.9. If in eq. 1.8, σ_p is replaced by $\sigma_p{}^+$ or $\sigma_p{}^0$, the corresponding resonance components are $\sigma_R{}^+$ and $\sigma_R{}^0$. Hence a series of parameters with increasing resonance components is $\sigma_R{}^0 < \sigma_R < \sigma_R{}^+$.

The other method, due to Swain and co-workers,[97] attempts to avoid the proliferation of scales of polar substituent constants by describing each constant in terms of a field constant, $\mathscr{F}$, and resonance constant, $\mathscr{R}$, according to eq. 11.5.

$$\sigma = f\mathscr{F} + r\mathscr{R} + h \tag{11.5}$$

The factors f, r, and h are weighting factors which differ for each reaction. This method appears to be a good one in principle, but has attracted criticism,[98] and has not been widely applied to electrophilic aromatic substitution. Somewhat similar is the Dewar–Grisdale treatment,[99] which also separates the substituent effects into field F and resonance M components. The effect of a substituent at position i on the reaction site j (σ_{ij}) is a function of both the distance between the two sites r_{ij}, and the amount of charge (q_{ij}) delocalized to position i from a $CH_2{}^-$ group placed at position j, according to eq. 11.6. The

$$\sigma_{ij} = F/r_{ij} + Mq_{ij} \tag{11.6}$$

intention was that values derived from substituted benzenes could be applied to polycyclic aromatics. However, for electrophilic substitution the method has proved rather unsatisfactory, probably because the method is based on π-distribution in the ground state rather than the transition state.[100] However, a modification of the method gives very good correlation of the effects of substituents in detritiation of polycyclic system.[101]

11.1.3 The Yukawa–Tsuno Equation

The ability of a substituent to stabilize the electron-deficient transition state by a mesomeric (conjugative) effect is a function of the electron demand of the reagent. For reaction of an aromatic with a less reactive electrophile, the transition state will be nearer to the Wheland intermediate, and consequently the bond to the electrophile will be stronger in the transition state. A greater proportion of the charge from the electrophile will therefore be transferred to the ring, and this in turn will elicit greater conjugative interaction with the electron-donating substituent.

The realization that the extent of resonance interaction in the transition state depends on the electron demand of the reagent led to the introduction of the Yukawa–Tsuno equation (eq. 11.7),[102] which introduces a parameter, r, that measures the importance of this interaction.

$$\log f = \rho[\sigma + r(\sigma^+ - \sigma)] \qquad (11.7)$$

It will be evident that $\sigma^+ - \sigma$ is a measure of the ability of a substituent to interact mesomerically with an electron-deficient nucleus, i.e. its *polarizability*. When $r = 0$, eq. 11.7 reduces to the Hammett equation (Chapter 1, eq. 1.7), whereas when $r = 1$, it reduces to the Hammett–Brown equation (eq. 11.2). Typical r values obtained in electrophilic substitutions are molecular halogenation 1.66, positive bromination 1.15, nitration 0.9, protiodesilylation 0.65, protiodegermylation 0.60, and protiodestannylation 0.44.

The scales of both r and $\sigma^+ - \sigma$ are arbitrarily chosen, and *any* two reactions with different amounts of resonance interactions in the transition states could have been used to provide the sets of sigma values. Indeed, a more recent version of the equation (eq. 11.8)[103] uses σ^0 values instead of σ values, and since the

$$\log f = \rho[\sigma^0 + r'(\sigma^+ - \sigma^0)] \qquad (11.8)$$

difference $(\sigma^+ - \sigma^0)$ describes a bigger resonance interaction than does $\sigma^+ - \sigma$, it follows that for a given reaction r' will be smaller than r. Because the set of σ values is larger than the set of σ^0 values, most electrophilic substitutions have been evaluated in terms of the earlier equation.

The Yukawa–Tsuno equation has been found to correlate rate data for many electrophilic substitutions with much greater precision than the Hammett–Brown equation. The ρ factors for these reactions are shown in parentheses (together with the associated r factors) in Table 11.4, which lists the ρ factors (calculated using the σ^+ values given in Tables 11.1 and 11.2) for all electrophilic aromatic substitutions. It is emphasized that most of the reactions for which *extensive* and *accurate* rate data are available require the Yukawa–Tsuno equation to correlate the data satisfactorily. It is probable that most electrophilic substitutions correlate best with the Yukawa–Tsuno equation, but for the majority of reactions too few data are available to permit a distinction to be made.

In the light of eq. 11.7, the deviations in the application of the Hammett–Brown equation (eq. 11.2), now become understandable. Two examples illustrate this.

First the variations in the ratios quoted in Table 11.3 are seen to result from the stronger capacity of the p-Ph and 3:4-benzo substituents compared with the p-Me substituent to interact mesomerically, the respective values of $\sigma^+ - \sigma$ for

Table 11.4 ρ Values for electrophilic substitutions

Reaction	ρ[a]
Arylation, MeCN, 40 °C	−0.8
Hydroxydeboronation, H_2O_2–H^+, 25 °C	−1.2
Protiodemagnesiation, Et_2O, 31.5 °C	−1.7
Carboxyamination, $EtCO_2N_3$, TFA–$PhNO_2$, 109 °C	−1.7
Chlorination (of 9-X-anthracenes), SO_2Cl_2, PhCl, 25 °C	−2.1
Bromination (of *p*-*N*,*N*-dimethylanilines), Br_2, H_2O, 25 °C	−2.2
Protiodestannylation (of $ArSnMe_3$), HCl–MeOH, 25 °C	−2.2
Ethylation, EtBr–$GaBr_2$, 25 °C	−2.4
Fluorination, F_2–N_2, −78 °C	−2.45
Protiodeplumbylation, $HClO_4$–EtOH, 25 °C	(−2.5, $r = 0.4$)
Protiodemercuriation, arylmercury(II) chlorides, aq. EtOH, 70 °C	−2.45
Bromodestannylation, Br_2, MeOH, 25 °C	−2.6
Protiodemercuriation, diarylmercurials, DMSO–dioxane, 32 °C	(−2.8, $r = 0.5$)
Iododemercuriation, ArHgCl, I_2, 25 °C	−2.9
Iododestannylation, I_2, MeOH, 25 °C	(−2.95, $r = 0.65$)
Protiodeiodination (of 4-hydroxybenzenes), aq. HOAc, 40 °C	−3.0
Mercuridestannylation, $Hg(OAc)_2$, THF, 20 °C	ca −3.0[b]
Mercuridemercuriation, ArHgCl, $HgBr_2$, 20 °C	−3.4
Chloromethylation, paraform–HCl, 85 °C	−3.5 to −4.9
Nitramine rearrangement, $HClO_4$, dioxane, 40 °C	−3.7
Bromodeboronation, Br_2, aq. HOAc, 25 °C	−3.8
Protiodestannylation [of ArSn(cyclohexyl)$_3$], $HClO_4$–EtOH, 50 °C	(−3.8, $r = 0.4$)
Protiodemercuriation, diarylmercurials, aq. DMSO, 30 °C	(−3.8, $r = 0.5$)
Protiodetritylation, HOAc–aq. H_2SO_4, 20 °C	−3.9
Nitration (gas-phase), $MeO^+(H)NO_2$,	ca −3.9
Mercuriation, $Hg(OAc)_2$, HOAc, 25 °C	−4.0
Chlorination (of 1,3-dimethoxybenzenes), SO_2Cl_2, PhCl, 25 °C	−4.0
Vinylation, 1-anisyl-2,2-diphenylvinyl bromide–$AgBF_4$–base, 120 °C	−4.1
Mercuriation, $Hg(OCOCF_3)_2$, 25 °C	−5.7 to −6.4
Protiodegermylation, $HClO_4$–MeOH, 50 °C	−4.4, $r = 0.6$)
Sulphonoxylation, (m-$NO_2C_6H_4SO_2O)_2$, ethyl acetate, 20 °C	−4.4
Amination, ArN_3, TFA, CH_2Cl_2, 25 °C	−4.5
Iodination, I_2–HNO_3–H_2SO_4, 25 °C	−4.5
Protiodeacylation, 89.8% H_2SO_4, 25 °C	−4.6
Chlorination (of *p*-X-acetanilides), Cl_2, HOAc, 25 °C	−4.7
Protiodeboronation, 74.5% aq. H_2SO_4, 60 °C	−5.2
Protiodesilylation, $HClO_4$–MeOH, 50 °C	(−5.3, $r = 0.65$)
Rhodiation, (octaethylporphyrinate)rhodium(III) chloride, 50 °C	(−4.5, $r = 0.3$)
Bromination (of *p*-X-anisoles), Br_2, H_2O, 25 °C	−5.6
Sulphonation, $H_2S_2O_7$, 25 °C	−6.1
Bromination, HOBr–$HClO_4$, 50% aq. dioxane, 25 °C	−6.2

(*continued*)

Table 11.4 (*continued*)

Reaction	ρ^a
Nitrosation (of 4-X-phenols), $NaNO_2$–aq. $HClO_4$, 0 °C	−6.2
Iodination, I_2–KIO_3–H_2SO_4, 25 °C	−6.3
Nitration, HNO_3–H_2SO_4, 25 °C	−6.5 ($-6.4, r = 0.9$)
Bromination (of *o*-X-anisoles), Br_2, H_2O, 25 °C	−6.5
Bromodesilylation, Br_2, aq. AcOH, 25 °C	($-6.8, r = 0.8$)
Nitrosation, $NaNO_2$–aq. $HClO_4$, 52.9 °C	−6.9
Thalliation, $Tl(OCOCF_3)_3$, TFA, 25 °C	ca −7.0
Chlorination, SO_2Cl_2, $MeNO_2$ 25 °C	−7.2
Nitrodemethylation (of polymethylbenzenes), HNO_3–Ac_2O, 25 °C	−8 to −12
Thalliation, $Tl(OAc)_3$, HOAc, 25 °C	−8.3 ($-8.6, r = 0.6$)
Protiodetritiation, TFA, 70 °C	−8.75
Acetylation, MeCOCl–$AlCl_3$, $C_2H_4Cl_2$, 25 °C	−9.1
Sulphonation, $H_3SO_4^+$, 25 °C	−9.3
Sulphonation, $ClSO_3H$, $MeNO_2$, 25 °C	ca −10
Chlorination, Cl_2, AcOH, 25 °C	−10.0
Bromination (of polymethylbenzenes), Br_2, H_2O, 25 °C	−10.7
Bromination, Br_2, H_2O, 25 °C	−11.6
Thiocarboxylation, RSCOCl–$AlCl_3$, 1,2-dichloroethane, 25 °C	ca −11.7[c]
Bromination, Br_2, AcOH, 25 °C	−12.1
Protiodealkylation	ca −12.5
Bromination, Br_2, TFA, 25 °C	−16.7
Acetoxylation, Ac_2O–HNO_3, 25 °C	ca −18

[a] Values in parentheses refer to correlation against the Yukawa–Tsuno equation.
[b] This is for correlation against σ values.
[c] This value was obtained by plotting against σ; the correlation against σ^+ values was poorer but gave a lower ρ value.

these substituents being 0.19, 0.18, and 0.14. The corresponding σ values are 0.02, 0.04, and −0.14, and since $r(\sigma^+ - \sigma)$ becomes increasingly important relative to σ as r increases, it follows that as r increases the partial rate factors for the former two substituents 'catch up' with those for *p*-Me.[104] Ultimately, in bromination in aqueous HOAc (which has a high r value), f_p^{Ph} becomes larger than f_p^{Me}.

Second, the behaviour of the *para*-halogens can be understood. Consideration only of their inductive effects would lead to an order of reactivities, H > I > Br > Cl > F, but superimposed on the inductive effect is the mesomeric effect, the two together leading to the order measured by σ values, H > F > I > Cl > Br, and this would be the order in a reaction for which $r = 0$. Their σ^+ values decrease in the order F > H > Cl > I > Br, and this order would apply in reactions for which

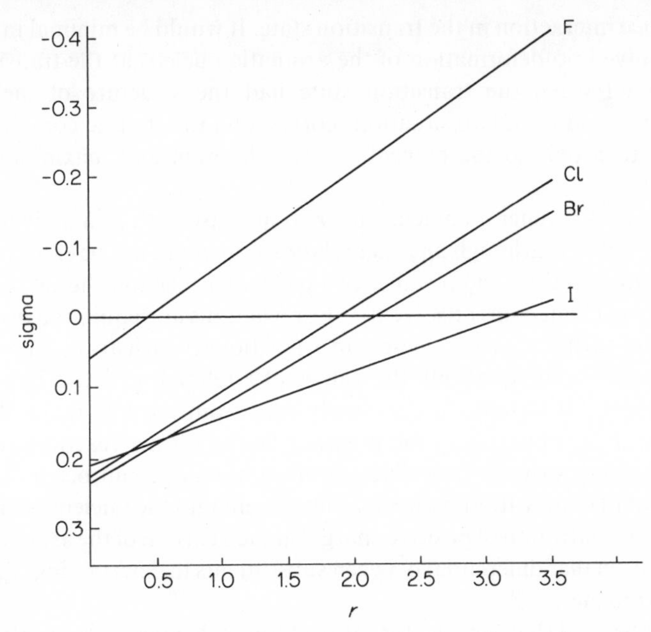

Figure 11.1. Variation in *p*-halogen substituent effect with *r*-value.

$r \approx 1$. As r increases, therefore, the order changes and various relative reactivities may arise, as illustrated diagramatically in Fig. 11.1; on this (points are omitted for clarity) the slopes of the lines are proportional to $\sigma^+ - \sigma$ and are therefore a measure of the conjugative electron-releasing ability of the halogen. (The order of the slopes also parallels the values $\sigma_p - \sigma_I$, which is another measure of this ability.) It is interesting that recent measurements of ^{13}C chemical shifts for various α-side-chain carbocations of the type ArC^+RR' require enhanced σ^+ values (designated σ^{c+} values) to correlate the data, and this corresponds to a reaction with an r factor of 3.4 (see Fig. 11.1), the ρ factors being in the range -15 to -18.[105] Under these circumstances, not only does *p*-F activate, but *p*-Cl and *p*-Br also do so, the corresponding σ^{c+} values being -0.40, -0.24, and -0.19.[105]

The physical interpretation of the parameters ρ and r in the Yukawa–Tsuno equation is as follows. The parameter ρ measures the extent to which the electron pair, used to form the new bond to the electrophile, has been removed from the aromatic system in the transition state. As the extent of the transfer is increased, the positive charge on the residual aromatic system is increased, and therefore the differential effects between substituents in stabilizing the transition state are increased; that is, an increase in electron transfer corresponds to an increase in selectivity, or ρ value. The parameter r is defined as a measure of the importance of resonance interaction in the transition state. It would be minimal in a reaction which involved no deformation of the aromatic nucleus at the transition state, and maximal when the transition state had the structure of the Wheland intermediate, and these two situations correspond to zero and complete transfer of the electron pair to the reagent, i.e. to minimum and maximum ρ values, respectively.

There is thus a general tendency for r to increase with ρ in a given aromatic system (see below), although an exact relationship might not be expected because it is not clear that the r and ρ values of a particular reaction should respond in a similar way to changes in the solvent.[51] For reactions and equilibria involving the formation of carbocations at side-chain α-positions (which are also correlated by the Yukawa–Tsuno equation), the values of r relative to ρ are larger than in electrophilic substitutions; this is clearly seen from the ^{13}C n.m.r. data noted above. This is attributable to the presence, in the former reactions, of a larger number of centres over which delocalization of charge can occur. Thus, for a given extent of conjugation between a substituent and the nucleus (same r), there is a lower concentration of positive charge on each carbon of the aromatic system, and hence a smaller differential between substituents in determining the energy of the transition state.[106]

It is emphasized that the ρ factor for reaction with a given electrophile will not necessarily be the same for reaction with differing systems. Not only may the extent of electron transfer from the *ipso* carbon to the electrophile in the transition state differ according to the reactivity of the aromatic (Section 11.1.5), but the ease of transmission of electrons from the substituent to the *ipso* carbon may differ according to the aromatic system (for a fuller discussion, see ref. 107). An example that shows this clearly comes from protiodetritiation by TFA at 70 °C of benzene and the 1-position of naphthalene. The conditions are identical, but substituent effects at the *para* (and *meta*) position are smaller in naphthalene (see Table 3.23) because delocalization of the charge in naphthalene is less favourable owing to bond fixation in the ground state [Section 3.1.2.13(iv)].

A further example concerns substituent effects between the 4- and 4′-positions in biphenyl. For detritiation (Section 3.1.2.10) and protiodesilylation (Section 4.7.1(iii)), the ρ factors are reduced ca 4-fold, and for reactions producing carbocations at the side-chain α-position, e.g. pyrolysis of 1-arylethyl acetates,[25] and solvolysis of 1-aryl-1-chloroethanes, 1-aryl-1-phenylchloromethanes, and 2-aryl-2-chloropropanes,[108,109] reduction factors of 3.2–3.5 apply. (The factors are smaller for these side-chain reactions presumably because, as argued above, a smaller fraction of charge is delocalized on to the aromatic ring.) It is noteworthy here also that the r factors are reduced ca 2-fold compared with the same reaction in benzene, because of the reduced conjugation between the aryl rings, which are at an angle of 45° to each other. [Note that in ref. 108 r' factors are used (eq. 11.8) and since these are larger than r factors, the attenuation factor is reduced.]

11.1.4 The Electronic Effect of Bulky Groups: Steric Hindrance to Solvation

In Section 1.4.3 it was shown that the intrinsic electron-releasing ability of alkyl groups parallels their size, because C–C hyperconjugation is greater than C–H hyperconjugation. The observation of the so-called Baker–Nathan order of electron release, e.g. *p*-Me > *p*-*t*-Bu, arises only because steric hindrance to solvation of the bulkier *tert*-butyl compound causes the transition state for reaction of the *tert*-butyl compound to be less stabilized than that for the methyl compound. Thus the Baker–Nathan order is found for reactions carried out in highly solvating media, and also in poorer solvating media where the solvating counter ion is especially bulky, e.g. SbF_6^-.

It follows that the electron-supplying effects of bulky groups is greatest under non-solvating conditions, and diminishes with increasing solvating power of the medium. This is clearly seen from the 'σ^+' values (i.e. log f/ρ values) obtained for these substituents in three reactions (Table 11.5):[38,40,110] pyrolysis of 1-arylethyl acetates (no solvent), protiodetritiation in TFA (poorly solvating solvent), and solvolysis of 2-aryl-chloropropanes in either aqueous ethanol or aqueous acetone (which are moderately good and excellent solvating solvents, respectively).

Table 11.5 Values of σ^+ derived from reactions with differing solvating conditions

Substituent	Solvolysis, aq. acetone	Solvolysis, aq. ethanol	Detritiation in TFA	Pyrolysis gas phase	$\sigma^+_{solv.} - \sigma^+_{pyrol.}$
p-Me	−0.311	—	−0.303	−0.29	−0.021
m-Me	−0.066	−0.077	−0.090	−0.098	0.032
p-*t*-Bu	−0.256	—	−0.312	−0.365	0.109
m-*t*-Bu	−0.059	−0.084	−0.175	−0.190	0.131
p-$SiMe_3$	0.020	—	—	−0.090	0.110
m-$SiMe_3$	0.010	—	—	−0.160	0.170
m-Ph	0.109	0.056	0.020	0.0	0.109
p-Cyclopropyl	−0.462	—	−0.47		(0.011)[a]
p-Cyclobutyl	−0.290		−0.345		(0.056)[a]
p-Cyclopentyl	−0.302		−0.35		(0.049)[a]
p-Cyclohexyl	−0.285	—	−0.338	−0.380	0.095[b]
p-Neopentyl	−0.240	—	−0.306	−0.350	0.110
p-Adamantyl	−0.250	—	−0.377	−0.425	0.175
p-Bicyclo[2.2.2]-octan-1-yl	−0.270	—	−0.391	—	—
p-*endo*-Norbornan-2-yl	−0.295	—	−0.358	−0.390	0.095
p-*exo*-Norbornan-2-yl	−0.309	—	−0.371	−0.420	0.111

[a]These are $\sigma^+_{solv.} - \sigma^+_{exchange}$ values.
[b]The $\sigma^+_{solv.} - \sigma^+_{exchange}$ value is 0.53.

These results demonstrate a number of important points:

(1) Whereas the σ^+ values for the *p*-methyl substituent are essentially constant under all three conditions, the σ^+ values for the other substituents show a regular and large increase (i.e. become more negative) with decrease in the solvating power of the medium. This increase is large for the bulky *para* substitutions, and larger still for the bulky *meta* substituents.

(2) The effects of the cycloalkyl substituents, although not measured for each in the gas phase, show the same trend, the value of $\delta\sigma^+$ between solvolysis and hydrogen exchange being substantially smaller for the cyclopropyl group than for the bulkier homologues.

(3) The results provide clear evidence that steric hindrance to solvation is more serious for *meta*- than for *para*-alkyl substituents, at least in the side-chain solvolytic reaction used to define the original σ^+ values. A consequence is that the original σ^+ value for *m*-Me of -0.066, still used by many workers, does not in fact correlate satisfactorily *any* electrophilic aromatic substitution! Hence the average value needed to correlate the data in hydrogen exchange (TFA and H_2SO_4), protiodesilylation, -germylation, and -stannylation, mercuriation, vinylation, acetylation, benzoylation, sulphonation, molecular and positive bromination, molecular and positive chlorination, and iodination, is -0.086, with higher values (up to -0.105) being needed for the reactions of higher ρ value. Likewise, the average value needed for the side-chain reactions 1, 2, 4, 9, 10, 13, 24–27, 29, and 30 in ref. 102, Table 1, is also 0.086.

The tendency for the σ^+ value to increase with increasing ρ value for electrophilic substitutions strongly indicates that the *m*-Me substituent is susceptible to polarizability effects to an extent not previously recognized, this property having been masked by solvation effects. Very strong evidence that this is the case comes again from gas-phase measurements: in the pyrolysis of 1-arylethyl methyl carbonates and 2-aryl-2-propyl acetates, reactions both of which have a more polar transition state than that in pyrolysis of 1-arylethyl acetates (which gave a σ^+ value of -0.098), a value of -0.135 was needed to correlate the effect of the *m*-Me substituent. For 1-arylethyl benzoates, a reaction with a transition state of intermediate polarity, a value of -0.115 was required.[6a,6b,7b] Evidently the more polar the transition state, and hence the greater the demand for resonance stabilization, the greater is the conjugative electron release from the *m*-Me substituent. Other *m*-alkyl groups could be expected to respond similarity, but no data are available.

(4) The anomalous behaviour of the $SiMe_3$ substituent is now readily explained in terms of steric hindrance to solvation. The solvolysis reaction predicted that in both *meta* and *para* positions this substituent should deactivate, whereas in most electrophilic substitutions activation is observed. It is noteworthy that not only is the difference between the solvolysis and gas-phase σ^+ values again greater for the *meta* position than for the *para* position, thereby paralleling the behaviour of the alkyl groups, but the values of $\delta\sigma^+$ are very similar. (Although silicon is larger

Table 11.6 Values of σ^+ derived from various reactions

Reaction	$\sigma^+_{p\text{-Me}}$	$\sigma^+_{p\text{-}t\text{-Bu}}$
Solvolysis of 2-aryl-2-chloropropanes	−0.311	−0.256
Protiodesilylation	−0.288	−0.259
Protiodegermylation	−0.294	−0.272
Bromodesilylation	−0.273	−0.236
Mercuriation	−0.340	−0.309
Bromination (H^+–HOBr)	−0.286	−0.256
Protiodetritiation (TFA)	−0.303	−0.312
Protiodetritiation (HOAc–H_2SO_4)	−0.320	−0.300
Bromination (Br_2–aq. HOAc)	−0.280	−0.240
Chlorination (Cl_2–HOAc)	−0.291	−0.260
Acetylation	−0.316	−0.310
Pyrolysis of 1-arylethyl acetates	−0.290	−0.365
Overall variation	0.067	0.129

than carbon, the methyl groups are further removed from the aromatic ring so that a comparable steric effect is to be expected.)

(5) The former interpretation of the Baker–Nathan effect, viz. that the variation in the *p*-*t*-Bu:*p*-Me activating ratio was due to variable C–H hyperconjugation operating on reagent demand, can be dismissed by examination of the values of σ^+ ($\log f/\rho$) for these two substituents from various reactions (Table 11.6). Clearly it is the *tert*-butyl group rather than the methyl group which produces the most variable activation, and this is due to steric hindrance to solvation.

11.1.5 The Reactivity of the Aromatic: the Reactivity–Selectivity Principle

According to the foregoing discussion, the structure of the transition state depends on the reactivity of the electrophile but not on that of the aromatic. This assumption cannot be strictly correct because there is in principle no difference between the electrophile and aromatic. According to the Hammond postulate,[111] in reactions of two aromatics with the same electrophile the transition state formed by the less reactive aromatic should more closely resemble the Wheland intermediate than that formed by the more reactive aromatic, just as the arguments presented earlier led to the view that as the reactivity of the electrophile is decreased, the transition state more closely represents the Wheland intermediate.

There are certain indications that this is the case; the large ρ value for bromination by bromine in aqueous HOAc suggests that there is a considerable proportion of unit positive charge on the aromatic ring in the transition state, yet the activation energies in this reaction for *N,N*-dimethylaniline and related compounds (activated ca 10^{18} times more than benzene) are close to zero.[112] This suggests that the reactivity of these aromatics is so great that the transition states now more closely resemble the reactants. (This implied shift of the transition state is, however, surprisingly large in the light of the observation that for hydrogen exchange of aromatics with electrophiles having a 10^{11}-fold difference in reactivity, the ρ factor was reduced by only ca 20%.[113])

Just as a given pair of aromatics show differing substrate and positional selectivities towards differing electrophiles, so a given electrophile will discriminate less between a pair of more reactive aromatics than between a pair of less reactive aromatics of comparable structure. Both of these results are consequencies of the *Reactivity–Selectivity Principle*, and are manifest in a breakdown of additivity. Additivity of substituent effects has been examined in protiodesilylation, mercuriation, protiodetritiation, nitration, molecular chlorination, bromination, and iodination of polymethylbenzenes. The results are complicated by buttressing effects when substitution occurs at a position next to two adjacent methyl groups, and also because the steric effect of two *o*-Me groups is probably greater than predicted from the results of one such group; for nitration the onset of encounter control prevents meaningful analysis of the results. For the first two reactions, which have small ρ factors, the observed reactivities are equal to or slightly greater than those predicted, but in the other four reactions, the observed reactivities for the more highly substituted compounds are lower than predicted. If steric effects alone were responsible, they would certainly be apparent in mercuriation, but they are not. It is therefore probable that as the aromatic becomes more reactive, the transition state is formed earlier in passage along the reaction coordinate with the consequence that the selectivity of the reagent towards different aromatics is reduced and the introduction of a further activating substituent has a smaller effect than expected.

A further consequence is that two deactivating substituents should produce greater deactivation than calculated, provided that they are of the $-I$, $-M$ type, and this prediction[114] has been confirmed in halogenation (Section 9.6.5). Two or more $-I$, $+M$ substituents tend to produce greater reactivity than predicted, attributable largely to greater conjugative electron release from each when placed in a less reactive compound. Examples of this behaviour are found in the protiodesilylation and protiodestannylation of pentafluorobenzene [Sections 4.7.1.(ii) and 4.9.1] and the nitration of polychlorobenzenes (Table 7.11).

Compounds containing an electron-withdrawing and an electron-supplying substituent are usually more reactive than predicted. There are very many examples of this behaviour and it may be attributed to the following. The effect of the electron-withdrawing substituent will be *less* than in benzene itself, because of the greater reactivity of the system in which it is placed. Conversely, the electron-supplying substituent will be *more* activating than when substituted in benzene. The net effect will therefore always be to make the observed reactivity *greater* than predicted. Quantitative examples may be found in hydrogen exchange (Table 3.12) and halogenation (Section 9.6.5).

The *Reactivity–Selectivity Principle* (and consequences flowing from it) can break down owing to various causes:

(1) In nitration, for example, *ipso* substitution followed by rearrangement can intervene with polysubstituted compounds.

(2) Comparison of aromatic systems with different transmission pathways, e.g. benzene and thiophene, can lead to seemingly anomalous behaviour if the ρ factor is taken merely as an indication of the reactivity of the electrophile; as noted above, it is also a measure of the ease of transmission of substituent effects through the aromatic system.

(3) It has been pointed out[115] that the principle follows from the Hammond postulate only if the following applies: in the reaction of two electrophiles A and B (the former being the more reactive and therefore giving the earlier transition state) with two different aromatics X and Y, the difference in the free energies of the products BX and BY should be significantly smaller than that between the products AX and AY. This may not always be the case.

Probably the clearest example of the principle comes not from electrophilic substitution, but from the related reactions whereby a carbocation is produced at the side-chain α-position, and in which the aromatic system is constant throughout. In the solvolysis of compounds $ArCX(Me)OC_6H_4NO_2$-4, a plot of $\log k$ for the unsubstituted compounds ($Ar = Ph$) against the ρ factors for 34 compounds (different X groups) was linear with only five deviant points (due to secondary effects).[116] A perfect linear correlation is also found between $\log k$ and ρ for pyrolysis of esters ArCH(Me)OR.[25]

Because the *Reactivity–Selectivity Principle* requires that the transition state for reaction of each compound containing a substituent will occupy a different position along the reaction coordinate from that for the unsubstituted compound, it is widely believed (see, e.g., ref. 115) that Hammett plots should therefore be curved. This is incorrect. The amount by which the transition state is transposed along the reaction coordinate will be *proportional to the reactivity of the compound and hence to the sigma value.* It follows, therefore, that a straight-line correlation must be obtained. The ρ value obtained will contain a small 'error' and will be reduced slightly from that which it should be insofar as it describes the position of the unsubstituted compound along the reaction coordinate. Since the movement of the transition state for compounds in reactions of high ρ factor is likely to be greater than for reactions of small ρ factor, it seems probable that the overall effect will be to condense the range of ρ factors from their 'true' values.

11.1.6 Theoretical Calculations of Aromatic Reactivities

Whereas Hammett-type equations merely predict relative reactivities in one reaction on the basis of data observed in other, similar, reactions, theoretical methods attempt to predict reactivities from basic principles. Like the treatments which assign a sigma parameter to each substituent to describe its electronic effect, the theoretical treatments predict that the order of activating or deactivating power of a substituent should remain constant irrespective of the nature of the reagent. Experimentally this is known not to be the case, and evidence makes it clear that it is necessary to consider both polarizability and solvent effects in a satisfactory treatment. The recognition of the extreme difficulty of this probably accounts for the reduced activity in this area in recent years.

As noted in Section 2.2, efforts have been concentrated upon predicting relative reaction rates rather than absolute values. Moreover, models for the transition state have to be used, such as the ground state, or, more commonly, the Wheland intermediate since this is more successful. This success stems from the endothermic nature of coordination of the electrophile with the aromatic due to the loss of aromatic resonance energy; hence, according to the Hammond postulate, the transition state will tend to be product-like.

There are two additional difficulties. First, whereas it is possible to calculate ground-state charge densities, there is no correlation between reaction rates and these partial charges,[117] reflecting the fact that the transition state is not reagent-like. (For the same reason, the frontier-orbital method, in which the index of reactivity is known as *superdelocalizability*, gives a poor indication of electrophilic aromatic reactivities.[118]) Unfortunately, there is no satisfactory method for calculating the effects of substituents in the Wheland intermediate. Second, the effect of steric hindrance cannot be predicted.

In view of these difficulties, most efforts have been directed towards predicting the relative reactivities of unsubstituted polycyclic aromatics, and the data from hydrogen exchange have attracted most interest because of the freedom from steric hindrance. Various methods are available for calculating reactivities, and comparison of various methods using a limited range of the hydrogen exchange data given in Table 3.17 shows that there is little difference in correlating abilities of the PPP (self-consistent field), CNDO/2 (all-valence electrons), Hückel (ω-technique, in which the coulomb integrals that are a measure of the electronegativities of the carbon atoms in the aromatic framework are allowed to vary from each other), and free-electron superdelocalizability methods, the correlation coefficients lying between 0.966 and 0.979.[119] *Ab initio* calculations could be expected to give better results, but at present it is not feasible to calculate results for molecules of the size of even the smallest aromatic. Satisfactory correlations of hydrogen-exchange rate data and basicity constants of polycyclic aromatics have also been obtained as expected in view of the similar pathways for protonation and exchange.[120]

Earlier it had been shown that data for nitration and chlorination of polycyclic aromatics could be correlated with reasonable precision against values of $-\Delta E/\beta$, where ΔE is the activation energy and β is the carbon–carbon resonance integral, if values of 6 and 12 kcal mol^{-1} were used, respectively, for β.[121]

Fortuitously, these values are almost exactly the same as the ρ values for the reactions, and since the normal value for β is ca 20 kcal mol^{-1}, the implication is that an electrophilic substitution for which the transition state was identical with the Wheland intermediate would have a ρ factor of ca -20. This seems reasonable in view of the fact that reactions which produce a full carbocation at the side-chain α-position in the (solvated) transition state have ρ-factors which are approximately one third of this value; ρ factors are known to be reduced approximately 2.5-fold per intervening carbon atom.

11.2 THE *ORTHO*:*PARA* RATIO

The factors which govern the relative reactivity of the *meta* and *para* positions in a monosubstituted aromatic compound are essentially confined to electronic effects. The *ortho*:*para* ratio is governed in addition by steric effects, and the superimposition of two groups of effects of different type presents difficulties in the elucidation of the significance of each individual effect. Although in certain cases it is evident that one particular factor is dominant in determining the ratio, in most cases it is at present not possible to separate the resultant effect into its components.

Factors which contribute to the determination of the *ortho*:*para* ratio are selectivity, steric hindrance, steric acceleration, interaction between the substituent and the reagent, electronic effects, and solvent and temperature effects.

11.2.1 Selectivity

If there is a linear free-energy relationship between substitution at the *ortho* and *para* positions, i.e. there is either no steric hindrance or a constant one, then $\log f_o$:$\log f_p$ will be constant.[122] Consider now substitution of toluene in two hypothetical reactions of ρ values -4 and -10, such that the foregoing applies and the value of the ratio is 0.8. It is simply calculated that in the reaction of ρ value -4 the *ortho*:*para* ratio will be 0.56, whereas in the reaction of ρ value -10 it will be 0.18, and this is simply a mathematical consequence of selectivity. Hence, although both reactions give different *ortho*:*para* ratios, there is no difference in steric effect between them. For a meaningful discussion of steric effects it is desirable to use $\log f_o$:$\log f_p$ values, though in many cases the partial rate factors are unavailable, in which case the *ortho*:*para* ratios may, with caution, be used.

11.2.2 Steric Hindrance and Steric Acceleration

The effects of steric hindrance and steric acceleration on the $\log f_o$:$\log f_p$ ratios have been described in Sections 2.5.1 and 2.5.2.

11.2.3 Interaction between Substituent and Reagent

Examples in which the substituent and the reagent interact, thereby facilitating *ortho* substitution, were discussed in the nitration of ethers and anilides, compounds that have an unshaired pair of electrons suitably placed in the side-chain [see Section 7.4.3.(x)]. The comparatively high *ortho*:*para* ratio (63:37) in nitration of benzyl acetate by acetyl nitrate[123] may have an analogous basis. Significant too may be the higher 2:4 ratio of nitro products obtained from 1-*p*-tosylamino-7-nitronaphthalene when nitration is carried out in a mixture of acetic and propanoic acids (2.5) than when fuming nitric acid is used (0.6).[124]

Other reactions in which high *ortho*:*para* ratios have been ascribed to *ortho* interactions are lithiation of compounds with unshared electron pairs in side-chains (Section 5.1), thalliation of anisole (Section 5.5), nitration of benzyltrimethylsilane [Section 7.4.3(xi)], hydroxylation of ethers (and also of toluene) (Sections 8.1.2 and 8.1.3), arylation of anisole (Section 8.3), and chlorination of phenol by 2,3,4,4,5,6-hexachloro-2,5-dienone (Section 9.2.5). The high ratio obtained for nitration of biphenyl under some conditions has been explained in a related way [Section 7.4.3(xv)]. It is probable that in a number of other instances such interactions partly determine the *ortho*:*para* ratio, but recognition of the effect is rendered difficult by the importance of other factors being unknown. In the following examples, however, *ortho* substitution is clearly facilitated.

First there is evidence that the high proportion of *ortho*-substituted product formed in the reaction between phenoxide ion and formaldehyde results from the formation of a complex between formaldehyde and the metal ion of a metal–phenoxide ion pair within which electron redistribution takes place (eq. 11.9).[125]

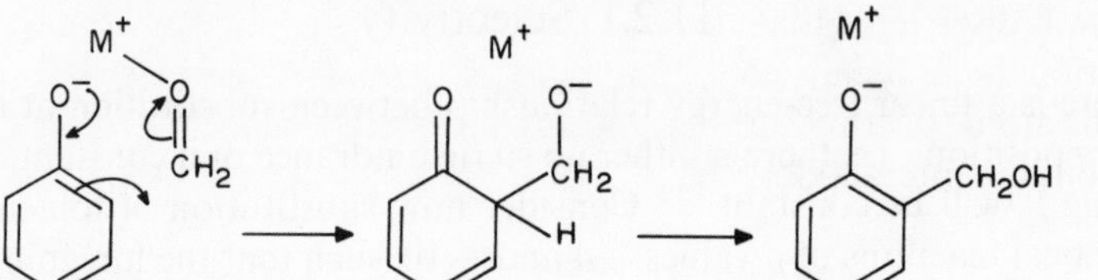

(11.9)

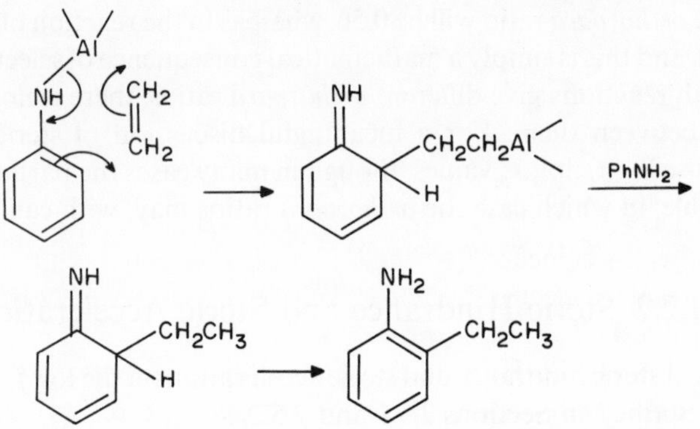

(11.10)

Second, the alkylation of primary and secondary aromatic amines with alkenes in the presence of aluminium anilide catalysts gives only *ortho* derivatives, attributed to the operation of a cyclic mechanism (eq. 11.10).[126]

An analogous explanation has been advanced for the predominance of *o*-alkyl- and 2,6-dialkylphenols formed by the alkylation of phenols with alkenes in the presence of aluminium phenoxide catalysts.[126]

11.2.4 Electronic Effects

Because so little is known about the magnitude of steric hindrance to *ortho* substitution, it is not always easy to evaluate the importance of polar factors in governing the *ortho*:*para* ratio when this is less than the statistical value of 2 ($\log f_o : \log f_p = 1$). When, however, the ratio is greater than this, then provided that it is not the result of either a facilitating *ortho* interaction or of steric acceleration, polar factors must be powerful enough to outweigh steric hindrance. The theory of the role of polar effects in determining the *ortho*:*para* ratio stemmed originally from studies of aromatic substitutions which gave rise to ratios greater than 2, and was then extended to other reactions which give lower ratios.

The *ortho*:*para* ratio in the nitration of nitrobenzene is exceptionally large (21), and it was first suggested that *ortho* substitution might be facilitated by a dipolar interaction in the transition state between the nitro substituent and the entering nitronium ion, the stereochemistry of the transition state (**7**) being suitable for the interaction.[127] This explanation was found to be inadequate when it was discovered that cyanobenzene also gives a high *ortho*:*para* ratio (8.5) in nitration; here, because of the linearity of the cyano group, the electron-rich nitrogen atom of this substituent is not stereochemically suited for the corresponding interaction.[128]

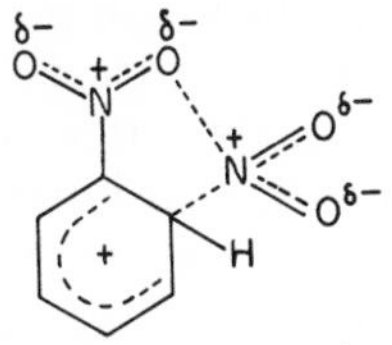

(**7**)

However, there is evidence that the high ratios obtained in these and related cases are the result of electronic effects that favour *ortho* over *para* substitution. The data in Table 11.7 are the *meta*:*para* and *ortho*:*para* ratios for nitration and chlorination (by $HOCl–H^+$) of benzenes containing electron-withdrawing substituents. First, high *ortho*:*para* ratios are obtained in both reactions and

Table 11.7 Orientation ratios in nitration and chlorination of PhX

	Nitration		Chlorination	
Substituent, X	*o*:*p*	*m*:*p*	*o*:*p*	*m*:*p*
NO_2	21.4	310	11.8	53.8
CO_2H	14.2	61.6	—	—
CN	8.6	40.4	8.0	25.4
CO_2Et	8.6	20.8	—	—
CF_3	2.0	30.4	3.8	19.6
CHO	2.0	8.0	5.2	11.0

show similar trends in each as the substituent is varied. The electrophiles in each reaction have different geometry, and it is therefore unlikely that specific steric interactions govern both sets of *ortho*:*para* ratios.[129] Second, for each reaction, the variation in *ortho*:*para* ratio follows closely that in the *meta*:*para* ratios, and, as pointed out earlier with reference to nitration,[130] this can hardly be coincidental. It is probable that the same factors determine both ratios.

In rationalizing these results in terms of electronic factors, it is first assumed that the transition state for these substitutions is similar in structure to the Wheland intermediate, represented as a hybrid of structures (**4**–**6**, Chapter 1). This assumption is reasonable, for the aromatics are each fairly unreactive so that, according to the Hammond postulate (Section 11.1.5), considerable progress should have been made along the reaction coordinate at the transition state. The simple Hückel treatment indicates that the charge in this ion is shared equally between three carbon atoms, as in **8**, but n.m.r. measurements of the charge distribution in the benzenonium ion show that the charge is more heavily concentrated at the *para* position, as shown in **9**.[131] Further, the calculated charge densities for this ion, obtained by a perturbation method allowing for electron interaction, give this same order of charge densities (*para* > *ortho* > *meta*).[132] If then, as is reasonable, this order is unchanged when the electrophile E replaces a proton in **9**, a substituent should have a more powerful effect on the stability of **9** when it is *para* to the entering reagent than when it is *ortho*, and a number of conclusions then follow:

H E 1/3+ 1/3+ 1/3+ (**8**)

H H + 0.26 0.09 0.30 (**9**)

(1) The transition state for the *para* substitution of the compounds in Table 11.7 will be of higher energy than that for *ortho* substitution, leading to the high *ortho*:*para* ratios. Moreover, the difference in transition-state energies should be greater as the electron-withdrawing capacity of the substituent is increased, and it follows, therefore, that the *ortho*:*para* ratios in Table 11.7 decrease approximately with decreasing σ^+ value of the substituent.

(2) Conversely, an electron-releasing group should cause the stabilities of the Wheland intermediates to fall in the order *para* > *ortho* > *meta*. For reactions in which the Wheland intermediate provides a satisfactory model for the transition state, the *ortho*:*para* ratio should be <2.

Since electron-releasing substituents activate the nucleus and therefore result in the transition state occurring earlier in the passage along the reaction coordinate, particularly when the electrophile is also of high reactivity, this theory is best examined using reactions involving unreactive electrophiles. The most satisfactory test is obtained from hydrogen exchange of toluene under various conditions because of the freedom from steric hindrance in this reaction, and because the selectivity of the reagent can be varied widely by varying the reaction conditions. It is found (Table 3.2) that the value of $\log f_o$:$\log f_p$ changes steadily from values >1 when the reagent is of low selectivity to values <1 as the selectivity is increased.[133] Data for hydrogen exchange of *tert*-butylbenzene (Table 3.2) and biphenyl (Table 3.14) show the same trend. In other reactions of high selectivity, such as molecular bromination and chlorination, values <1 are also obtained for toluene.

(3) When both the electrophile and the aromatic are of high reactivity, the transition state should involve less deformation of the aromatic system than has occurred in the Wheland intermediate, and the relative rates of substitution at *ortho* and *para* positions should be governed approximately by the relative electron densities at these positions in the unperturbed molecule. For substituents such as methyl it has been argued that the *ortho* position should be more negatively polarized than the *para* position,[129] and it is therefore understandable that toluene gives *ortho*:*para* ratios >2 in reactions with unselective reagents. The high ratio in nitration of benzeneboronic acid in acetic anhydride has likewise been attributed to the larger effect of the $+I$ substituent (the boron anion, structure **63**, Chapter 7) on the *ortho* than the *para* position.[134]

(4) The gradual decrease in the $\log f_o$:$\log f_p$ ratio with increase in f_p (a measure of the selectivity of the electrophilic reagent) for hydrogen exchange of toluene is now seen to result from opposing effects: the high ratios, which occur when f_p is small, are determined essentially by ground-state electron densities, whereas the low ratios, which occur when f_p is large, are governed by the relative stabilities of the Wheland intermediates. The similar trends shown by *tert*-butylbenzene in hydrogen exchange[133] and by toluene in halogenation (Fig. 9.1) argue against the importance of steric hindrance in these reactions, even when the ratios are small. On the other hand, the data for halogenations of *tert*-butylbenzene do not follow

this trend: the random scatter of the points in the plot of log f_o against log f_p for these reactions indicates that steric hindrance is the overriding factor in determining the log f_o:log f_p ratio. Again, a plot of log f_o (ordinate) against log f_p (abscissa) for a large number of reactions of toluene is a smooth curve, convex upwards, as expected on the basis of the above arguments, provided that reactions in which there is reason to believe that steric hindrance is important are omitted.[129]

(5) There is a further consequence of the charge distribution shown in the intermediate **9**. Since the free energy of activation for substitution at a given position should be proportional to the charge at that position, it follows that the logarithms of the partial rate factors will also be proportional to the charge. It may therefore be predicted that provided steric hindrance is either absent or trivial, the ratio log f_o:log f_p should be $0.26/0.30 = 0.87$, and this should be found for substituents exhibiting substantial electronic effects, in fairly unselective reactions, i.e. those with transition states resembling the Wheland intermediate.[135,136] By contrast, substituents that have a small electronic effect usually do so because there is a balance between inductive and conjugative effects; as a result, the theoretical ratio will not be observed for these because of the importance of the short range (inductive-field) effects as described in (3) above. The log f_o:log f_p ratios for a number of substituents in reactions which conform to these requirements are given in Table 11.8, and it is seen that values are very close to those theoretically predicted,[135,136] as is also the average value. One may therefore predict σ^+ (*ortho*) values for conditions free from steric hindrance according to eq. 11.11.[136] An alternative equation, viz. $\sigma^+ (ortho) = 0.07 + 0.95\sigma^+ (para)$,[137] is unsatisfactory in that it does not predict the *ortho* values given in the same paper, predicts *o*-F to activate, and predicts *o*-Me to activate insufficiently.

$$\sigma^+ (ortho) = 0.87\sigma^+ (para) \qquad (11.11)$$

(6) For substituents that are of the $-I$, $+M$ type, two considerations apply. If the conjugative effect is large then in reactions with large demands for conjugative stabilization of the transition state, the ratio of ca 0.87 will be obtained. In reactions of low conjugative demands then a much lower ratio will be obtained. The SMe, OPh, and SPh substituents show this particularly well. They each give the theoretical ratio in protiodetritiation, whereas in protiodesilylation the ratios become 0.70, 0.48, and 0.11, respectively.[135] If, on the other hand, the conjugative effect is small, the ratio will be governed largely by the inductive effect and will be low in all reactions. Thus, for the halogens, the log f_o:log f_p ratios in, e.g., hydrogen exchange, nitration, and chloromethylation lie in the order F < Cl < Br, whereas if they resulted from steric hindrance the order should be reversed.

(7) The other rate ratio that is commonly measured is the α:β ratio for substitution in naphthalene. It follows from the arguments outlined above that

Table 11.8 Values of $\log f_o$:$\log f_p$ for various substituents

Substituent	Reaction	$\log f_o$:$\log f_p$
Me	Protiodesilylation, 50 °C	0.95
	Protiodegermylation, 50 °C	0.95
	Protiodetritiation, TFA, 70 °C	0.88
	Nitration, $AcONO_2$, 25 °C	0.905
	Chlorination, Cl_2, HOAc, MeCN, $MeNO_2$, 25 °C	0.90 (av.)
Et	Chlorination, Cl_2, HOAc, 25 °C	0.91
i-Pr	Protiodetritiation, TFA, 70 °C	0.86
t-Bu	Protiodetritiation, TFA, 70 °C	0.87
Cyclobutyl	Protiodetritiation, TFA, 70 °C	0.875
Cyclopentyl	Protiodetritiation, TFA, 70 °C	0.87
Cyclohexyl	Protiodetritiation, TFA, 70 °C	0.885
CH_2Ph	Protiodetritiation, TFA, 70 °C	0.81
$(CH_2)_2Ph$	Protiodetritiation, TFA, 70 °C	0.83
CH_2SiMe_3	Protiodetritiation, TFA, 70 °C	0.81
$(CH_2)_2SiMe_3$	Protiodetritiation, TFA, 70 °C	0.91
$(CH_2)_3SiMe_3$	Protiodetritiation, TFA, 70 °C	0.88
$(CH_2)_4SiMe_3$	Protiodetritiation, TFA, 70 °C	0.86
OH	Protiodesilylation, 50 °C	0.885
OMe	Protiodesilylation, 50 °C	0.81
	Protiodegermylation, 50 °C	0.84
	Protiodetritiation, TFA, 70 °C	0.92
	Chlorination, Cl_2, HOAc, 25 °C	0.885
SMe	Protiodetritiation, TFA, 70 °C	0.88
OPh	Protiodetritiation, TFA, 70 °C	0.855
	Nitration, HNO_3–Ac_2O, 0 °C	0.88
SPh	Protiodetritiation, TFA, 70 °C	0.88
	Nitration, HNO_3–Ac_2O, 0 °C	0.85
Ph	Protiodetritiation, TFA, 70 °C	0.87
C_6H_4-Ph-4	Protiodetritiation, TFA, 70 °C	0.92
(Fluorene)	Protiodetritiation, TFA, 70 °C	0.885
2-F-C_6H_4	Protiodetritiation, TFA, 70 °C	0.88
3-F-C_6H_4	Protiodetritiation, TFA, 70 °C	0.94
NO_2	Nitration, HNO_3–H_2SO_4, 25 °C	0.98
CO_2Et	Nitration, $AcONO_2$, 18 °C	0.85
CO_2H	Protiodesilylation, 50 °C	0.825
SO_2^-	Protiodesilylation, 50 °C	0.83
	Nitration, HNO_3–H_2SO_4, 25 °C	0.96
SO_2Et	Nitration, HNO_3–Ac_2O, 25 °C	0.96
Average		0.885

$\log f_\alpha$:$\log f_\beta$ could reasonably be expected to be constant (assuming that conjugative interactions at the two sites are not significantly different).[2] Values that have been obtained in hydrogen exchange are 1.48 ± 0.07, so that values which differ appreciably from this may be attributed to steric effects.[2] Thus the high values of 3.14 (protiodegermylation), 2.72 (protiodesilylation) and 2.09 (bromodesilylation) are clear indications of steric acceleration, whilst the low value of 1.33 in sulphonation clearly is due to steric hindrance. Values of 1.57 (nitration) and 1.61 (bromination with Br_2–HOAc) are rather high; the nitration data are pre-g.l.c results and may merit reinvestigation.

11.2.5 The Effects of Solvent and Temperature

There have been few investigations of these effects. A systematic survey of the effect of solvent on the uncatalysed chlorination of toluene at 25 °C showed the *ortho*:*para* ratios to lie in the region of 60:40 and 40:60 for hydroxylic and non-hydroxylic media, respectively.[138] This appeared to be due to the electrophilic reagent being a complex between chlorine and the solvent, so that the selectivity depends on the nature of the solvent.

Raising the temperature of a reaction should reduce the selectivity, and hence lower the *ortho*:*para* ratio, given that $\log f_o$:$\log f_p$ is constant, but this aspect does not appear to have been examined. Raising the temperature in the nitration of biphenyl under various conditions produced a substantial lowering of the ratio, but the reason for this not known.[139]

11.3 STRAIN EFFECTS

The consequences of strain effects have been described in Section 2.6. The relationship between partial rate factors and strain may be seen from consideration of the data in Table 11.9 relating to molecules **10** and **11**.[140]

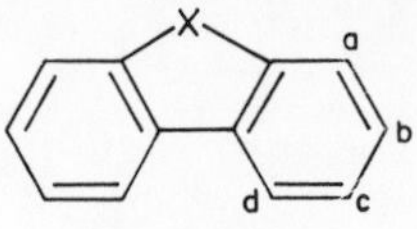

(10)
X=CH_2, O, S, NH

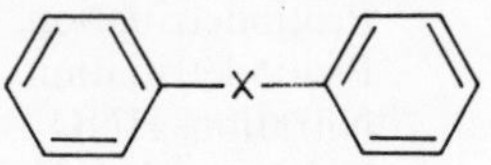

(11)

Positions a and d in **10** are α to the five-membered central ring whereas positions b and c are β to it. Positions a and d should therefore have reduced reactivity if the central ring is strained (Section 2.6). Positions a and c in **10** correspond to the *ortho* and *para* positions in **11** (and are each *meta* to the bridge joining the two rings, so the effect of this on each position should be the same). It follows, therefore, that the strain in the central ring should cause the ratio $\log f_a$:$\log f_c$ for **10** to be reduced relative to the $\log f_o$:$\log f_p$ ratio for **11**, determined under the same conditions. The data for detritiation in Table 11.8 show that this is indeed the case and, moreover, the ratio is reduced more for

Table 11.9

		Compounds **10**		Log f_o:log f_p	
Reaction	X	Log f_a:log f_c	Log f_d:log f_b	Compounds **11**	Biphenyl
Detritiation	CH_2	0.63	0.885	0.81	0.90
	O	0.62	0.85	0.855	
	S	0.785	0.92	0.88	
Desilylation	O	−0.02	−0.05	0.48	1.75
	S	0.077	2.46	0.11	
Nitration	CH_2	—	0.9	0.74	1.02
	O	0.85	>0.65	0.88	
	NH	0.92	—	1.03	
Chlorination	CH_2	—	0.835	—	0.83
	NAc	0.715	1.0	0.865	

fluorene (**10**, X = CH_2) and dibenzofuran (X = O), than for dibenzothiophene (X = S) which has a less strained central ring. The data for nitration, desilylation, and chlorination of these compounds show similar behaviour.

Positions b and d in **10** correspond to the *ortho* and *para* positions in biphenyl, each being *meta* to the group X. It would be expected, therefore, that the log f_d:log f_b ratio should be less than the log f_o:log f_p ratio in biphenyl. The results are less unambiguous here because of the large difference in reactivity between compounds **10** and biphenyl, which causes conjugative interactions to be significantly different in the two molecules. Nevertheless, except in chlorination (where the conjugative effects are greatest), the former ratio is smaller.[140]

11.4 *IPSO* FACTORS

The ease of replacement of one substituent by another depends on a number of factors:

(1) Electron supply to the *ipso* site by other substituents in the aromatic ring.

(2) Steric hindrance to the incoming electrophile. This appears to be particularly important in causing cleavages by the nitro group to take place very much slower than cleavages by the nitroso group (see Chapter 10). It causes cleavage of the *tert*-butyl group to be comparable to, or even slower than, cleavage of the methyl group.

(3) Ease of cleavage of the bond between the *ipso* carbon and the leaving group.

(4) Reactivity of the electrophile.

Ipso factors should therefore ideally be defined for benzene containing no substituents other than that which is being replaced. This has not always been

Table 11.10 *Ipso* factors for electrophilic aromatic substitution

Reaction	Leaving group	f_1	Section/Ref.
Protiodeplumbylation	Pb(cyclohexyl)$_3$	$\sim 2 \times 10^{12}$	Section 4.7.1
Protiodestannylation	Sn(cyclohexyl)$_3$	$\sim 3.5 \times 10^9$	Section 4.7.1
Protiodegermylation	$GeEt_3$	$\sim 3.6 \times 10^5$	Section 4.7.1
Protiodesilylation	$SiMe_3$	$\sim 10^5$	Section 4.7.1
Protiodeboronation	$B(OH)_2$	~ 4	141
Bromodestannylation	$SnMe_3$	2×10^{12}	Section 10.42
Bromodesilylation	$SiMe_3$	2×10^8	Section 10.38
Chlorodesilylation	$SiMe_3$	10^3–10^4	Section 10.38
Nitrosodecarboxylation	CO_2H	13	Section 10.27
Nitrodemethylation	Me	4.7	Section 10.23
Bromode-*tert*-butylation ($HOBr$–H^+)	*t*-Bu	1.4	Section 10.25
Chlorodebutylation (Cl_2–HOAc)	*t*-Bu	1.0	Section 10.25
Bromodemethylation (Br_2–aq. HOAc)	Me	0.29	142
Bromode-*tert*-butylation (Br_2–aq. HOAc)	*t*-Bu	0.255	142
Nitrodesilylation	$SiMe_3$	0.66	Section 10.37
Diazodesilylation	$SiPh_3$	<1	Section 10.38
Nitrosodeiodination	I	0.18	Section 10.60
Nitrosodebromination	Br	0.079	Section 10.60
Nitrosodechlorination	Cl	0.061	Section 10.60
Diazodeiodination	I	0.149	Section 10.59
Diazodebromination	Br	0.0089	Section 10.59
Diazodechlorination	Cl	0.0070	Section 10.59
Protiodecarboxylation	CO_2H	0.002	143
Bromodecarboxylation	CO_2H	0.01–0.0014	Section 10.30
Bromodebromination (Br_2–aq. HOAc)	Br	$<4 \times 10^{-7}$	142

possible, so that some of the values in the compilation of all known *ipso* factors (Table 11.10) may subsequently be revised. The values for bromine exchange were not obtained from labelling experiments, but rather from the rate of formation of the intermediate adducts, which were isolated.

REFERENCES

1. M. J. S. Dewar and R. J. Sampson, *J. Chem. Soc.*, 1956, 2789; 1957, 2946, 2952.
2. R. Taylor and G. G. Smith, *Tetrahedron*, 1963, **19**, 937.
3. D. E. Pearson, J. F. Baxter, and J. C. Martin, *J. Org. Chem.*, 1952, **17**, 1511.
4. C. W. McGary, Y. Okamoto, and H. C. Brown, *J. Am. Chem. Soc.*, 1955, **77**, 3037.
5. L. M. Stock and H. C. Brown, *Adv. Phys. Org. Chem.*, 1963, **1**, 35.
6. H. B. Amin and R. Taylor, *J. Chem. Soc., Perkin Trans. 2*, (a) 1978, 1090; (b) 1978, 1095; (c) 1979, 624.
7. (a) E. Glyde and R. Taylor, *J. Chem. Soc., Perkin Trans. 2*, 1975, 1463; (b) H. B. Amin and R. Taylor, *J. Chem. Soc., Perkin Trans. 2*, 1979, 228.

8. H. B. Amin and R. Taylor, *Tetrahedron Lett.*, 1978, 267.
9. E. Glyde and R. Taylor, *J. Chem. Soc., Perkin Trans. 2*, 1977, 1541.
10. M. A. Hossaini and R. Taylor, *J. Chem. Soc., Perkin Trans. 2*, 1982, 187.
11. E. Glyde and R. Taylor, *J. Chem. Soc., Perkin Trans. 2*, 1973, 1632.
12. R. Taylor, *J. Chem. Soc. B*, 1971, 622.
13. R. Taylor, *J. Chem. Soc.*, 1962, 4881; *J. Chem. Soc. B*, 1968, 1397; 1971, 2382; *J. Chem. Soc., Perkin Trans. 2*, 1975, 277; E. Glyde and R. Taylor, *J. Chem. Soc., Perkin Trans. 2*, 1975, 1783; H. B. Amin and R. Taylor, *J. Chem. Soc., Perkin Trans. 2*, 1975, 1053; R. A. August, C. Davis, and R. Taylor, *J. Chem. Soc., Perkin Trans. 2*, 1986, 1265.
14. I. N. Juchnovski and I. G. Binev, *Tetrahedron*, 1977, **33**, 2993.
15. Values of σ taken from O. Exner, in *Correlation Analysis in Chemistry*, Eds N. B. Chapman and J. Shorter, Plenum Press, London, 1978, Table 10.1.
16. H. C. Brown, G. C. Rao, and M. Ravindranathan, *J. Am. Chem. Soc.*, 1970, **92**, 829.
17. A. J. Hoefnagel and B. M. Wepster, *J. Am. Chem. Soc.*, 1973, **95**, 5357.
18. H. C. Brown and Y. Okamoto, *J. Am. Chem. Soc.*, 1958, **70**, 4979.
19. R. Taylor, *J. Chem. Soc., Perkin Trans. 2*, 1978, 755.
20. P. A. S. Smith, J. H. Hall, and R. O. Kan, *J. Am. Chem. Soc.*, 1962, **84**, 485.
21. K. V. Seshadri and R. Ganesan, *Tetrahedron*, 1972, **28**, 3827.
22. A. Cornelis, S. Lambert, and P. Laszlo, *J. Org. Chem.*, 1977, **42**, 381.
23. W. Hanstein and T. G. Traylor, *Tetrahedron Lett.*, 1967, 4451; W. Hanstein, H. J. Berwin, and T. G. Traylor, *J. Am. Chem. Soc.*, 1970, **92**, 829, 7476.
24. (a) R. Taylor, *J. Chem. Soc. B*, 1971, 1450; (b) J. M. Brittain, P. B. D. de la Mare, and J. M. Smith, *J. Chem. Soc., Perkin Trans. 2*, 1981, 1629.
25. R. Taylor, unpublished work.
26. P. G. Sennikov, S. E. Skobeleva, V. A. Kuznetsov, A. N. Egorochkin, P. Riviere, J. Stage, and S. Richelme, *J. Organomet. Chem.*, 1980, **201**, 213.
27. R. Baker, C. Eaborn, and R. Taylor, *J. Chem. Soc. B*, 1972, 97.
28. A. Daneshrad, C. Eaborn, and D. R. M. Walton, *J. Organomet. Chem.*, 1975, **85**, 35; C. Eaborn, A. R. Hancock, and W. A. Stanczyk, *J. Organomet. Chem.*, 1981, **218**, 147.
29. C. Eaborn, T. A. Emokpae, V. I. Sidorov, and R. Taylor, *J. Chem. Soc., Perkin Trans. 2*, 1974, 1454.
30. V. F. Traven, B. S. Korolev, T. V. Pyatkina, and B. I. Stepanov, *J. Gen. Chem. USSR*, 1975, **45**, 943.
31. D. D. Davis, *J. Organomet. Chem.*, 1981, **206**, 21.
32. G. D. Hartman and T. G. Traylor, *J. Am. Chem. Soc.*, 1975, **97**, 6147.
33. T. G. Traylor and J. C. Ware, *J. Am. Chem. Soc.*, 1967, **89**, 2304.
34. F. Fringuelli, G. Marino, and A. Taticchi, *J. Chem. Soc. B*, 1971, 2304; 1972, 158, 1738.
35. Calculated from the data of P. P. Alikhanov, *J. Gen. Chem. USSR*, 1977, **47**, 339, 550; *J. Org. Chem. USSR*, 1977, **13**, 691.
36. N. A. Clinton, R. S. Brown, and T. G. Traylor, *J. Am. Chem. Soc.*, 1970, **92**, 5228.
37. M. A. Hossaini and R. Taylor, *J. Chem. Soc., Perkin Trans. 2*, 1982, 187.
38. P. B. Fischer and R. Taylor, *J. Chem. Soc., Perkin Trans. 2*, 1980, 781.
39. M. M. J. LeGuen and R. Taylor, *J. Chem. Soc., Perkin Trans. 2*, 1976, 559.
40. E. Glyde and R. Taylor, *J. Chem. Soc., Perkin Trans. 2*, 1977, 678.
41. H. V. Ansell and R. Taylor, *J. Chem. Soc., Perkin Trans. 2*, 1978, 751.
42. L. B. Jones and J. P. Foster, *J. Org. Chem.*, 1970, **35**, 1777.
43. J. K. Kochi and G. S. Hammond, *J. Am. Chem. Soc.*, 1953, **75**, 3452.
44. J. A. Landgrebe and R. H. Rynbrandt, *J. Org. Chem.*, 1966, **31**, 2585.
45. O. Buchman, M. Grosjean, and J. Nasielski, *Helv. Chim. Acta*, 1964, **47**, 2037.
46. A. J. Cornish and C. Eaborn, *J. Chem. Soc., Perkin Trans. 2*, 1975, 875.
47. P. Gassman and T. L. Guggenheim, *J. Org. Chem.*, 1982, **47**, 4002.

48. Y. Kusuyama and Y. Ikeda, *Bull. Chem. Soc. Jpn.*, 1973, **46**, 204.
49. H. Cerfontain and Z. R. H. Schaasberg-Nienhuis, *J. Chem. Soc., Perkin Trans. 2*, 1976, 1780.
50. A. C. Boicelli, R. Danieli, A. Mangini, A. Ricci, and G. Pirazzini, *J. Chem. Soc., Perkin Trans. 2*, 1974, 1343.
51. R. Taylor and G. G. Smith, *Tetrahedron*, 1963, **19**, 937.
52. R. Taylor, *J. Chem. Soc., Perkin Trans. 2*, 1973, 253.
53. W. A. Sheppard, *J. Am. Chem. Soc.*, 1970, **92**, 5419.
54. A. Streitwieser and H. S. Klein, *J. Am. Chem. Soc.*, 1964, **86**, 5170.
55. G. A. Olah *et al.*, *J. Am. Chem. Soc.*, 1987, **109**, 3708.
56. C. J. Byrne, D. A. R. Happer, M. P. Hartshorn, and H. K. J. Powell, *J. Chem. Soc., Perkin Trans. 2*, 1987, 1649.
57. C. J. Byrne, D. Christoforou, D. A. R. Happer, and M. P. Hartshorn, *J. Chem. Soc., Perkin Trans. 2*, 1988, 147.
58. N. V. Kondratenko, G. P. Syrova, V. I. Popov, Yu. N. Sheinker, and L. M. Yagulpolskii, *J. Gen. Chem. USSR*, 1971, **41**, 2075.
59. K. Kamiyama, H. Minato, and M. Kobayashi, *Bull. Chem. Soc. Jpn.*, 1973, **46**, 2255.
60. Yu. A. Ustynyuk, O. A. Subbotin, L. M. Buchneva, V. N. Gruzdneva, and L. A. Kazitsyna, *Proc. Acad. Sci. USSR*, 1976, **227**, 175.
61. L. I. Belenkii, V. S. Bogdanov, and I. M. Karmanova, *Bull. Acad. Sci. USSR*, 1979, 1599.
62. R. Taylor, G. J. Wright, and A. J. Homes, *J. Chem. Soc. B*, 1967, 780.
63. J. Vaughan and G. J. Wright, *J. Org. Chem.*, 1968, **33**, 2580.
64. M. M. A. Stroud and R. Taylor, *J. Chem. Res. (S)*, 1978, 425.
65. Y. El-din Shafig and R. Taylor, *J. Chem. Soc., Perkin Trans. 2*, 1978, 1263.
66. J. M. Blatchly and R. Taylor, *J. Chem. Soc.*, 1964, 4641.
67. H. V. Ansell, M. M. Hirschler, and R. Taylor, *J. Chem. Soc., Perkin Trans. 2*, 1977, 353.
68. K. C. C. Bancroft, R. W. Bott, and C. Eaborn, *J. Chem. Soc.*, 1964, 4806.
69. R. Baker, C. Eaborn, and J. A. Sperry, *J. Chem. Soc.*, 1962, 2382.
70. R. Baker, R. W. Bott, C. Eaborn, P. M. Greasley, *J. Chem. Soc.*, 1964, 627.
71. C. Eaborn, A. Fischer, and D. R. Killpack, *J. Chem. Soc. B*, 1971, 2142.
72. H. V. Ansell, P. J. Sheppard, C. F. Simpson, M. A. Stround, and R. Taylor, *J. Chem. Soc., Chem. Commun.*, 1978, 586; *J. Chem. Soc., Perkin Trans. 2*, 1979, 381.
73. H. V. Ansell and R. Taylor, *J. Org. Chem.*, 1979, **44**, 4946.
74. W. J. Archer and R. Taylor, *J. Chem. Soc., Perkin Trans. 2*, 1981, 1153.
75. P. Golborn, C. Eaborn, R. E. Spillett, and R. Taylor, *J. Chem. Soc. B*, 1968, 1112.
76. A. P. Neary and R. Taylor, *J. Chem. Soc., Perkin Trans. 2*, 1983, 1233.
77. A. P. Laws, A. P. Neary, and R. Taylor, *J. Chem. Soc., Perkin Trans. 2*, 1987, 1033.
78. M. C. A. Opie, G. J. Wright, and J. Vaughan, *Aust. J. Chem.*, 1971, **24**, 1205; H. V. Ansell and R. Taylor, *Tetrahedron Lett.*, 1971, 4915.
79. K. C. C. Bancroft and G. R. Howe, *J. Chem. Soc. B*, 1970, 1541; 1971, 400.
80. M. M. J. Le Guen and R. Taylor, *J. Chem. Soc., Perkin Trans. 2*, 1974, 1274.
81. M. M. J. Le Guen, Y. El-din Shafig, and R. Taylor, *J. Chem. Soc., Perkin Trans. 2*, 1979, 803.
82. W. J. Archer, Y. El-din Shafig, and R. Taylor, *J. Chem. Soc., Perkin Trans. 2*, 1981, 675.
83. H. V. Ansell, M. S. Newman, and R. Taylor, unpublished work.
84. W. J. Archer, R. Taylor, P. H. Gore, and F. S. Kamounah, *J. Chem. Soc., Perkin Trans. 2*, 1980, 1828.

85. A. Streitwieser, A. Lewis, I. Schwager, R. W. Fish, and S. Labana, *J. Am. Chem. Soc.*, 1970, **92**, 6525.
86. M. M. Hirschler and R. Taylor, *J. Chem. Soc., Chem. Commun.*, 1980, 967.
87. A. P. Laws and R. Taylor, *J. Chem. Soc., Perkin Trans. 2*, 1987, 591.
88. R. Taylor, *J. Chem. Soc., Perkin Trans. 2*, 1975, 1287.
89. A. P. Laws and R. Taylor, *J. Chem. Soc., Perkin Trans. 2*, 1987, 1683.
90. E. M. Arnett and R. D. Bushick, *J. Org. Chem.*, 1962, **27**, 111.
91. L. I. Zakharkin, V. N. Kalinin, and A. P. Snyaku, *J. Gen. Chem. USSR*, 1971, **41**, 1521.
92. A. D. Allen, M. Rosenbaum, N. O. L. Sato, and T. T. Tidwell, *J. Org. Chem.*, 1982, **47**, 4234; A. G. Harrison, R. Houriet, and T. T. Tidwell, *J. Org. Chem.*, 1984, **49**, 1302.
93. Y. Okamoto and H. C. Brown, *J. Am. Chem. Soc.*, **80**, 4976.
94. R. W. Taft, *J. Phys. Chem.*, 1960, **64**, 1805.
95. H. van Bekkum, P. E. Verkade, and B. M. Wepster, *Recl. Trav. Chim. Pays-Bas*, 1959, **78**, 815.
96. E. R. Vorpagel, A. Streitwieser, and S. D. Alexandratos, *J. Am. Chem. Soc.*, 1981, **103**, 3777.
97. C. G. Swain and E. C. Lupton, *J. Am. Chem. Soc.*, 1968, **90**, 4328; C. G. Swain, S. H. Unger, N. R. Rosenquist, and M. S. Swain, *J. Am. Chem. Soc.*, 1983, **105**, 492; C. G. Swain, *J. Org. Chem.*, 1984, **49**, 2005.
98. W. F. Reynolds and R. D. Topsom, *J. Org. Chem.*, 1984, **49**, 1989; A. J. Hoefnagel, W. Oosterbeck, and B. M. Wepster, *J. Org. Chem.*, 1984, **49**, 1993; M. Charton, *J. Org. Chem.*, 1984, **49**, 1997; R. W. Taft, J.-L. Abboud, and M. J. Kamlet, *J. Org. Chem.*, 1984, **49**, 2001.
99. M. J. S. Dewar and P. J. Grisdale, *J. Am. Chem. Soc.*, 1962, **84**, 3539, 3548; M. J. S. Dewar, R. Golden, and J. M. Harris, *J. Am. Chem. Soc.*, 1971, **93**, 4187.
100. C. Eaborn and A. Fischer, *J. Chem. Soc. B*, 1969, 152.
101. D. A. Forsyth, *J. Am. Chem. Soc.*, 1973, **95**, 3594.
102. Y. Yukawa and Y. Tsuno, *Bull. Chem. Soc. Jpn.*, 1959, **32**, 971.
103. Y. Yukawa, Y. Tsuno, and M. Sawada, *Bull. Chem. Soc. Jpn.*, 1966, **39**, 2274.
104. C. Eaborn and R. Taylor, *J. Chem. Soc.*, 1961, 1012.
105. D. P. Kelly, M. D. Jenkins, and R. A. Mantello, *J. Org. Chem.*, 1981, **46**, 1650.
106. R. O. C. Norman and G. K. Radda, *Tetrahedron Lett.*, 1962, 125.
107. R. Taylor, in *The Chemistry of Heterocyclic Compounds*, Ed. S. Gronowitz, Wiley, Chichester, 1986, Vol. 44, Pt. 2, pp. 10–15.
108. Y. Tsuno, W.-Y. Chong, Y. Tawaka, M. Sawada, and Y. Yukawa, *Bull. Chem. Soc. Jpn.*, 1978, **51**, 596.
109. T. Inukai, *Bull. Chem. Soc. Jpn.*, 1962, **35**, 400.
110. M. A. Hossaini and R. Taylor, *J. Chem. Soc., Perkin Trans. 2*, 1982, 187.
111. G. S. Hammond, *J. Am. Chem. Soc.*, 1955, **77**, 334.
112. R. P. Bell and E. N. Ramsden, *J. Chem. Soc.*, 1958, 161.
113. R. Taylor and T. J. Tewson, *J. Chem. Soc., Chem. Commun.*, 1973, 836.
114. R. Taylor, *Specialist Periodical Report on Aromatic and Heteroaromatic Chemistry*, Chemical Society, London, 1973, Vol. 1, p. 188.
115. A. Pross, *Adv. Phys. Org. Chem.*, 1977, **14**, 69.
116. R. Taylor, *J. Org. Chem.*, 1979, **44**, 2024.
117. N. S. Isaacs and D. Cvitas, *Tetrahedron*, 1971, **27**, 439.
118. K. Fukui, T. Yonezawa, and C. Nagata, *Bull. Chem. Soc. Jpn.*, 1954, **27**, 423; *J. Chem. Phys.*, 1957, **27**, 1247.
119. L. von Szentpaly, *Chem. Phys. Lett.*, 1981, **77**, 352.

120. G. Dallinga, A. A. Verrijn Stuart, P. J. Smit, and E. L. Mackor, *Z. Elektrochem.*, 1957, **61**, 1019; E. L. Mackor, P. J. Smit, and J. H. van der Waals, *Trans. Faraday. Soc.*, 1957, **53**, 1309.
121. See M. J. S. Dewar and D. S. Urch, *J. Chem. Soc.*, 1958, 3079, and earlier papers in this series.
122. R. Taylor, *Chimia*, 1968, **22**, 1.
123. E. Hayashi and Y. Nishi, *J. Pharm. Soc. Jpn.*, 1960, **80**, 841.
124. E. R. Ward and J. E. Marriott, *Chem. Ind.* (*London*), 1962, 507.
125. H. G. Peer, *Recl. Trav. Chim. Pays-Bas*, 1959, **79**, 851; 1960, **79**, 825.
126. G. G. Eke, J. P. Napolitano, A. H. Filbey, and A. J. Kolka, *J. Org. Chem.*, 1957, **22**, 639, 642.
127. G. S. Hammond and M. F. Hawthorne, in *Steric Effects in Organic Chemistry*, Ed. M. S. Newman, Wiley, New York, 1956, p. 180.
128. G. S. Hammond and K. J. Douglas, *J. Am. Chem. Soc.*, 1959, **81**, 1184.
129. R. O. C. Norman and G. K. Radda, *J. Chem. Soc.*, 1961, 3610.
130. P. B. de la Mare and J. H. Ridd, *Aromatic Substitution*, Butterworths, London, 1959, p. 82.
131. G. A. Olah, *Acc. Chem. Res.*, 1971, **4**, 240; G. A. Olah, R. H. Schlosberg, R. D. Porter, Y. M. Mo, D. P. Kelly, and G. D. Mateescu, *J. Am. Chem. Soc.*, 1972, **94**, 2034.
132. C. MacLean and E. L. Mackor, *Mol. Phys.*, 1961, **4**, 241; J. P. Colpa, C. MacLean, and E. L. Mackor, *Tetrahedron*, 1963, **19** (Suppl. 2), 65; W. J. Herre and J. A. Pople, *J. Am. Chem. Soc.*, 1972, **94**, 6901. In contradiction to these calculations and results in ref. 131, ^{13}C n.m.r. data indicate that in the *static* benzenonium ion, the charge is more heavily concentrated at the *ortho* position (structure 7, Chapter 2).
133. R. Baker, C. Eaborn, and R. Taylor, *J. Chem. Soc.*, 1961, 4927.
134. D. R. Harvey and R. O. C. Norman, *J. Chem. Soc.*, 1962, 3822.
135. F. P. Bailey and R. Taylor, *J. Chem. Soc. B*, 1971, 1146.
136. H. V. Ansell, M. M. J. Le Guen, and R. Taylor, *Tetrahedron Lett.*, 1973, 13.
137. V. A. Koptyug and V. I. Buraev, *J. Org. Chem. USSR*, 1980, **9**, 1599.
138. L. M. Stock and A. Himoe, *Tetrahedron Lett.*, 1960, (13), 9.
139. R. Taylor, *Tetrahedron Lett.*, 1966, 6093.
140. R. Taylor, *J. Chem. Soc. B*, 1968, 1559.
141. H. G. Kuivila and K. V. Nahabedian, *J. Am. Chem. Soc.*, 1961, **83**, 2159.
142. E. Baciocchi and G. Illuminati, *J. Am. Chem. Soc.*, 1967, **89**, 4017.
143. J. L. Longridge and F. A. Long, *J. Am. Chem. Soc.*, 1968, **90**, 3092.

Index

References to substituents are given with respect to the parent aromatic, i.e. chloro substituent effects are to be found under chlorobenzene. Compounds involved in *ipso* substitutions are referred to only with respect to the parent aromatic, devoid of the *ipso* substituent. For example, bromodesilylation of 3-(chlorophenyl)trimethylsilane is listed under chlorobenzene. Trivial names are not used for alkylbenzenes. Reagents listed are those not covered by the subheadings in the List of Contents.